# Economicity, Safety and Reliability for Motor Vehicles Congress (and SIAR International Automotive and Transport Engineering Congress) 2023

IOP Conference Series: Materials Science and Engineering
Volume 1303

Bucharest, Romania
2-4 November 2023

ISBN: 978-1-7138-9567-1
ISSN: 1757-8981

**Printed from e-media with permission by:**

Curran Associates, Inc.
57 Morehouse Lane
Red Hook, NY 12571

**Some format issues inherent in the e-media version may also appear in this print version.**



Institute of Physics
Dirac House, Temple Back
Bristol BS1 6BE UK

Phone: 44 1 17 929 7481
Fax: 44 1 17 920 0979

techtracking@iop.org

**Additional copies of this publication are available from:**

Curran Associates, Inc.
57 Morehouse Lane
Red Hook, NY 12571 USA
Phone: 845-758-0400
Fax: 845-758-2633
Email: curran@proceedings.com
Web: www.proceedings.com

# TABLE OF CONTENTS

IOP Conf. Series: Materials Science and Engineering 1303 (2024) 011001 doi:10.1088/1757-899X/1303/1/011001

# ESFA 2023

**2-4 november 2023**

## ECONOMICITATEA, SIGURANȚA ȘI FIABILITATEA AUTOVEHICULELOR
## ECONOMICITY, SAFETY AND RELIABILITY FOR MOTOR VEHICLES

**The XXXIII[rd] SIAR International Automotive and Transport Engineering Congress**

**The X[th] ESFA Congress hosted to the National University of Science and Technology POLITEHNICA Bucharest**

**Editors:** **Mihail-Daniel IOZSA**
**Marius-Florin TOMA**
**Gheorghe FRĂȚILĂ**

**National University of Science and Technology POLITEHNICA Bucharest** is the oldest and most prestigious engineer school in Romania. Its traditions are related to the establishment, in 1818, by Gheorghe Lazăr, of the first higher technical school with teaching in Romanian, at the Saint Sava Abbey in Bucharest. In 1832, it was reorganized into the St. Sava College.
www.upb.ro

**Automotive Engineering Department** was founded in 1959 within the Faculty of Transports, from Polytechnic Institute of Bucharest. The department ensures the development of two bachelor's study programs and three master's study programs
www.autovehiculerutiere.pub.ro

**Society of Automotive Engineers of Romania** is the professional organization of automotive engineers, independent, non-profit legal entity, active member of FISITA (Fédération Internationale des Sociétés d'Ingénieurs des Techniques de l'Automobile – International Federation of Automotive Engineering Societies) and EAEC (European Automotive Engineers Cooperation).
www.siar.ro

**ESFA congresses** have been organized by the Automotive Engineering Department since 1981, being one of the scientific events with a high visibility among specialists in the field of Automotive Engineering. The ESFA congress is part of a series of SIAR, congresses that are organized annually, by rotation, in the universities that train engineers in the field of Automotive Engineering and in which there are sections with SIAR members.
www.siarcongress.eu

IOP Conf. Series: Materials Science and Engineering 1303 (2024) 011001 doi:10.1088/1757-899X/1303/1/011001

## Preface

The ESFA Congress is a part of a series of SIAR (Society of Automotive Engineers from Romania) congresses which are organized annually, by rotation, in Romanian universities that train engineers in the field of Automotive Engineering (Bucharest, Braşov, Pitesti, Craiova, Cluj, Chisinau, Timișoara). This congress is the 33rd edition of the SIAR Congress.

The ESFA Congress are hosted by National University of Science and Technology POLITEHNICA Bucharest, the oldest and most prestigious engineer school from Romania. These congresses are organized by the Automotive Engineering Department since 1981, being one of the scientific events with great visibility among specialists in the field of Automotive Engineering.

According to the organizers' expectations, more than 150 Romanian specialists and from other countries, renowned companies involved in various activities specific to the field of automotive engineering, participated in this congress.

The Congress provided optimal conditions for the presentation of the latest achievements of scientific and technological research in the automotive industry and automotive components, in the field of vehicle maintenance and road transport, and facilitated the exchange of information and experience between participants.

One of the main purposes of the congress was to promote and present the most modern products and current technologies in the field of automotive engineering from the most prestigious romanian companies and abroad. For this purpose, exhibitions (indoor and outdoor), demonstrations, experimental tests, appropriate software presentations etc. were organized within the congress.

On the occasion of the SIAR congress, the final stage of the *Student Competition in Automotive Engineering "Prof. university Eng. Constantin GHIULAI"* is organized annually by SIAR, with two sections: Vehicle Dynamics (nineth edition) and Automotive CAD - CATIA (sixth edition).

About of 70 students from almost all Romanian university centers which have studied programs in the field of Automotive Engineering participated in the final stage of this competition.

ESFA 2023 Proceedings Editors

## Honorary Committee

IOP Conf. Series: Materials Science and Engineering 1303 (2024) 011001 doi:10.1088/1757-899X/1303/1/011001

**Scientific and Technical Committee**

## CONGRESS MANAGEMENT

***Congress Chairman***
*Mihail-Daniel IOZSA*
National University of Science and Technology POLITEHNICA Bucharest

***President of Society of Automotive Engineers of Romania (SIAR)***
*Nicolae BURNETE*
Technical University of Cluj-Napoca

***President of Scientific & Technical Committee***
*Gheorghe FRĂȚILĂ*
National University of Science and Technology POLITEHNICA Bucharest

***SIAR General Secretary***
*Minu MITREA*
„Ferdinand I" Military Technical Academy of Bucharest

***Website Coordinator***
*Dinu COVACIU*
Transilvania University of Braşov

***Exhibition***
*Marius-Florin TOMA*
*Cristian-Alexandru RENȚEA*
National University of Science and Technology POLITEHNICA Bucharest

***Student Contests***
*Dan-Alexandru MICU*
*Adrian-Aristide ASIMOPOLOS*
National University of Science and Technology POLITEHNICA Bucharest

## ORGANIZING COMMITTEE

Automotive Engineering Department, Faculty of Transports,
National University of Science and Technology POLITEHNICA Bucharest

*Mihail-Daniel IOZSA*
*Marius-Valentin BĂȚĂUȘ*
*Marius-Florin TOMA*
*Gheorghe-Valerian CROITORESCU*
*Alexandru DOBRE*
*Nikolaos-Cristian NUȚU*
*Cornelia STAN*
*Alexandra-Raluca MOISESCU*
*Ștefan VOLOACĂ*
*Dan-Alexandru MICU*
*Ana-Maria MANEA*
*Cristian-Alexandru RENȚEA*
*Iordan FILERU*
*Adrian-Aristide ASIMOPOLOS*
*Alexandru ANCUȚA*

*National University of Science and Technology POLITEHNICA Bucharest -* ***Rectorat***

IOP Conf. Series: Materials Science and Engineering 1303 (2024) 011001 doi:10.1088/1757-899X/1303/1/011001

**Brief overview of ESFA2023 Congress**

**Plenary session - keynote speakers**

- Cornel STAN, West Saxon University of Zwickau, Germany, « *Automobile of the future: a complex mobility* system» (on-line presentation)
- Adrian-Constantin CLENCI, Pitești University Center and Ioan-Mircea OPREAN, Bucharest University Center, National University of Science and Technology POLITEHNICA Bucharest, Romania, « *The internal combustion engine in the context of sustainable road mobility* »
- Stefan KANIA, AVL Austria, « *Software-defined vehicle using thermal management as an example"* »
- Corina SANDU, Mechanical Engineering Department, Virginia Tech, USA, « *Advances in mobility of autonomous vehicles in off-road conditions* » (video recording)
- Robert BERCU and Marian POPESCU, FEV Romania, « *Euro 7 – Emissions & monitoring regulations – Future challenges* »
- Ioana TIRCOMNICU UNGA, INAS, Romania, « *Simulation in automotive: better, faster, cheaper*»

**The final stage of the SIAR International Contest for Automotive Engineering Students** *"Professor Eng. Constantin GHIULAI"*:

- *Vehicle Dynamics* section – 9th edition;
- *Automotive CAD* section– 6th edition.

**Oral presentations of scientific papers by sections:**

- *Green vehicles*
- *Smart vehicles*
- *Advanced powertrains*
- *Advanced engineering methods and tools*
- *Fuel economy and pollution control*
- *New materials and technologies*
- *Vehicle dynamics, safety and comfort*
- *Vehicle design and testing*
- *Terrain vehicles*
- *Traffic and road transport systems*

**Exhibition**

During the ESFA congress, indoor and outdoor exhibition stands of the sponsors and the main collaborators of the Automotive Engineering Department were organized within the Faculty of Transports.

**Aknowledgements**

The organizing committee would like to thank to National University of Science and Technology POLITEHNICA Bucharest, as well as the companies that provided support for the organization of the ESFA 2023 congress.

BOSCH Tehnică pentru o viață
REGISTRUL AUTO ROMÂN
infoprodigital AUTOMOTIVE
FEV
MAN MHS Truck & Bus Group
AUTOKLASS Official home of Mercedes-Benz
INAS power to change
AKA
AVL
ed EUROPEAN DRINKS
MOLSON COORS beverage company
ZAREA

IOP Conf. Series: Materials Science and Engineering 1303 (2024) 011002 doi:10.1088/1757-899X/1303/1/011002

## Peer Review Statement

All papers published in this volume have been reviewed through processes administered by the Editors. Reviews were conducted by expert referees to the professional and scientific standards expected of a proceedings journal published by IOP Publishing. The papers were reviewed for the scientific content, the novelty of the informations, language correctness and the paper format.

1. Type of peer review: Double anonymous
2. Conference submission management system: Morressier
3. Number of submissions received: 109
4. Number of submissions sent for review: 102
5. Number of submissions accepted: 55
6. Acceptance Rate (Submissions Accepted / Submissions Received × 100): 50.4 %
7. Average number of reviews per paper: 2.00
8. Total number of reviewers involved: 35
9. Contact person for queries:
   Name: Dinu COVACIU
   Affiliation: Transilvania University of Brasov, Romania
   Email: dinu.covaciu@unitbv.ro

# Experimental research regarding the autonomy of electric cars in real driving conditions

**V Nișulescu[1], F Ivan[2], M F Toma[1], G Bancă[1] and C Rențea[1]**

[1]University Politehnica of Bucharest, Romania.
[2]University of Pitești, Romania.

E-mail: dsggearbox@yahoo.com

**Abstract**. Currently, almost all car manufacturers have in the production scheme at least one vehicle model with a hybrid or electric energy source. This fact is generated by the high interest in the market related to the subsidies offered to the purchase by the states but also by the high performance in terms of pollution, noise, and economy, which they can achieve. The only performance indicator which for the moment is not at a satisfactory level is represented by autonomy. At this point, the electric cars have an autonomy between 250 and 800 km driven after a single charge of the traction battery. It should be mentioned though that, the autonomy of electric vehicles declared by European manufacturers is obtained in laboratory conditions, these being carried out in accordance with the provisions of the European Union regulations. Obviously, there are significant differences between the electric autonomy obtained in laboratory conditions and the autonomy obtained in real driving conditions. These differences are related to the ideal test conditions achieved at the laboratory level and which cannot be fully found in real driving conditions. Starting from these premises, in a first part of the paper, are presented the theoretical arguments regarding the determination of autonomy in real driving conditions with a driving profile specific to the RDE testing and are described the technical details of the electric vehicle under investigation. In the second part of the paper, are presented the results of the experimental research for the three driving zones specific to the RDE testing: driving in the urban zone, driving in the rural zone, and driving on the motorway. The downward evolution of the state of charge of the traction battery for each driving zone is shown graphically. Following the processing, analysis and interpretation of the data acquired from the car electronic control units, are formulated conclusions regarding the causes that influence the electric autonomy.

## 1. Introduction

The specific autonomy of electric vehicles defined by the distance traveled after a single charge of the traction battery, represents the main comparative aspect considered by those interested before purchasing a new car [1]. Knowing the growing trend of potential clients' interest regarding the cars equipped with electric power source, the manufacturers invest heavily in the research of traction batteries to meet the expectations of customers [2]. Even if the materials and manufacturing procedures applied in the case of traction batteries are very advanced and in continuous evolution, the autonomy available for series road vehicles satisfies partially the need for mobility. If in the previous period the development trend of electric vehicles in terms of autonomy was mainly oriented towards urban use, currently the competitiveness is very high for use in extra-urban conditions and the motorway, where the speed is high or very high [3].

To give good results in terms of autonomy for such speed conditions, high power batteries, specific tires and new materials are used to lighten the vehicle's own weight, knowing that the electrical power consumed is dependent on the weight and the load applied to the vehicle [4]. Also, an important factor is the way of testing to obtain the electrical autonomy declared by the manufacturers

[5]. Currently, at the European level, most electric road vehicles are tested in order to obtain the autonomy in accordance with the Commission Regulation (EU) 2017/1151 of 1 June 2017 supplementing the Regulation (EC) No 715/2007 of the European Parliament and of the Council relating to type-approval of the vehicles, this means that the speed profile which the electric vehicle must respect is the one established by the World-wide harmonized Light duty Test Cycle (WLTC) [8].

The WLTC cycle is composed of four test speed zones, low speed, medium speed, high speed, and very high speed. Thus, the electric vehicle is subject to testing conditions at the laboratory level, which faithfully simulates a mode of exploitation as close as possible to the real one [6,7]. Obviously, the WLTC test gives a clear image regarding the reduction of the state of charge of the traction battery, but the ideal conditions obtained at the laboratory level are very difficult to reproduce in real driving. Starting from these premises, the authors proposed detailed research concerning the reduction of the state of charge of the traction battery in real driving conditions, using the running profile specific to the RDE test (Real Driving Emissions), applicable to vehicles equipped with thermal engines in the case of obtaining the type approval.

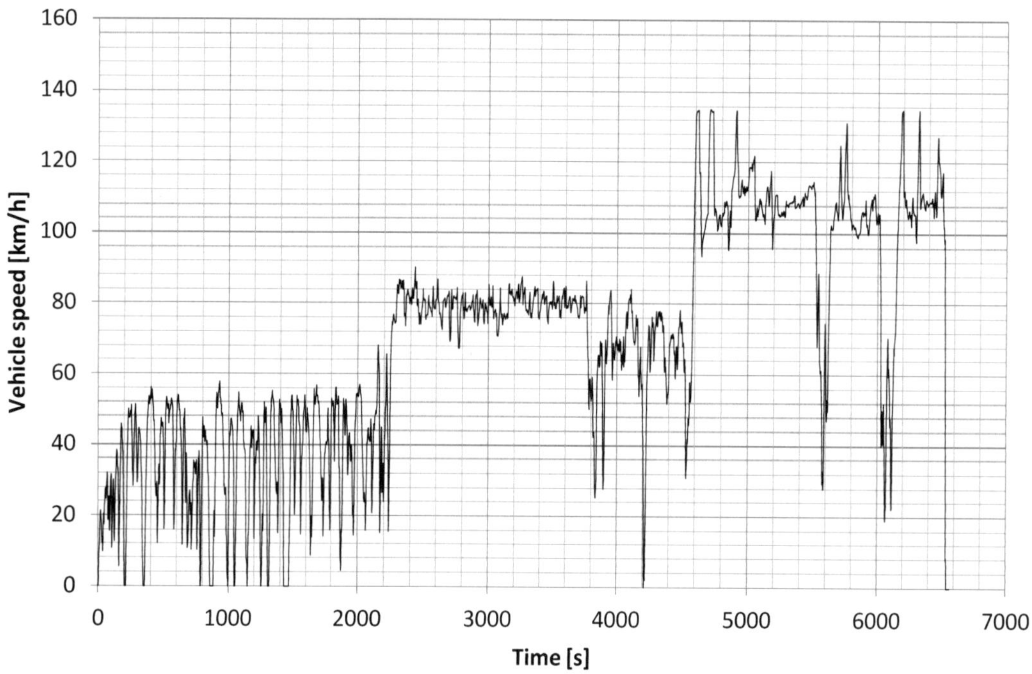


**Figure 1.** Vehicle speed profile specific to testing in real driving conditions [8]

An example of a speed profile obtained as a result of testing the vehicle in real driving conditions, is indicated in Figure 1. The procedural aspects regarding the testing in real driving conditions are described in the same regulation indicated above, among the most important ones that can be mentioned are: test duration between 90 minutes and 120 minutes, defining three vehicle speed zones specific to urban, rural and motorway driving, the distance traveled by the vehicle for each driving zone greater than 16 km, 29 - 44% of urban share and 23 - 43% of rural and motorway shares, urban speed zone is characterized by vehicle speeds lower than or equal to 60 km/h, rural speed zone is characterized by vehicle speeds higher than 60 km/h and lower than or equal to 90 km and the speed range of the motorway driving shall properly cover a range between 90 and at least 110 km/h. Also, the vehicle speed shall normally not exceed 145 km/h. This maximum speed may be exceeded by a tolerance of 15 km/h for not more than 3 % of the time duration of the motorway operation. Thus, can be seen that the test in real driving conditions is not very restrictive but covers the entire operating range of a vehicle, and the behavior of the vehicle in relation to the real operating conditions can be easily highlighted being able to draw clear conclusions regarding the proposed objectives.

## 2. Experimental

To carry out the experimental research, a medium-class car with 5 seats equipped with an electric traction motor and reducer was tested. The research was carried out with the vehicle running on the streets in real driving conditions. Before carrying out the tests, the itinerary to be followed was established, which would meet the specific requirements of the test in real driving conditions. To validate the experimental results, repeated tests were carried out, on different days and in different traffic conditions, thus removing possible erroneous assessments regarding the evolution of the state of charge of the traction battery.

The most important technical characteristics specific to the vehicle under research are:

- Maximum speed: 140 km/h;
- Own weight: 1500 kg;
- Front traction;
- Maximum power: 65 kW;
- Maximum torque: 200 Nm;
- Traction battery technology: Li-Ion, 400V.
- Traction battery power: 22 kWh;
- Type of charging applied to the traction battery: single-phase slow charging, power 2.3 kW/h.

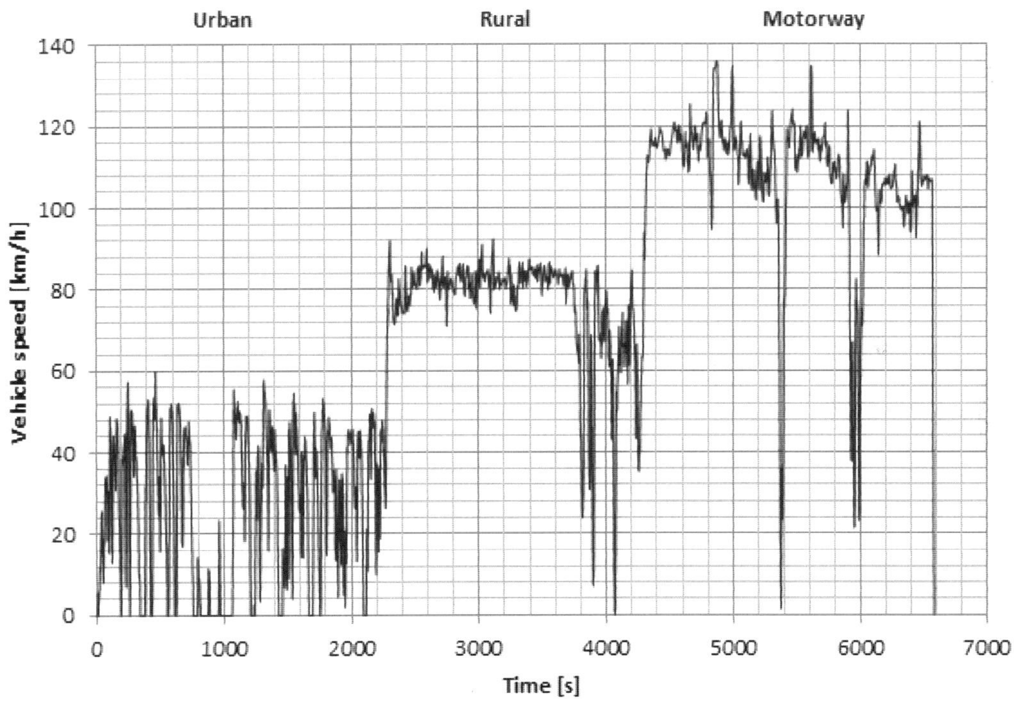


**Figure 2.** Vehicle speed profile obtained after the testing in real driving conditions.

As previously specified, the electric car was subjected to repeated tests to ensure the similarity of the results, and one of the representative tests and its results was selected to be presented in this paper. The speed profile obtained as a result of the testing in real driving conditions is shown in figure 2, where can be observed the limits of the three driving zones traveled by the car, urban driving, rural driving and motorway driving.

Regarding the specific speed of each driving zone, it can be observed that the vehicle speed is within the limits specified in the test procedure. Also on the graph, in the case of the three driving zones, there are moments in which the speed of the car is 0 km/h, the stopping being necessary because of the situations encountered in traffic, these being even an important feature of the testing in real driving conditions.

**Table 1.** The characteristics of the testing

| | Rural | Urban | Motorway | Trip |
|---|---|---|---|---|
| **Duration (s)** | 2276 | 2022 | 2269 | 6567 |
| **Distance (km)** | 30.97 | 39.31 | 68.11 | 138.39 |
| **Average velocity (km/h)** | 48.98 | 69.99 | 108.06 | 75.86 |

In table 1, are described the characteristics obtained after testing and it can be observed that the total duration of the testing is approximately 110 minutes. In the present case, because the research was done on a car equipped with an electric power source and the objective being the analysis regarding the evolution of the state of charge of the traction battery, the percentages allocated to each driving zone were not important, a special attention being attributed to speed compliance assigned to the testing zone and respectively to the comparable test time for any of the three zones.

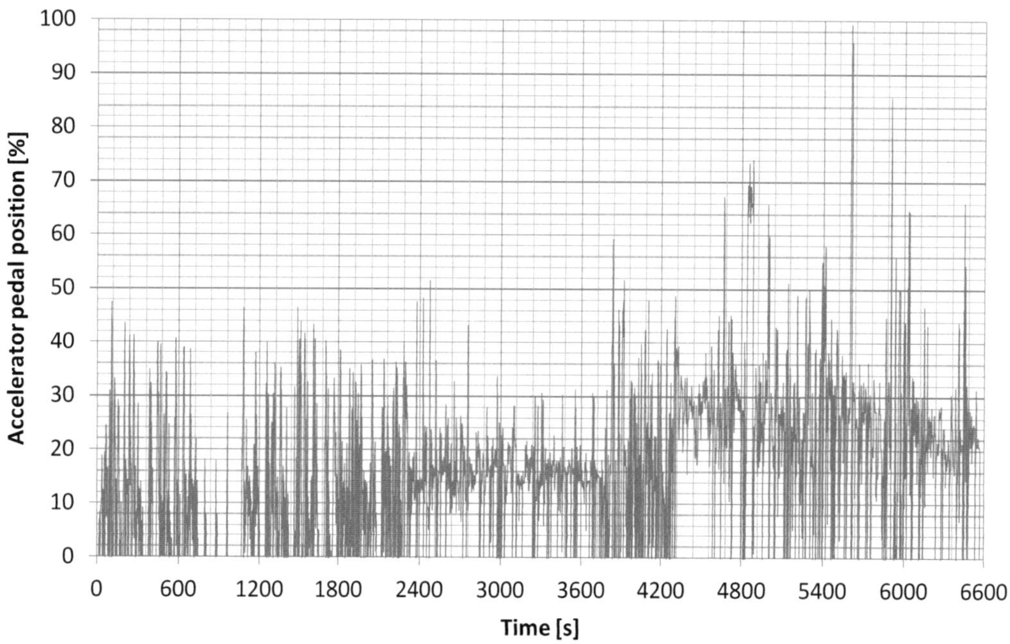


**Figure 3.** The variation of the accelerator pedal position during testing

The dynamics of the testing is highlighted based on the variation of the accelerator pedal position shown in figure 3. Can be observed that, in the case of all driving zones the position of the accelerator pedal shows permanent variations that are explainable and specific to real driving conditions. It is also observed that in the case of cars equipped with the electric energy source, the reaction to moderate variations of the accelerator pedal is immediate and the car speed required in relation to the demand is ensured quickly.

There are very few zones on the graph where the accelerator pedal was pressed between 60% and 100% for the vehicle to meet the speed request and these are found in the motorway driving zone.

## 3. Results and discussions

To highlight the research results, were created graphs regarding the evolution of the state of charge of the traction battery for each driving zone specific to the test in real driving conditions. Thus, using this information, it is possible to precisely analyze the way in which the state of charge of the traction battery decreases and, respectively, the dependence between it and the dynamic test conditions of the car.

To correctly analyse the decrease mode of the state of charge of the traction battery and respectively the zones with the best economic performance characterized by low consumption of electrical energy, over the evolution of the state of charge was integrated the allure of the speed achieved by the car to fulfil the real test conditions.

Obviously, the accuracy of the results and their interpretation was possible by realizing corroboration between the values indicated in the graphic form and the data acquired in real time during the tests from the main electronic control units of the electric traction system.

The evolution of the state of charge of the traction battery in the case of urban driving is described in figure 4. The analysis of the speed profile realised by the car highlights the fact that, in the case of urban driving the maximum speed is less than 60 km/h and there are multiple moments when the car stops. Also, a peculiarity resulting from urban driving is the multiple movements from the place and sudden accelerations followed by long decelerations to comply with the real driving conditions caused by traffic.

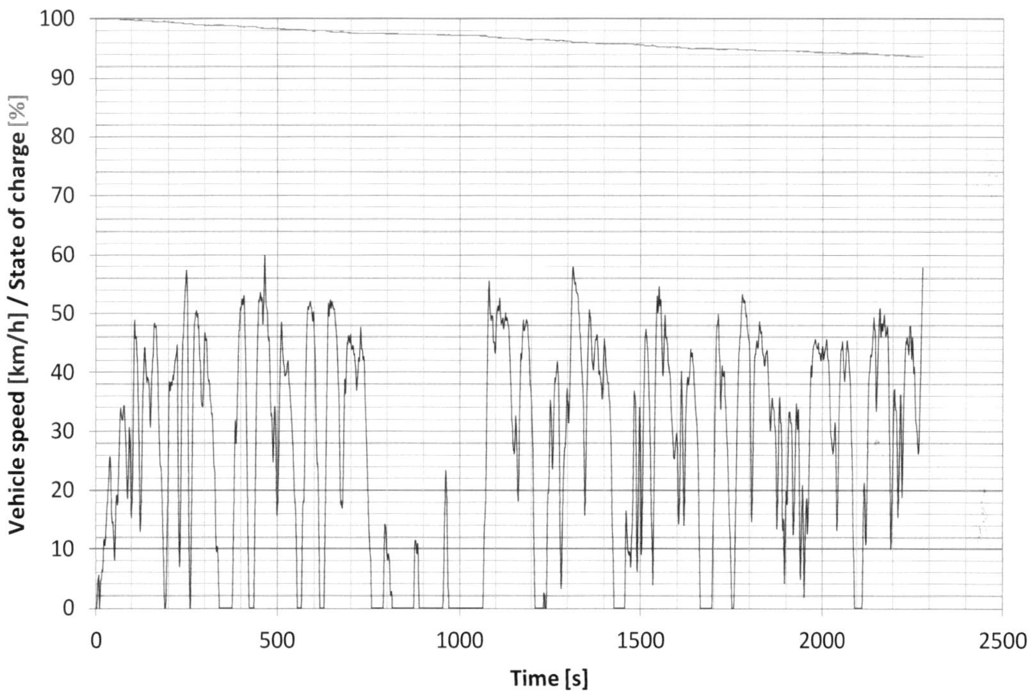


**Figure 4.** Evolution of the state of charge of the traction battery in the case of urban driving

Regarding the evolution of the state of charge of the traction battery after a distance traveled of about 31 km in a time of about 38 minutes, can be seen that it has a very smooth but practically insignificant downward trend.

This is explained by the fact that the dynamic demands in the case of urban driving are moderate, as well as by the fact that in the urban zone we find sufficient deceleration and recuperative braking regimes that ultimately allow the recovery of electrical energy, this being reintroduced into the traction battery.

So, with reduced consumption of electrical energy and multiple possibilities for its recovery, the urban driving zone indicates that the use of electric vehicles in these conditions is very useful.

Figure 5 shows the evolution of the state of charge of the traction battery in the case of rural driving. In the rural zone, the speed profile is characterized by a variation between 60 km/h and 90 km/h, but obviously due to the real road and traffic conditions, certain moments can be encountered when the car speed reaches 0 km/h.

Also, in the case of rural driving, sudden accelerations and decelerations can be encountered, generated because of road and traffic conditions, which, obviously, can have a beneficial role in energy recovery so necessary in such conditions.

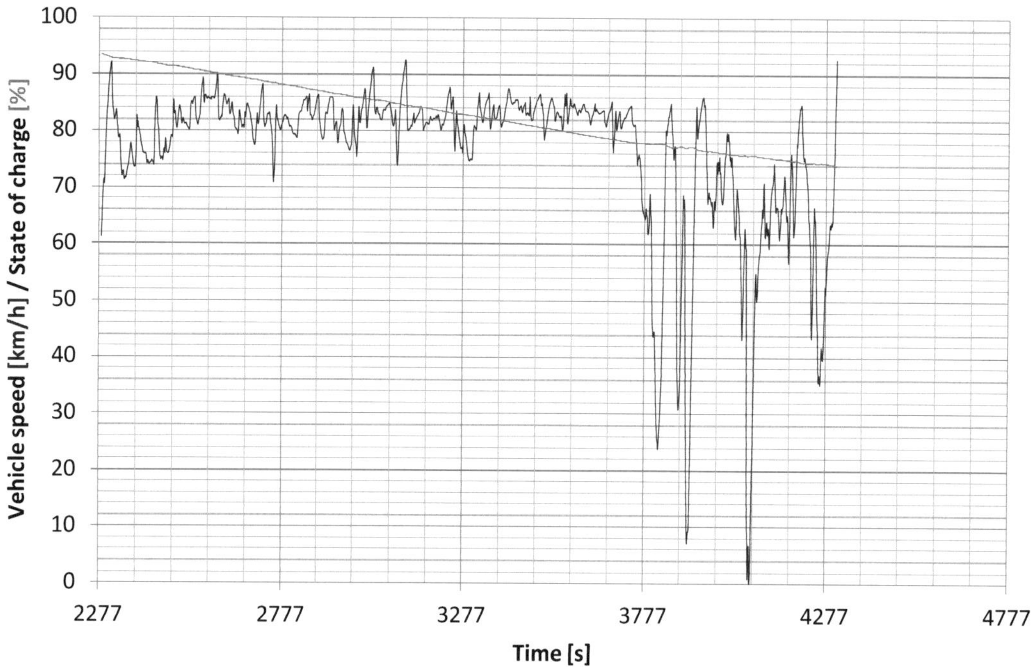


**Figure 5.** Evolution of the state of charge of the traction battery in the case of rural driving

The analysis of the evolution of the state of charge of the traction battery after a distance traveled of approximately 40 km in 34 minutes, indicates that even in these operating conditions the tendency to decrease the state of charge of the traction battery is not a fast and significant one.

It is true, because of the increase in travel speed and the reduced number of moments of deceleration practically without multiple possibilities of energy recovery, an accelerated trend can be noted, but the behavior of the traction battery is very good compared to the driving conditions.

The evolution of the state of charge of the traction battery in the case of driving on the motorway is described in figure 6. In driving conditions on the motorway, the speed profile achieved by the car is between 90 km/h and 135 km/h.

Even in these conditions of operation at high speed, moments can be encountered with deceleration regimes that can obviously favor recovery but considering the high consumption of electrical energy because of the practical speed requirement, the contribution of these deceleration regimes is insignificant. Also, there are moments when the speed of the car can be low or close to 0 km/h also because of the road and traffic conditions.

Regarding the reduction of the state of charge as a result of the distance traveled of approximately 68 km in 38 minutes, the graphic analysis shows clearly the influence of the travel speed on the state of charge of the traction battery.

Thus, under these operating conditions, the state of charge decreases rapidly, it being highlighted that the consumption of electrical energy in conditions of running at high speed is very significant.

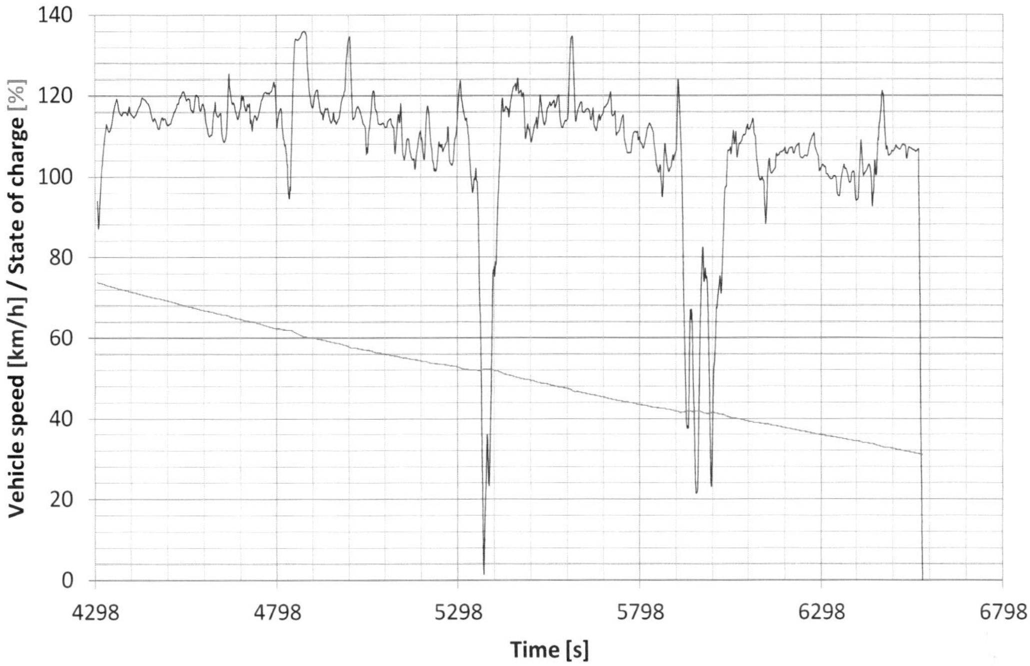


**Figure 6**. Evolution of the state of charge of the traction battery in the case of motorway driving

As a result of the graphic analysis of the three driving zones in real driving conditions, it is evident that the state of charge of the traction battery is maintained very well in the case of urban and rural driving as a result of the reduced consumption of electricity and the multiple moments available for energy recovery, but it is clear that using the car at high speed on the motorway has a major impact on maintaining the state of charge of the traction battery, in such conditions it quickly depletes.

## 4. Conclusions

As a result of the test of the middle-class car in real driving conditions, it was possible to highlight the way in which the state of charge of the traction battery decreases in relation to the specific conditions of the test. In the case of urban driving, the behavior of the traction battery is very good. Even if urban driving involves many starts from a standstill and sudden accelerations, these are practically supplemented by multiple moments of recuperative braking that help maintain the state of charge of the traction battery. Also, the low speed of up to 60 km/h with which the vehicle moves in urban conditions is a benefit because in such situations the consumption of electrical energy is reduced.

Regarding the use of the car in the rural zone, the decrease in the state of charge of the traction battery has a behavior almost like that of urban driving. Although in rural conditions the speed of the vehicle has significant values between 60 km/h and 90 km/h, not having enough moments of recuperative braking, the mode of reducing the state of charge of the traction battery is smooth as a result of the moderate consumption of electrical energy. Another positive contribution to maintaining the state of charge of the traction battery is the speed applied to the car in rural conditions, which is mostly constant without charge variations on the electric traction motor.

Driving the car in motorway conditions, with speed variations between 90 km/h and 135 km/h, has a negative impact on the state of charge of the traction battery. In such test conditions, the reduction of the state of charge has a rapid evolution as a result of the high dynamic demands generated by the high travel speed, the consumption of electrical energy directly dependent on this, but also the non-existence of the possibilities of energy recovery which in the other conditions of running could have an important positive contribution.

Considering the above, it can be concluded that the cars equipped with the electric energy source have a very good efficiency when driving in urban and rural conditions. In the case of these types of driving, the behavior of the traction batteries characterized by the state of charge is very good. Even if the traction battery behaves satisfactorily for motorway driving conditions, current electric cars clearly represent an alternative to those equipped with thermal engines, the distance covered by them after a single charge of the traction battery being dependent on the power of the electric motor and of the traction battery, basically the lower the power of the electric motor and the higher the power of the battery, the better the autonomy will be.

**Author's ORCID ID**
C Rențea: https://orcid.org/0009-0007-8536-8638
M F Toma: https://orcid.org/0000-0002-8588-1964

**References**

[1] Buchmann I 2001 *Batteries in a Portable World*, 2nd end. Cadex Electronics Inc. ISBN: 978-0968211823
[2] Paul D and Stănescu O 2015 *Electric, hybrid and fuel cell cars*, Politehnica, Timișoara, ISBN: 978-606-35-0011-4
[3] Jeong N T and Yang S M 2016 *Urban driving cycle for performance evaluation of electric vehicles.* Int. J. Automot. Technol 17(1), 145-151
[4] Bamban V et al 2015 *Analysis of range extender electric vehicle performance using vehicle simulator*. Energy Procedia 68, 409-418
[5] Shi Q et al 2011 *The study of a new method of driving cycles construction*. Procedia Eng. 16,79-87
[6] Ababei S and Culea G 2003 *Electrical measurements and data acquisition*, Tehnica-Info, ISBN 9975-63-214-9
[7] Barlow T J et al 2009 *A reference book of driving cycles for use in the measurement of road vehicle*. Published Project Report PPR354, TRL Limited
[8] EUR-Lex Homepage, *http://data.europa.eu/eli/reg/2017/1151/oj*, last accessed 2023/06/20

# Considerations on reducing emissions by increasing the use of electric vehicles in urban area

**A Vilcan[1], V Nicolae[2], I Martin[1] and M Istrate[1]**

[1] AV TRANSPORT PLANNING SRL, Ionel Perlea 11, Apt. 15, S1, Bucharest, Romania
[2] National University of Science and Technology *Politehnica* Bucharest – Pitesti University Center

E-mail: viorel.nicolae@upit.ro

**Abstract.** The pace between city development, increasing number of working places and emissions control is a challenge nowadays in urban environment. The gradual increase of electric cars ownership is essential for reducing the emissions level in urban areas. The city policy is of high importance for encouraging the use of electric cars, especially for trip purposes to work and business. The paper deals with the effects of increasing the fleet of electric vehicles on the emissions level in urban areas. The effect of considering a certain share of electric car trips to work and business is analyzed. Considering this, the concept of "green city corridors" can be introduced and considered as a priority for the transport policy in urban areas.

## 1. General considerations

This paper is based on the estimation of the impact of electric vehicles use, with the help of the Pitesti Multi-modal Urban Transport Model. For this it is foreseen that 0.5% of the total car trips are carried out with electric cars. Based on this consideration, and by applying the Pitesti transport model for the MA and PM peak hours, reduction of the CO2 emissions is estimated for year 2030.

Thus, results are considered for both AM and PM peak hours, for time horizon 2030.

## 2. Electric vehicles in Romania and in Argeș county

Based on the official information, the registered road vehicles in Romania and in Argeș county – Pitesti is the capital city of the county, at 31.12.2022 are as shown in the table below.

The total share of electric cars in total car fleet is as follows:

- 0.314% at national level and 0.193% in Argeș county.

**Table 1.** Registered road vehicles in Romania and in Argeş county – Pitesti is the capital city of the county, at 31.12.2022.

| Category | No | Age (years) | | | | | |
|---|---|---|---|---|---|---|---|
| | | 0--2 | 3--5 | 6--10 | 11--15 | 16--20 | >20 |
| **National level** | | | | | | | |
| **Cars** | 7865186 | 246401 | 447330 | 808640 | 1543357 | 2384002 | 2435456 |
| **Argeş county** | | | | | | | |
| **Cars** | 247126 | 6796 | 13253 | 28028 | 55663 | 76895 | 66491 |

| Category | Fuel | | | | Age classes | | | |
|---|---|---|---|---|---|---|---|---|
| | Gasoline | Diesel | Electric | Hybrid | 0--4 | 5--8 | 9--12 | Over 12 |
| **National level** | | | | | | | | |
| **Cars** | 3711866 | 3946451 | 24691 | 127331 | 535924 | 641583 | 799620 | 5888059 |
| **Argeş county** | | | | | | | | |
| **Cars** | 132598 | 106679 | 476 | 2238 | 15646 | 20296 | 30086 | 181098 |

## 3. Basic forecasts 2030

The basic 2030 forecasts are based on the foreseen socio-economic pattern in the urban area, thus population, workplaces, employment, car ownership and activities. The figures below show the Volume/Capacity ratio and LOS – Level of Service in junctions for AM and PM peak hours, in a normal working day, without the implementation of the TMS.

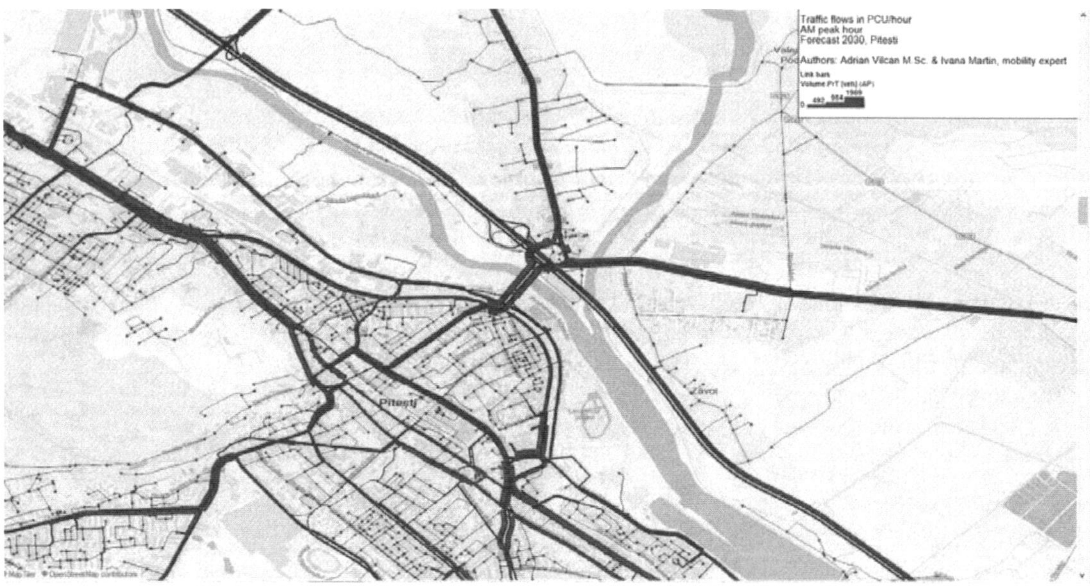


**Figure 1**. Traffic flows in PCU/hour, AM peak, Pitesti, forecast 2030

IOP Conf. Series: Materials Science and Engineering 1303 (2024) 012002 doi:10.1088/1757-899X/1303/1/012002

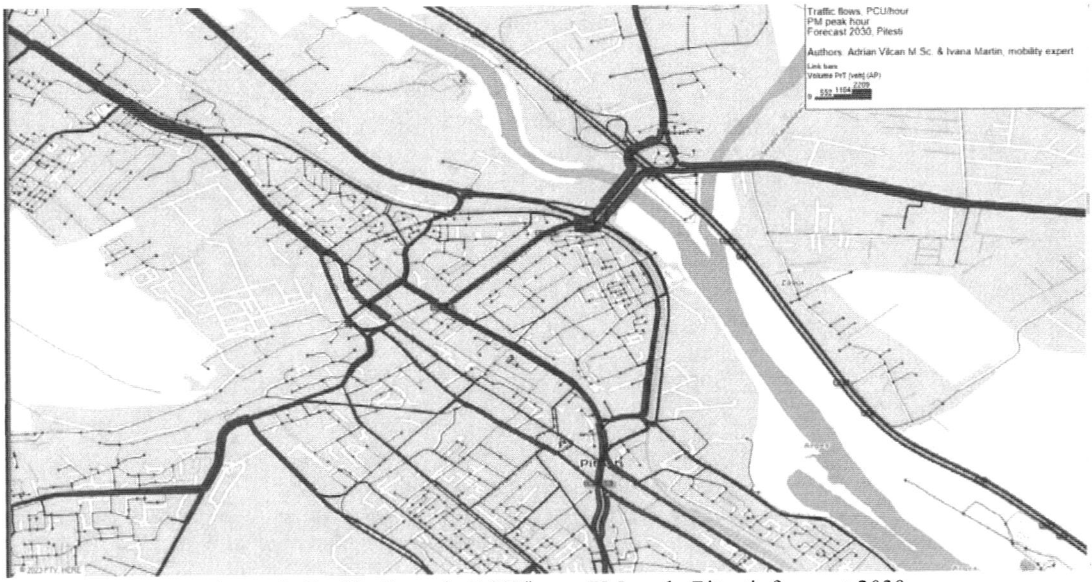


**Figure 2**. Traffic flows in PCU/hour, PM peak, Pitesti, forecast 2030

Based on the above estimations, the emissions are calculated for the following scenarios:

- Scenario A: all cars are considered as conventional fuel cars;
- Scenario B: 0.193% of cars are considered electric cars, and the rest as conventional fuel cars.

Results are shown for both AM and PM peak in the following tables.

**Table 2.** Emissions 2030 AM Scenario A

| | | | | | | | | | | |
|---|---|---|---|---|---|---|---|---|---|---|
| | | | | | | **AutoB** | **AutoM** | **OGV1** | **OGV2** | **PSV** |
| | | **Gas** | **Factor** | | $CO_2$ | 7308 | 3576 | 294 | 4359 | 626 |
| | | $CO_2$ | 1 | | $N_2O$ | 1 | 0 | 0 | 0 | 0 |
| | | $N_2O$ | 298 | | $CH_4$ | 3 | 0 | 0 | 0 | 0 |
| | | $CH_4$ | 23 | | | | | | | |
| | | | | | **tonnes per year at AM** | | | | | |
| | | | | | $CO_2$ech | **16,623.97** | | | | |
| | | | | | HGV | 4654 | | | | |
| **Turns** | | **Gas** | **Factor** | **tonnes/year** | | | | | | |
| Veh-h | 486.96 | $CO_2$ | 1 | 479.90 | | | | | | |
| l/h | 1.2 | $N_2O$ | 298 | 16.53 | | | | | | |
| Liter fuel | 584.35 | $CH_4$ | 23 | 0.17 | | | | | | |
| | | | | **496.59** | | | | | | |
| | | | | | $CO_2$ech | **17,120.57** | | | | |

IOP Conf. Series: Materials Science and Engineering 1303 (2024) 012002 doi:10.1088/1757-899X/1303/1/012002

**Table 3.** Emissions 2030 AM Scenario B

| | | | | | | | | | | |
|---|---|---|---|---|---|---|---|---|---|---|
| | | | | | | **AutoB** | **AutoM** | **OGV1** | **OGV2** | **PSV** |
| | | **Gas** | **Factor** | | **$CO_2$** | 7293 | 3569 | 294 | 4359 | 626 |
| | | $CO_2$ | 1 | | **N2O** | 1 | 0 | 0 | 0 | 0 |
| | | $N_2O$ | 298 | | **CH4** | 3 | 0 | 0 | 0 | 0 |
| | | $CH_4$ | 23 | | | | | | | |
| | | | | | **tonnes per year at AM** | | | | | |
| | | | | | **$CO_2$ech** | **16,602.25** | | | | |
| | | | | | HGV | 4654 | | | | |
| **Turns** | | **Gas** | **Factor** | **tonnes/year** | | | | | | |
| Veh-h | 486.0162 | $CO_2$ | 1 | 478.97 | | | | | | |
| l/h | 1.2 | $N_2O$ | 298 | 16.49 | | | | | | |
| Liters fuel | 583.2194 | $CH_4$ | 23 | 0.17 | | | | | | |
| | | | | **495.63** | | | | | | |
| | | | | | **$CO_2$ech** | **17,097.88** | | | | |

Thus, the reduction of $CO_2$ emissions is of 22.69 tons per year at AM peak (0.133% of the total).

Emissions at PM peak hour are estimated as presented below.

**Table 4.** Emissions 2030 PM Scenario A

| | | | | | | | | | | |
|---|---|---|---|---|---|---|---|---|---|---|
| | | | | | | **AutoB** | **AutoM** | **OGV1** | **OGV2** | **PSV** |
| | | **Gas** | **Factor** | | **$CO_2$** | 7753 | 3755 | 365 | 5449 | 656 |
| | | $CO_2$ | 1 | | **$N_2O$** | 1 | 0 | 0 | 0 | 0 |
| | | $N_2O$ | 298 | | **$CH_4$** | 3 | 0 | 0 | 0 | 0 |
| | | $CH_4$ | 23 | | | | | | | |
| | | | | | **tonnes per year at AM** | | | | | |
| | | | | | **$CO_2$ ech** | **18,481.56** | | | | |
| **Turns** | | **Gas** | **Factor** | **tonnes/year** | | | | | | |
| Veh-h | 574.677653 | $CO_2$ | 1 | 566.34 | | | | | | |
| l/h | 1.2 | $N_2O$ | 298 | 19.50 | | | | | | |
| Liters fuel | 689.613184 | $CH_4$ | 23 | 0.20 | | | | | | |
| | | | | **586.05** | | | | | | |
| | | | | | **$CO_2$ ech** | **19,067.61** | | | | |

**Table 5.** Emissions 2030 PM Scenario B

| | | | | | | | | | | |
|---|---|---|---|---|---|---|---|---|---|---|
| | | **Gas** | **Factor** | | **$CO_2$** | 7738 | 3748 | 365 | 5449 | 656 |
| | | $CO_2$ | 1 | | **$N_2O$** | 1 | 0 | 0 | 0 | 0 |
| | | $N_2O$ | 298 | | **$CH_4$** | 3 | 0 | 0 | 0 | 0 |
| | | $CH_4$ | 23 | | | | | | | |
| | | | | | **tonnes per year at AM** | | | | | |
| | | | | | **$CO_2$ech** | **18,458.59** | | | | |
| | | | | | | | | | | |
| **Turns** | | **Gas** | **Factor** | **tonnes/year** | | | | | | |
| Veh-h | 573.5685 | $CO_2$ | 1 | 565.25 | | | | | | |
| l/h | 1.2 | $N_2O$ | 298 | 19.46 | | | | | | |
| Liters fuel | 688.2822 | $CH_4$ | 23 | 0.20 | | | | | | |
| | | | | **584.92** | | | | | | |
| | | | | | **$CO_2$ech** | **19,043.51** | | | | |

Thus, the reduction of $CO_2$ emissions is of 24.10 tons per year at PM peak (0.126% of the total).

At both peak hours reduction of $CO_2$ emissions is 46.79 $CO_2$ each tons/year.

The reduction might be low, but we need to keep in mind that the share of the EV in the total fleet is very low in the present: 0.193%.

## 4. Main conclusions

It is estimated that for EV share of 0.193%, the reduction of $CO_2$ emissions is 46.79 tons per year. If we extrapolate this result at a full day, the estimation is that the total yearly reduction of $CO_2$ emissions would be 234 tons per year.

The share of EV will increase in the coming years, as shown by global trends and the trends at EU level and enhancing measures to promote the acquisition of electric cars would be useful to be considered at the local level.

Moreover, from the mobility demand data it can be identified what segments of population are performing more than one trip per peak hour, to encourage the acquisition and use of the EV for these categories of users, in order to get a higher impact on the reduction of $CO_2$ emissions.

However, combining these measures with promoting the use of Hybrid cars and public transport vehicles, as well as of the electric freight vehicles for delivery, would contribute to improving the urban environment overall.

**Author's ORCID ID**

V Nicolae: https://orcid.org/0000-0002-2536-3372

**References**

[1] AV Transport Planning SRL / PANTEIA bv /Synergetics Corporation SRL, 2017, *SUMP – Sustainable Mobility Plan Pitesti*, city of Pitesti

[2] Vilcan A 2018 AV Transport Planning SRL, *Traffic study for the implementation of the Traffic Management System in Pitesti city*

[3] Institute of Transportation Engineers, 2000, *Highway Capacity Manual,* ISBN 0-309-06681-6

[4] Institute of Transportation Engineers, 1999, *Transportation Planning Handbook*, ISBN: 0-935403-33-7, Publication No. TB-011A, 1000/TA/0402

# Research concerning the autonomy of the electric vehicles, simulated and measured, in the case of driving at the low and the medium speed, specific to the WLTC test.

**C Minzatu, V Nişulescu, M F Toma, C Renţea and Gh Bancă**

Politehnica University of Bucharest, Romania

E-mail: corneliu.minzatu@renault.com

**Abstract.** The most important attributes expected from today's electric vehicles are the electricity consumption and the autonomy, which must be able to unequivocally replace the results achieved by the vehicles equipped with an internal combustion engine. These expectations come equally from the car manufacturers and from the end users. Based on these premises, in a first part of the paper, are presented the theoretical aspects concerning the test procedure for determining the autonomy of electric vehicles, applicable at the European level, with the highlighting of the low and medium driving zones, specific to the WLTC test. Also, are indicated theoretical details regarding the model of the electric vehicle tested and simulated, using the AMESIM modelling and simulation platform. In the second part are presented the comparative results obtained by measuring and simulating regarding the electric autonomy, respectively the influence that the driving at the low and the medium speed has, upon of the state of charge depletion of the traction battery.

## 1. Introduction

The electric autovehicles have become becoming more popular in more recent years, numerous governments around of the world are committing to decrease carbon emissions by switching away of fossil fuels. One of the most critical aspects of the electric vehicle's performance is the autonomy, which is affected by several factors. In the last few years, the simulations have become a popular tool for the prediction of the electric vehicle autonomy, their accuracy has often been doubted. In contrast, the testing under the real conditions has been presented as being more accurate, but it can be expensive and time-consuming.

This paper proposes to compare the simulated and the measured performance of the electric vehicle autonomy at the low and the medium speeds, with a special focus on the Worldwide Harmonized Light Vehicles Test Cycle (WLTC). In this way, we hope to identify the essential differences between the two approaches and to provide information for improving the predictions of autonomy of the electric autovehicles.

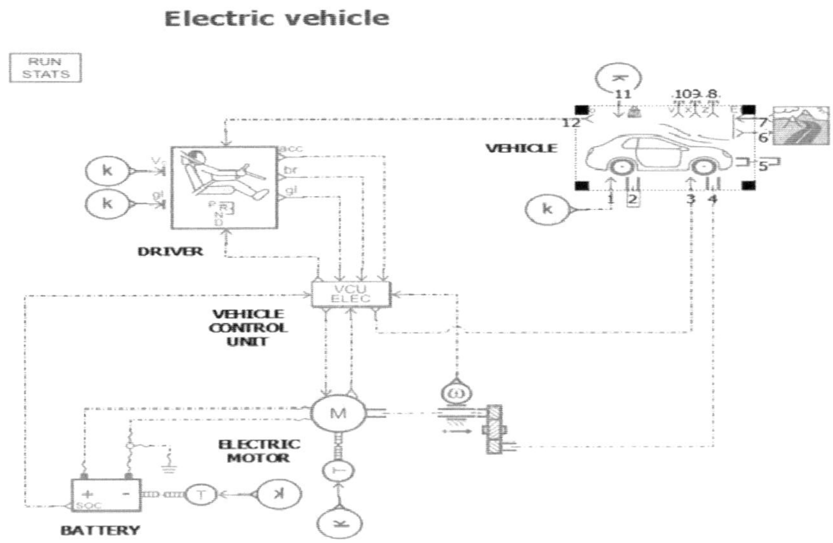


**Figure 1.** Example of an electric vehicle simulation model using the AMESIM modelling and simulation platform.

Regarding the applicable legislation for the test for determining the electricity consumption and European-level autonomy, it is applicable to European Commission Regulation 2017 / 1151 of 1st of June of 2017 completing the Regulation of the European Parliament and of the Council 715 / 2007 on the type-approval of the vehicles. The speed profile which the electric vehicle under the test must follow is that specified by the WLTC, described in the Figure 2[8].

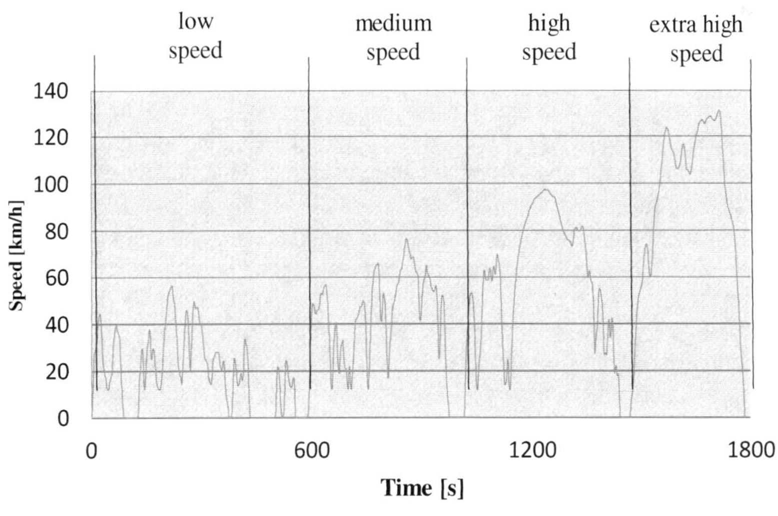


**Figure 2.** Worldwide Harmonized Light Vehicles Test Cycle

A quick analysis of the WLTC cycle speed profile shows that it is distinguished by an enhanced dynamicity over previously applicable test cycles and has four different speed zones containing a different percentage of speed type, i.e., the low range 600 seconds, medium range almost 400 seconds, high range almost 500 seconds and very high range almost 300 seconds. The main characteristics of the cycle include: driving time 30 minutes, route length about 23 kilometres, maximum speed 131 km/h, average speed about 47 km/h, stationary percentage 13%, the consideration of the optional equipment respectively the changing points of the shift gears where there is applicable, which may be defined individually on the basis of clearly defined elements taking taking into account the geometrical characteristics and dynamic performance of the vehicles tested [1].

Another component that needs to be developed is the charging infrastructure for the traction batteries, which unfortunately, both at the European and the worldwide level, is not able to manage the amount of the manufactured autovehicles. With this in mind, the development for the electric autovehicles, respectively the increase of their number, is strictly dependent by the degree of the development of the infrastructures for the electric charging [2].

The autonomy has two main characteristics with which it is closely related, respectively the dependence of the traction battery power and the conditioning generated by the comfort systems which are a part of the respective vehicles [5]. Also, a characteristic of the traction battery with an effect on the autonomy is the health status which should be maintained at a high level for as long as possible [4].

WLTC introduces also three distinct testing categories depending on the mass/power ratio: Class I up to 22 kW/tonne, characterised by a maximum test speed of 70 km/h, Class II up to 34 kW/tonne, characterised by a maximum test speed of 90 km/h, and Class III from 35 kW/tonne, where the maximum test speed is approximately 131 km/h. At present, most vehicles meet the specific test conditions of Class III, where the power is higher than 35 kW/tonne. Thus, for each specific speed in the cycle, the average speed and maximum speed are defined [7].

WLTC focuses mainly on fuel and electricity consumption measurement and of the emissions from passenger cars and of the light commercial vehicles. It takes into consideration various parameters such as the vehicle mass, the aerodynamic resistance, and the rolling resistance to simulate the real-world driving conditions with more accuracy than the previous test cycle (NEDC).

However, it is important to note that the regulations and the testing standards may evolve over that time. It is possible that there may have been some updates or some developments in testing procedures since my last update. I recommend that you consult the latest guidelines or the last regulations of the relevant authorities or organizations for the latest information on WLTC and any specific test classes related to the power-weight ratio for all vehicles. In addition to the technical benefits conferred by the simulation platforms, the development costs are also significantly reduced, making the simulation a key component with fundamental implications for the industrial research and development [6].

## 2. Experimental

As mentioned above, one of the main goals of today's electric vehicles is electric range, i.e. the number of kilometres travelled on a single charge of the traction battery, which has certainly increased in recent years, but which does not fully meet all of society's needs.

To do this, we looked at how the charge level of the traction battery depletes according to the specific driving zones of the WLTC cycle, characterised by different speed regimes and different dynamic loads.

The experimental studies were carried out by driving the vehicle on the dynamometer at a speed corresponding to the load profile described by the WLTC cycle. Prior to the test, the traction battery was charged to 100% and, after charging, the vehicle was preconditioned and maintained at a controlled temperature for approximately 12 hours.

The vehicle tested is a medium-class vehicle whose general architecture is shown in the diagram below and whose technical specifications are as follows:

- Own weight: 1120 kg ;
- Inertia: 1245 kg ;
- Front wheel drive;
- Electric engine type: synchronous with wound rotor;
- Max power: 64 kW ;
- Maximum torque: 205 Nm ;
- Traction battery power: 21 kWh;
- Technology of traction battery: Lithium-Ion, 400V.

The following data were used for inertia simulation and mill stand load requirements:
- Power and load carried by the running stand at 80 km/h: 6.8 [kW] ;
- Rolling resistance coefficients: a = 6,9 [N]; b = 0,0459 [N/(km/h)2].

Information on compliance with test conditions, i.e. the required speed profile, was acquired at a frequency of 10 Hz [3]. This type of information validates the quality of the test and allows deviations from the speed profile of up to 2 km on uphill or downhill gradients.

Figure 1 shows the model of the electric vehicle made for the purpose of simulation, the visually identifiable elements being:
- an element that simulates the simulated vehicle,
- a component called "Driver" that simulates the driver's behaviour and controls the vehicle in the WLTC simulation,
- a component that simulates the electric motor mounted on the simulated vehicle
- an element simulating the electric vehicle electronic control unit .
- a component simulating the transmission
- a component simulating the traction battery.

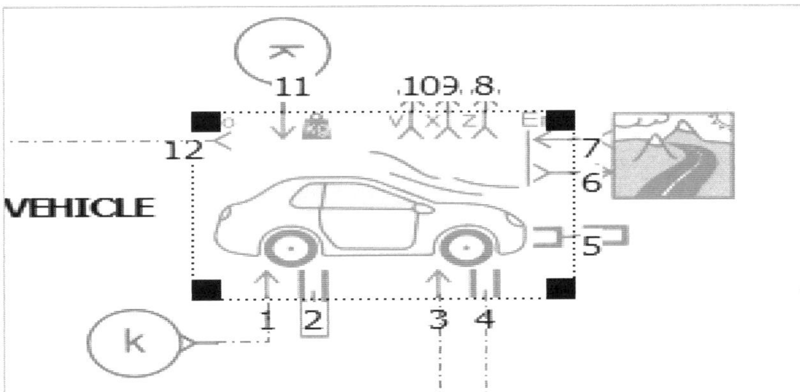

**Figure 3.** Electric vehicle component

The simulation model's main input quantities are the components listed above, but it also uses other sub-components to obtain information on inertia, torque, engine speed and temperatures, which form the basis of the internal calculation algorithms [9]. Figure 3 shows the vehicle component modeled in the simulation, the role of the ports present on the vehicle is as follows:
- ports 1 and 3, through which torque signals from front and rear axle braking are received;
- the ports 2 and 4, which receive drive torque information and return axle-specific angular velocity information;
- the port 5, is strictly mechanical through which the external forces applicable to the car are input and through which speed and acceleration information is returned;
- the ports 6 and 7, through which information is received on wind speed, as well as ramp and roll gradient, respectively.

Some of the relationships used by the model to establish the simulated routes are presented below:
- for the wheel radius:

$$R_R = \frac{1}{2} \cdot D_r + 0.01 \cdot H \cdot l \quad (1)$$

where $D_r$- the wheel rim diameter $[m]$ ; $H$ - the height of the tire [%]; $l$ - the width of the tyre $[m]$.

- for the total mass of the vehicle which also considers the effect of the wheel inertia in the linear motion:

$$m_v = m + 4 \cdot \frac{J_r}{R_{r2}} \cdot l \qquad (2)$$

Where $m$ - the total mass of the vehicle $[kg]$; $J_r$ - the wheel inertia $[kgm^2]$; $l$ - the wheel radius $[m]$.

- the driving on the roller stands, without the longitudinal slip of the tire, the motor force is determined with the following relation:

$$F_m = \frac{T_2 + T_4}{R_d} \qquad (3)$$

where $T_2$ și $T_4$ - the input torque in ports 2 and 4, specific to the front and rear axles $[Nm]$; $R_d$ - the dynamic radius of the wheel$[m]$.

The simulated resistive forces from the vehicle model are determined with the following relationships:

- the force generated by the resistance of the ramp $F_p[N]$

$$F_p = m \cdot g \cdot \sin(\arctan(0.01 \cdot \alpha)) \qquad (4)$$

where $m$ - the total mass of the vehicle $[kg]$; $g$ - the gravitational acceleration 9,81 $[m/s^2]$; $\alpha$ – the inclination of the rolling road [%].

- the force generated by the aerodynamic resistance $F_a[N]$

$$F_a = \frac{1}{2} \cdot \rho_a \cdot C_x \cdot S \cdot (v + v_v)^2 \qquad (5)$$

where $\rho_a$ - the air density defined in the sub-model $[kg/m^3]$; $C_x$ - the aerodynamic coefficient [ ]; $S$ - the section of the vehicle $[m^2]$; $v$ - the speed of the vehicle $[m/s]$; $v_v$ - the wind speed $[m/s]$.

- the force generated by the rolling resistance $F_{rul}[N]$

$$F_{rul} = m \cdot g \cdot (f + k \cdot v + wind \cdot v^2) \qquad (6)$$

where $f, k$ *și* $wind$ are defined as the constant values.

## 3. Results and discussions

Results of the experimental research, it was possible to highlight the graphs with the level of the charge of the traction battery specific to each speed zone included in the WLTC test. Also, based on the data acquisition with the specific equipment, it was possible to draw some conclusions on the way in which the traction battery energy is consumed and on the way in which the energy is recovered.

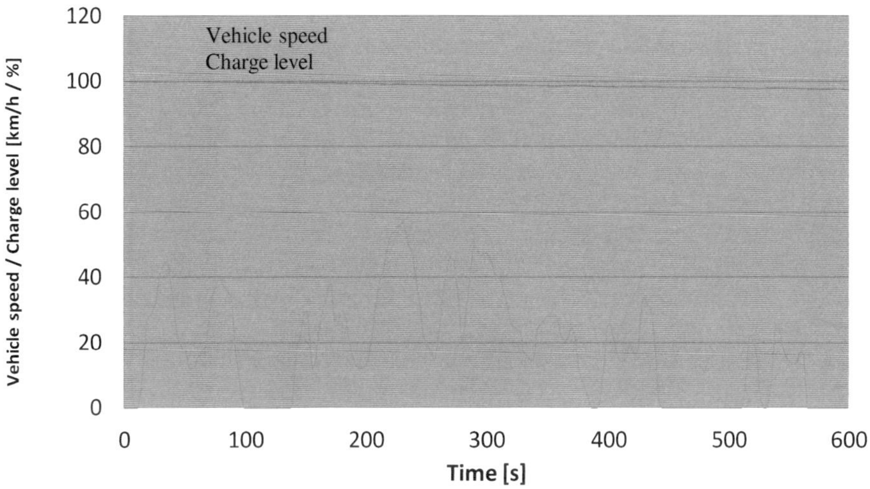


**Figure 4.** Traction battery charge level, low speed zone

The low speed driving zone is shown in the figure above and is characterized by a very low load level. As we can see, this type of driving simulates the specifics of a busy city, with an average speed of around 19.0 km/h and a top speed of around 55.6 km/h. In these situations, the traction battery functions very well, with low energy consumption and multiple regenerative braking, where obviously the impact on the traction battery's charge level and therefore on range is not very significant.

Experimental research indicates that these types of low-speed, low-load driving achieve the best performance in terms of traction battery range, which is also why certain types of electric vehicles are intended for use in heavy urban traffic. Also, the medium speed driving described in Figure 5 is not problematic in terms of range, as even in this situation we can speak of low loads and moderate speed, with an average speed of 38.6 km/h and a maximum speed of about 75.7 km/h.

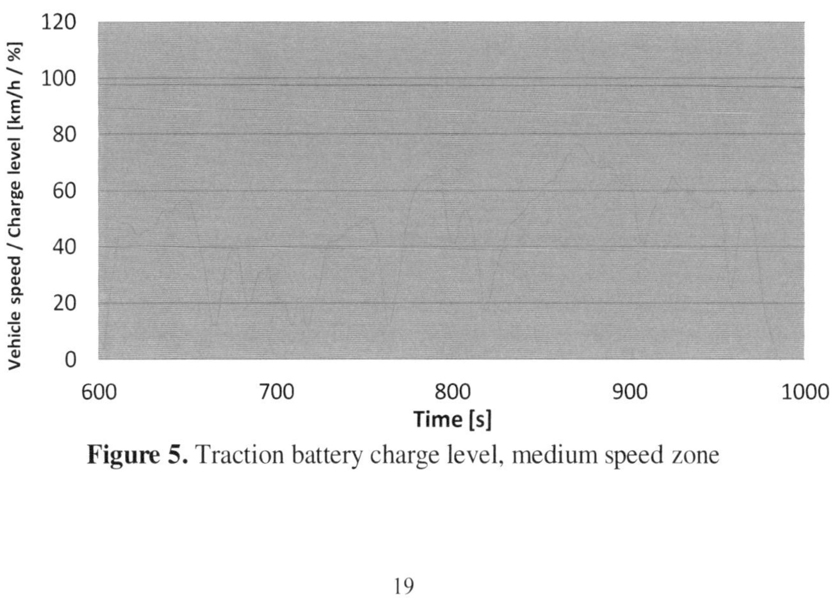


**Figure 5.** Traction battery charge level, medium speed zone

IOP Conf. Series: Materials Science and Engineering 1303 (2024) 012003 doi:10.1088/1757-899X/1303/1/012003

## 4. Conclusions

Experimental research indicates that the state of charge, i.e. the range of the traction batteries, is very satisfactory under low and medium speed driving conditions in urban areas with heavy road traffic. In addition, maximum performance in terms of recovery of electrical energy by regenerative braking is also achieved under the same conditions.

The life of a battery is also measured in the number of times that the battery is recharged, from empty to full, instead of the time or distance travelled by the vehicle. During the usage of the vehicle, one of the best options to prolong the life of a traction battery is to prevent any sudden speed changes. An advantage in this sense is the low-speed driving mode and the medium speed driving mode.

Instead of the sudden stop or braking to a stop in a short time, a better idea is to let the vehicle to run downhill or to a stop, of course as far as traffic conditions allow. Running the electric vehicle at low to medium speeds generates the kinetic energy, and this extra energy, as mentioned before, is redirected to the traction battery.

Also, a very high influence in the establishment of a very good behavior of the traction battery autonomy at low and medium driving speeds conditions is the driving mode chosen, lately, by all electric vehicles manufacturers, being available also the ECO driving mode, which gives a slow progressive acceleration.

The development of the market for vehicles equipped with electric propulsion systems is driving the need to increase the electrical power of traction batteries, which must fulfil new functions such as dynamics and economy to meet the most demanding needs of contemporary society.

**Author's ORCID ID**
M F Toma: https://orcid.org/0000-0002-8588-1964
C Rențea: https://orcid.org/0009-0007-8536-8638

**References**

[1] Jeong NT and Yang SM 2016 *Urban driving cycle for performance evaluation of electric vehicles*. Int. J. Automot. Technol 17(1), 145-151
[2] Buchmann I 2001 *Batteries in a Portable World*, 2nd end. Cadex Electronics Inc. ISBN 978-0968211823
[3] Ababei S and Culea G 2003 *Electrical measurements and data acquisition*, Tehnica-Info, ISBN 9975-63-214-9
[4] Barlow T J et al 2009 *A referance book of driving cycles for use in the measurement of road vehicle*. Published Project Report PPR354, TRL Limited
[5] Shi Q et al 2011*The study of a new method of driving cycles construction*. Procedia Eng. 16, 79-87
[6] Paul D and Stănescu O 2015 *Electric, hybrid and fuel cell cars*, Politehnica, Timișoara ISBN: 978-606-35-0011-4
[7] Bamban V et al 2015 *Analysis of range extender electric vehicle performance using vehicle simulator*, Energy Procedia 68, 409-418
[8] EUR-Lex Homepage, *http://data.europa.eu/eli/reg/2017/1151/oj,* last accessed 2023/05/31.
[9] *Help AMESim Rev12*

IOP Conf. Series: Materials Science and Engineering 1303 (2024) 012004 doi:10.1088/1757-899X/1303/1/012004

# Research concerning the autonomy of the electric vehicles, simulated and measured, in the case of driving at the high and the very high speed, specific to the WLTC test.

**C Minzatu, V Nișulescu, Gh Frățilă, C Stan and Gh Bancă**

Politehnica University of Bucharest, Romania

E-mail: corneliu.minzatu@renault.com

**Abstract**. In the current research, the autonomy and the energy consumption are the most important attributes expected from the electric road vehicles. Because the electric car manufacturing industry is in a continuous progress, this offers its customers a wide range of constructive variants through which their operational requirements can be met. It also has high expectations of the electric car manufacturers in the current energy context, to develop new high-performance models that can be considered as a real alternative to the current internal combustion engine solutions. So, the research part continues, and is a priority for all car manufacturers. Moreover, the computer simulation is also included as an intermediate stage before the final projects are validated. In the first part, there are detailed theoretical aspects of the electric vehicle to be tested and the test procedure applicable to it, specific to the WLTC cycle, for the high and very high-speed test area. Also, is described the vehicle model simulated using the AMESIM modelling platform. In the second part, are presented the experimental and simulated results performing a comparative analysis. Following the analysis, it has been possible to draw graphs showing the evolution of the state of charge and the mode of exhaustion of the autonomy, in the case of the two driving zones analyzed, respectively at high and very high speed.

## 1. Introduction

Vehicles equipped with an electric power source attract a great deal of interest today, so electric vehicle manufacturers are proposing various design variants that can meet even the highest expectations [1].

Another component that needs to be developed is the charging infrastructure for the traction batteries, which, unfortunately, both in Europe and worldwide, is not able to handle the number of vehicles produced. In this context, the development of electric vehicles, namely the increase in their number, is strictly dependent on the degree of development of the charging infrastructure [2].

Two performance criteria are perhaps among the most important expectations that users have of the electric vehicles, respectively the autonomy and the power consumption. The autonomy represented by the total number of the kilometers travelled after a single traction battery charge is different from one manufacturer to another and obviously lower or higher depending on the destination of the vehicle [3]. The autonomy has two main characteristics with which it is closely related, respectively the dependence on the traction battery power and the conditioning generated by the comfort systems that are part of the respective vehicles [5]. Also, a characteristic of the traction battery with an effect on the autonomy is the health status which should be maintained at a high level for as long as possible [4].

The simulation of the autonomy of the electric autovehicle being researched using the AMESim platform is based on the legislation applicable to the electric vehicles by the type-approval authorities, respectively European Commission Regulation 2017 / 1151 of 1st of June of 2017 supplementing European Parliament and Council Regulation No 715 / 2007 on type approval of vehicles [8].

Consequently, the speed profile to be met by the electric vehicle during simulation is that set by the World Harmonised Light Duty Cycle (WLTC),described in the figure 1.

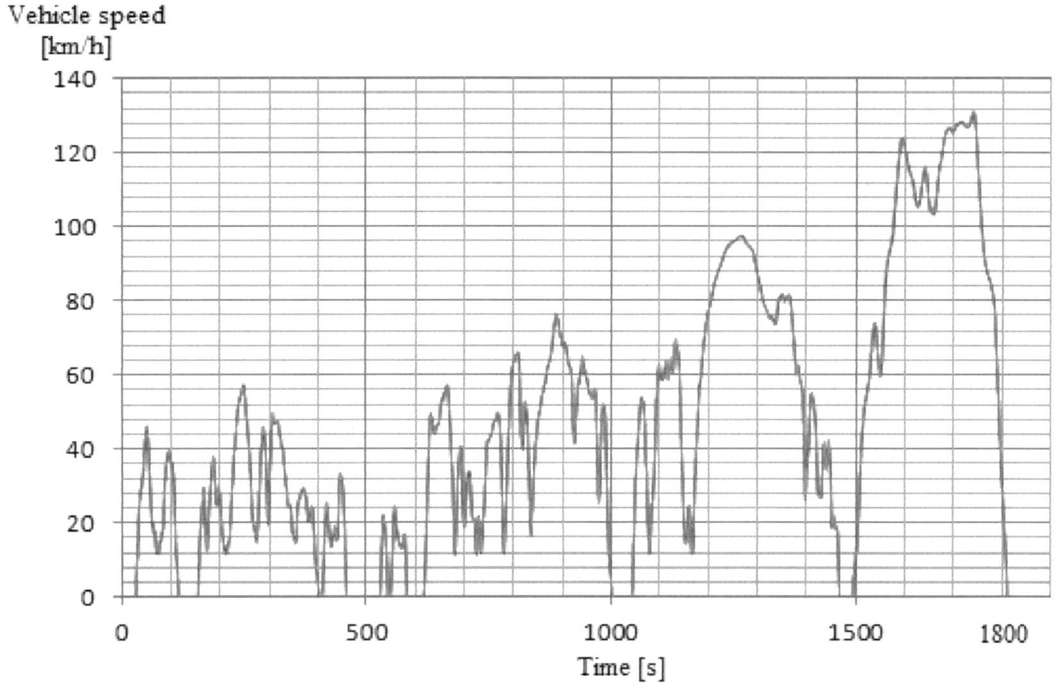


**Figure 1.** World Harmonised Light Duty Cycle (WLTC)

The speed profile for the WLTC cycle has four zones characterized by a different dynamicity and a different percentage of driving time, respectively a low speed of 600 seconds, a medium speed of 400 seconds, a high speed of 500 seconds and a very high speed of 300 seconds. It can therefore be seen that the vehicle is put through a wide range of the dynamic stresses in order to reproduce as faithfully as possible all the situations it may encounter in the exploitation. The driving time per cycle is 1800 seconds, the distance travelled approximately 24 kilometres, with a maximum speed of 130 km/h, an average speed of approximately 48 km/h and a stationary percentage of 14%.

The AMESim platform is used for the simulation, as it is one of the most favoured modelling and simulation platform by the vehicle manufacturers before the implementation phase. The AMESim platform has a wide range of simulation functions used by the vehicle manufacturers and other industrial fields such as machine building, shipbuilding, aviation, etc., and it can satisfy a very wide range of them in terms of the virtual and the calculation model design as well as the simulation of efficiency through the assisted tests.

The platform can be used to simulate the types of the vehicles, the engines, the mechanical and the automatic gearboxes, the axles, the differentials, the mechanical and the electrical components of the listed systems, the emission treatment systems, the injection systems, the cooling systems, the lubrication systems, etc. Another feature is the integration of the specific test methods defined by the legislative regulations on the type or the system approval, which really bring great advantages in the research phases prior to the development and the implementation of the components or the systems [7].

In addition to the technical benefits conferred by the simulation platforms, the development costs are also significantly reduced, making the simulation a key component with a fundamental implication in the industrial research and in the industrial development [6].

## 2. Experimental

The AMESim platform was used for modelling and simulation. The test was carried out on a mid-size vehicle equipped with an electric motor and gearbox. All data of the simulated and modelled vehicle are the same as the data of a real vehicle, which has been tested on a dynamometer to validate the simulation results.

The model used aims to determine, on the basis of mathematical algorithms, the mode of discharge of the state of charge of the traction battery and the electrical energy consumption according to the established relationships, which are implemented according to the operating conditions imposed by the WLTC, but in this study only the aspects linked to the electric vehicle's range are highlighted, energy consumption being the subject of another study.

The vehicle tested is a medium-class vehicle whose general architecture is shown in the diagram below and whose technical specifications are as follows:

- Own weight: 1120 kg ;
- Inertia: 1245 kg ;
- Front wheel drive;
- Electric engine type: synchronous with wound rotor;
- Max power: 64 kW ;
- Maximum torque: 205 Nm ;
- Traction battery power: 21 kWh;
- Technology of traction battery: Lithium-Ion, 400V.

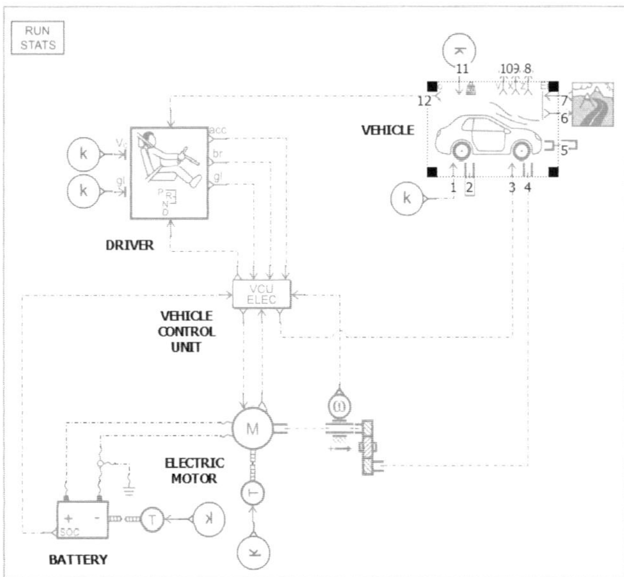


**Figure 2.** AMESim model for an electric vehicle

The following data were used for inertia simulation and mill stand load requirements:

- Power and load carried by the running stand at 80 km/h: 6.8 [kW] ;
- Rolling resistance coefficients: a = 6,9 [N]; b = 0,0459 [N/(km/h)2].

Figure 2 shows the model of the electric vehicle made for the simulation, the visually identifiable elements being:

- an element that simulates the simulated vehicle,

- a component called "Driver" that simulates the driver's behaviour and controls the vehicle in the WLTC simulation,
- a component that simulates the electric motor mounted on the simulated vehicle
- an element simulating the electric vehicle electronic control unit .
- a component simulating the transmission
- a component simulating the traction battery.

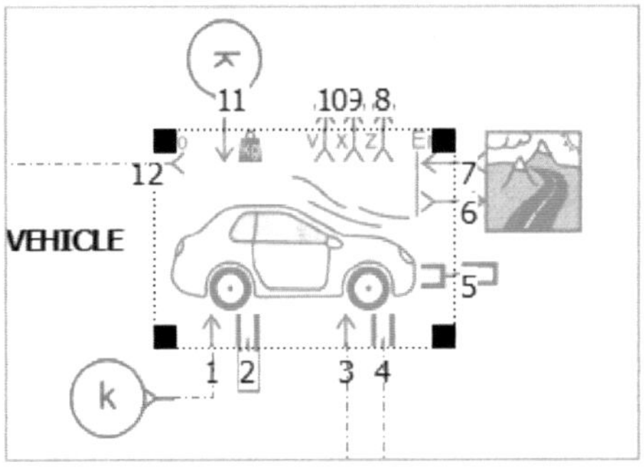


**Figure 3.** Electric vehicle component

The simulation model's main input quantities are the components listed above, but it also uses other sub-components to obtain information on inertia, torque, engine speed and temperatures, which form the basis of the internal calculation algorithms [9]. Figure 3 shows the vehicle component modeled in the simulation, the role of the ports present on the vehicle is as follows:

- The ports 1 and 3, through which are received the torque signals resulting from the braking of the front and for the rear axle;
- The ports 2 and 4, through these are received the informations about the drive torque and return the angular velocity information, specific to each axle;
- The port 5, is strictly mechanical, through which the external forces applicable to the vehicle are entered and through which information on the speed and the acceleration is returned;
- The ports 6 and 7, through which the informations are received about the wind speed, as well as respectively the ramp and the rolling slope.

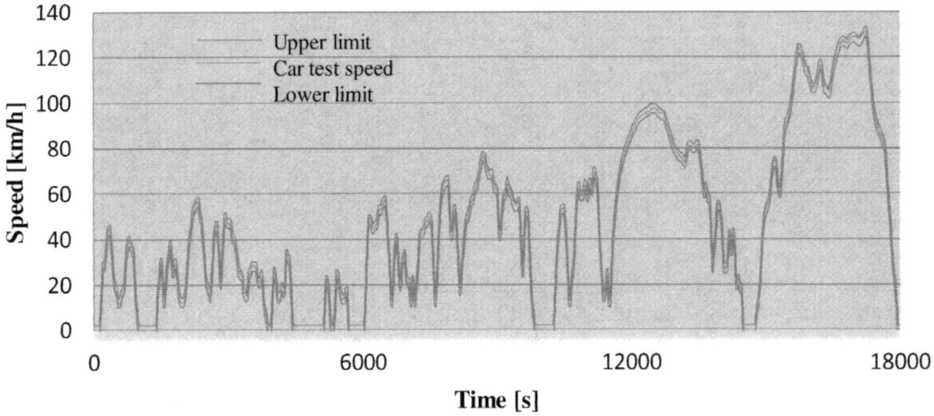


**Figure 4.** WLTC speed profile followed by the tested vehicle.

IOP Conf. Series: Materials Science and Engineering 1303 (2024) 012004 doi:10.1088/1757-899X/1303/1/012004

Information on whether the test conditions correspond to the required speed profile was recorded at a frequency of 10 Hz [3]. This type of information validates the quality of the test and makes it possible to deviate from the upper or lower limit speed profile by up to 2 km.

In Figure 4, the speed profile was maintained throughout the test period, indicating that the vehicle was capable of meeting the dynamic demands to which it was subjected. In order to correctly analyse how the reduction in the state of charge of the traction battery, which implies a reduction in range, was achieved, several journeys were made and average values were calculated for each of the four specific zones of the WLTC cycle.

Based on the information provided by the electronic control units controlling the traction battery, it was possible to analyse the energy balance, i.e. the ratio between the energy supplied by the traction battery to the traction system and the energy recovered by the traction battery after regenerative braking.

The traction battery in the study vehicle can only withstand a slow and fast AC charge. The charging mode used on the traction battery prior to testing was slow to ensure a correct and full charge up to 100% [2].

## 3. Results and discussions

Based on the simulation it was possible to draw up graphs regarding the evolution of the state of charge of the traction battery for each high and very high-speed zone specific to the simulated running on the WLTC cycle. Therefore, it was analysed the simulated behaviour of the state of charge in the traction battery in relation to the load and the speed of the electric motor, obviously generated by the travel speed and the cycle dynamics. The simulation was run until the traction battery state of charge was fully depleted. Based on this information, it was also possible to highlight the speed zones where the electric vehicle performs best in terms of economy, characterised by a low electricity consumption.

In figure 5 it is represented the variation of the regenerative load over time of the vehicle submitted to the test at the high and at the very high speed specific to the simulated running on the WLTC cycle, for one test cycle. It is observed how the SOC state of charge decreases very much over the time, despite the regenerative braking with a very high inertia at the high and at the very high speed, after the studies revealing that the temperature also plays a very important role in the evolution of the regenerative state of the charge of the electric vehicles.

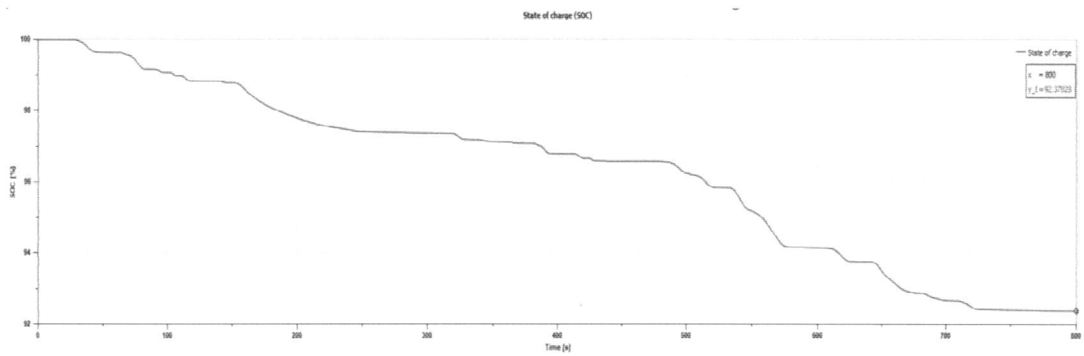


**Figure 5.** Variation of the regenerative state of charge for the tested car at the high and very high speeds for one test cycle in WLTC

In fact, as a result of the studies, it has been shown that a very important role regarding the depletion of the state of charge of the traction batteries fitted in the electric vehicles, in the case of the high and the very high-speed running, is represented, besides the temperature, by the number of the charging and of the complete discharge cycles of the traction battery.

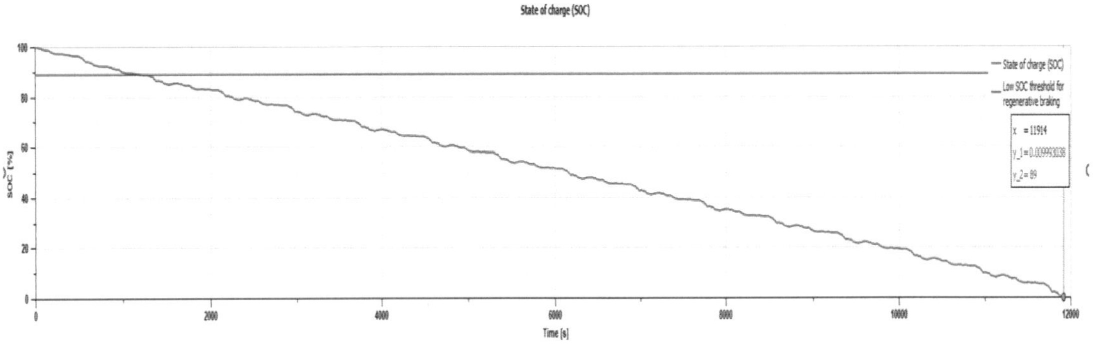


**Figure 6.** Variation of the regenerative state of charge for the tested vehicle at the high and at the very high speeds for 15 test cycles in WLTC

Figure 6 shows the variation of the regenerative load over the time of the vehicle tested at the high and at the very high speeds specific to the driving simulated on the WLTC cycle for 15 test cycles. It is observed how the SOC state of charge decreases very much over the time, despite the regenerative braking with very high inertia at the high and at the very high speeds, and the studies show that the temperature also plays a very important role in the evolution of the regenerative state of charge of the electric vehicles. Also, it was preferred to have several test cycles, considering that these tests were carried out under different temperature and under different climatic conditions, and in certain speed ranges, at the high and at the very high speeds of the vehicle being tested. This makes that the accuracy of the results of these tests are much closer to the truth, considering that there are many factors influencing these measurements.

In Figure 7, in such situations, if we speak of an average speed of 55.7 km/h and a maximum speed of about 96.9 km/h, the energy recovered is usually practically insignificant, with the state of charge and thus the range decreasing gently and gradually.

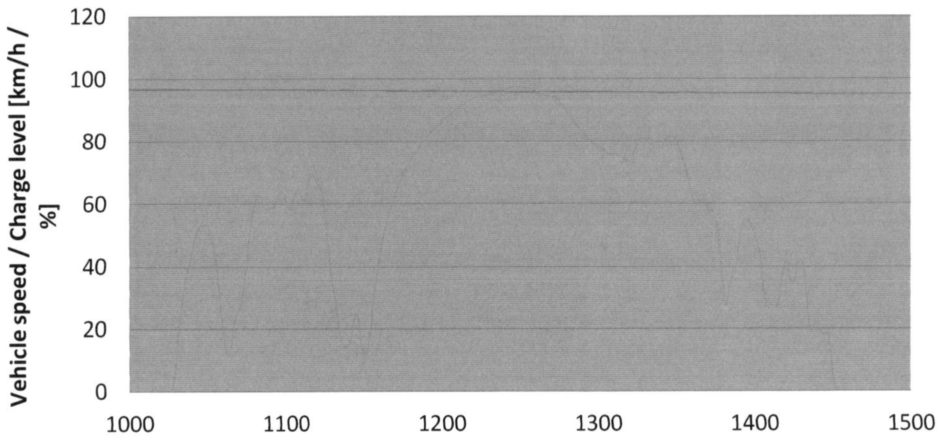


**Figure 7.** Traction battery charge level, at the high-speed zone

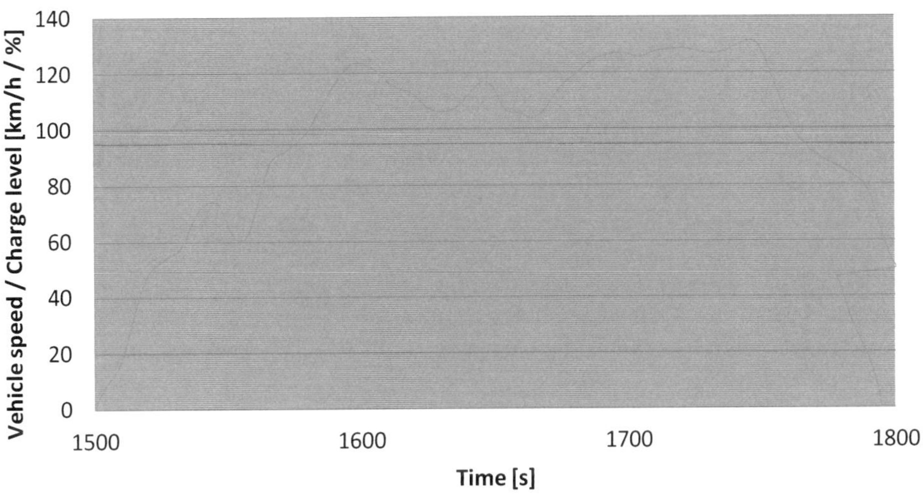


**Figure 8.** Traction battery charge level, at the very high-speed zone

The very high-speed running zone, graphically represented in the figure above, has a special character in conditions to maintain the state of charge of the traction battery. At these speeds, the average speed is 93 km/h and the top speed 132.2 km/h. As a general rule, in this zone, the vast majority of production electric vehicles reach their maximum design speed, characterized by a maximum load where energy recovery is non-existent.

At this driving speed, the charge level of the traction battery drops rapidly and range is immediately affected. This is the worst case scenario for the energy stored in the traction battery, due to the very high consumption generated by the electric traction motor.

The tests applied to the studied electric vehicle were repeated until the total autonomy of the traction battery was exhausted, respectively until the speed limit could no longer be respected by the vehicle.

## 4. Conclusions

As a result of the simulated drive, it was found that, when the vehicles are used at the high and at the very high speeds, characterized by the high and by the very high dynamic loads, the maintenance of the state of charge is difficult to achieve and the effect on the electric autonomy is very high.

With direct reference to the high to the very high-speed range of the simulated WLTC cycle, this can be assimilated to the use of the vehicles in the extra-urban and the highways. Under such conditions, basically without any regenerative braking, the recovered energy contribution of the traction battery is non-existent, so the impact on the electric autonomy is very high.

The simulated evolution of the state of charge of the traction battery with the effect in the electric autonomy, when running on the WLTC cycle, indicates that the electric vehicles with an electric power source have a very good utility in the urban mode and are satisfactory for the extra-urban or for the highway mode. The evolution of the technologies used to produce the traction batteries and their increased power can supplement the use of the vehicles in terms of their autonomy in the extra-urban or in the highway mode.

It was also found from the studies that the temperature, the climate conditions and the number of charging and the discharging cycles of the traction battery are very important for the depletion of the state of charge of the traction batteries of the electric vehicles in the case of the high and of the very high-speed driving.

During high-speed driving, the state of charge of the traction batteries decreases gradually and the contribution of the recovered electrical energy is negligible. Such a situation leads to a rapid reduction in range due to the increased power consumption of the electric traction motor.

The most unfavorable situation for the depletion of the traction battery's state of charge is when vehicles are driven at very high speeds. Under these conditions, with a high load and high electric motor speed, consumption is very high, electrical energy recovery is practically non-existent and this type of driving leads to an immediate depletion of the traction battery's state of charge.

It should be noted that the main research objectives in the field of traction batteries should focus on increasing performance in terms of load levels at high and very high speeds. Of course, this performance is not only related to the traction battery, but further research is also needed in the area of optimising the design of electric motors, transmission or passenger comfort system, which play an extremely important role in such situations.

The development of the market for vehicles equipped with electric propulsion systems is driving the need to increase the electrical power of traction batteries, which must fulfil new functions such as dynamics and economy to meet the most demanding needs of modern society.

**References**

[1] Buchmann I 2001 *Batteries in a Portable World*, 2nd end. Cadex Electronics Inc. ISBN 978-0968211823
[2] Paul D and Stănescu O 2015 *Electric, hybrid and fuel cell cars*, Politehnica, Timișoara ISBN: 978-606-35-0011-4.
[3] Jeong N T and Yang S M 2016 *Urban driving cycle for performance evaluation of electric vehicles*. Int. J. Automot. Technol 17(1), 145-151
[4] Bamban V et al 2015 *Analysis of range extender electric vehicle performance using vehicle simulator*. Energy Procedia 68, 409-418
[5] Shi Q et al 2011 *The study of a new method of driving cycles construction*. Procedia Eng. 16,79-87
[6] Ababei S and Culea G 2003 *Electrical measurements and data acquisition*, Tehnica-Info, ISBN 9975-63-214-9.
[7] Barlow T J et al 2009 *A referance book of driving cycles for use in the measurement of road vehicle*. Published Project Report PPR354, TRL Limited
[8] *EUR-Lex Homepage*, http://data.europa.eu/eli/reg/2017/1151/oj, last accessed 2022/08/20.
[9] *Help AMESim Rev12*

IOP Conf. Series: Materials Science and Engineering 1303 (2024) 012005 doi:10.1088/1757-899X/1303/1/012005

# Review of PEDELECS as an alternative to conventional means of urban transportation

**V Teodorașcu, N V Burnete and N Burnete**

Technical University of Cluj-Napoca, Faculty of Automotive Engineering Mechatronics and Mechanical Engineering, Department of Automotive Engineering and Transports, Cluj-Napoca, Romania

E-mail: teodorascu.ov.vlad@student.utcluj.ro

**Abstract**. Sustainable transportation solutions are more crucial than ever because of the pressing need to increase resource efficiency while lowering greenhouse gas emissions. The most recent advancements in the e-cycle sector have made a significant contribution to this goal and have attracted the interest of numerous businesses that offer mobility services. This paper discusses the main topics surrounding pedal electric cycles (PEDELECS) with an accent on a niche application, namely e-cargo cycles. The paper highlights not just the defining characteristics of these categories, but also other relevant aspects such as barriers to market penetration, general legislation, benefits for specific applications, as well as the significance of incentives, local infrastructure, and urban policies. Based on the available literature, it can be concluded that PEDELECS have a considerable potential to contribute to sustainability goals in urban areas due to their numerous benefits (functionality without emissions, less space occupied on roads, ability to access destinations with increased precision, ability to travel in car-restricted areas, lower costs compared to conventional vehicles, added health benefits to the users, versatility etc.). However, there are also significant challenges and barriers that must be overcome before they can see widespread adoption.

## 1. Introduction

Urban transportation accounts for more than 23% of the greenhouse gas emissions in the European Union (EU) [1], making it one of the primary sectors of concern in the effort to reduce environmental impacts in cities. City logistics plays a significant role in promoting sustainability by addressing challenges related to emissions, noise, and traffic congestion. The idea of "green logistics" in urban settings has increased the demand for electric vehicles, cargo bikes, and other forms of electric mobility as cleaner options for transportation. This resulted in a greater demand for freight and passenger transportation that could fit into the limited space of densely populated cities, specifically through micromobility solutions that address the aforementioned challenges.

One potential solution to the issue of urban mobility is the widespread use of electric cycles (e-cycles), especially electric cargo (e-cargo) cycles, for personal as well as business needs as a method to mitigate the negative effects of conventional transportation systems. In terms of transport capacity, range, and cost [2], e-cargo cycles can be placed somewhere between e-cycles and conventional vehicles. The electric motor of e-cargo cycles, which allows for greater payload capacities, longer trips, and overcomes the serious drawback of driver fatigue, is one of their primary advantages over their non-electrified counterparts [3]. Additionally, since they don't generate local pollutants, e-cargo cycles can travel through city centers and other areas where cars aren't allowed.

A very important characteristic of pedal electric cycles (PEDELECS) is the range, which can be influenced by multiple factors such as battery capacity, route characteristics, payload etc. The progress in battery technology coupled with an optimized control has resulted in increasingly greater ranges, that allows PEDELECS to compete with conventional vehicles for many applications. Furthermore, due to their fast-travelling capabilities in otherwise congested areas, micromobility vehicles are seen as a useful and efficient solution in an urban setting. This is supported by the fact that, on average, 40-60% of all car journeys within cities are less than 5 km [4,5]. Furthermore, e-cargo cycles are a valid and useful solution for businesses relying on delivery, due to their lightweight build, a cargo capacity exceeding 100 kg, and the ability to even transport children [6,7]. To highlight the acceptance of cargo bikes as a substitute for motor vehicles in the United States of America, the study of Riggs [6] showed that 66% of the survey respondents who had previously used a car as their primary mode of transportation switched to using a cargo-cycle after purchasing one.

To put the average distance and duration of most car journeys into perspective, a study [7] that focused on the impacts of e-micromobility in urban areas compiled various reports of distance, duration, and energy consumption for e-bikes and e-scooters in different cities. The result showed that the average travel distance and time varies significantly between cities (1 to 11 km per trip). With respect to the average distance travelled, Şengül and Mostofi [7] reported between 0.72 and 2.4 km for e-scooters and between 3 and 4.5 km for e-bikes. This difference in travel distance is correlated with the maximum vehicle speed and the power of the electric motor, which allow e-bikes to cover the same distance in a shorter time (due to the lower speed limit of e-scooters) [8]. In another study [9] participants reported that for longer recreational trips, they preferred e-cycles over conventional bicycles. To highlight the day-to-day aspects of using an e-bike as well as the underlying factors for choosing this method of transportation Sun et al. [10] considered two sample groups from the Netherlands and China. In terms of trip distance, the study showed that participants from the Netherlands travel by e-bike between 10 and 20 km, while participants from China travel by e-bike for approximately 10 km. The participants reported that they preferred the e-bike over car and public transportation because it is less affected by congestion, and therefore provides a more reliable travel time, it has better road accessibility, and it requires less time searching for parking, especially in busy urban areas.

Compared to other literature reviews [7,11,12], this paper seeks to combine all aspects of PEDELECS and as such, bring together the pros and cons for e-micromobility adoption and their potential to contribute to sustainability goals. The highlighted aspects include the integration of PEDELECS in sustainable mobility concepts and the challenges that e-micromobility adoption faces from an industry, individual, and infrastructure perspective.

## 2. Typology and characteristics

As previously mentioned, PEDELEC stands for "pedal electric cycle", which is a low-powered electric cycle providing pedal assistance via an electric motor. With the continuous integration of modern technologies in the urban transportation business, there is no surprise that PEDELECS gained the attention of both e-cycle manufacturers and users. Due to the broad definition of PEDELECS, many types of electrically assisted cycles can fall under the term, ranging from pedal assist e-cycles, throttle-controlled e-cycles, S-PEDELECS (PEDELECS designed for speed), mid-drive e-cycles and hub-drive PEDELECS, all having cargo variants or available upgrades that boost their versatility. With respect to the number of wheels, the available configurations can be listed as: bicycles (bikes), tricycles (trikes) and quadricycles (quads).

A specific sub-category of PEDELECS are e-cargo cycles, which encompass a broad range of options adapted for carrying passengers and/or heavy loads. As a result of their increasing popularity, there is a wide variety of e-cargo cycle topologies available on the market. In Gruber and Narayanan [13] five different e-cargo cycle types are listed based on vehicle construction type (Figure 1). With respect to their maximum speed, some e-cycles can reach up to 40 km/h when the electrical assist is enabled. The other defining characteristics are the maximum load capacity (usually 50 – 250 kg and in exceptional cases up to 500 kg [14,15]), the cargo space (up to 2.5 m$^3$ [16,17]), and the range (currently it usually varies between 40 and 80 km [11,15]).

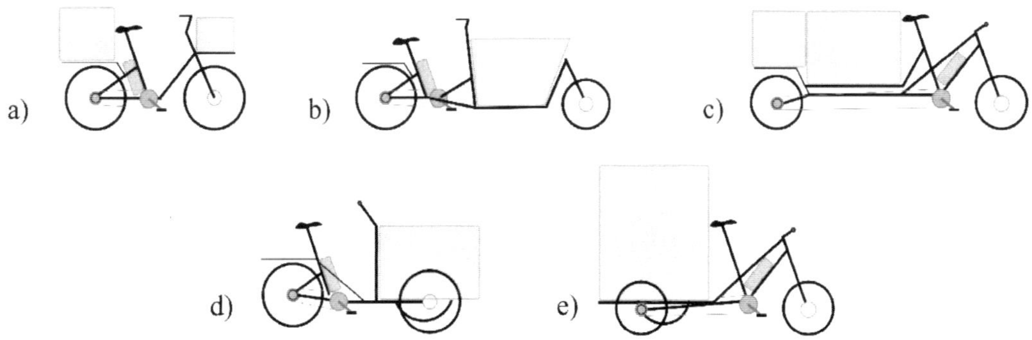


**Figure 1.** Different topologies for cargo cycles
a – Delivery bicycle, b – Long John bicycle, c – Longtail bicycle, d – Front load tricycle, e – Heavy load tricycle

### 2.1. *Purpose*

In terms of purpose, e-cargo bikes show high versatility and can be used for personal transportation (with or without passengers), service trips (service providers with or without equipment travelling to a customer's location) as well as delivery trips [13,18]. Delivery trips can further branch in five market segments: home delivery services, postal services, internal transport for companies with large facilities, parcel services and courier services.

PEDELECS are suitable for both personal and business uses in areas with a high-density population, having the added benefit of being able to use dedicated cycle lanes and access areas in which conventional vehicles are prohibited. Furthermore, by taking advantage of the cycling infrastructure, PEDELECS are able to perform tasks with times comparable to those of conventional vehicles but with added advantages such as reduced cruising times and resource requirements, as well as avoiding congestion [7].

### 2.2. *Cargo space*

Carirns and Solman [19] estimated in their study that 10 – 30% of urban delivery trips can be completed using e-cargo cycles as an alternative to vans and other light commercial vehicles. Considering that delivery trips may constitute up to 25% of traffic in large cities, if cargo/e-cargo cycles replaced delivery and service vehicles for suitable trips, this could be reduced by 1.5 – 7.5%. It is important to note that the delivery and transport efficiency of e-cargo cycles are especially dependent on their load capacity and/or cargo space. Athanassopoulos et al. [20] conducted a study in Hamburg that used Life-cycle assessment to analyze three alternatives: electric cars and two types of e-cargo tricycles (with and without parcel storage containers). The payload capacity of the e-cargo tricycles were 306 kg and 220 kg, respectively, with a cargo space of 2.2 m$^3$ and 1.5 m$^3$. The study also involved an urban parcel delivery company, which tested various e-cargo cycle integration strategies, with the goal of reducing emissions while maintaining operational efficiency. For a different project, "Ich ersetze ein Auto" [21], 40 couriers received e-cargo bicycles with a payload capacity of 90 kg and a cargo space of 0.2 m$^3$. During the test period (21 months), more than 127,000 shipments were carried out (with a total distance of more than 500,000 km). The study showed that 90% of participants see e-cargo cycles as a viable option for courier deliveries. Another study [22] used two e-cargo tricycles with a payload capacity of 280 kg and a cargo space of 1.5 m$^3$ for last-mile delivery in an effort to increase the energy efficiency of urban freight. With respect to the cargo space, the study concluded that the limited size of the shipments as well as the limited number of packages were useful for the pilot analysis, but it is not viable for long-term use (which contradicts the findings of other projects, see section 3).

Another important aspect is the efficiency of a delivery trip, which is determined by travel distance, traffic density and capacity of delivery (cargo volume occupied by a delivery payload) which would lead to the conclusion that a higher capacity of delivery results in a higher efficiency. However, studies have shown that e-cargo cycles' delivery trip efficiency decreases with the increase in drop size [23,24].

## 3. Challenges

Even though PEDELECS have many advantages, their integration in the urban ecosystem comes with challenges. A first important aspect is the average consumer decision to adopt a micromobility solution over the conventional means of transportation, which can be influenced by multiple aspects: price and time savings, social perception, safety, and other. This is supported by the study of Bretones and Marquet [25] which showed that beyond the monetary costs and other functional aspects, e-micromobility adoption relates more to symbolic factors that can be associated with subjective factors (such as self-identity, sense of belongingness, and pro-environmental attitudes). On the other hand, the adoption of PEDELECS for businesses is connected with efficiency and capacity. To analyze these aspects, an Italian pilot project "Pro-e-bike" [26], which involved thirty-nine companies that integrated the use of e-scooters and e-bikes for the delivery of goods in urban areas. Aside from mitigating $CO_2$ emissions, the study showcased some of the concerns that big companies have about the widespread adoption of e-bikes. These concerns stem from the perception that e-bikes are inefficient and that the load capacity is suboptimal. However, the study demonstrated that choosing the right type of PEDELEC for the task (e-bikes for letter distribution and e-cargo bikes for bigger packages) leads to high efficiency, thus proving that they can be a reliable alternative to traditional combustion engine vehicles. The main characteristics of the PEDELECS used in the study as well as the main results of each operator are summarized in Table 1.

**Table 1.** Main e-vehicle characteristics and results reported in [26].

| | **Operator 1** | **Operator 2** | **Operator 3** | **Operator 4** |
|---|---|---|---|---|
| **Type of vehicle** | Electric delivery scooter | Long John bike | Long John bike | Electric delivery bike |
| **Market price (€)** | € 5,000 - 6,000 | € 1,200 | € 1,700 | € 2,000 |
| **Consumption (Wh/km)** | 30 - 40 Wh/km | 10 Wh/km | 10 Wh/km | 4 Wh/km |
| **Autonomy (km)** | 100 - 130 km | 55 km | 45 km | 75 km |
| **Battery capacity (Wh)** | 4,000 Wh | 540 Wh | 432 Wh | 300 Wh |
| **Charging time (h)** | 6 - 8 h | 4 h | 4 h | 3 h |
| **Payload (kg)** | 175 kg | 80 kg | 80 kg | 25 kg |
| **Cargo space ($m^3$)** | 0.160 $m^3$ | 0.250 $m^3$ | 0.250 $m^3$ | 0.070 $m^3$ |
| **Monthly usage** | 15.3 days/month | 8.8 days/month | 5 days/month | 21.3 days/month |
| **Daily parcel deliveries** | 114/day | 6/day | 3/day | 57/day |
| **Daily mass delivered** | 50.8 kg/day | 31.6 kg/day | 21.7 kg/day | 33.6 kg/day |
| **Daily volume delivered** | 0.035 $m^3$ | 0.531 $m^3$ | 0.533 $m^3$ | 0.134 $m^3$ |
| **Daily covered distance** | 39 km/day | 18 km/day | 6 km/day | 24 km/day |
| **Average speed** | 16 km/h | 17 km/h | 16 km/h | 20 km/h |

## *3.1. Costs*

### *3.1.1. Purchase price*

For the average consumer, one of the first aspects affecting the modal switch or the overall adoption of PEDELECS is the purchase price. This can be due to the overwhelming variety and versatility that e-bikes and other micromobility options offer, which come with a proportional price range that varies depending on the manufacturer, the region, and the archetype. The price barrier is supported by multiple studies, which identified the purchase price as a frequent concern of the average consumer [27–32]. Although there are currently more expensive e-cargo bikes available, many e-cycle models are less expensive to buy when compared to traditional motorized vehicles [14,33,34]. In this regard, a study done by Cappelle [35] surveyed local suppliers regarding the catalogues, prices, and reasoning behind choosing certain brands. It was found that the price of PEDELECS from local suppliers (in Flanders) ranged between 695€ and 3600€, with an average of 1,691€. Similar prices were reported in [26] namely 1,200€ for light front-load e-bikes, and 5,000-6,000€ for electric scooters.

### *3.1.2. Energy costs*

Another aspect that can hinder adoption of PEDELECS is the energy cost. This includes energy consumption per travelled kilometer, as well as the cost per kilowatt-hour during charging.

With respect to the energy consumption per travelled kilometer, with the same amount of energy (1kWh) e-scooters can travel up to 128 km, compared with conventional vehicles, such as gasoline (less than 1.6 km) and electric (6.4 km) cars [7]. The high range values are supported by the low consumption values reported in the literature. For example, the study of Martínez-Navarro et al. [36] analyzed the energy consumption of e-scooters and determined a value of 0.012 kWh/km. Other studies found different values: Brdulak et al. [37] reported a value of 0.04 kWh/km, while Gubman et al. [38] reported 0.01 kWh/km. Although e-cargo cycles are larger, and heavier, it was found that they require a similar amount of energy, with values between 0.009 and 0.018 kWh/km [26,36,39–41]. However, there are studies that reported higher values, such as the study of Navarro et al. [22] which tested smart urban logistics solutions in the cities of Barcelona and Valencia. The test involved the use of electric tricycles for last mile delivery in transshipment terminals. The study showed that in Barcelona, over the course of one month two tricycles performed a total of 740 shipments with 1.38 shipments/km and a total energy consumption of 32.3 kWh, resulting in an average of 0.030 kWh/km. In contrast, in Valencia, the tricycles performed 1194 shipments with 1.40 shipments/km, a total energy consumption of 21.6 kWh, and an average of 0.011 kWh/km.

The charging costs depend on multiple factors such as the charging location (home/charging facilities), energy cost in the respective country and the characteristics of the PEDELEC model (battery size, charging efficiency etc.). To better understand these aspects, Table 2 shows the costs of charging an electric bike in the EU countries.

With respect to costs for businesses, Nocerino et al. [26] compared the costs associated with an electric micromobility fleet to those of a traditional fleet for deliveries and freight transport operators. The authors reported similar costs for the two fleets present, with a slight economical advantage for the electric fleet due to the sustainability aspect, which brings the added benefits of reduced environmental impact and increased publicity for the company, that further translates into marketing benefits and opportunities. It must be noted that, even though there are significant lower costs associated with the e-fleet in case of energy costs (↓ 87%) and congestion charges (↓ 100%) these are offset by a higher annual depreciation (↑ 37%), platform renting (↑ 38%) and salaries (↑ 6%).

**Table 2.** Charging cost of an electric bike in EU countries.

| Country | Energy cost [42] (€/kWh) | Minimum battery capacity (kWh) | Maximum battery capacity (kWh) | Minimum cost/charge (€) | Maximum cost/charge (€) |
|---|---|---|---|---|---|
| **Italy** | 0.3192 | | | 0.0798 | 0.3192 |
| **Belgium** | 0.3791 | | | 0.094775 | 0.3791 |
| **Bulgaria** | 0.0956 | | | 0.0239 | 0.0956 |
| **Czechia** | 0.2077 | | | 0.051925 | 0.2077 |
| **Denmark** | 0.3636 | | | 0.0909 | 0.3636 |
| **Germany** | 0.2333 | | | 0.058325 | 0.2333 |
| **Estonia** | 0.2085 | | | 0.052125 | 0.2085 |
| **Ireland** | 0.3823 | | | 0.095575 | 0.3823 |
| **Greece** | 0.4556 | | | 0.1139 | 0.4556 |
| **Spain** | 0.2966 | | | 0.07415 | 0.2966 |
| **France** | 0.1723 | | | 0.043075 | 0.1723 |
| **Croatia** | 0.117 | | | 0.02925 | 0.117 |
| **Cyprus** | 0.2551 | | | 0.063775 | 0.2551 |
| **Latvia** | 0.2448 | 0.25 | 1 | 0.0612 | 0.2448 |
| **Lithuania** | 0.2007 | | | 0.050175 | 0.2007 |
| **Luxembourg** | 0.1696 | | | 0.0424 | 0.1696 |
| **Hungary** | 0.0853 | | | 0.021325 | 0.0853 |
| **Malta** | 0.1201 | | | 0.030025 | 0.1201 |
| **Netherlands** | 0.3197 | | | 0.079925 | 0.3197 |
| **Austria** | 0.209 | | | 0.05225 | 0.209 |
| **Poland** | 0.1002 | | | 0.02505 | 0.1002 |
| **Portugal** | 0.1972 | | | 0.0493 | 0.1972 |
| **Romania** | 0.2683 | | | 0.067075 | 0.2683 |
| **Slovenia** | 0.1533 | | | 0.038325 | 0.1533 |
| **Slovakia** | 0.1315 | | | 0.032875 | 0.1315 |
| **Finland** | 0.1806 | | | 0.04515 | 0.1806 |
| **Sweden** | 0.2489 | | | 0.062225 | 0.2489 |

#### *3.1.3. Maintenance costs*
A significant benefit of e-bikes, along with fuel cost savings, is the lower maintenance cost [43]. However, it is difficult to estimate an annual value, since it varies depending on the type of e-bike, its features, and the time of service (for example, long-term use leads to a bigger necessity of servicing). Furthermore, the maintenance cost is influenced by the service costs, which are country dependent.

### *3.2. Safety*
Safety is widely perceived as the biggest negative aspect of PEDELECS by both users and non-user of electric micromobility services [44–53]. This perception is also one of the main factors affecting

market penetration and modal shift to PEDELECS. Most participants identify the infrastructure as the main cause for their perception. This was confirmed by the fact that the majority of electric micromobility users reported that they feel safer when using dedicated lanes because they are separated from motor traffic [48,50,54]. Participants also suggest that, for the overall safety perception to increase it is necessary to encourage the use of protective gear, together with the promotion of clear and more advanced safety regulations and training on road norms and rules [51,53].

In terms of road safety there are also differences between PEDELEC types. Riggs [6] noted that some PEDELEC types, such as e-cargo bikes, have the added benefit of being more visible on the road, due to their size. This mitigates their risk of collisions in comparison to a conventional bicycle.

When it comes to accidents involving PEDELECS, there are many possible causes, such as: underestimation of the e-bike speed by other traffic participants, e-bike speed control difficulties [55], errors, aberrant and/or aggressive behaviours [56] (it must be noted that the authors also concluded that PEDELEC users with a driver's license for cars are less likely to have an accident compared to those without one), traffic signal violation, illegal use of motor vehicle lanes, over-speeding, overloading, violation of manned vehicle regulations, and cycling in reverse [12] (considered to have the highest probability for traffic accidents) etc.

### *3.3. Incentives*

Incentives provide the necessary means to overcome the cost of barrier associated with the purchasing of an electric bike and specifically, incentives for travel that are managed in organizations can make a difference for people who might be risk-averse or tentative about trying out a PEDELEC.

In their study Fyhri et al. [57] aimed to observe the travel behavior and greenhouse gas emissions reduction of participants in an incentive program with e-bikes in Oslo. The study used an app that tracked the behavior of 619 participants out of which 153 were already using e-bikes. The study showed that the participants who used the incentive program to purchase an e-bike increased their cycling activity (by kilometers traveled) from 17% to 52% and that each e-bike user, saved between 87 and 144kg of $CO_2$ emission per year compared to the control group.

A study of de Kruijf et al. [58] describes the implementation of an incentive program in the province of North-Barbant in the Netherlands. To reduce car congestion and stimulate the use of e-bikes Participants in the project "B-Riders" received incentives based on the e-bike usage while commuting. The financial incentive varied with the time of day/traffic (0,15 €/km during peak hours and 0,08 €/km during off-peak hours) and was limited to 1,000 € for one participant. Results showed that the program led to a strong modal shift of the participants. More specifically, participants used their car (from 62% to 28%) and conventional bicycle (from 33% to 1%) less. Additionally, a follow-up survey conducted one month after the program ended, revealed that participants used their e-bikes for 68% of all commute trips. E-bikes substitution of cars and conventional cycling has also seen significant increases. The authors noted that e-bikes were used 80% of the time over short distances (0–5 km). Although their use decreased with the distance, it still accounted for 63% of trips longer than 20 km. The last survey (conducted six months after the program ended) showed that the use of e-bikes increased even more, while car use decreased to 24%. Additionally, the use of e-bikes for commute trips rose to 73%. The increase in e-cycling is most noticeable in the distance range of 0–20 km.

Sundfor and Weber [59] analyzed the e-bike subvention program implemented by the municipality of Oslo in which the applicants could obtain a €500 subsidy when purchasing a new e-bike. For the participants that have been granted subvention, the authors noted a significant increase in cycling activity compared to the control groups. It was concluded that the modal shift as well as increased physical activity are both effects of the incentive program.

Finally, it is important to note that incentives used to increase the penetration of PEDELECS can also come in the form of infrastructure adjustments such as access to bike lanes [60] and bike storage [24].

### *3.4. Legislation*

Other factors that must be taken into account are the regulations and the restrictions concerning the use of PEDELECS on public roads. The implementation of policies is different among countries and can include the maximum speed of the vehicle, the user age restrictions, the vehicle pricing policy, etc.

In Poland, e-scooters are allowed to be used only on cycle lanes or paths, while their speed is limited to 20 km/h. However, if no cycle infrastructure is available, e-scooters are allowed to use roads with a speed limit of 30 km/h. When travelling with an e-scooter on roads that have a speed limit above 30 km/h and no cycle lanes, e-scooters may be used on pedestrian roads or pavement. Furthermore, the minimum age requirement for accessing public roads is 10 and all users under 18 years are required to have a licence [61]. In contrast, the use of micromobility services in Italy involves additional requirements (possessing a license, restrictions such as age limit, and limitation of usage to specific hours of the day [62]). In the city of Chicago [63] the maximum speed of travel for e-micromobility vehicles is limited to 15 mph (~24km/h), but there are also other restrictions such as: with the exception of children under 12 years of age, traveling on pavement is prohibited; the user of shared electric scooters must be at least 18 years old with the exception of people over 16 years of age riding with a guardian.

In other areas there are also equipment restrictions, as showed in a study covering ten cities in the United States of America [64]. Users of e-scooter must wear protective gear and stickers or lights in order to be more visible during the night. Additionally, it is against the law to use electronic devices while driving or transport more than one person (unless the vehicle is designed to do so). Additionally, e-scooters are only permitted on bike lanes and sidewalks, and their top speed is limited to 15 mph (25 km/h).

As presented in Leger et al. [50] the minimum age required to use an e-bike in British Colombia and Ontario is 16 years old, while the maximum speed is higher, 32 km/h. Additionally, neither a license nor registration are required, but safety equipment is mandatory. The legal minimum age to use an e-bike in Manitoba and Quebec is 14. Similarly, the legal age to operate an e-bike in Alberta is 12. Additionally, neither a license nor registration are required, although helmets are.

It is important to notice that there are inconsistencies in the regulations. This was also the conclusion of Pimentel and Lowry [65] which showed that in Oregon, the speed of e-scooters is limited to 25 km/h and traveling on pavements is prohibited, but at the same time, micromobility vehicles are required to travel with the traffic speed (for example if traffic flows at 40 km/h the e-scooter should travel at 40 km/h or more). Another inconsistency is related to different regulation in the same state. For instance, wearing a helmet is a requirement in King County, Washington, but not in other areas of the state.

### *3.5. Urban policies*

Urban policies play a major role in the modal shift from conventional vehicles to e-micromobility solutions. As noted by many researchers, implementing vehicle access restrictions is of major importance [11,14,22,66]. For example, Choubassi et al. [66] concluded that the implementation of no-truck zones plays a significant role in the adoption of e-cargo cycles. Furthermore, according to, Lenz and Riehle [14] implementing a tax for access to city centers with conventional vehicles also supports PEDELEC adoption, mainly e-cargo cycles. Parking policies are another important aspect that can boost the use of PEDELECS over conventional vehicles. For example, higher parking fines and limited parking spaces for conventional vehicles further support the penetration of PEDELECS [14,66,67].

### *3.6. Infrastructure*

Public acceptance and willingness to shift from traditional means of transportation to PEDELECS is strongly dependent on the benefits that this transition can bring to individuals and businesses. The benefits of these electric vehicles are in close connection with the context they are used in (ownership, purpose, location etc.). One important aspect to be accounted for is the available transport infrastructure. Infrastructure dependent factors can be summarized in several categories:

- Cycling infrastructure (bike lanes and traffic control):
  - Faxér et al. [68] showed that infrastructure has a major impact on the safety perception. No participant of the study expressed safety concern when using the bike lane, but eight out of ten participants said that they feel uncomfortable sharing the road with cars;
  - Rudolph and Gruber [69] highlighted the need for a state-of-the-art infrastructure in order to properly use e-cargo cycles. The larger width and access requirements (wider cycle lanes and parking places) for e-cargo cycles can make them difficult to function in current infrastructural designs;
- Parking/Urban morphology:
  - Rudolph and Gruber [69] highlighted the need for adapting of current parking availability to include the use of e-cargo cycles (due to their larger size and higher usage frequency);
  - Schliwa et al. [70] also highlighted the importance of parking for sustainable city logistics that involve the use of e-cargo cycles.
- Road safety regulations (details in chapter 3.5):
  - User age requirements [54,62,63];
  - Speed limitations [54,63,65,71];
  - Safety equipment [54,65];
  - Area restrictions [62,63,65];
- Charging stations:
  - Nocerino et al. [26] mentioned the lack of adequate charging stations as a problem in the adoption of e-cargo cycles together with the limited autonomy and the technical issues that may arise (engine or battery issues).

In their effort Blazejewski et al. [72] listed several benefits of using e-cargo cycles that boost the efficiency of business operations, and that encourage decarbonization, such as:

- by using the cycling infrastructure, e-cargo cycles have the ability to travel through city centers with ease;
- e-cargo cycles take up less road space than light commercial vehicles;
- e-cargo cycles can be parked right outside the destination of the trip;
- the fact that, even though training can be beneficial for new users, e-cargo cycles and other PEDELECS do not require a driver's license;
- using e-cargo bikes instead of light commercial vehicles leads to a reduction in $CO_2$ emissions, congestion, and noise pollution.

Considering these benefits and the infrastructure aspects listed above, as well as the characteristics of e-cargo cycles, it can be concluded that the infrastructure requirements are significantly lower (both in terms of costs and space) compared to those of conventional vehicles, but absolutely necessary for micromobility to be successful.

#### *3.6.1. Docked vs dock-less solutions*

Docked bike sharing schemes are an emerging shared mobility service relying on networks of fixed stations that enable users to pick up, use and drop their bike (to one of the available stations) when they complete their travel. The docking stations are built to regulate customer behavior and to facilitate payment. One of the most crucial elements for the success of such programs is the installation of bike stations [73–75]. To ensure financial success docked bike sharing stations are mostly placed in areas with high usage expectations. However, a study from Beairsto et al. [76] shows that the stations should also be located in less accessible places to reduce accessibility inequalities. To solve this issue Chen et al. [75] proposed a framework for determining the appropriate location for docked bike sharing stations.

Given the challenges that traditional station-based bike-sharing systems encountered (convenience, accessibility, space, the need for public subsidies and limited time availability) as their popularity grew, the spread of bike-sharing systems has been limited. However more recently the functions of docking stations have been integrated into smartphone apps providing payment options (scanning a quick-response (QR) code or using near field communication (NFC) systems). Additionally, the bikes have GPS sensors to provide fleet tracking and management. Bike sharing systems are a shared mobility solution that is widely viewed as a healthy, efficient, and flexible method of transportation [77]. Dock-less sharing systems present more convenience and flexibility through the use of smartphone applications and embedded GPS systems as well as internet for locating and unlocking bikes. The location of a bike in a dock-less system depends on where the previous user left it, thus increasing flexibility through the multiple locations that the bikes can be found [78]. In their study Chen et al. [78] identified several common points between the two types of bike-sharing systems:

- Service usage:
  - Mode shift mainly from public transit and walking;
  - Limited car use (when bike-sharing systems and public transit are integrated);
  - Morning and afternoon usage peaks during weekdays;
  - Short trip duration;
- Last mile implications:
  - General cycling benefits such as interaction with a built environment, increased physical activities and improved mood;
- Accessibility and distribution:
  - Concentrations of bicycles in popular areas;
  - Traditionally, more disadvantaged groups have less access;

### *3.6.2. Parking*

In terms of parking, PEDELECS bring multiple benefits such as being able to park virtually at any trip destination and a shorter trip with the amount of time saved for parking cruising. In this regard, for a study done in the Netherlands, Moolenburgh et al. [79] reported a 30% decrease in the round-trip time. This decrease has been mainly attributed to the shorter routes for cycles compared to the routes for conventional vehicles. In the conventional car trips cruising for parking can take up a significant portion of the trip especially in densely populated areas. In their study Dalla Chiara and Goodchild [80] showed that parking cruising time is approximately 2.3 minutes per trip representing, in their case, 28% of the total trip time. Furthermore, the authors found that, by using e-cargo cycles instead of conventional vehicles, the total operation time decreased by 40%. It must be noted that this reduction in operation time is partly due to the reduction in parking cruising time.

## 4. Integration with sustainable mobility concepts

When compared to other sustainable mobility concepts, PEDELECS have the advantage of being an electric form of transportation as well as actively being included into urban mobility plans to further cut greenhouse gas emissions. Of the numerous projects and innovative solutions that seek to boost integration of PEDELECS in urban mobility, the following are mentioned [81,82]:

- **Cleanergy 4 Micromobility** - proposes cableless and renewable energy docks for e-scooters. The project's primary goal is to build docks for e-scooters to address the current problems with e-scooter expansion in metropolitan areas. These docks for e-scooters should include cableless locks, helmet storage, solar panel strips, and, in order to ensure renewable energy to the docking stations, an off-grid battery storage system.
- **ECOSWAP** – Through an ecosystem for swapping out batteries, the concept promotes the use of electric motorbikes by providing a digital platform that enables the use of the same battery-

swapping stations by electric motorbikes and a dedicated business model that is put into practice and is approved by users.

- **SEVES** – secondary use of EV batteries as an energy storage solution. To accelerate the electrification of all urban transportation systems, the project objective is to develop an energy storage solution based on re-used batteries from electric vehicles (BEV) as an energy storage system.
- **Solar-powered micromobility docking stations to encourage intermodality** – as a market-ready solution for cities. The project proposes the usage of Solum Helios autonomous docking station e-micromobility vehicles as well as the usage of solar-charging systems for electric batteries. The docking stations can be installed anywhere with minimal disturbance to the environment, and they allow users a secure way of charging their e-scooters and e-bikes.
- **BICIFICATION** – supporting model shifts and bicycle use through gamification and rewards. The project aims to encourage the modal shift towards green mobility solutions through a reward-based gamification solution. The project involves approximately 1,500 users who are eligible to receive monetary rewards (which can be used in local stores) from local authorities. BICIFICATION also provides real-time data (via an open data platform) that includes: cycling routes, kilometers travelled, case studies, heat maps, reports on the amount of reduced $CO_2$ emissions, and rewards granted to the users.
- **SmartHubs -** The project's goal is to create and validate effective and economically feasible solutions for mobility hubs. To improve the implementation of smart hubs, the SmartHubs projects requires testing and developing a planning tool for decision and support.
- **HALLO** – hubs for last mile delivery solutions. The project's objective is to create shared urban consolidation and distribution centers through studies based on a series of pilots in Barcelona and Stockholm to reduce environmental and traffic problems in urban areas by. To make the replication of these strategies easier, the initiative is also developing a roadmap that outlines potential location planning, business models, and issues that may arise during implementation.
- **Inclusiv_eBike -** The project has the objective to develop a new electric vehicle that can improve the inclusivity of transitioning to personalised, self-driving, micromobility services. Additionally, the vehicle aims to further reduce energy consumption and increase the availability of automated micromobility.

## 5. Conclusions

This review has focused on evaluating PEDELECS and their ability to change the status quo in urban mobility. Based on the available literature, PEDELECS have a considerable potential to contribute to sustainability goals in urban areas due to their numerous benefits, such as:

a) functionality without emissions,
b) less space occupied on roads (due to their smaller size),
c) ability to access destinations with increased precision,
d) ability to travel in car-restricted areas (such as city centers),
e) low purchasing, maintenance, and usage costs (compared to conventional vehicles),
f) added health benefits to the users, and g) versatility (PEDELECS are suitable for personal and business usage alike – such as last mile delivery services and freight transport).

However, there are also significant challenges and barriers that must be overcome, before they can see widespread adoption, such as:

a) a proper local infrastructure such as dedicated bike lanes, docking stations and proper parking spaces;
b) public acceptance, which is mainly influenced by the safety perception associated with micromobility solutions, but also by concerns associated with the social and socioeconomic status;

c) aspects that can influence travel decisions such as journey purpose, distance to destination and weather conditions, as well as
d) consistent and clear regulations.

Nonetheless, there are significant ongoing efforts to support the integration of PEDELECS in the urban mobility environment. These efforts include incentive programs, innovative solutions for adoption and availability, pilots, dedicated infrastructure investments and mobility projects that benefit from the integration of personalized and accessible solutions. Finally, PEDELECS are no panacea for urban mobility issues, but if properly used, they can bring significant improvements.

**Author's ORCID ID**
V Teodorașcu: https://orcid.org/0009-0008-1659-0216
N Burnete: https://orcid.org/0000-0001-6647-459X
N V Burnete: https://orcid.org/0000-0001-5447-4728

## 6. References

[1] European Commission 2013 Communication from the Commission to the European Parliament, the Council, the European economic and social committee and the committee of the regions, COM(2013)913, Bruxelles (2013)
[2] Gruber J, Kihm A and Lenz B 2014 A new vehicle for urban freight? An ex-ante evaluation of electric cargo bikes in courier services *Research in Transportation Business & Management* **11** 53–62
[3] Transport for London 2009 Cycle freight in London: A scoping study (London)
[4] Cuenot F, Fulton L and Staub J 2012 The prospect for modal shifts in passenger transport worldwide and impacts on energy use and CO2 *Energy Policy* **41** 98–106
[5] Basel-Stadt M, Stadt Bern V, Stadt Luzern T, Stadt St.Gallen T, Stadt Winterthur T and Stadt Zürich T 2017 *Städtevergleich Mobilität*
[6] Riggs W 2016 Cargo bikes as a growth area for bicycle vs. auto trips: Exploring the potential for mode substitution behavior *Transportation Research Part F Traffic Psychology and Behaviour* **43** 48–55
[7] Şengül B and Mostofi H 2021 Impacts of e-micromobility on the sustainability of urban transportation—a systematic review Applied Sciences (Switzerland) **11**
[8] Cherry C R, Yang H, Jones L R and He M 2016 Dynamics of electric bike ownership and use in Kunming, China *Transport Policy* **45** 127–35
[9] Castro A, Gaupp-Berghausen M, Dons E, Standaert A, Laeremans M, Clark A, Anaya-Boig E, Cole-Hunter T, Avila-Palencia I, Rojas-Rueda D, Nieuwenhuijsen M, Gerike R, Panis L I, de Nazelle A, Brand C, Raser E, Kahlmeier S and Götschi T 2019 Physical activity of electric bicycle users compared to conventional bicycle users and non-cyclists: Insights based on health and transport data from an online survey in seven European *Transportation Research Interdisciplinary Perspectives* **1** 100017
[10] Sun Q, Zhao J, Spahn A and Verbong G 2023 Pathway towards sustainability or motorization? A comparative study of e-bikes in China and the Netherlands *Global Environmental Change* **82** 102735
[11] Narayanan S and Antoniou C 2022 Electric cargo cycles - A comprehensive review *Transport Policy 2022* **116**
[12] Ma C, Yang D, Zhou J, Feng Z and Yuan Q 2019 Risk Riding Behaviors of Urban E-Bikes: A Literature Review *International Journal of Environmental Research and Public Health 2019*, Vol. 16, Page 2308 16 2308
[13] Gruber J and Narayanan S 2019 Travel Time Differences between Cargo Cycles and Cars in Commercial Transport Operations *https://doi.org/10.1177/0361198119843088* **2673** 623–37
[14] Lenz B and Riehle E 2013 Bikes for Urban Freight? *https://doi.org/10.3141/2379-05* 39–45

[15] Schier M, Offermann B, Weigl J D, Maag T, Mayer B, Rudolph C and Gruber J 2016 Innovative two wheeler technologies for future mobility concepts 2016 *11th International Conference on Ecological Vehicles and Renewable Energies*, EVER 2016
[16] Urban Arrow Tender - The Professional Electric Cargo bike (available at: https://urbanarrow.com/business-bikes/tender/)
[17] Chariot FS1 - Cargo Cycling (available at: https://www.cargocycling.com/cargo-cycles/chariot-fs1/)
[18] Wrighton S and Reiter K 2016 CycleLogistics – Moving Europe Forward! *Transportation Research Procedia* **12** 950–8
[19] Carirns S and Solman L 2019 Potential for e-cargo bikes to reduce congestion and pollution from vans in cities
[20] Athanassopoulos T, Dobers K and Clausen U 2016 Reducing the Environmental Impact of Urban Parcel Distribution *Lecture Notes in Logistics 2016* 159–81
[21] Gruber J and Kihm A 2016 Reject or Embrace? Messengers and Electric Cargo Bikes *Transportation Research Procedia 2016* **12** 900–10
[22] Navarro C, Roca-Riu M, Furió S and Estrada M 2016 Designing New Models for Energy Efficiency in Urban Freight Transport for Smart Cities and its Application to the Spanish Case *Transportation Research Procedia 2016* vol **12**
[23] Marujo L G, Goes G V., D'Agosto M A, Ferreira A F, Winkenbach M and Bandeira R A M 2018 Assessing the sustainability of mobile depots: The case of urban freight distribution in Rio de Janeiro *Transportation Research Part D: Transport and Environment 2018* **62** 256–67
[24] Sheth M, Butrina P, Goodchild A and McCormack E 2019 Measuring delivery route cost trade-offs between electric-assist cargo bicycles and delivery trucks in dense urban areas *European Transport Research Review* **11**
[25] Bretones A and Marquet O 2022 Sociopsychological factors associated with the adoption and usage of electric micromobility. A literature review *Transport Policy2022* **127**
[26] Nocerino R, Colorni A, Lia F and Luè A 2016 E-bikes and E-scooters for Smart Logistics: Environmental and Economic Sustainability in Pro-E-bike Italian Pilots Transportation *Research Procedia vol* **14**
[27] Glavić D, Trpković A, Milenković M and Jevremović S 2021 The E-Scooter Potential to Change Urban Mobility—Belgrade Case Study *Sustainability 2021,* **13** 5948
[28] An K, Chen X, Xin F, Lin B and Wei L 2013 Travel Characteristics of E-bike Users: Survey and Analysis in Shanghai *Procedia - Social and Behavioral Sciences 2013* **96** 1828–38
[29] Ye Y, Xin F and Wei L 2014 Characteristics of the Electric Bicycle: A Comparative Analysis with Bicycles and Public Transit CICTP 2014: Safe, Smart, and Sustainable Multimodal Transportation Systems - *Proceedings of the 14th COTA International Conference of Transportation Professionals* 3450–8
[30] Zuev D 2018 Urban mobility in modern China: The growth of the E-bike Urban Mobility in *Modern China: The Growth of the E-bike* 1–110
[31] Kaplan S, Wrzesinska D K and Prato C G 2018 The role of human needs in the intention to use conventional and electric bicycle sharing in a driving-oriented country *Transport Policy 2018* **71** 138–46
[32] Alamelu R, Sivasundaram Anushan C and Selvabaskar S G 2015 Preference of e-bike by women in India–a niche market for auto manufacturers *Business: Theory and Practice* **16** 25–30
[33] Leonardi J, Browne M and Allen J 2012 Before-After Assessment of a Logistics Trial with Clean Urban Freight Vehicles: A Case Study in London *Procedia - Social and Behavioral Sciences* **39** 146–57
[34] Ormond P A, Telhada J and Afonso P 2019 Evaluating the Economic and Environmental Impact of the Urban Goods Distribution by Cargo Cycles—A Case Study in São Paulo City *Proceedings of the World Conference on Transport Research*—WCTR

[35] Cappelle J, Timmermans J-M, Lataire P and Mierlo J Van 2007 The Pedelec Market in Flanders EET2007 European EleDrive Conference
[36] Martínez-Navarro A, Cloquell-Ballester V A and Segui-Chilet S 2020 Photovoltaic electric scooter charger dock for the development of sustainable mobility in urban environments *IEEE Access* **8** 169486–95
[37] Brdulak A, Chaberek G and Jagodziński J 2020 Determination of electricity demand by personal light electric vehicles (PLEVs): An example of e-motor scooters in the context of large city management in Poland *Energies* **13**
[38] Gubman J, Jung A, Kiel T and Strehmann J 2019 Shared E-Scooters: Paving the Road Ahead
[39] Koning M and Conway A 2016 The good impacts of biking for goods: Lessons from Paris city *Case Study on Transport Policy* **4** 259–68
[40] WIRED Electric Scooters Are Better for Cities Than Cars (available at: https://www.wired.com/story/e-scooter-micromobility-infographics-cost-emissions/)
[41] Affonso P D O J, Afonso P and Telhada J 2018 Evaluating the economic and environmental impact of the urban goods distribution by cargo cycles – a case study in São Paulo City
[42] Eurostat Statistics | Electricity prices for household consumers - bi-annual data (available at: https://ec.europa.eu/eurostat/databrowser/view/nrg_pc_204/default/table?lang=en)
[43] Choubassi C, Seedah D P K, Jiang N and Walton C M 2016 Economic Analysis of Cargo Cycles for Urban Mail Delivery *https://doi.org/10.3141/2547-14* **2547** 102–10
[44] Bieliński T and Ważna A 2020 Electric Scooter Sharing and Bike Sharing User Behaviour and Characteristics *Sustainability 2020* **12** 9640
[45] Jones T, Harms L and Heinen E 2016 Motives, perceptions and experiences of electric bicycle owners and implications for health, wellbeing and mobility *Journal of Transport Geography* **53** 41–9
[46] Sellaouti A, Arslan O and Hoffmann S 2020 Analysis of the use or non-use of e-scooters, their integration in the city of Munich (Germany) and their potential as an additional mobility system *2020 IEEE 23rd International Conference on Intelligent Transportation Systems, ITSC 2020*
[47] Ling Z, Cherry C R, MacArthur J H and Weinert J X 2017 Differences of Cycling Experiences and Perceptions between E-Bike and Bicycle Users in the United States *Sustainability 2017* **9** 1662
[48] Rayaprolu S and Venigalla M 2020 Motivations and Mode-choice Behavior of Micromobility Users in Washington, *DC Journal of Modern Mobility Systems* **1** 110–8
[49] Elias W and Gitelman V 2018 Youngsters' Opinions and Attitudes toward the Use of Electric Bicycles in Israel *Sustainability 2018* **10** 4352
[50] Bateman L B, Fouad M N, Sullivan A, Heider L and Oates G R 2021 Barriers and facilitators to bikeshare programs: A qualitative study in an urban environment *Journal of Transport & Health* **21** 101062
[51] De Ceunynck T, Wijlhuizen G J, Fyhri A, Gerike R, Köhler D, Ciccone A, Dijkstra A, Dupont E and Cools M 2021 Assessing the Willingness to Use Personal e-Transporters (PeTs): Results from a Cross-National Survey in Nine European Cities *Sustainability 2021* **13** 3844
[52] Eccarius T, Leung A, Shen C W, Burke M and Lu C C 2021 Prospects for shared electric velomobility: Profiling potential adopters at a multi-campus university *Journal of Transport Geography* **96** 103190
[53] Esztergár-Kiss D and Lopez Lizarraga J C 2021 Exploring user requirements and service features of e-micromobility in five European cities *Case Studies on Transport Policy* **9** 1531–41
[54] Leger S, McLaughlin D and Tracksdorf K 2018 Leading the Charge on Canadian E-bike Integration: A Discussion on the Emerging and Unchartered Role of Micromobility
[55] Haustein S and Møller M 2016 E-bike safety: Individual-level factors and incident characteristics *Journal of Transport & Health* **3** 386–94
[56] Yao L and Wu C 2012 Traffic Safety for Electric Bike Riders in China *https://doi.org/10.3141/2314-07* 49–56

[57] Fyhri A, Sundfor H B and Weber C 2016 Effect of subvention program for electric bicycle in Oslo on bicycle use, transport distribution and CO2 emissions
[58] Kruijf J, Ettema D, Kamphuis C B M and Dijst M 2018 Evaluation of an incentive program to stimulate the shift from car commuting to e-cycling in the Netherlands *Journal of Transport & Health* **10** 74–83
[59] Sundfør H B and Fyhri A 2022 The effects of a subvention scheme for e-bikes on mode share and active mobility *Journal of Transport & Health* **26**
[60] Assmann T, Lang S, Müller F and Schenk M 2020 Impact assessment model for the implementation of cargo bike transshipment points in urban districts *Sustainability (Switzerland)* **12**
[61] Eltis Poland proposes new legislation for e-scooters and other personal transport devices (available at: https://www.eltis.org/in-brief/news/poland-proposes-new-legislation-e-scooters-and-other-personal-transport-devices)
[62] Campisi T, Akgün N and Tesoriere G 2020 An Ordered Logit Model for Predicting the Willingness of Renting Micro Mobility in Urban Shared Streets: A Case Study in Palermo, Italy *Lecture Notes in Computer Science (including subseries Lecture Notes in Artificial Intelligence and Lecture Notes in Bioinformatics)* vol 12250 LNCS
[63] City of Chicago 2021 E-SCOOTER PILOT EVALUATION CHICAGO
[64] Sun Q, Feng T, Kemperman A and Spahn A 2020 Modal shift implications of e-bike use in the Netherlands: Moving towards sustainability? *Transportation Research Part D: Transport and Environment* **78**
[65] Pimentel R W and Lowry M B 2020 IF YOU PROVIDE, WILL THEY RIDE? MOTIVATORS AND DETERRENTS TO SHARED MICRO-MOBILITY
[66] Choubassi C, Seedah D P K, Jiang N and Walton C M 2016 Economic Analysis of Cargo Cycles for Urban Mail Delivery *https://doi.org/10.3141/2547-14* **2547** 102–10
[67] Tipagornwong C and Figliozzi M 2014 Analysis of Competitiveness of Freight Tricycle Delivery Services in Urban Areas *https://doi.org/10.3141/2410-09* **2410** 76–84
[68] Faxér A Y, Olausson E, Olsson L, Smith G and Pettersson S 2018 Electric cargo bike with a twist - A field test of two innovative bicycle concepts *31st International Electric Vehicle Symposium and Exhibition, EVS 2018 and International Electric Vehicle Technology Conference 2018, EVTeC 2018*
[69] Rudolph C and Gruber J 2017 Cargo cycles in commercial transport: Potentials, constraints, and recommendations *Research in Transportation Business and Management* **24**
[70] Schliwa G, Armitage R, Aziz S, Evans J and Rhoades J 2015 Sustainable city logistics - Making cargo cycles viable for urban freight transport *Research in Transportation Business and Management* **15**
[71] Bieliński T and Ważna A 2020 Electric scooter sharing and bike sharing user behaviour and characteristics *Sustainability (Switzerland)* **12**
[72] Blazejewski L, Davies N and Sherriff G 2020 Delivering The Last Mile Scoping the Potential for E-cargo Bikes Healthy Active Cities
[73] Frade I and Ribeiro A 2015 Bike-sharing stations: A maximal covering location approach *Transportation Research Part A: Policy and Practice* **82** 216–27
[74] Lin J R and Yang Ta-Hui T H 2011 Strategic design of public bicycle sharing systems with service level constraints *Transportation Research Part E: Logistics and Transportation Review* **47** 284–94
[75] Chen W, Chen X, Cheng L, Chen J and Tao S 2024 Locating new docked bike sharing stations considering demand suitability and spatial accessibility *Travel Behaviour and Society* **34** 100675
[76] Beairsto J, Tian Y, Zheng L, Zhao Q and Hong J 2022 Identifying locations for new bike-sharing stations in Glasgow: an analysis of spatial equity and demand factors *Annals of GIS* **28** 111–26

[77] Shen Y, Zhang X and Zhao J 2018 Understanding the usage of dockless bike sharing in Singapore *International Journal of Sustainable Transportation* **12** 686–700
[78] Chen Z, van Lierop D and Ettema D 2020 Dockless bike-sharing systems: what are the implications?
[79] Moolenburgh E A, Van Duin J H R, Balm S, Van Altenburg M and Van Amstel W P 2019 Logistics concepts for light electric freight vehicles: a multiple case study from the Netherlands *Transportation Research Procedia* **46** 301–8
[80] Dalla Chiara G and Goodchild A 2020 Do commercial vehicles cruise for parking? Empirical evidence from Seattle *Transport Policy* **97** 26–36
[81] EIT Urban Mobility - *Mobility for more livable urban spaces* - Highlights report 2022
[82] EIT Urban Mobility - *Mobility for more livable urban spaces* - Highlights projects 2020-2021

IOP Conf. Series: Materials Science and Engineering 1303 (2024) 012006 doi:10.1088/1757-899X/1303/1/012006

# Analysis of Energetic and Traction Performances for an Electric Vehicle in Real Driving Conditions

**A A Ancuța, Ș Voloacă, G Frățilă, G Danciu**

Department of Automotive Engineering, Faculty of Transport, University „Politehnica" of Bucharest

E-mail: alex.ancuta96@yahoo.com

**Abstract.** In continuation of the studies related to the performances of electric vehicles and their optimization solutions, in this work a simulation model will be created in accordance with all the demands that appear on the vehicle during driving and the model will be adapted for urban driving conditions on a certain length of time, so that the performance of the electric propulsion system can be determined, such as: the state of charge of the battery, the consumption of electricity, the characteristics of the electric motor, as well as the variation of the autonomy of the vehicle. The simulation model made will later be correlated with a series of experimental data determined in real driving conditions so that the accuracy of the made model can be checked and validated so that it can be used on different types of electric vehicles with different performances and destinations.

## 1. Introduction

The issue related to the spread of electric vehicles worldwide and the permanent need to determine solutions to optimize dynamic and energetic performances represent aspects of heightened interest in terms of recent concerns and each completed study constitutes an additional step in the deepening of certain notions necessary to complete a numerical simulation model that can be used on several types of electric vehicles depending on destination and interest. However, in order to be able to validate the developed model, different experimental determinations are needed in certain real driving conditions through which, after validating the model, concrete valid optimization solutions can be identified and proposed.

So this paper proposes the analysis of certain performances for another electric vehicle, different in terms of characteristics from those studied in previous works, in which the already existing simulation model presented in the paper [1] suffered some changes to become more accurate and to be able to provide some more conclusive results. In addition to the model architecture, which contains the vehicle dynamic model, the driver model, the transmission model, the electric motor model and of course the battery model, this paper will also present some experimental results measured under real driving conditions for a short distance and a short time in the urban area, but in which the basic parameteres like speed and torque varied greatly. Using the data obtained through experimental measurements, the simulation model will be parameterized according to the characteristic sizes of the analyzed electric vehicle so that the main purpose of the work is the validation of the results obtained through simulation by comparison with the real, measured ones and obviously the resolution of certain inconsistencies or errors that may appear and can influence the accuracy of the model.

Finally, the performances of the studied vehicle will be presented under the conditions of the considered real cycle.

## 2. Theoretical aspects and data used for realize the study

To carry out the study, experimental data were recorded when driving in the urban area at a temperature close to that of the ambient environment in a short cycle of 320 seconds, over a distance of less than 5 km. The vehicle used was the Peugeot e-208 model, a vehicle with reduced overall dimensions, hatchback bodywork with 5 seats, extremely suitable for driving in the urban area. The e-208 model has a total weight of 1910 kg and is equipped with an electric motor on the front axle which develops a electric power of 100 kW, a maximum torque of 260 Nm and a maximum speed of 14000 rpm and the efficiency $\eta_m$=0.98. It uses a lithium-ion battery which presents the following characteristics: nominal battery voltage 400 V, usable energy 46.3 kWh and battery capacity of 115 Ah [3] and the efficiency of the battery was considered 1. Other important data about the vehicle used in the study are maximum speed: V=150 km/h, the front track: $E_1=1500\ mm$, height: $H_a=1450\ mm$, cross-sectional area: $A=2.145\ m^2$, drag coefficient: $c_x=0.29$, total reduction coefficient: $i_t$=9.701 and transmission efficiency $\eta_t$=0.98 [2] and tire rolling radius: $r_r$=0.303 m, calculated using the wheel size type of e-208: *205/45R17* [3]. All these values will be used in the simulation model in the same way as in last work [1].

An Exxotest VCI-Muxdiag type device was used to realize the measurements and the program used to process the experimental data and provide the communication interface between the electric vehicle and the measuring device was the INCA program. An important observation is that the response time of the program was set to *500 ms*.

Following the registrations, a series of values for the vehicle speed, torque variation of the electric motor during the actual cycle and the position of the accelerator pedal were saved. These values were later exported to Excel, where the graphs related to the variation of those parameters over time were generated, which will be used to validate the simulation results under the same conditions.

In the following paragraphs you can find the graphs obtained from the measurements recorded in real driving conditions:

### 2.1. *Vehicle speed profile*

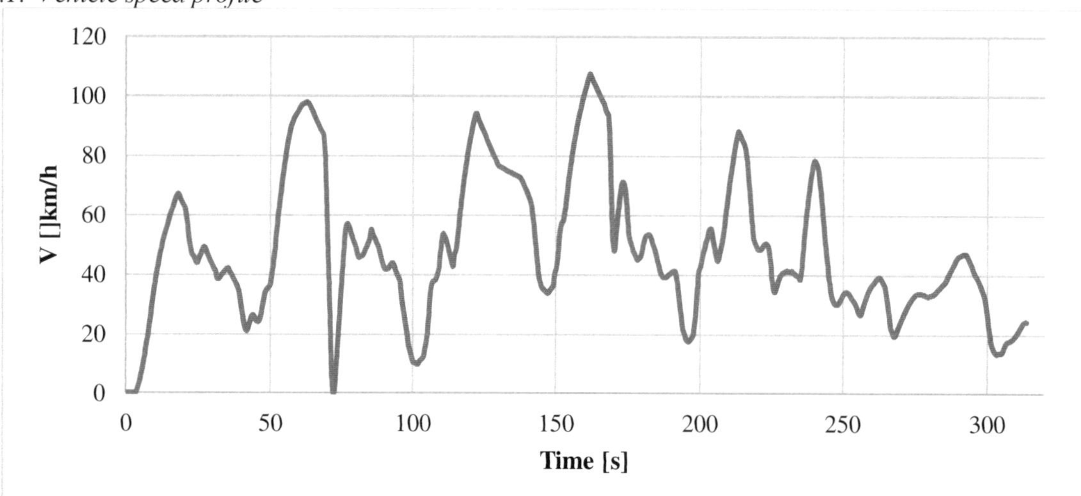


**Figure 1.** Speed profile obtained during the considered driving cycle for Peugeot e-208.

From figure 1 it can be seen that during the approx. 313.5 seconds of driving, the vehicle speed had a large variation and there were also moments with important breakings that were recorded. The maximum speed reached by the vehicle on this proposed cycle was ~105.5 km/h.

### *2.2. Electric motor torque profile*

In figure 2 it can be seen the variation of the electric motor torque during the approx. 313.5 seconds of driving. The negative part of the obtained diagram shows us according to speed variation, the braking moments during the measurements. The maximum torque reached by the vehicle on this proposed cycle was 259.3 Nm in the second 72 and the minimum value -160.7 Nm in the second 143. To do the simulations, vehicle data will be entered into the already existing model: driving time, response time and distance will be considered as input data in the conditions in which the measurements were made and also the speed profile obtained in figure 1 will be entered.

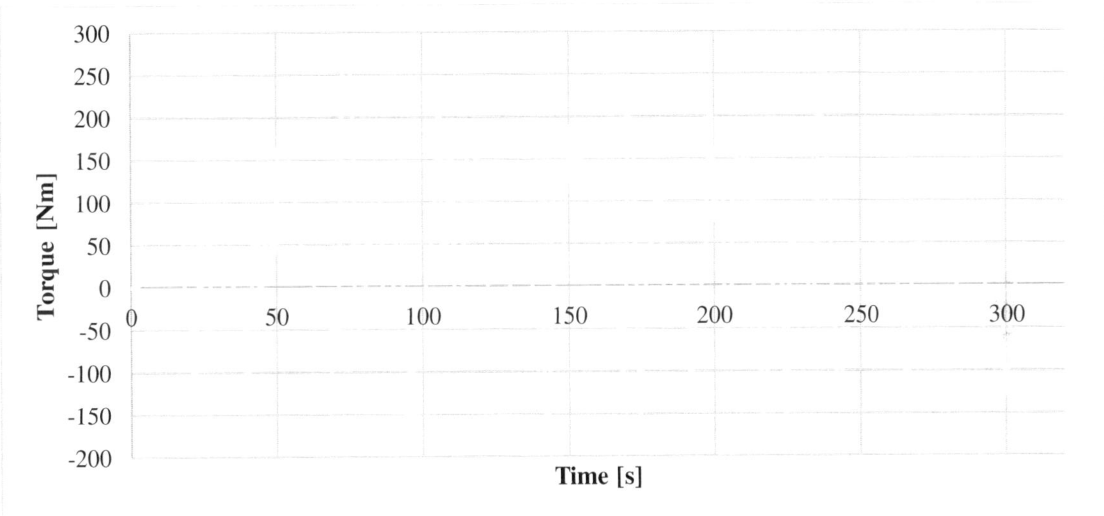


**Figure 2.** Torque profile obtained during the considered driving cycle for Peugeot e-208.

## 3. Model used for simulation

As specified in the first part of the paper the model used for the simulation is based on the structure already presented in [1]. The dynamic model of the vehicle under horizontal driving conditions, the model of the driver, the electric motor and the transmission with a single gear, the lithium-ion battery was considered for which an already existing block in the Simscape library of the Simulink program was used. Figure 3 shows the simulation model used:

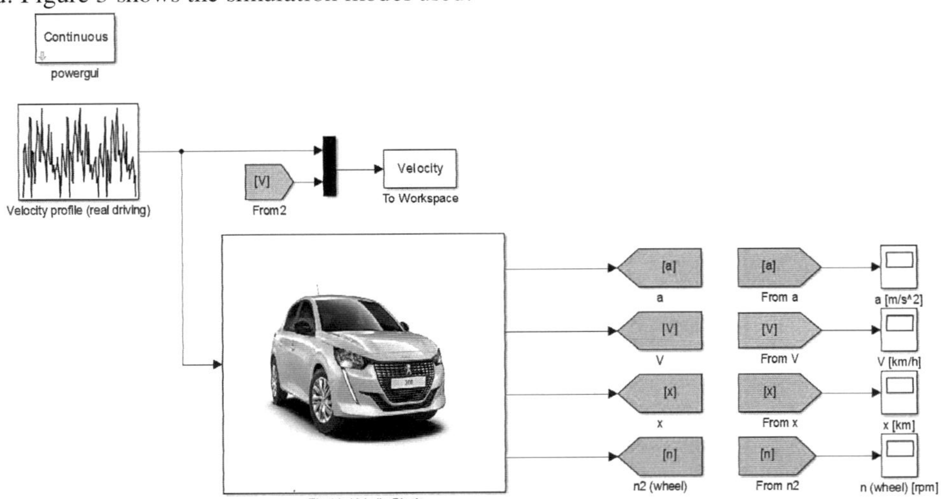


**Figure 3.** Simulation model for electric vehicle.

Additionally, a branch was added to simulate the power consumed when running on the considered cycle and the energy corresponding to this power in order to be able to compare the results with those obtained by measurements. The following calculation relationship was used to determine the power [4]:

$$P = T \cdot \omega \; [kW] \quad (1)$$

Where ***T*** is electric motor torque, [Nm] and ***ω*** is angular speed, [$s^{-1}$] which is determined using the next relation [5]:

$$\omega = \frac{2\pi \cdot n}{60} \; [s^{-1}] \quad (2)$$

In order to obtain the energy profile by simulation, the power profile obtained with relation 1 was integrated as a function of time. So, in the next figure is presented the simulation model obtained:

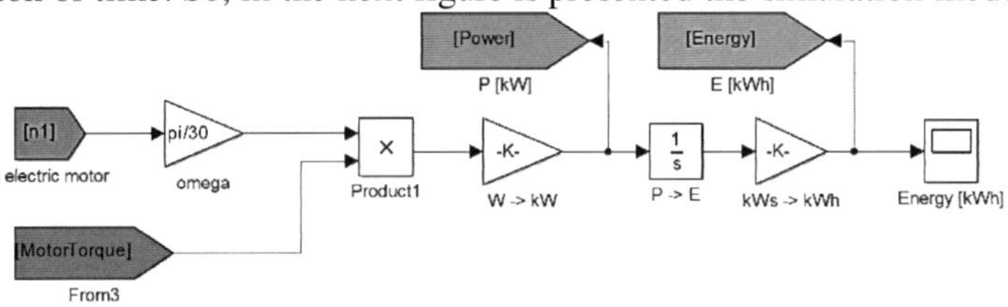


**Figure 4.** Simulation model for power and energy profiles.

As for the battery once the power was determined by simulation using the model from figure 4, it was considered that its nominal voltage remains constant throughout the simulation cycle ($U_{nom}$ = 400 V) and the variation of the battery discharge current was determined by the ratio between the power profile and voltage *(P=f(t)/U)* so as to identify some performances of the battery during the driving cycle. The model used for the battery can be found in figure 5:

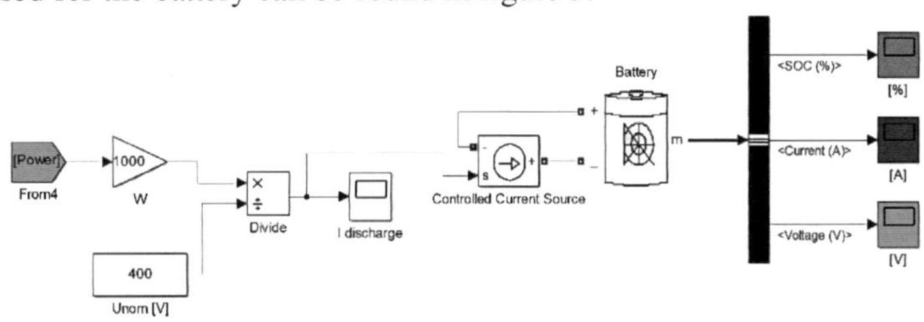


**Figure 5.** Simulation model for lithium-ion battery.

Finally, in order to verify the model and to correlate the parameters obtained through simulation with those obtained through experimental measurements a model was developed where the measured profiles were entered from Excel using a „RepeatingSequence" type block and through the „Mux" block compared to the simulated profiles. That model took into account the units of measurement of all parameters so that the results are conclusive; it can be found below:

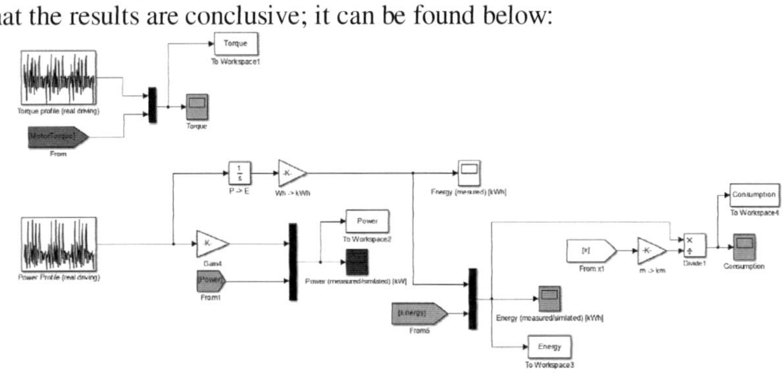

**Figure 6.** Simulation model for comparing the simulated results with the measured results.

The final model obtained in MatLAB-Simulink is presented in figure below. To simplify the connections between the electric vehicle body and the other branches of the model we used the „GoTo" and „From" commands to retrieve and connect certain parameters such as: power, torque of the electric motor and energy consumed:

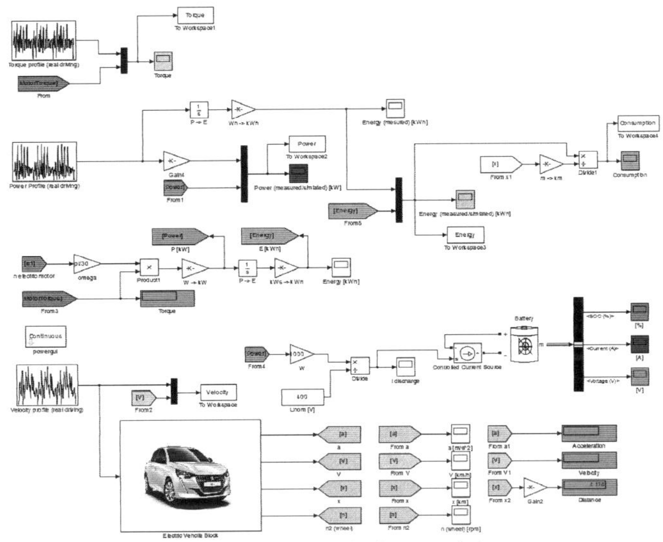

**Figure 7.** Simulation model.

## 4. Results

The figures below show the results obtained from the simulations compared to the measured results:

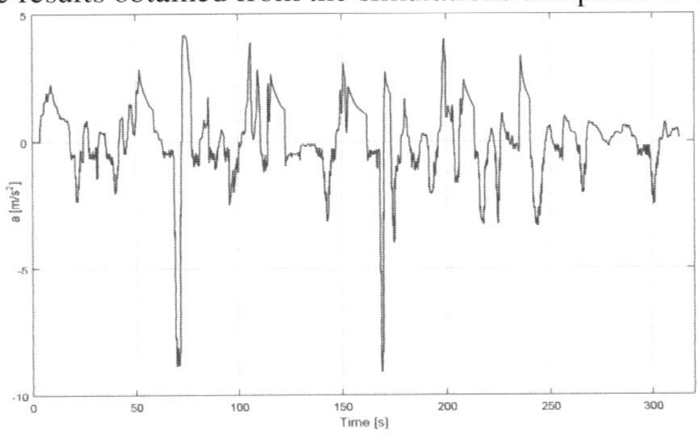


**Figure 8.** Acceleration profile during the simulation.

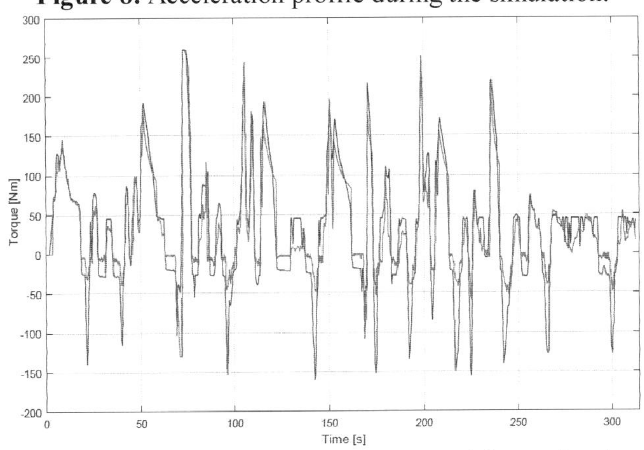


**Figure 9.** Torque profile (blue line – real torque, red line – simulated torque).

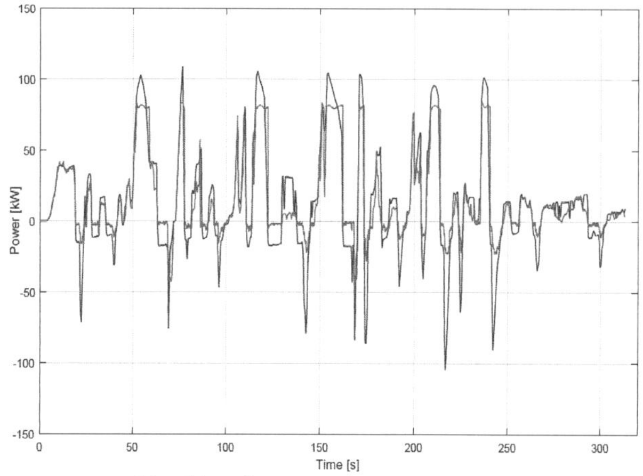


**Figure 10.** Power profile (blue line – real power, red line – simulated power).

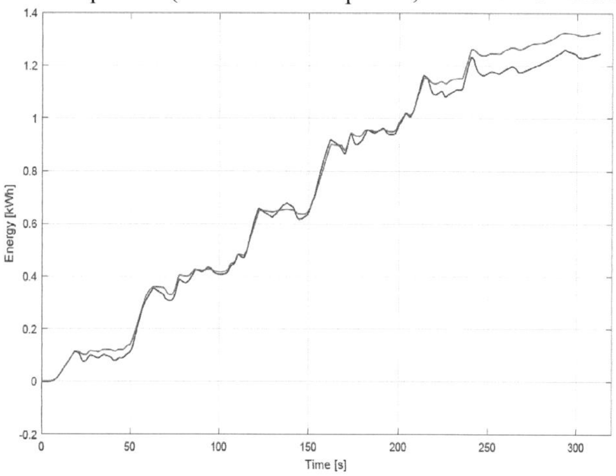


**Figure 11.** Energy profile (blue line – real energy, red line – simulated energy).

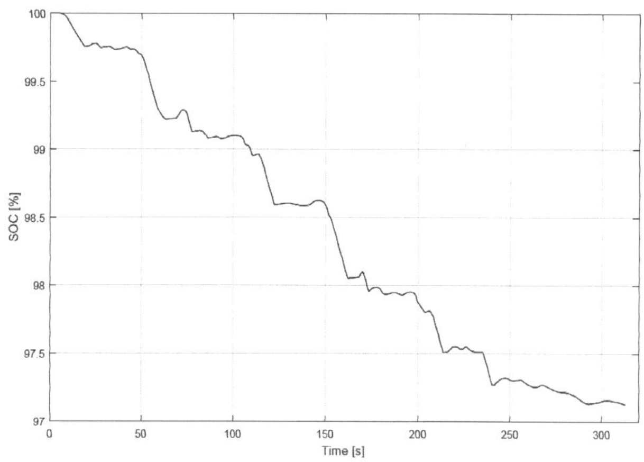


**Figure 12.** Battery SOC variation.

## 5. Comments of the results

- Regarding the speed obtained by simulation compared to the profile measured in real conditions, it is observed that these two coincide and the simulated speed follows exactly the real profile. The maximum speed reached by the electric vehicle when driving in real conditions was **106.9** km/h while the maximum value resulting from the simulation was **107.4** km/h;

- The acceleration profile of the electric vehicle was also obtained by simulation (figure 8) and the resulting maximum value was **4.168** m/s$^2$ in the 73rd second at which point the maximum torque value is also reached on the considered cycle as can be seen in figure 9;
- In the upper part of the diagram (figure 9) the simulated torque follows the real profile in a very large proportion but in the negative part, where the regenerative brakes are important, it is constant that the torque values are lower from the simulation which can be caused by the settings from „Braking Command" where the set lower limit is not enough for braking to occur exactly as in reality. The maximum value of the real torque was **259.6** Nm, and the measured one of **260** Nm, which causes an error of 0.15% between the two;
- Regarding the simulated power (figure 10), it shows some limitations of the maximum points in the positive part of the diagram, never managing to reach the nominal value of 100 kW as it happens in the case of the measured profile. This can also be determined by the „Motor Command" settings but also by the type of solver used which can introduce different errors in this regard. The maximum value reached by simulation was **83.25** kW compared to **100** kW in the real case, thus resulting in an error of 16.5%, which determines the identification of problems and the optimization of the simulation model;
- In figure 11 can see the variation of the energy consumed on the considered cycle in the two analyzed cases, which resulted in a real value of **1.248** kWh and a simulated value of **1.328** kWh so that an error can be calculated: $\varepsilon = \frac{|1.248-1.328|}{1.248} \cdot 100 = 6.41\%$. In this case, an energy consumption of **7.63** kWh/100 km was recorded by simulation under the imposed driving conditions compared to the measured one of **7.167** kWh/100 km;
- The state of charge of the battery changed by **2.87**% in the 320 seconds during the simulation, the distance traveled being **4,172** km, which means that if the vehicle traveled 100 km, the state of charge would reach **68.8**%. Thus, if we compare the theoretical autonomy mentioned by the manufacturer of **340** km [3], when running in the regime of the considered cycle, the vehicle could cover a maximum of **300** km.

## 6. Conclusions

The simulation model developed and obtained in this work and in the previous paper [1] can be used to determine the performances of an electric vehicle according to certain imposed driving conditions.

Use of measured data as input led to results that are extremely close to reality, which proves that the model is valid and ready for future optimizations that lead to even more accurate results.

Finally in order to eliminate the errors that were detected in the torque and power profiles obtained by simulation, other types of solvers will be tried and the modification of certain basic settings will be considered so that the model is no longer limited by the program.

It is very important to know that the Results obtained by simulation are very useful as a validation key.

## Acknowledgments

I would like to thank first of all to my colleague Mr. Renţea Cristian Alexandru for his support in the optimization of the simulation model and for the suggestions offered, to Mr. engineer Vişan Cristopher for his help on the part of data acquisition and measurements and last but not least to Mr. Professor Fraţilă Gheorghe, my PhD supervisor for all the advices and thoughts offered throughout the activities and of course to my mentor Mr. Professor Danciu Grigore for the support provided in my all work.

## References

[1] Ancuţa A A 2023 „*Analysis of Thermal Influence on the Operation of a Li-Ion Battery Used by an Electric Vehicle*" – published in 2023 46th International Spring Seminar on Electronic Technology (ISSE, 10-14 May, Timişoara, Romania) and accessible in IEEE Xplore, DOI:

https://doi.org/10.1109/ISSE57496.2023.10168455, ISBN 979-8-3503-3484-5, ISSN 2161-2536.
[2] https://www.automobilecatalog.com/performance/2021/2917250/peugeot_e208.html#gsc.tab=0
[3] https://ev-database.org/car/1168/Peugeot-e-208, last access 10.05.2023
[4] Racicovschi V, Danciu Gr, Chefneux M 2007 Automobile Electrice şi Hibride (in English „Electric and Hybrid Vehicles") – Editura Electra, 2007; ISBN 978-973-7728-98-2.
[5] Andreescu C, Stan C and Manea A M, - *DINAMICA AUTOVEHICULELOR PE ROȚI – Aplicații numerice* (*in english „VEHICLE DYNAMICS. Applications)* – Publishing by A.G.I.R, Bucharest, 2022, ISBN 978-973-720-885-9.

# Equalizing the battery charges of rechargeable accumulators

**O I Petrov**

Department of Transport, Technical University of Moldova, Chisinau, 9/8 Student Street, MD 2045, Republic of Moldova

E-mail: oleg.petrov@tran.utm.md

**Abstract.** Charge imbalances in high-voltage batteries, including battery cell arrays, such as lithium-ion in hybrid vehicles, tend to develop and increase over time as the batteries are charged and discharged, or even when the battery remains charged but not in use. This reduces the efficiency of the battery charging and discharging modes, and limits battery life and capacity. The article describes a method and device for performing the charge equalization operation when charging a storage battery. The elements of the charger circuit have been selected, the parameters of which are sufficient for charging with equalizing the charge of the Audi Q5 hybrid Quattro car. Calculations of the pulse mode of charging the battery of the specified car, as well as the time spent on charging with equalizing the battery charges are given. The charging time of the considered battery with the elimination of a 15% imbalance of the batteries will last up to 139 minutes (2.31 hours).The results of calculating the operating parameters of the device indicate that the device for equalizing the charge of the batteries is operable, can be implemented on a modern element base at a relatively low cost, and is capable of charging the battery in a time acceptable for the operation of the car.
**Keywords:** charging and equalizing the charges of rechargeable batteries, batteries of accumulators, Battery Management System, constant current, constant voltage, electrolytic capacitors, electronic switches, resistive voltage divider.

## 1. Introduction

Most new electric vehicles are equipped with lithium-ion batteries. Batteries account for about 40% of the cost of electric vehicles [1]. Since the voltage of individual battery cells varies from 1V to 4.5V, the cells are connected in series to increase the battery voltage level. Batteries of electric vehicles consist of elements connected in series to power electric drives of vehicles with voltage up to 400 V. At the same time, the characteristics of these elements must be strictly the same. However, it is known that no two batteries are the same, even if they are made by the same manufacturer and have the same model.

Small changes in the characteristics of each cell will still take place, be it SoC (the state of charge), be it ESR (Equivalent Series Resistance), capacity or temperature characteristics. When cells are connected to each other to form a battery, changes in characteristics result in an imbalance in cell voltages. For example, if one cell has a slightly lower capacity, it will continue to discharge, and each time it is charged, it will experience a slight overvoltage. Over time, this will lead to a decrease in its capacity and service life. This phenomenon makes equalizing the charge of individual battery cells an indispensable procedure. The process of balancing voltages and SoCs between cells when they are connected and fully charged is called cell balancing. The authors of Bortecene Yildirim and others 2019 [2] investigated the following types of battery cell balancing:

*a. Passive balancing*
*b. Active balancing*
*c. Runtime balancing*
*d. Lossless balancing*

***Passive Balancing*** This is a simple form of balancing by switching a resistor between cells. Energy losses for heating balancing resistors (usually from 30 to 40 ohms) allow cells with an increased charge to be discharged. This type is acceptable when balancing requirements are low. However, as cells age, the amount of balancing required to optimize energy increases. As a result, the amount of energy lost to heat increases. It can also increase charging time when trying to reach the maximum SoC for the entire battery cell package.

***Active Balancing*** The idea here is to redistribute energy throughout the cells. Power from the cells with the highest SoC is transferred to the cells with the lowest SoC. This is the ideal approach to cell balancing. However, this means that the system must be able to move energy between cells in a battery. Ideally between any two cells. Lots of wires and lots of switches means more weight, complexity and cost.

***Runtime Balancing*** Each cell is connected to a separate low power DC/DC converter, then each converter is connected in series. This then allows full control of the power given and received by each cell, depending on their capabilities.

This system can provide a much higher level of reliability, but requires a significant increase in cost.

***Lossless Balancing*** This approach turns the cells on and off while charging. This means we have a lot of switches and these switches must be rated for full current. Bortecene Yildirim and others 2019 [2] showed that dissipative balancing is the simplest and most economical method, but its efficiency is low and it is not suitable for high power applications. The main disadvantage of capacitive balancing is voltage balancing, which limits the speed and accuracy of balancing. Finally, live balancing has fast and efficient balancing features and is also suitable for used batteries, but with added complexity and increased cost.

Thus, each balancing method has its advantages and disadvantages, and their choice depends on the cost of the system and the application. However, capacitive balancing can be improved to improve balancing accuracy and reduce power loss.

This article is devoted to the development and research of a new Device and a method for charging and equalizing the level of charge in a battery of accumulators (application for invention number 2434 dated 24.05.23, AGEPI R. Moldova), based on the principle of charging a battery of batteries by transferring energy in small portions of storage devices due to multiple connecting them in parallel to the batteries in the second stage (charging at constant voltage).

In the process of research, methods of mathematical, structural and simulation modeling of instantaneous circuits of the charging device and aligning the level of battery charging were used. The calculations were performed using the Microsoft Office Excel 2007 program.

The list of designations of the elements presented in Figure 1 and 2:

1 – a device for charging and aligning the charge level of batteries;
2 – high-voltage battery;
$B_1$, $B_2$, ..., Bn – accumulators of the high-voltage battery;
$C_1$, $C_2$, ..., Cn – electrical energy storage devices, in particular capacitors;
$R_1$, $R_2$, ..., Rn – voltage divider resistors;
$S_1$, $S_2$ - electronic switches for switching on the battery of batteries and energy storage devices;
$BS_1$, $BS_2$, ..., BSn – electronic switches for connecting batteries to drives;
$RS_1$, $RS_2$, ..., RSn – electronic switches for connecting voltage divider resistors to energy storage devices;
D – sensor, voltage divider;
BSM - battery control unit;
DC-DC – power supply for charging batteries;
GND – ground;
K – the point of connection of the power source to the device for charging and equalizing battery charges.

The main effective electrical quantities in the device are:

$EB_1$, $EB_2$, ..., $EBn$ – EMF of high-voltage batteries;
$\boldsymbol{i}_{Bch}$ – battery charging current;
$\boldsymbol{i}_{Bcv}$ - is the equalizing current of accumulator charging from storage devices;
$\boldsymbol{i}_{Cch}$ – storage current charging;
$\boldsymbol{i}_{Scv}$ – equalizing current of storages and voltage divider for equalization of voltage of storages;
$U_{CC}$ – is the voltage of the DC-DC power supply at the stage of charging the batteries with direct current;
$U_{CV}$ – is the voltage of the DC-DC power supply at the power stage of charging batteries at a constant voltage;
$U_C$, $U_C$‘ – the voltage of the charged energy storage devices and the residual voltage after the transfer of energy to the batteries.The level of these voltages is measured using a sensor D (voltage divider);
$\boldsymbol{U_{R1}}$, $\boldsymbol{U_{R2}}$, ... , $\boldsymbol{U_{Rn}}$ – voltage of voltage divider resistors;
$\boldsymbol{U_{C1}}$, $\boldsymbol{U_{C2}}$, .., $\boldsymbol{U_{Cn}}$ – energy storage voltage;

$E_{B1}$, $E_{B2}$, .., $E_{Bn}$ – EMF of the batteries.

**2. Battery leveling device**

The proposed Device (1) for equalizing the level of charge of batteries (2) of rechargeable batteries ($B_1$, $B_2$, ..., Bn) when charging from a direct current source (DC-DC) operates under the control of the Battery Management System (BMS) and consists of: **n** energy storage devices connected in series in a circuit, which can be capacitors ($C_1$, $C_2$, ..., Cn); electronic switches ($S_1$) and ($S_2$) for connecting to a direct current source (DC-DC) circuits: **n** rechargeable batteries ($B_1$, $B_2$, ..., Bn) and n energy storage devices ($C_1$, $C_2$, ..., Cn), respectively; **n** series connected via electronic switches ($RS_1$, $RS_2$, ..., RSn) into a chain of resistors ($R_1$, $R_2$, ..., Rn), one output of each of the resistors ($R_1$, $R_2$, ..., Rn) is connected by an electrical jumper to the corresponding output of the drive energy ($C_1$, $C_2$, ..., Cn), in turn, one output of each of the energy storage devices ($C_1$, $C_2$, ..., Cn) is connected by an electrical jumper through electronic switches ($BS_1$, $BS_2$, ..., BSn) with the corresponding output connected in series in rechargeable battery circuit ($B_1$, $B_2$, ..., Bn); to the analogue input of the control unit (BMS) through a voltage divider (D) connect the input output of the drive $C_1$ to control the voltage of the drives. The connection diagram of the device is shown in fig. 1 where **n** and **i** are positive integers, **n** is greater than 1 and $1 \leq i \leq$ **n**.

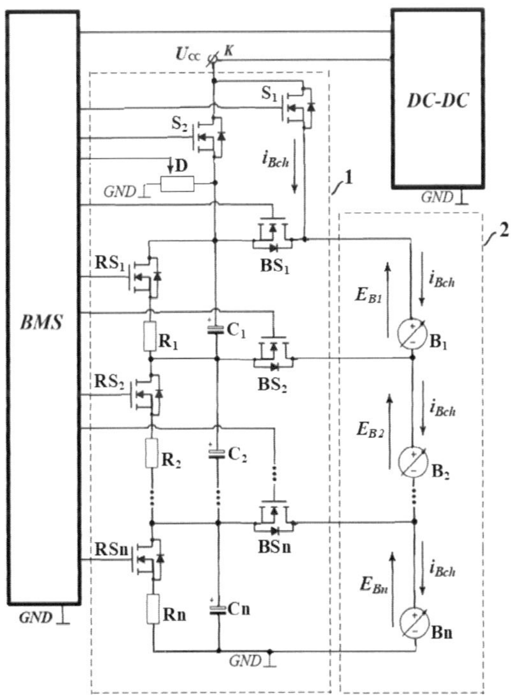


**Figure 1** Scheme of a device for equalizing the level of battery charge

**3. Battery equalization method**

One of the most common methods for charging a lithium-ion battery is the hybrid charging approach known as CC-CV. In this approach, the battery is first charged in 2 stages: charging at constant current in the CC phase, after which the battery enters the CV phase with a given constant voltage, resulting in a continuous decrease in the charging current. The battery voltage in the CV phase is increased to the maximum safe threshold. This stage ends when either the final value of the descending current, or the target power is reached [3].

In the CC charge phase, it is safe for the battery to handle higher charge currents between 0.5C and 3C. CC charging continues until the battery voltage has reached the "full" or floating voltage level, at which point, the constant voltage phase begins.

Constant voltage (CV) charge: The constant voltage (CV) threshold for Lithium cells is usually between 4.1V and 4.5V per cell. The charger Battery Management System (BMS) monitors the battery voltage during CC charging [4].

For an Audi Q5 hybrid Quattro battery pack with a capacity of 5000mA/h, the nominal charge current in the first stage can be 1000-2500mA, and the boost charge current can be in the range of 2.5-5A. In this CC mode, the battery voltage is brought to the level of the end-of-charge voltage. As soon as the desired voltage is reached, charging begins in CV mode, in which the charging current is gradually reduced. When the current drops to a minimum, charging is completed and the current is turned off.

The proposed technical solution is used under the control of the BMS program. The switching scheme of electronic switches in accordance with the phases of the state of the device is presented in Table. 1.

At the first stage of charging batteries ($B_1$, $B_2$, ..., Bn) from a power source (DC-DC) in constant current mode ***Ucc***, use:

- phase 1.0 of the circuit state (see Fig. 2, A) of the proposed device, is characterized by the fact that the energy storage circuit ($C_1$, $C_2$, ..., Cn) is connected in parallel to the battery circuit ($B_1$, $B_2$, ..., Bn) using electronic switches ($BS_1$, $BS_2$, ..., BSn) and charge the battery ($B_1$, $B_2$, ..., Bn) with a constant charging current $\boldsymbol{i_{Bch}}$, and energy storage devices ($C_1$, $C_2$, ..., Cn) with the charging current $\boldsymbol{i_{Cch}}$. Charge up to the battery charge level SoC = 80% [5].

At the second stage of charging batteries ($B_1$, $B_2$, ..., Bn) from a power source (DC-DC) in the constant voltage mode Ucv, in order to prevent the flow of charging current through the series circuit of batteries ($B_1$, $B_2$, ..., Bn), they are charged by multiple parallel connection to the appropriate energy storage devices ($C_1$, $C_2$, ..., Cn), which, before each connection, are charged to the voltage set by the battery manufacturer:

- phase 1.1 of the state of the circuit (see Fig. 2 C) of the proposed device, is characterized by the fact that for 2τ (τ is the time constant of the storage device), the voltage Ucv is applied to the chain of energy storage devices ($C_1$, $C_2$, ..., Cn) for additional drive charging ($C_1$, $C_2$, ..., Cn) up to SoC = 99% of drives;

- phase 1.2 of the state of the device circuit (see Fig. 2 C) is characterized by the fact that for a time of 1τ a resistor ($R_1$, $R_2$, ..., Cn) is connected in parallel to each storage device ($C_1$, $C_2$, ..., Rn) and cause the flow of an equalizing current $\boldsymbol{i_{Scv}}$ to equalize the voltage (***$UC_1$, $UC_2$, .., Ucn***) on each drive to the voltage specified by the manufacturer, since a chain of series-connected resistors of the same resistance ($R_1$, $R_2$, ..., Rn) divides the source voltage (DC-DC ) for voltages specified by the battery manufacturer ($B_1$, $B_2$, …, Bn);

- phase 1.3 of the state of the circuit of the proposed device, is designed to measure the total voltage (***$UC_1$, $UC_2$, .., Ucn***) of drives ($C_1$, $C_2$, ..., Cn), which is read from the sensor D and entered into the memory of the BMS device.

- phase 1.4 of the state of the circuit (see Fig. 2 D) of the proposed device, serves to charge batteries ($B_1$, $B_2$, ..., Bn) with an equalizing current $i_{Bcv}$, which flows when the storage devices ($C_1$, $C_2$, ..., Cn) and batteries are connected in parallel ($B_1$, $B_2$, … ,Bn). The peculiarity of this charging method is that when the EMF $E_B$ of the batteries (B1, B2, ..., Bn) is equal to the voltage $U_C$ of the storage devices ($C_1$, $C_2$, ..., Cn), the equalizing current iBcv between these pairs will not flow, and for those batteries , for which the EMF $\boldsymbol{E_B}$ has not reached the voltage level determined by the manufacturer, the equalizing current $i_{Bcv}$ from the storage devices will flow and their voltage will decrease, therefore the total voltage (***$U_{C1}$, $U_{C2}$, .., Ucn***) of the storage devices ($C_1$, $C_2$, ..., Cn) is also at the end charging will go down;

- phase 1.5 of the state of the circuit of the proposed device, is designed to measure the residual voltage $U_{CV}'$ of storage devices ($C_1$, $C_2$, ..., Cn) which is read from the sensor D and entered into the memory of the BMS device.

- phase 1.6 of the circuit state is designed to compare the residual voltage $U_{CV}'$ with the value stored in the memory of the BMS device. If the voltage values $\boldsymbol{U_{CV}' = U_{CV}}$ are equal, the second stage of charging the batteries (B1, B2, ..., Bn) is stopped, and if the residual voltage $\boldsymbol{U_{CV}'}$ is less, that is, part of the storage energy is spent on charging the batteries, then go to phase 1.1 and repeat the charging process.

Thus, the charging of each battery separately will be performed by current pulses of the corresponding storage device until the battery EMF is equal to the storage voltage. According to the maximum power transfer theorem, the load will receive maximum power from the source when the load resistance (Ra) is equal to the internal resistance (Rc) of the source.

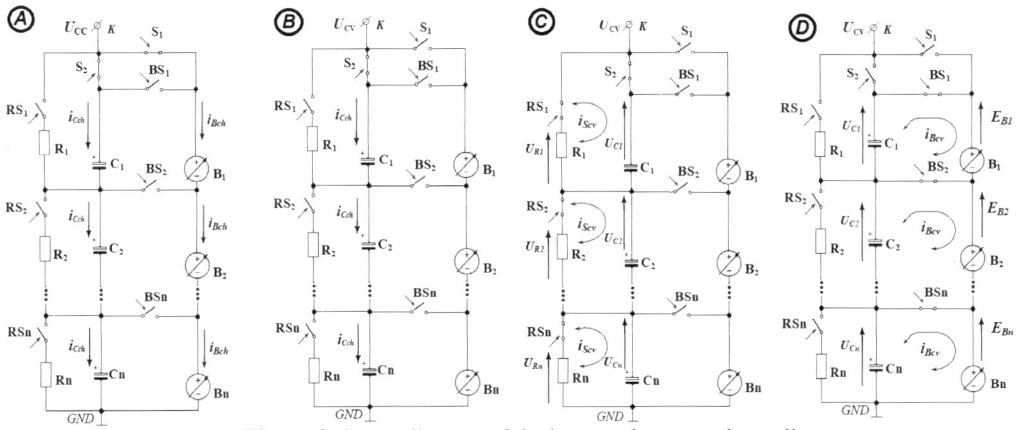


**Figure 2.** Status diagram of the battery charger and equalizer

**Table 1.** Schemes for switching on electronic switches for charging and equalizing the charge of batteries at the second stage of battery charging (2) (see Figure 2)

| Circuit state phase devices (1) | Setting the state of the battery charger circuit (2) | | Functions of the way to equalize the level of charge of the batteries (B1, B2, ..., Bn) in the battery after setting the state of the device circuit |
|---|---|---|---|
| | Electronic switch **on** | Electronic switch **off** | |
| **1.0** | $S_1$, $S_2$ | $BS_1$, $BS_2$, … , BSn; $RS_1$, $RS_2$, … , RSn | Battery 1 and energy storage devices ($C_1$, $C_2$, ..., Cn) are charged at a voltage $\boldsymbol{U_{CC}}$ with a constant current from a DC-DC source |
| **1.1** | $S_2$ | $S_1$; $BS_1$, $BS_2$, … , BSn; $RS_1$, $RS_2$, … , RSn | Energy storage devices ($C_1$, $C_2$, ..., Cn) are charged at a constant voltage $\boldsymbol{U_{CV}}$ from a DC-DC current source |
| **1.2** | $S_2$; $RS_1$, $RS_2$, … , RSn | $S_1$; $BS_1$, $BS_2$, … , BSn | The voltages $\boldsymbol{U_C}$ of the energy storage devices $C_1$, $C_2$, ..., Cn are equalized by the voltage across the resistors ($R_1$, $R_2$, ..., Rn) of the voltage divider, within 1τ of the capacitor charging time constant |
| **1.3** | | $S_1$; $S_2$; $BS_1$, $BS_2$, … , BSn; $RS_1$, $RS_2$, … , RSn | The BMS control unit reads the voltage of the energy storage devices $\boldsymbol{U_C}$ from the voltage divider D and assigns it to the variable $\boldsymbol{U_{CV}}$ |
| **1.4** | $BS_1$, $BS_2$, … , BSn; | $S_1$; $S_2$; $RS_1$, $RS_2$, … , RSn | Up to 20% of the energy of the energy storage devices ($C_1$, $C_2$, ..., Cn) is transmitted by a current pulse to the batteries ($B_1$, $B_2$, ..., Bn) during the time necessary to equalize the voltage of the battery and the storage device |
| **1.5** | | $S_1$; $S_2$; $BS_1$, $BS_2$, … , BSn; $RS_1$, $RS_2$, … , RSn | The BMS control unit reads from the voltage divider D the voltage of the energy storage devices $\boldsymbol{U_{CV}}$' |
| **1.6** | | $S_1$; $S_2$; $BS_1$, $BS_2$, … , BSn; $RS_1$, $RS_2$, … , RSn | The BMS control unit checks $\boldsymbol{U_C}$<$\boldsymbol{U_C}$' and, if the condition is met, then the cycle is repeated, if not, then the charging operation is completed with battery equalization (B1, B2, … ,Bn) |

**4. Calculation of the parameters of the charging process with battery voltage equalization**

To analyze the effectiveness of the proposed device and confirm the possibility of implementing the invention, we will carry out an approximate calculation of the elements of the device and the parameters of the process of charging the Audi Q5 hybrid Quattro battery [6].

To fulfil this condition, the battery equalization device must have a storage device, in particular a capacitor with an internal resistance (Rc) equal to the internal resistance of the battery, which for the Audi Q5 hybrid Quattro battery is Ra = 20 mΩ (see Table 2).

**Table 2**. Parameters high-voltage battery Audi Q5 hybrid Quattro

| SPECIFICATION | TECHNICAL INFORMATION |
|---|---|
| 266-volt, 72-cell, lithium-ion hybrid battery system | A38 |
| High voltage battery | |
| Rated battery voltage | 266 V |
| Battery EMF, EBA | 3.7 V |
| Number of cells (in series) batteries | $n_A$ = 72 accumulators |
| The capacity of one accumulators of the batteries | 5.0 Ач = 5000 мA/h |
| Working temperature | +15 - +55°C |
| Battery capacity | $E_{AB}$ = 13 kWth (4 680 · $10^3$ J) |
| Energy capacity of one accumulator in a battery | $E_A = 65 \cdot 10^3$ J |
| Useful battery capacity | 0.8 kWth |
| Cell (accumulator) internal resistance | *Rc* = 20 mΩ |

In our case, an electrolytic capacitor manufactured by Jb Capacitors Company (Table 3) with an internal resistance Rc = 19 mΩ is suitable [7]. However, commercially available capacitors have a wide range of parameters: the capacitance of two identical capacitors can differ by +/- 20%.

**Table 3**. Parameters of electrolytic capacitors [7]

| SPECIFICATION | TECHNICAL INFORMATION |
|---|---|
| Manufacturer | Jb Capacitors Company |
| Capacitor type | Aluminum electrolytic JNK |
| Capacity | *Cc* = 27 000 uF +/- 20% |
| Working voltage | 10 V DC (for 4.2 V reserve 135%) |
| Case dimensions | Ø30 x 35 mm |
| Resource | 20 000 h |
| Working temperature | -40° ~ +105°C |
| Resistance Type ESR 20°C, 120Hz | *Rc* =19 mΩ = 0.019 Ω |

Therefore, in charging mode 1.1, the constant voltage from the DC-DC source will be distributed unevenly across the capacitor circuit.
To eliminate this disadvantage of capacitors, in the proposed technical solution, after charging the capacitors, they are briefly connected to the voltage divider of the DC-DC source and the voltages are equalized, since the divider is created from the same resistors ($RS_1$, $RS_2$, ..., RSn), the deviation of the electrical resistance of which is not more than +/-1%.
For example, for a high-voltage Audi Q5 hybrid Quattro battery, a metal film resistor with a resistance of Rr = 5 Ohm was chosen with a manufacturing accuracy of +/- 1% (Table 4).
For electronic switches $S_1$, $S_2$, $BS_1$, $BS_2$, … , BSn; $RS_1$, $RS_2$, … , RSn accept NPN Silicon Epitaxial Transistor Bipolar FGA25N120ANTDTU1200 V, 25 A NPT Trench IGBT Transistor Table 5. Using ON Semiconductor's proprietary trench design and advanced NPT technology, the 1200V NPT IGBT offers superior conduction and switching performances, high avalanche ruggedness and easy parallel operation.

**Table 4**. Parameters of voltage divider resistors [8]

| SPECIFICATION | TECHNICAL INFORMATION |
|---|---|
| Tip: | WIREWOUND RESISTOR |
| Manufacturer | VISHAY INTERTECHNOLOGY INC. LLC |
| Model: | RS0055R000FE73 |
| Putere nominală | 5 Watts |
| Rezistență, Rsh: | 5 Ohm |
| Tolerance: | 1% |

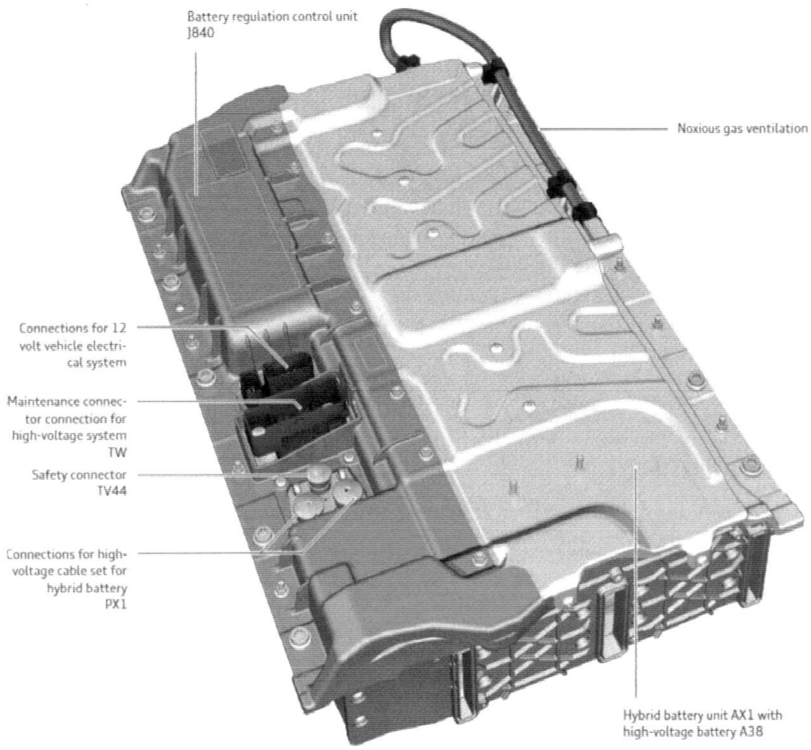


**Figure 3.** Audi Q5 hybrid Quattro high-voltage battery

The first stage of charging ends when the battery charge reaches SoC=80%. At the first stage of charging the battery (2), its batteries ($B_1$, $B_2$, ... , Bn) for example lithium-cobalt oxide batteries ($LiCoO_2$) and storage devices connected to them in parallel, in this case electrolytic capacitors) ($C_1$, $C_2$, ... , Cn) will be charged to a voltage of $U_C$ = 3.9 V.

**Table 5.** Parameters of electronic switches [9]

| SPECIFICATION | TECHNICAL INFORMATION |
|---|---|
| Marking: | GA25N120 |
| Manufacturer: | ON Semiconductor, USA |
| Hull: | TO-3P |
| Collector−Emitter Voltage: | $V_{CC}$ = 1200 Vdc |
| Collector Current | $I_C$ = 25 A |
| Maximum power dissipation: | 125 W |
| Operating temperature range: | -55 to +150°C |

DoD (Depth of Discharge) is a numerical compliment at DoD = 20%, the value of SoC = 80% [10]. If there is an imbalance of +/-10% battery charges, individual batteries of the battery will be charged to a voltage of $U_C$ = 3.85 V. (Fig. 4. at DoD = 30%)

At the second stage of charging, the proposed technical solution is used as follows: in charging mode 1.1, turn off the electronic switches ($BS_1$, $BS_2$, ..., BSn; $RS_1$, $RS_2$, ..., RSn) and to power the capacitor bank ($C_1$, $C_2$, ..., Cn) turns on the electronic switch $S_2$ and supplies a constant voltage $\boldsymbol{U_{CV}}$ from a DC-DC source to a chain of $\boldsymbol{n_C}$ = 72 capacitors. Each capacitor must receive a voltage ***Uash (i)*** = 4.20 V. To charge $\boldsymbol{n_A}$ = 72 lithium-ion batteries of an Audi Q5 hybrid Quattro battery, the voltage $\boldsymbol{U_{CV}}$ of the DC-DC source, for charging at constant voltage, must be equal to:

$$U_{CV} = U_a(i) \cdot n_A \tag{1}$$
$$U_{CV} = 4.20 \cdot 72 = 302 \text{ (V)}$$

The total resistance of internal resistances ***Rc*** of series-connected capacitors ($C_1$, $C_2$, … ,Cn) with a number ***nc***, equal to the number of batteries $n_A$ and will be:

$$R_{CS} = Rc \cdot n_A \tag{2}$$
$$R_{CS} = 0.019 \cdot 72 = 1.368 \text{ (Ohm)}$$

Therefore, when connected to the $U_{CV}$ voltage, the initial charging current can be:

$$i_{Sch} = U_{CV} / R_{CS} \tag{3}$$
$$i_{Sch} = 302 / 1.367 = 221 \text{ (A)}$$

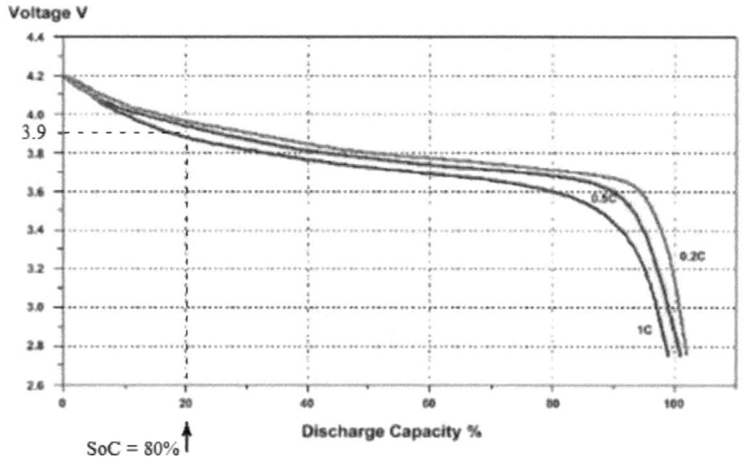


**Figure 4.** Voltage and remaining energy of lithium-ion batteries [10]

In order to prevent the capacitors from burning out from such a current, in the proposed technical solution, the capacitor banks are charged together with the batteries at the first stage of charging at a constant current value, therefore, at the end of the first stage, the capacitors will also be charged to a voltage of ***Ua' (i)*** = 3.9 V, and some due to imbalance will be charged up to UDC = 3.85 V. For capacitors, this voltage value is 91.67% (from 4.2V), which is equivalent to a charge time of $2.5\tau$ for capacitors [13].

Therefore, at the second stage of battery charging, to charge the energy storage (capacitors), there will be enough time $3\tau$ for the capacitor voltage to be 4.18 V (99.5% of 4.2 V) [13]. The total charge of the capacitor ***C(i)*** is calculated by the formula:

$$E = U^2 \cdot C / 2 \tag{4}$$

Let's take the EMF level at the beginning of the second stage of charging the battery ***Ea''*** and at the end of ***Ea''*** based on the materials of Isabelle Sourmey 2022 [11]. In this case, one storage capacitor ***C(i)*** can transfer energy to the accumulator ***B(i)*** at the beginning of the second stage with ***Uash (i)***:

$$\Delta Ecstart = (Uash\ (i)^2 - E1a^2) \cdot C / 2 \tag{5}$$
$$\Delta Ecstart = (4.2^2 \cdot 3.85^2) \cdot 0.027 / 2 = 0.038 \text{ (J)}$$

At the end of the second stage, this capacitor can transfer energy to the battery:

$$\Delta Ecend = (Uash\ (i)^2 - E2a^2) \cdot C / 2 \tag{6}$$
$$\Delta Ecend = (4.2^2 \cdot 4.10^2) \cdot 0.027 / 2 = 0.0112 \text{ (J)}$$

For one cycle of connecting the capacitor to the battery during the second stage, on average, you can transfer:

$$\Delta Ec = (\Delta Ecstart + \Delta Ecend) / 2 \quad (7)$$

$$\Delta Ec = (0.038 + 0.0112) / 2 = 0.0246 \text{ (J)}$$

Charging batteries at the second stage, using the proposed technical solution, with an imbalance of +/- 10%, means that individual batteries after the first stage can have a charge of about 80 - 85% [12], and those that did not reach 20 -25% that is, such batteries need to be added with the help of capacitors another $\Delta Ep$ = 23% of energy to charge up to $E_A$ = 65 · 103 J:

$$E_A''(i) = E_A(i) \cdot \Delta Ep / 100 \quad (8)$$

$$E_A''(i) = 65 \cdot 10^3 \cdot 23 / 100 = 15.015 \cdot 10^3 \text{ (J)}$$

Number of cycles to transfer with capacitors more $\Delta Ep$ =23% of energy to charge up to $E_A$ = 4680 · $10^3$ J:

$$N\,cycle = E_A''(i) / \Delta Ec \quad (9)$$

$$N\,cycle = 15.015 \cdot 10^3 / 0.0246 = 610.4 \cdot 10^3$$

Let us determine the charging time of the capacitors (**$C_1$, $C_2$, ..., Cn**) at the second stage. The duration of the process of charging the capacitor depends on two values: on the electromotive force of the source - $U_{CV}$ and on the time constant -.
The interval time for the capacitor for full charging is also known as transient response time $\tau$. We can find the value from the product of the resistance and capacitance [13]. Hence,

$$\tau = R \cdot C \quad (10)$$

Where:

$\tau$ = time constant, measured in seconds (s);
$R$ = resistance, measured in ohm (ohm);
$C$ = capacitance, measured in Farads (F).

For capacitors (Table 3) time constant:

$$\tau_C = R_C \cdot C_C \quad (11)$$

$$\tau_C = 0.019 \cdot 0.027 = 0.513 \cdot 10^{-3} \text{ (c)}$$

Charging capacitors ($C_1$, $C_2$, ..., Cn) to a value of 99.3% capacity, that is, up to a voltage of 4.2 V, to transfer this charge to the batteries, with an average capacitor voltage of $U_C$ = 3.93V, will require at least $3\tau$ of time. This means that a single drive charge will require:

$$t_C = -\ln(1 - UD_C / Uash(i)) \cdot 3\tau'_C \quad (12)$$

$$t_C = -\ln(1 - 3.85 / 4.20) \cdot 3 \cdot 0.513 \cdot 10^{-3} = 3.82 \cdot 10^{-3} \text{ (s)}$$

The total time spent on multiple refuel top-up ling of the energy storage (capacitor) before transferring energy to the batteries will be:

$$\sum t'_C = t'_C \cdot N\,cycle \quad (13)$$

$$\sum t'_C = 3.82 \cdot 10^{-3} \cdot 610.4 \cdot 10^3 = 2332 \text{ (s)} = 38.87 \text{ (min)} = 0.65 \text{ (hour)}$$

In the process of charging the capacitors, in state phase 1.1, the circuits of the device (1) after $2\tau$ pass into the state phase 1.2 (capacitor charging with voltage equalization). This phase lasts for another $1\tau$, that is, one third of $t'_C$, which is:

$$td'' = 3.82 \cdot 10^{-3} / 3 = 1.27 \cdot 10^{-3} \text{ (s)} \quad (14)$$

The charging time of the capacitors for power transfer in the second stage is much less than the time required to charge the lithium-ion battery. To charge a lithium-ion battery to a level of 99%, on average, it will take $t_{2A}$ = 100 minutes (6000 s). These data are accepted from Rahul Bollini 2022 [14], and from Battery University BU-409 [15].

After each top-up of the drive to transfer energy to the batteries, time will be spent on charging and equalizing the voltage of the batteries when the device is in phase 1.4 of the circuit state:

$$t_{2A}' = t_{2A} / N\,cycle \quad (15)$$
$$t_{2A}' = 6000 / 610.4 \cdot 10^3 = 9.83 \cdot 10^{-3}\ (s)$$

Therefore, if we neglect the time spent on switching, then the total operating time of the device for charging the Audi Q5 hybrid Quattro battery from 80% to 99% of capacity, with a battery charge imbalance of 20% (see Table 2) will be:

$$\sum t_{CA} = t_C' + t_{2A}' \cdot N\,cycle \quad (16)$$
$$\sum t_{CA} = (3.82 \cdot 10^{-3} + 9.83 \cdot 10^{-3}) \cdot 610.4 \cdot 10^3 = 8332\ (s) = 139\ (min) = 2.31\ (hour)$$

This confirms the possibility of charging the batteries from the capacitors at the second stage and performing the equalization of the battery charges. The increase in the charging time of the batteries of hybrid vehicles in the second stage of charging, caused by the equalization of the level of charge of the battery cells, does not exceed 40%. The comparison is based on Rahul Bollini 2022 [14] and Battery University BU-409 [15], where the indicated time was estimated at 100 minutes.

## 5. Calculation of the efficiency of the proposed equalization charger

Using a voltage divider across resistors to equalize the voltage of the capacitors before charging the batteries will cause energy wastage. To calculate the losses, the operation of the proposed device was simulated in relation to the Audi Q5 hybrid Quattro high-voltage battery (Table 2). The calculation of the instantaneous capacitor charging circuit at the moment of time corresponding to the charge time of the time constant $4\tau_C$ at the moment the voltage divider is turned on to equalize the voltages of the storage capacitors is performed. The charge time corresponding to the time constant $4\tau_C$ was chosen because the capacitors have already passed the time $2\tau$ when charging together with the batteries. This circuit will make it possible to select the value of the resistances of the resistors to equalize the voltages of the capacitors.

For a more compact representation of the calculation data, 72 batteries connected in series into a battery are divided into groups: ***nga*** = 9 groups, ***nag*** = 8 batteries in each group and, accordingly: ***ngc*** = 9 groups of capacitors, ***ncg*** = 8 capacitors.

The rated voltage ***Ua*** = 4.2 V is taken in accordance with the battery charging voltage of the Audi Q5 hybrid Quattro battery (Table 2).

The disbalance of storage capacitors capacitances is set by the coefficient of difference between the internal resistances of the group capacitors: ***Disb of Resrg(i).***

To simplify the calculation of the internal resistance of capacitors (energy storage) ***Resr'(i),*** we express in terms of data on the values (expressed as a percentage) of voltage (Voltage) and current (Current), depending on the charge of the capacitor [13]:

$$Resr'(i) = Resr(i) \cdot Voltage/Current \quad (17)$$

The equivalent resistance of the shunt resistor of the voltage divider and the internal resistance of the storage capacitor is determined by the formula:

$$Requiv'(i) = Resr'(i) \cdot Rsh/(Resr'(i)+Rsh) \quad (18)$$

The equivalent resistances of each group of series-connected resistors and capacitors are determined by the formula:

$$Rgequiv'(i) = Re'(i) \cdot nag \quad (19)$$

The value of the nominal EMF of the capacitor as the product of the battery charging voltage determined by the manufacturer and the correction factor [11] is determined by the formula:

$$EMFc' = Uc \cdot Voltage/100 \quad (20)$$

The value of the rated charging current of the capacitor as the product of the charging current at the battery charging voltage determined by the manufacturer and the correction factor [11] is determined by the formula:

$$Ic' = Idc\text{-}dc \cdot Current/100 \tag{21}$$

The value of the charging current of capacitors with voltage equalization from a DC-DC source is determined by the formula:

$$I_{dc\text{-}dc}' = U_{dc\text{-}dc} / \sum_{i=1}^{i=ngc} Rgequiv'(i) \tag{22}$$

Electromotive force of a group of capacitors:

$$EMFcg' = Ecg \cdot Voltage/100 \tag{23}$$

Power loss to shunt resistors:

$$\Sigma Ploss\ instant' = \sum_{i=1}^{i=ngc} (Ucg'(i) \cdot Ishg'(i)) \tag{24}$$

Energy loss for shunt resistors:

$$\Sigma Eloss\ instant' = \sum Ploss\ instant' \cdot Ncycle \cdot td' / 3600 \tag{25}$$

Equivalent resistance of a group of shunted capacitors:

$$Resrg'(i) = \Sigma Rcesrg \cdot (Voltage / Current) \cdot Disb\ of\ Resrg(i) \tag{26}$$

Equivalent resistance of a group of shunted capacitors:

$$Rgequiv'(i) = Rshg \cdot ncg \cdot R'esrg(i) / (Rshg \cdot ncg + Resrg'(i)) \tag{27}$$

Capacitor group voltage:

$$Ucg'(i) = Rgequiv'(i) \cdot Idc\text{-}dc' \tag{28}$$

Capacitor group charging current:

$$Icg'(i) = Ucg'(i) / Resrg'(i) \tag{29}$$

The current of a group of resistors shunting capacitors:

$$Ishg'(i) = Ucg'(i) / (Rshg \cdot ncg) \tag{30}$$

Power loss on groups of shunt resistors:

$$Pc(i) = Ucg'(i) \cdot Icg'(i) \tag{31}$$

Energy loss on groups of shunt resistors:

$$Plossg'(i) = Ucg'(i) \cdot Ishg'(i) \tag{32}$$

Voltage applied to shunted capacitor:

$$Uc'sh(i) = Ucg'(i) / ncg \tag{33}$$

Voltage applied to capacitors:

$$Uc'sh\ less\ (i) = Udc\text{-}dc \Big/ \sum_{k=1}^{k=ngc} Resrg'(k) \cdot Resrg'(i) / ncg \tag{34}$$

**Table 6.** Initial parameters of the energy storage voltage equalization mode

| *Initial data* | | *Data adapted to Time Constant τ = 4* | |
|---|---|---|---|
| *Uc = Ea charging ,(V)* | **4.2** | *t'd (c) equalizat. time* | 0.00127 |
| *ncg* | 8.00 | *N cycle* | 610400 |
| *ngc* | 9.00 | *Voltage (%)* | 98.16 |
| *Resr nom, (Ohm)* | 0.019 | *Current (%)* | 0.02 |
| *ΣRcesrg, (Ohm)* | 0.15 | *EMF' c (V)* | 4.12 |
| *Rsh nom, (Ohm)* | 5.00 | *I'c (A)* | 0.04 |
| *Rshg, (Ohm)* | 40.00 | *I'dc-dc (A)* | 0.88 |
| *EMFcg, (V)* | 33.60 | *EMF'cg (V)* | 32.98 |
| *U dc-dc nom, (V)* | 302.40 | *ΣP'loss instant(Wt)* | 254.03 |
| *I dc-dc by* τ=0, (A) | 221.05 | *ΣE'loss instant(Wth)* | 54.70 |

**Table 7.** Expected parameters of the energy storage voltage equalization mode

| *Disb of Resrg(i)* | **1** | **1,2** | **1** | **0,8** | **1** | **1** | **0,8** | **1** | **1,2** |
|---|---|---|---|---|---|---|---|---|---|
| *Resrg' (i),(Ohm)* | 916.73 | 1100.08 | 916.73 | 733.39 | 916.73 | 916.73 | 733.39 | 916.73 | 1100.08 |
| *Rgequiv'(i),(Ohm)* | 38.33 | 38.60 | 38.33 | 37.93 | 38.33 | 38.33 | 37.93 | 38.33 | 38.60 |
| *Ucg' (i), (V)* | 33.62 | 33.86 | 33.62 | 33.28 | 33.62 | 33.62 | 33.28 | 33.62 | 33.86 |
| *Icg' (i), (A)* | 0.04 | 0.03 | 0.04 | 0.05 | 0.04 | 0.04 | 0.05 | 0.04 | 0.03 |
| *I'shg (i), (A)* | 0.84 | 0.85 | 0.84 | 0.83 | 0.84 | 0.84 | 0.83 | 0.84 | 0.85 |
| *Pc (i), (Wt)* | 1.23 | 1.04 | 1.23 | 1.51 | 1.23 | 1.23 | 1.51 | 1.23 | 1.04 |
| *Plossg' (i), (Wt)* | 28.27 | 28.66 | 28.27 | 27.68 | 28.27 | 28.27 | 27.68 | 28.27 | 28.66 |
| *Uc' sh (i) (V)* | **4.20** | **4.23** | **4.20** | **4.16** | **4.20** | **4.20** | **4.16** | **4.20** | **4.23** |
| *Uc' sh less (i) (V)* | **4.20** | **5.04** | **4.20** | **3.36** | **4.20** | **4.20** | **3.36** | **4.20** | **5.04** |

The calculation was made for the case of using capacitors with a typical capacitance deviation of +/- 20% (parameters in Table 3), used as energy storage devices for pulsed energy transfer to batteries and divider resistors (parameters in Table 4). The resistors of the charging voltage divider $U_{CV}$ used to equalize the voltage of the capacitors made it possible to equalize the voltage to the limits that allow charging the batteries (see Fig. 5. and Table 7).

A new technical solution to the problem of charging a battery of batteries with the equalization of battery charges using energy storage devices is proposed. A device and a method for charging batteries connected in series are described, which excludes the flow of current through batteries that have reached the level of full charge, when charging batteries that have not reached this level.

By calculating the instantaneous mode scheme at the second stage of voltage equalization and charging of storage devices with the expected spread of storage parameters, it is shown that charging lithium-ion batteries using the proposed device and the proposed method of high-voltage battery Audi Q5 hybrid Quattro allows charging lithium-ion batteries with current pulses , formed by capacitive storage devices at the second stage, at a constant value of the charging voltage close to the nominal one, preventing the flow of current through the charged batteries of the battery and without exceeding the permissible level of the charging voltage of the batteries.

The charging time of the considered high-voltage battery with the elimination of 15% of the imbalance of the batteries will last up to 139 minutes (2.31 hours).

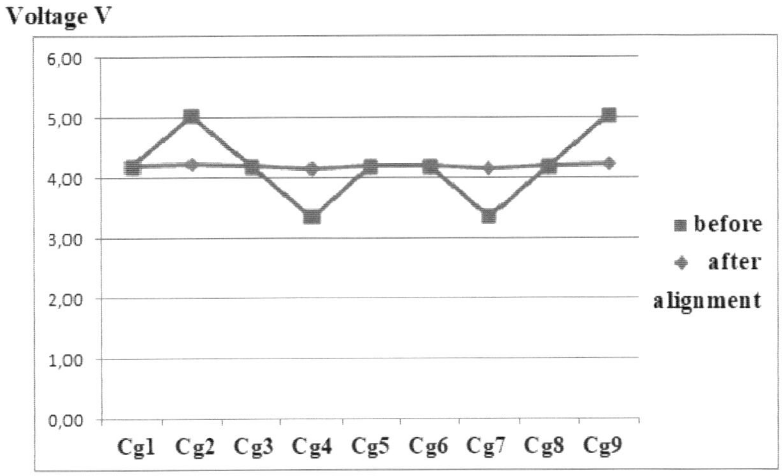


**Figure 5.** Capacitor voltage in device drive groups.

To equalize the charges of the Audi Q5 hybrid Quattro high-voltage battery using capacitive storage devices in this example, additional energy costs $\Sigma Eloss'$ = 54.70 Wth will be required, which is 4.2% of the full battery charge $E_{AB}$ = 1300 Wth. The expected energy loss when using a particular Audi Q5 hybrid Quattro high-voltage battery is 4.2% of a full battery charge.

## 6. Discussion

The proposed device allows you to perform individual charging of each battery at the final stage of this process using a simpler circuit compared to other methods with individual control of the voltage of each individual battery. The performed study, despite the calculations of the parameters of the device and its modes of operation, in relation to the real-life high-voltage battery Audi Q5 hybrid Quattro, requires additional studies of equalizing processes when equalizing the voltages of capacitive energy storage devices, as well as the processes of charging batteries from the energy storage devices of the proposed device. Theoretical studies of the operation of the device will require the use of special programs and a lot of time. It is advisable to carry out further studies of the proposed device experimentally on the operating model of the device in the modes of charging real batteries using modern measuring equipment at the appropriate stand. The creation of a test bench for studying the charging modes of lithium-ion batteries will not only provide new scientific results and bring the proposed technical solution to practical application, but will also allow it to be used for laboratory work in such disciplines as "Electrical and electronic equipment of vehicles", "Electric and hybrid cars".

## 7. Conclusions

The proposed device for aligning the level of battery charging during battery charging can be created on the basis of commercially available electronic components and used to equalize the charges of high-voltage batteries used in the automotive industry, as well as to accumulate energy from renewable energy sources.

The use of the proposed device and the method of equalizing the level of charge of the batteries in the battery allows you to fully use the useful capacity of the batteries.

Using the proposed device increases the charging time in the second stage by 40%, so the device can be used with each charge.

The proposed device and method for leveling the charge level of the batteries in the battery differs in that it allows:

- will increase the service life of batteries by eliminating the flow of current through fully charged batteries, that is, to prevent them from overcharging and overheating;
- will reduce energy losses for heating balancing resistors, since they are turned on for a short time in the phase of capacitor voltage equalization;

- does not require the connection of sensors to control the voltage of each battery cell in the battery separately.

**Conflicts of Interest:** The author declare no conflict of interest. The funding sponsors had no role in the design of the study; in the collection, analyses, or interpretation of data; in the writing of the manuscript, and in the decision to publish the results.

**References**

[1] Andrew Ryan. Tackling the rising cost of electric vehicle batteries, *Fleet News,* 20/07/2022 Available online: https://www.fleetnews.co.uk/electric-fleet/costs-and-incentives/tackling-the-rising-cost-of-electric-vehicle-batteries
[2] Bortecene Yildirim and others. Evaluation and Comparison of Battery Cell Balancing Methods ©2019 IEEE, Available online: https://eprints.ncl.ac.uk/file_store/production/261931/F4ADEC30-2EDD-4881-A81A-AC0BFDE3D9EB.pdf
[3] Kailong LIU, Kang LI, Qiao PENG, Cheng ZHANG A brief review on key technologies in the battery management system of electric vehicles Front. Mech. Eng. 2019, 14(1): 47–64 Available online: https://doi.org/10.1007/s11465-018-0516-8
[4] BU-409: Charging Lithium-ion. "Batteries in a Portable World - A Handbook on Rechargeable Batteries for Non-Engineers", 4th edition of Battery University, 25-Oct-2021, Available online: https://batteryuniversity.com/article/bu-409-charging-lithium-ion
[5] Isabelle Sourmey, Charging your lithium-ion batteries: 5 expert tips for a longer lifespan. Saft Connected Energy Division- January 13, 2022. Available online:https://www.saft.com/energizing-iot/charging-your-lithium-ion-batteries-5-expert-tips-longer-lifespan
[6] SSP 489 Audi Q5 hybrid Quattro, 2007. Available online: https://procarmanuals.com/vag-ssp-489-audi-q5-hybrid-quattro/ (in Deutsch)
[7] Jb Capacitors Company, 2023. Available online: https://www.jbcapacitors.com/pdf/JNK-5000H-at-105-Snap-in-Aluminum-Electrolytic-Capacitor.pdf
[8] Chipdip, Armenia, Sales office of VISHAY INTERTECHNOLOGY INC. LLC, USA, 2023, Available online: https://www.chipdip.am/product/rs0055r000fe73-wirewound-resistor-5-ohm-5w-1 (in Russian).
[9] NdFT Company, 2023. Available online: https://ndft.com.ua/product-category/semiconductor-devices/tranzistory-i-drugie-aktivnye-komponenty/silovye-tranzistory-bipolyarnye/ (in Russian).
[10] Rob McEwen Voltage Vs remaining energy of Lithium Ion batteries, 2020, Available online: https://meloyelo.nz/knowledge-base/voltage-vs-remaining-energy-of-lithium-ion-batteries/
[11] ALL ABOUT BATTERIES- Isabelle Sourmey - Application Engineer at Saft Connected Energy Division- January 13, 2022, Available online: https://www.saft.com/energizing-iot/charging-your-lithium-ion-batteries-5-expert-tips-longer-lifespan.
[12] Charging Lithium-Ion and LiPo Batteries the Right Way, by Maker.io Staff. 10/13/2021, Available online: https://www.digikey.com/en/maker/blogs/2021/charging-lithium-ion-and-lipo-batteries-the-right-way.
[13] Simple Equation for Capacitor Charging With RC Circuits Copyright © 2023 Wira Electrical. Available online: https://wiraelectrical.com/equation-for-capacitor-charging/.
[14] Rahul Bollini, Understanding Charge Discharge Curves of Li-ion Cells, EVreporter Magazines Jan 2022, Available online: https://evreporter.com/understanding-charge-discharge-curves-of-li-ion-cells/.
. [15] The material on Battery University BU-409: Charging Lithium-ion Figure 3: Volts/capacity vs. time when charging lithium-ion. 2023. Available online: https://batteryuniversity.com/article/bu-409-charging-lithium-ion

IOP Conf. Series: Materials Science and Engineering 1303 (2024) 012008 doi:10.1088/1757-899X/1303/1/012008

# A Duster Plug-in Hybrid Electric Vehicle with Wireless Charging System

**D G Marinescu[1], A Marinescu[2], V Nicolae[1], L Calin[3], S Parlac[1], I Vieru[1] and A Stan[1]**

[1]National University of Science and Technology *Politehnica* Bucharest – Pitesti University Center

[2]Technical Sciences Academy of Romania

[3] Romanian Automotive Register

E-mail: viorel.nicolae@upit.ro

**Abstract.** The paper presents the *GASTER Electricway 4WD* concept, Plug-in Hybrid Electric Vehicle (PHEV) type, in development within the *Automotive Engineering* Research Centre of the University of Pitesti, aiming to create a green and versatile version of Romanian *Dacia Duster*. The car has been built by implementing a hybrid propulsion system with electrified rear axle. This hybrid system is PHEV, parallel type with two shafts, organized in a motorized solution E4WD (Electricway 4WD). It is composed of a thermal power train with a 1.6 litre engine fuelled with Compressed Natural Gas (CNG) and a CVT (Continuously Variable Transmission), placed in the front of the vehicle. The electric powertrain, located at the rear side of the vehicle includes an electric motor coupled with a mechanical transmission. The traction battery is Lithium-ion NMC type. The charging of the traction battery can be performed with the on board charger of by cable from a Wallbox or public station, in the *Standard* version, and Plug-in bidirectional or Wireless from renewable sources in the *Premium* version.

## 1. Introduction

The paper reveals for the first time the details of the *GASTER Electricway 4WD PHEV* concept car project (Figure 1).

**Figure 1**. *GASTER Electricway 4WD* PHEV concept.

It is a Plug-in Hybrid Electric Vehicle Concept in development within the *Automotive Engineering* Research Center of the National University of Science and Technology *Politehnica* Bucharest – Pitesti University Center in a partnership project between *Automotive Engineering* Research Center and *Renault Technologie Roumanie*.
The vehicle is built on the platform of the SUV *Dacia DUSTER II,* made by the Automobile Dacia - Group Renault at their plant located near Pitesti city, Romania.

## 2. Main project objectives

The *GASTER Electricway 4WD PHEV* concept has been developed aiming to create a green SUV part time all-wheel drive type. This will be used for research of the means to reduction of pollutant emissions under 100g CO2/km, using a hybrid system with engine fueled by Compressed Natural Gas (CNG) coupled by an intelligent transmission CVT and a part time all-wheel drive by rear axle electrified.

**Table 1.** Main *GASTER Electricway 4WD PHEV* performances.

| | |
|---|---|
| CO2 Emissions | under 100 g/km |
| Maxim speed in electric mode | 80 km/h |
| Autonomy in electric mode | 50 km |
| Autonomy in thermic mode with CNG | 300 km |
| Reserve autonomy in thermic mode fueled with Petrol | 300 km |

## 3. Architecture

The *GASTER Electricway 4WD* concept has an architecture similar with *GRAND HAMSTER Electricway 4WD* [1]. It is realised with new components and represents a Plug- in evolved solution of, in agreement with Euro 6 pollution requirements. The architecture of the *GASTER Electricway 4WD* concept is presented in Figure 2. It is a parallel system type, with torque addition and with two shafts.

The organization of this hybrid propulsion equipment *Plug-in* type is done by dividing it in two parts, (motorized solution E- 4WD): the thermal powertrain in the front side and the electric powertrain in the rear side.

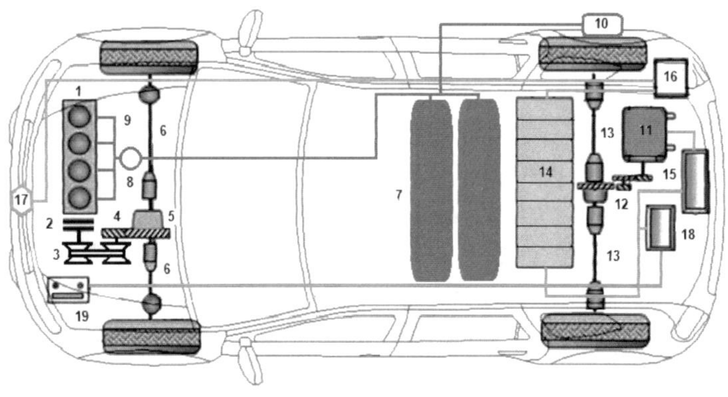


**Figure 2.** The architecture of the *GASTER Electricway 4WD PHEV* concept.

The thermal powertrain with engine (1) mainly fuelled by CNG (8,9) coupled by a CVT transmission (2,3,4,5) is mounted in the front side of the vehicle. The electric powertrain with electric motor (11) added is mounted in rear side. The CNG fuel tank with 2 cylinders (7) is located under the rear seats.

For safety, it has been provided a supplementary gasoline fuel tank (not presented in the figure), located on the rear side of the vehicle. The refill sockets of CNG and gasoline (10) are placed on the rear right side. The lithium-ion battery (14) is installed above the rear axle. The Traction Inverter Module for the electric motor (15), the on board battery charger (16), the auxiliary 12 V battery charger (18) are mounted in the rear side. The refill socket of the traction battery (17) and the auxiliary 12 V battery (19) are placed on the front side.

The architecture of the thermal management of vehicle includes 4 complexes hydraulic circuits for colling/heating of the components: one colling high temperature circuit of the thermic engine, one colling low temperature circuit of the electric components, one heating circuit of the CNG reducer and one climate circuit in the cabin.

## 4. Thermal powertrain

The thermal powertrain mounted in the front side is created with elements from Renault-Nissan, not yet introduced in the production of the Dacia vehicles. It includes an engine, 1,6 SCe (Figure 3, right) fuelled for the first time by Compressed Natural Gas (CNG) (Figure 4) coupled with an automatic gearbox X – *Tronic* CVT DK0 type, V- Belt (Figure 5).

**Figure 3.** The new thermal powertrain of the *GASTER Electricway 4WD PHEV* with engine fueled by CNG and coupled with an automatic gearbox, CVT Type

The engine of the *GASTER Electricway 4WD PHEV* is an 1,6 SCe, four-cylinder, H4M type, 84 kW @ 5500 rpm, liquid cooled [2] fuelled for the first time by Compressed Natural Gas (CNG) Figure 4.

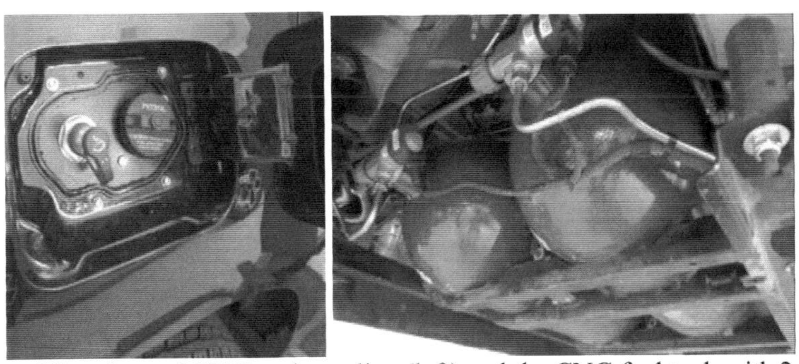

**Figure 4.** The refill sockets of CNG and gasoline (left) and the CNG fuel tank with 2 cylinders (right) of the *GASTER Electricway 4WD PHEV* concept.

Considering the engine`s adaptation to CNG fueling and its integration in a PHEV, other modifications were operated, as follows:

- Integration of the specific components of the LANDI RENZO OMEGAS EVO natural gas system that meets the strictest European emissions standards Euro 6. The system features OBDII connectivity via the K and CAN protocols and offers thorough diagnostics on the connection with petrol injectors and the functioning of individual gas injectors [3].
- Replacement of the original AC compressor with an electro compressor, fuelled by the traction battery.
- Building a new exhaust gas system, adapted to the architecture of the central and rear areas under the floor.
- Building a new gasoline tank for stopping the gasoline vapours during the CNG and electric mode functioning.
- Building a *ThermoSTART* preheating system for reducing the CO2 emission at cold starting of the engine fuelled by CNG, powered with energy from the traction battery or from a charging point in the charging mode.

The second main novelty of the concept is the thermal engine coupling H4M type with a Nissan CVT (Continuously Variable Transmission), new *X Tronic* generation [4]. The transmission features a ratio coverage of 7.0 using a sub planetary gear to switch between low and high ranges, allowing smaller pulleys to reach higher transmission ratio coverage. This raises responsiveness on start-ups and at low speeds while allowing lower engine rpm at high speeds, resulting in higher fuel economy. This generation *X Tronic CVT* model achieves an improvement in fuel economy of up to 10% compared to previous CVTs for comparable vehicles. Combined with Adaptive Shift Control (which adapts shifting patterns to match each driver's style and the road), this transmission delivers responsive and powerful acceleration. It also helps keep the engine from revving too fast at high speeds and minimize noise.

**Figure 5.** The cockpit of the *GASTER Electricway 4WD PHEV* concept with Continuously Variable Transmission.

## 5. Rear axle and electric powertrain

The electric powertrain with an electric motor induction type and a reducer- differential transmission of the *GASTER Electricway 4WD PHEV* (figure 6, center) is mounted in the rear side in an innovative solution, with an *H* type axle and twisted travers (figure 6, left). This rear axle is an adapted solution of the *GRAND HAMSTER Electricway 4WD* [1].

The electric motor, liquid cooled, has the maximum power peak 32 kW, able to achieve a maximum speed of 80 km/h with zero $CO_2$ emissions. This electric motor also handles the harvest of energy during braking and deceleration in order to recharge the battery and increase the range for 100% electric driving.

In order to adapt the rear axle to the new weight distribution, the rear suspension has been revised and includes two pneumatically supplementary air suspensions (figure 6, right).

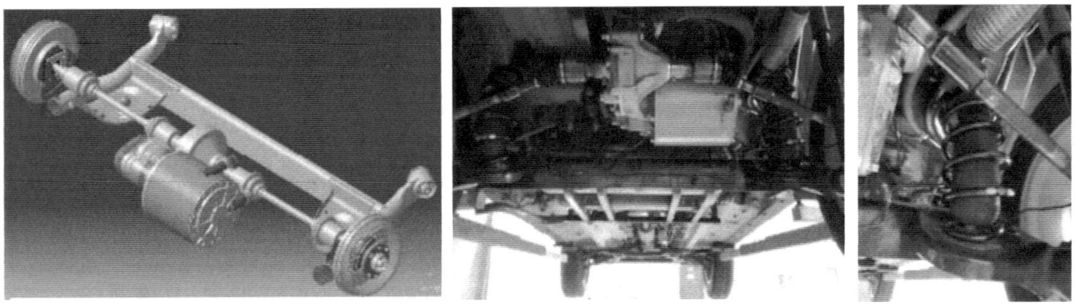

**Figure 6.** The rear axle "H" type (left), the electric powertrain (center) and the pneumatically supplementary air suspensions (right) of the *GASTER Electricway 4WD PHEV* concept.

## 6. The Traction Battery

The Traction Battery System includes the battery pack, the Battery Management System and the Battery Monitor System.

The battery pack is Lithium Nickel-Manganese-Cobalt (NMC) technology 400V, 9,8 kWh from the Renault CAPTUR PHEV. It has 96 cellules, series coupled [2]. This battery 1 equipped with external switch 2 is placed on the floor of the rear side of vehicle (Figure 7).

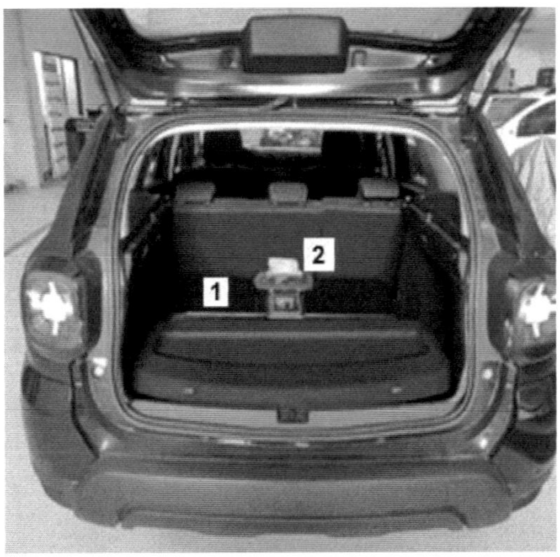


**Figure 7**. The Traction Battery of the *GASTER Electricway 4WD PHEV* concept installed above the rear axle.

**Table 2.** The Traction Battery characteristics.

| | |
|---|---|
| Technology | Lithium Nickel-Manganese-Cobalt (NMC) |
| Voltage | 400 V |
| Capacity | 9,8 kWh |
| Number of cells/modules | 96 / 8 |
| Management | BMS Orion 2, 96 channels |
| Cooling | liquid |
| Mass | 115 kg |

The battery is controlled by an Orion 2 Li-Ion Battery Management System [5]. This BMS protects and monitors the battery pack by monitoring several sensors and uses several outputs to control charge and discharge into the battery. The BMS measures inputs from cell voltage taps, the total pack voltage tap, a hall effect current sensor and thermistors. Using the programmed settings, the BMS then controls the flow of current into and out of the battery pack through broadcasting charge and discharge current limits via the CANBUS. The BMS relies on the user to provide external controls that respect the current limits set by the BMS to protect the batteries as the BMS does not have integrated switches. During and immediately after charging, the BMS will balance the cells using internal shunt resistors based on the programmed settings.

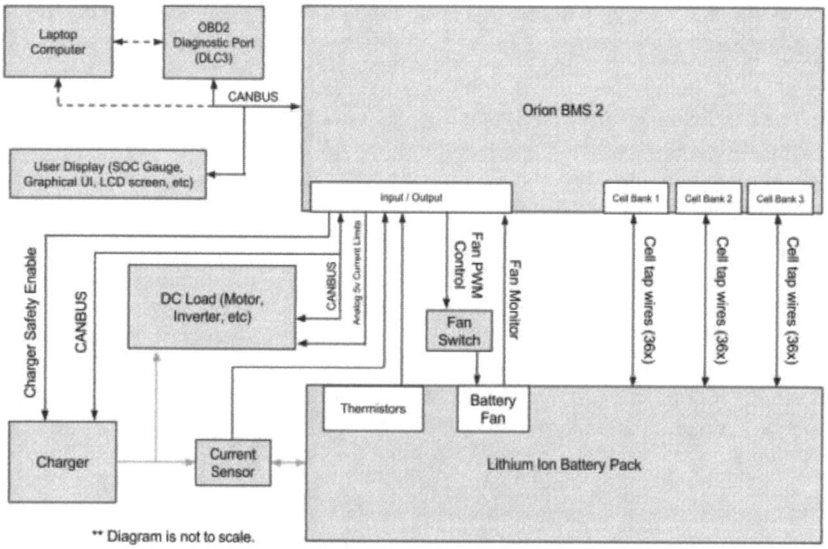


**Figure 8**. The diagram of the ORION 2 Battery Management System adapted of the *GASTER Electricway 4WD PHEV* Traction Battery [5].

## 7. The *GASTER Electricway 4WD* Batteries Charging Systems

The GASTER Electricway 4WD PHEV concept is equipped with two batteries: the 12 V auxiliary battery that feeds the low-voltage electrical network and the 400 V traction battery of the electric propulsion system. The 12V auxiliary battery is charged by a DC/DC converter of 1000.W TDC IT 320 – 12 type, air cooling, with CAN protocol.

Two types of charging systems are provided for the traction battery: The *Standard* system, by conduction (plug-in) and the *Premium* system, by induction (wireless) and plug-in (bidirectional).

### *7.1 The* Standard *charging system*

The *Standard* type achieves plug-in charging via a 3.3 kW AC, air cooled On Board Charger (OBC), EVPT23-32010 type. It's equipped with CAN communication interface to realize real-time communication with BMS. The charger with high efficiency, high stability and high protection grade (IP67) realizes an intelligent temperature compensation function in the charging process, preventing the damage to battery caused by charge-off or charge due, greatly extending the lifespan of battery.

The socket integrated on the front of the vehicle (figure 9, right) is a standard *Mennekes* socket, Type 2 with charging protocol for communication between vehicle and charging station.

**Figure 9.** The socket 1 Phase, *Mennekes* Type 2 of *the GASTER Electricway 4WD* PHEV for charging the traction battery in *Standard* Mode.

To monitor the state of operation of the plug-in charger, an operation indicator is mounted near to the socket and the plug is removed from the socket with a release button (figure 9, left).
In *Standard* mode the 9.8 kWh, 400 V battery is charged by a 3.3 kW, 10 A OBC from a Wallbox in at least 180 minutes.

*7.2 The* Premium *Wireless charging system and standardization*
For the *GASTER Electricway 4WD PHEV Premium* charging version, a bi-directional charger has been designed to enable G2V/V2G/V2L [6] combined with a wireless charging system. This solution is described in Figure 9.

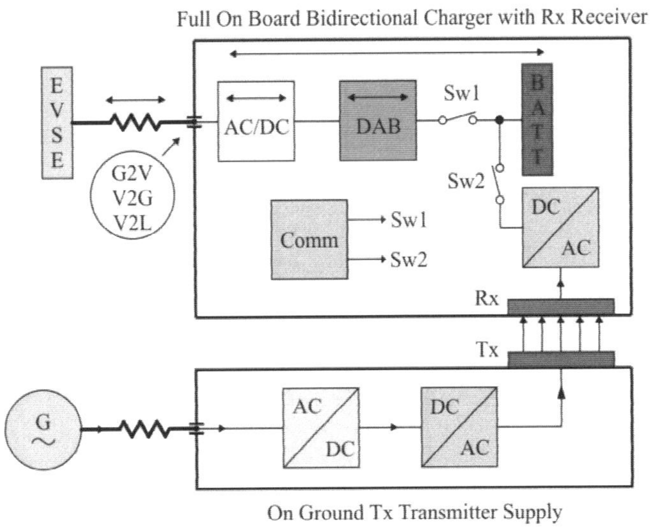


**Figure 10.** GASTER E 4 WD PHEV *Premium* charging system.

The system contains an integrated bi-directional charger for controlled charging/discharging of the battery in the V2G/G2V/V2L system [6] and a system for receiving the energy transmitted wirelessly via the Rx coil located in the lower rear part of the PHEV (Figure 11). The V2L (Vehicle to Load) scheme is a function that allows stored energy to be used to power external devices such as camping equipment, power tools and a variety of consumer devices. Because the battery capacity is relatively small, powering appliances in a home (fridge, air conditioner, etc.) can only be done for a short time. Switching from the plug-in to the wireless system is done from the vehicle on-board via the static switches Sw1 and Sw2 operated via the *Comm* block. Table 3 shows the position of the SW switches according to the chosen mode.

**Table 3.** Switch position Sw1 and Sw2.

| | Sw1 | Sw2 |
|---|---|---|
| G2V, V2G, V2L | on | off |
| Wireless | off | on |

The wireless system is powered either from the grid or from a range extender G which can be a photovoltaic, wind, portable electric or CNG-powered motor generator.

Contactless charging (WPT-Wireless Power Transfer) of EV or PHEV batteries is currently a consolidated technique through standards created by IEC [7] and SAE [8]. The situation was completely different in 2014-2017 when the *DACIA Electron EV* concept, the first Romanian electric car [9] on a *DACIA Sandero* vehicle base was developed. The theoretical and practical experience gained then was used in the *GASTER Electricway 4WD PHEV*.

These standards offer solutions for three power classes: 3.7, 7.2 and 11 kW, generically referred to as WPT1, WPT2 and WPT3, and three working distances depending on the ground clearance of the vehicle in question, defined as follows: Z1 =100-150, Z2=140-210 and Z3=170-250 mm. The standards impose a single operating frequency (85 kHz) for the transfer of electrical energy between the Tx-Rx coils of the inductive coupler associated with the system interoperability conditions according to ISO 19163 [10]. The three WPT power classes and Z distances must be interoperable so that both residential and public charging stations can be used regardless of the power level for which they have been sized.

The Table 4 shows the interoperability conditions according to the power class and the Table 5 shows the interoperability conditions according to the ground clearance of the studied vehicle. For *GASTER Electricway 4WD PHEV* class Z2 is chosen which covers all payload regimes at which it could be used: 140 mm for maximum load up to 210 mm for unloaded vehicle. The choice of power class for the 9.8 kWh battery depends on the duration of the charging process desired by the user.

**Table 4.** The interoperability at different power classes.

| | | Rx | | |
|---|---|---|---|---|
| | | WPT 1 | WPT 2 | WPT 3 |
| Tx | WPT 1 | Yes | Yes | Yes |
| | WPT 2 | Yes | Yes | Yes |
| | WPT 3 | Yes | Yes | Yes |

**Table 5.** The interoperability at different distances Tx-Rx.

| | | Rx | | |
|---|---|---|---|---|
| | | WPT 1 | WPT 2 | WPT 3 |
| | Z1 | Yes | No | No |
| Tx | Z 2 | Yes | Yes | No |
| | Z 3 | Yes | Yes | Yes |

The 9.8 kWh battery at 400 V has a capacity of 24.5 Ah. When charging at power class WPT1 (3.7 kW), the charging current is 9.25 A (C-Rate = 0.38) and the charging time is 160 min, and at power class WPT2 (7.2 kW) the charging current is 18 A (C-Rate = 0.73) and the charging time drops to 78 min. As a result, it is considered that the Tx/Rx sizing in class WPT2 is acceptable depending on the battery cooling conditions.

*7.3 The sizing the inductive coupler*

As shown above, the *GASTER Electricway 4WD PHEV Premium* is equipped with an Rx receiver sized for WPT2/Z2 according to J2954 [8] shown in Figure 11.

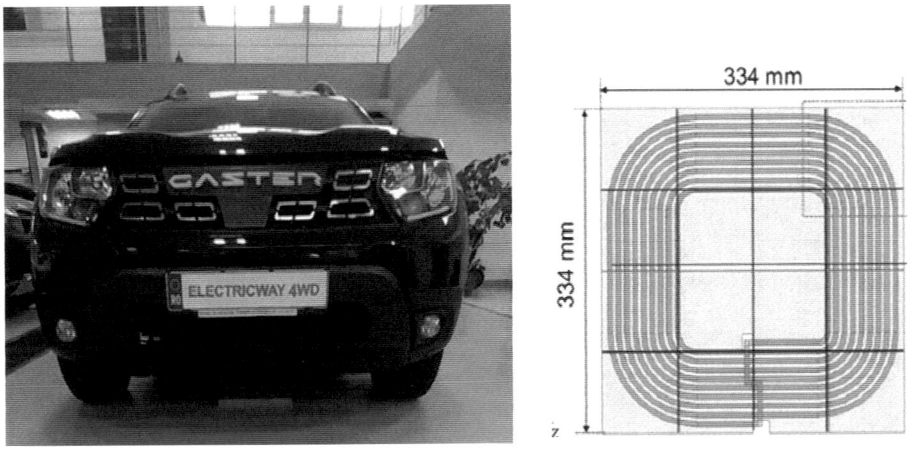


**Figure 11.** The Tx transmitter located on the ground (left) and the Rx dimensions according to [8] (right).

The Rx is square-shaped with rounded corners, has 8 turns and 4 mm pitch. Under the coil is a magnetic field concentrator made of ferrite and a 0.5 mm Al plate that prevents the residual magnetic field from penetrating the vehicle's steel underbody. The whole assembly is placed in a rugged housing made of a dielectric material.

**Conclusions**

The *GASTER Electricway 4WD* is a Plug-in Hybrid Electric Vehicle Concept in development within the *Automotive Engineering* Research Center of the National University of Science and Technology *Politehnica* Bucharest – Pitesti University Center in a partnership project between *Automotive Engineering* Research Center and *Renault Technologie Roumanie*.

It represents an experimental vehicle for studying the ways of reducing the pollutant emissions of DACIA DUSTER car under 100g $CO_2$/km, using a hybrid system with engine fueled by Compressed Natural Gas (CNG) coupled by an intelligent transmission CVT and a part time all-wheel drive by rear axle electrified.

For reducing the pollutant emissions, we building a new gasoline tank that stopping the gasoline vapors during the CNG and electric mode functioning.

The starting of the thermal engine fueled by GNC is assisted by the *ThermoSTART* system, an auxiliary system for preheating. This system uses both the thermal energy from the previous use of the engine and the electricity from the network (through the plug-in system).

The battery charging system depends on the *GASTER Electricway 4WD PHEV* version: the *Standard* version has only Plug-in charging and the *Premium* version has both charging/discharging via a bi-directional Plug-in charger with V2G/G2V/V2L and wireless charging from various renewable sources.

Our academic research on this subject will continue with the following stages:

- Performing the road tests in order to determine all the performances in real conditions.
- Simulation of the driving, consumptions and performances of the car with the *CRUISE* software from AVL Austria.

**Acknowledgments**

We express our gratitude to *RENAULT TECHNOLOGIE ROUMANIE* and *AVL* for their support in this project.

**Author's ORCID ID**
V Nicolae: https://orcid.org/0000-0002-2536-3372

**References**

[1] Marinescu D G, Tabacu I, Tabacu S, Serban Fl, Nicolae V and Vieru I., 2013 *A Plug-in Hybrid Diesel-Electric Vehicle*, EVS26 -Worldwide International Battery, Hybrid and Fuel Cell Electric Vehicle Symposium & Exhibition, Los Angeles, California, May 6-9 2012. *Lecture Notes Electr. Eng. J.*, vol. 191, no. 3, pp.449 -461
[2] www.renault.com *Renault E Tech Hybrid*
[3] www.landirenzo.com *Component and installation LANDI RENZO OMEGAS EVO handbook*
[4] https://www.nissan-global.com
[5] www.orionbms.co *Orion 2 Li-Ion Battery Management System Operation Manual*
[6] Marinescu A and Golovanov N 2023 *VGI: smart grid integration of electric vehicles*, JESI, Journal of Engineering Sciences and Innovation, Volume 8, Issue 1 / 2023, pp. 69-76
[7] *** IEC 61980; *Electric Vehicle Wireless Power Transfer (wpt) Systems—Part 1: General Requirements.* International Electrotechnical Commission: Geneva, Switzerland, 2020
[8] *** "SAE J2954 2020 *Surface vehicle standard, wireless power transfer for light-duty plug-in/electric vehicles and alignment methodology*
[9] Marinescu A, Dumbrava I, Vintila A, Marinescu D G, Nicolae V, Neagu D and Radu A 2017 *The Way to Engineering EV Wireless Charging: DACIA Electron*, Electric Vehicles International Conference & Show , Bucharest
[10] *** ISO 19163 2017 *Electrically Propelled Road Vehicles—Magnetic Field Wireless Power Transfer—Safety and Interoperability Requirements*. ISO: Geneva, Switzerland

# The design and development of an Integrated Propulsion System – Phase 1: the wheel architecture

**V Croitorescu, D Preda and L Giurca**

POLITEHNICA Bucharest National University of Science and Technology, Scientific Research and Continuous Training Center for Sustainable Automotive Technologies, Romania

E-mail: valerian.croitorescu@upb.ro

**Abstract**. The paper presents the design and development of an integrated powertrain system where the suspension system is also included in the wheel together with the electric machine. Based on a patent of one of the authors, the solution was virtually developed and tested. Different advancements were performed during the development phases for choosing the final architecture. Currently, the integrated propulsion system is manufactured and implemented on a L6E road vehicle to be tested and optimized.

## 1. Introduction

The future of the sustainable transportation excellent includes the integrated propulsion systems as one of the pioneers in the road vehicles new era. The road vehicles powertrains current trends clearly express the concerns about the harmful emissions and the electric vehicles as one viable solution. While many electric vehicles powertrains use the individual components operating independently, the integrated propulsion systems approach the interconnectivity between the individual components, achieving the potential of the electric machine, the battery package, the power electronics and the control systems to simultaneously operate in perfect unison. The integrated propulsion systems have several advantages including the energy efficiency and the range increase, together with the dynamic behavior improvement and adaptive power management. The innovative level for the integrated propulsion systems is brought by their modularity together with the multi-choice connectivity between the integrated components. While the multiple design solutions are available, the space utilization demands are strictly defined by each of them. Most of the solutions approaches the integration of the individual components within the vehicle-axle, resulting an electric driven axle (e-axle), and within the wheel, resulting an electric-driven-wheel (e-wheel). These two different possible solutions are very well researched by most of the road vehicles` after-market suppliers, being already tested and validated. The many challenges for these solutions come during the implementation level, where the plug-and-play solution is highly demanded.

The e-axle is an integrated propulsion system that brings in the same unit the electric machine, the power electronics, the transmission and sometimes also the battery package. The e-axle have compact design, being a great space-saving solution that can easily be installed in a road vehicle chassis and can be connected to the suspension system. If the battery package is part of the same unit where the electric machine is installed, the cooling systems are also integrated.

The e-wheel is an integrated propulsion system that brings the electric machine inside the wheel. Being known as in-wheel or hub-motor, this solution is innovative by eliminating the traditional components that belong to an axle or to a drivetrain (shafts, differential etc.), freeing space and winning less vehicle weight in comparison with traditional electric or conventional powertrains. However, this solution is quite complex, because of the wheel assembly and of the difficulties that are brought by the unsprung weight that can affect handling and comfort.

This paper approaches the solution where the electric machine together with the suspension system are hosted inside the wheel. The objectives of this solution include improving efficiency, handling and comfort. One of the authors` patent represents the main starting point of this research.

## 2. The state of the art

The approached state of the art consists of presenting the already developments for road vehicles suspension systems and in-wheel motors used for road vehicles. Several innovations were presented during the last years for both the road vehicles suspension systems and for the in-wheel electric machines for road vehicles use.

The suspension system is the main system that offers ride comfort, handling and stability at the same time. It consists of an elastic component and a damping component. Several solutions for the suspension systems are known, including the mechanical, the adaptive, the hydraulic or the pneumatic suspension. Although the excitation forces can be transmitted in a vertical, longitudinal or lateral direction current suspension systems mainly isolate forces in the vertical direction. In many situations, multi-directional insulation can significantly improve safety and comfort. The already made research shows that the in-wheel suspension is advantageous, bringing added value to conventional suspension systems. In [18] the in-wheel suspension consists of suspensions arms that are equidistant set around the central wheel hub to provide shock absorption even if the rim is rigid. In [1] some other in-wheel suspension systems solutions were studied, and one prototype was tested. In [5], different in-wheel suspension systems were virtually tested, also considering the energy consumption increase and the difficulties in correct steering, while a new concept was presented and tested.

In addition to the suspension systems components, the tire was in-deep research subject for its elastic behavior as part of the wheel. Some well-known tire manufacturers developed airless tires, with promising partnerships for reaching series manufacturing [13, 17]. In addition to these solutions, one tire manufacturer together with one rim manufacturer presented the flexible wheel [9, 12]. The body airless tire was studied by [3, 4, 7], different design solutions being analyzed during different operation scenarios and for different manufacturing materials.

The in-wheel motors are already well-implemented in different transport solutions, from scooters to utility and construction vehicles. The hub-motor architecture was implemented around 1895 by Lohner-Porsche, its advancements being continuously tested since then. Many after-market road vehicles suppliers are producing in-wheel motors as part of their strategy to contribute to electric vehicles development [11, 14, 15, 16, 19]. As the competition between suppliers is continuously increasing, similar solutions to the hub-motors are also implemented, one of them consisting of the near wheel motor [10]. Based on the existing solutions, researchers studied different design improvements [8] and virtually tested the e-wheel dynamics.

## 3. The proposed and applied technologies

The proposed e-wheel architecture should integrate the suspension system and the electric machine. Firstly, based on the presented state of the art and on the patent [2] obtained by one of the authors, five different e-wheel architectures (figure 1) were investigated where the main challenge is the suspension system integration.

The first solution (figure 1.a) consists of a wheel with three Ω type springs (that can be changed to U shape springs), with variable length and an angle of 120 degrees between them. Each spring is mechanically fixed to the wheel rim and to the center of the wheel hub. Inside each spring, a hydraulic shock absorber can be fixed. This solution demands a big diameter wheel (above 18 inches), is relatively easy to build, but with increased difficulty in manufacturing the Ω type springs.

The second solution (figure 1.b) consists of a wheel with curved elastic spokes and additional rubber springs, being fixed to the rim and to the wheel hub. This solution demands a big wheel diameter (above

18 inches) and it has few components in comparison with the other five studied solutions. This solution is part of a patent registered in China by the same author and patent owner, under the number CN209159289.

The third solution (figure 1.c) consists of a wheel where the suspension is based on rubber elements and lamellar elastic springs inserted into them, being rigidly fixed to the rim and to the wheel hub. The elastic and absorption behaviors are achieved separately. This solution has no limitations for the wheel diameter, but the manufacturing approach includes the development of a personalized vulcanization mold for each targeted wheel diameter.

The fourth solution (figure 1.d) consists of a wheel with radial elastic elements fixed at the same time to the wheel rim and to the wheel hub. The radial elastic elements must be manufactured using the same material and needs to develop a personalized vulcanization mold for the outer band.

The fifth solution (figure 1.e) consists of a wheel with articulated elements. This solution has the highest modularity from all five studied solutions. Its modularity includes the possibility to easily change the elastic elements, to adapt them following the requests in terms of weight, elastic characteristics, size and volume, by keeping the same mechanical design and structure.

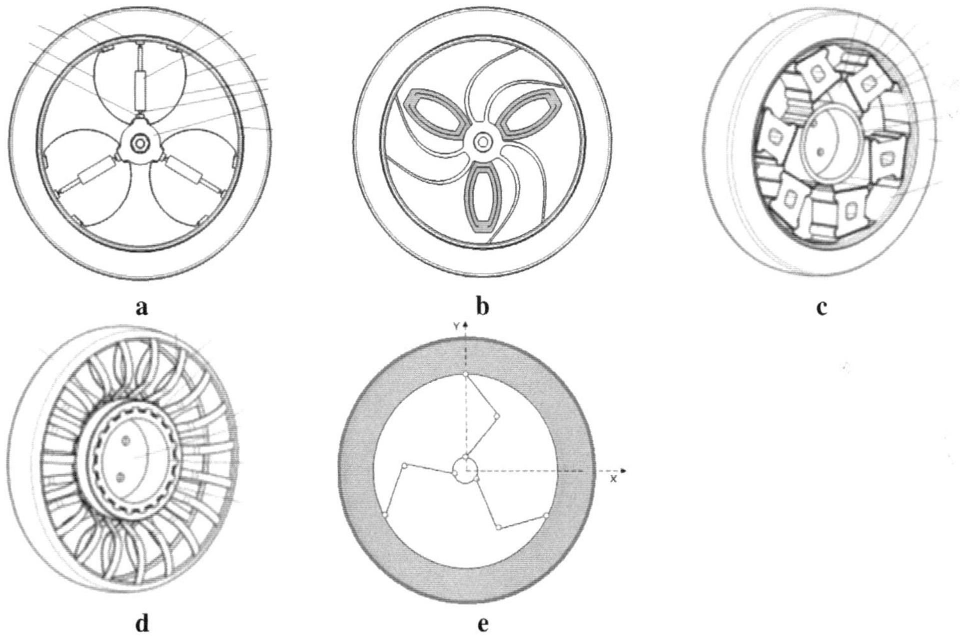


**Figure 1.** The five studied e-wheel architectures [2].

The fifth solution was adopted to be investigated in deep. Therefore, starting from its basic design, three different design proposals derived from it were researched, but only one was continued for production and implementation.

The first proposal (figure 2.a) for the chosen solution consists of three different group of levers. Each group have two levers connected to two other levers with connecting rods. This mechanism is blocked when the elements are rigid. To ensure an elastic behavior, the elements have deformation within the elastic limit. Unfortunately, this kind of mechanism offers a reduction of few millimeters for the unsprung mass, being considered not a viable solution for the electric vehicle proposed for testing.

The second proposal (figure 2.b) for the chosen solution consists of six levers connected two by two by a shaft on which a helical spring is mounted. One end of the helical spring is connected to the outer lever, while the other end of the helical spring is connected to the inner lever. While crossing an obstacle, the angle between the levers varies (figure 2.b.2).

a b.1. b.2.
b

**Figure 2.** Two of three of the proposed designs for the e-wheel architecture chosen solution [2].

The mathematical model for this design proposal was created and run using a virtual spring design and calculation software [21]. The helical spring wire diameter was considered equal with 6 mm, while the external diameter for the helical spring was considered equal with 30 mm. The helical spring was considered to have ten coils. Using these values, the mathematical model calculates the helical spring life cycle and the torque depending on its deformation (figure 3). Unfortunately, the chosen spring with the presented characteristics breaks. Increasing the helical spring wire diameter represents a solution but is not acceptable being very hard to be further implemented and not appropriate for the unsprung mass.

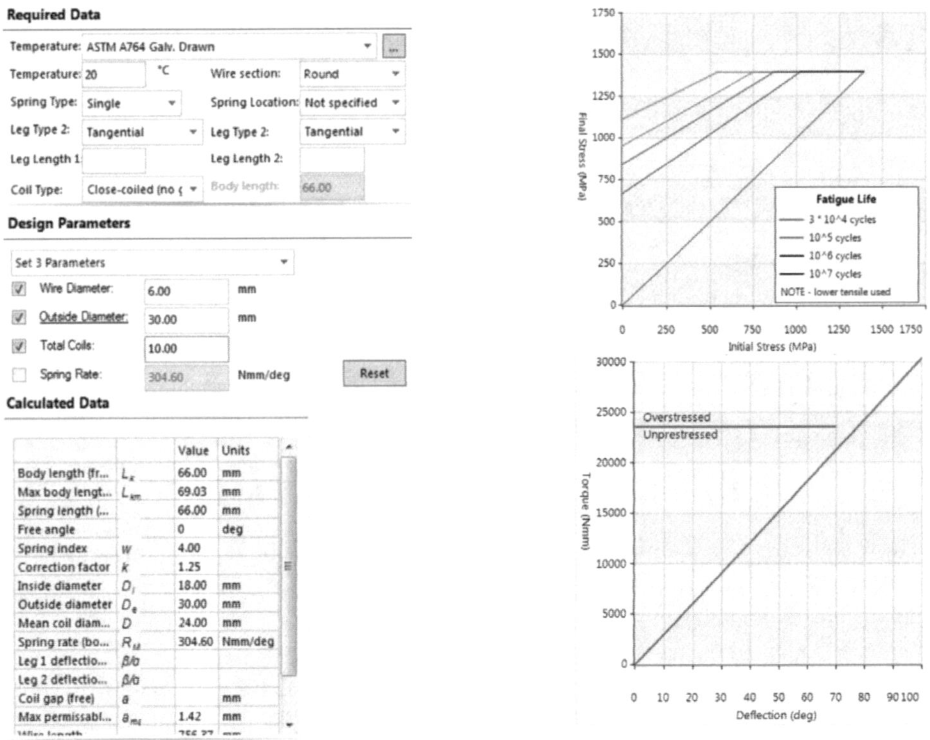


**Figure 3.** The helical spring caculation for the second design proposal.

The third proposal (figure 4) for the chosen solution keeps the same design approach as the second proposal, but with an upgrade. The previous proposal with only one helical spring on each median axis has a low level of redundancy. If the spring breaks, the function of the suspension is completely compromised. The upgrade consists of using six helical springs instead of three, being aligned symmetrically regarding the wheel median vertical axis. The existence of two helical springs for each group has high level of redundancy allowing the system to work for a limited time even if one spring is broken.

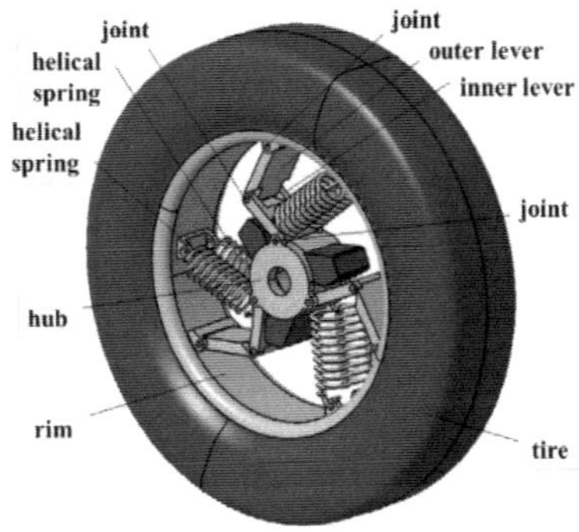


**Figure 4.** The final accepted design.

The mathematical model for this design proposal was also created and run using the same virtual spring design and calculation software [21]. Preliminary sizing steps were performed with several iterations. Based on the obtained results, together with the requirements for the outer diameter value of about 40 mm and the maximum length of ten coils, it was determined that the diameter of the wire should be over 5.5 mm. Several suppliers were contacted to purchase the most suitable helical spring already manufactured. For each of the most suitable received helical springs where the characteristics are meeting all the constraints, the same virtual spring design and calculation software was used.

The chosen helical spring has the following characteristics: the wire diameter value of 5.8 mm, outside diameter value of 41.30 mm, the length of nine coils, the maximum length value of 102 mm.

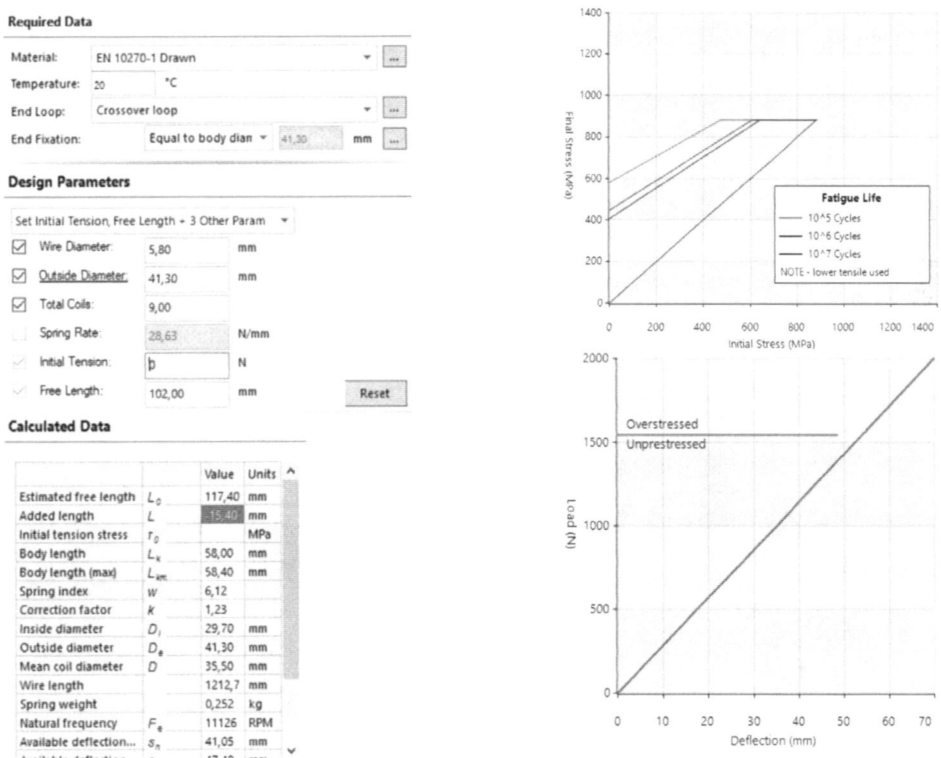


**Figure 5.** The chosen helical spring calculation.

In addition to the suspension system that is integrated into the wheel, the electric machine should also be integrated. Currently, integrating the electric machine into the already chosen design represents a big challenge. The electric machine size represents a major constraint. However, there are different possible design solutions for coupling the electric machine even if its size is larger than the available hub size.

The first available solution (figure 6.a) consists of using two different electric machines that are coupled directly next to the drive wheels. The coupling can be done either by using a spline or grooves on the output shafts of the electric machines. When choosing the electric machines for this solution, the main constraint is to have the output shaft with spindle or with grooves, while the other requirements are strictly related to size, power and easy integration.

The second available solution (figure 6.b) consists of using the already existing electric machine that are installed and are in use on the electric vehicle proposed for testing. As a result of this solution, no important changes are required, only flanges are needed for each of the drive wheels to be connected to the existing electric machines.

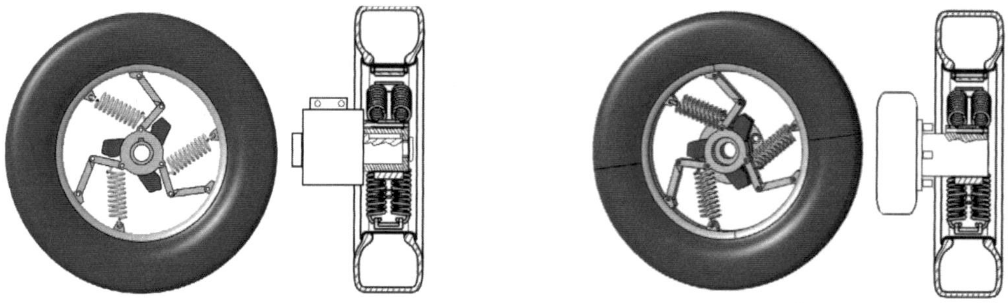

**a** **b**

**Figure 6.** Investigated solutions for coupling the electric machine to the wheel.

Based on the comparisons made for the two investigated solutions, adding the size requirement of the hub to achieve its modularity, the wheel architecture is chosen, being presented in figure 7. The flange is part of the wheel hub and must be customized in a modular way for being easily implemented to the existing electric vehicle proposed for testing and to other configurations. The e-wheel will be assembled with the electric drive motor by four different nuts specially configured to obtain the minimum size of the wheel hub.

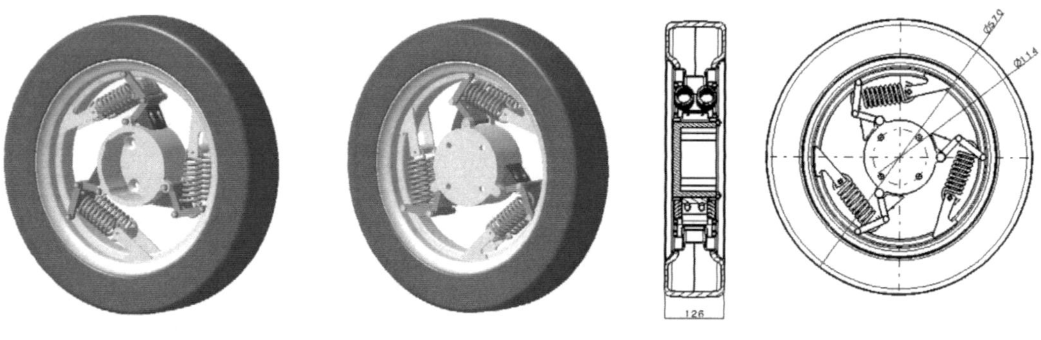


**Figure 7.** The wheel architecture.

## 4. The prototype

Each e-wheel component was built with respect to the determined sizes and then assembled. The rims (figure 8.a) were obtained from existing wheels, type 4Tx16, from which the original hubs were extracted. The new hub (figure 8.b) consists of 18 independent plates assembled using nuts, with respect to the sizing of the electric machine. The springs mountings (figure 8.c) were realized by cutting from metal sheets, then they were welded to the rims. The outer and inner levers (figure 8.d) were

IOP Conf. Series: Materials Science and Engineering 1303 (2024) 012009 doi:10.1088/1757-899X/1303/1/012009

manufactured using computer numerical control machines. The springs (figure 8.e) were bought from a supplier. The prototype was produced (figure 8.f) and the tire was mounted (figure 8.g).

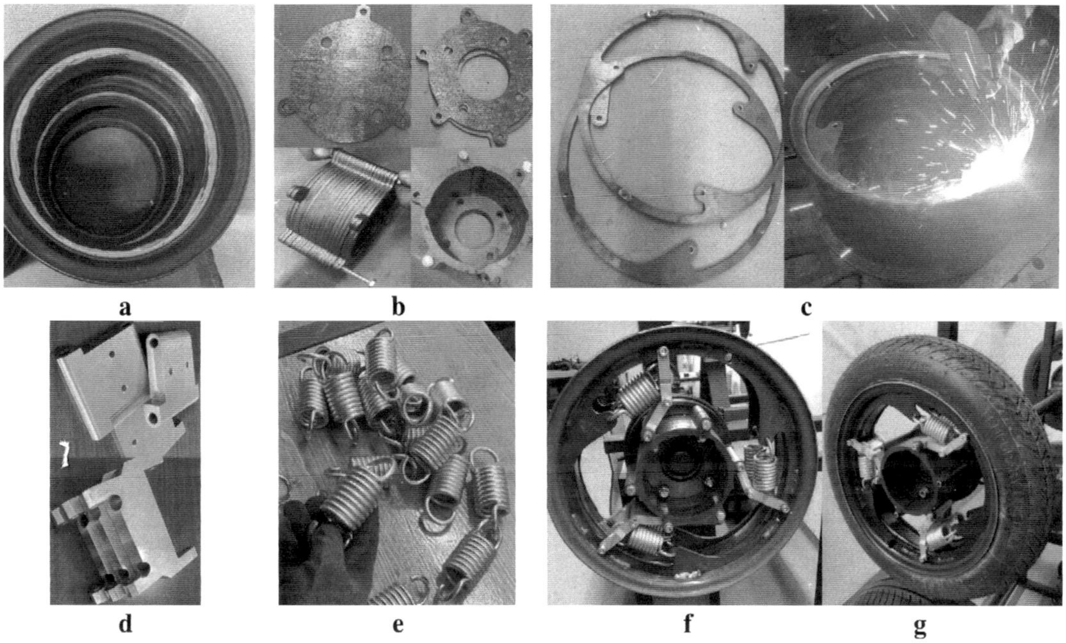


**Figure 8.** The e-wheel prototype with each of its components.

## 5. The implementation

The e-wheel prototype was implemented on an existing L6E category electric vehicle (figure 9) for being on-road tested. Further results are pending to be obtained for optimizing the e-wheel.

**Figure 9.** The e-wheel prototype used by a L6E category electric vehicle.

## 6. Conclusions

The paper presents the stages of research and development for one of the phases of obtaining an integrated propulsion system. This first phase was dedicated to the wheel architecture. Several solutions for the suspension system integrated in the wheel and for the electric machines integrated in the wheel were studied. In-wheel suspension systems represent an innovative solution, being under development by many car manufacturers suppliers. They offer several benefits including more free space for the chassis, less unsprung weight, better handling. In-wheel electric machines are widely used for different electric vehicle categories. They offer several benefits including flexible vehicle design, reduced energy losses, simplified drivetrain. Starting from one author patent, different wheel architectures were virtual tested. Based on the obtained results, one solution was adopted, and its prototype was produced. Currently, the prototype is tested on a L6E category electric vehicle.

The studies of integrated propulsion systems represent a current and a future concern for most technological companies involved in powertrain electrification. The achieved solution presented in this paper has great potential for being replicated and produced in series.

**Acknowledgments**
Authors express their gratitude for the assistance and the continuous support received from all the team members of the research and development project "Cercetarea, dezvoltarea si lansarea in productie a Sistemului de Propulsie Integrat (SPI) pentru vehicule electrice", developed entirely in Romania.

**Author's ORCID ID**
V Croitorescu: https://orcid.org/0000-0003-1987-5088

**References**
[1] Azimi M 2009, Design, Prototyping, and Testing of an In-Wheel Suspension System, Master Thesis, University of Waterloo, Ontario, Canada
[2] Giurca LG 2016, Sistem de propulsie integrat pentru vehicule electrice sau hibride - Brevet de inventie RO131493, publicat in BOPI nr. 1/2016, Oficiul de Stat pentru Inventii si Marci, Bucuresti
[3] Mazur V and Rykov S 2021, The Ability of Airless Tires to Self-Purify, *IOP Conference Series: Earth and Environmental Science*. 666. 042012. 10.1088/1755-1315/666/4/042012
[4] Mazur VV 2021 Theoretical Study of the Force Heterogeneity of Airless Tires Made of Elastic Polyurethanes, Proceedings of the 7th International Conference on Industrial Engineering (ICIE 2021)
[5] Shashavali S , Raghunatha Reddy C and Yadiki GK 2016 Design and Analysis of Four Wheeler Airless Tire, *International Journal of Advanced Technology and Innovative Research*, ISSN 2348–2370 Vol.08, Issue.22, December-2016, Pages:4298-4305
[6] Skarka W, Kocoń W, Skoberla R and Bzymek A 2019 Innovative Suspension System Integrated to the Wheel Rim for Lightweight Vehicle, *Transdisciplinary Engineering for Complex Socio-technical Systems*, K. Hiekata et al. (Eds.), doi:10.3233/ATDE190147
[7] Sriwijaya R, Tjandra K and Hamzah R 2021 The Effect of Bank Angle on the Pressure Distribution and Deflection of Non-pneumatic Tires. *IOP Conference Series: Materials Science and Engineering*. 1194. 012008. 10.1088/1757-899X/1194/1/012008.
[8] Winai C and Prasert H 2014 Design of Efficient In-Wheel Motor for Electric Vehicles, *Energy Procedia*, Volume 56, Pages 525-531, ISSN 1876-6102, https://doi.org/10.1016/j.egypro.2014.07.188.
[9] https://www.automotiveworld.com/news-releases/maxion-flexible-wheel-michelin-acorus-technology/
[10] https://www.borgwarner.com/technologies/p2-hybrid-architecture
[11] https://www.bosch-mobility.com/en/solutions/powertrain/vehicle-platforms/pre-integrated-system-solutions/
[12] https://www.caranddriver.com/features/a21946899/how-michelin-is-trying-to-reinvent-the-wheel/
[13] https://www.drive.com.au/news/airless-tyres-could-be-a-reality-this-decade/
[14] https://in-wheel.com/en/technology-2/#in-wheel-motor-technology
[15] https://lightyear.one/articles/lightyear-and-elaphe-develop-the-most-efficient-production-powertrain-in-the-world
[16] https://www.lordstownmotors.com/pages/tech
[17] https://newatlas.com/automotive/hankook-i-flex-airless-tire-hyundai-l7-concept/
[18] https://www.proteanelectric.com/innovation/#patent-portfolio

[19] https://www.schaeffler.com/en/media/press-releases/press-releases-detail.jsp?id=87902208
[20] https://www.softwheel.technology/technology/
[21] http://springcalculator.com/

# The design and development of an Integrated Propulsion System – Phase 2: the functional electric behaviour strategy

**V Croitorescu, B Duran and L Giurca**

POLITEHNICA Bucharest National University of Science and Technology, Scientific Research and Continuous Training Center for Sustainable Automotive Technologies, Romania

E-mail: valerian.croitorescu@upb.ro

**Abstract**. The paper presents the integrated propulsion system implementation steps on a L6E road vehicle. The implementation followed several technical levels which are properly described within the paper. The functional behaviour in terms of electric propulsion was defined and developed. All the dedicated components were investigated, set and customized, to equip a L6E vehicle. Currently the electric machine, the battery pack, the user interface, the speedometer, the pedals, the power electronics etc. are implemented on a real existing L6E road vehicle, after they were defined and virtually tested.

## 1. Introduction

The current needs of automotive industry in electrification are great. Combining the requirements of world governments with the economic and environmental concerns, the electric propulsion systems designed for road vehicles are one of the possible key factors to push the electrification forward.

When talking about reducing harmful emissions, most automotive engineers think to the hybrid or electric powertrains, since the internal combustion engine road vehicles represent a significant source of carbon dioxide emissions at the tailpipe. In addition, even though the hybrid vehicles have less local carbon dioxide emissions while operating, and the electric vehicles produce zero tailpipe emissions, the electricity they use is produced in a polluting energy plant.

The governments around the world are implementing increasingly stringent standards and regulations on road transport emissions. They are also providing financial support to customers for the purchase of electric vehicles. Most countries have a strict deadline for implementing the restriction on sales of road vehicles equipped only with internal combustion engines. The car manufacturers are facing the biggest challenge since the turn of the twentieth century, to electrify their entire manufacturing fleet. The customers are more inclined to use electric vehicles. Expanding charging infrastructures convincingly offers a convenient alternative to fuel stations. The advancements in battery technology are growing dynamically, both for the chemistry used and for battery packages modularity. The energy efficiency of the electric propulsion systems is better than of the internal combustion engine. In addition, by using electric propulsion systems the dependence on fossil fuels is reduced and the running costs are lower because the electricity is often cheaper.

With the benefit of the current framework, the development of different designs for the propulsion systems is very appropriate. When these propulsion systems are `integrated`, even modular, their development potential increases, and the users are the ones who impose the limits. The road electric vehicles integrated propulsion systems are those that have arrangements of their systems such as the

IOP Conf. Series: Materials Science and Engineering 1303 (2024) 012010 doi:10.1088/1757-899X/1303/1/012010

electric machine, the transmission and sometimes the controller, the battery package and the power electronics in the same unit. These systems benefit from customized design, quite complex due to the strict requirements regarding sizing, weight, price and easy integration. The various components that the integrated propulsion systems include bring together advanced technologies that focus on energy efficiency, performance and durability.

This paper presents the implementation of an electric integrated propulsion system on a L6E vehicle and focuses on the functional behavior of the mandatory electric systems. This paper is part of a series of papers that address the integrated propulsion systems for electric vehicles.

## 2. The approach: from the concept idea to requirements and testing

The integrated propulsion system was defined by the specific goals and requirements: the performance objectives, the energy efficiency, the costs. All these three targets include different steps. Starting with the components` selection, the electric machine, the battery package, the power electronics, the controller were defined. Continuing with the integration and packaging, the physical layout was designed together with the integration plan of the components to the vehicle chassis. The control system needs advanced algorithms to manage the power flow between the components and optimize their performance, while the thermal management and the regenerative braking control are mandatory. The virtual testing represents the first testing level, while using a customized test bench for validation the results help to ensure that the integrated propulsion system meets the standards and the regulations.

## 3. The powertrain components

### *3.1. The electric machine*

The electric machine (figure 1) for this solution is a brushless direct current motor (BLDC) in-wheel-hub motor, manufactured by QS Motors [7], having the nominal power of 2 kW, and the peak power of 4 kW, using customizable voltage of 48 / 60 / 72 Volts for a rotary velocity of 665 / 833 / 1000 min-1, the maximum torque of 160 Nm, the declared efficiency between 85 to 92 %, maximum current of 30 Amps, peak current of 80 Amps (reaching also 100 Amps for a short period of time), the angle between the Hall transducers of 120 deg. It weighs 15 kilograms. The electric machine has one axle, being dedicated to being mounted inside wheels or nearby the wheels. Following the already developed wheel architecture, the electric machine will be coupled to the wheel.

**Figure 1.** The BLDC electric machine [7].

### *3.2. The energy storage system*

The energy storage system is customized for a second vehicle designed especially for the in-wheel electric machine solution, operating at 48 Volts. It consists of four different LiFePO4 in-series batteries, of 12.8 Volts, 50 Ah and 640 Wh each. The main parameters of the batteries are presented in table 1. Each battery has integrated battery management system, including a state of charge indicator. Its manufacturer declares that its life cycle is expected to be more than 2000 cycle of charge-discharge at maximum of 80 % deep of discharge. The temperature of the battery is very important to its state of charge and its end of life. Therefore, the manufacturer clearly presents that the appropriate charging temperature should be between 0 and 45 degC (figure 2).

**Table 1.** The electric, mechanical and functional parameters of the batteries.

| Electric parameters | | Functional/Mechanical parameters | |
|---|---|---|---|
| Nominal voltage [V] | 12.8 | Dimensions (Lxlxh) [mm] | 229 x 138 x 214 |
| Nominal capacity [Ah] | 50 | Weight [kg] | 8.12 |
| Energy [Wh] | 640 | Terminal type | M6 |
| Resistance [mΩ] | ≤ 30 | Battery housing | ABS, UL-94 V-0 |
| Efficiency [%] | 99 | Housing protection | IP56 |
| Self-discharge | Max. 2%/month | Modules in series (S) / parallel (P) | 4 S / 20 P |
| Charge voltage [V] | 14.4 – 14.8 | Cell Type-Chemistry | LiFePO4 Cylindrical Cell |
| Charge current [A] | 25-50 | Life cycle [cycles] | > 2000cycles @1C,100% DOD |
| BMS charge cut-off [V] | 15.6 | BMS temperature cut-off [degC] | 80 |

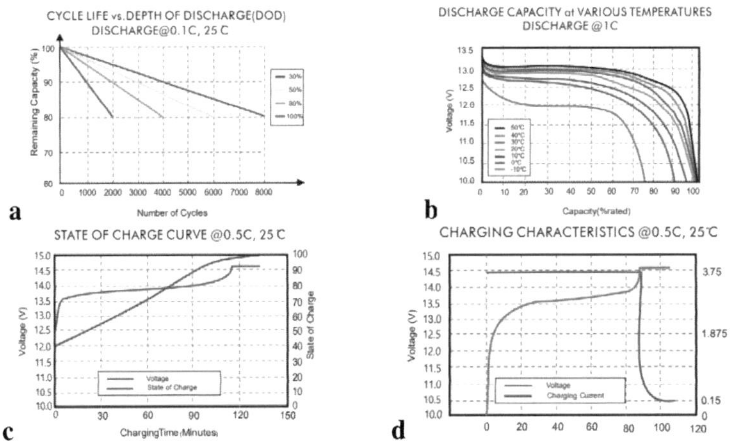


**Figure 2.** The batteries performances [2].

### *3.3. The battery package charger*

The battery package charger is very important to the presented electric powertrain design solution. Not only the choice of charging technology, but also the selection of the correct charging method is a feature that must be considered during the charging procedure. The most popular charging strategies to recharge Li-Ion batteries are the constant current / constant voltage (CC / CV) and pulse current charging methods. However, these methods do not consider the internal processes of the battery that influence the charging capacity and its aging. As a result, some promising charging strategies that are based on more complete Li-Ion battery models are under research. But many solutions are using too many components that effect the sustainability of the charger. Therefore, to minimize the number of the components and to reduce the size and the costs of the battery charger, the inverter and the electric machine were used. After studying several strategies, together with the requirement to keep the electricity consumption low, a charger manufactured by Yewy was chosen (figure 3.a). It offers a power of 900 W for a nominal voltage of 48 Volts and a constant current of charging of 15 Amps. It weighs around 4 kilograms. It benefits of three-stage intelligent charging (figure 3.b): Constant Current (CC), Constant Voltage (CV), turbulence, being able to operate fully automatic, blocking high voltage, high current, ensuring full charging without overcharging. Its safety design is based on power safety regulations, with short circuit, overcurrent, overvoltage, reverse connection protection, automatic current monitoring, charging mode switching.

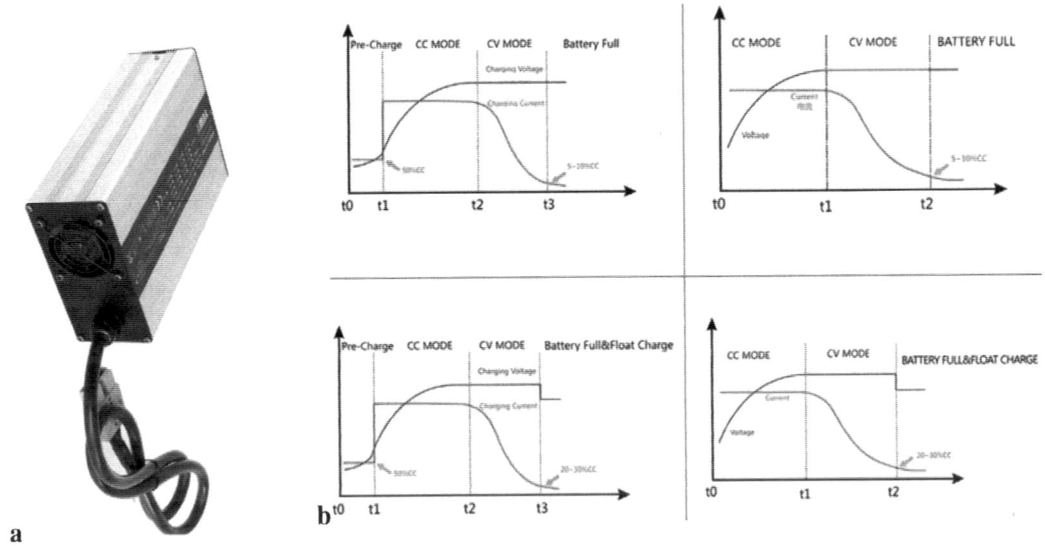


**Figure 3.** The battery charger (a) and the charging modes (b) [9].

*3.4. The throttle pedal*

The throttle pedal (figure 4.a) represents one of the most important devices for an electric powertrain, being responsible to connect the driver demands to the electric machine management system and to the controller. The chosen throttle pedal is manufactured from metal on a solid structure. It is water resistant, and it can work at voltages from 5 to 96 Volts. Its operation consists of transforming the mechanical displacement into an electric signal based on the Hall principle.

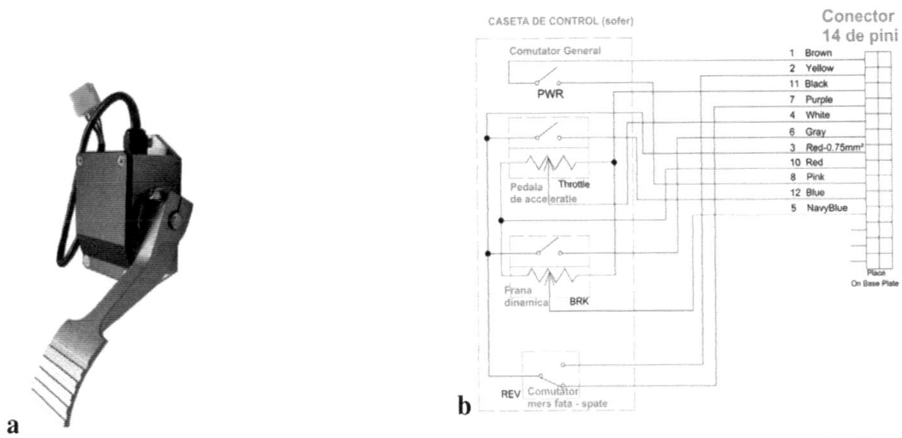


**Figure 4.** The throttle pedal (a) and its wiring diagram (b) [3, 4].

*3.5. The controller and driver system design and development*

The controllers (figure 5.a) were chosen to be used for both electric machines, one controller for one electric machine. The electric wiring (figure5.b) was customized for following the chosen architecture. Both controllers are based on sinusoidal wave form, dedicated to operating together with BLDC motors with three hall sensors. The controllers are monitoring the battery voltage to stop driving when it is too high or to progressively cut back motor drive power as it is too low. During regenerative braking, the controllers are monitoring the battery recharging voltage also. The controllers are responsible to configure the electric machine speed between 20% and 100%. The operating frequency for the controllers is between 10 to 20 kHz.

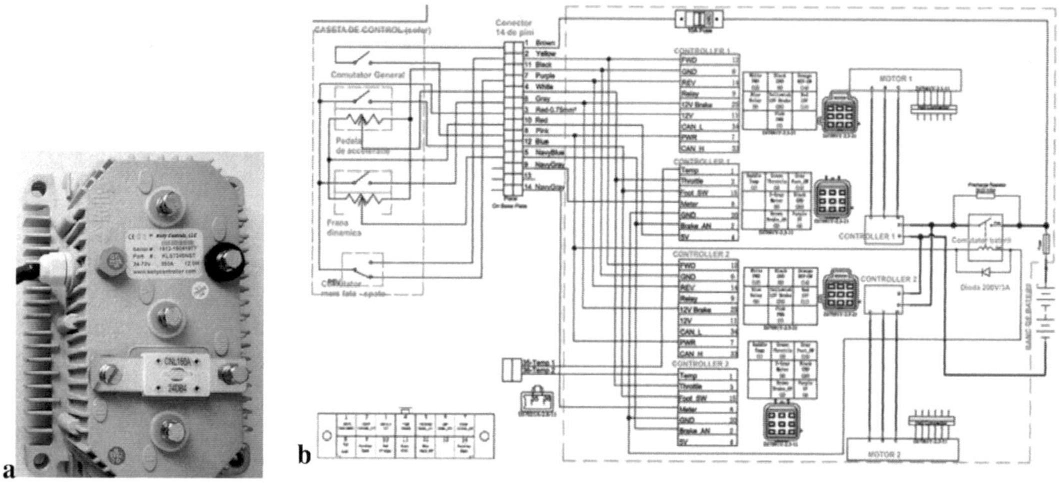


**Figure 5.** The controller (a) and the wiring diagram (b) [8].

The controllers' behavior is strict related to the driver system design and development. The electric machine driver system represents the set of electronic components that converts the energy from the batteries into a sequence of electrical power pulses applied to the electric machine to control its behavior. The electric machine driver system development needs parameters like the electric machine power, the peak power, the regenerative braking power.

For input, the electric machine driver system uses: (i) the signals received from the encoder that is located inside the electric machine, respectively from the Hall transducers that indicate the position of the magnets inside the motor, (ii) the throttle pedal position, that is used for switching on and off the internal transistors to run the electric machine proportionally to its position.

The electric machine driver system consists of the transistors and the integrated circuit. The transistors must be selected to support the electric machine power, as the lower their internal resistance is, the lower the dissipated current will be. Six different groups of transistors have been used (SW1...SW6). They can continuously switch a voltage of 60 Volts and a current of 70 Amps. For achieving a control load of 300 Amps, the optimal solution consists of connecting several transistors in parallel (figure 6). The obtained values are presented in table 2.

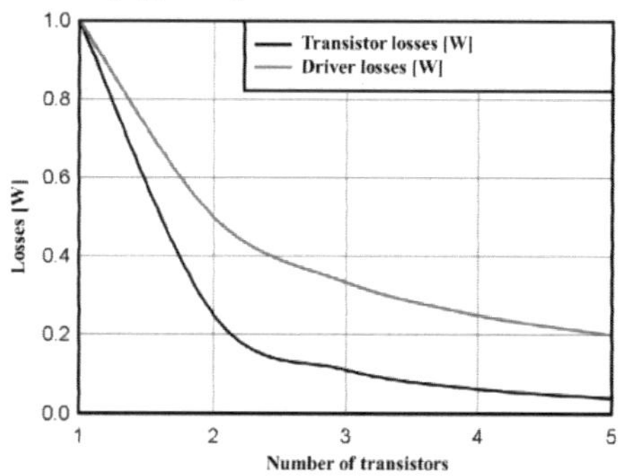


**Figure 6.** Power losses evolution for the transistor and for the driver.

**Table 2.** The power losses evolution while connecting transistors in parallel.

| Transistors in parallel | Power losses / transistor [W] | Power losses / driver [W] |
|---|---|---|
| 1 | 16.900 | 101.4 |
| 2 | 4.225 | 50.7 |
| 3 | 1.880 | 33.8 |
| 4 | 1.060 | 25.35 |
| 5 | 0.680 | 20.28 |

IOP Conf. Series: Materials Science and Engineering 1303 (2024) 012010 doi:10.1088/1757-899X/1303/1/012010

The integrated circuit receives and interprets the information from the hall transducers located in the electric motor. The DC electric machine with permanent magnets is using three phases A, B and C. The motor starts when the driver provides voltage on two of the three phases, accordingly with the information received from the encoder.

The group of transistors are switching in turn in a certain sequence, being divided into UP and DOWN. The positive voltage is provided on one phase when the UP transistors are open, while the negative voltage is provided when the DOWN transistors are open. When applying a positive voltage to terminal B and a negative voltage on terminal C, the electric motor receive control. The Hall transducers are sending the information for continuously run of the motor, while the driver system is applying a certain voltage to another terminal. The control circuit for the electric machine with permanent magnets is presented in table 3.

**Table 3.** The control circuit for the electric machine with permanent magnets.

| **Terminal A [type of voltage]** | **Terminal B [type of voltage]** | **Terminal C [type of voltage]** |
|---|---|---|
| Unpowered | Positive | Negative |
| Negative | Positive | Unpowered |
| Negative | Unpowered | Positive |
| Unpowered | Negative | Positive |
| Positive | Negative | Unpowered |
| Positive | Unpowered | Negative |

For each of the phases the terminals will receive positive voltage, negative voltage, or will be unpowered. This voltage sequence will allow the magnetic field produced by the coil to be attracted by the next magnet and to be repel by the previous one, resulting the torque to start the motor. This sequence is given by an integrated circuit, which converts the information received by the Hall transducers into the power sequence for the electric machine. The electric machine rotary velocity is controlled by each terminal voltage and the electricity supply sequence. The pulse width modulation is responsible for control it, the limits being between 0 and 100 %. For example, for a 48 Volts supply and a pulse length of 50%, the resulting voltage is 24 Volts. The pulse width is controlled by the acceleration pedal.

The output driver is responsible to send information to the electronic control unit about the battery voltage, the electric machine rotary velocity, the current energy consumption, the error messages.

### *3.6. The user interface – the speedometer*

The user interface consists of the speedometer that is providing information about the battery voltage, the trip distance, the trip time, the vehicle speed, the absorbed current. Its programming is easy to handle, all its features being intuitive to set. Its operation voltage is between 48 to 96 Volts.

**Figure 7.** The speedometer [1, 5, 6].

## 4. Implementing and testing the powertrain components on the prototype L6E vehicle

The above presented powertrain components were implemented and integrated to a prototype L6E vehicle. The implementation was followed by the in-real-life testing activities of the subassemblies and of the new vehicle. The results allow to evaluate the performances the achieved integrated propulsion system has. Using two electric machines, one for each wheel of the driven axle, the expected results are expected to be better than to an already existing similar solution.

Firstly, the e-wheel with integrated suspension system was tested using a customized dedicated test bench. Then, the e-wheel with integrated suspension system was tested on the prototype L6E vehicle, especially developed for testing this solution. The prototype L6E vehicle consists of its own chassis, a steering system, a braking system, two different axles, where the rear axle is driven by the two BLDC electric hub motors, coupled directly to the e-wheel, while the front axle is steered. The battery package (figure 8.a) is placed above the rear axle, directly mounted to the vehicle chassis, being protected by a customized housing that also hosts the power electronics (figure 8.b). The L6e vehicle has only one left seat but it can be equipped with the second seat, placed on the vehicle right side. The interface with the driver is assured by the speedometer.

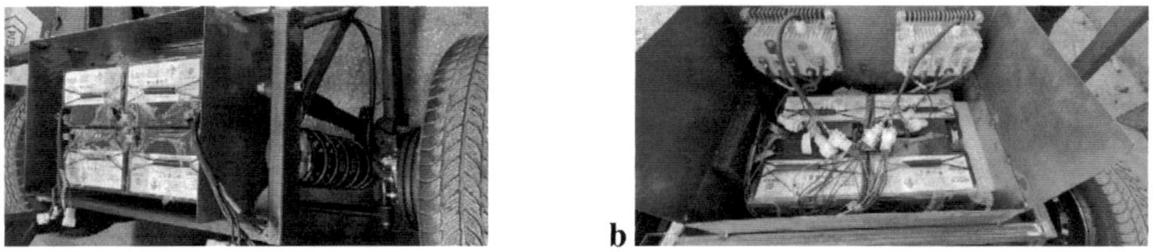

a b

**Figure 8.** The battery package and the power electronics implementation.

Before starting the on-road testing, the battery package was charged (figure 9.a), while the speedometer was set to run for the first time. In addition, the tire pressures were set.

a b c d

**Figure 9.** Charging the battery package (a), setting up the speedometer (b), setting the tires pressure (c) and beginning the on-road testing (d).

## 5. Conclusions

The paper presented the technical level that were achieved for designing and developing an electric powertrain, together with choosing each component and implementing them to the prototype L6E vehicles. The current solution of the integrated propulsion system for electric vehicles, where the suspension system is in the wheel, clearly improves the road behavior in terms of maneuverability and dynamic performances using two electric machines, each driving a wheel. The electric machines are coupled to the wheels, as one of the parts of the suspended mass, belonging to the rear rigid axle. This design considerably reduces the energy consumption because of the simplified kinematics, the lighter rear axle where no conventional suspension is needed and the torque distribution between the driven wheels.

The functional behavior strategy for this vehicle was defined. Different configurations for the driver system were investigated. The number of transistors and their arrangement were used to calculate the power losses. All the components that interact and are connected to the power electronics were considered as contributors to the power losses. The L6E prototype is currently under testing and its optimization is mandatory.

**Acknowledgments**
Authors express their gratitude for the assistance and the continuous support received from all the team members of the research and development project "Cercetarea, dezvoltarea si lansarea in productie a Sistemului de Propulsie Integrat (SPI) pentru vehicule electrice", developed entirely in Romania.

**Author's ORCID ID**
V Croitorescu: https://orcid.org/0000-0003-1987-5088

**References**
[1] http://www.cnqsmotor.com/en/article_read_610.html
[2] https://www.e-acumulatori.ro/utility/file/download?file=download_files%2Fhd12-8-503317.pdf&hash=4da4000b98cfd3af5169df55c6eb9d8c
[3] https://www.sayoon.net/en/ProductView.Asp?ID=302
[4] https://www.sayoon.net/uploadfile/PDF12/JKH_E_SAYOON.PDF
[5] http://www.siaecosys.com/upfile/201908/2019082255661793.pdf
[6] http://www.siaecosys.com/upfile/201908/2019082255697717.pdf
[7] https://www.qsmotor.com/13-inch-scooter-motor/
[8] http://www.qs-motor.com/product/kelly-controller-kls7245n-30v-72v-270a-sealed-sinusoidal-bldc-motor-controller-for-4000w-5000w-motor/
[9] https://www.yewycharger.com/battery-charger/electric-golf-carts-battery-charger/48v-battery-charger-for-golf-cart.html

IOP Conf. Series: Materials Science and Engineering 1303 (2024) 012011 doi:10.1088/1757-899X/1303/1/012011

# The design and development of an Integrated Propulsion System – Phase 3: the testing bench

**V Croitorescu, A Ionescu, D Preda, B Duran, L Giurca**

POLITEHNICA Bucharest National University of Science and Technology, Scientific Research and Continuous Training Center for Sustainable Automotive Technologies, Romania

E-mail: valerian.croitorescu@upb.ro

**Abstract**. The paper presents the testing bench dedicated to the integrated propulsion system (IPS). The IPS consists of a wheel that carries the electric machine and the suspension inside. Its behaviour needs to be tested while being used independently on a test bench and then on road tests. Because no already existing testing bench was available for this kind of testing, a dedicated testing bench was designed and developed. Its main role is to investigate the functional behaviour of the wheel, together with the integrated suspension system behaviour. The tests were made following different scenarios, where different tire dimensions, different environment properties, different road profiles etc. were reproduced.

## 1. Introduction

The innovation fits very well with the road vehicles design during the current progress of automotive industry, while simplifying the solutions and reducing the costs are mandatory. Both are conditioned by the research efforts made through virtual and physical testing, while the equipment used for testing define the value of the research results. Consequently, when the system to be tested has higher innovation level, the test benches to be used are difficult to be find. When it becomes impossible to find suitable test benches for new and innovative systems, there is a need to develop dedicated customizable test benches.

The testing bench for an e-wheel with integrated suspension system and the electric machine coupled to the hub is a very complex and highly targeted equipment. Many electronic components, power electronics, control systems are part of this type of testing bench, and their integration goes hand in hand with the adaptability to work together properly.

The testing bench for in-wheel suspension systems must reproduce real-world conditions, including road bumps, potholes and various road surface irregularities. It must have multi-axis testing capability, including vertical displacement, in addition to yaw, pitch/tilt and roll movements. The testing bench must have another crucial capability: to adjust different loading conditions for the evaluation of different types of in-wheel suspension, depending on the types of the vehicles for which it is dedicated. For a more comprehensive evaluation of an e-wheel with integrated suspension system, it is mandatory to allow dynamic testing: to replicate the acceleration and deceleration, braking and regenerative braking, steering behaviors. Different suspension configurations should be tested without limits.

The data logging and analysis systems will collect, store and analyze the information from different transducers, which can be also linked with various simulation software. By integrating a simulation software as part of the testing bench, the virtual testing will enable design refinement before the manufacturing processes begin, and well in advance of physical testing. In addition, the adaptive testing should use real-time feedback from the transducers used and to adjust the parameters for different scenarios.

The modularity and the flexibility of the testing bench make it easy to reconfigure and adapt to different configurations, sizes and requirements to meet.

The safety represents a top priority for the testing bench. All component parts and all the operation procedures must be provided carefully. The mechanisms of the testing bench must have emergency stop mode, actuated manually or automatically.

This paper tries to cover most of the above-mentioned tasks for developing a testing bench dedicated for an e-wheel with integrated suspension and the electric machine coupled to its hub.

## 2. The state of the art

The current literature presents several solutions for different test benches for being used either for e-wheels or for suspension systems. The technology of e-wheels is evolving rapidly and when a solution like the in-wheel integrated suspension comes as an addition, the difficulty of finding physically testing solutions increases.

The road vehicles tire stiffness represents an important characteristic for evaluating the e-wheel with integrated suspension system. The force ratios under vertical and horizontal loads on the tire can be investigated by different testing benches. Investigating the tire vertical behavior can be done using testing benches that are also used for testing the suspension systems. The test rig presented in [1] allows to test independently the shock absorber and the spring, or if needed, to test all the suspension system components simultaneously. Testing the tire behavior can also be done independently, this test rig being able to apply a displacement of controlled frequency and amplitude. Special attention must be given to the rim deformation that can influence the measurements.

The industry dedicated to developing testing benches facilities helps testing independently the springs, the shock absorbers and the other components that are part of the suspension system. Testing different suspension system components is possible by using the test rigs from [9] and from [10]. The multiple design configurations allow easy adjustment of the system frequency and choosing different types of actuators. The extended features of this already existing testing rig dedicated very well cover the detailed investigation of all the components of the automotive suspension systems.

The dynamic wheel force measurement represents an important task, used to determine the dimensions of a testing bench supporting load bearing or of the springs and dampers. One such technique is presented by [11], where several requirements are applied to maintain the driving dynamics, including not to change the unsprung mass or the stiffness of the wheel.

In [2], a test bench for suspension systems is presented, where its development started from the quarter-car model, where the controller and its functional behavior were preliminary simulated. The different road surfaces were reproduced by a linear motor placed under the wheel of the quarter-car model. The same approach of reproducing the road profile is presented by [4], where the vertical excitation of the wheel is achieved by an electrical linear motor connected to the wheel`s base plate. A testing rig suspension system based on a quarter car is also presented in [6], where the road surface is reproduced by using a second wheel with a cam on its external profile, while the wheel of the quarter car model is considered as follower. The road surface can be reproduced also by using the vertical movement of the suspension system produced by a cam, as being presented in [8], where the test rig includes an induction motor and a gearbox for reducing the axle speed.

In [7], an automotive suspension testing rig is presented as having the characteristics to be used also as an educational, training and research tool. Its design allows investigating the impact of various road surface bumps and humps for both an active and a passive suspension system. In [3], the vehicle wheel load estimation is presented by using the suspension mechanism. The developed virtual model allows to easily understand where the most significant loads are met in relation to the components of the suspension system.

The in-wheel electric motor may suffer from common faults, both mechanical and electrical. The operating conditions of an in-wheel electric motor was studied by [5], which developed a dedicated test bench to also simulate the road shocks and the vertical load.

Based on the existing solutions, the testing bench to be developed will be open for different improvements and optimization criteria.

## 3. The proposed design and the applied technologies

The main component for the test bench is the system to be tested and validated, which is the e-wheel that integrates the suspension system and has the electric machine coupled to its hub. The e-wheel dimensions represent one of the applied limitations before starting the test bench design. The test bench must allow to test different e-wheel sizes (especially the diameter). Therefore, it was designed from the beginning to be configurable. The proposed solution for the test bench is based on combining different already developed solutions with new approaches for achieving the current e-wheel needs. The test bench development consists of the definition the 3D design (figure 1) and its validation, followed by the manufacturing processes for each of the components, in addition to choosing the appropriate systems that are part of it.

The frame for the test bench was designed to be made of rectangular tubes, on which an electric motor (figure 2) is fixed. Along with the electric motor, a reduction gearbox (figure 2) was needed to ensure the rotary velocity below 1400 rotations/minute.

**Figure 1.** The test bench design

a b

**Figure 2.** The electric motor with the reducing gearbox (a) and its operating principle (b) [12]

The electric motor has its own controller, which regulates its rotary velocity. The output shaft of the electric motor reducer is coupled to an axial radial/ thrust bearing containing two bearings. The axial radial/ thrust bearing supports the e-wheel with its integrated suspension. Another radial axial/ thrust bearing, like the first one, is mounted to the frame parallel to the first one. This second axial radial/ thrust bearing drives an upper cam that simulates the road profile. This second axial radial bearing is fixed to the frame by means of oval holes that allow the upper cam wheel to be mounted so that it exerts on the e-wheel a force equivalent to the weight of the car distributed on the rear wheel.

## 4. The prototype

The prototype of the test bench (figure 3) was built with respect to the e-wheel characteristics in terms of size, testing approach and costs. Each component was built, manufactured or purchased. The dimensions from figure 3 correspond to testing the 14 inches e-wheel.

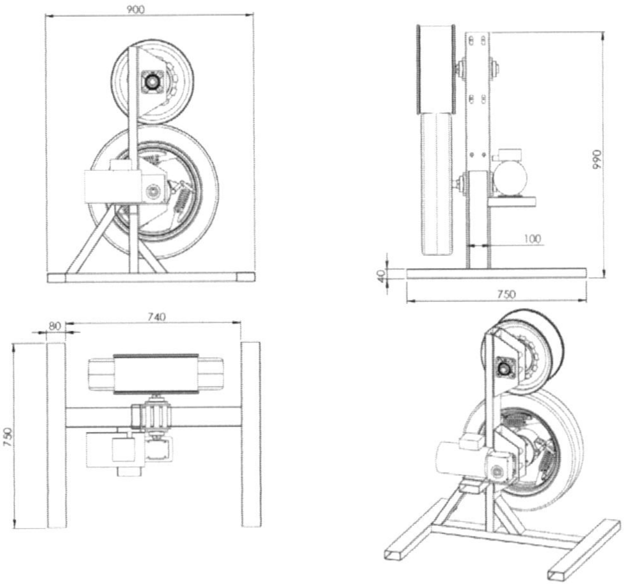


**Figure 3.** The test bench dimensions for the 14 inches e-wheel

The manufacturing processes consist of several operations, both technological and control operations. The main technological operations include the mechanical processing, the welding, the assembly of the mechanical components, the assembly of the electrical and electronic components, the configuration step.

The frame was built from rectangular tubes (figure 4.a), that were cut based on the established dimensions, then being welded or assembled by screws and nuts, with respect to the design, and prepared to be finally painted. The electric motor and its reducing gearbox (figure 2) were purchased. They were mounted to the frame, using a dedicated metal support (figure 4.d). and an axial radial bearing (figure 4.c). Another metal support was manufactured for the second axial radial bearing that is needed for the upper wheel. Two linchpins (figure 4.b) were also manufactured for making possible the mounting of the e-wheel to the metal support and of the second wheel with the cam that simulates the road profile (figure 4.e).

Before the final assembly of the test bench, the dimensional control of each component was carried out. In addition, a variable wheel load force adjustment was set, while several cams have been made to simulate various road profiles. The assembly for the testing bench includes the e-wheel with its integrated suspension system, the upper wheel that simulates the road profile, the electric motor that drives the e-wheel, together with its reducing gearbox and controller (figure 5).

a b
c d e

**Figure 4.** The manufacturing processes for: (a) the frame, (b) the linchpins, (c) the axial radial bearing, (d) the metal support, (e) the upper wheel with the cam

**Figure 5.** Some of the assembly steps for the test bench

## 5. The testing scenarios and the obtained results

The dynamic test bench (figure 6.a) has the role to investigate the behavior for the e-wheel with integrated suspension. The position for the upper wheel allows to choose between various heights that are coming from the road, easily simulating the road profile. A three-axis displacement transducer was used to measure the electromechanical parameters of the e-wheel.

First testing scenario consists of measuring the oscillation on the vertical axis relative to the road for a road profile with height irregularities of maximum ten millimeters (10 mm) (table 1, figure 6.b). For the second testing scenario, the road profile height irregularities were increased to 15 mm (table 2, figure 6.c), while for the third testing scenario they were increased to 20 mm (table 3, figure 6.d).

**Table 1.** The displacement of the wheel on the vertical axis relative to the road profile of 10 mm

| Time [s] | z [mm] | Time [s] | z [mm] | Time [s] | z [mm] | Time [s] | z [mm] |
|---|---|---|---|---|---|---|---|
| 0.0473882 | 0.4200079 | 9.6468699 | -0.6559449 | 19.246534 | 0.3789712 | 28.846533 | -2.6407261 |
| 1.0073884 | 0.7994604 | 10.606871 | 0.6901649 | 20.206533 | 0.2538803 | 29.806172 | -0.050637 |
| 1.9673882 | 1.4593887 | 11.56687 | 0.9452886 | 21.166535 | 4.4709373 | 30.766173 | 1.2566905 |
| 2.9273892 | -0.309827 | 12.526869 | 0.4008937 | 22.126535 | -1.2461983 | 31.726173 | -1.5011406 |
| 3.8873882 | -0.6157556 | 13.486871 | 1.3331156 | 23.086533 | 0.1502482 | 32.686172 | 0.1099331 |

| | | | | | | | |
|---|---|---|---|---|---|---|---|
| 4.8473877 | -0.4846868 | 14.446871 | -1.6564088 | 24.046533 | 0.4404008 | 33.646172 | -0.7214967 |
| 5.8073879 | -1.386307 | 15.406871 | 0.8798955 | 25.006534 | -0.2886623 | 34.606172 | -0.345035 |
| 6.7673879 | -0.6970087 | 16.366871 | -0.9753715 | 25.966534 | 0.4196007 | 35.566172 | 2.5066967 |
| 7.7273879 | 0.8788394 | 17.32687 | -1.1094482 | 26.926532 | 1.7585709 | 36.526172 | 2.0105774 |
| 8.6868699 | -0.331484 | 18.28687 | 0.8576493 | 27.886533 | -0.3162503 | | |

**Table 2.** The displacement of the wheel on the vertical axis relative to the road profile of 15 mm

| Time [s] | z [mm] | Time [s] | z [mm] | Time [s] | z [mm] | Time [s] | z [mm] |
|---|---|---|---|---|---|---|---|
| 0.04429 | 0.612882 | 13.48268 | 0.489077 | 26.9224 | 0.48316 | 40.36203 | 0.702092 |
| 1.003993 | 0.024304 | 14.44268 | 0.361433 | 27.8824 | -2.26482 | 41.32203 | 2.821445 |
| 1.963994 | 0.033806 | 15.4025 | 0.194411 | 28.8424 | -0.16681 | 42.28203 | -0.86388 |
| 2.923995 | -0.08537 | 16.36251 | -0.0304 | 29.8024 | 2.98648 | 43.24203 | -4.59013 |
| 3.883995 | 0.451246 | 17.32263 | 3.169335 | 30.7624 | -1.65273 | 44.20203 | 3.424962 |
| 4.843995 | 0.021495 | 18.28263 | -1.98349 | 31.72203 | 4.36871 | 45.16203 | 0.673999 |
| 5.803994 | 0.667506 | 19.24263 | 0.504617 | 32.68203 | -0.47678 | 46.12203 | 4.737623 |
| 6.763995 | 1.063161 | 20.20263 | 3.273795 | 33.64203 | -0.25178 | 47.08203 | 2.314242 |
| 7.723563 | 0.164879 | 21.1624 | 2.89712 | 34.60203 | -3.13816 | 48.04201 | -2.50379 |
| 8.682997 | 2.090292 | 22.1224 | 1.826002 | 35.56203 | -1.34926 | 56.87911 | -0.09161 |
| 9.642997 | -0.15343 | 23.0824 | -1.2246 | 36.52203 | -2.03299 | 57.84757 | 0.061119 |
| 10.60268 | -0.10035 | 24.0424 | -1.0652 | 37.48203 | -2.84267 | 58.80757 | 0.028667 |
| 11.56268 | 1.715568 | 25.0024 | 0.163069 | 38.44203 | -1.90174 | 59.76223 | -0.06514 |
| 12.52268 | -1.90075 | 25.9624 | -0.51381 | 39.40203 | -4.19783 | 60.72223 | 0.065137 |

**Table 3.** The displacement of the wheel on the vertical axis relative to the road profile of 20 mm

| Time [s] | z [mm] | Time [s] | z [mm] | Time [s] | z [mm] | Time [s] | z [mm] |
|---|---|---|---|---|---|---|---|
| 0.037892 | -0.67728 | 22.10309 | -0.27353 | 44.18268 | -5.3052 | 67.45766 | 0.403645 |
| 0.996623 | 0.046905 | 23.06309 | -0.62133 | 45.14268 | -6.36975 | 68.41766 | 0.113275 |
| 1.955888 | -0.0126 | 24.02309 | -2.61437 | 46.10268 | 2.661116 | 69.37766 | -0.04465 |
| 2.915888 | 0.085171 | 24.98309 | -3.83298 | 47.06268 | 3.481288 | 70.33766 | -3.89297 |
| 3.87589 | -0.1903 | 25.94309 | -1.63167 | 48.01664 | -0.8537 | 71.29766 | 2.93115 |
| 4.835888 | -0.03454 | 26.90308 | -1.92528 | 48.97664 | 3.699876 | 72.25766 | 2.93838 |
| 5.795588 | 0.072542 | 27.86308 | 3.391936 | 51.13421 | 6.248023 | 73.21766 | -0.17561 |
| 6.754655 | 0.230999 | 28.82266 | -5.92999 | 52.0942 | -0.03973 | 74.17409 | 0.629756 |
| 7.713888 | -0.19927 | 29.78266 | 3.599284 | 53.0542 | 8.896811 | 75.13409 | 1.147321 |
| 8.673528 | 0.042747 | 30.74266 | 5.967854 | 54.0142 | 5.176649 | 76.09409 | 1.652118 |
| 9.632527 | -0.27713 | 31.70268 | -3.22369 | 54.9742 | 2.273607 | 77.05409 | -0.11272 |
| 10.59113 | 0.239647 | 32.66268 | -0.20222 | 55.9342 | 2.461061 | 78.01409 | 0.547186 |
| 11.5482 | -0.42081 | 33.62268 | 3.852866 | 56.8942 | 3.732144 | 78.97409 | -0.06629 |
| 12.50753 | -1.4082 | 34.58268 | 2.250919 | 57.8542 | -0.78691 | 80.01262 | 0.237801 |
| 13.46723 | 0.182626 | 35.54268 | 1.142612 | 58.81766 | -4.94543 | 80.97262 | 0.054849 |
| 14.42653 | -0.5031 | 36.50268 | -0.0822 | 59.77766 | 0.320404 | 81.93262 | 0.076311 |
| 15.38653 | 5.431729 | 37.46268 | 3.835338 | 60.73766 | -0.37631 | 84.33517 | -0.06368 |
| 16.34553 | 2.099473 | 38.42268 | 2.73874 | 61.69766 | 0.397499 | 85.29517 | 1.591817 |
| 17.30553 | 3.507032 | 39.38268 | -0.03287 | 62.65766 | 1.655737 | 86.25517 | -0.6745 |
| 18.26476 | 4.195607 | 40.34268 | -2.32201 | 63.61766 | -0.24855 | 87.21517 | -0.02042 |
| 19.22325 | 2.908476 | 41.30268 | -4.44628 | 64.57766 | -0.40419 | 88.17517 | 0.032801 |
| 20.18309 | 2.871919 | 42.26268 | -2.83292 | 65.53766 | -0.55922 | 89.13772 | 0.236015 |
| 21.14309 | 0.185894 | 43.22268 | 0.075703 | 66.49766 | -0.06188 | 90.09772 | 0.275128 |

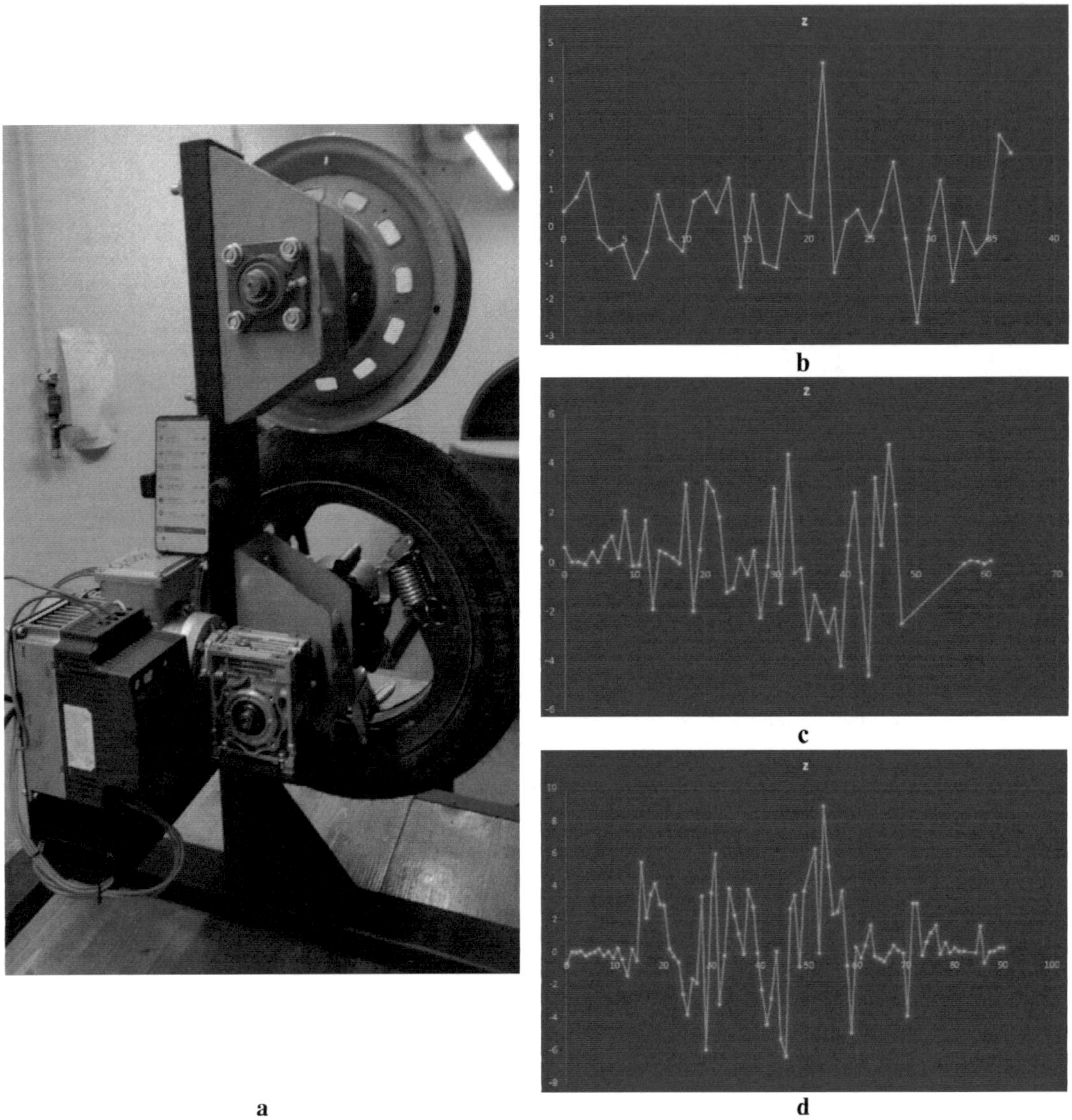

**Figure 6.** The test bench under operation (a) and the oscillation on the vertical axis relative to the road for a rod profile of: (b) 10 mm, (c) 15 mm, (d) 20 mm

The system was tested for five hours continuously for each road profile and no damage to the component parts was found. During the testing scenarios, certain sounds were met which are mainly due to the rectangular profile chosen for the road profile, which is considered the most unfavourable in terms of mechanical demands. The wheel displacement along vertical axis for each of the various road profiles are within acceptable limits, the shock of bumps being easily dissipated by the integrated tire-suspension assembly. The temperatures of the most important components of the e-wheel reached values within acceptable limits, which indicates the absence of unwanted frictional contacts.

## 6. Conclusions

The test bench development stages included the definition for the final design by virtual modeling and validation. The development stages also included the manufacturing operations for the frame, the metallic support, the linchpins, the axles for mounting the wheels, the upper wheel together with the cam. To ensure the enough clearance for the e-wheel, the frame was configured to allow different tire dimensions and easily modify the distance between the mounting axis that correspond to the e-wheel and to the upper cam wheel. Several cam sizes were developed for simulating various road profiles. A variable load force adjustment was created. The drive system and the controller were set for operation, while the data collection was achieved using the mobile phone application of the transducer.

The resulted test bench allows testing the wheel with integrated suspension simulating the entire load and speed range. In addition to the functional behavior tests for the e-wheel with integrated suspension system, another important test consisted of the endurance test of the mechanical elements of the e-wheel, which are directly related to the functional safety of the system.

**Acknowledgments**
Authors express their gratitude for the assistance and the continuous support received from all the team members of the research and development project "Cercetarea, dezvoltarea si lansarea in productie a Sistemului de Propulsie Integrat (SPI) pentru vehicule electrice", developed entirely in Romania.

**Author's ORCID ID**
V Croitorescu: https://orcid.org/0000-0003-1987-5088

**References**

[1] E. C. Acosta, J. J. C. Aguilar, J. A. C. Carrillo, J. M. V. García, J. P. Fernández and M. G. A. Vargas, (2020) Modeling of Tire Vertical Behavior Using a Test Bench, IEEE Access, vol. 8, pp. 106531-106541, 2020, doi: 10.1109/ACCESS.2020.3000533
[2] K. Hyniova, (2016) On Testing of Vehicle Active Suspension Robust Control on An One-Quarter-Car Test Stand, International Journal of Mechanical Engineering, Volume 1, 2016, ISSN: 2367-8968
[3] M Kluziewicz, M Maniowski, (2016) Calibration procedure of measuring system for vehicle wheel load estimation, Scientific Conference on Automotive Vehicles and Combustion Engines (KONMOT 2016), IOP Conf. Series: Materials Science and Engineering 148 (2016) 012010 doi:10.1088/1757-899X/148/1/012010 2016
[4] Guido Koch , Enrico Pellegrini, Sebastian Spirk and Boris Lohmann, (2010) Design and Modeling of a Quarter-Vehicle Test Rig for Active Suspension Control, Technical Reports on Automatic Control Vol. TRAC-5, July 21, 2010
[5] Liu, B.; Xue, H.; Ding, D.; Sun, N.; Chen, P., (2023) In-Wheel Motor Fault Diagnosis Using Affinity Propagation Minimum-Distance Discriminant Projection and Weibull-Kernel-Function-Based SVDD. Sensors 2023, 23, 4021. https://doi.org/10.3390/s23084021
[6] Anirban C. Mitra, Tanushri Soni, and G. R. Kiranchand, (2016) Optimization of Automotive Suspension System by Design of Experiments: A Nonderivative Method, Hindawi Publishing Corporation Advances in Acoustics and Vibration Volume 2016, Article ID 3259026, 10 pages http://dx.doi.org/10.1155/2016/3259026
[7] Salah Mohammad, (2017) A Laboratory Automotive Suspension Test Rig: Design, Implementation and Integration, Jordan Journal of Mechanical and Industrial Engineering, Volume 11 Number 2, April. 2017 ISSN 1995-6665 Pages 67 -71
[8] Mahmoud Omar, M.M. El-kassaby, Walid Abdelghaffar, (2017), A universal suspension test rig for electrohydraulic active and passive automotive suspension system, Alexandria Engineering Journal, Volume 56, Issue 4, 2017, Pages 359-370, ISSN 1110-0168, https://doi.org/10.1016/j.aej.2017.01.024
[9] https://idea4t.com/test-systems/components/durability-test-laboratory/
[10] https://labtech.org/education/automotive/brake-steering-suspension-systems/
[11] https://www.rdu-gmbh.de/en/messraeder-zur-dynamischen-radkraftmessung/
[12] https://www.resitec.pt/zArchives/Pages/29/Files/catalogo-de-redutores.pdf

IOP Conf. Series: Materials Science and Engineering 1303 (2024) 012012 doi:10.1088/1757-899X/1303/1/012012

# Proposed solution for the electrification of a motor vehicle formerly powered by an internal combustion engine (case study: Oltcit)

**L Racila[1], I Dumitru[2], A Oprica[2] and L Matei[2]**

[1] University of Craiova, Faculty of Mechanics, Department of Applied Mechanics and Civil Constructions, Craiova, Romania
[2] University of Craiova, Faculty of Mechanics, Department of Vehicle, Transportation and Industrial Engineering, Craiova, Romania

E-mail: mateiclucian@gmail.com

**Abstract.** A vehicle powered by electricity is a kind of alternative fuel-powered automobile that uses electrically powered engines and controllers for them in place of a conventional internal combustion engine. Instead of burning a carbon-based fuel, power is generated using battery packs. This not only results in monetary savings but also has an extremely reduced negative influence on the surrounding natural environment. The high price that a buyer must pay to get one of the electric cars that are now on the market is one disadvantage. In this article, we explore the possibility of developing an electric vehicle with the use of an electrical conversion kit. The fundamentals and constituent parts of electric car systems will be broken down, and the procedure for transitioning from an internal combustion engine vehicle to an electric vehicle will be analyzed in this paper.

## 1. Introduction

The widespread perception that oil is a resource that will be almost depleted within the next half-century is one of the primary factors contributing to the ever-increasing cost of a barrel of oil [1]. Electric vehicles and hybrid electric vehicles, which major manufacturers are now proposing, indicate a shift in the transportation paradigm for cities. In addition, a number of organizations and energy experts are proposing new regulations that will stimulate research, development, and demonstration projects aimed at boosting electric vehicles.

On the other hand, low-cost solutions that make use of dependable components that are already on the market might also be presented [2, 3]. Consequently, the transformation of cars powered by internal combustion engines into electric vehicles is an appealing option for the interim time of coexistence.

Electric vehicle conversion is the term used to describe this procedure the transformation procedure [4]. This is a difficult procedure from an engineering standpoint, and it requires understanding of multiple domains and mainly in mechanics but also the electronics domain is important to complete. The electric motor and its associated controller, an energy storage system, and its associated charging system are the primary systems that need to be integrated.

Batteries are the option that has been suggested the most often for the energy storage system. As a result, a mechanism for charging the batteries is required as well. The use of static power converters has historically been the technology of choice for the implementation of battery charging systems, but there are a variety of topologies that also may be employed [5, 6]. Because of this different charging algorithms will need to be utilized in order so that charging technology behind batteries will work.

Charging systems for batteries may also be used to convert electric vehicles and hybrid vehicles into an energy storage system [7]. This system both receives and distributes energy, which contributes to the stability of the power grid. Therefore, in addition to the process of charging batteries, the flow of energy may also occur in the other direction, depending on the features of the power grid and the advantages gained by the driver.

This article, discusses the process of converting an internal combustion engine vehicle into an electric vehicle [8]. All of the essential components, including the powertrain power electronics converter and the battery charging system, has their designs and prototype implementations described.

## 2. Electric vehicle development steps

This build started off with a manual OLTCIT 11R vehicle, which can be seen in the figure 1 bellow. We decided to go with this vehicle because it was readily accessible, its bed provided a convenient spot for us to store batteries, and its manual gearbox makes it possible for us to more easily convert it to one with an electric engine.

Based on the weight of the vehicle, the weight of the planned electrical components, dimensions, and the intended range, we were able to determine the amount of power that would be necessary to achieve a variety of possible peak speeds as well as the range that would be associated with each of those speeds.

**Figure 1.** The OLTCIT 11R vehicle.

The electric vehicle design was calibrated in three stages:

- Electric traction motor (ME), which is of the PMSM type (permanent magnet synchronous motor)
- Command and control system (SCC), which consists of the BorgWarner – Sevcon Gen. 4 unit and the start/stop components
- Energy storage system (SSE), which consists of 6 12V Pb batteries/accumulators, providing a total voltage of 72 V.

### *2.1. Electric traction motor*

The electric traction motor is a three-phase permanent magnet synchronous motor (PMSM) with 10 poles, 8000 rpm, and a nominal power of 13 kW (rated current of 157 A).

However, the electric traction motor is capable of assuring a peak output of around 48 kW for approximately 1 minute maximum (at a current of max. 600 A). This peak output is crucial for accelerating the vehicle quickly, especially during overtaking or merging vehicle flow. Beyond this one-minute period, the motor's output gradually decreases to avoid overheating and potential damage.

Nonetheless, the electric traction motor provides sufficient power for regular driving conditions and ensures a smooth and efficient performance of the electric vehicle.

The ME1905 electric motor is produced by Motenergy, its overall dimensions being given as follows in the figure 2:

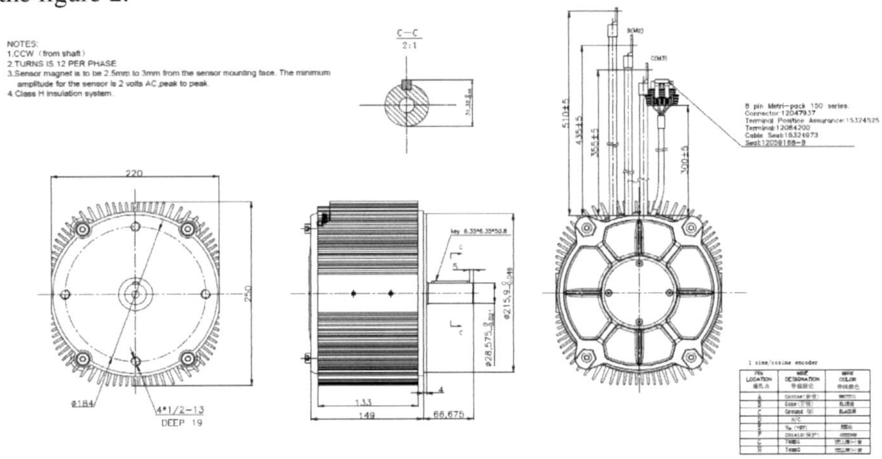

**Figure 2.** Electric motor main dimensions.

The operating curves of the motor (input power, output power, voltage, current, efficiency) are shown in the following figure 3. The input power curve shows the amount of power supplied to the motor, while the output power curve represents the power produced by the motor. The voltage and current curves depict the electrical characteristics of the motor, showcasing how these values change with varying load conditions. Lastly, the efficiency curve illustrates the motor's ability to convert input power into useful output power, providing insights into its overall performance. Together, these curves offer a comprehensive understanding of the motor's behavior under different operating conditions.

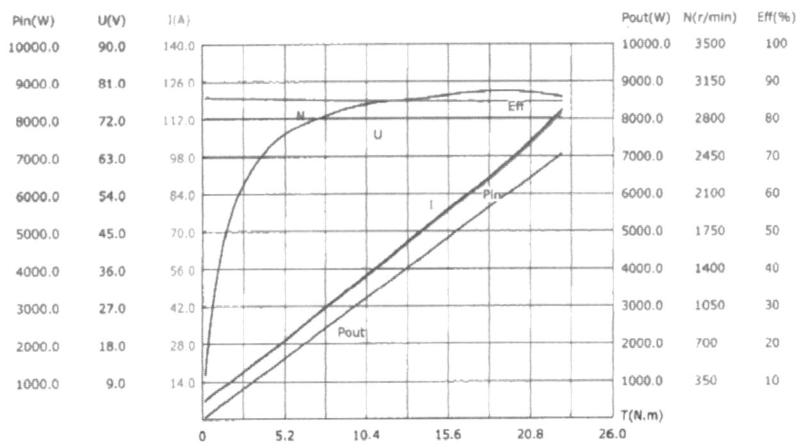


**Figure 3.** Electric motor load characteristics curve.

As can be seen, the output power of the motor is around 7 kW, and the current draw is approximately 100 Amperes, when the supply voltage is 72 Volts, which is the nominal voltage of the battery pack that is being utilized. The torque is around 22 Nm, which is adequate torque for an Oltcit car to operate.

Controlling the position of the motor is accomplished by use of a sinusoidal encoder, also known as a sin/basket encoder, which is connected to the Sevcon command and control system.

*2.2. Command and control system*

The command-and-control system (SCC) is produced by BorgWarner, being a Sevcon Gen. 4 system, see figure 4. It is a highly advanced and efficient system designed for electric vehicles. The SCC seamlessly integrates with the vehicle's drivetrain, providing precise control over power delivery and optimizing overall performance. With its cutting-edge technology, the SCC ensures smooth acceleration, regenerative braking, and enhanced energy management, making it a reliable choice for electric vehicle manufacturers.

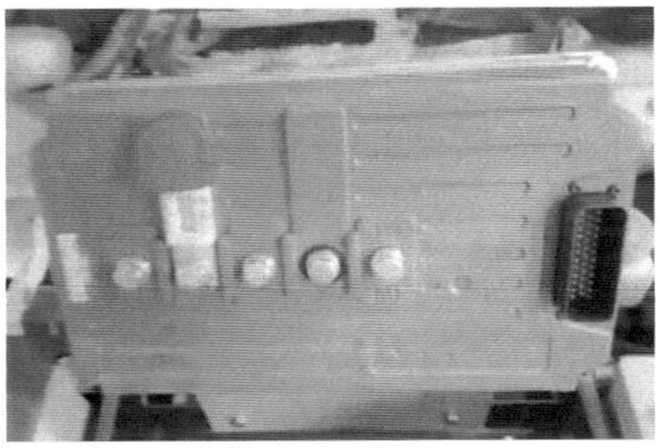

**Figure 4.** Sevcon Gen. 4 system.

The Sevcon has two parts, a low-voltage one for command and control, and a high-voltage one, which is responsible for powering the three-phase electric traction motor. Three-phase voltage is obtained by means of an inverter built into the SCC. The command-and-control part is interfaced through a 35-pin Ampseal jack. This allows for easy connection and disconnection of the Sevcon with other electrical systems. The high-voltage section, on the other hand, is connected directly to the motor through heavy-duty cables. By separating the command-and-control functions from the power delivery, the Sevcon ensures efficient and safe operation of the electric traction motor, making it an essential component in electric vehicle technology.

*2.3. Energy storage system*

The battery system (SSE) consists of 6 Varta batteries with Pb, AGM type, capable of providing a maximum current of 680 amperes. The rated voltage is 12 V, and the capacity of one battery is 60 Ah. The 6 batteries inserted will provide a voltage of 72 V. This battery system is commonly used in large-scale applications such as backup power systems for commercial buildings or renewable energy storage systems. With a maximum current of 680 amperes, it can support high-power devices and equipment. The capacity of 60 Ah per battery ensures a long-lasting and reliable power supply, while the combined voltage of 72 V allows for efficient energy distribution.

**Figure 5.** Battery system.

## 3. Connections and testing of the EV

The battery pack is directly connected to the Sevcon command and control system, the positive terminal of the package having a 500 A fuse embedded in the SCC structure. The delimitation of the force part from the control part is done by means of a power contactor, the control of this contactor being made in low voltage.

**Figure 6.** On/off command for the EV.

**Figure 7.** Safety fuse.

On the power/control side of the contactor (powered at low voltage from SCC Sevcon) there is a 5 A protective fuse, emergency stop button and key contact that activates the force part of the contactor. The emergency stop button and key contact provide an additional layer of safety by allowing immediate shutdown of the force part of the contactor in case of any unforeseen circumstances. When activated, the force part of the contactor allows the flow of high voltage power from the SCC Sevcon to the connected system. This setup ensures that the control of the contactor is securely separated from the high voltage power, preventing any accidental or unauthorized activation.

Furthermore, the emergency stop button and key contact are strategically placed in easily accessible locations, ensuring that they can be quickly and effortlessly reached in case of an emergency. The emergency stop button is designed to be highly visible and easily identifiable, with a bright red color and a prominent label. This ensures that anyone in the vicinity can immediately recognize and activate it in the event of a hazardous situation. Additionally, the key contact adds an extra layer of security by requiring a physical key to be inserted and turned in order to activate or deactivate the force part of the contactor.

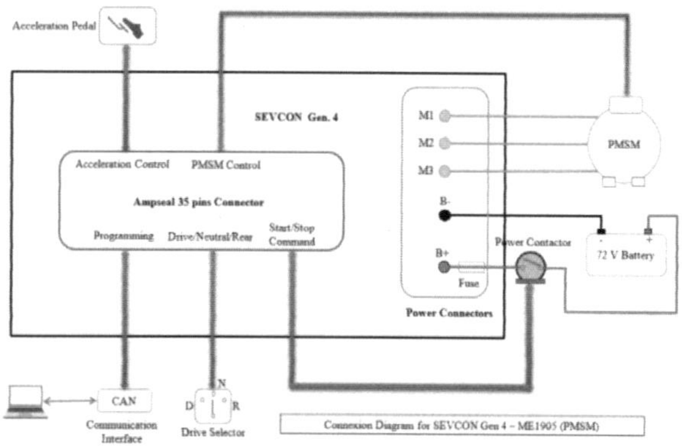


**Figure 8.** General connection system.

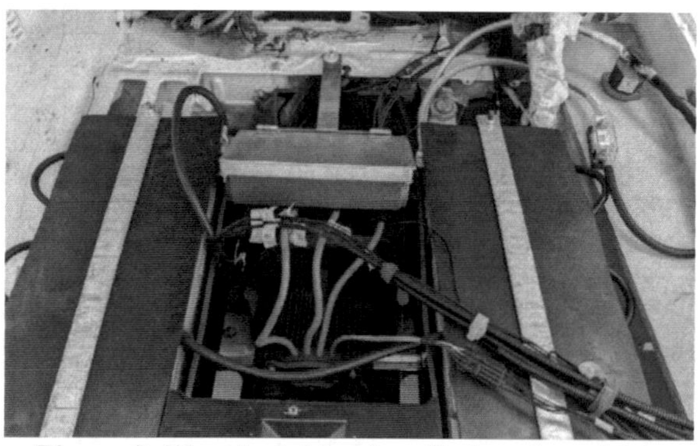

**Figure 9.** Electrical drive installed on the vehicle

The Sevcon controller is programed using BorgWarner DTV software. With the BorgWarner DTV software, users can easily adjust parameters such as acceleration, deceleration, and torque settings to optimize the performance of the Sevcon controller. Additionally, the software provides real-time data analysis and diagnostic tools, enabling efficient troubleshooting and maintenance of the controller. Overall, the integration of BorgWarner DTV software enhances the functionality and reliability of the Sevcon controller in various applications.

The general schematic for the EV connections can be seen in the figure 8. It is important to mention that the schematic for EV connections may vary depending on the specific model or manufacturer, so it may not be accurate to say that this is a general schematic that applies to all EVs. Further more the entire electrical drive can be seen in figure 9, installed on the vehicle.

## 4. Conclusion

The conversion process involved removing the engine and fuel system and replacing them with an electric motor and battery pack. The paper also discussed the challenges encountered during the conversion, including modifying the vehicle's wiring and ensuring compatibility with existing systems. Overall, the successful conversion demonstrated the feasibility of transforming conventional vehicles into electric ones, highlighting the potential for reducing emissions and dependence on fossil fuels.

The paper conversion proved that as these technologies continue to evolve, the process of converting vehicles to electric power will become even more efficient and cost-effective. Furthermore, the integration of smart charging systems and renewable energy sources, such as solar panels, were identified as potential enhancements to further reduce the environmental impact of electric vehicle conversions.

Additionally, it is crucial to address any potential challenges and barriers that may hinder the transition, such as limited range and charging time. By addressing these issues and investing in research and development, the potential for electric vehicle conversions to revolutionize transportation and reduce greenhouse gas emissions is immense.

In conclusion, the paper emphasized the importance of continued research and development in the field of electric vehicle conversions to accelerate the transition towards sustainable transportation.

## References

[1] Pedrosa D, Monteiro V, Goncalves H, Martins J S and Afonso J L 2014, October A Case Study on the Conversion of an Internal Combustion Engine Vehicle into an Electric Vehicle. *2014 IEEE Vehicle Power and Propulsion Conference (VPPC)*.

[2] Silva J E D and Urbanetz Junior J 2019 Converting a Conventional Vehicle into an Electric Vehicle (EV). *Brazilian Archives of Biology and Technology*, *62*(spe).

[3] Eydgahi A and Long E (n.d.). 2011 Converting an Internal Combustion Engine Vehicle to an Electric Vehicle, *ASEE Annual Conference & Exposition Proceedings*.

[4] Pote B 2023 March 6 *Are electric cars a solution to tackle climate change?* Bon Pote. https://bonpote.com/en/are-electric-cars-a-solution-to-tackle-climate-change/

[5] *Smart charging system of the electric vehicle CEPIUM*. (n.d.). CORE Reader. https://core.ac.uk/reader/55621320

[6] Shi T, Zhao F, Hao H and Liu Z 2018 November Development Trends of Transmissions for Hybrid Electric Vehicles Using an Optimized Energy Management Strategy. *Automotive Innovation*, *1*(4), 291–299. https://doi.org/10.1007/s42154-018-0037-5

[7] Zainuri F, Sumarsono D A, Adhitya M and Siregar R 2017 Design of synchromesh mechanism to optimization manual transmission's electric vehicle. *AIP Conference Proceedings*.

[8] Siregar A, Sumarsono Z and Adhitya M 2017 Design a New Generation of Synchromesh Mechanism to Optimization Manual Transmission's Electric Vehicle. *Proceeding of the 15th International Conference on QIR (Quality in Research)*, *1*(1), 410–418.

IOP Conf. Series: Materials Science and Engineering 1303 (2024) 012013 doi:10.1088/1757-899X/1303/1/012013

# Assessment Possibilities for Trajectory Fusion of Two Vehicles in the Case of Automated Vehicles

**N Cordoş, I Barabás, A Todoruţ, I Duma, A Molea, D Moldovanu, N Burnete, M Jurchiş and M Ghereş**

Department of Automotive Engineering and Transports, Technical University of Cluj-Napoca, RO

E-mail: nicolae.cordos@auto.utcluj.ro

**Abstract.** Because of the increasing integration of automation and cutting-edge technology into modern transportation systems, the development of efficient and dependable Automated Vehicles (AVs) has become an important area of research. One of the most important characteristics of AVs is their ability to perceive and navigate complex environments safely. The process of combining trajectory data from multiple vehicles, known as trajectory fusion, has the potential to improve AV perception. The feasibility of implementing trajectory fusion in the context of two autonomous vehicles coexisting in the same operational environment is investigated in this study. The primary goal of this research is to investigate the effectiveness of trajectory fusion in increasing the environmental awareness of Automated Vehicles (AVs), while also investigating its implications for the decision-making mechanisms that govern these vehicles. The MATLAB Simulink simulation platform was essential in carrying out this investigation. The intricate dynamics of trajectory fusion within the context of dual AVs were meticulously analyzed and evaluated using this software.

## 1. Introduction

The evolution of automated vehicles (AVs) has transformed transportation networks, promising increased safety, efficiency, and convenience. However, the success of AVs deployment is dependent on their capacity to effectively perceive and navigate complicated situations. The idea of trajectory fusion, which combines the trajectory data from many vehicles to produce a more complete depiction of the environment, is essential to obtain this capability. This paper delves into the assessment possibilities of trajectory fusion within the context of two automated vehicles, employing MATLAB Simulink simulations to explore its implications and advantages. In order to study the consequences and benefits of trajectory fusion, this research uses MATLAB Simulink simulations to examine the assessment possibilities of the technique in the setting of two automated vehicles.

### 1.1. State of the art

A potential answer to the problems brought on by separate vehicle sensors is trajectory fusion. It can lessen sensor limits, enhance perceptual precision, and make it possible to have a more comprehensive grasp of the world. Trajectory fusion offers the potential to improve AVs' situational awareness and decision-making abilities by incorporating data from nearby vehicles [1].

Cooperative perception is a concept that is introduced through AV cooperation through trajectory fusion. By utilizing one another's observations, AVs can efficiently increase the range of their sensing capabilities. This collaborative strategy shows promise for overcoming difficulties like occlusions, sensor errors, and ambiguous situations [2].

Simulation is a crucial stage in assessing trajectory fusion. A flexible environment for modeling and evaluating intricate interactions between AVs is offered by MATLAB Simulink. Simulations enable researchers to evaluate the performance of trajectory fusion algorithms under various circumstances by generating controlled and reproducible scenarios [3].

Simulations allow for a more in-depth examination of the impact of trajectory fusion on AV performance. Researchers can test the resilience and scalability of trajectory fusion algorithms by varying parameters such as sensor noise, communication range, and environmental complexity [4]. This method allows for the investigation of many scenarios and provides insights into potential problems.

The paperwork [5] seeks to answer the following fundamental problems about trajectory fusion: (1) Explain how trajectory fusion improves AV vision and decision-making. (2) What are the difficulties in establishing trajectory fusion for two AVs? Some studies aim to provide empirical data and insights into the possible benefits and complexity of trajectory fusion using MATLAB Simulink simulations.

The subsequent sections will delve into the methodology, detailing the simulation setup, trajectory fusion algorithms, and assessment metrics employed. By focusing on a scenario involving two AVs, the study [6] strikes a balance between realism and controllability, enabling a comprehensive analysis of trajectory fusion's effects.

The findings of the study [7] have ramifications for the development of cooperative AV systems, contributing to a better understanding of the possibilities of trajectory fusion. As the AV landscape evolves, trajectory fusion could be critical in improving AV perception and decision-making capabilities in complex contexts.

The paper [8] addresses the challenges of autonomous driving in complex environments. The authors present unique trajectory planning algorithms that take into account real-world circumstances, including interactions with traffic dynamics and human drivers. The project aims to improve the adaptability and safety of highly autonomous vehicles by stressing maneuver-based methods. The research advances the science of autonomous driving by providing insights into successful trajectory planning approaches for navigating real roads with complex traffic circumstances and driver interactions through a thorough analysis.

The aim of this paper is to demonstrate the significance of trajectory fusion for enhancing the perceptual capabilities of AVs and paving the way for further advancements in AV technology. The simulation is used to demonstrate the practical development of a track-level fuser and the use of the object track data format. The driving scenario and models derived from the "Automated Driving Toolbox", which seamlessly interact with the tracking and track fusion models derived from the “Sensor Fusion and Tracking Toolbox”, within the simulated environment using MATLAB Simulink. This dynamic fusion demonstrates the promise of sophisticated fusion approaches for improving the efficacy and performance of automated driving systems.

## 2. Materials and methods

In the analyzed scenario (Figure 1), two cars are driving down a road with constant speed of 12 m/s (approx. 43 km/h). A lower speed than the locality's legal speed was considered. Vehicle 1 is positioned in front and is outfitted with two forward-facing sensors: a short-range radar (SRR) and a vision sensor. Long-range radar (LRR) is installed in Vehicle 2, which is trailing Vehicle 1 at a distance of 10 meters. Other cars are parked along the right side of the road. Notably, a pedestrian is standing in the space between two of the parked vehicles.

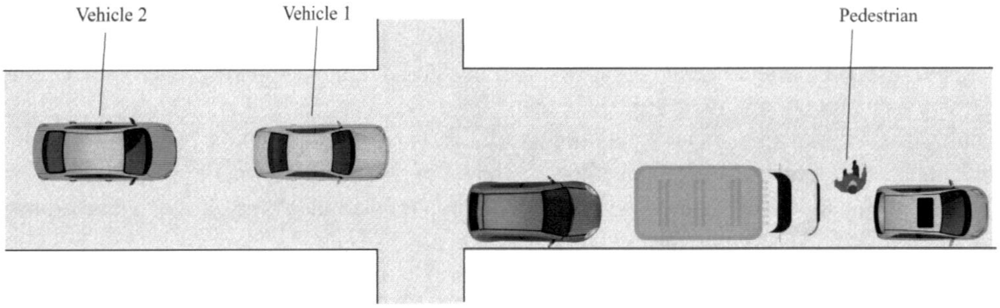


**Figure 1**. The analyzed scenario.

The action takes place at the X-coordinate of 60 meters when we see a symbolic representation of the pedestrian's presence. Through the use of various sensor setups, inter-vehicle distances, and the presence of pedestrians, this scenario captures the complexity present in real-world surroundings. Such situations demonstrate the necessity of efficient sensor fusion and trajectory planning techniques.

Due to the close proximity of Vehicles 2 and 1 (10 m), Vehicle 2's radar sensor coverage is blocked by Vehicle 1. As a result, tracks transmitted from Vehicle 1 serve as the basic initialization for the majority of the tracks that the track fuser on Vehicle 2 maintains.

In Vehicle 1, a pair of sensors send data to separate local trackers. These trackers monitor objects using local sensor detections and then feed these confined trajectories to the Vehicle's track fusion system. Vehicle 2, on the other hand, is equipped with a single sensor that relays detections to its dedicated local tracker. The local track fusion algorithm within Vehicle 2 then uses the local trajectories solely from Vehicle 2 as input. A basic functional diagram of these sensors on vehicles can be seen in Figure 2.

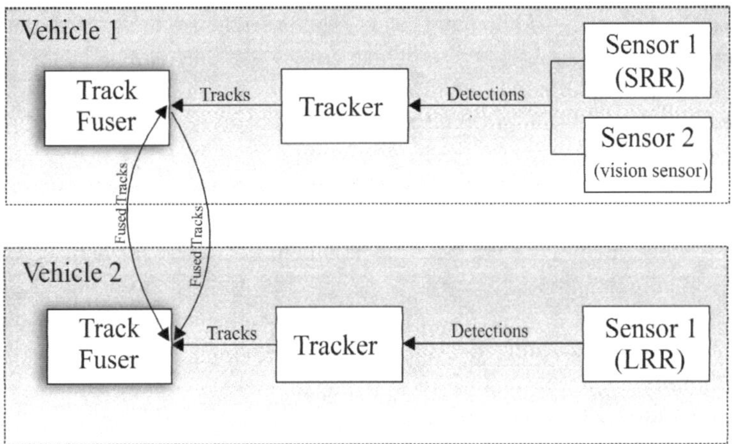


**Figure 2**. Functional diagram of the vehicle's sensors.

Each Vehicle's track fuser joins the tracks it receives from the track fuser of the other Vehicle with the tracks it has locally. Each track fuser broadcasts its fused tracks after each update, which feeds into the subsequent update of the track fuser on the opposite Vehicle.

The fused tracks from one Vehicle are updated by the fused tracks on the other Vehicle in this architecture. The first vehicle is then transmitted these fused tracks. When updating the track fuser with tracks from another vehicle, the rumor propagation must be prevented.

Vehicle 2 transmits track information about the item to Vehicle 1, which is the only one really tracking it. The track fuser on Vehicle 1 must thus be aware that the tracks it receives about this item from Vehicle 2 do not, in fact, include any new information that has been updated by a different source.

To distinguish between tracks that include new information and tracks that simply repeat information, Vehicle 2 must be specified as an external source to the track fuser on Vehicle 1, and Vehicle 1 as an external source to the track fuser on Vehicle 2. Only tracks that are modified by a track fuser based on information from an internal source must be defined as self-reported. As a result, the track fuser in each Vehicle can disregard updates from tracks that bounce back and forth between track fusers without containing any new information.

## 3. Results and interpretation

When the simulation first starts (Figure 3), the behavior takes the following course: Vehicle 1 begins by recognizing the presence of parked automobiles along the right side of the street. The linked tracks for these stationary Vehicles are then put through a verification process to confirm their existence and characteristics.

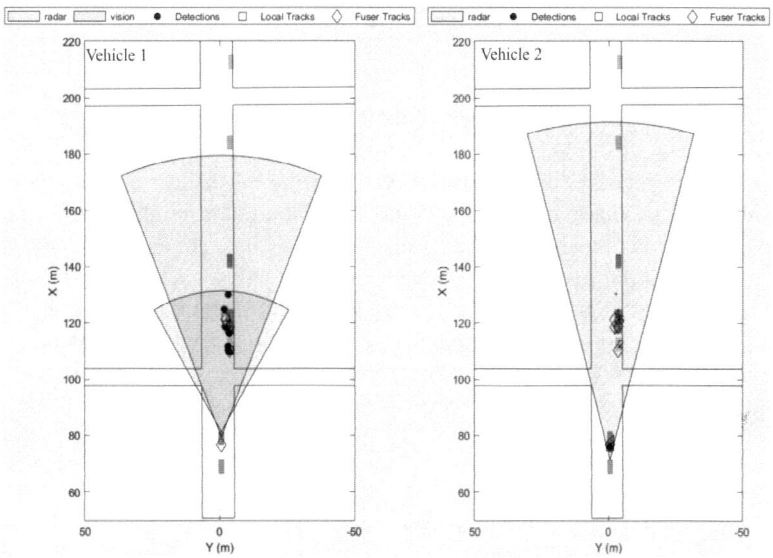


**Figure 3**. The initial position of the Vehicle's.

In the initial phase, Vehicle 2's tracker dedicates its resources entirely to detecting and monitoring the immediate presence of Vehicle 1, which is positioned directly ahead. As this occurs, Vehicle 1's track fuser enters a confirmation process to ensure the accuracy of the identified tracks. Once the confirmation is successfully accomplished, these verified tracks are then shared, a process initiated by Vehicle 1's track fuser. They are subsequently absorbed and seamlessly integrated into the track fuser of Vehicle 2. This orchestrated synchronization within the realm of track fusion bestows upon Vehicle 2 an advanced level of comprehension concerning the existence and dynamic behavior of these tracks.

This interplay showcases the collaborative aspect of track fusion, where one vehicle's tracking information benefits another, enhancing overall situational awareness and object-tracking proficiency. The intricate synchronization within this fusion process ensures that the data shared between vehicles is not only accurate but also contributes to a more comprehensive understanding of the environment, enabling safer and more efficient automated vehicle operations. The scenario progresses when Vehicle 2 successfully detects and tracks the nearby parked cars, fusing their tracks with those of Vehicle 1. This shows the Vehicle's expanding capabilities. Vehicle 2 successfully locates and tracks the pedestrian at the 4-second mark, instantly fusing the corresponding track at the 4.4-second mark into the simulation (Figure 4).

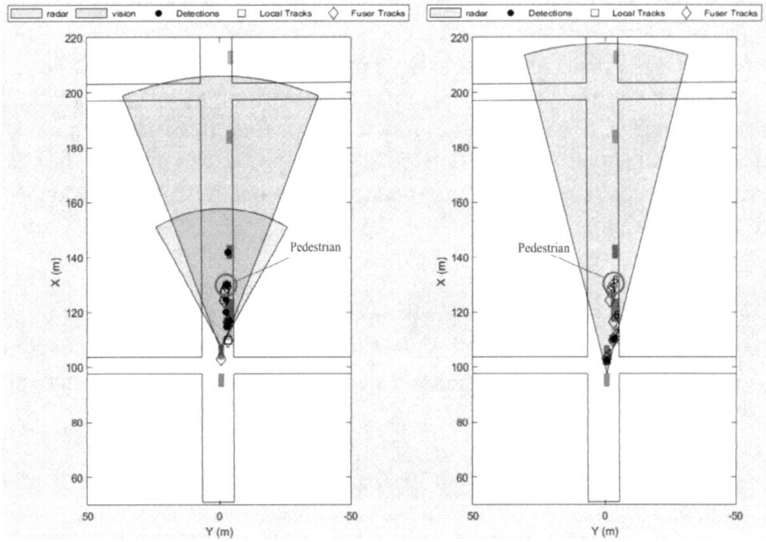


**Figure 4.** The fusion phase of the sensors within the simulation.

However, there is a further delay of about two seconds before the pedestrian is autonomously detected and tracked by Vehicle 2's internal sensors (Figure 4). This delay could have negative effects on pedestrian safety, especially if the pedestrian starts to cross during this time. This emphasizes the significance of quick detection and fusion to enable quick responses and improve safety in dynamic circumstances. Track fusion often offers advantages in terms of faster and more accurate object detection because it combines information from multiple sensors, providing a more comprehensive view of the surroundings. Comparing the scenario with fusion tracks and the one with no fusion track, it can help to evaluate the effectiveness of track fusion in improving detection time and overall safety in automated vehicles.

Finally, in the final phase of the simulation (Figure 5), it is critical to further explore how the vehicles behave when they pass the objects, causing them to leave their field of view. The fused tracks connected to these objects are in this case discarded by both trackers.

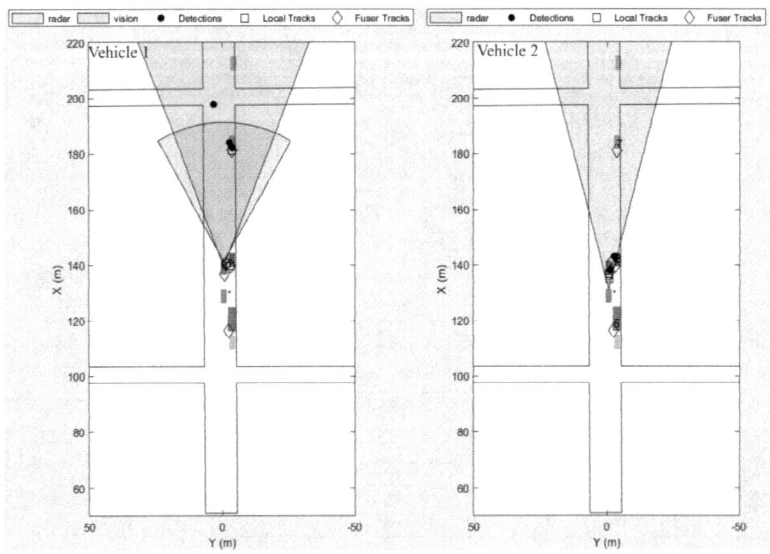


**Figure 5**. The final phase of the simulation.

This intentional track-dropping highlights that the exchanged fused tracks between the two Vehicles comply to a system that maintains accuracy and validity in the trajectory information instead of perpetuating rumors. This method confirms that the trajectory fusion system maintains information precision without unintentionally distributing speculative data amongst Vehicles and underlines the strict dedication to reliability.

## 4. Conclusions

The simulation emphasized the real-world advantages of track-to-track fusion and highlighted its potential to raise safety and situational awareness in the context of automotive applications.
In the complex world of automotive applications, it effectively illustrated how this fusion method has the ability to drastically raise safety and increase situational awareness. The simulation demonstrated the crucial role that track-to-track fusion plays in the development of automated Vehicles by linking theory and application, and it offered a promising trajectory development towards safer and more perceptive driving situations. The results of the simulation thus confirm the importance of fusion techniques in promoting the integration of autonomous cars into real-world environments while enhancing their general operational efficacy.

The significance of trajectory fusion for improving the perceptual capacities of AVs is shown in this research, which opens the door for further developments in AV technology. The simulation clearly highlighted the concrete real-world advantages inherent in track-to-track fusion by utilizing the strong MATLAB Simulink framework. The knowledge gathered from this work can be used to build trajectory fusion techniques that can be incorporated into the next generation of automated cars, ultimately leading to more secure and dependable autonomous transportation systems.

**Author's ORCID ID**
N Cordos: https://orcid.org/0000-0002-9877-6530
A Todorut: https://orcid.org/0000-0003-3368-3701
I Duma: https://orcid.org/0000-0003-2303-4636
A Molea: https://orcid.org/0000-0003-3848-2722
D Moldovanu : https://orcid.org/0000-0003-3039-1965
N Burnete: https://orcid.org/0000-0001-6647-459X
M Gheres: https://orcid.org/0000-0001-8194-5126

## References

[1]. Pérez M D, Olivier K, Salvador D and Phillipe M 2021 *Multisensor-based predictive control for autonomous parking* in IEEE Transactions on Robotics **38** DOI:10.1109/TRO.2021.3091285.
[2]. Chen Q, Ma X, Tang S, Guo J 2019 *F-cooper: "Feature-based cooperative perception for autonomous Vehicle edge computing system using 3D point clouds* in Proceedings of the 4th ACM/IEEE Symposium on Edge Computing, DOI:10.1145/3318216.3363300.
[3]. Kavas T O, Gelbal S, Cantas MR, Aksun B and Guvenc L 2022 *V2X Communication between Connected and Automated Vehicles (CAVs) and Unmanned Aerial Vehicles (UAVs)* in Sensors **18** DOI: 10.3390/s22228941.
[4]. Kunyang C, Ting Q, Bingzhao and G Hong C 2023 *Consensus-Based Distributed Cooperative Perception for Connected and Automated Vehicles* in Trans. Intell. Transport. Sys. **24** DOI:10.1109/TITS.2023.3264608.
[5]. Yifang M, Zhenyu W, Hong Y and Lin Y 2020 *Artificial Intelligence Applications in the Development of Autonomous Vehicles: A Survey* in IEEE/CAA J. Autom. Sinica, vol. **7**, no. **2**, pp. 315-329 DOI: 10.1109/JAS.2020.1003021.
[6]. Shahian JB, Theja T and Sabri C 2019 *Real-time hybrid multi-sensor fusion framework for perception in autonomous vehicles* in Sensors **19** DOI: 10.3390/s19204357.

[7]. Burger C, Thomas S and Martin L 2020 *Interaction aware cooperative trajectory planning for lane change maneuvers in dense traffic* in IEEE 23rd International Conference on Intelligent Transportation Systems (ITSC), DOI:10.1109/ITSC45102.2020.9294638.

[8]. Glaser S, Vanholme B, Mammar S, Gruyer D and Nouvelière L 2010 *Maneuver-Based Trajectory Planning for Highly Autonomous Vehicles on Real Road With Traffic and Driver Interaction* in IEEE Transactions on Intelligent Transportation Systems, vol. **11**, DOI: 10.1109/TITS.2010.2046037.

IOP Conf. Series: Materials Science and Engineering 1303 (2024) 012014 doi:10.1088/1757-899X/1303/1/012014

# Overview of the use of oxygen sensors in automotive applications

**D Tutunea, I Dumitru and O V Stănciuc-Oţăt**

Faculty of Mechanical Engineering, University of Craiova, Romania

E-mail: dragostutunea@yahoo.com

**Abstract**. In the process of combustion, the air-fuel ratio is important for control strategies to reduce fuel economy and improve engine performance. Oxygen sensors are key elements in electronic fuel injection systems to indicate the oxygen content and send electrical signals to the electronic control unit. Starting with the first generation of oxygen sensors, which had limited applications, all automotive manufacturers now use wideband sensors to detect oxygen concentration with rapid reaction and linear output characteristics. Also, these new sensors provide accurate information for a wide range of air coefficients ($\lambda$ = 0.65-2.4), improving combustion efficiency. In this paper, several types of narrowband and wideband sensors were analyzed, measuring the activation temperature for each sensor until a reliable signal was obtained.

## 1. Introduction

Today the increase in energy demands in correlation with the shortages in the supply of fossil fuels force the European countries to improve strategies for emissions control of combustion gases. The automotive sector uses many types of control sensors to asses engine performance by using various fuels and using different configurations targeting a better engine control. Adopting of a sophisticated engine management system in time will reduce fuel, costs and finally will protect the environment [1].

The lambda sensor plays an important role in the in electronic fuel injection system to measure and monitor the quantity of oxygen in the exhaust gases and send this information into electric signals to ECU. Control strategies for gasoline engines are based on the accurate estimation of the complete/incomplete combustion and the oxygen content to regulate air-fuel ratio (A/F) [2]. The base of lambda technology began back in 1889 when Walther Nernst discovered the "Nernst equation", a thermodynamic relationships used to measure the oxygen content. The oxygen sensor was later used by the company Bosch in 1976 and where U-shaped tubes without any heaters. The disadvantage of earlier sensors was the time necessary to reach over 300°C (almost 2 minutes) and in this time the electronic fuel injection enters in open loop control (1st and 2nd generation). The third generation of lambda sensors as emissions become more restrictive were mounted inside the sensor a heating element to reduce the time and to increase the heating rate at the start of the engine. The fourth generation of the sensor reduce the heating time up to 15 degrees and are called narrow-domain oxygen sensors using a ceramic molding technology (Figure 1). The voltage change near stoichiometric ratio is recorded and if the voltage is over this range is less responsive. This narrow measurement has limitations and this type of sensor cannot meet the requirements of modern engine control modules. The five generation of lambda sensor measures a wide range of air fuels ratio with a linear signal output and is composed by two parts (sensor battery and pump oxygen battery). In Table 1 is presented the performances of various types of oxygen sensors.

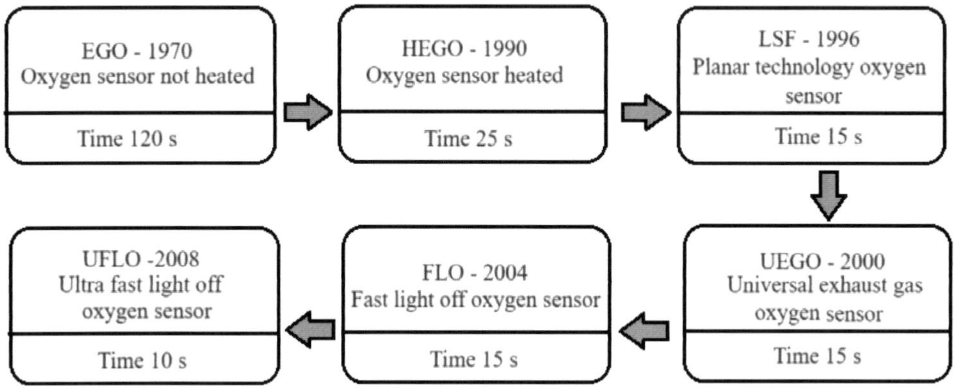


**Figure 1.** Time evolution of oxygen sensors [3].

**Table 1.** Generations of oxygen sensors.

| Lambda - oxygen generation | Parameters |
|---|---|
| 1st and 2nd generation | No heating elements, open loop control at cold starting |
| 3rd generation | Heating elements, response time < 1min |
| 4th generation | Heating elements, response time < 15 s, fast heating speed, exhaust emissions can be controlled |
| 5th generation | Linear output, response time < 10 s, rapid reaction, accuracy of measurements |

The sensors used in the automotive industry are either based on an oxygen ion conductor (for example, ZrO2) or a mixed ionic electronic conductor (SrTiO3, CeO2, TiO2). The major disadvantage of the ZrO2 sensor is its low sensitivity at oxygen partial pressure, which is why it is used in gasoline engines. Also, this sensor is complicated to manufacture with a high cost of production and is instable to temperature fluctuations. As a solution, the industry adopted TiO2-based sensors, which function more properly due to the fact that this sensor does not need a reference gas. The small dimension, low production cost, better durability, and high operation temperature [4,5].

Currently, the automotive industry uses dual-cell lambda sensors (considered wide-range sensors). This sensor is very complex, with two cells (the pumping cell and the Nernst sensing cell). The working principle consists in the passing of the exhaust gases into the Nernst cell, with the pumping cell pumping oxygen into the Nernst cell, recording a wide range of A/F from 10 to 32. In this study, the activation temperature of internal heaters was investigated to analyze their response in time for narrowband and wideband oxygen sensors.

## 2. Closed loop control vs open loop control

The electronic fuel injection system used today compares the air/fuel ratio commanded by the electronic control unit of the engine with the air/fuel ratio read by the lambda sensor. If the values recorded by the sensor do not match the ECU, change the quantity of fuel or air to equalize the two. A closed-loop control lambda sensor provides feedback to the ECU about the oxygen content, and this information is used to operate the engine by maintaining an ideal air/fuel ratio, resulting in maximum efficiency and low emissions. If the oxygen content is too high, the ECU will decrease the air content by adjusting the throttle. If the oxygen content is too low, the ECU will reduce the quantity of fuel by commanding the injectors. This system is complex and may prioritize emissions reduction over the high performance of the engine (Figure 2). The open-loop control lambda sensor is not working or sending the wrong values, and the ECU calculates the air/fuel ratio on the basis of pre-programmed values from the engine manufacturer. Also, the ECU will ignore the signal received from the lambda sensor when the vehicle is accelerating (the mixture is enriched to provide power) or decelerating (the fuel is reduced to improve economy).

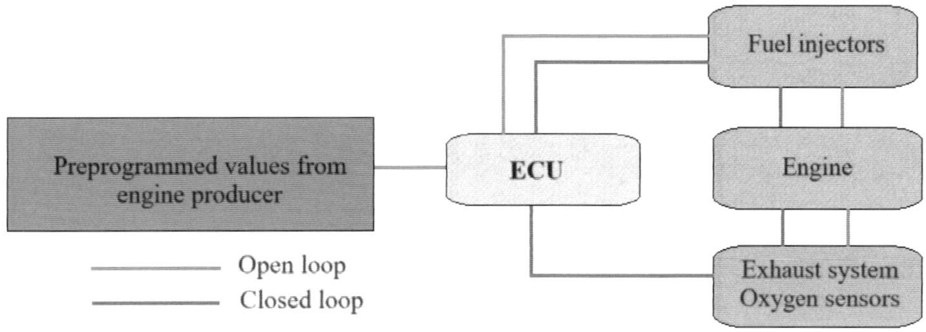


**Figure 2.** Open vs closed loop.

## 3. Heated exhaust gas oxygen - HEGO

In colder conditions and at the start of the engine, there is insufficient heat available in the exhaust gases to warm the oxygen sensor at around 300°C to activate the sensor and decrease unwanted emissions. Also, at long periods of idling, the exhaust gas temperature decreases significantly. To reduce this time and to ensure that the aftertreatment gas system is efficient, engine manufacturers used heated lambda sensors called HEGO. The implementation of this system requires additional wires for the heating element mounted inside the sensor. At earlier types, the sensor had only three wires (two for the heater, one for the signal, and the ground from the sensor metal body). In the present, lambda sensors use four wires (two for the heater, one for the signal, and a separate ground wire). Table 2 presents the evolution of the lambda sensor.

**Table 2.** Lambda sensor wiring colours.

| | | |
|---|---|---|
| | | Zirconia Binary |
| 1 | 1 x wire | Black = signal |
| 2 | 2 x wire | Black = signal, Grey = ground |
| 3 | 3 x wire | Black = signal, White = heater, White = heater |
| 4 | 4 x wire | Black = signal, Grey = ground, White = heater, White = heater |
| | | Titania |
| Type 1 | 3 x wire | Red = heater (+), White = heater (-), Black = signal (-) |
| Type 1 | 4 x wire | Red = heater (+), White = heater (-),Yellow = signal (+), Black = signal (-) |
| Type 2 | 4 x wire | Grey = heater (+), White = heater (-), Yellow = signal (+), Black = signal (-) |
| | | Wideband sensor |
| 1 | 5 x wire | Black - ground wire, Gray - reference voltage wire White - signal wire, Purple - heater ground wire, Green - heater monitor wire |
| 2 | 6 x wire | Black - ground wire, Gray - reference voltage wire, White - signal wire Purple - heater ground wire, Green - heater monitor wire , Brown - heater ground wire |
| 3 | 7 x wire | Signal Output - gray or black wire, Sensor Ground - black wire, Sensor Power - red wire, Heater Ground - black or white wire, Heater Control - brown or tan wire, Pump Current - blue wire, Pump Ground - black wire |

## 4. Methods and materials

To test the heating time necessary to activate the oxygen sensor, two experiments were realized in the laboratory of engines from the Faculty of Mechanics Craiova. In the first experiment, several oxygen sensors (Table 2) were tested by supplying the heaters with energy sources and recording the time when the sensor was activated by measuring their temperature and analyzing their signal with an oscilloscope (Figure 3). The second experiment recorded the heating time of the oxygen sensor in different engine operating conditions. First, the engine is started at idle speed, the oxygen sensor is supplied with energy to activate the heaters, the exhaust temperature is recorded with a digital thermometer, the speed of the engine is recorded with a digital tachometer, and the lambda signal is analyzed with a multimeter and an oscilloscope (Figure 4). The following equipment was used to carry out the experiments:

- Single cylinder gasoline engine (Table 3);
- Digital tachometer type AXIOMET AX-2901;
- Digital multimeter type DT9205A;
- Digital thermomether TM 902C;
- Portable IR thermometer Optris LS LT;
- MINI DS211 Portable LCD Digital Oscilloscope;

**Table 3.** Types of oxygen sensors.

| Number | Type | Number of wires |
|---|---|---|
| 1. | Bosch LSH-24 | 3 wires |
| 2. | Delphi ES 10456 | 3 wires |
| 3. | Bosch 15725 - 106719 | 3 wires |
| 4. | Bosch LSF 4.2 | 4 wires |
| 5. | Bosch LSU 52 | 4 wires |
| 6. | 98AB-9F472-BB (Ford) | 4 wires |
| 7. | Bosch LSU 49 | 5 wires |
| 8. | LHA (XFI) | 5 wires |
| 9. | X Autohaux | 6 wires |

**Table 4.** Engine specifications.

| | |
|---|---|
| Brand | Gospsodarul Profesionist |
| Configuration | Single cylinder |
| Type | Four stroke, air cooled |
| Displacements | 163 $cm^3$ |
| Bore | 68 mm |
| Stroke | 45 mm |
| Power | 5.5(4.1)/3600 CP(kW)/rpm |
| Torque | 8.9/2800 N.m/rpm |
| Weight | 15 kg |

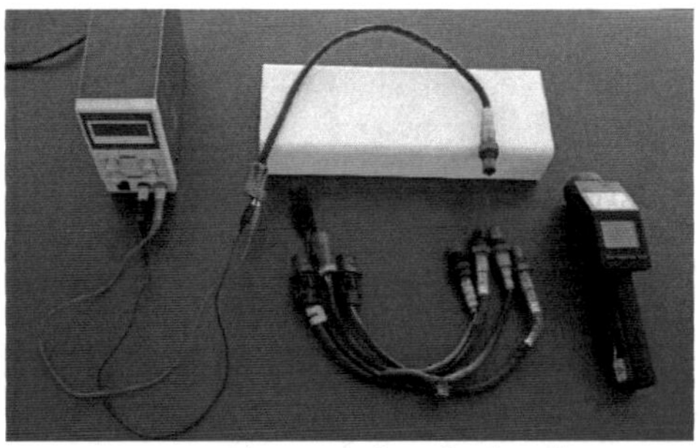

**Figure 3.** Testing oxygen sensor and measuring the heating temperature with Portable IR thermometer Optris LS LT

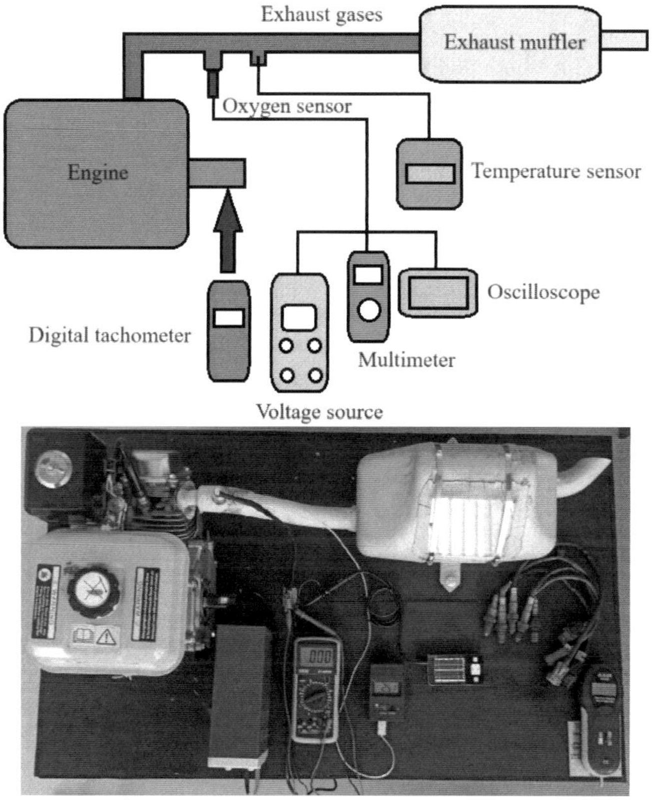


**Figure 4.** Testing oxygen sensor on single cylinder gasoline engine

## 5. Results and discussion

In the first test, several oxygen sensors were tested by supplying their heaters with energy to obtain on the oscilloscope a reliable signal and to measure the time necessary to activate the sensor (Table 5).

The older generation oxygen sensor presents a higher time for the internal coil to warm up to heat the ceramic element to the desired degree. The new generation of wideband sensors with 5 or 6 wires has an activation temperature of around 10 seconds, which helps reduce emissions during idling and cold starts. However, the new lambda sensors (5, 6, or 7 wires) need to be heated at higher temperatures (600–800 °C) than the narrowband sensors, which have an operation range of 300–350 °C.

**Table 5.** Experimental results for the first test.

| Types of oxygen sensor | Number of wires | Time to produce a signal [s] | Temperature measured with Portable IR thermometer [°C] |
|---|---|---|---|
| Bosch LSH-24 | 3 wires | 25 | 354 |
| Delphi ES 10456 | 3 wires | 26 | 355 |
| Bosch 15725 - 106719 | 3 wires | 28 | 362 |
| Bosch LSF 4.2 | 4 wires | 17 | 360 |
| Bosch LSU 52 | 4 wires | 16 | 352 |
| 98AB-9F472-BB (Ford) | 4 wires | 18 | 361 |
| Bosch LSU 49 | 5 wires | 10 | 480 |
| LHA (XFI) | 5 wires | 11 | 484 |
| X Autohaux | 6 wires | 10 | 510 |

The second test uses a single-cylinder engine to study the response time of oxygen sensors when they are subjected to heating from exhaust gases (Table 6). The engine was warmed up until the oil temperature reached 80 °C (a thermally stable regime). The sensor was mounted on the exhaust pipe, and heaters were supplied with energy. The heat generated by the exhaust gases improves with a range of 2–6 seconds in the activation time for oxygen sensors. Through the exhaust gas temperature sensor, the heaters must be carefully controlled by reducing their power when the load and engine speed increase. However, a fast change in the sensor temperature will cause a fracture in the sensing element.

**Table 6.** Experimental results for the second test.

| Types of oxygen sensor | Number of wires | Time to produce a signal [s] | Temperature of the exhaust gases measured with Digital thermometer TM 902C (at 300 rpm, oil temperature 80°C) [°C] |
|---|---|---|---|
| Bosch LSH-24 | 3 wires | 19 | 450 |
| Delphi ES 10456 | 3 wires | 21 | 456 |
| Bosch 15725 - 106719 | 3 wires | 23 | 461 |
| Bosch LSF 4.2 | 4 wires | 14 | 430 |
| Bosch LSU 52 | 4 wires | 12 | 428 |
| 98AB-9F472-BB (Ford) | 4 wires | 15 | 445 |
| Bosch LSU 49 | 5 wires | 8 | 441 |
| LHA (XFI) | 5 wires | 9 | 439 |
| X Autohaux | 6 wires | 7 | 447 |

## 6. Conclusion

The present study proposes two methods to study and test oxygen sensors in terms of their response to time. To achieve this goal, an experimental stand was designed in the laboratory using a single-cylinder engine and various types of narrowband and wideband lambda sensors. From the experiments, narrowband sensors with 3 or 4 wires require a longer time to generate a stable signal with the increase in gas emissions. The new wideband sensors present an activation time of around 10 seconds but need to be heated to higher temperatures. These exhaust gas temperatures can be reached only with supercharged engines. To increase the lifetime of the oxygen sensors, it is necessary to have an active heating control to avoid thermal cycling.

**Author's ORCID ID**

O V Otat: https://orcid.org/0000-0002-1939-7533

**References**

[1] Aravind S, Ragupathi P, Sathish K D, Vignesh G, 2020 *A Numerical Investigation of Automotive Lambda Sensor to Improve the Life Span of the Sensor using CFD*, IOP Conf. Series: Materials Science and Engineering 923 012003

[2] Lee JH 2003 *Review on zirconia air-fuel ratio sensors for automotive applications*, Journal Of Materials Science 38 pp 4247 – 4257

[3] *Technical report – FAE lambda sensor. Evolution and types*, 17 April 2023 Company, Report.

[4] Yousef S.H. Najjar, Abad Alrahman W.H. Almardini, Amro J.A. Sawan, 2019 *Comparison of performance using wide-band versus narrow-bansd oxygen sensors in gasoline engines*, Case Studies in Thermal Engineering 16 100552

[5] Schaefer F, Egger S, Steiner A, Eichel R A., 2012 *Online characterization of operational parameters in an SOFC system with anode-exhaust gas recirculation by oxygen sensors*, Fuel Cells

# Kinematic and dynamic analysis of a planetary gear box in steady-state working regime

**T Ciobotaru[1*], L S Grigore[1], L Barothi[1], O Alexa[1*]**

[1] Military Technical Academy „Ferdinand I", Bucharest, Romania

[*] E-mail: cticusor2004@yahoo.com; octavian.alexa@gmail.com

**Abstract**. The paper presents a method for analysis of planetary gear boxes in order to calculate the speed and torque of the epicyclic gear trains elements. Based on these findings, the power flow within the gear box is determined. Additionally, the mechanical efficiency is estimated in order to assess the overall efficiency of the transmission.

## 1. Introduction

Planetary gear boxes consist of epicyclic gear trains, clutches and brakes connected so a structure with 2DOF or 3DOF results. Usually, 2DOF epicyclic gear trains are used, but also epicyclic gears with more than 2DOF may be used too (i.e. Ravigneaux mechanisms).

The assessment of the planetary gear box imposes the determination of kinematics (speed, overall ratio), power flows and overall efficiency [1], for all the stages the gear box could realize. The above mentioned data calculated for steady state working regime could be used for evaluation of the solution analysed, and as reference base for analysis of transitory regimes (such as shifting).

Kinematic issues of planetary gear box are, at least partially, solved using graphical methods (such as speed vector planes [2], [3]) or by using variants of bond graph theory (an exhaustive presentation of various variants as well as of their capabilities is presented in [4]). The above mentioned methods for analysing the planetary gear boxes involves the usage of specialized knowledge and present real difficulties in the case of structures with more than 2 epicyclic gear trains. Analytical methods for studying planetary gear train structures are detailed in [5] without direct application to planetary gearboxes. A useful approach is applied in [6], consisting in combining analytical method with block diagrams; this approach allowed to analyse the working possibilities of power-split infinitely variable transmissions.

Frequently, the gear trains design parameters have to be optimized; for ordinary gear trains, there are optimization methods which allow the calculation of gear teeth numbers, as it is presented in [7]. For the planetary gear boxes, no reference was identified for the calculation of planetary gear design parameters which provide imposed gear box ratios.

An optimal design math model of epicyclic gear train based on the planet gear design features is detailed in [8]; the model is solved using the differential evolution algorithm programmed in Visual Basic 6.0, proving the interest in using the computer for analysing the planetary gears. For Ravigneaux mechanisms, a variant of the optimisation algorithm is detailed in [9].

The inverse kinematic planetary gear unit is presented in [10] without taking into consideration the aspects regarding the planetary gear boxes.

The method presented below is simple, uses basic knowledge regarding the epicyclic power trains, may be used for complex structures of gear boxes and permits an exhaustive characterization of kinematics and power flows of planetary gear boxes, including the calculation of overall mechanical efficiency. This method is based on analytical approaches detailed in [11], [12] and [13], and provide a comprehensive analysis as well as the solving of kinematic inverse problem.

In order the present the proposed methods, a hydro-mechanical transmission type ZF-ECOMAT 7 HP 602 was selected.

The transmission of the series ZF-ECOMAT 7 HP 600 is largely used in trucks, buses as well as in military vehicles.

The transmission consists of:

- torque converter (TC) of which stator could became freed do to one-way clutch FW; the turbine of torque converter may be connected directly to impeller by coupling the clutch ATC;
- planetary gearbox which includes four epicyclic gear trains, four brakes and tree clutches.

The kinematic diagram of the transmission type ZF-ECOMAT 7 HP 602 is shown in Figure 1.

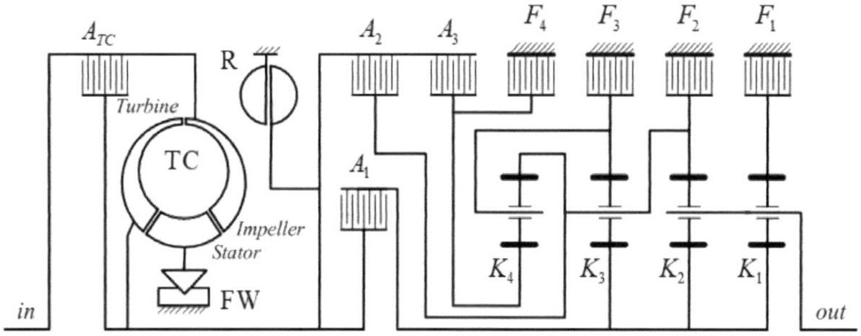


**Figure 1** Diagram of ZF 7 HP-602 transmission

## 2. Kinematics

The proposed method consists of several successive steps.

The first step is devoted to transforming the kinematic diagram into nodal diagram applying the following rules:

- each brake and clutch are represented by kinetic nodes, with the following notation: $F_1$, $F_2$, $F_3$, $F_4$ (brakes) $A_1$, $A_2$ and $A_3$ (clutches);
- each epicyclic gear train is represented by a generalized node (a circle) with the following notations:
  - constant of the mechanism indexed $j$, noted $K_j$ : $K_j = \frac{z_2}{z_1}$, where indexes 2, 1, 0 refer respectively to the ring gear, sun gear and planet carrier, while $z$ represents the tooth number;
  - sun gear: 1 (inside the circle); $1+3\cdot(j-1)=3\cdot j-2$ (outside the circle);
  - ring gear: 2 (inside the circle); $2+3\cdot(j-1)=3\cdot j-1$ (outside the circle);
  - planet carrier: 0 (inside the circle); $0+3\cdot(j-1)=3\cdot j-3$ (outside the circle).
- the input shaft is noted $a$, and the output shaft is noted $b$.

The results of this step are presented in Figure 2.

The second step of the method deals with the calculation of the speed of all central elements of the transmission, indexed 1 to 12. The kinematic equations may be grouped as following:

- Willis equations written for the four epicyclic mechanisms [12] (four equations):

$$\omega_{3j-2} - K_j\omega_{3j-1} - (1-K_j)\omega_{3j-3} = 0; \quad j = 1\ldots4 \tag{1}$$

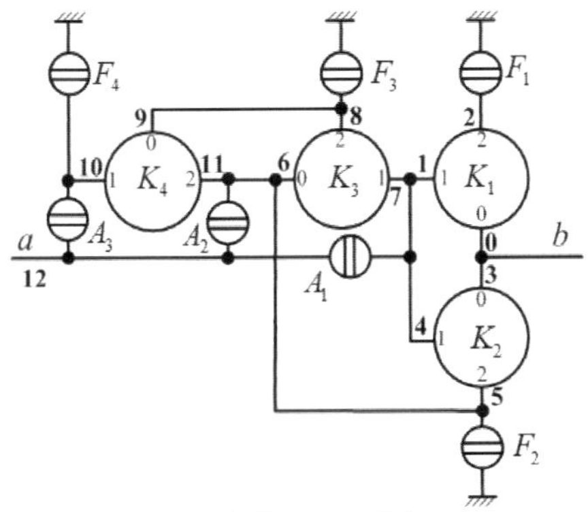


**Figure 2** Nodal diagram of the planetary gearbox (*a* – input; *b* – output)

- equations for permanently connected central elements, six equations may be written:

$$\omega_0 = \omega_3;\quad \omega_1 = \omega_4;\quad \omega_1 = \omega_7;$$
$$\omega_6 = \omega_{11};\quad \omega_8 = \omega_9;\quad \omega_6 = \omega_5 \qquad (2)$$

The equation for the input element is:

$$\omega_{12} = 1. \qquad (3)$$

In equation (3) it was adopted the value 1 for the input speed in order to facilitate comparative analysis of the speeds of the output elements in all stages of the gearbox.

The equation types (1), (2) and (3) totalize 11 equations; according to Figure 2, there are 13 central elements, indexed from 0 to 12. The two necessary equations for assembling a determined system of equations results from the conditions imposed for obtaining the desired stage of the gearbox: the coupling of two friction elements (clutch, brake); the coupling sequences are detailed in Table 1.

**Table 1** Stages of the planetary gearbox: coupled friction elements, ratios

| Stage (gear) | Coupled elements | | Corresponding equations | | Ratio relation | |
|---|---|---|---|---|---|---|
| 1st | $A_1$ | $F_1$ | $\omega_1 = \omega_{12};\, \omega_2 = 0$ | (4) | $K_1 + 1$ | (5) |
| 2nd | $A_1$ | $F_2$ | $\omega_1 = \omega_{12};\, \omega_8 = 0$ | (6) | $K_2 + 1$ | (7) |
| 3rd | $A_1$ | $F_3$ | $\omega_1 = \omega_{12};\, \omega_5 = 0$ | (8) | $(1+K_2)(1+K_3)(1+K_2+K_3)^{-1}$ | (9) |
| 4th | $A_1$ | $F_4$ | $\omega_1 = \omega_{12};\, \omega_{10} = 0$ | (10) | $\dfrac{K_2+K_3+K_4+K_2K_3+K_2K_4+1}{K_2+K_3+K_4+K_2K_4+1}$ | (11) |
| 5th | $A_1$ | $A_2$ | $\omega_1 = \omega_{12};\, \omega_{11} = \omega_{12}$ | (12) | 1 | |
| 6th | $A_2$ | $F_4$ | $\omega_{11} = \omega_{12};\, \omega_{10} = 0$ | (13) | $\dfrac{K_2+K_4+K_2K_4+1}{K_2+K_3+K_4+K_2K_4+1}$ | (14) |
| 7th | $A_2$ | $F_3$ | $\omega_{11} = \omega_{12};\, \omega_8 = 0$ | (15) | $(K_2+1)\cdot(K_2+K_3+1)^{-1}$ | (16) |
| R | $A_3$ | $F_2$ | $\omega_{10} = \omega_{12};\, \omega_5 = 0$ | (17) | $-(K_2+K_4+K_2K_4+1)\cdot(K_3)^{-1}$ | (18) |

Adding to the equations (1), (2) and (3) one of the equations (4), (6), (8), (10), (12), (13), (15) or (17), a system of 13 linear equations resulted; the system of equations was solved symbolically using the matrix method detailed in [11]. Using symbolic calculus in PTC Mathcad Prime it was possible to obtain the relations which allow the calculation of the ratio of all gearbox stages.

The 3rd step of the proposed method consists of the estimation of numerical values of epicyclic gear trains constants ($K_1 \ldots K_4$) based on the existing gear ratios (values provided by the manufacturer). It may be observed that there are seven such relations involving only four unknown parameters; consequently, an overdetermined non-linear system resulted, usually without unique accurate solution. Nevertheless, it is noticed that relations (5), (7), (9) and (11) contain all the unknown parameters and, consequently, may be used for the estimation of the epicyclic gear constants by imposing the actual values of the gearbox ratios (inverse kinematic problem).

The calculated constants are the following: $K_1 = 4.600$; $K_2 = 2.430$; $K_3 = 2.440$; $K_4 = 2.404$.

Due to the presentation of data regarding the gearbox ratios with only two digits, it is presumed that the calculated values for epicyclic gear train constants will generate error in the case of calculation of estimated gearbox ratios. The results of the calculations are presented in Table 2.

**Table 2** Error resulted by using calculated values of $K_j$

| | 1st gear | 2nd gear | 3rd gear | 4th gear | 5th gear | 6th gear | 7th gear | R |
|---|---|---|---|---|---|---|---|---|
| Actual ratio | 5.60 | 3.43 | 2.01 | 1.42 | 1.00 | 0.83 | 0.59 | -4.84 |
| Calculated ratio | | | | | 1 | 0.827 | 0.584 | -4.786 |
| Error (%) | | | | | 0 | 0.34 | -0.76 | 1.12 |

The magnitude of the errors is low enough to allow further usage of the estimated values for $K_j$ for calculation of central element speeds; the results of such calculations are presented in Table 3.

**Table 3** Calculation results of central elements speed

| Unit | Element | 1st gear | 2nd gear | 3rd gear | 4th gear | 5th gear | 6th gear | 7th gear | R |
|---|---|---|---|---|---|---|---|---|---|
| | 0 | 0.179 | 0.292 | 0.292 | 0.704 | 1.000 | 1.209 | 1.711 | -0.209 |
| $K_1$ | 1 | 1.000 | 1.000 | 1.000 | 1.000 | 1.000 | 1.717 | 3.440 | -0.717 |
| | 2 | 0.000 | 0.138 | 0.138 | 0.640 | 1.000 | 1.099 | 1.336 | -0.099 |
| | 3 | 0.179 | 0.292 | 0.292 | 0.704 | 1.000 | 1.209 | 1.711 | -0.209 |
| $K_2$ | 4 | 1.000 | 1.000 | 1.000 | 1.000 | 1.000 | 1.717 | 3.440 | -0.717 |
| | 5 | -0.159 | 0.000 | 0.000 | 0.583 | 1.000 | 1.000 | 1.000 | 0.000 |
| | 6 | -0.159 | 0.000 | 0.000 | 0.583 | 1.000 | 1.000 | 1.000 | 0.000 |
| $K_3$ | 7 | 1.000 | 1.000 | 1.000 | 1.000 | 1.000 | 1.717 | 3.440 | -0.717 |
| | 8 | -0.635 | -0.410 | -0.410 | 0.411 | 1.000 | 0.706 | 0.000 | 0.294 |
| | 9 | -0.635 | -0.410 | -0.410 | 0.411 | 1.000 | 0.706 | 0.000 | 0.294 |
| $K_4$ | 10 | -1.777 | -1.395 | -1.395 | 0.000 | 1.000 | 0.000 | -2.404 | 1.000 |
| | 11 | -0.159 | 0.000 | 0.000 | 0.583 | 1.000 | 1.000 | 1.000 | 0.000 |
| input | 12 | 1.000 | 1.000 | 1.000 | 1.000 | 1.000 | 1.000 | 1.000 | 1.000 |

**Table 4** Speed variation due to up-shifting

| | Shifting sequence | | | | | |
|---|---|---|---|---|---|---|
| Element | 1 to 2 | 2 to 3 | 3 to 4 | 4 to 5 | 5 to 6 | 6 to 7 |
| 0 | -0.069 | -0.121 | -0.086 | -0.208 | -0.170 | -0.350 |
| 1 | -0.388 | -0.414 | -0.294 | -0.296 | -0.170 | -0.496 |
| 2 | 0.000 | -0.057 | -0.040 | -0.189 | -0.170 | -0.318 |
| 3 | -0.069 | -0.121 | -0.086 | -0.208 | -0.170 | -0.350 |
| 4 | -0.388 | -0.414 | -0.294 | -0.296 | -0.170 | -0.496 |
| 5 | 0.062 | 0.000 | 0.000 | -0.172 | -0.170 | -0.289 |
| 6 | 0.062 | 0.000 | 0.000 | -0.172 | -0.170 | -0.289 |
| 7 | -0.388 | -0.414 | -0.294 | -0.296 | -0.170 | -0.496 |
| 8 | 0.246 | 0.170 | 0.120 | -0.122 | -0.170 | -0.204 |
| 9 | 0.246 | 0.170 | 0.120 | -0.122 | -0.170 | -0.204 |
| 10 | 0.689 | 0.578 | 0.410 | 0.000 | -0.170 | 0.000 |
| 11 | 0.062 | 0.000 | 0.000 | -0.172 | -0.170 | -0.289 |
| 12 | -0.388 | -0.414 | -0.294 | -0.296 | -0.170 | -0.289 |

From the Table 3 results that the maximum speed occurs for sun gear of $K_1$, $K_2$ and $K_3$ epicyclic gear trains; this aspect doesn't imply negative behaviour of gearbox due to reduced rotation inertia of sun gears. The maximum relative speed of planet gear type elements occurs in 7[th] gear for $K_1$ and $K_2$ sun gears.

Another kinematics issue refers to the speed variation of the central elements during the shifting; this issue is important in order to assess the inertia moments of central elements, a special attention being paid to ring gear and to carrier due to their bigger inertia. It is assumed that during the shifting process the output speed of the gearbox doesn't change (the vehicle speed is constant); consequently, at an up-shift, the input speed of gearbox decreases with the ratio of the gearbox speed ratios existent before and after the shifting process. The relative reduction of the input shaft speed at up-shifting is presented in Table 4.

From the above data, it may observe that a maximum variation of speed occurs for the sun gear of $K_4$ epicyclic train. For the ring gears, the maximum speed variation occurs also for $K_4$. Considering the low input speed used in the first stage (usually, the drivers uses the first stage only for starting the acceleration process and prefers to shift up without reaching the maximum speed of the engine), the magnitude of the speed variation decreases accordingly with the low input speed.

## 3. Power flow

The next step of gearbox analysis deals with the calculation of torques acting on outer elements.

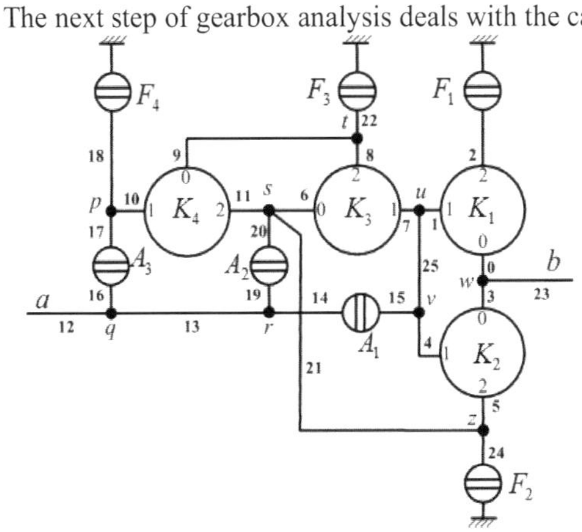


**Figure 3** Nodal diagram for torque calculation

It is noticed that the ramified nodes included in the nodal diagram (Figure 2) obey to two main laws:

- energy conservation – the sum of power flows acting on the ramified node is null;
- torque equilibrium – the sum of torques acting on the ramified node is null.

Due to the fact that the ramified nodes imply torque distribution, it is necessary to index all the ramified nodes, as well as braches connected to it; the results of these operations are presented in Figure 3.

For each of the four epicyclic gear trains, the following two equations regarding torque distribution may be written [11]:

$$K_j \cdot T_{3j-2} - T_{3j-1} = 0; \quad M_{3j-3} + M_{3j-2} + M_{3j-1} = 0, \quad j = 1 \ldots 4, \tag{19}$$

in total, eight equations.

The torque equilibrium law for ramified nodes indexed *p, q…w, z* conducts to the following equations respectively:

$$\begin{gathered} T_{10} + T_{17} + T_{18} = 0; T_{12} + T_{13} + T_{16} = 0; T_{13} + T_{14} + T_{19} = 0; T_6 + T_{11} + T_{20} + T_{21} = 0; \\ T_8 + T_9 + T_{22} = 0; T_1 + T_7 + T_{25} = 0; T_4 + T_{15} + T_{25} = 0; T_0 + T_3 + T_{26} = 0; T_5 + T_{21} + T_{24} = 0 \end{gathered}, \tag{20}$$

in total, nine equations. The input element is indexed 12; we assume that the input torque is:

$$T_{12} = 1. \tag{21}$$

In order to obtain a stage of the gearbox, two friction elements have to be coupled and only these friction elements could transfer torque; all other friction elements cannot transfer torque, and, consequently the torques of their input/output shafts (modelled as branches) are null. These considerations lead to the specific equations written (for clutches and brakes respectively) for each stage:

$$\begin{array}{l}
1^{st}\text{ stage: } T_{14}=T_{15};\ T_{16}=0;\ T_{17}=0;\ T_{19}=0;\ T_{20}=0;\ T_{22}=0;\ T_{18}=0;\ T_{24}=0; \\
2^{nd}\text{ stage: } T_{14}=T_{15};\ T_{16}=0;\ T_{17}=0;\ T_{19}=0;\ T_{20}=0;\ T_{2}=0;\ T_{18}=0;\ T_{22}=0; \\
3^{th}\text{ stage: } T_{14}=T_{15};\ T_{16}=0;\ T_{17}=0;\ T_{19}=0;\ T_{20}=0;\ T_{2}=0;\ T_{18}=0;\ T_{24}=0; \\
4^{th}\text{ stage: } T_{14}=T_{15};\ T_{16}=0;\ T_{17}=0;\ T_{19}=0;\ T_{20}=0;\ T_{2}=0;\ T_{22}=0;\ T_{24}=0; \\
5^{th}\text{ stage: } T_{14}=T_{15};\ T_{16}=0;\ T_{17}=0;\ T_{19}=T_{20};\ T_{2}=0;\ T_{18}=0;\ T_{22}=0;\ T_{24}=0; \\
6^{th}\text{ stage: } T_{14}=0;\ T_{15}=0;\ T_{16}=0;\ T_{17}=0;\ T_{19}=T_{20};\ T_{2}=0;\ T_{22}=0;\ T_{24}=0; \\
7^{th}\text{ stage: } T_{14}=0;\ T_{15}=0;\ T_{16}=0;\ T_{17}=0;\ T_{19}=T_{20};\ T_{2}=0;\ T_{18}=0;\ T_{22}=0; \\
\text{R stage: } T_{14}=0;\ T_{15}=0;\ T_{16}=T_{17};\ T_{19}=0;\ T_{20}=0;\ T_{2}=0;\ T_{18}=0;\ T_{22}=0.
\end{array} \quad (22)$$

in total eight equations for each stage.

Assembling for each stage the equation types (19), (20), (21) and one of the equations groups (22), a system of 26 equation results. It became possible to calculate the power flow of element $j$ using the relation:

$$P_j = \omega_j \cdot T_j \quad (23)$$

The ratio relations presented in Table 1 allow identifying the epicyclic gear trains which take part in the power transmission. Consequently, for the first two stages the power flows through a single epicyclic gear train used as ordinary gear. For the 5th stage, the planetary gearbox works "frozen", all its elements having the same speed.

The results obtained for power flow in remaining stages are presented in Figure 4 to Figure 8.

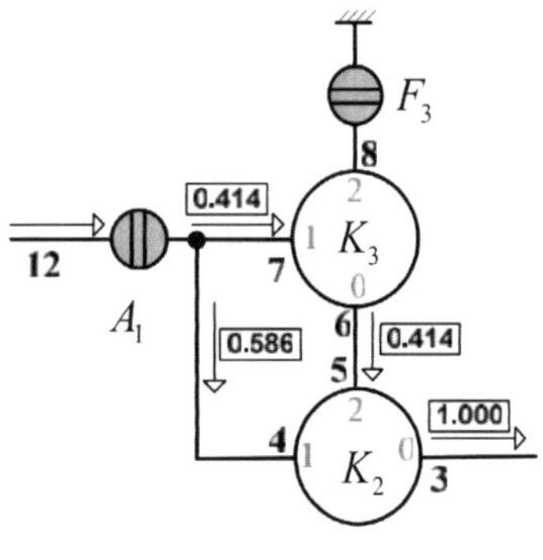


**Figure 4** Power flow – 3rd stage

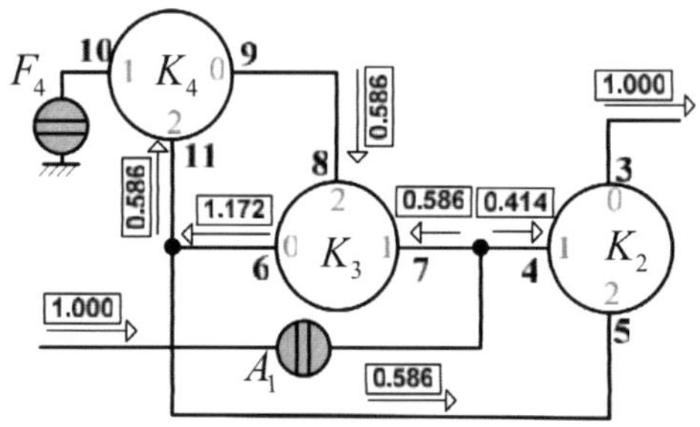


**Figure 5** Power flow – 4th stage

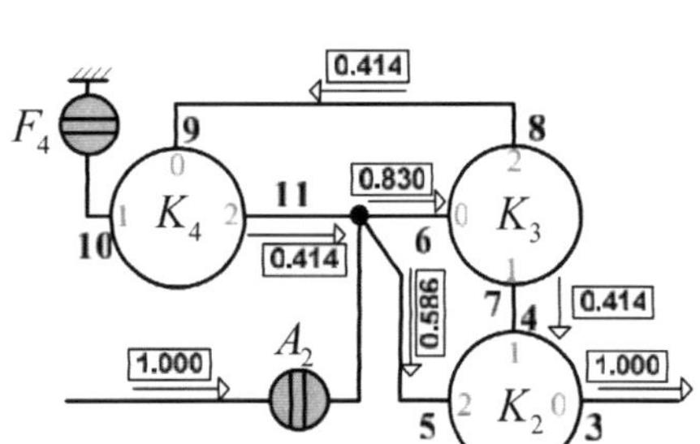


**Figure 6** Power flow – 6th stage

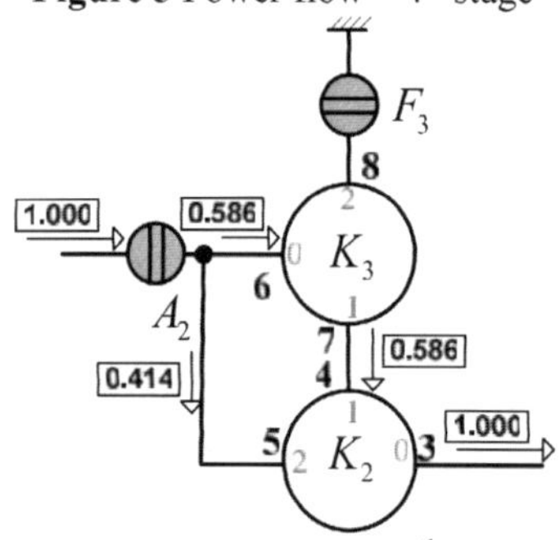


**Figure 7** Power flow – 7th stage

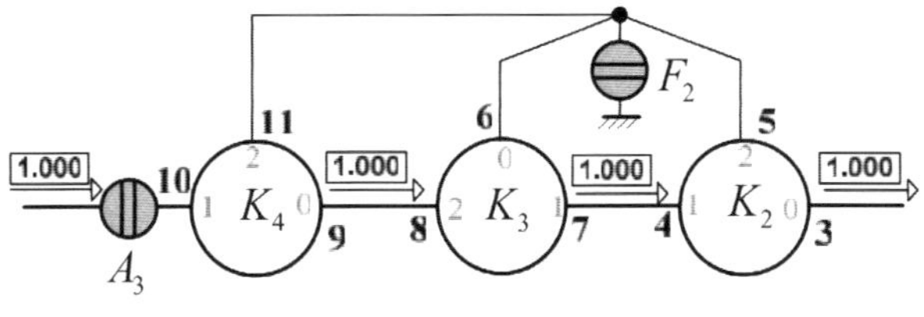


**Figure 8** Power flow – Reverse stage

It may observe that in Reverse stage, despite the three epicyclic gear trains participate in power transmission, actually there is a serial configuration consisting of three gears which transmit the same amount of power.

Analysing the power flow in the 4th stage, it could observe that element 6 transmits a power flow greater than the input power (with 17.2%); that means the mechanisms $K_3$ and $K_4$ faces a closed loop power flow.

## 4. Overall mechanical efficiency of the gearbox

The efficiency of the epicyclic gear train is analysed by various papers in order to emphasise the influence of the mechanism constants, torque, speed, function (differential/summing) on the efficiency [14], [15], [16].

A common denominator of these papers is the need to previously know the power flow through each epicyclic gear train, as well as the difficulties in "assembling" the individual power losses into overall efficiency of the gearbox.

Instead, a generalized method, presented in [13], was adopted in order to reduce the volume of calculations as well as to avoid the possible error propagation from previous calculations.

In general, for a planetary gearbox, the ratio of the stage, denoted $i_s$, is a rational function of epicyclic gear trains constants $K_j$:

$$i_s = f_s(K_j) \quad (24)$$

For the planetary gearbox analysed, the above relation may be written as:

$$i_s = f_s(K_j), \quad s = 1,...,8; \quad j = 1,...,4 \quad (25)$$

where the Reverse stage was considered as the 8th stage.

For the epicyclic gear train with the carrier blocked, the mechanical efficiency, noted $\eta_0$, may be calculated using the same formulas as for ordinary gears (having fixed rotational axis); for the purpose of the present paper, it was adopted the value: $\eta_0 = 0.965$ [11].

The overall efficiency in the stage $s$ of the gearbox may be calculated using the following relation:

$$\eta_s = \frac{f_s(\gamma_{s,j} \cdot K_j)}{f_s(K_j)}, \quad s = 1,...,8; \quad j = 1,...,4 \quad (26)$$

where the functions $f_s(K_j)$ are presented in Table 1 (the relations (5), (7), (9), (11), (14), (16) and (18), respectively).

The factors $\gamma_{s,j}$ are calculated using the Kreines relation detailed in [13]:

$$\gamma_{s,j} = \ln\left[u_{s,j} \cdot \exp\left(\eta_0\right)\right] \quad j = 1...4; \quad s = 1...8 \quad (27)$$

where:

$$u_{s,j} = \operatorname{sign} \frac{K_j}{f_s(K_j)} \cdot \frac{\partial f_s(K_j)}{\partial K_j}, \quad j = 1...4; \quad s = 1...8 \qquad (28)$$

The derivatives from the relations (28) were calculated symbolically using PTC Mathcad Prime, forwarded by numerical calculation of efficiency considering the calculated values of ratios presented in Table 1.

Mechanical efficiency in the 5th stage is assumed to be 1 because the power is transmitted without any transformation of torque and speed through the gearbox.

The obtained results are presented in Table 5.

**Table 5** Overall efficiency of the planetary gearbox

| | 1st gear | 2nd gear | 3rd gear | 4th gear | 5th gear | 6th gear | 7th gear | R |
|---|---|---|---|---|---|---|---|---|
| $u_{s,1}$ | +1 | NA | NA | NA | NA | NA | NA | NA |
| $u_{s,2}$ | NA[a] | +1 | +1 | +1 | NA | +1 | +1 | +1 |
| $u_{s,3}$ | NA | NA | +1 | +1 | NA | −1 | −1 | −1 |
| $u_{s,4}$ | NA | NA | NA | −1 | NA | +1 | NA | +1 |
| $\eta_s$ | 0.971 | 0.975 | 0.980 | 0.981 | 1 | 0.985 | 0.975 | 0.918 |

[a] denotes the epicyclic gear trains that are not involved in power flow

It may observe the following aspects:

- the low efficiency occurs for stages (1, 2, 3 and R) in which the power transmits using a unique circuit, with lowest value for Reverse stage in which three epicyclic gear trains are connected in series (see Figure 8);
- the highest efficiency is recorded in 6th gear due to the circulation of power flows through three epicyclic gear trains (see Figure 6) without closed loop circulation as in the case of 4th stage (see Figure 5).

It is mentioned that this method is only a rough estimate which allows identifying situations in which the overall efficiency of the gear box is very low due to closed loop power flow circulation.

## 5. Conclusion

The proposed methodology for analysing the planetary gearboxes has the following main advantages:

- allows the fast calculation of the gear ratios, and power flow for multi-path power flow (power split) transmissions, with direct applicability for analysing the automatic transmission;
- allows the extensive usage of computers and appropriate commercial software for performing a comprehensive analysis of planetary gear boxes, resulting time saving and high accuracy;
- allows the fast evaluation by computation of the overall efficiency and power loss of the gearbox;
- may serve as first estimation for more accurate calculations of efficiency taking into consideration the influence of torque and speed on power losses;
- may serve as basis for assessing the dynamic behaviour of the gearbox during stage shifting process.

**Acknowledgments**

This work was supported by a grant of the Romanian Ministry of Research, Innovation and Digitization, CNCS - UEFISCDI, project number PN-III-P1-1.1-PD-2021-0079, within PNCDI III, and has as scope the setup of general methods for planetary gear boxes analysis.

**Author's ORCID ID**

T Ciobotaru: https://orcid.org/0009-0003-0129-5867

**References**

[1]. Del Castillo J 2002 The analytical expression of the efficiency of planetary gear trains. *Mechanism and Machine Theory - MECH MACH THEOR*. 37. 197-214. 10.1016/S0094-114X(01)00077-5

[2] Oprean I M 1999 *Transmisii automate pentru automobile*, Editura Printech, Bucureşti, ISBN 973-9475-61-2 12-90

[3] Marciniec A, Sobolak M and Połowniak P 2022 Graphical method for the analysis of planetary gear trains, *Alexandria Engineering Journal*, Volume 61, Issue 5, Pages 4067-4079, https://doi.org/10.1016/j.aej.2021.09.036

[4] Xue H L, Liu G and Yang X H 2016 A review of graph theory application research in gears, *Proc IMechE Part C: J Mechanical Engineering Science*, **230(10)** 1697–1714

[5] Mathis R and Mond Y 1999 Une theorie unifiee des trains epicycloidaux, *C. R. Acad. Sci. Paris*, t. **327**, Serie II b, p. 1115-1121, Machines, mecanismes, robotiques/Machines, mechanisms, robotics

[6] Preda I, Ciolan G and Covaciu D 2017 Design-Parameters Setup for Power-Split Dual-Regime IVT. IOP *Conference Series: Materials Science and Engineering*. 252. 012064. 10.1088/1757-899X/252/1/012064

[7] Preda I and Ciolan G 2013 Algorithm to define the speed ratios of the tractor complex geartrains. *INMATEH-Agricultural Engineering*. 41. 77-84

[8] Wang Z K and Zheng Y L 2012 The program and optimal design of planetary gear based on differential evolution algorithm. *Advanced Materials Research*. Vol. 482–484. Trans Tech Publications. P. 2438–2441

[9] Simionescu P A, Beale D G and Dozier G V 2004 Teeth-number synthesis of an automatic planetary transmission using evolutionary computation, *Proceedings of the ASME 2004 International Design Engineering Technical Conferences and Computers and Information in Engineering Conference*. Vol 2: 28th Biennial Mechanisms and Robotics Conference, Parts A and B. Salt Lake City. Utah. USA. P. 291–298

[10] Mathis R and Rémond Y 2002 Solving the kinematic inverse problem for epicyclic gear trains. *VDI BERICHTE*. 1665. 479-490

[11] Ciobotaru T and Alexa O 2019 *Ingineria autovehiculelor militare cu senile, vol. 3, Transmisia. Frânele. Agregatul Energetic*, Editura Academiei Tehnice Militare, Bucureşti, ISBN 978-973-640-286-9

[12] Ciobotaru T, Frunzeti D and Jäntschi L 2010 Method for analyzing epicyclic gearboxes. *Int.J Automot. Technol*. **11**, 167–172, https://doi.org/10.1007/s12239-010-0022-4

[13] Ciobotaru T, Frunzeti D, Rus I and Jäntschi L 2010 Method for analysing multi-path power flow transmissions. *Proceedings of the Institution of Mechanical Engineers, Part B: Journal of Engineering Manufacture*.;**224**(9)1447-1454. doi:10.1243/09544054JEM1668

[14] Esmail E L, E. Pennestrì E and Hussein Juber A H 2018 Power losses in two-degrees-of-freedom planetary gear trains: A critical analysis of Radzimovsky's formulas, *Mechanism and Machine Theory* **128** 191–204 https://doi.org/10.1016/j.mechmachtheory.2018.05.015

[15] Mathis R and Rémond Y 2009 Kinematic and dynamic simulation of epicyclic gear trains, *Mechanism and Machine Theory* **44** 412–424, doi:10.1016/j.mechmachtheory.2008.03.004

[16] Mantriota G and Pennestrì E 2003 Theoretical and Experimental Efficiency Analysis of Multi-Degrees-of-Freedom Epicyclic Gear Trains, *Multibody System Dynamics* **9** 389–408

IOP Conf. Series: Materials Science and Engineering 1303 (2024) 012016 doi:10.1088/1757-899X/1303/1/012016

# Aspects of combustion in diesel engine at hydrogen use-a theoretical approach

**A Cernat[1], N Neguresu[1], C Pana[1] and C Nutu[2]**

[1]Department of Thermotechnics, Engines, Thermal and Frigorific Equipment, University Politehnica of Bucharest, Bucharest, Romania
[2]Departmentof Automotive, University Politehnica of Bucharest, Bucharest, Romania

E-mail: cernatalex@yahoo.com

**Abstract**. The use of alternative fuel may be a viable solution in order to ameliorate the engine performance especially in terms of pollutant emissions. Among the alternative fuels that can be use to fuel internal combustion engines hydrogen can be a viable alternative fuel especially due to the advantage of reducing the carbon emission at its use as alternative fuel even for partial substitution of classic fuel. Hydrogen has good combustion properties like higher Lower Heating Value, large inflammability limits, higher combustion speed, which may has a benefic influence on combustion process. The use of hydrogen to diesel engine bring few important issues that must be solved in order to assure the normal engine operation, starting with the fuelling system and engine with the control of the combustion process. The paper presents some results obtained during the theoretical modeling of the in-cylinder process at a diesel engine fuelled with classic fuel and hydrogen. The diesel fuel is energetically substituted by hydrogen in percent's of 25% and 30%. The influences of hydrogen use on in-cylinder maximum pressure, maximum pressure rise rate, heat release rate, combustion temperature, indicated thermal efficiency and nitrogen oxides and smoke emission levels are shown and analyzed.

## 1. Introduction

The substitution of the classic fuels with alternative ones may be a viable method in order to ameliorate the level of the engine gas emissions. From this point of view hydrogen can be use fuel to fuel diesel engines with the advantage of reducing the carbon emission at its use as alternative fuel even for partial substitution of classic fuel.

Hydrogen has good combustion properties like superior Lower Heating Value, large inflammability limits, superior flame speed comparative to diesel fuel, which may has a good influence on combustion process. Hydrogen use as alternative fuel in diesel engine is starting to come to the attention of the specialists in the field, which are trying to continuously improve the energetic and functional-constructive formula of the hydrogen fuelled diesel engine. For diesel engines, hydrogen fuelling brings few important issues that can be solved in order to assure the engine running, starting with the adequate fuelling system use for the combustion process management. Because of a low value of the hydrogen cetane number the use of classic fuel pilot injection to initiate the ignition of the air-fuel mixture is required. If diesel fuel is used as substitute fuel, the engine being fuelled with diesel and hydrogen in dual fuel mode, the substitution of diesel with hydrogen must be established taking into consideration energetically and mechanical operation aspects.

Some researchers use hydrogen as fuel for engines, good perspectives results being shown. Shahpouri [1] develops o study based on a simulation with one-dimensional model used for hydrogen fuelling to predict the level of the pollutant gases to an ICE. Shahpouri observes the increase of the in-cylinder temperature, which may affect the $NO_x$ level in the exhaust gases, which may require a special tune for amelioration of the $NO_x$ and soot emissions level. In general, at hydrogen fuelling, the NOx emission level is higher, depending on fuels cyclic quantities, but emission levels for soot, PM, CO, unburn HC and $CO_2$ are decreased by hydrogen. The thermal efficiency is increased [1]. Anticaglia [2] develops a one dimensional CFD model to study the possibilities to increase of the brake thermal efficiency and to decrease the NOx emission level at internal combustion engines fuelled by hydrogen. Anticaglia [2] presents the influences of the hydrogen on combustion process reflected in the thermal efficiency rise. Hydrogen use to the advantages of higher break mean effective pressure at lean mixture use [2]. Krebs [3] develops a multi-zone model for hydrogen combustion modelling and shows the increase of maximum pressure during combustion. Also the peak value of the pressure tends to appears sooner per cycle. The rise of combustion temperature peak appears also at the increase of the mixture combustion speed as presence of hydrogen [3]. Also the combustion starts earlier but the combustion duration and speed is influenced by the air-fuel ratio[3]. The increase of the in-cylinder water mole fraction at hydrogen combustion leads to the decrease of the $NO_x$ level in the exhaust gases, but the decrease in also related to the air-fuel ratio [3]. The maximum of $NO_x$ appears sooner per cycle since hydrogen ratio increase, especially for lean dosages area, but for stoichiometric conditions the influences are not so obvious [3]. Michl [4] uses a model to study the influence of hydrogen combustion, that substitute the diesel fuel, on heat transfer and he conclude that during hydrogen combustion peak values for temperature appears for short periods of time, the heat release being higher and shorter. Maghbouli [5] shows that hydrogen use as gaseous fuel in the engine admission in percent of 1% to 7% volume, assures the rise of pressure and temperature, for 7% of hydrogen, the rate of the pressure rise and the rate of the heat release being accelerated, knocking combustion being favoured. Ragupathy [6] uses a basic hydrogen fuel and observe the decrease of the injection duration decrease and the heat release during combustion is higher. Hydrogen combustion favours the rise of the in-cylinder pressure, the maximum value of the pressure per cycle occurs farther from Top Dead Centre (TDC) versus classic fuelling. A similar tendency appears for rate of heat release, HRR, the peak of heat release rate being increased once with the rise of the hydrogen content and the angle of maximum HRR appears closer to TDC, the reduction of the combustion duration being shown [6]. Molina [7] shows that higher percentages of the hydrogen assure the acceleration of the combustion flame speed for all range of air-to-fuel ratio, the peak pressure and the HRR being increased. The pressure diagrams allure tends to be very different al hydrogen use, the peak pressure appears closer to TDC with almost 5 CAD [7]. The maximum pressure is increased from 110 bar till 148 bar at hydrogen maximum share and the pressure rise rate is increased from 2 to 3.4 bar/crank angle degree, (°CA). The total duration of the burning is shortened, and the rapid stage of combustion appears with almost 8 degrees sooner for 50% hydrogen and with 6 degrees for 100% hydrogen. The main phase of combustion occurs with 12 degrees sooner for 100% hydrogen versus diesel fuel [7]. The maximum of HRR appears with 10 degrees sooner at hydrogen maximum share comparative to classic fuelling [7] The indicated efficiency decreases with almost 2.5% for $\lambda$=1.5 for maximum hydrogen content comparative to classic fuelling. But for area of $\lambda$=1.5…2 the indicated efficiency increase with 2.9% at hydrogen use and the lowest indicated specific fuel consumption being achieved at hydrogen fuelling [7]. In the exhaust gases, the $CO_2$ level is lower at hydrogen fuelling especially at lean mixture use [7]. Cameretti [8] shows that hydrogen assures a more rapid combustion process, the heat release rate peak being increased and achieved sooner with 5 °CA per cycle versus classic fuelling. Also, the combustion law at hydrogen use is superior to methane or to diesel fuel [8]. The maximum pressure increases from 52 bar to 56 bar at hydrogen use but the angle of peak pressure appears at a similar position versus Top Dead Centre [8]. Hydrogen may influence the final combustion stage, an increased second peak in the curve of the HRR may appears at hydrogen use comparative to classic fuelling, leading to a superior and more vertical combustion law [8]. This aspects concord with the increase of the combustion temperature from 2400 K for classic fuelling till 2600 K at hydrogen use [8]. The moment of peak pressure tends appears sooner during combustion, with 3 crank angle degrees sooner per cycle for hydrogen fuelling [8]. The level of the HC emission level drops from 14900 ppm to ~1300 ppm [8]. The level of the $NO_x$ emission is superior, 2.6 ppm for $H_2$, comparative to classic fuelling, 0.4 ppm and the solution of EGR use is proposed [8]. The $CO_2$ emission level is around 32000 ppm for diesel fuel, but at $H_2$ fuelling the level drops till 17900 ppm and remains steady comparative to classic fuelling [8]. Monemian [9] uses different percent of hydrogen to fumigate a diesel engine and observes large amounts of hydrogen can assure to the reduction of the combustion duration and of the dispersion between values of the indicated mean effective pressure (IMEP) [9]. Hydrogen provides a large share of the premixed burning phase share, to the increase of the maximum pressure and of the heat release rate, the peak angles being closer to TDC [9]. The indicated efficiency is increased at hydrogen use up till 2.2% [9]. The $NO_x$ emission level slightly decrease for small amount of hydrogen, till 10%, but for large hydrogen percent a slightly increase in NOx emission level appears [9]. Soot, HC and $CO_2$ levels in the exhaust gases are lower for $H_2$ versus diesel fuel, the soot emission being drastically reduced [9]. After its experiment Monemian develops a modelling process and comperes experimental result with the model result and the similar influences of hydrogen on engine performances being obtained [9]. The model results show that a large $H_2$ quantity assure the rise of the peak temperature from around 1740 K to 1850 K [9]. Santoso [10] shows that hydrogen use in large quantities, around 49 l/min, affects the thermal efficiency. Small hydrogen flows into the inlet air assure the increase of the combustion pressure above 70 bar for 21 l/min and higher hydrogen flows 36…49 l/min leads to lower maximum pressures, around 62 bar [10]. Also, for lower hydrogen flow the maximum pressure rise rate is higher [10]. The HRR is decreased for large $H_2$ flows, and its peak values appears later during combustion process [10]. Simulation results show similar tendency the combustion being retarded [10]. Helldorff [11] uses a CFD model to investigate the hydrogen combustion in diesel engine and its main conclusion shows that hydrogen peroxide product assures to the reduction of the burning duration, rise of heat release rate and maximum pressure. Almutairi [12] shows the rise of the in-cylinder temperature, HRR and $NO_x$ emission level because of the higher flame temperature at hydrogen-air mixture combustion.

The paper presents some results obtained during the theoretical modelling of the in-cylinder process at a CFR engine fuelled with diesel fuel and hydrogen. The classic fuel is energetically replaced by hydrogen in percent's of 25% and 30%. The influences of hydrogen use on in-cylinder maximum pressure, maximum pressure rise rate, heat release rate, combustion temperature, indicated thermal efficiency and NO and smoke emission levels are shown and analysed. Hydrogen use may lead to the increase of the combustion parameters, like gases pressure, rate of pressure rise, rate of heat release, gases temperature, thermal efficiency and the NO and smoke emission levels can be ameliorated.

## 2. In-cylinder processes modelling

The processes are simulated with a one-zonal model developed by authors and applied for a CFR diesel engine which has the characteristics: 85 mm bore, 115 mm stroke, 13 compression ratio, 0.652 $dm^3$ swept volume, 13 °CA start of injection, standard diesel fuel consumption 13 $cm^3$ per minute classic fuel consumption, operating at 900 rev/min. The modelling is achieved for classic fuel use, diesel fuel, and for diesel fuel and hydrogen use when the energetic replacement ratio of diesel fuel by hydrogen, XE, is 25% and 30%.

In-cylinder fluid state is evaluated from the relations of energy and mass conservation, applied for intake, compression, burning, expansion and exhaust.

The equation of energy balance [13, 14]

$$\frac{dU}{d\alpha} = \frac{dQ_{comb}}{d\alpha} - \frac{dL}{d\alpha} - \frac{dQ_{wall}}{d\alpha} \tag{1}$$

uses the internal energy, $dU/d\alpha$, the heat released during burning, $dQ_{comb}/d\alpha$, the mechanical work, $dL/d\alpha$, the heat transferred to the walls, $dQ_{wall}/d\alpha$ for a $d\alpha$ calculation step, [13, 14].
Internal energy variation [13, 14]:

$$\frac{dU}{d\alpha} = n_M \cdot c_v - \frac{dT}{d\alpha} \tag{2}$$

depends on the in-cylinder mass, $n_M$, determined from the fuel molecular mass, $M_f$, and cyclic fuel dose $m_f$ [13, 14]:

$$n_M = (\lambda \cdot L_t + \frac{1}{M_f}) \cdot (\mu_{ch} + \gamma) \cdot m_f \tag{3}$$

where the fuel quantity:

$$m_f = \frac{\rho_0 \cdot V_s \cdot \eta_v}{\lambda \cdot L_t \cdot M_a} \tag{4}$$

uses the density of the inlet fluid $\rho_0$ determinate from the state equation at $p_0$, $T_0$ condition, [13, 14].
During combustion, the simulation two burning mechanisms, preformed and diffusive, are taken into consideration by Vibe laws [13, 14]:

$$\xi = R_c \times \left[1 - e^{-6{,}9 \times x_r^{m_p^*+1}}\right] + (1 - R_c) \times \left[1 - e^{-6{,}9 \times x_d^{m_d^*+1}}\right] \tag{5}$$

where the fraction of the heat released during the premixed burning stage is $R_c$, the kinetic parameters are $m_p$* and $m_d$* and the relative angles for the beginning and share of the premixed and diffusive phases, $x_r$ and $x_d$ [13, 14].
In order to estimate the temperature variation, the mechanical work [13, 14]:

$$\frac{dL}{d\alpha} = p \cdot \frac{dV}{d\alpha} \tag{6}$$

the heat lost to the walls [13, 14]:

$$\frac{dQ_{wall}}{d\alpha} = K_{conv} \cdot A_{wall} \cdot (T - \overline{T_{wall}}) \cdot \frac{1}{6 \cdot n} \tag{7}$$

the combustion heat release [13, 14]:

$$\frac{dQ_{comb}}{d\alpha} = m_f \cdot Q \cdot \frac{d\xi}{d\alpha} \tag{8}$$

are used. The Q is the reaction heat which is equal with the LHV of the fuel for $\lambda>1$. A double Vibe law defines the combustion characteristic $d\xi/d\alpha$:

$$\frac{d\xi}{d\alpha} = R_c \cdot f_1 + (1 - R_c) \cdot f_2 \tag{9}$$

where $f_1$ and $f_2$ are burning parameters of the first two stages of the burning process [13, 14].

Finally based on previous relation and the energy balance equation, the in-cylinder temperature variation is evaluated by the relation [13, 14]:

$$\frac{dT}{d\alpha} = \frac{k-1}{n_M \cdot R_M} \cdot \left(\frac{dQ_{comb}}{d\alpha} - \frac{Q_{wall}}{d\alpha} - \frac{dL}{d\alpha}\right) \tag{10}$$

the current temperature being evaluated from the previous value, in the Δα=1...3 °CA:

$$T_2 = T_1 + \left(\frac{dT}{d\alpha}\right) \cdot \Delta\alpha \tag{11}$$

The determination of the global temperature provide the determination of the pressure, [13, 14]:

$$p = \frac{n_M \cdot R_M \cdot T}{V} \tag{12}$$

Also, highly important process of expansion is evaluated in similar mode, the use of energy balance equation for the expansion evolution, [13, 14]:

$$\frac{dT}{d\alpha} = \frac{1}{N_C \cdot C_V} \cdot \left(\frac{Q_{wall}}{d\alpha} + \frac{dL}{d\alpha}\right) \tag{13}$$

with the amount of gas in the cylinder:

$$N_C = (N_g + N_f) \cdot \overline{m_f} \tag{14}$$

offers the in-cylinder pressure variation in this specific process, [13, 14].

In terms of pollutant emission, the smoke and $NO_x$ are important for diesel engine, an estimation of their levels being of interest.

The Hiroyasu formula is used for estimation of the soot mass formation speed [14-16]:

$$m_{sf}\left[\frac{mg}{s}\right] = 2.5 \cdot 10^{-3} \cdot m_{tf} \cdot p^{0.5} \cdot \exp\left(-\frac{52298}{R \cdot T}\right) \tag{15}$$

and the Morel-Karibel formula is used for evaluation of the soot oxidation speed [14-16, 23]:

$$m_{so}\left[\frac{mg}{s}\right] = 9.55 \cdot 10^{-4} \cdot m_{soot} \cdot p^{0.5} \cdot \exp\left(-\frac{5000}{T}\right) \tag{16}$$

the fuel mass, $m_{tf}$ ,the combustion pressure and the temperature, p and T, being taking into consideration. From the ratio of the mass of the oxidized soot subtracted from the mass of soot formed, the estimated mass of soot is obtained, as an estimation of the soot comparative level [14-16]. The Schroer formula is used to estimate the NO emission level [13, 14]:

$$NO\ [ppm] = C_1 \cdot (\lambda - 0.89)^{2.11} \cdot n^{-0.39} \cdot Q^{2.49} \cdot m_a^{-5.01} \cdot \left(\int_0^Q T^{11} \cdot p^9 \cdot dQ\right)^{0.28} \tag{17}$$

where: $C_1=2.4 \cdot 10^{-11}$ , $[(1/s)^{0.39} \cdot g^{5.01} \cdot ppm / kJ^{2.77} \cdot K^{3.08} \cdot bar^{2.52}]$- constant , λ- coefficient of air excess [-], n- speed [1/min], Q- heat quantity [kJ/cycle], $m_a$-inlet mass [g], T-temperature [K], p-pressure [bar], [13, 14].

## 3. Results and discussions

The pressure diagrams for classic fueling, XE=0, and for diesel fuel and $H_2$ use, XE=0.25 and 0.30 are presented in the figure 1.

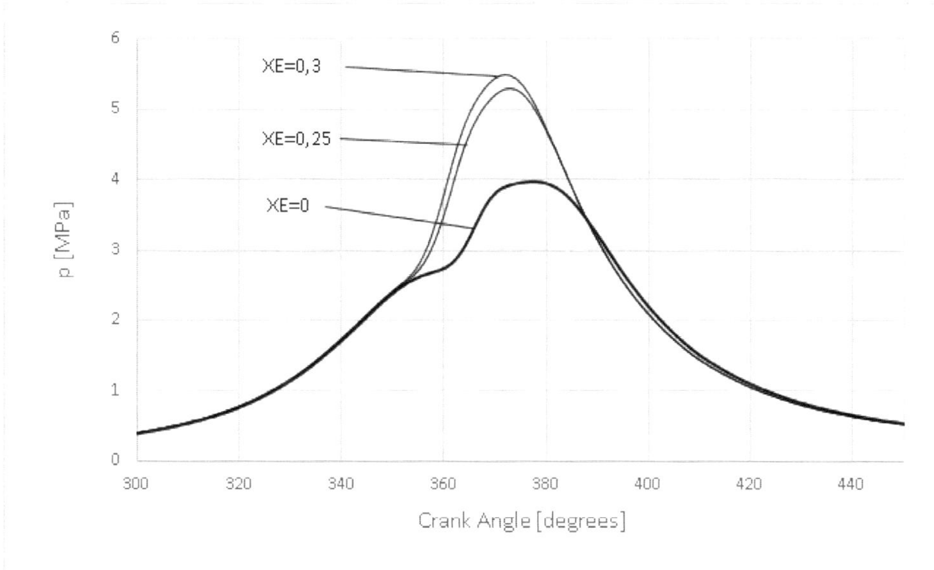


**Figure 1.** The indicated diagrams for XE=0 and $H_2$ use, XE=0.25, XE=0.30.

The raised LHV and burning speed of hydrogen associated with a shorter duration of the burning process provides to the increase of the maximum pressure during combustion, the peak values being 5.29 MPa for XE=0.25 and 5.48 MPa for XE=0.3, with 33.2% and 38%, respectively higher comparative to XE=0. The maximum rate of pressure rise (MPRR) increase from 0.147 MPa/°CA till 0.251....0.269 MPa/°CA, in the normal rage for diesel engine, but higher with 70.7% at XE=0.25 and with 82.3% at XE=0.30 versus XE=0. The reduction of the burning duration at hydrogen use is in corelation with the increase of the MPRR. The acceleration of the burning process, especially of the rapid phase leads to a closer angle of peak pressure to TDC. The angle of peak pressure is 377°CA for diesel fuel use, 373°CA for XE=0.25, 372°CA for XE=0.30 which represents an acceleration of rapid phase combustion at hydrogen use. Similar results and tendency variations for in-cylinder pressure were obtained by other researchers [3, 5, 6, 7, 8, 9, 10, 11].

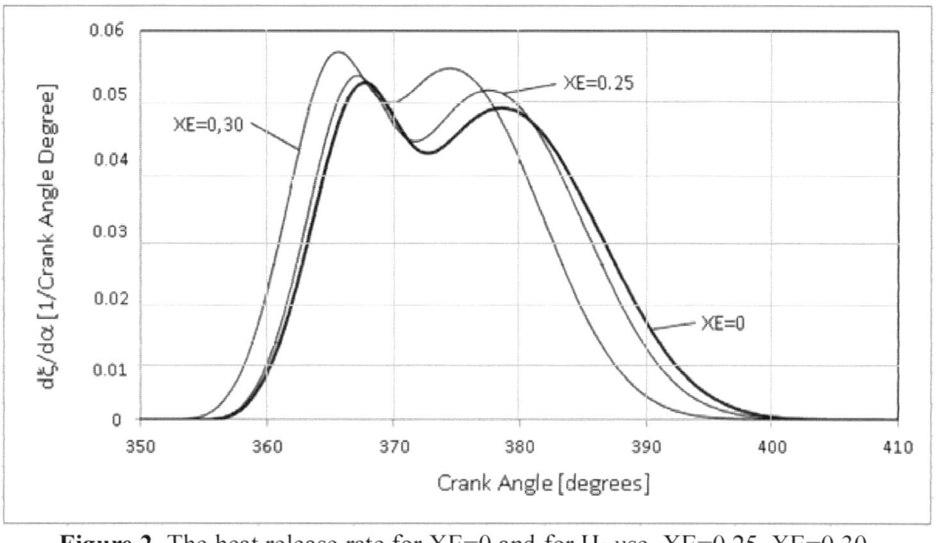


**Figure 2.** The heat release rate for XE=0 and for $H_2$ use, XE=0.25, XE=0.30.

The burning speed and LHV of $H_2$ superior to diesel fuel assures to the increase of the HRR with 6%, figure 2, the first HRR peak being closer to TDC due to the increase of the share of rapid combustion phase. Similar results and tendency variations for heat release rate were obtained by other researchers [5, 6, 7, 8, 9, 11, 12]

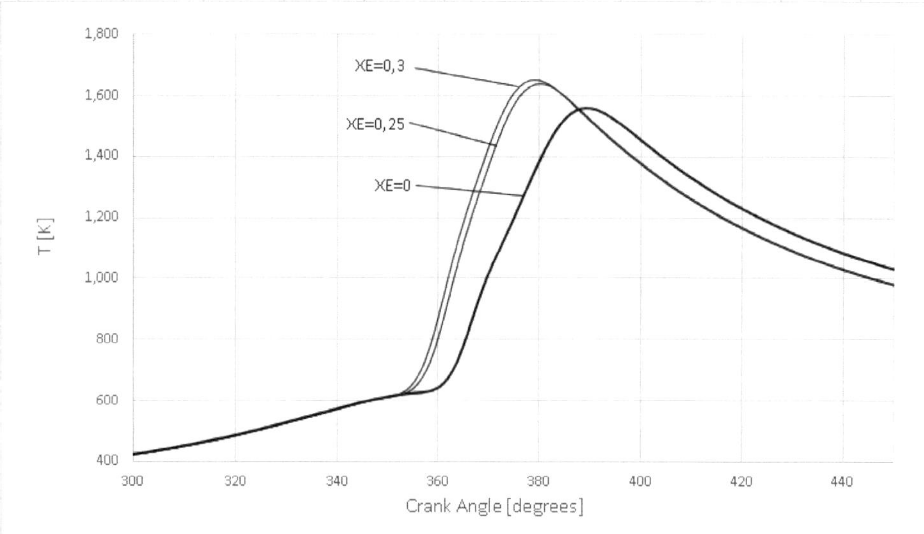


**Figure 3.** In-cylinder global temperature at XE=0 and for $H_2$ use, XE=0.25, XE=0.30.

A higher heat release rate at hydrogen use assures to the rise of the in-cylinder global temperature once with the rise of the hydrogen cyclic percent, figure 3. The maximum temperature at diesel fuel use is 1559K, the value being increased with 5.1% at XE=0.25 and with 5.8% at XE=0.30 at dual fuelling. In correlation with the pressure variation, figure 1, the angle of maximum temperature appears closer to TDC once hydrogen quantity increase, closer with 10 degrees for XE=0.25 and with 11 degrees for XE=0.30 which may represent the increase of isocore combustion share. Similar results and tendency variations for in-cylinder temperature were obtained by other researchers [1, 4, 5, 8, 9, 12]

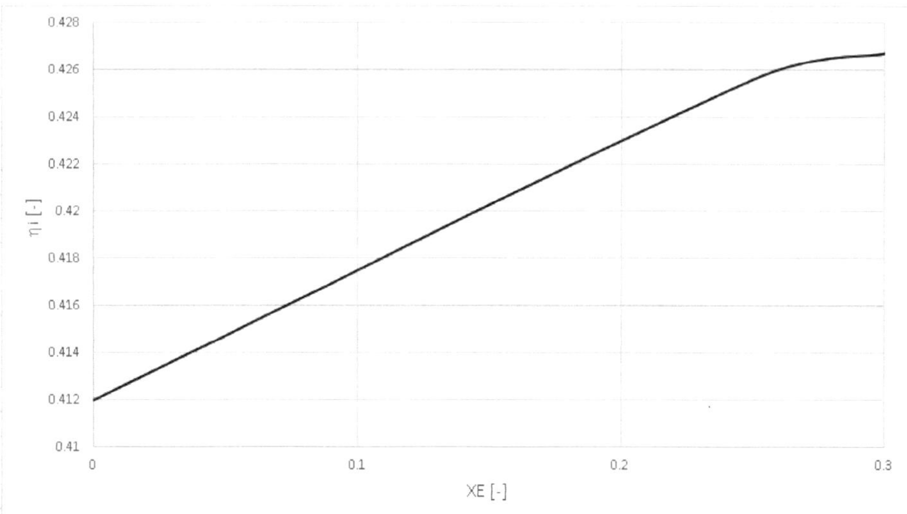


**Figure 4.** Indicated efficiency versus XE.

A raised LHV of air-fuel mixture, a decreased duration of the burning process duration at $H_2$ use leads to the improve of the engine efficiency is improved by hydrogen use, the indicated efficiency increases till 0.425 for XE=0.25 and 0.426 for XE=0.30, values that are superior to the diesel fuel use with 3.15% and 3.39%, respectively, figure 4. Similar results and tendency variations for indicated efficiency were obtained by other researchers [2, 7]

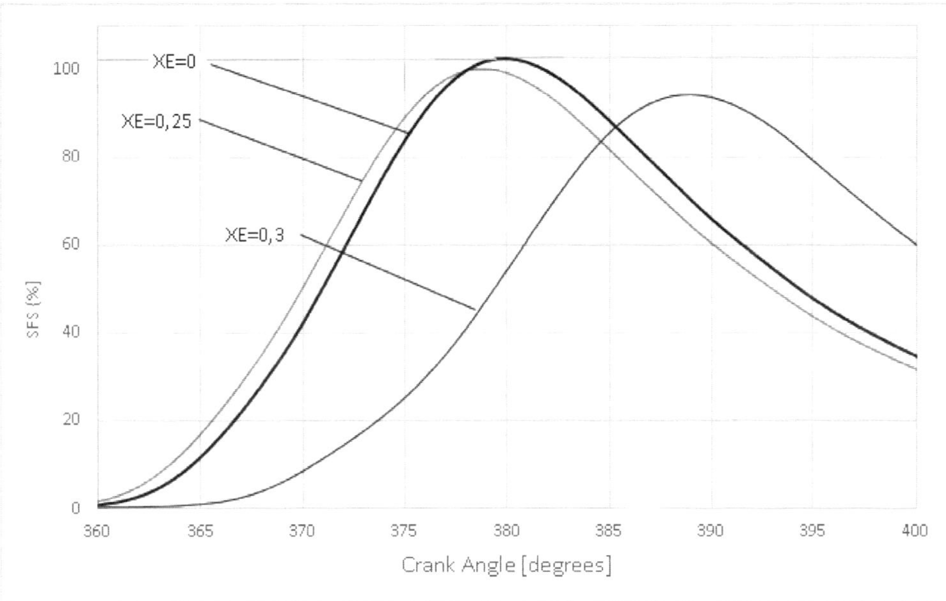


**Figure 5.** Speed of soot forming at different XE.

Lower cyclic amounts of diesel assures the decrease of the carbon content which affects the soot forming speed, figure 5. The rise of the $H_2$ flow leads to the reduction of the soot forming speed, the peak of soot forming speed being with 2.8% less for XE=0.25 and 8.17% for XE=0.3, comparative to diesel fuel fuelling. A faster burning process, reflected is acceleration of the heat release rate, figure 2, is also visible in allure of the soot forming speed, figure5.

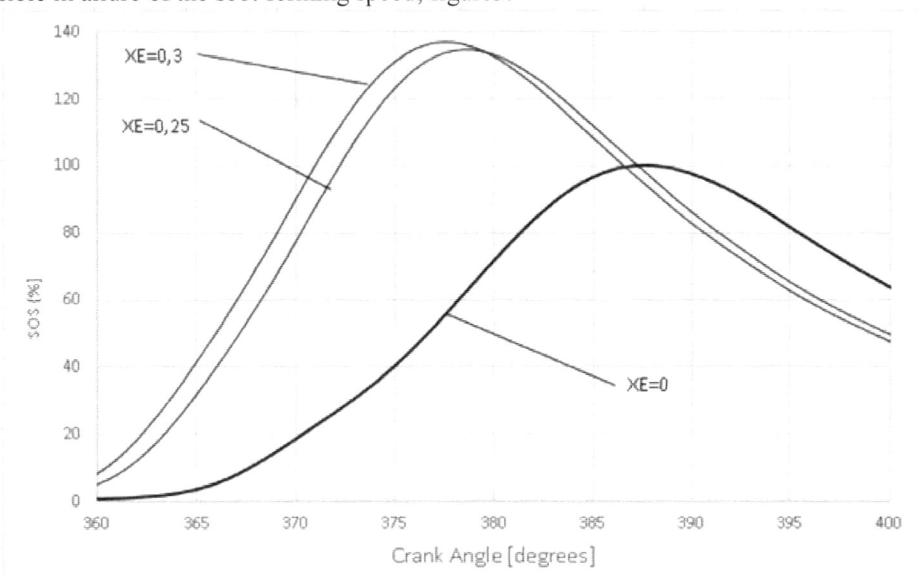


**Figure 6.** Speed of soot oxidation speed of the soot at different XE.

The speed of soot oxidation speed increase with 35.4% for XE=0.25 and with 36.5% for XE=0.30, versus XE=0. Due to hydrogen higher burning speed, the presence of hydrogen in the air-fuel mixture accelerate the soot oxidation, figure 6. A faster burning process, reflected in HRR acceleration, figure 2, is also visible in allure of the soot oxidation speed, figure5.

By integrating the speeds of soot formation and soot oxidation, associated laws for the formation and oxidation of soot are obtained. The soot balance, realized as the difference between the obtained laws, allows the evaluation of the smoke emission, which is reduced by 3.44% at XE=0.25 and by 5.92% at XE=0.30, figure 7.

The level of the soot starts to decrease at $H_2$ use, in conditions that hydrogen provides lower soot quantities which are faster oxidation by the hydrogen-diesel fuel flames. Soot reduction was achieved also by others researchers [1, 9].

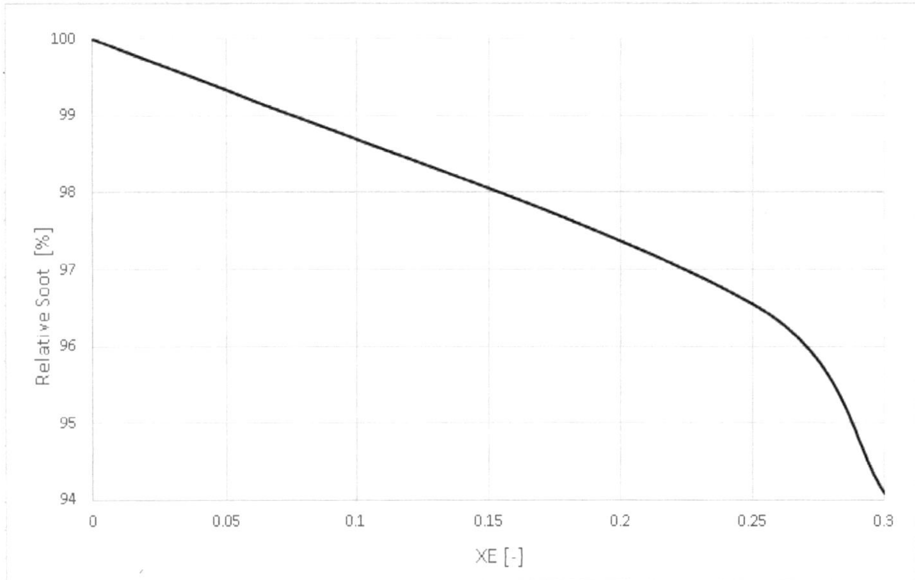


**Figure 7.** Relative soot emission level at different energetic substitute ratios.

Combustion of the air-hydrogen mixture leads to the rise of the HRR, pressure and temperature of the in-cylinder gases which increase the peak of the NO formation speed, but in condition of a lower NO quantities due to reduction of the classic fuel amount that which burn into the rapid stage of combustion once with the decrease of the diesel cyclic dose at the XE increase. The NO forming speed, figure 8, is with 64% for XE=0.25 and with 94% for XE=0.30, in correlation with the rise of burning rate, figures 2, 5 and 6.

From the integration of the NO formation speeds, laws of variation of the NO emission are obtained, noting that NO is formed by 37.2% less at XE=0.25 and by 33.1% less at XE=0.3, figure 9. The reduction in the emission of nitrogen oxides can be attributed to the reduction in the amount of oxygen available per cycle, along with the increase in the dose of hydrogen, noting that the lambda is reduced from 1.987 for the classic fuel to the value 1.561 for XE=0.25 and to the value 1.493 for XE =0.3. Similar results of nitrogen oxides reduction were obtained by other researchers [1-3, 9].

From the point of view of reduction perspectives of soot and nitrogen oxides the use of small quantities of $H_2$ injected into the engine inlet can represent a viable solution.

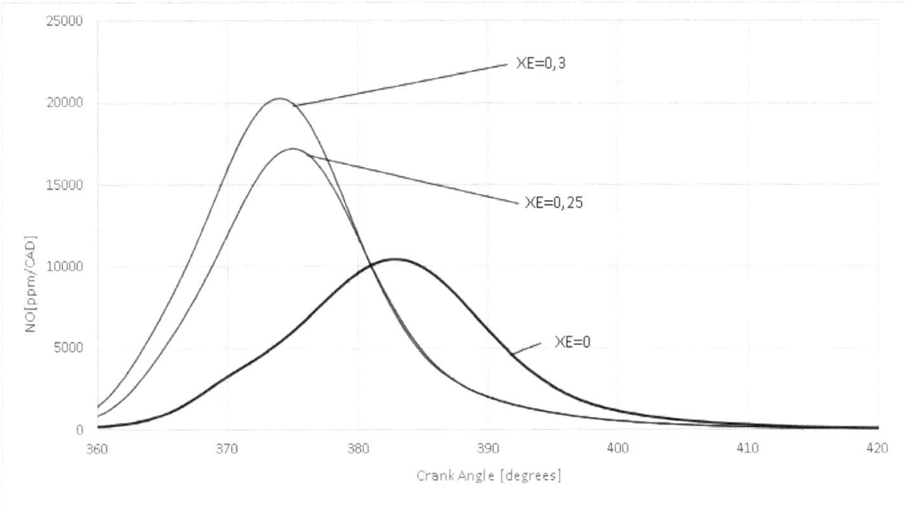


**Figure 8.** The NO formation speed at different energetic substitute ratios.

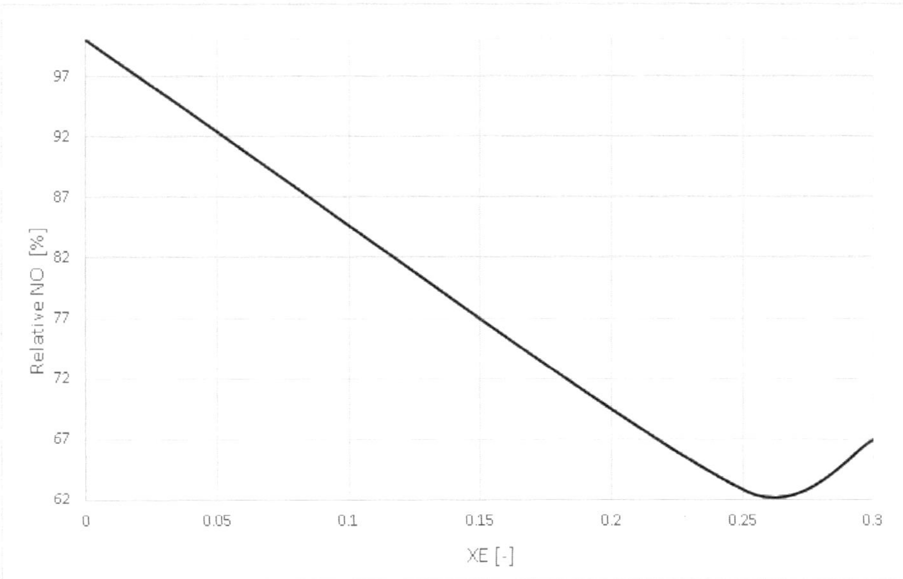


**Figure 9.** The NO in-cylinder level at different XE.

## 4. Conclusions

Following the analysis of the modelling results, it can be stated that:

-hydrogen use in percent's of 25% and 30% assures the rise of peak pressure with 33.2% and 38%, respectively higher comparative to the value allocated to diesel fuel; the maximum of the pressure rise rate increases from 0.147 MPa/°CA till 0.251 MPa/°CA....0.269 MPa/°CA, in the normal rage for diesel engine, but higher with 70.7% at XE=0.25 and 82.3% at XE=0.30 versus XE=0 and can be related with the reduction of the burning duration due to higher burning speed of air-$H_2$ mixture; the angle of maximum pressure is 377°CA for diesel fuel use, and get closer to TDC at 373°CA for XE=0.25 and 372°CA for the XE=0.30.

- the heat release rate increases with 6% and the first pick occurs closer to TDC which is in correlation with the faster burning of air-$H_2$ mixture and increased share of rapid combustion phase;

- increase of the in-cylinder global temperature once with the rise of $H_2$ percent, the value being increased with 5.1% at XE=0.25 and with 5.8% at XE=0.30 at dual fuelling.

- the indicated efficiency increases till 0.425 for XE=0.25 and 0.426 for XE=0.30., values that are superior to the diesel fuel use with 3.15% and 3.39%, respectively;

-the increase of $H_2$ percent leads to the reduction of the soot forming speed, the peak of soot forming speed being with 2.8% lower for XE=0.25 and with 8.17% for XE=0.3, comparative to diesel fuel fuelling, in correlation with the decrease of the diesel fuel which assure the decrease of the carbon percent in the air-fuel mixture leading to the decrease of the speed of soot formation; the speed of soot oxidation increases with 35.4% for XE=0.25 and with 36.5% for XE=0.30 versus diesel fuel; the soot balance shows that the smoke emission is reduced by 3.44% at XE=0.25 and by 5.92% at XE=0.30, in conditions that hydrogen provides lower soot quantities which are faster oxidation by the hydrogen-diesel fuel flames;

- the increase of the maximum speed of NO formation is favoured by $H_2$ amount increase, but in condition of a lower NO quantities due to reduction of the classic fuel amount which combust into the rapid stage once with the decrease of the diesel percent at the increase of $H_2$ quantity; the NO forming speed is higher with 64% for XE=0.25 and with 94% for XE=0.30, in relation with the rise of the burning rate in general; the laws of NO variation show that NO is formed by 37.2% less at XE=0.25 and by 33.1% less at XE=0.3; the reduction in the emission of nitrogen oxides can be attributed to the reduction in the amount of oxygen available per cycle, along with the increase in the dose of hydrogen, noting that the lambda is reduced from 1.987 for the classic fuel to the value 1.561 for XE=0.25 and to the value 1.493 for XE =0.3.

- thermal efficiency increase, soot and NO reduction perspectives can define the use of $H_2$ injection into the inlet air as a viable solution.

**Acknowledgments**
The authors address special thanks to AVL GmbH Graz Austria for providing the necessary equipment's. This work was supported by a grant of the Ministry of Research, Innovation and Digitization, CCCDI - UEFISCDI, project number PN-III-P2-2.1-PED-2021-0427, within PNCDI III.

**Author's ORCID ID**
A Cernat: https://orcid.org/0000-0003-0551-0758
C Pana: https://orcid.org/0000-0002-4629-6551
N Negurescu: https://orcid.org/0000-0003-3107-6836
C Nutu: https://orcid.org/0000-0001-9085-6053

**References**

[1] Shahpouri S, Gordon D, Hayduk C, Rezaei R, Koch C R and Shahbakhti M 2023 Hybrid emission and combustion modeling of hydrogen fueled engines *International Journal of Hydrogen Energy* doi.org/10.1016/j.ihydene.2023.03.153

[2] Anticaglia A, Balduzzi F, Ferrara G, De Luca M, Carpentiero D, Fabbri A and Fazzini L 2023 Feasibility analysis of a direct injection H2 internal combustion engine: Numerical assessment and proof-of-concept *International Journal of Hydrogen Energy* doi.org/10.1016/j.ihydene.2023.04.339

[3] Krebs S and Biet C 2021 Predictive model of a premixed, lean hydrogen combustion for internal combustion engine *Transportation Engineering* **5** 100086 doi.org/10.1016/j.treng.2021.100086

[4] Michl J, Neumann J, Rottengruber H and Wensing M 2016 Derivation and validation of a heat transfer model in a hydrogen combustion engine *Applied Thermal Engineering* **98** pp 502-512doi.org/10.1016/j.applthermaleng.2015.12.062

[5] Maghbouli A, Yang W, An H, Shafee S, Li J and Mohammadi S 2014 Modeling knocking

combustion in hydrogen assisted compression ignition diesel engines *Energy* **76** pp 768-779 doi.org/10.1016/j.energy.2014.08.074

[6] Ragupathy K 2017 Modeling and Analysis of the Diesel Engine with Addition of hydrogen-hydrogen-oxygen gas *Thermal Science* **21** pp 465-471 DOI:10.2298/TSCI17S2465R

[7] Molina, Novella R, Soriano G J and Girona O M 2021 New Combustion Modelling Approach for Methane-Hydrogen Fueled Engines Using Machine Learning and Engine Virtualization *Energies* **14** (20) 6732 pp 1-21 doi.org/10.3390/en14206732

[8] Cameretti M C, De Robbio R, Ezio M and Marco P 2022 CFD Study of Dual Combustion in a Research Diesel Engine Fueled by Hydrogen *Energies* **15** (15) 5521 pp 1-21 doi.org/10.3390/en15155521

[9] Emad M and Alascdair C 2020 Hydrogen Fumigation on HD Diesel Engine: An Experimental and Numerical Study *Diesel and Gasoline Engines* editor Richard Viskup chapter DOI:10.5772/intechopen.89425 Book DOI:10.5772/intechopen.75259 ISBN 978—1-78985-447-3, EBOOK (PDF) ISBN 978-1-78985-448-0 PRINT ISBN 978-1-78985-248-6 pp 200

[10] Santoso W B, Bakar R A and Nur A 2013 Combustion characteristics of the diesel-hydrogen dual fuel engine at low loads *Energy Procedia* **32** pp 1-10

[11] Helldorff H and Micklow G 2022 Chemical Kinetic Modeling of Hydrogen –Diesel Co-combustion in Compression Ignition Engines *SAE Int.J.Fuels Lubr.* **15** (3) pp 235-254 SAE International US doi.org/10.4271/04-15-03-0011

[12] Almutairi F S, Ranga Dinesh, J A van Oijen 2023 Modelling of the hydrogen-blended dual-fuel combustion using flamelet-generated manifold and preferential diffusion effects *International Journal of Hydrogen Energy* **48** (4) pp 1602-1624

[13] Popa M, Negurescu N and Pana C 2003 Diesel Engines. Processes, MatrixRom, Bucharest.

[14] Cernat A, Negurescu N, Pana C and Nutu C 2023 Theoretical aspects of the use of hydrogen in the diesel engine *IOP Conference Series: Materials Science and Engineering* **1290** pp 1-14

[15] Jung D and Assanis D N 2004 Modeling of Direct Injection Diesel Engine Emissions for a quasi-dimensional Multi-zone Spray Model, *International Journal of Automotive Technology*, volume **5**, number 3, pp 165-172

[16] Haiyan M 2001 Fundamental Studies of Combustion Phenomena in Dual Fuel Engines, PhD-Thesis, School of Mechanical and Manufacturing Engineering, The University of New South Wales

# Aspects of the combustion variability analysis at an automotive engine fuelled with hydrogen

**R Georgescu[1], C Pana[1], N Negurescu[1], A Cernat[1], C Nutu[2] and C Sandu[1]**

[1]Department of Thermotechnics, Engines, Thermal and Frigorific Equipment, University Politehnica of Bucharest, Bucharest, Romania
[2]Department of Automotive, University Politehnica of Bucharest, Bucharest, Romania

E-mail: raresgeorgescu@yahoo.com

**Abstract**. Over the last decades, the use of the alternative fuels was one of the main research activities for specialists in the field of internal combustion engines. The development of the modern automotive engines is constantly challenged by the more severe emission legislation. The engine emissions levels and the fuel efficiency are directly influenced by the engine operation, reproduction of the combustion phases from one cycle to other, cyclic dispersion during combustion process being important. In general, the use of alternative fuels in internal combustion engines provides an improvement of the energetic and pollution performance, or just a slight improvement of them, but the study of the combustion process must be completed with aspects regarding the cyclic variability. In particular, using this alternative fuel, a study of cyclic variation of the combustion process would be necessary in order to establish if the normal operation of the engine can be ensured. The paper presents some aspects of the analysis of the cyclic variability at a spark ignition engine fuelled with gasoline and hydrogen. During the engine operation at the regime of 2500 rev/min speed and 55% engine load, a number of 250 consecutive combustion cycles was recorded for classic fuel use and for hydrogen use. The coefficient of cyclic variation (CCV) or the coefficient of variation (COV) is determined for different combustion parameters such as maximum pressure, maximum pressure rise rate and mass fraction burned, defined by angles at which the conventional fractions of 10%, 50% and 90% of the heat of reaction is released. Thus, the values of the COV for maximum pressure $(COV)_{pmax}$, maximum pressure rise rate $(COV)_{dp/d\alpha}$, angles of 10, 50 and 90% heat release as $(COV)_{10\%}$, $(COV)_{50\%}$ and $(COV)_{90\%}$ were calculated and compared with the admissible limit of 10%. The combustion variability analysis establishes the limits of the normal operation of the spark ignition engine fuelled with gasoline and hydrogen compared with the classic fuelling method.

## 1. Introduction

Hydrogen is a crucial component in alternative fuel research and is considered to be the most environmental friendly energy source. Recent research indicates that hydrogen fuel has great potential to be used in internal combustion engines and can achieve zero carbon emissions without radical changes [1]. The pollutant emissions of the hydrogen fueled engine, compared with fossil fuel, are quite reduced, especially carbon emission, but the problem is the relatively high $NO_x$ emission due to the high flame temperature of hydrogen, so it`s necessary to study the low-temperature combustion and lean-burn regimes. Because of the difficulty of storing large quantities on-board and the power loss at stoichiometric ratio, using pure hydrogen to fuel the engine is not feasible, so hydrogen blended fuels can be used as an alternative in the transition phase for automotive use.

According to [2], cyclic variations or cycle-to-cycle variation (CCV), have a direct effect on vehicle drivability. CCV can deteriorate the engine`s output torque, resulting in combustion instability, engine noise and vibration. The main method to evaluate cyclic variation is the coefficient of variation in indicated mean effective pressure ($COV_{IMEP}$). The drivability problems of vehicles start to be noticed when $COV_{IMEP}$ exceeds 10% for SI engines [3]. [4] and [5] also stated that cyclic variability affects driving comfort, reduces the thermal efficiency and increases pollutant emissions.

COV for maximum pressure is defined with the following formula [6]:

$$COV_{Pmax} = \frac{\sqrt{\frac{1}{N_c - 1} \sum_{k=1}^{N_c} \left(p_{\,max_k} - p_{\,max_m}\right)^2}}{p_{\,max_m}} \tag{1}$$

$$p_{\,max_m} = \frac{\sum_{k=1}^{N} p_{\,max_k}}{N_c} \tag{2}$$

where (1) is the standard deviation and (2) is the avarage value for maximum pressure.

[7] studied cyclic variation for a SI engine fuelled with gasoline-hydrogen blends of 2.14%, 5.28% and 7.24% under lean-burn settings. An increase of air-fuel ratio also increased the COV and is more noticeable at air-fuel ratios beyond 1.1, showing that cyclic variations increase when fuel mixtures become leaner. The addition of hydrogen reduces COV, the minimum cyclic variation being observed for the largest hydrogen fraction. The COV is basically unaffected for hydrogen fractions above 5.28%, for any air-fuel mix.

Using a DI gasoline engine with PFI hydrogen injection, [8] investigated the COV of IMEP. With 11.67% hydrogen addition, the maximum COV reached 1.07% compared with almost 20% for gasoline use only.

A single cylinder gasoline engine was modified to operate on hydrogen with manifold injection [9]. When running only $H_2$, the engine produced 29% less maximum indicated power, BTE increased with 2%. $COV_{IMEP}$ was around 0.85% for hydrogen engine at $\lambda = 0.75$ and 1.3% for gasoline engine at $\lambda=1.3$. The cyclic variations in all combustion periods for hydrogen engine were lower than for gasoline engine. The authors concluded that even at lean mixtures, hydrogen produces a more stable engine operation than gasoline at rich mixtures.

Dual fuelling SI engines with gaseous renewable fuels is an efficient approach because it combines the advantages of both fuels. Hydrogen DI, compared with PFI, has the advantage that reduces abnormal combustion phenomena like preignition, backfire and knock, but in terms of emissions, a higher level of $NO_x$ and an increase in cyclic variation was noted. Using 18% EGR flow rate, showed a reduction of $NO_x$ with 77.8% [2].

[10] studied the effect of hydrogen DI on combustion stability and showed that for hydrogen fractions under 5.3%, the COV drops from 3.3% to 0.51%, with the excess air-fuel ratio decreasing from 1.8 to 1. The authors agreed that an increase of hydrogen fraction will reduce the COV for a given excess air ratio. At stoichiometric conditions, a slight COV increase was observed because of the concentration of hydrogen at the top of the combustion chamber which creates a lean mixture near the cylinder walls that affects the combustion stability. [11] directly injected 10% hydrogen in a SI under lean burn conditions with 15% throttle opening. The results showed that an increase in excess air ratio reduced the mean value of the maximum cylinder pressure and increased the COV because the mixture becomes leaner and deteriorates the flame propagation speed, reducing the maximum cylinder pressure. Notable cycle-to-cycle variations occur when the engine is operating with very lean mixtures, but with hydrogen, these variations are much reduced compared to other hydrocarbon fuels [12].

According to [2], [4] and [8], the main sources of cyclic variation in a SI engine are: (a) turbulence formation in the cylinder flow field, (b) variation in fuel-air ratios, (c) the homogenous nature of fuel mixtures, (d) characteristics of spark discharge and flame development, (e) the quantity of leftover and recirculated gases in the cylinder. A high turbulence flow can affect the fuel concentration in the cylinder. A lower fuel-air ratio than desired ratio for optimum operation, slows down the flame propagation process, which is important during the initial flame development phase which directly affects the cyclic dispersion during combustion. During operation of a SI engine, only a few cycles are obtained at optimum spark timing, so faster combustion cycles need advanced spark timing and slow combustion cycles need late spark timing. This fluctuation of spark timing also add up to cyclic variation. The air-fuel mixture within the cylinder is not an homogenous mix with the recirculated exhaust gases and this fact affects the combustion process but because of the higher burning velocity of hydrogen than that of gasoline, the cyclic variation can be reduced. There are also other methods to reduce the cyclic variation such as: using a correct air/fuel ratio, controlled spark timing, using fuel with high RON.

Recent studies try to focus on creating a predictive model for a premixed hydrogen in an ICE [13]. Besides efficiency, they also predicted the $NO_x$ emission and abnormal combustion phenomena [14] used a simple first-law zero-dimensional model that predicted reasonably well the trends obtained by experiment, though the simplifications limit its predictive capability. [4] model incorporates cycle-to-cycle variability and takes into consideration the effect of hydrogen addition and was validated by direct comparison with experimental results. The authors discovered that for lean mixtures, when the fraction of hydrogen increases, COV decreases to a minimum value, which depends on the fuel-air equivalence ratio. At fuel ratio near 0.9 and 70% $H_2$ by volume are the minimum values of COV.

Set side by side with pure gasoline, hydrogen addition can make the ignition more stable and accelerate combustion rate to improve the brake thermal efficiency even more under lean burn condition. Additionally, it reduces the CO and HC emissions because of more complete combustion, but generates more $NO_x$ due to the higher combustion temperature [15] [16], [17], [18], [20], [21], [22]. Decreasing of spark advance can improve the $NO_X$ [19].

The purpose of this paper is to determine the effects of fuelling a 4-cylinder spark ignition engine with different fractions of gaseous hydrogen on energetic performance and cyclic variability over 250 cycles. The engine was fueled at first with gasoline to establish the reference at the engine load $\chi$=55%, engine speed of n=2500 min-1 and different air-fuel ratios ($\lambda$).

## 2. Experimental test bench

The test bench is equipped with a 1.5l, 4 cylinder turbocharged Daewoo A15MF engine fueled only with gasoline at the engine load $\chi$=55% and speed of n=2500 min-1. Parameters such as: ambient temperature, intake air temperature, exhaust gases temperature, water temperature, air consumption, boost pressure and pollutant emissions. The in cylinder pressure was measured across 250 engine cycles and the mean values of these measurements were used to determine maximum pressure ($p_{max}$), maximum pressure rise rate $(dp/d\alpha)_{max}$, heat release rate ($dQ/d\alpha$) and total heat released (Q). To determine the cycle variation during combustion process, the coefficient of variability (COV) for maximum pressure $(COV)_{pmax}$, maximum pressure rise rate $(COV)_{dp/d\alpha}$, angles of 10, 50 and 90% heat release as $(COV)_{10\%}$, $(COV)_{50\%}$ and $(COV)_{90\%}$ were calculated and compared with the admissible limit of 10%.

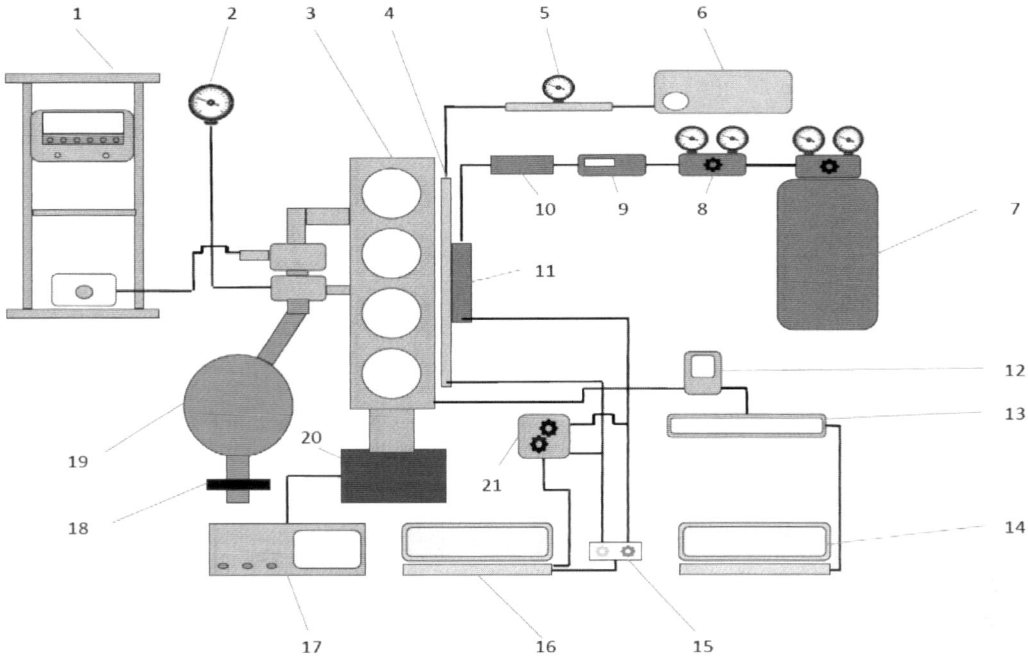


**Figure 1.** Test bench schema of the SI engine fueled with gasoline and hydrogen

The main components of the test bench: 1 - gas analyser, 2 – inlet supercharger pressure gauge, 3 - A15MF engine, 4 – gasoline injection system, 5 – gasoline flowmeter, 6 – gasoline fuel tank, 7 – hydrogen bottle with pressure reducer, 8 – second pressure reducer for hydrogen, 9 – flowmeter for hydrogen, 10 – flame arrestor, 11 – hydrogen injection system, 12 - electric charge amplifier, 13 - data acquisition system, 14 – PC equipped with data acquisition board and software, 15 – remote for fine tune of fuel injectors, 16 – PC, 17 – dyno control cabinet, 18 – differential pressure gauge for intake air, 19 – intake air tank, 20 – eddy current dynamometer, 21 – open electronic control unit.

## 3. Results

The energetic performances of the engine are presented in the next graphs at fuelling only with gasoline (in blue) and at fueling with 5.82% hydrogen-gasoline blend (in red), at air-fuel ratio $\lambda$=0.89.

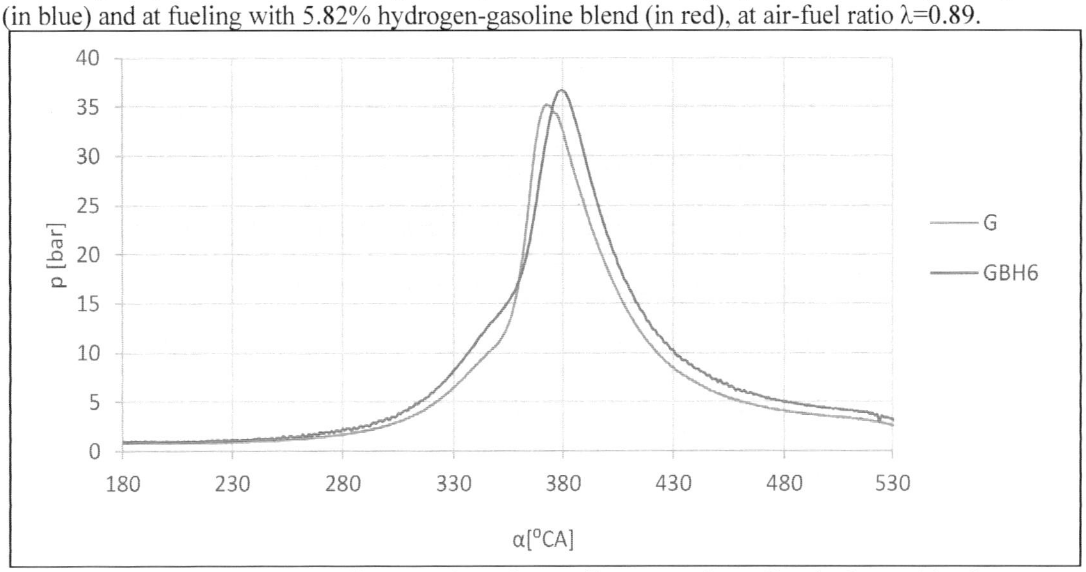


**Figure 2.** Pressure diagrams for only gasoline use and gasoline-hydrogen blend.

The pressure diagrams for both fuels are presented in fig. 2. Because of the small fraction of hydrogen used, the engine adjustments were kept the same. For the blended fuel, the initial combustion phase was shorter and move to the left of the TDC because of the high flame speed of hydrogen. The maximum pressure is higher for the blended fuel 36.65 bar vs gasoline use 35.16 bar.

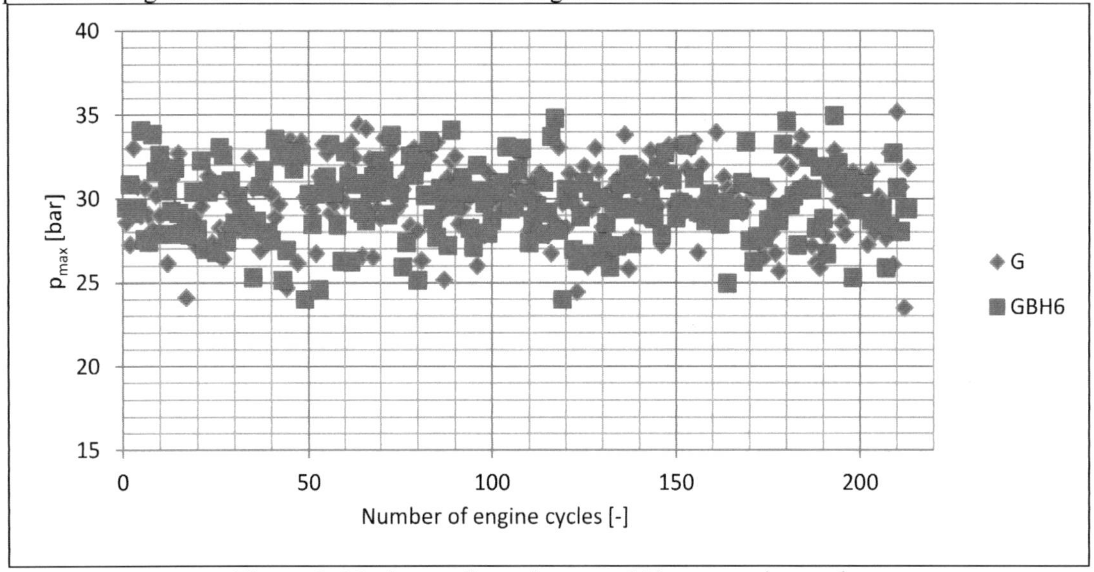


**Figure 3.** Maximum dispersion over 250 consecutive cycles

In figure 3, the maximum pressure values for 250 consecutive cycles are presented. The graphic shows an improvement of dispersion in values at fueling with gasoline-hydrogen blend vs only gasoline. The COV for maximum pressure (figure 4) is determined with formula (1).

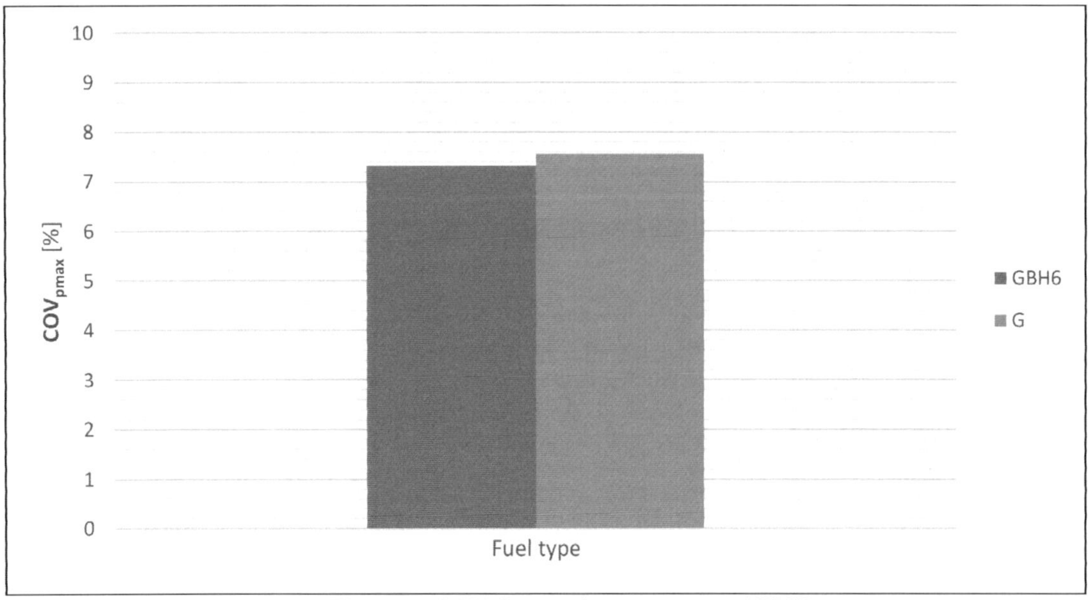


**Figure 4.** COV for maximum pressure $(COV)_{pmax}$

After the calculation, a small improvement from 7.56% to 7.32% was noted (fig. 4) when using 5.82% hydrogen-gasoline fuel blend.

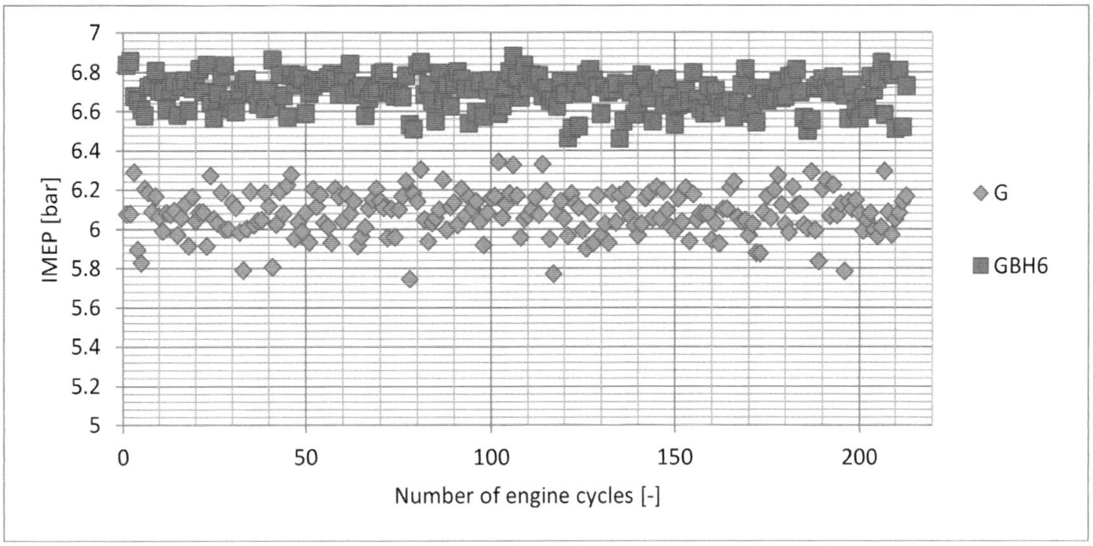


**Figure 5.** IMEP dispersion over 250 engine cycles

The dispersion for the indicated mean effective pressure is presented in figure 5. For λ=0.89, the dispersion of IMEP for 250 engine cycles for the hydrogen-gasoline blend is much reduced comparing to the gasoline use. This effect was obtained primarily because hydrogen burned more completely than gasoline, due to the excellent diffusivity properties and short quenching distance that allowed the flame to travel near the cylinder wall, achieving complete combustion.

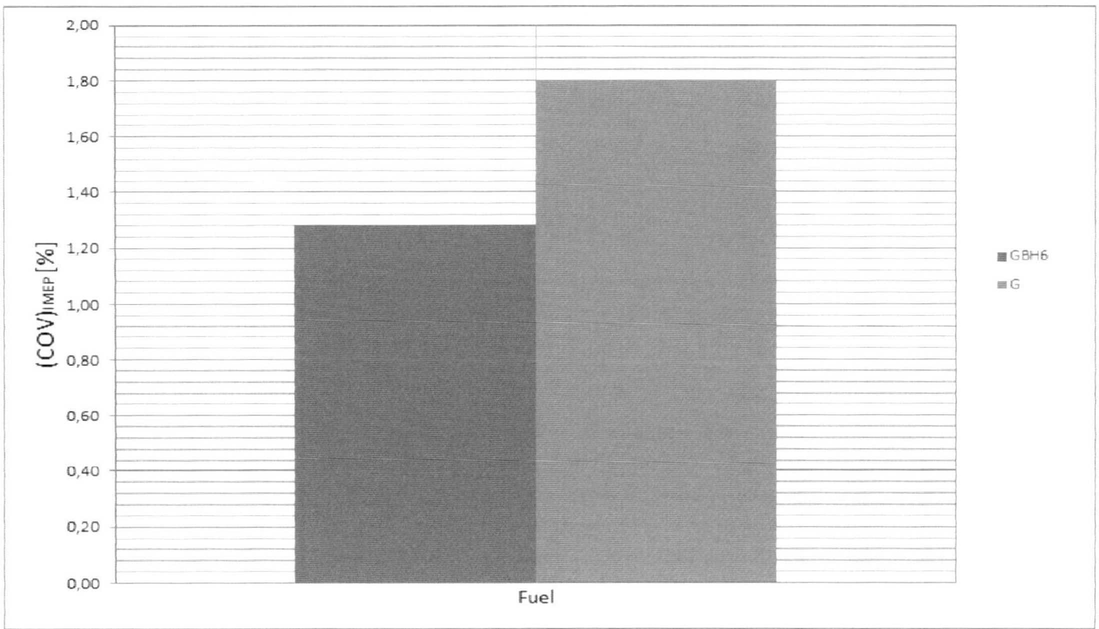


**Figure 6.** COV for IMEP over 250 engine cycles

The addition of hydrogen slightly improved the stability of combustion also as pointed out in fig. 6. The $COV_{IMEP}$ for gasoline-hydrogen fuel is 1.28 vs 1.80 for gasoline use.

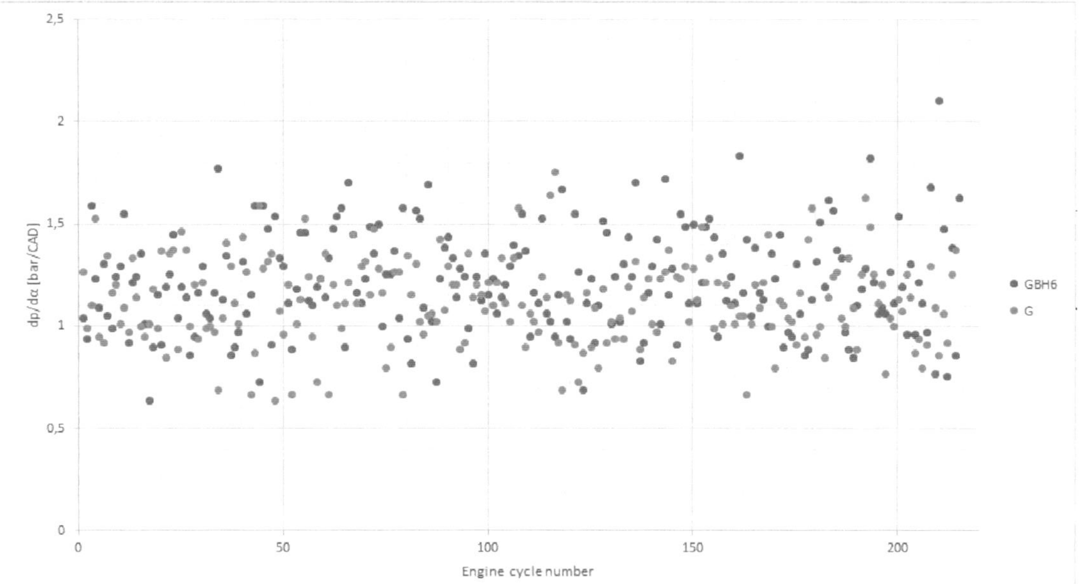


**Figure 7.** Maximum pressure rise rate over 250 engine cycles

At all air-fuel ratios, higher maximum pressures are observed for gasoline-hydrogen fuel. The maximum pressure rise rate $(dp/d\alpha)_{max}$ was calculated using data from the in-cylinder transducer. In figure 7, hydrogen-gasoline blend has overall higher pressure rise rates (red dots) than regular gasoline fuel even at rich air-fuel ratio like $\lambda$=0.89 used for this graphic.

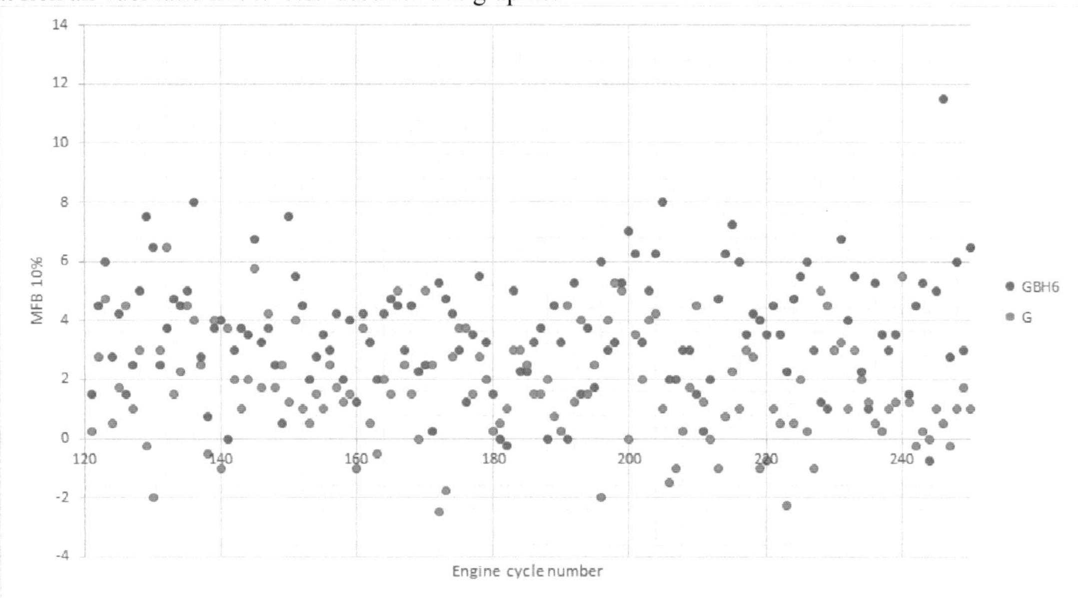


**Figure 8.** Angle of 10% MFB

The mass fraction burned (MFB) for 10% heat release is represented in figure 8 where a high dispersion for both fuels can be observed with a slight improvement for the gasoline-hydrogen fuel.

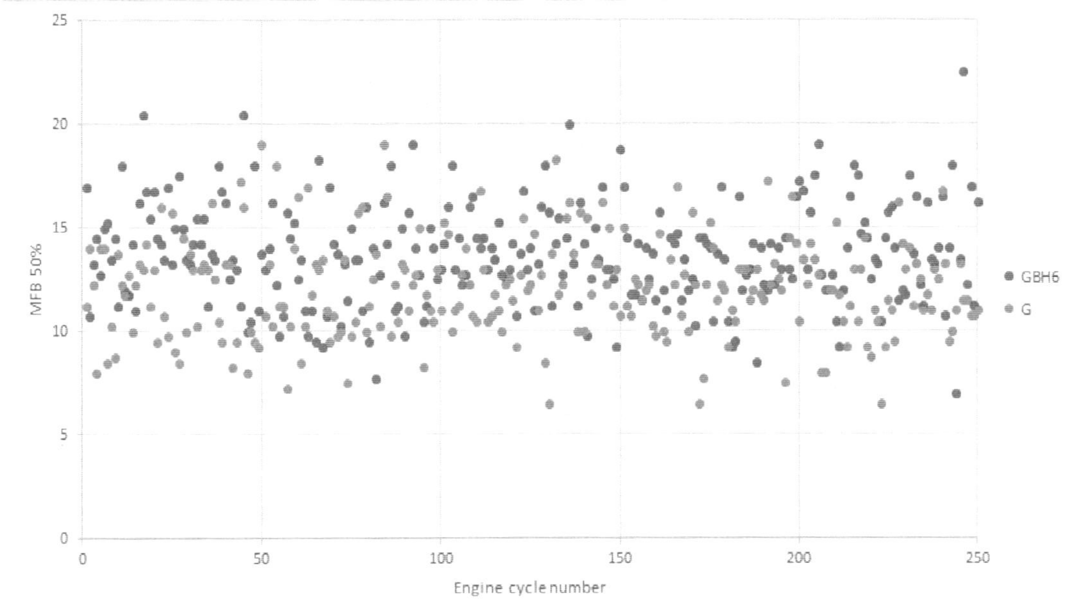


**Figure 9.** Angle of 50% MFB

For the mass fraction burned for 50% heat release, the addition of hydrogen lightly improves the combustion stability as shown in figure 9.

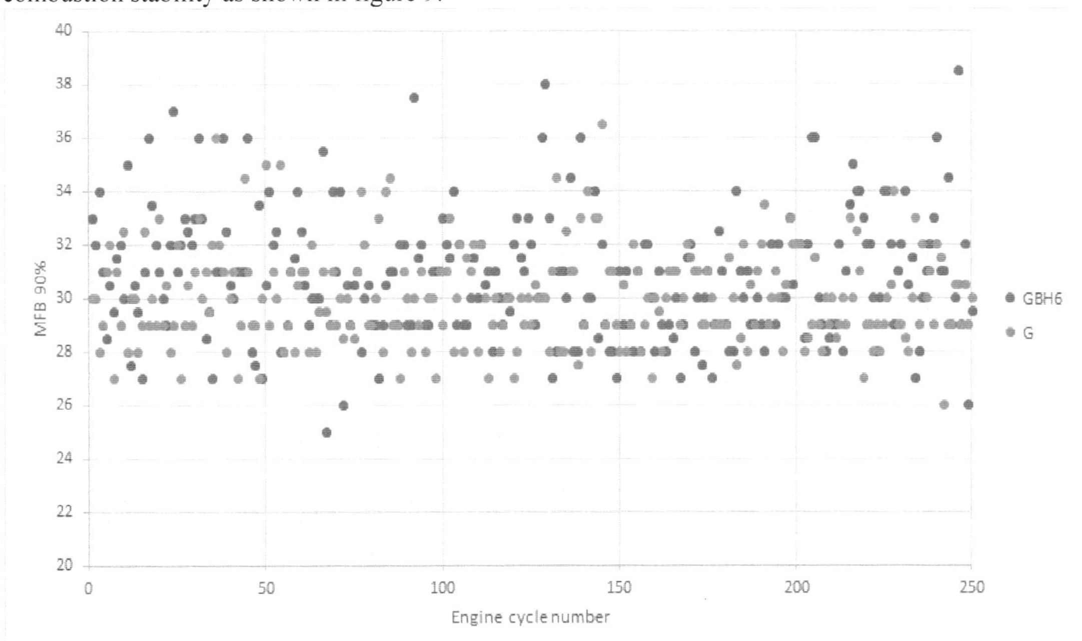


**Figure 10.** Angle of 90% MFB

Gasoline-hydrogen blend releases 90% of the heat faster that regular gasoline in the first and main phase of the combustion.

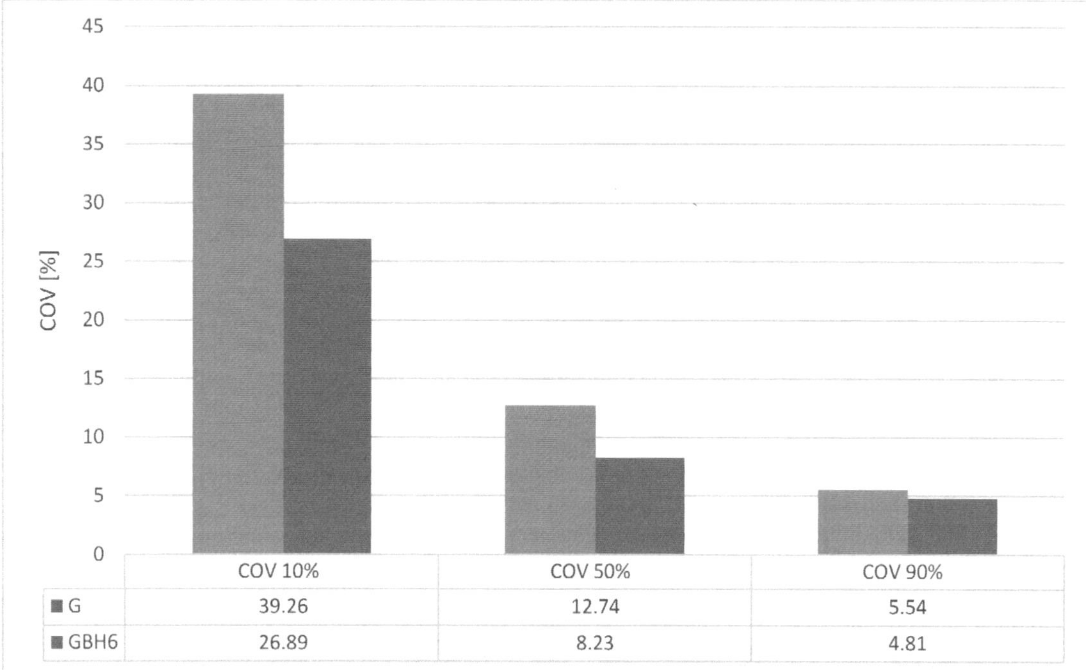


**Figure 11.** COV for 90% MFB

Calculating the coefficient of variation for angle 10% of the MFB, $COV_G$=39.26% and $COV_{GBH6}$=26,89%, where G = gasoline and GBH6 = gasoline-hydrogen 5.82%, showed that the addition of hidrogen to the blend have a drastic impact on the initial burning phase, accelerating the combustion because of the high burning speed of hydrogen. The COV for angle 50% also indicates more stability for the main combustion phase for the use of gasoline-hydrogen blend 8.23% vs 12.74%. As for the final phase of the combustion, hydrogen has little to no impact 4.81% for the gasoline-hydrogen blended fuel vs 5.54% for regular gasoline fuel.

The reduction of cyclic dispersion during the main phase of combustion is influenced by the lower cyclic dispersion of the initial phase at hydrogen use. The duration of the initial phase of combustion is influenced by the initial pressure, initial temperature, dosage, but as the reaction speed depends to a greater extent on the concentration than on pressure and temperature, the dosage has a greater influence, locally, in the cylinder, the presence of hydrogen influencing this aspect. Superior combustion properties of hydrogen compared to gasoline, such as lower heating value (119600 kJ/kg versus 42690 kJ/kg), higher flammability limits ($\lambda$=0.13…10.8 versus $\lambda$=0.709…1.10), higher laminar flame speed (2.37 m/s versus 0.37...0.43 m/s), low minimum ignition energy (0.018 mJ versus 0.2...0.3 mJ), high diffusivity (0.61 $cm^2/s$ versus 0.05 $cm^2/s$), the last two favor the acceleration of specific processes of the initial phase, increasing the reaction speed, reducing the duration of the initial phase, an aspect that leads to the reduction of cyclic variability during the initial phase, which correlates with the tendency to reduce the total duration of combustion and cyclic dispersion during the main phase. From the analysis of the mediated heat release laws, it can be noted that the duration of the initial phase is reduced by 7 $^0$CA at hydrogen use, and the heat release rate reaches positive values faster per cycle, by about 1$^0$CA, compared to the classic fuelling. The reduction of the duration of the initial phase correlates with the reduction of the total duration of combustion by 26 $^0$CA, an aspect that correspond with the tendency to reduce cyclic dispersion and COV values at hydrogen use comparative to gasoline. In general, the reduction of the duration of the initial phase is consistent with a reduced cyclic dispersion during it. Thus, the lower COV of the beginning of combustion, COV10%, leads to the reduction of cyclic variability during the main phase, COV50%, and of the end of combustion COV90%, in the case of hydrogen use.

## Conclusions

Hydrogen has a significant impact on combustion, especially on the flame development and propagation period. At the studied regime, the addition of 5.82% hydrogen showed an improvement of the combustion. The variation in maximum pressure improves slightly from 7.56% to 7.32% for the blended fuel and a small increase of maximum pressure is noted, 36.65 bar for hydrogen-gasoline fuel vs 35.16 bar for gasoline use. The variation of the indicated mean effective pressure decreased from 1.80% for gasoline to 1.28% for the blended fuel. In the first phase of combustion, both fuels have high coefficients of variation, but the addition of hydrogen enhanced the flame development phase with roughly 31%. The COV for 50% heat release also improved from 12.74% to 8.23% for the blended fuel. Regarding the final phase of the combustion, hydrogen has a small impact on variation, under 1%.

Hydrogen improves combustion and reduces cycle variability because of the high laminar flame speed which shortens the combustion time. Although hydrogen addition requires engine modifications, it is considered a good alternative fuel for spark ignition engines to reduce pollutant emissions maintaining at the same time engine performance.

The research will continue with a study on the engine adjustments when fueling with different fractions of hydrogen and the effect on performance and pollutant emissions.

## Acknowledgments

The authors express their gratitude to AVL List GmbH Graz, Austria, for providing the necessary equipment's. This work was supported by a grant of the Ministry of Research, Innovation and Digitization, CCCDI - UEFISCDI, project number PN-III-P2-2.1-PED-2021-0427, within PNCDI III. This work was supported by a grant of the Ministry of Research, Innovation and Digitization, CNCS - UEFISCDI, project number PCE 5/2022, PN-III-P4-PCE-2021-0777, within PNCDI III.

## References

[1] Sun Z, Hong J et al 2023 Hydrogen engine operation strategies:Recent progress, industrialization challenges and perspectives *Int. J. of Hydrogen Energy* **48**, 366-392

[2] Purayil S T P et al 2023 Review of hydrogen-gasoline SI dual fuel engines: Engine performance and emission *Energy Reports* **9,** 4547-4573

[3] Heywood J B 1998 Internal combustion engine fundamentals, New York:McGraw-Hill Book Company

[4] Martinez-Boggio S D, Medina A et al 2016 Quasi-dimensional modelling of cyclic variability in gasoline-hydrogen spark engines *Int. J. of Thermodynamics*

[5] Martinez-Boggio S D, Curto-Risso P L et al 2016 Simulation of cycle-to-cycle variations on spark ignition engines fueled with gasoline-hydrogen blends *Int. J. of Hydrogen Energy* **41**, 9087-9099

[6] N. Neguresсu, C. Pană, M.G. Popa 2013 Spark ignition engines. Processes, Matrix ROM, Bucharest

[7] Ceviz M A, Sen A K et al 2012 Engine performance, exhaust emissions and cyclic variations in a lean-burn SI engine fuelled by gasoline-hydrogen blends *Appl. Therm. Eng.* **36**, 314-324

[8] Ji C et al 2018 Performance of a hydrogen-blended gasoline direct injection engine under various second gasoline direct injection timings *Energy Convers. Manag.* **171,** 1704-1711

[9] Gurbuz H et al 2013 An experimental study on performance and cyclic variations in a spark ignition engine fuelled with hydrogen and gasoline *J. of Thermal Science and Technology,* 33-41

[10] Niu R, Yu X et al 2016 Effect of hydrogen proportion on lean burn performance of a dual fuel SI engine using hydrogen direct-injection *Fuel* **186**, 792-799

[11] Yu X, Wu H et al 2016 Research on cycle-by-cycle variations of an SI engine with hydrogen direct injection under lean burn conditions *Appl. Therm. Eng.* **109**, 569-581.

[12] Kim J M et al 1995 Performance characteristics of hydrogen fueled engine with the direct injection and spark ignition system SAE Paper 952498

[13] Krebs S, Biet C 2021 Predictive model of a premixed, lean hydrogen combustion for internal combustion engines *Transport Engineering* **5**, 100086
[14] Urroz J C, Dieguez P M et al 2023 Gaseous fueling of an adapted commercial automotive spark-ignition engine: Simplified thermodynamic modelling and experimental study running on hydrogen, methane, carbon monoxide and their mixtures *Fuel* **337**, 127178
[15] Assad M, Penazkov O 2023 Comprehensive analysis of the operation of an internal combustion engine fueled by hydrogen-containing mixtures *Energy Reports* **9**, 4478-4492
[16] Wang Shuofeng, Ji Changwei 2012 Cyclic variation in a hydrogen-enriched spark-ignition gasoline engine under various operating conditions *Int. J of Hydrogen Energy* **37**, 1112-1119
[17] Wang Shuofeng, Ji Changwei et al 2010 Experimental study on improving gasoline engine lean operating performance with hydrogen addition *J. of Engineering Thermophysics* **31**, 2145-2148
[18] Ji Changwei, Yang J-X et al 2013 Effect of standard hydroxygen addition on the performance of a gasoline engine *Journal of Beijing University of Technology* **39**, 1378-1383
[19] Ji Changwei, Zhang Bob et al 2012 Emission performance of a hydrogen-enriched gasoline engine *Journal of Beijing University of Technology* **38**, 151-155
[20] Wang Shuofeng, Ji Changwei et al 2012 Experimental study on the idle performance of a spark-ignition engine fueled with hydrogen-gasoline blends *Chinese Society for Internal Combustion Engines* **30**, 451-455
[21] Su Teng, Ji Changwei et al 2017 Reducing cyclic variation of a gasoline rotary engine by hydrogen addition under various operating conditions *Int. J. of Hydrogen Energy* **42** 25428-25435
[22] Dhyani V, Subramanian K A 2017 Effect of Operating Parameters on Combustion Anomalies and Cyclic Variations in a Hydrogen Fuelled Spark Ignition Engine *Conference NCICEC-2017*
[23] Yu Xiumin, Li Guanting et al 2019 A comparative study on effects of homogeneous or stratified hydrogen on combustion and emissions of a gasoline/hydrogen SI engine *Int. J. of Hydrogen Energy* **44**, 25974-25984

IOP Conf. Series: Materials Science and Engineering 1303 (2024) 012018 doi:10.1088/1757-899X/1303/1/012018

# Design of Advanced Power Transmissions

**P Pavlov and A Dobreva**

Department of Machine Science, Machine Elements, Engineering Graphics and Physics, University of Ruse "Angel Kanchev", 7017 Ruse, Bulgaria

E-mail: adobreva@uni-ruse.bg

**Abstract**. During the last five years, significant progress has been made in the development of automotive transmissions, and especially in increasing the gear ratios and improving the efficiency. With the help of software products and programs, new transmission design schemes have been constantly developed, which are characterized by improved operational parameters. Different optimization methods have been considered about the implementations of main objectives: power transmissions' component design and efficiency improvement. Conclusions have been made concerning the opportunities for reducing the energy consumption by engineering tools inside the power transmissions.

## 1. Introduction

The largest concentration of carbon monoxide, nitrogen oxides, ozone and other chemical compounds has been registered in the central part of the cities. In [1], the authors' team creates an extremely significant mathematical model of four analyzed situations in order to suggest important recommendations for reducing pollution in an urban environment. The author teams of [2, 3, 4 & 5] investigate and propose several significant measures that can be implemented aiming to reduce energy consumption and deleterious emissions, the most important of which is related to the renewal of automotive vehicles and innovation of their components.

Besides, using new automobiles and new technologies to design gearboxes can contribute to reduce noise pollution in urban residential areas. In [6, 7 & 8] the authors evaluate noise pollution from car traffic. The authors identify opportunities for solving the problems of traffic and pollution in the central part of the cities by applying extremely interesting and relevant traffic simulations.

As summarized briefly and unambiguously in [9], sources of air pollution are motorcycles, cars, trucks, airplanes, ships, locomotives, etc. It is emphasized that technologies which are applied in engines such as computer control and direct gasoline injection can improve fuel economy and reduce harmful emissions.

Transmission technologies such as the application of seven or more speeds, dual-clutch transmissions (DCTs) and continuously variable transmissions (CVTs) also improve fuel economy and reduce emissions. As it has been mentioned, better transportation planning for passengers and cargo also leads to similar beneficial results, Figure 1.

## 2. Theoretical background

The level of harmful emissions and the energy efficiency of cars are highly dependent on the operational parameters and the quality of the engines. The authors of [10 & 11] conducted an up-to-date experimental study on the function of air temperature, the formation of the air/fuel mixture, the modification of the cooling system and several design parameters of the engine.

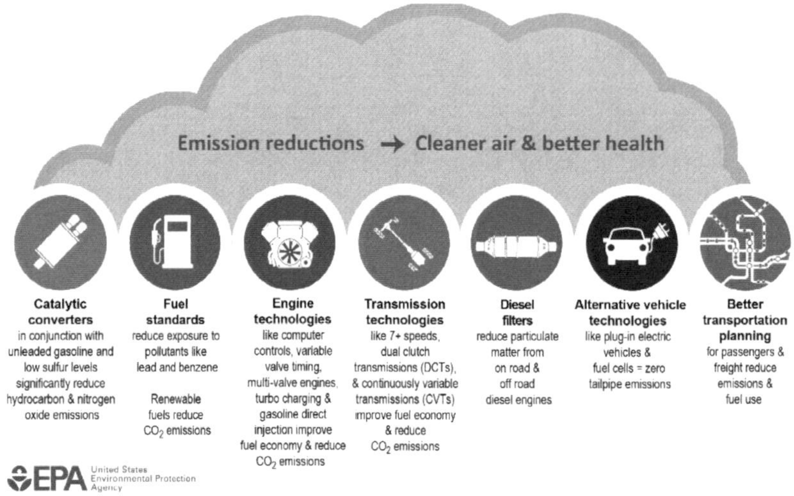


**Figure 1.** Solutions for Transportation air pollution, [9]

In order to increase the combustion quality of the engine, as a result of the research implemented, measures of particular importance have been suggested: application of different injection strategies; optimization of combustion components, etc. [12].

Along with the development of new technologies in the area of design and production of engines, the application of new methods and approaches concerning the layout of automotive transmissions is also growing extremely rapidly. According to [10, 11] over the past few years, continuously variable transmissions (CVT) and dual-clutch transmissions (DCT) have also increased their market share significantly.

Over the past few years, transmission design has evolved significantly, impacting both performance and fuel economy. Modern, well-performing transmissions have been shown to be one of the most cost-effective technologies for improving automotive fuel economy, [13].

According to the authors' team, the determination of the transmission ratios takes into account the decisions on the number of planetary gears, the number of clutches and the number of kinematic combinations of different transmission elements. A transmission design procedure is described in [14] in which planetary gears, joints, clutches, brakes and motors have been represented with appropriate mathematical constraints. Suitable analytical equations for these constraints have been solved in order to determine several possible design solutions.

An interesting solution for improving the quality of the automotive industry has been proposed in [15]. The authors determine the time required for each operation and the corresponding components in order to realize a change in the production process. Innovative methods based on time, visual control and visual inspection have been applied.

Based upon the results of the theoretical analysis implemented, the authors consider that it is possible to make significant changes in the design and calculation procedures for transmissions and gearboxes in order to improve their functional parameters and their energy efficiency.

## 3. Methods and prerequisites for solving the problem

Based on the experience of the authors' team in the field of design theory, systems approach and new product development, a methodology for improving the operational parameters of transmission and gearbox components is proposed.

The requirements for clarity in the layout of a new design version of the gears correspond to the necessity for unambiguously executable manufacturing and assembly processes.

Improved and advanced procedures for evaluation and comparative analysis of different design solutions for a certain problem, involving several main stages, have been described in details in [16-18].

Methodology suggested in [18] includes: analysis of elements in the power flow of the vehicle and the possibilities to reduce mechanical losses in these components; new and modified generally valid assessment criteria; a new approach for calculating the cardan drive in order to reduce its internal dynamic load; a method for calculating internal meshing with increased efficiency; specific criteria for the analysis and evaluation of design solutions that have a relatively equal degree of significance.

The main results of these studies described the prerequisites created for conducting simulation and experimental research of gears and other power transmissions.

Besides, in a previous study [19], one of the authors considered the strength calculations of gearbox components by several design methods - conventional standard calculations, automated design with a CAD system, and strength simulation with an appropriate CAD system.

Due to these reasons, the author's team considers to introduce two new stages to the methodology presented in [18]: improvement of the algorithm for determining the transmission ratio, which will enable the improvement of the transmission elements. This new stage of the methodology will be part of the refinement of the algorithm for switching gears and the selection of optimal gear ratios for future transmissions. The second new methodology stage is to be a precise calculation of the meshing according to the contact strength criterion based upon a comparative analysis of a database of accumulated calculations.

## 4. Results and Discussion

Calculations of geometric, kinematic parameters and strength parameters of an exemplary five-speed automobile transmission (Figure 2) have been performed.

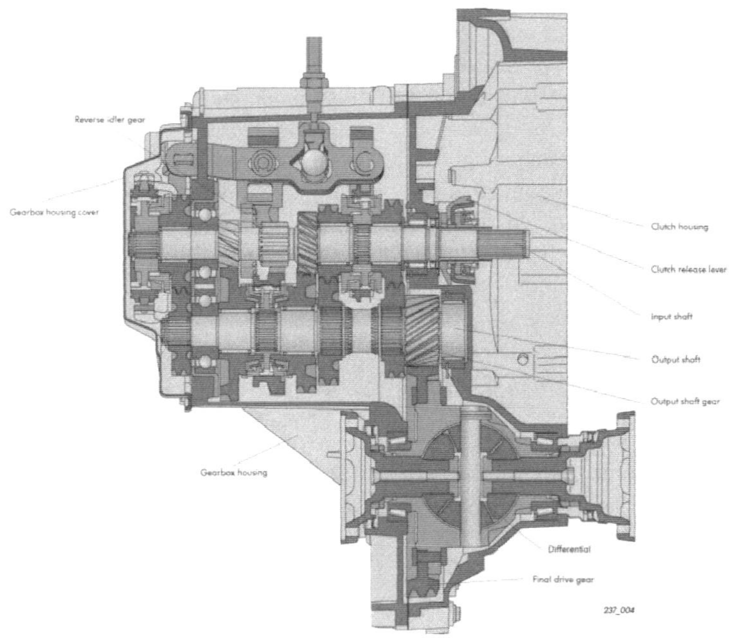


**Figure 2.** General view of the structure of a manual gearbox 02T [19]

The modules of the gear sets are: m = 2.25 mm or and m=1.75 mm. The number of teeth and gear ratio of each stage are indicated in Table 1.

The results of the calculations lead to the following conclusions: the values for contact stresses obtained are about 5% lower when working with SolidWorks in comparison with the calculations implemented according to ISO 6336, [20]. The corresponding coefficients and assumptions when working with the CAD system have been compared with the analytical results implemented. It has been found that the system accepts a greatest possible meshing width depending on the module compared to the recommendations of the international standard.

**Table 1.** Teeth numbers and gear ratios

| Gear | | Gear ratio |
|---|---|---|
| Gear ratio - $z_2 : z_1$ | 1st gear | 38 : 11 = 3,455 |
| | 2nd gear | 43 : 22 = 1.955 |
| | 3rd gear | 41 : 32 = 1,281 |
| | 4th gear | 38 : 41 = 0,927 |
| | 5th gear | 37 : 50 = 0,740 |
| | Reverse gear | 35 : 24 x 24 : 11 = 3,182 |

## 5. Conclusions

A theoretical investigation in the area of advanced power transmissions is presented. Based upon the results of the theoretical analysis done, the authors consider that it is possible to make significant changes in the design and calculation procedures for transmissions and gearboxes in order to improve their functional parameters and their energy efficiency.

The author's team introduces two new stages to the transmission design methodology: improvement of the algorithm for determining the transmission ratio and more precise calculation of the meshing according to the contact strength criterion.

The results obtained can be used at an early stage of design process of transmissions, allowing the designer to choose the optimal parameters in terms of maximum efficiency.

**Acknowledgements**
The authors' team expresses gratitude and acknowledgements for the support of the scientific fund at the University of Ruse, Bulgaria and for the support of the University of Pitesti, the University of Craiova and the Transilvania University of Brașov for their collaboration and experience exchange.

**Author's ORCID ID**
P Pavlov: https://orcid.org/0000-0003-4155-0861
A Dobreva: https://orcid.org/0009-0007-1643-1293

## References

[1] Tarulescu S, Tarulescu R and Soica A 2008 *Mathematical Model of Pollution Compounds Calculus in Function of Traffic Capacity from Urban Areas*, WSEAS transactions on environment & development, pp. 350-359
[2] Tarulescu S, Tarulescu R, Soica A and Leahu C Smart Transportation $CO_2$ Emission Reduction Strategies. *IOP Conf. Ser.: Mater. Sci. Eng*. **252**, 012051, (2017), pp. 1-10
[3] Chiru A, Covaciu D, Florea D and Timar J 2009 *Noise Mapping for Urban Road Traffic and its Effect on the Local Community*, The 3rd International Conference on Computational Mechanics and Virtual Engineering, Brasov, pp. 1-8
[4] Covaciu D, Florea D and Timar J 2015 *Estimation of the noise level produced by road traffic inroundabouts*, Applied Acoustics, Volume **98**, pp. 43-51
[5] Tarulescu S and Tarulescu R 2016 *Urban Transportation Solutions for the CO2 Emissions Reduction Contributions*, 12th International Congress of Automotive and Transport Engineering *(CONAT)*, doi: 10.1007/978-3-319-45447-4_49
[6] Titu A M, Boroiu A A, Mihailescu S, Pop A B and Boroiu A 2022 *Assessment of Road Noise*

*Pollution in Urban Residential Areas - Case Study in Pitesti, Romania.* Appl. Sci. 2022, 12, 4053.
[7] Boroiu A A, Neagu E, Boroiu A and Vieru I 2018 *Proposals for the Reorganization of Road Traffic in the Central Area of Pitesti Municipality based on Microsimulation Performed Traffic Analyses.* IOP Conference Series: Materials Science & Engineering, Vol. **444**/072021, pp. 1-11, doi:10.1088/1757-899X/444/7/072021
[8] Boroiu A A, Tabacu I, Ene A, Neagu E and Boroiu A 2017 *Computing Programs for Determining Traffic Flows from Roundabouts* IOP Conf. Series: Materials Science and Engineering **252 (2017) 012053**, pp. 1-7, doi:10.1088/1757-899X/252/1/012053
[9] Meyer M D and Elrahman O A 2019 *Transportation and Public Health An Integrated Approach to Policy, Planning, and Implementation*, doi:10.1016/B978-0-12-816774-8.05001-5
[10] Tarulescu S and Tarulescu R 2016 *Researches on Combustion Quality for a Single Cylinder Diesel Engine,* International Congress of Automotive and Transport Engineering, doi: 10.1007/978-3-319-45447-4_48, pp. 1-9
[11] Covaciu D, Florea D and Timar J 2015 *Estimation of the noise level produced by road traffic inroundabouts,* Applied Acoustics, Volume **98**, pp. 43-51
[12] Tarulescu S and Tarulescu R 2016 *Urban Transportation Solutions for the $CO_2$ Emissions Reduction Contributions*, 12th International Congress of Automotive and Transport Engineering (CONAT), doi: 10.1007/978-3-319-45447-4_49
[13] Kim N, Rousseau A, and Lohse-Busch H 2014 *Advanced Automatic Transmission Model Validation Using Dynamometer Test Data*, SAE Technical Paper 2014-01-1778, doi: 10.4271/2014-01-1778.
[14] Raghavan M, Bucknor N, Maguire J, Hendrickson J et al. 2007 *The Algebraic Design of Transmissions & EVTs*, SAE Technical Paper 2007-01-1458, doi: 10.4271/2007-01-1458.
[15] Crăciunoiu N, Oprica A and Pascu C 2021 *Aspects about Deployment of Lean Principles for Improving the Production Process Quality in Automotive Industry*. RoJAE, vol. **27** no. 2 , pp. 45 – 88, http://siar.ro/wp-content/uploads/2021/08/RoJAE-27_2.pdf#page=5.
[16] Dobreva A and Dobrev V 2007 *Research of Technical Parameters of Transmissions for Vehicles and Agricultural Machines.//* UPB: Scientific Bulletin, Series D: Mechanical Engineering, **69**, pp 103 – 109.
[17] Dobreva A 2013 *Theoretical Investigation of the Energy Efficiency of Planetary Gear Trains*, Springer Verlag Dordreeht, pp 289-298.
[18] Dobreva A and Dobrev V 2018 *Innovative Methodology for Decreasing Mechanical Losses in Vehicles*, Proc. of the 4th International Cong. of Aut. & Trans. Eng., Springer Verlag, pp 234 – 242.
[19] Dimitrov Y and Pavlov P 2020 *Strength Research of a Gear from a Car Gearbox*. Proc. of University of Ruse, Vol **59**, 4.1, pp. 45 – 50.
[20] ISO 6336-2:2019. Calculation of load capacity of spur and helical gears, Part 2: Calculation of surface durability (pitting)

# Hydrogen fuelling a passenger car spark ignition engine for a cleaner environment

**C Nutu[1], C Pana[2], N Negurescu[2], A Cernat[2] and R Georgescu[2]**

[1]University Politehnica of Bucharest, Department of Automotive Engineering.
[2]University Politehnica of Bucharest, Department of Thermotechnics, Engines, Thermal Equipment and Refrigeration instalations.

E-mail: nikolaoscristian87@gmail.com

**Abstract.** The tightening of the restrictions on polluting emissions in the last decade and the possibility of the disappearance of the internal combustion engine in the field of transport must motivate the researchers in the field of internal combustion engines to find new solutions to fuel the spark ignition engine, aiming to reduce the negative impact of polluting emissions on the environment. Unconventional fuels represent a very good choice to fuel the spark ignition engines, based on their excellent combustion properties and hydrogen stands out among them. This paper presents experimental results of fuelling a car spark ignition engine with mixtures of gasoline and hydrogen. The experimental investigations have shown an improvement in the combustion process of the engine, a reduction in the energetic fuel consumption, an increase in the brake thermal efficiency and a decrease in the pollutant emissions level.
**Keywords**: emissions, hydrogen, efficiency, combustion.

## 1. Introduction

The remarkable performances achieved by internal combustion engines in recent years have been possible both due to the sustained interest of researchers in the field and due to the requirements given by international regulations regarding polluting emissions and increasing efficiency, especially in the current context of fossil fuel saving requirements. With the increasing restrictions related to pollution standards in accordance with the regulations of the European Commission on the use of internal combustion engines for cars, new viable methods must be found to supply them with alternative fuels so that the thermal efficiency increases and the level of polluting emissions is sharply reduced compared to of the classic case of fossil fuel supply. A viable solution for meeting these goals is to fuel IC engines with hydrogen. Hydrogen may be used both as a single fuel and as an additional fuel, ensuring the reduction of the global consumption of fossil fuels and the reduction of the level of carbon dioxide emission, which has a real potential greenhouse effect. Hydrogen fueled engines produce a lower level of particulate and smoke emissions without requiring major changes to their design. Hydrogen fueling solutions can be successfully applied both to new internal combustion engines in the design stage and to internal combustion engines already in operation. Thus, by optimizing the operating regimes and by optimizing the settings of internal combustion engines fueled by hydrogen, older engines can be brought to levels closer to modern ones in terms of emissions and in terms of brake thermal efficiency and specific energy consumption.

Hydrogen is an element which can be found in the Universe in large quantities and on Earth it is found in the air, as a part of water and as a part of hydrocarbons. The principal methods of hydrogen production are: methane steam reforming, partial oxidation of hydrocarbons, biomass gaseification, coal and waste, thermal cracking and electrolysis of water [2, 3, 4, 5]. The elements of difficulty in using hydrogen to fuel spark ignition engines are represented by the calibration of fueling systems and their adjustments to obtain the best results in terms of the level of pollutant emissions, thermal efficiency and in terms of specific fuel consumption. In a study carried out in papers [6, 7, 8, 9], very good results were obtained in terms of the performance and level of pollutant emissions of the hydrogen fueled spark ignition engine. A key property of in-cylinder combustion that has contributed to improved energy and pollution performance is the higher burn rate of hydrogen compared to gasoline. Thus, through its higher burning speed, hydrogen ensured the placement of the moment of maximum pressure near the TDC of the engine and in the case of combustion of very poor air-fuel mixtures, obtaining the pronounced reduction of nitrogen oxides.

A major problem for modern spark-ignition engines equipped with direct fuel injection systems is the level of particulate emission that is higher than in the case of engines equipped with indirect fuel injection systems. The level of particle emission can be reduced in the case of hydrogen fueling in a dual system, the gasoline being partially energetically substituted with hydrogen, because first of all it reduces the content of carbon atoms in the mixture and decreases the risk of thermal cracking of the fuel droplets and secondly, the higher combusting speed of hydrogen reduces the risk of fossil fuel droplets depositing on the internal walls of the combustion chamber [10].

### *1.1. The properties of hydrogen.*

The properties of hydrogen are presented in the table 1, presented comparatively with those of gasoline and diesel fuel.

**Table 1**. The properties of hydrogen, gasoline and diesel fuel [1].

| Property | Hydrogen | Gasoline | Diesel fuel |
|---|---|---|---|
| Molar weight [$kg\cdot kmol^{-1}$] | 2,016 | 114 | 204 |
| Density [$kg/m^3$] at 0°C, 760 mmHg | 0,0899 | 735-760 | 825-870 |
| Lower heating value [kJ/kg] | 119617 | 42690 | 42600 |
| Autoignition temperature [K] | 858 | 753-823 | 473-493 |
| Research Octane Number RON [-] | 130 | 90-98 | 30 |
| Cetane Number CN [-] | - | - | 45-55 |
| Flame speed [m/s] $\lambda$=1,20°C, 760 mmHg | 2.37 | 0.25-0.35 | - |
| Air-Fuel Ratio as kg air/kg fuel [-] | 34.32 | 14.5 | 14.76 |
| Minimum ignition energy [mJ] | 0.018 | 0.2-0.3 | - |
| Flame temperature [K] | 2300 | 2411 | 2327 |
| Diffusivity [$cm^2/s$] | 0.63 | - | - |
| Flammability limits $\lambda$s.....$\lambda$i | 0.136-10.12 | 0.709-1.1 | 0.34-1.68 |
| (%) vol | 4.1-75.6 | 1.48-2.3 | 0.7-5 |

It can be observed in the table above that the density of hydrogen is very low compared to fossil fuels and therefore hydrogen quickly take all the available space and this aspect can affect the intake process in the case of hydrogen injecting in the intake plenum. To not affect the rated power of the engine a good choice is to use hydrogen as an additive fuel, the gasoline being partially substituted. The higher combusting speed of hydrogen than that of gasoline is an advantage when used in a SI engine, the first combustion phase becoming shorter, thus it can be obtained an increasing in the pressure inside the cylinder. Because hydrogen has an extremely low ignition energy, it develops a trend to auto-ignite from hot surfaces like the EX valve, spark plug, or hot gases. The laminar combusting speed of hydrogen is 7...10 times higher than gasoline combusting speed and this is a good advantage to reduce the initial combustion phase, and also the turbulent burning speed inside the cylinder is increasing, leading to a reduction in the rapid combustion phase inside the cylinder. To avoid knock when using stoichiometric mixtures a good solution is to make the mixture lean. To avoid the increase in the intensity of knock EGR cooling is preferred and this also helps to reduce the dilution of intake air.

Due to hydrogen's lower heating value higher than value of hydrocarbons, we can obtain an increase in the BTE of the engine and reduce the BSFC. Burning hydrogen produces an adiabatic flame of approximately 2300 K [10] and may increase the temperature in the combustion chamber, which can increase the level of NO emission [10]. The smoke emission level of the engine fuelled with hydrogen decreases, especially for the engines with direct gasoline injection when the engines are bi-fuelled with hydrogen, the second fuel being provided to the intake plenum of the engine and gasoline directly injected, because by making the speed of combustion higher the phenomenon of the occurrence of thermally cracking of fuel drops on the colder areas of the combustion chamber decreases [10, 11].

### *1.2. Fuelling methods*

The SI engine may be fueled with $H_2$ by indirect injection or direct injection. Because the port injection is common among SI engines, it can be applied without difficulty for engines already in operation. Also, the standard fuelling system can still be used on the vehicle and the engine may run in dual mode. The use of direct hydrogen injection requires serious engines modifications, particularly the mounting of injectors directly inside the combustion chamber. By injecting $H_2$ directly affecting the filling of the cylinder with air is avoided. In the papers [12] and [13], the authors $H_2$ fueled the SI engine by indirect injection, gasoline being provided by the classic fuel system. The burning process was optimal positioned near the TDC of the engine due to greater combustion speed of $H_2$ [12, 13].

## 2. Experimental investigations

The A15MF supercharged SI engine has been used to carry out the experiments, fuelled with gasoline and hydrogen. The engine ran at 2500 revolutions per minute and 55% engine load with $\lambda$=1, and the gasoline was 5% substituted with $H_2$.

$$xc = \frac{m_{H_2} H_{i_{H_2}}}{m_{H2} H_{i_{H2}} + m_{gas} H_{i_{gas}}} \cdot 100 \tag{1}$$

**Table 2.** The A15MF engine.

| | |
|---|---|
| Nr. of cylinders | 4 |
| Displ. [cm$^3$] | 1498 |
| Compr. ratio [-] | 9.5 |
| Bore [mm] | 76.5 |
| Stroke [mm] | 81.5 |
| Conrod length [mm] | 123.5 |
| Rated power [kW] / speed [rpm] | 63/4800 |
| Rated torque [Nm] / speed [rpm] | 136/3400 |

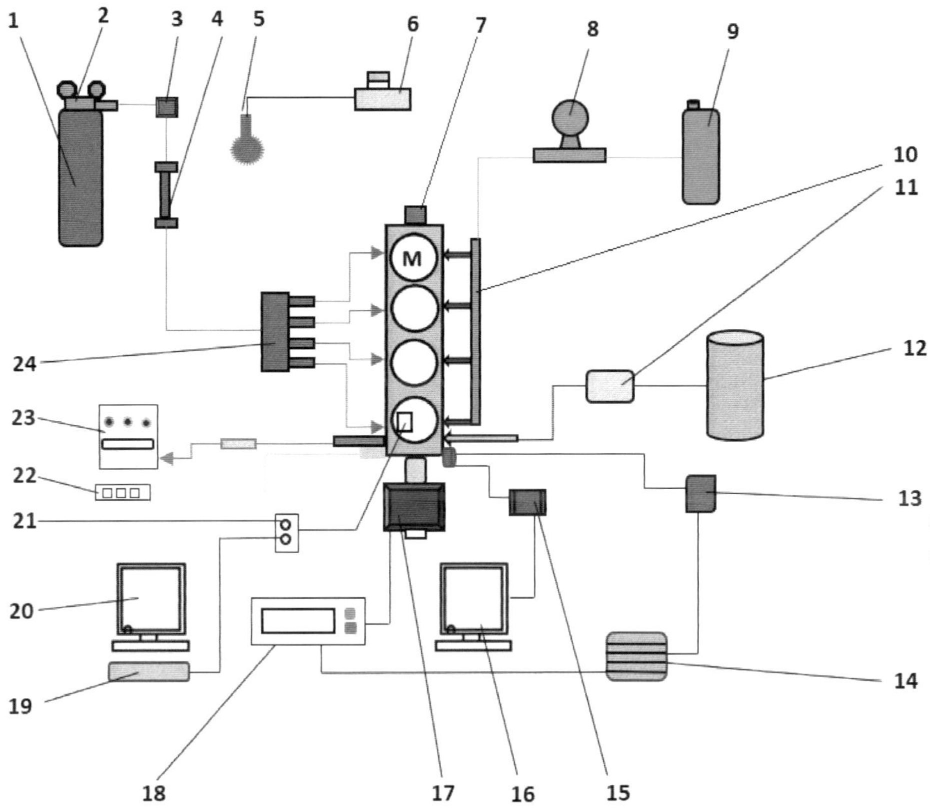


1 – hydrogen storage tank; 2 – pressure regulator; 3 – hydrogen flow meter; 4 – flame trap; 5 – hydrogen sensor; 6 – acoustic and light warning device; 7 – crankshaft position sensor; 8 – fuel mass flow meter; 9 – fuel tank; 10 – fuel injectors; 11- airflow meter; 12 – air reservoir; 13 – throttle body actuator; 14 – power unit for throttle body actuator; 15 – hydrogen injectors ECU; 16 – PC for hydrogen ECU; 17 – eddy current dyno; 18 – brake control panel; 19, 20 – PC for data acquisition system; 21 – load amplifier; 22 – display for cooling system, oil and air temperatures; 23 – gas analyser; 24 – hydrogen injectors.

**Figure 1.** The test bed diagram.

*2.1. The working procedure.*
To carry out the experiments at the beginning was determined the reference, supplying the engine only with gasoline, afterwards the gasoline was 5% partially substituted with hydrogen, using the indirect injection method, aiming to conserve the rated power and torque.

*2.2. Experimental results.*
In the figure below is presented the max pressure which increased for the case of fuelling with $H_2$ because of a higher flame speed in the homogeneous mixture of hydrogen and air. The maximum rate of pressure rise also was higher, and is presented relative to the reference case, figure 3.

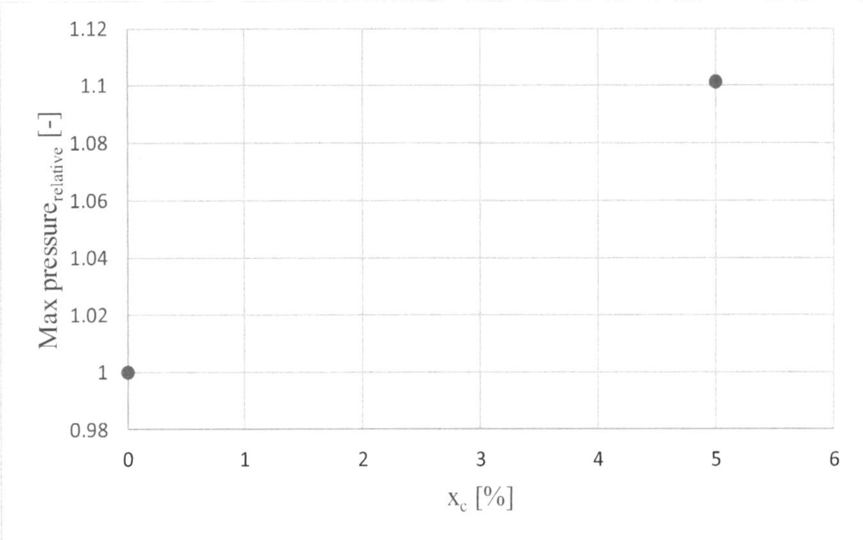


**Figure 2.** Max pressure presented relative to reference case.

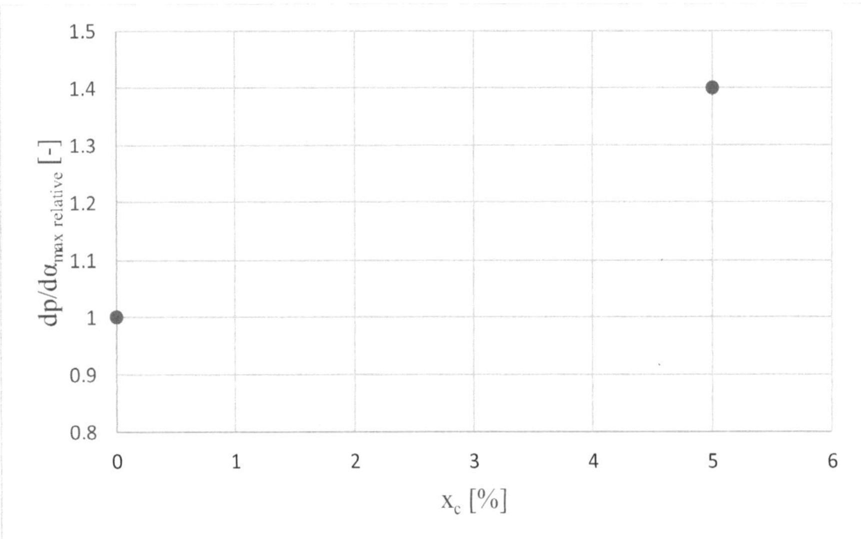


**Figure 3.** Max rate of pressure rise, presented relative to the reference case.

The BSEC decreased when the engine was supplied with $H_2$, due to higher lower heating value of $H_2$. In the figure 4 is shown the BSEC.

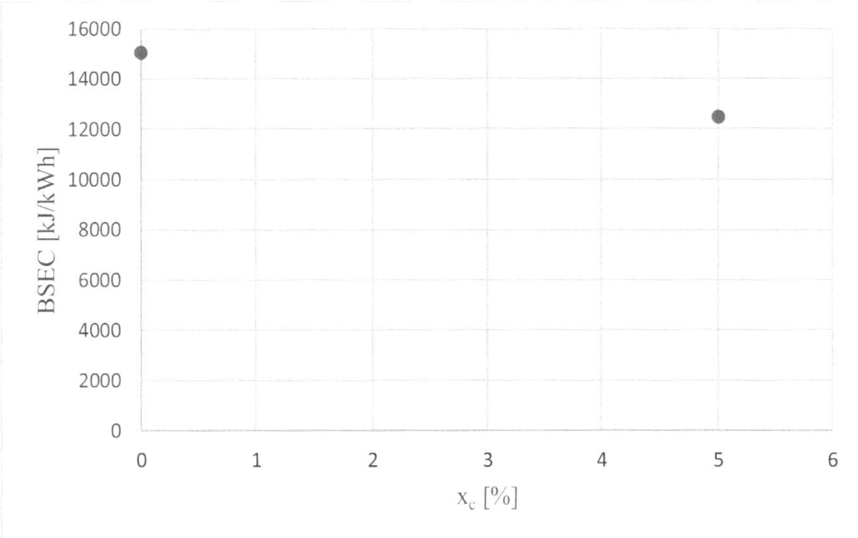


**Figure 4.** The BSEC.

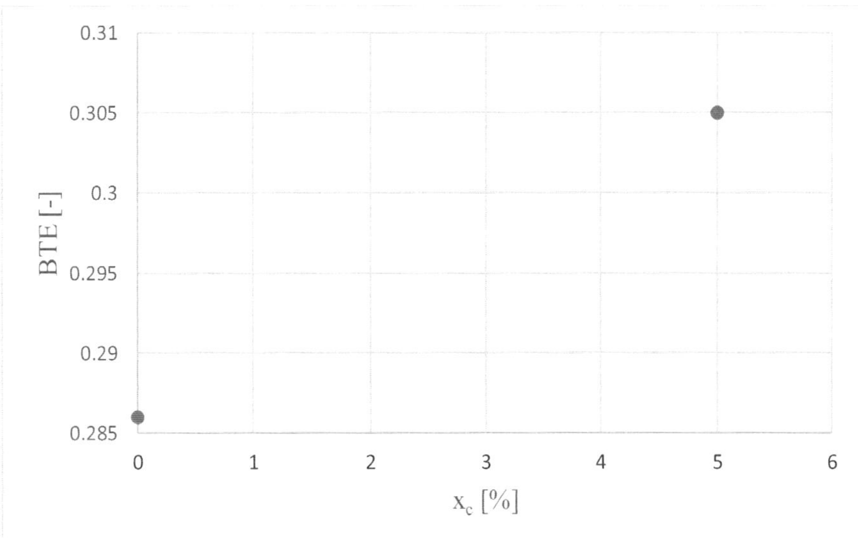


**Figure 5.** The BTE.

The BTE increased in the case of fuelling with $H_2$, the figure 5 is showing the BTE for the presented cases, relative to standard engine case.

The NO emission level decreases in the case of supplying the engine with $H_2$, figure 6, similar results are presented for the $CO_2$ and unburnt hydrocarbons emissions level, figures 7 and 8.

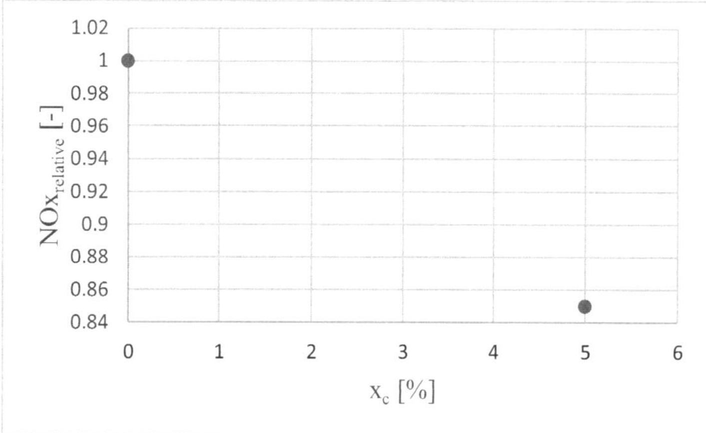


**Figure 6.** The NO emission relative to the reference case.

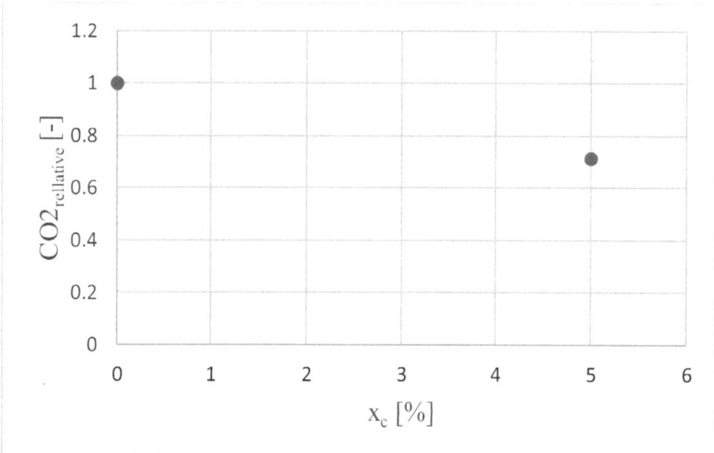


**Figure 7.** The $CO_2$ emission presented relative to the reference case.

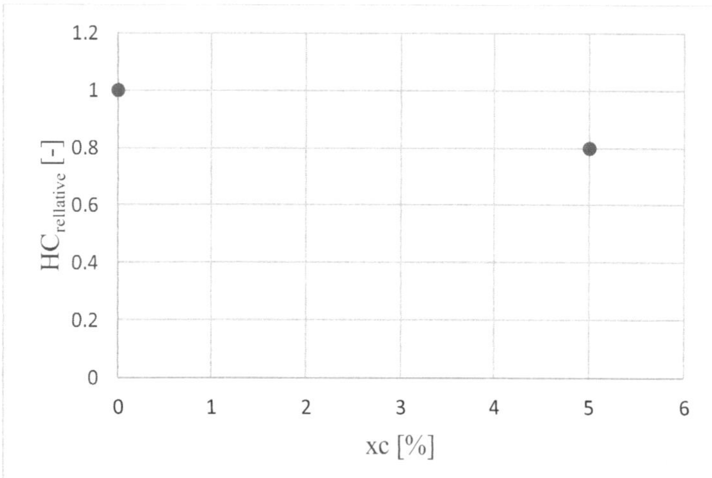


**Figure 8.** The unburnt hydrocarbons relative to the reference case.

## 3. Conclusions

1. The in cylinder pressure and its maximum rate of rise increased when the engine was fuelled with hydrogen with 9% and 29%.
2. The BSEC decreased with approximately 20% for $H_2$ fuelling.
3. The BTE increased with 6% in the case of hydrogen fuelling.
4. The NO emission level was lower with 18% when the engine was fuelled with $H_2$.
5. The $CO_2$ emission level and the UHC emission level decreased with 40% and 25% respectively.

**Acknowledgment**
This work was supported by a grant of the Ministry of Research, Innovation and Digitization, CCCDI - UEFISCDI, project number PN-III-P2-2.1-PED-2021-0427, within PNCDI III.

**Author's ORCID ID**
C Nutu: https://orcid.org/0000-0001-9085-60530758
C Pana: https://orcid.org/0000-0002-4629-6551
N Negurescu: https://orcid.org/0000-0003-3107-6836
A Cernat: https://orcid.org/0000-0003-0551-0758

## References

[1] Popa M G, Negurescu N and Pana C 2013 *Motoare cu aprindere prin scanteie. Procese,* MatrixRom, Bucuresti.
[2] Wróbel K, Wróbel J, Tokarz W, Lach, J, Podsadni K and Czerwi´nski A 2022 Hydrogen Combustion Engine Vehicles: A Review, *Energies*, **15**, 8937.
[3] Ogden JM 1999 Prospects for Building A Hydrogen Energy Infrastructure. *Annu. Rev. Energy Environ*, **24**, 227–279.
[4] Surygała J 2008 *Wodór Jako Paliwo Wydawnictwa Naukowo-Techniczne* PWN: Warszawa, Poland.
[5] Barreto L, Makihira A and Riahi K 2003 The hydrogen economy in the 21st century: A sustainable development scenario. *Int. J. Hydrogen Energy*, **28**, 267–284.
[6] Ji C, Wang S and Zhangn B 2010 Combustion and emissions characteristics of a hybrid Hydrogen gasoline engine under various loads and lean conditions *Inter. J. Hydr. Ener.* **35**.
[7] Ji C and Wang S 2009 Effect of hydrogen addition on combustion and emissions performance of a spark ignition gasoline engine at lean conditions *Inter. J. Hydr. Ener.* **34**.
[8] Negurescu N, Pana V, Popa M G and Cernat A 2011 P*erformance comparison between hydrogen and gasoline fuelled Spark Ignition engine*, paper ThSci2011.036, journal Thermal Science, Vol. 15, No. 4, 12.
[9] Negurescu N, Pana N, Popa M G and Cernat A 2011 *Experimental and Theoretical Researches of Hydrogen Use in Spark Ignition Engine*, Journal of Environmental Protection and Ecology, Vol **12**, No4A.
[10] Aghahasani M, Gharehghani A,Mahmoudzadeh Andwari A, Mikulski M, Pesyridis A, Megaritis T and Könnö J 2022 *Numerical Study on Hydrogen–Gasoline Dual-Fuel Spark Ignition Engine*, *Processes*, ***10***, 2249.
[11] Niu R, Yu X, Du Y, Xie H, Wu H and Sun Y 2016 *Effect of hydrogen proportion on lean burn performance of a dual fuel SI engine using hydrogen direct-injection. Fuel*, **186**, 792–799.
[12] Wang S and Ji C 2012 Cyclic variation in a hydrogen-enriched spark-ignition gasoline engine under various operating conditions*Inter. J. Hydr. Ener.* **37** 1112-19.
[13] Kiyoshi I, Takashi S, Yasuhiro U, Kazuo Y and Toshihisa O 1997 *Investigation of cyclic variation of IMEP under lean burn operation in spark-ignition engine* SAE Paper 972-830.

IOP Conf. Series: Materials Science and Engineering 1303 (2024) 012020 doi:10.1088/1757-899X/1303/1/012020

# The effects of LPG use over the performance of a Diesel engine

**C V Molnar, S Izvorean, A T Borborean and V Stoica**

Politehnica University Timisoara, Faculty of Mechanical Engineering, 1 Mihai Viteazu Blvd., Timisoara Romania

E-mail: andrei.borborean@upt.ro

**Abstract.** This study is based on the effects of using an auxiliary fuel such as gas on the performance of diesel engines. More attempts to use gas as a secondary fuel in a diesel engine were made over the years, resulting in significant increases of the engine performance. Other effects of dual fuel use, are lower fuel consumption, and an improved combustion, that has a direct effect on pollutant emission. For this investigation a 4-cylinder diesel engine with mechanical injection, equipped with a speed regulator disable system. An auxiliary fuel (gas) is introduced through the intake manifold of the engine using a volume base measurement system. According to the literature, by increasing the main injection advance, we can see further improvement in the engine performance, this being a result of better timing of the burn process. The regular burn process uses only a part of the oxygen in the cylinder, by introducing the gas in the intake, it uses the rest of the oxygen in the cylinder in the burn process. This in theory will improve the efficiency of the engine and thus reduces the emissions.

## 1. Introduction

The use of gas as an auxiliary fuel for diesel internal combustion engines has become an increasingly widespread topic nowadays, from the point of view of environmental protection.

There are various methods to introduce the gas as a dual fuel into the engine. Many measurements were made using the gas injection on the intake manifold and they are described in the literature. In order to make an installation as simple as possible from a constructive point of view, we opted for the gas introduction on the intake, without using injectors. This solution offers a lower precision in adjusting a gas dose, but sufficient to obtain some relevant experimental results.

Diesel engines, often known for their reliability and the high powers they can develop, are also some of the most polluting sources for the environment due to the amount of polluting emissions from the composition of the exhaust gas. Specialized literature tells us that by introducing a quantity of gas into the cylinder of diesel engines, a proportional quantity of used fuel should be replaced [1]. This phenomen must be studied in detail, because we also know from specialized literature that gas has a delay in ignition, so an analysis of this phenomenon could show us if we obtain a certain efficiency by using gas as a dual fuel in a diesel engine in what concerns power, fuel consumption or reducing polluting emissions.

It is also considered from the literature that the use of gas for diesel engines would have a beneficial effect in changing polluting emissions by reducing the amounts of $CO_2$ that would be considered the main factor of global warming, but also an increase in the level of hydrocarbons [2].

Diesel engines are much more efficient than spark ignition engines, but a major problem with them is related to high CO2 emissions. These emissions can be reduced by using a fuel with a low carbon content such as NG, LPG or biogas. For compression ignition engines, the main fuel cannot be completely replaced, which means that there is a minimum dose of diesel fuel required for its good operation [3].

According to the measurements performed on our test bench, following the analysis performed on the measured data, is observed an increase in amount of HC and a decrease in the amount of $CO_2$ and NOX which is inversely proportional to the amount of gas introduced in the combustion process. Research studies [4] and [5] suggest that by increasing the dose of gas introduced into the intake manifold, the dose of air is reduced. This has an influence on the HC pollutant emissions from the exhaust gas which decreases in direct proportion with Lambda value.

According to the literature, by introducing the gas on the intake manifold, it replaces part of the air which entering in combustion chamber. Inside the cylinder, the gas is mixed with the diesel fuel, resulting the pre-mix phase, which leads to an ignition of the fuel before the top dead center, implicitly to an increase in engine power [6]. Following the measurements on the test bench, we observe that by keeping the same engine speed and changing the dose of gas introduced into the combustion chamber, the specific consumption of diesel fuel decreases and result a significant increase in engine power. Into the references [7] and [8] are presented the effects of dual fuel supply on the duration of the internal combustion process. With increase in dose of gas introduced into the engine, ignition delay also increases, resulting an almost complete combustion of the fuel mixture

## 2. Experimental test bench

The experimental test bench is composed of a D110 diesel engine connected to a dynamometer which measures the force and torque to the engine crankshaf. The D110 engine is a compression ignition engine, equipped with direct injection, which develops a maximum power of 48 kW (65 Cp) and a maximum torque of 289.5 Nm at 1800 rpm. This experimental test bench is equipped with Arduino-controlled electronic equipment for measuring engine parameters and a dual fuel supply system on the intake manifold.

In the following figure are represented the component elements of the dual fuel supply installation. From the fuel tank (1) the LPG or bio-gas is released by a pressure regulator (2), then it is transported through gas pipes in a gas meter (3) from which through a uni-directional valve (4) it is introduced in the intake manifold (5). In the intake manifold is made the mixture between LPG or biogas and air, which is then introduced into the engine (6).

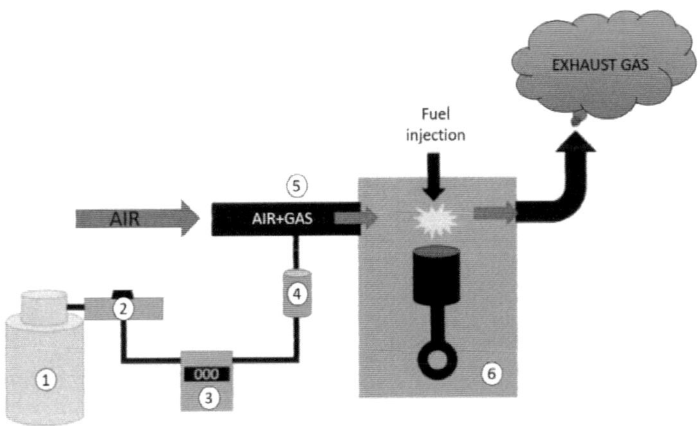


**Figure 1**. LPG supply installation.

The hydraulic dynamometer is a brake that takes the mechanical energy produced during engine operation. During operation, quantities of water are introduced inside the hydraulic dynamometer depending on the required load. Transformation of mechanical energy into heat takes place at the friction contact between the water and the brake rotor. The principle of force measurement by hydraulic dynamometer is based on the measurement of the opposition force to the rotation of the dynamometer housing. This being done on the test bench by using a scale built with load cells connected to a HX711 conversion module. The measured weight is processed by an Arduino Nano microcontroller.

The electronic equipment used to measure the motor parameters is Arduino based. The monitored parameters are as follow: engine speed, coolant temperatures, intake and exhaust manifold temperatures, the load measured by the hydraulic dynamometer, but also fuel consumption reported per hour of operation. All these functional parameters are transferred to an interface made in LabView. Using calculation formulas we obtain data such as torque, power, and specific consumption of liquid fuel.

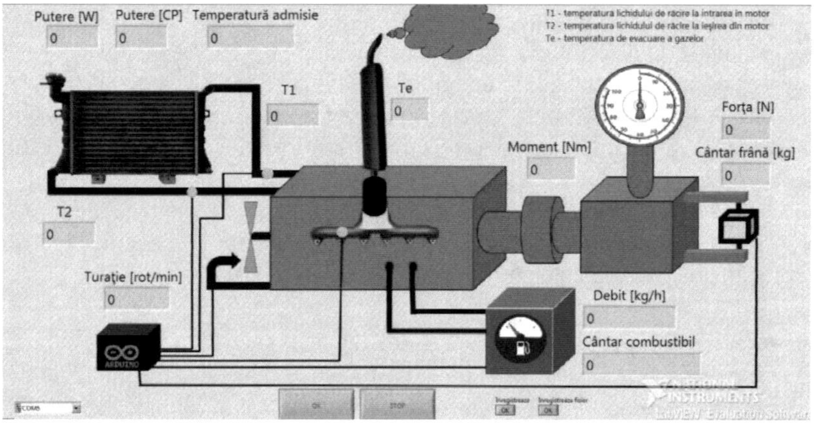


**Figure 2.** General overview of the test bench and the monitored parameters.

## 3. Measurement procedure

To perform the measurements, we opted for the adjustment of a constant engine speed regime, of approximately 1402 rpm, with a deviation of 1.03%.

Also, to keep the same constant engine speed, we used a locking device for the injection pump regulator. The operating principle is based on locking the injection pump rack to obtain a constant engine speed and a variable load. The fuel used for dual fueling is LPG and consists of a mixture of mostly propane and butane. In this sense, according to the literature, gas mixtures with propane content give the best contribution to the performance of dual fueled diesel engines. [9]

During the measurements, 3 different operating cases were studied. In first case we analyzed the operation of the engine only with diesel fuel, and in the following two cases we studied the operation of the engine by reducing the dose of diesel fuel and increasing the dose of gas, but also keeping the dose of diesel at the value of approximately 1.05 kg/h with a deviation of 8.4% and the increase in the gas dose.

The pollutant emissions measurements were made with Capelec CAP 3201. For a better understanding of the values on the following graphs, it is necessary to know the emission values for diesel operation only.

**Table 1.** Emission Values for Diesel Operation Only.

| Diesel Only Emissions | |
|---|---|
| CO [% vol.] | 0.02 |
| CO2 [% vol.] | 3.8 |
| HC [ppm vol.] | 76 |
| O2 [% vol.] | 14 |
| Lambda [-] | 3.53 |
| NOx [ppm vol.] | 479 |
| K [1/m] | 0.13 |

On the side of polluting emissions, after the tests, we can see an improvement due to the use of gas as an auxiliary fuel. The gas introduced into the intake manifold replaces part of the air but also of the diesel fuel introduced into the cylinder during compression. This fact has an observable contribution, modifying the combustion process. Following the analysis of the measured results, we observed a significant decrease in NOX and CO2 from the exhaust gas, with the increase the dose of gas. In addition to the positive values obtained on the emissions side, we also observe an increase in HC which is not beneficial. Although it is assumed that the gas mixed with diesel should provide a better combustion, the increase in HC emissions, are the result of the incomplete combustion of the gas [10].

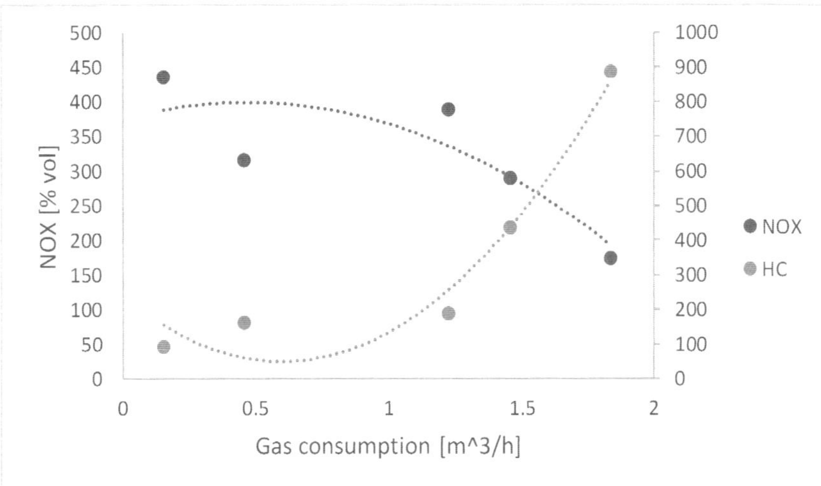


**Figure 3**. Variation of NOx and HC as a function of Gas consumption.

Decreasing the diesel dose requires an increase of the gas dose in order to keep the power and the speed constant.

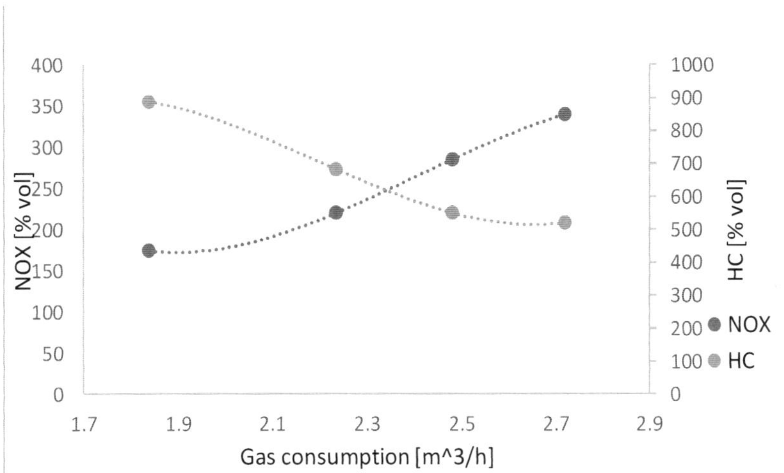


**Figure 4.** Variation of NOx and HC in function of gas consumption when we keep a constant Diesel dose and increase gas dose.

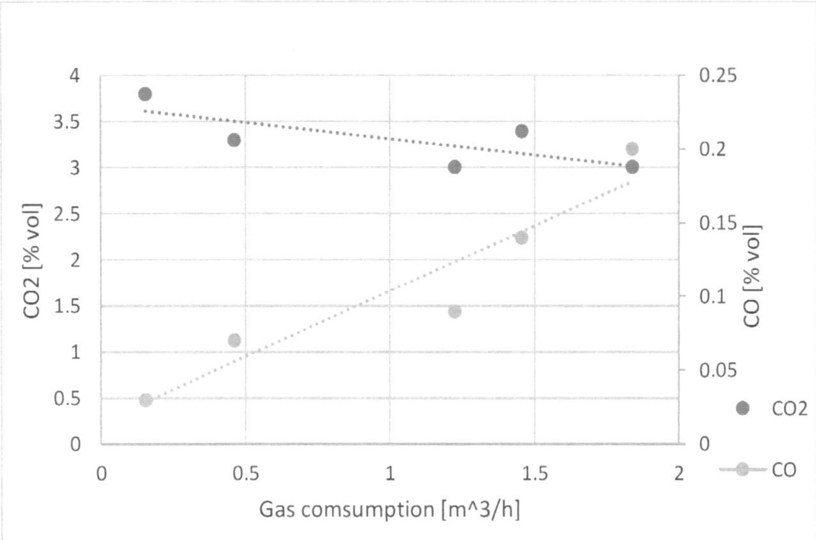


**Figure 5.** Variation of $CO_2$ and CO in function of Gas consumption when we decrease Diesel dose and increase Gas dose

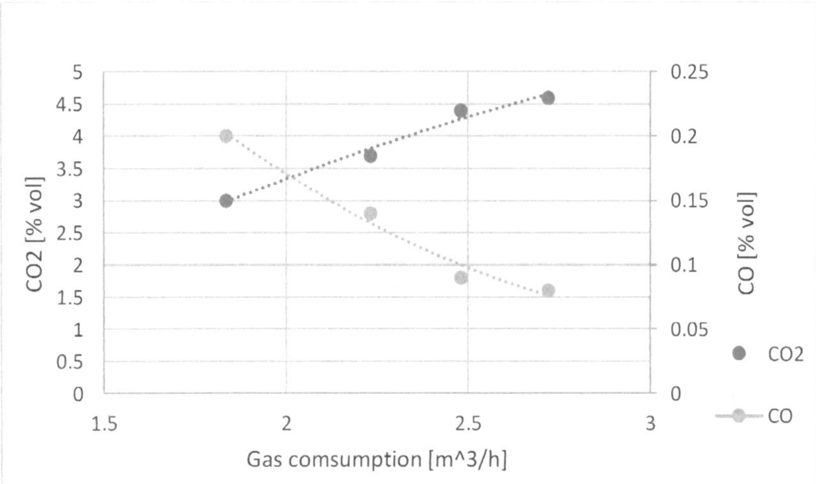


**Figure 6.** Variation of $CO_2$ and CO in function of Gas consumption when we keep a constant Diesel dose and increase Gas dose.

On the engine efficiency side, no significant improvement can be observed. The engine efficiency is around 26% during operation with only diesel, and by adding gas it has a deviation of 10% from its initial value. The addition of gas does not have a positive contribution to the engine's performance, because with large doses of gas introduced, a slight decrease of the engine efficiency is observed due to irregular operation.

Regarding to the price of the fuel required for the operation of the engine under the conditions listed above, the mixture of gas and diesel brings a significant benefit because with small doses of gas introduced, a significant amount of diesel is substituted, and the total price of the two fuels is lower than the one required to operate only with diesel.

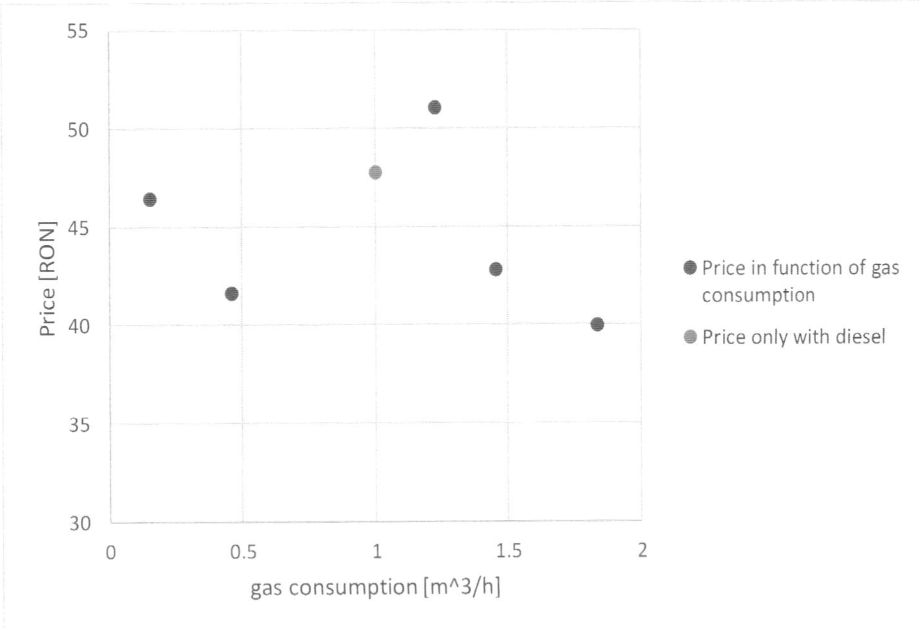


**Figure 7.** Differences between Diesel and LPG+Diesel price for 1h engine operation.

## 4. Conclusion

We consider that the use of gas as a dual fuel for diesel engines has a quite noticeable impact in the case of polluting emissions from the exhaust gas. The decrease in emissions specific to diesel engines can be seen through the decrease in NOX, but on the other hand, an increase in HC due to unburned gas is also obtained. Also, the reduction of CO2 is important because it is considered the main factor of global warming.

On the other hand, we do not see an improvement in terms of engine efficiency. At low rates of diesel substitution, the engine performance remains almost identical to the case of operating only with diesel, but when substituting a larger amount of liquid fuel, a slight decrease in efficiency is observed. For this reason, we believe that there should be a minimum dose of diesel for perfect operation in order to correlate engine performance with the improvement of polluting emissions.

The analysis that was made on the financial side shows us that it is profitable to use gas as a dual fuel for diesel engines, but the calculations were made only on the basis of current LPG prices. If another gas with the same thermal properties were used, such as biogas, which has a lower price, it would result in a significant price decrease compared to the price of diesel.

## Author's ORCID ID

A Borborean https://orcid.org/0000-0001-6818-9549
V Stoica https://orcid.org/0000-0002-7945-7817

## References

[1] Sharma A, Ansari N A, Pal A, Singh Y and Lalhriatpuia S 2019 Effect of biogas on the performance and emissions of diesel engine fuelled with biodiesel-ethanol blends through response surface methodology approach, *Renewable Energy,* vol. 141

[2] Esenay A and Nafiz K 2022 The effects of hydrogen enriched natural gas under different engine loads in a diesel engine, *International Journal of Hydrogen Energy,* vol. 47

[3] Jiří V, Bortel I, Takáts M and Diviš M 2017 Emissions and performance of diesel–natural gas dual-fuel engine operated with stoichiometric mixture, *Fuel,* vol. 208

[4] Yoon S H and Lee C S 2011 Experimental investigation on the combustion and exhaust emission characteristics of biogas-biodisel dual-fuel combustion in a CI engine, *Fuel Processing Technology,* vol. 92

[5] Sunyoup L, Changgi K, Seokhwan L, Jeongwoo L and Junghwan K 2022 Experimental investigation on combustion and particulate emissions of the high compressed natural gas reactivity controlled compression ignition over wide ranges of intake conditions in a multi-cylinder engine using a two-stage intake boost system, *Fuel Processing Technology,* vol. 228

[6] Nayak S K, Hoang A T, Niˇzeti´c S, Nguyen X P and Le T H 2022 Effects of advanced injection timing and inducted gaseous fuel on performance, combustion and emission characteristics of a diesel engine operated in dual-fuel mode," *Fuel,* vol. 310

[7] Di Blasio G, Belgiorno G and Beatrice C 2017 Effects on performances, emissions and particle size distributions of a dual fuel (methane-diesel) light-duty engine varying the compression ratio, *Applied Energy,* vol. 204

[8] Sahoo B B, Sahoo N and Saha U K 2009 Effect of engine parameters and type of gaseous fuel on the performance of dual-fuel gas diesel engines—A critical review, *Renewable and Sustainable Energy Reviews,* vol. 13

[9] Ulishney C J and Dumitrescu C E 2022 Effect of gas composition on the performance and emissions of a dual-fuel diesel-natural gas engine at low load conditions, *Fuel,* vol. 324

[10] Imamoto T, Kawahara N and Tomita E 2022 PREMIER combustion characteristics of a pilot fuel-ignited dual-fuel biogas engine with consideration of cycle-to-cycle variations," *Fuel,* vol. 314

# Analysis of metal content in the lubricating oils used in truck diesel engines

**R Niculescu[1,2], M Nastase[2], D Negrea[2], C Zaharia[1] and A Clenci[1,2]**

[1]University of Science and Technology "POLITEHNICA" Bucharest, University Centre Pitesti, Department Automobiles and Transport, 1, Tg. din Vale Street, Piteşti, 110040, Argeş, Romania
[2]University of Science and Technology "POLITEHNICA" Bucharest, University Centre Pitesti, Regional Research & Development Center for Innovative Materials, Products and Processes for the Automobile Industry (CRC&D-Auto), 11, Doaga Street, Piteşti, 110440, Argeş, Romania

E-mail: rodica.niculescu@upit.ro

**Abstract.** The lubricating oil of an internal combustion engine (I.C.E.) performs several functions with a major contribution to the reliability of the system, which involves proper operation during the operating period. Changing the engine oil early or late could generate additional costs and/or damage the engine's performance. Thus, it becomes necessary to periodically analyse the oil used in the engine. In this work, the authors propose to analyse the metal content of the used engine oil and, based on the results obtained, to issue maintenance strategies. Samples taken from a number of 43 DAF trucks equipped with the same type of engine, MX - 11 330 H2, EURO 6, which use the same assortment of 10W30 oil, were analysed.

## 1. Introduction

Lubricating oils are decisive in reducing the process of friction between an engine moving element and serve to diminish their wear in the period of technical systems operation, which surely conduct to increase total efficiency, along with energy and fuel economy. Engine oils are the most universal of all lubricants and are commonly used for gasoline and diesel engine vehicles. It is well known that engine oils are composed of the base oil and a maximum of 20% additives [1]. The base oils can be mineral oils, obtained in the petroleum distillation process, or synthetic oils.

In the period of engine operation, different situations/changes can occur, which facilitate the appearance of contamination, gradually leading to the diminishing of oil quality and its properties. In conclusion of that, the specialists in the domain recommend technical overhaul and the lubricating agent need to be eliminated from the engine and replaced with a fresh one.

The oils used in powered vehicles can intermediate the appearance of a big environmental problem, especially when is about logistics of the waste disposal. It is interesting that this process is based on the reality that the engine oil is found in small quantities and utilized by a huge number of car owners which are in distinctives sites. Thus, according to [1], in Europe, the quantity of engine oils is considered large being between 1.7 and 3.5 million tons and varies every year. For that, it can be said that these quantities of used engine oils imply important environmental and economic consequences.

The chemical composition of used oils is based on important factors, such as the base oil and additives used, the conditions to which the oil was subjected during the engine's operation. Thus, the analysis of the variation of the physico-chemical properties of the engine oil during its use is a subject of interest for the reliability of vehicles. Used oils may contain, among other elements, hydrocarbons, water, glycol and heavy metals [2–4].

High concentration of heavy metals can be present in engine oils and this fact implies many discussions and therefore interests. The biggest issue is highlighted when it is about the adverse effects on the human health and environment. There are different sources of metals in used oils, first it is well known that they can be native parts of crude oil and from the additives [5]. Secondly, metals can appear from the action of wear and corrosion of the metal surfaces in contact during operating period and also from the deterioration process of used oil, like oxidation process, additives transformation in solid residues, etc.

It's well known the possibility that the engine oil to have microscopic metal shavings on occasion and it is a truth that their appearances should be impossible to see with the naked eye. If the metal shavings in the oil have dimensions that make them easily visible, they become very dangerous for the engine.

During the cold start of the engine, it is known that the lubrication is deficient and particles resulting from the dry friction of the engine parts may appear in the oil. However, the accumulation of metal shavings that happens unexpectedly can reveal a sign of wear inside the engine [3].

So, without appropriate lubrication, the engine metal elements begin to rub producing wear, allowing shavings to consume from the parts. The next step is that these metal shavings are left in the oil and this process can lead to significant issues as they are circulated through the engine [6]. The metals which can be found in engine oil are iron, copper, bronze or brass, zinc, nickel, molybdenum or chromium, manganese, aluminium, etc. [7,8].

This paper aims to complete with one step the research gap in the domain by analysing the concentrations of selected heavy metals (Cu, Fe, Zn) in used oils from the same diesel engine which equips trucks of different mileage and different year of manufacture (2016 – 2021). Moreover, another objective of this paper is to make preliminary diagnose of internal combustion engines to emit preventive maintenance strategies in the period of operation.

The conclusions of the study can also help to make an important and effective decision regarding the timing of the technical overhaul, but also regarding the waste oil management strategies [9].

## 2. Methodology and experimental setup

The determination of the metals content in the oil used in diesel truck engines is carried out by X-ray fluorescence (XRF) spectrometry. This is a non-destructive analytical technique used for the quantitative and qualitative analysis of the composition of materials. Briefly, this analysis consists of irradiating a sample (solid or liquid) with high-energy X-rays. When an atom is hit with such an X-ray, an electron on an inner layer (K or L) is displaced. The stability of the atom will be restored by filling the vacant place with an electron coming from a higher energy layer. Following this relocation of the electron, its energy decreases by releasing a fluorescent X-ray (fluorescent light). The energy of this ray is, obviously, equal to the energy difference between the two states of the electron. The basis of XRF analysis is, in fact, the measurement of this energy [10].

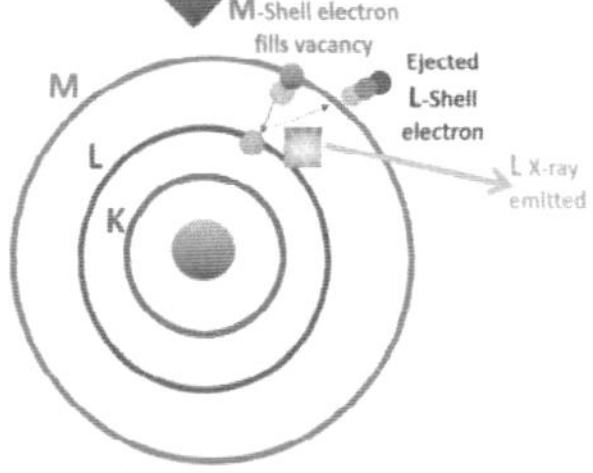


**Figure 1.** X-ray fluorescence (XRF) [10].

The equipment used in this research is Spectro Midex M which is shown in figure 1, a multifunctional X-ray micro fluorescence (XRMF) analysis system. Samples in solid or liquid form are very small and are not destroyed by analysis, as stated previously. The detector of the device measures the X-photons emitted from the sample under analysis, these being counted with a impulses counter. The number of impulses is entered into the data processing system of the device under the name intensity I. The value of the intensity thus measured is associated with the concentration of the respective element in the sample [11] .

The spectra resulting from the analysis of the sample show a graphic representation of the intensity peaks. Each peak indicates an element, and the height of the peak indicates the concentration of the element in the analyzed sample [10].

**Figure 2.** Spectro Midex M, a multifunctional X-ray micro fluorescence (XRMF) analysis system
Source: [12].

Taking into account the above, 43 samples of used engine oil collected under similar conditions from DAF trucks that have the same type of engine with characteristics presented in table 1 were analyzed. All the trucks were manufactured between years 2016 and 2021.

**Table 1.** DAF truck engine characteristics.

| | |
|---|---|
| Engine type | Diesel, EGR, DPF, SCR/Paccar MX-11330 H2 |
| Cylinder capacity | 10800 cm$^3$ |
| Engine power@speed | 449 HP@1600 rpm |
| Emissions standards | EURO 6 |
| Mileage | 100000 - 800000 km |
| Type of the engine oil | DAF Xtreme FE 10W30 (approved by the manufacturer) |

Among the component elements detected in the analyzed samples, in this work we focused on Fe, Cu and Zn. Iron is part of the alloys from which the cylinders, the crankshaft, are made and, through their wear, iron particles can be found in the engine oil. Iron can also come from the corrosion of parts in contact with water. Water can get into the engine through various ways, such as: the fuel, the environment, from the cooling system through leaks. Copper particles can get into the engine oil through the wear of bearings and heat exchangers. Also, copper is used as a lubrication additive in systems subject to severe demands. Zinc is found in EP (extreme pressure) type additives, used in lubricants intended for systems that work under extreme pressure conditions. It is also used in bearing alloys and is a common material for metal coatings [13].

A qualitative analysis was made of the resulting spectra for the 43 samples comparing with the spectrum of the unused oil. Considering the relationship between the number of impulses seen in the spectrum and the concentration of the component in the analyzed sample, as presented above, we calculated the difference between the number of impulses for each component studied in the sample (used oil) and the unused oil, noting with DIF_Fe, DIF_Cu respectively DIF_Zn. Afterwards, the average value was determined for the samples coming from trucks with similar mileage.

## 3. Results and discussion

Relying on the methodology presented, the analysis was made for the iron content presented in figure 3. Figure 3a shows DIF_Fe, and figure 3b shows the average DIF_Fe by mileage categories.

Figure 3b shows that the average iron content decreases with the increase in the number of kilometers traveled by the truck, i.e., with the total mileage of the truck. In our opinion, when the truck is new, that is, it has low mileage, the grinding of the parts is intense, the particles of material that break due to the friction of the moving parts are increased. A slight increase is observed for trucks with a mileage of over 700,000 km.

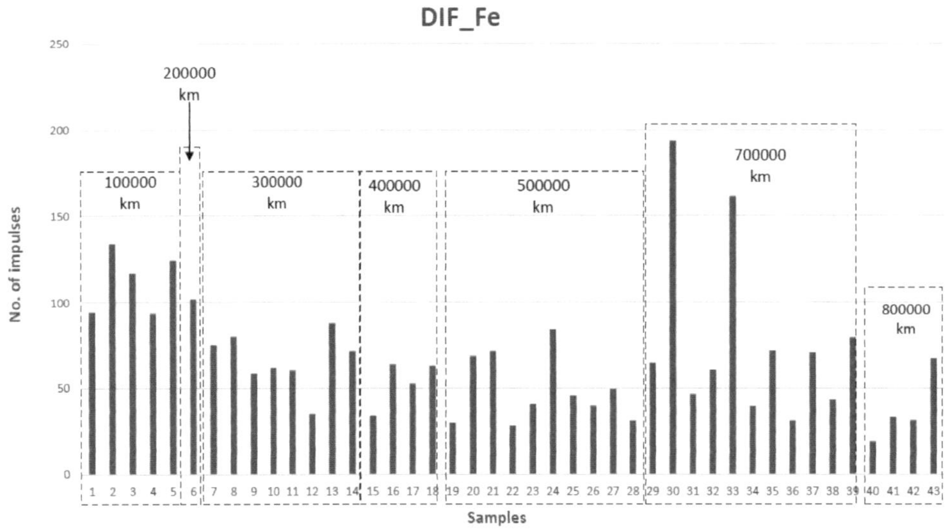


a) DIF_Fe

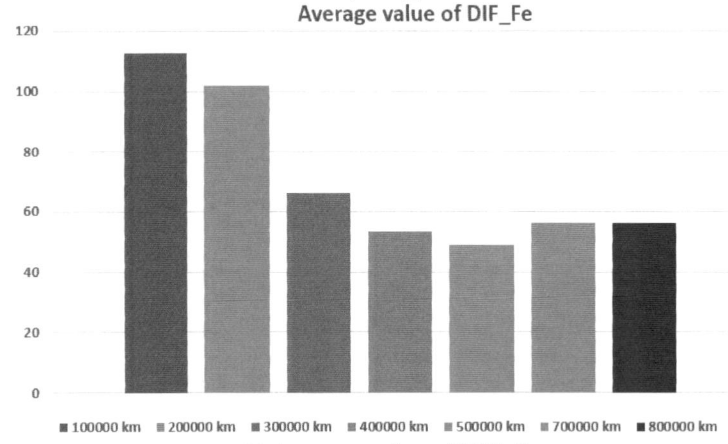


b) Average value of DIF_Fe

**Figure 3.** Iron content.

In the category of trucks with a mileage of 700,000 km, there are two samples (Sno_30 and Sno_33) that exceed the average for this segment of trucks (fig 3a). In order to have a better idea of the trend regarding the iron content in the samples, when calculating the average value of DIF_Fe, the two samples were eliminated.

The high iron content in the two samples leads to an increase in viscosity and sediment level, which is in agreement with the results regarding the analysis of the physico-chemical properties of the same samples of used motor oil presented by the authors in the work published in [4].

Figure 4 shows the spectra of the two samples overlapped on the spectrum of fresh engine oil in which the high Fe content is clearly highlighted.

For the trucks from which the Sno_30 and Sno_33 samples were collected, it is recommended to check the oil level (the amount in the engine). The degree of wear of parts such as: segments, cylinders, bearings must also be checked.

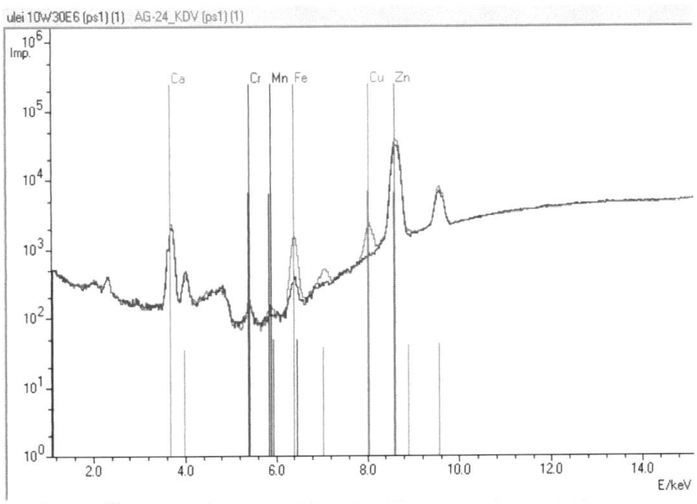


Proba Sno_30

| | |
|---|---|
| DIF_Fe | 193.95 |
| DIF_Cu | 305.089 |
| DIF_Zn | 1411.7 |
| DIF_Mn | 0.59527 |

Viscozitate cinematica la 40°C 94.53 mm$^2$/s [4]

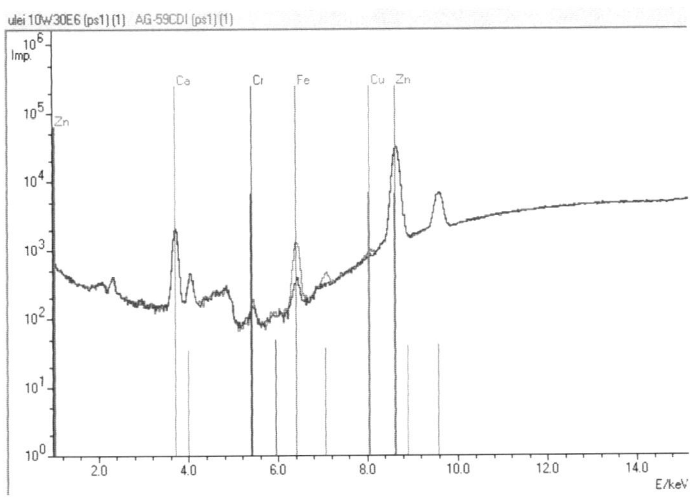


Proba Sno_33

| | |
|---|---|
| DIF_Fe | 161.4281 |
| DIF_Cu | 31.013 |
| DIF_Zn | 217.5 |

Viscozitate cinematica la 40°C 81.33 mm$^2$/s [4]

**Figure 4.** Spectra of the Sno_30 and Sno_33 samples overlapped on the spectrum of fresh engine oil.

Figure 5 shows the copper content. In Fig. 5b, which shows the DIF_Cu average by mileage category, it can be seen that the Cu content, as in the case of iron, decreases with the increase in mileage. The explanation can be that in the first hundred thousand kilometers traveled, the engine is in run-in period. From figure 5a showing DIF_Cu, it can be seen that the Sno_30 sample has an increased copper content. In the calculation of the average value of DIF_Cu, this value was excluded (as in the case of iron). Therefore, the maintenance recommendations previously made for the truck from which this sample was collected become mandatory.

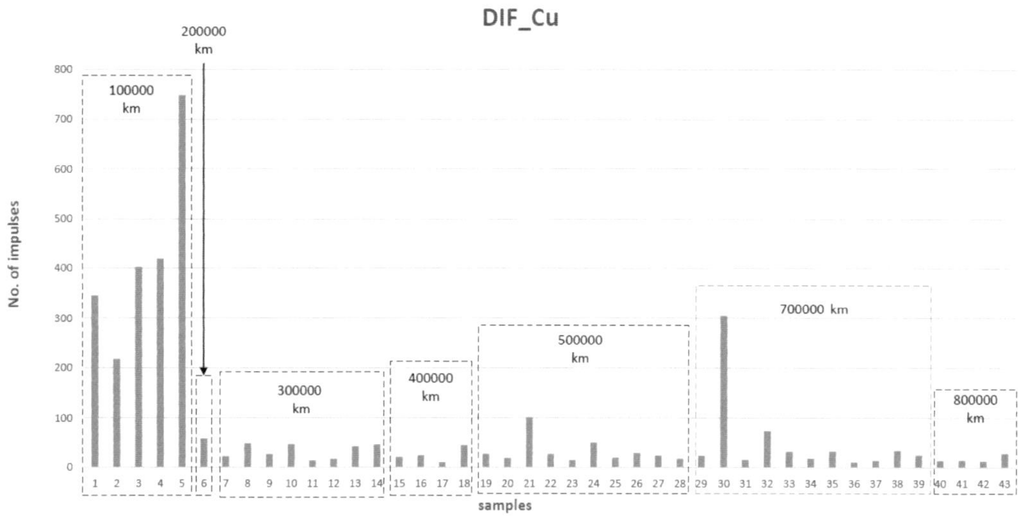


a) DIF_Cu

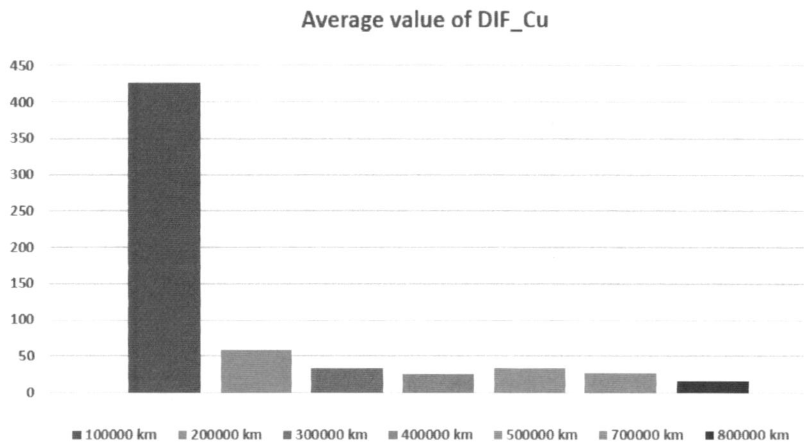


b) Average value of DIF_Cu

**Figure 5.** Copper content.

Figure 6 shows the zinc content. In fig. 6b, which shows the average value of DIF_Zn by mileage category, it can be seen that the content of Zn, as in the case of iron and copper, is decreasing with the increase of mileage. The explanation can be that in the first hundred thousand kilometers traveled, the engine is in run-in period. For trucks with high mileage, over 500,000 km, the zinc content begins to increase slightly. The degree of wear of parts such as: segments, cylinders, bearings make their deterioration process accelerate.

From figure 6a, which shows DIF_Zn, it is also noticeable that the sample Sno_30 shows an increased content of zinc. In the calculation of the average value of DIF_Zn, this value was excluded. The maintenance recommendations made previously for the truck from which this sample was collected are kept.

Figure 6 shows that some samples were excluded, because negative DIF_Zn values were obtained, which makes us believe that there were measurement errors for these cases.

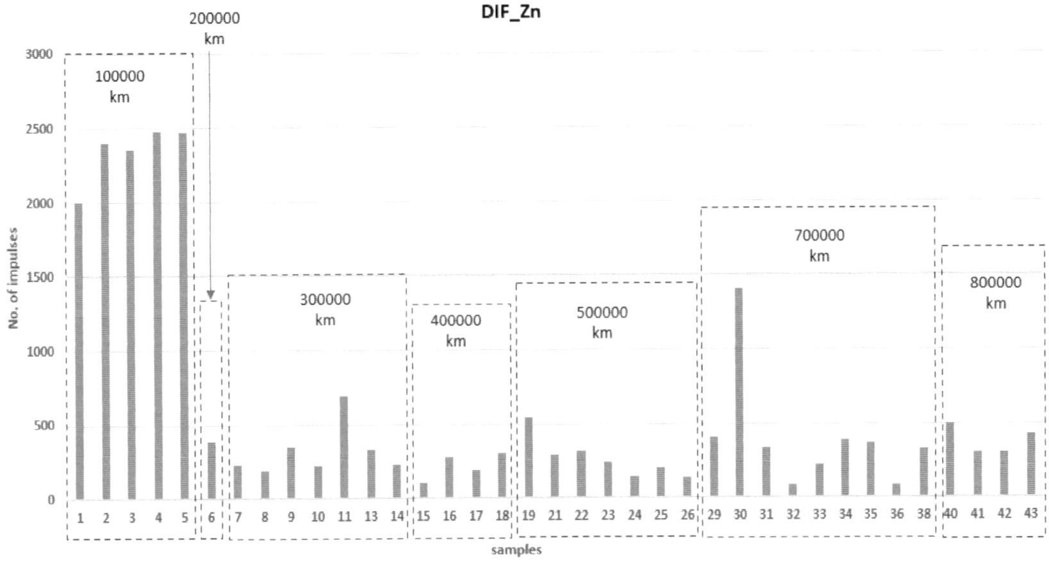


a) DIF_Zn

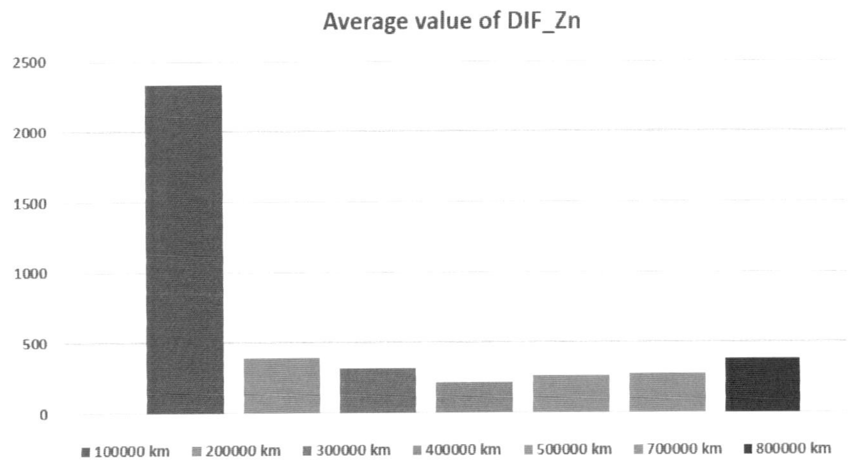


b) Average value of DIF_Zn

**Figure 6.** Zinc content.

Although, the work does not propose to study the manganese contamination of the samples, nevertheless, out of the 43 samples, manganese was found in three samples: Sno_2, Sno_5, Sno_30. The first two are in the 100,000 km category, the third in the 700,000 km category. Manganese can come from the intense abrasion process during break-in (Sno_2, Sno_5) or for sample Sno_30 it can be caused by insufficient lubrication (small amount of oil, considerable oil loss), or from the addition of an engine oil that contains an additive on manganese base. Figure 7 shows the spectra of samples a) Sno_2 and b) Sno_5 overlapped on the spectrum of fresh oil.

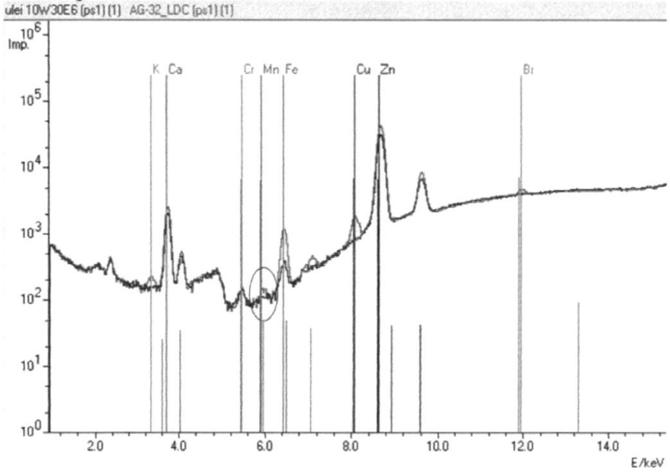


a) Sno_2

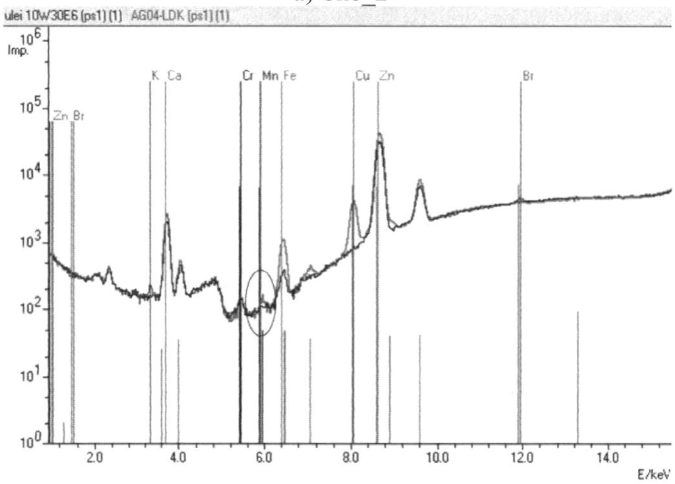


b) Sno_5

**Figure 7.** Spectra of samples a) Sno_2 and b) Sno_5.

## 4. Conclusions

It is well known the close connection between the reliability of the vehicle engine and the lubrication performance, the subject being of major interest both for manufacturers and for users, especially in the case of vehicle fleet owners.

Thus, vehicle manufacturers recommend precise periods of technical overhaul. Only that, in the period between technical overhaul, the conditions of vehicle operation, vary widely and sometimes unexpectedly and can affect engine oil quality in different ways. From here, two directions of discussion and analysis can be distinguished: on the one hand, the oil deteriorates rapidly, and compliance with the overhaul period can lead to early wear of the engine with considerable economic implications. On the other hand, the oil is not damaged, and it can be used even after the overhaul period, with economic consequences, as well, but also in terms of the environment, i.e. reducing the amount of waste. Considering what has been said, an analysis of the oil used in the engine can lead to the right decisions.

Metal contamination of engine oils during operation can occur for various reasons and can produce major negative effects on the engine. Determining the content of iron, copper and zinc can lead to a quick diagnosis of the engine. These metals are part of the alloys from which different engine components are made, but also of the composition of the various lubricant additives.

From the analysis presented in this paper, it can be seen that the oil from engines with a mileage of less than 100,000 km, which are therefore in the break-in period, have a higher metal content than in any other category of mileage. It has been observed that the amount of metals in the oil decreases with the increase in engine speed. The oils that come out of this trend lead to the idea that the engines that worked with these oils have problems, so they should be diagnosed and put in a maintenance process. It must be checked if the amount of oil in the engine is within the prescribed limits, what is the degree of wear of the parts (cylinders, segments, bearings, valves, etc.), if the operator has not topped up with another type of oil that has different basic elements, especially additives.

Consequently, this paper demonstrates that it is possible to diagnose engines by analyzing the physico-chemical properties of the oil used in the engine and issuing effective maintenance strategies. Moreover, the results of this paper regarding contamination with metals are in agreement with the results obtained in the paper [3,4] regarding the physical properties of the same oil samples.

**Author Contributions:** The authors have equal contribution to the work presented in this article

**Author's ORCID ID**
R Niculescu: https://orcid.org/0000-0002-5833-7752

**References**

[1] Szyszlak - Bargłowicz J, Zając G and Wolak A 2021 Heavy metal content in used engine oils depending on engine type and oil change interval *Arch. Environ. Prot.* **47** 81–94.

[2] Hamawand I, Yusaf T and Rafat S 2013 Recycling of waste engine oils using a new washing agent *Energies* **6** 1023 – 49.

[3] Zaharia C, Niculescu R, Năstase M, Clenci A and Iorga - Simăn V 2022 Engine oil analysis to evaluate the degree of its wear during the period of operation of the vehicle, *IOP Conf. Ser. Mater. Sci. Eng.* **1220** 012037

[4] Zaharia C, Niculescu R, Năstase M and Clenci A 2022 Diagnosing the truck engine operation by evaluating engine oil wear and issuing maintenance recommendations EAEC – MVT

[5] E.L.& D.B. Scott A. Stout 2018 Metal concentrations in used engine oils: Relevance to site assessments of soils *Environ. Forensics.* **19** 191–205.

[6] MECHANIC BASE, (n.d.). https://mechanicbase.com/engine-oil/how-much-metal-in-oil-is normal/.

[7] Artur Wolak WG and Grzegorz Zając 2019 Determination of the content of metals in used lubricating oils using AAS *Pet. Sci. Technol.* **37** 93–102.

[8] Saif-ur-Rehman Kashif ZF, Aneesa Zaheer and Fariha Arooj 2018 Comparison of heavy metals in fresh and used engine oil *Pet. Sci. Technol.* **36** 1478 – 81.

[9] Pinheiro C T, Quina M J and Gando - Ferreira L M 2021 Management of waste lubricant oil in Europe: A circular economy approach, *Crit. Rev. Environ. Sci. Technol.* **51** 2015 – 50.

[10] ThermoFischer Scientific https://www.thermofisher.com/blog/ask-a-scientist/what-is-rf-x-ray-fluorescence-and-how-does-it-work/.

[11] Lorentz J and Horea Iustin N 2009 Chimie Analitică şi Instrumentală *Acad. Pres Acad.* 3–249.

[12] Spectro Analytical Instruments - AMETEK, (n.d.). https://www.environmental-xpert.com/products/spectro-midex-m-20095.

[13] Posible origin of metals in lubricating oil http://noria.mx/lublearn/analisis-de-elementos/.

# Surface Defects Vibration Measurements of Automotive Tapered Roller Bearings

**B A Sabo**[1] and **Ș Voloacă**[2]

[1] Timken Romania S.A., Ploiesti, Romania
[2] University POLITEHNICA of Bucharest, Romania

E-mail: bogdan.sabo@timken.com

**Abstract**. The paper investigates vibration testing methodologies for automotive tapered roller bearings. The European Union regulations have indirectly led to improvements in the NVH performance of the bearings that affect both the industrial and automotive markets. The goal of this study was to examine the influence of localized small defects on the vibrations' magnitudes within tapered roller bearings. A Rockwell indenter was used to create three different surface defects. These three levels, representing different indentation pressures, mimic defects that could occur during manufacturing or bearing installation. Indents were placed on inner raceways, the outer raceways, and the roller bodies. After the assembly of the components, different vibration testing methodologies were applied to quantify those types of defects. This paper presents a new envelope analysis approach to monitoring for defects. The paper presents the current Fast Fourier Transform (FFT) testing approach, whereby the machine sensitivity is representative of a typical quality control bearing thresholds. This paper also presents a new, more robust method and analyzed for its suitability in the detection of localized defects.

## 1. Introduction

Rolling element bearings play a critical role in the functionality of machinery. Almost all rotating applications contain bearings, and rolling element bearings are thus found in almost all industries. Like all componentry where friction is present, noise is generated. This noise can be exacerbated by geometric imperfections, such as those that can be generated due to manufacturing issues, part assembly, or in application bearing operation.

A new fault detection methodology lends itself to improved noise and vibrating monitoring of the imperfections. To reduce the number of bearings with imperfections, it's necessary to understand the type of vibration signals produced during the bearing's lifetime. Thus, engineers can explore new approaches that involve optimizing the bearing's geometric design, material selection, lubrication, and manufacturing process to minimize unwanted noise and vibrations.

Nowadays, requirements and standards regarding noise and vibration are more strictly applied in industrial settings. To address those requirements in the manufacturing process, vibration measurement and data interpretation were introduced. Condition monitoring serves as a detector of early stages of bearing damage, enabling forecasted repairs to be scheduled by the maintenance team. This approach helps to reduce downtime and optimize maintenance practices. Also, for quality control, at the end of each assembly line, vibration measurement equipment is installed to detect manufacturing imperfections before the bearing is sent to the customer.

High volume production, in any size range, necessitated the development of new detection methods. The envelope detection method is a qualitative technique used by many industries, including the aerospace [1]. The technique of envelope analysis is detailed by S. Kim [2], with the aim of analyzing roller bearing elements in mechanical transmissions.

Konstantin-Hansen and Herlufsen [3] describe envelope analysis as the Fast Fourier Transform of a modulated signal. This methodology can be used for investigation of machinery faults possessing a modulating effect with frequency characteristics. It can be used on gearboxes, motors, turbines, etc., and is also a great tool to find local defects such as cracks and spalls on bearing functional surfaces.

Relative motion of the rollers over a defect generates impacts that excite bearing periodic frequencies. Each bearing component, when in relative motion, generates a different periodic defect frequency. Those frequencies are calculated by using geometric characteristics and the rotating speed in the application. The ball pass frequency outer race (BPFO), ball pass frequency inner race (BPFI), ball spin frequency (BSF), and fundamental train frequency (FTF) can be calculated using modules available online or from the bearing producers.

Desheng Li and Young Sup Kang [4] published a simulation model for tapered roller bearing vibration introduced by a geometrical imperfection on the outer ring of the assembly. The model was correlated with measurements.

Doguer and Strackeljan [5] presents a focus on bearing fault simulation to improve the understanding of fault-induced vibration and early-state fault detection. The authors propose a multi-body system model of a rolling element bearing and introduce fault elements to simulate different fault geometries.

Patel, Pandey, and Tandon [6] developed a model to analyze the vibrations of deep groove ball bearings. The model takes into consideration multiple defects on the surfaces of the inner and outer races. Results from the model are compared to experimental measurements obtained from both undamaged and defective deep groove ball bearings. The Runge-Kutta method is used to solve the motion equations. This method calculates the vibration levels both in time and frequency domains for the shaft, balls, and housing.

R.M. Stoica [7] created a simple mechanical transmission integrating an electric motor, a rigid coupling, a driven shaft, two housing bearings, and one monoaxial accelerometer that was used to collect data in three directions in the proximity of the bearing.

Adamczak, Domagalski, Wrzochal, Piotrowicz, and Wnuk [8] describe a new industrial test rig designed specifically to measure the frictional moment of rolling bearings during the production stage. The frictional moment is a critical parameter that influences the quality and suitability of bearings for various applications including the automotive industry, machine industry, and household appliances. By measuring the frictional moment, information can be obtained about the bearing's structure, the quality of elements, its cleanliness, and the specific properties of the lubricant used. Reducing friction in rolling bearings is crucial to increase bearing efficiency, reduce energy consumption, and protect the environment.

The envelope analysis was determined in [9] using the kurtogram of bearing fault detection. With the help of a kurtogram, the frequency band of the raw vibration signal with the highest impulsivity and its corresponding central frequency and bandwidth can be determined.

Liu J, Wang L, and Shi Z [10] focused on studying local defects in cylindrical roller bearings (CRB), specifically on the contact surfaces. The authors investigate the acceleration peak that occurs when the rolling element passes over the defect.

H. Arslan and N. Aktürk [11] developed a shaft bearing model to study the vibrations of an angular contact ball bearing with and without defects. The assembly of the shaft and bearing is modeled to be a mass-spring system that exhibits nonlinear characteristics under dynamic conditions. The equations of motion for the shaft and rolling elements are derived in both the radial and axial directions.

In [12], an experimental investigation focuses on analyzing the vibration signature generated by rolling elements as they enter and exit a defect in the outer raceway of a bearing. The study aims to measure and analyze the vibration responses of the bearing housing and the displacement between the raceways.

## 2. Bearing component defect determination

To observe the actual irregularities that influence the dimension allowance and to make it possible to reject the faulty bearings, The Timken Company often uses a testing machine at the end of the line inspection. The measurements are performed with a Vertical Sound Test machine.

The current methodology quantifies the level of the vibration by measuring the dynamic force level of each tested bearing while operating at a specific preload and speed. The level of the measured dynamic force is transformed using a Fast Fourier Transform (FFT).

This paper proposes a new testing methodology to identify potential defects. It is based on the envelope analysis method. Envelope analysis is an established technique that identifies the faults of the bearings' rolling elements. It uses a handpass-filtered signal with a rectifier and smoothing circuit. From the frequency spectrum of the signal, the filter extracts the resonance excited by the bearing fault. For improved results, in this study a new transducer was added in addition to the force transducer and mounted in the radial direction: the accelerometer.

### *2.1. Bearing periodic defect frequencies determination*

In rotating machinery systems using bearings, each component rotates at a different speed. Consequently, each component generates noise and vibration at a different frequency. The bearing vibration behaviour depends on shaft speed and the bearing geometry.

With this information, is possible to calculate the bearing periodic frequencies. For this study, tapered roller bearing (TRB) assemblies were used. The following equations detail the formulas used to determine the bearing defect frequencies of the TRBs.

$$F_{\text{inner race}} = \frac{\left(\frac{S}{60}\right) \times Z \times \sin\alpha}{\sin\alpha + \sin\beta}\ [Hz] \qquad (1)$$

where: $F_{\text{inner race}}$ or BPFI – Ball Pass Frequency Inner

$$F_{\text{outer race}} = \frac{\left(\frac{S}{60}\right) \times Z \times \sin\beta}{\sin\alpha + \sin\beta}\ [Hz] \qquad (2)$$

where: $F_{\text{outer race}}$ or BPFO – Ball Pass Frequency Outer

$$F_{\text{roller}} = \frac{\left(\frac{S}{60}\right) \times \sin\alpha \times \sin\beta}{\sin\gamma \times (\sin\alpha + \sin\beta)}\ [Hz] \qquad (3)$$

where: $F_{\text{roller}}$ or BSF – Ball Spin Frequency

$$F_{\text{train outer}} = \frac{\left(\frac{S}{60}\right) \times \sin\alpha}{\sin\alpha + \sin\beta}\ [Hz] \qquad (4)$$

where: $F_{\text{train outer}}$ FTF – Fundamental Train Frequency (cage speed) when outer ring rotates.

$\alpha$ – 1/2 included outer ring angle
$\beta$ – 1/2 included inner ring angle
$\gamma$ – 1/2 included roller angle
Z – number of rollers
S – bearing rotating speed

### *2.2. Testing preparation*

Regardless of the methodology, the study was performed on a tapered roller bearing part number produced by The Timken Company. The bearing's boundary dimensions and characteristics are presented in figure 1.

IOP Conf. Series: Materials Science and Engineering 1303 (2024) 012022 doi:10.1088/1757-899X/1303/1/012022

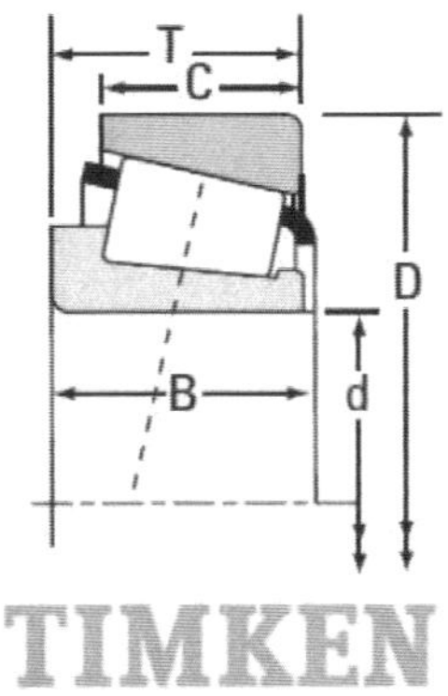


| Dimension | Value |
|---|---|
| d | 80 mm |
| D | 140 mm |
| B | 33 mm |
| C | 28 mm |
| T | 35.25 mm |
| BPFI | 10.29 Hz |
| BPFO | 7.7 Hz |
| BSF | 3.29 Hz |
| FTFI | 0.43 Hz |

**Figure 1.** Tapered roller bearing (TRB) specific dimensions.

The Rockwell hardness test method, defined by ASTM E-18, is the most used hardness testing method for metals. It is preferred over other methods due to its ease of use and high accuracy. This hardness test measures the permanent depth of indentation caused by the force/load applied by the indenter. The process involves applying a preliminary test force (preload or minor load) to the sample using a diamond cone indenter. This preload breaks through the surface to spoil the surface finish.

Three levels of localized defect were intentionally created for each part. The diamond indenter was operated at three pressure levels on the inner ring raceway, the outer ring raceway, and the roller body. The values of force applied to the bearing components were 60 kgf, 100 kgf and 150 kgf, and were defined by defect magnitude level: A, B, and C. The localized defects were quantified using a profile gauge from the metrology department.

The process used for indenting the bearing components is presented in figure 2 — the inner ring raceway (left), the outer ring raceway (centre), and the roller body (right).

**Figure 2.** Identing process for inner and outer ring and roller body:
(from left to right: the inner ring raceway, the outer ring raceway, and the roller body).

The principle of vibration identification with raised material typically involves analyzing the oscillations occurring on the surface of an object or structure that has raised features or protrusions.

The profile was measured with a Jenoptik HOMMEL T8000 tester. It is designed to measure and analyze the surface roughness parameters of various types of surfaces for quality control, process optimization, and inspection tasks across various industries. Figure 3 presents how the profile was measured for the outer ring imperfection.

**Figure 3.** Profile measurements with Jenoptik HOMMEL T8000 tester.

Using this tester, it was possible to determine the height of the raised material. Compared to the normal profile of the raceway, after the artificial defects were created, the resulting values of the heights were obtained (figure 4): the smallest one of 7.3 microns for an indentation obtained with 60 kgf (left side), the next-highest one of 12.5 microns for 100 kgf (center), and the highest one of 15.7 microns for 150 kgf (right side).

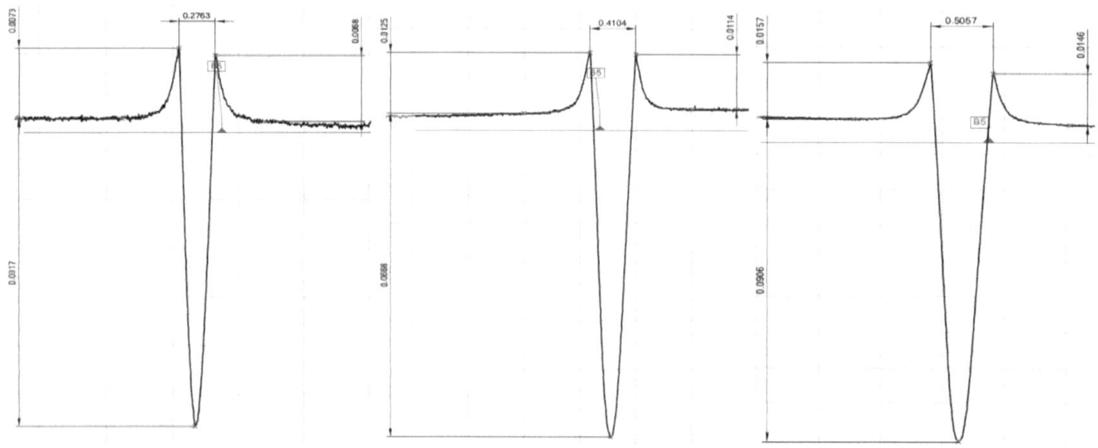


a) Defect level A b) Defect level B c) Defect level C

**Figure 4.** Localized defect dimensions (units: mm).

*2.3. Vibration measurement equipment*

The noise, vibration, and harshness (NVH) laboratory equipment available at The Timken Company-Romania facility consists of an acquisition system from Bruel and Kjaer (B&K) and measurement equipment from PCB Piezotronics. The data acquisition system used to acquire the data was the B&K Type 3050 with six input channels. The measuring equipment used included the PCB force transducer 208C03, which measures the bearing axial axis with a nominal sensitivity of 8.91 mV/lb, and the PCB accelerometer, Type 352C34, which measures the bearing radial axis with a nominal sensitivity of 99.8 mV/g. All the acquired data was processed with NVH dedicated software from Bruel and Kjaer – B&K Connect. Figures 5 and 6 present the measurement equipment used, the B&K Connect system, and the Vertical Sound Test machine.

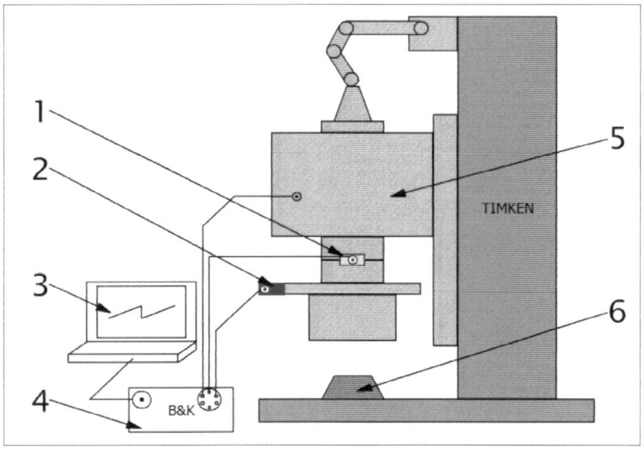


**Figure 5.** Testing bench for tapered roller bearings:
1. Axial force transducer; 2. Radial accelerometer; 3. Workstation; 4. DAQ;
5. Vertical sound test machine; 6. Tested tapered roller bearing.

**Figure 6.** Force transducer and accelerometer position on the Vertical Sound Test machine.

## 3. Experimental results

The tests were performed on nine assemblies with the same characteristics, on bearings produced by The Timken Company. Each tested bearing assembly had a defect produced by different applied magnitudes on every bearing component.

The recorded data were obtained from two measurement points (Figure 6). The first point is represented by a force transducer and the second one by an accelerometer.

The force transducer measures the vibration in the axial direction of the bearing. The accelerometer measures the vibration in the radial direction of the bearing.

The measurements were made at a turning speed of 1800 rpm. The peak amplitude of the spindle is identified for frequencies of 30 Hz and 60 Hz.

The results are presented in the following figures for all the components of the tested bearings. The top row of the figures shows the radial signal from the accelerometer and the bottom row shows the axial signal from the force transducer. The recordings began with the defect with the highest magnitude, defect level C, and concluded with the smallest one, defect level A.

### *3.1. The inner ring vibration measurement*

Figure 7 presents the envelope frequency spectrum for the inner ring defect with level C magnitude. The envelope spectrum results indicate that the radially placed accelerometer seems to be more sensitive to the raceway imperfections. The envelope spectrum indicates that the force transducer placed in the axial direction identified the defects, but the amplitude of the defect harmonics is smaller. Additionally, other frequencies with similar amplitudes are present there.

The plotted results identify the first, second, and third BPFI at frequencies of 309, 618, and 927 Hz. The amplitude recorded by the accelerometer for 309 Hz was 798 mm/s$^2$. The second defect-specific frequency (618 Hz) had an amplitude of 435 mm/s$^2$ and the third (927 Hz) is at 254 mm/s$^2$. When the results from the force transducer are analyzed, the following amplitudes can be viewed in the graph: 2.78 N, 1.83 N, and 2.92 N.

Also, as mentioned before, the spindle-specific frequencies of 30 Hz and 60 Hz can be observed in the measurements made by the force transducer and accelerometer.

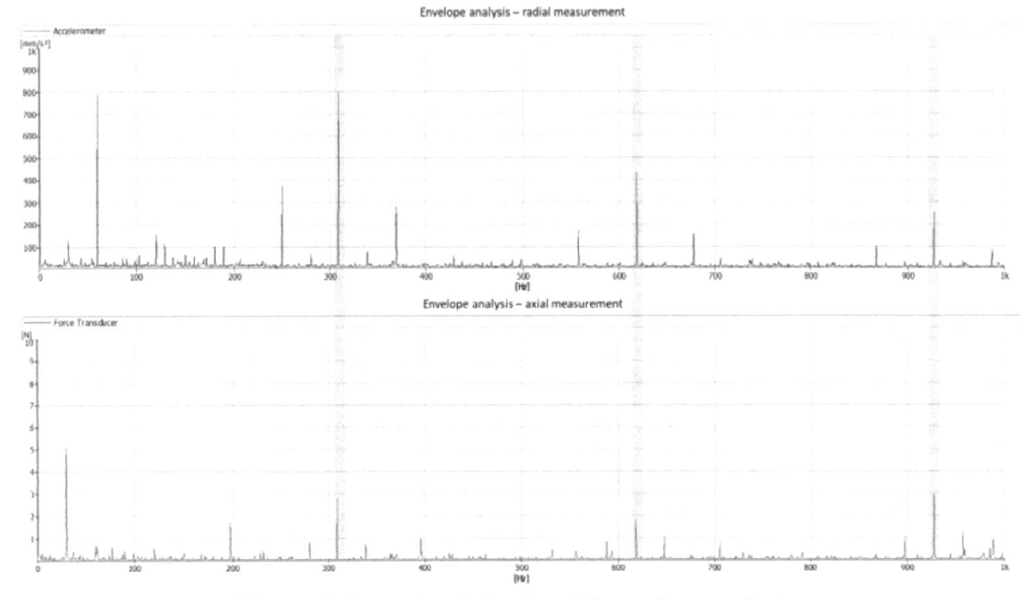


**Figure 7.** Inner ring defect level C envelope analysis.

Figure 8 presents the acquired data from the force transducer and accelerometer measurement points for an inner ring defect with level B amplitude. The acceleration data also identifies the BPFI with reduced amplitude, and the first, second, and third defect harmonics are present. When the force transducer data is analyzed, the amplitude decreases substantially compared to the C magnitude. The amplitude recorded by the accelerometer for 309 Hz is 312 mm/s$^2$, with 182 mm/s$^2$ for the second defect-specific frequency at 618 Hz and 93 mm/s$^2$ for the third at 927 Hz. When the results from the force transducer are analyzed, the following amplitudes can be viewed in the graph: 0.54 N, 0.86 N, and 2.92 N.

From these results, one can conclude that the optimal measuring method for identifying this type of defect is the accelerometer.

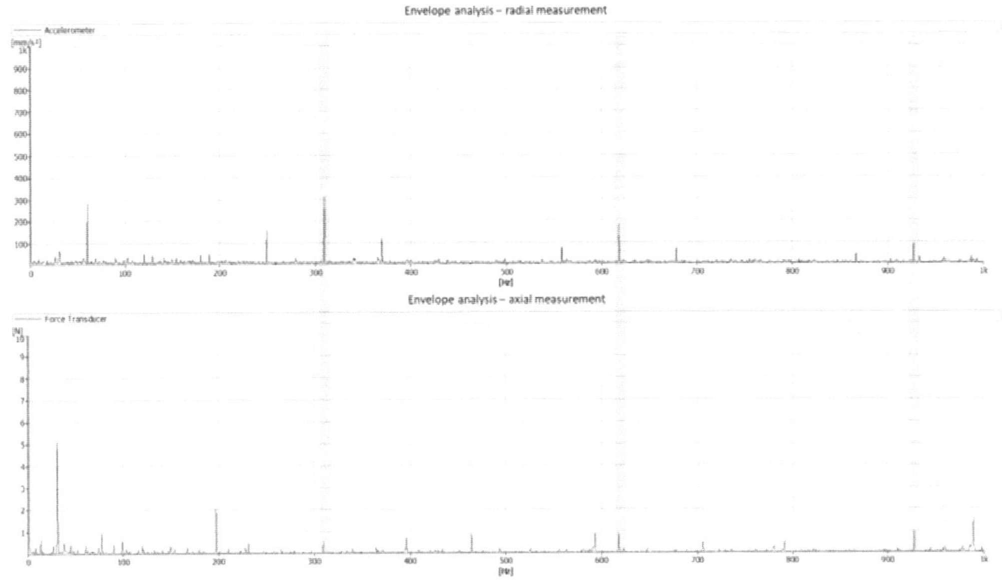


**Figure 8.** Inner ring defect level B envelope analysis.

In figure 9 are presented the results for the inner ring with a level A magnitude defect. From the envelope spectrum, the conclusion is that the defect magnitude is not big enough to be detected by vibration measurement with the accelerometer.

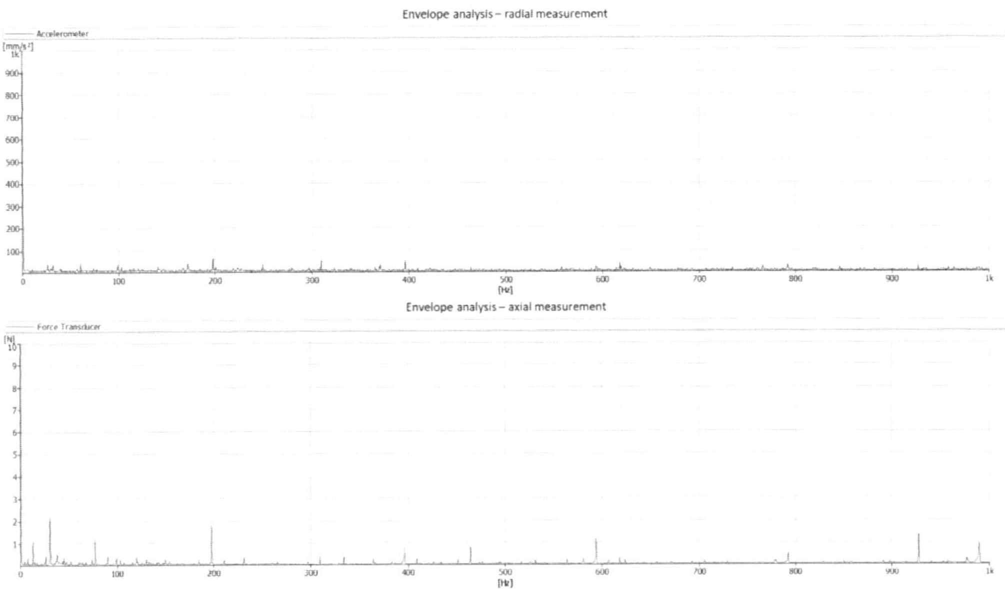


**Figure 9.** Inner ring defect type A envelope analysis.

### *3.2. The outer ring vibration measurement*

When the data for the outer ring is analyzed, the amplitudes of the vibrations from the same defects are higher compared to the inner ring. BPFO defect harmonics can be identified in figure 10. The identified specific harmonics of the defect are the first, second, third, and fourth. The amplitude recorded by the accelerometer for 231 Hz is 1158 mm/s$^2$, 694 mm/s$^2$ for the second-defect specific frequency (462 Hz), 502 mm/s$^2$ for the third (693 Hz), and 235 mm/s$^2$ for the fourth (924 Hz). When the results from the force transducer are analyzed, the following amplitudes can be viewed in the graph: 1.25 N, 1.78 N, 0.45 N, and 0.43 N.

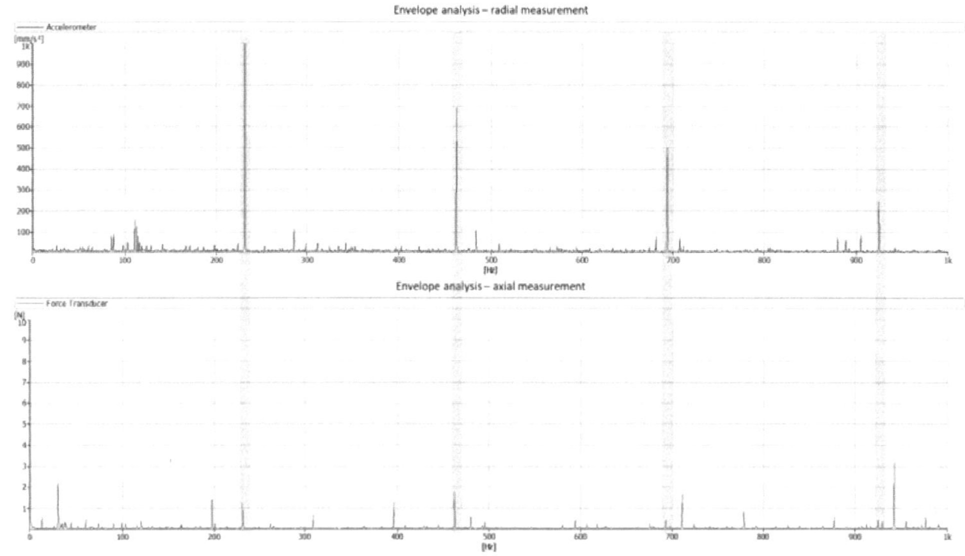


**Figure 10.** The outer ring defect level C envelope analysis.

For the outer ring with a magnitude B defect (figure 11), the force transducer measurements are not relevant. The defect amplitudes are too small and can't be clearly identified on the envelope spectrum. When discussing recorded measurement values, the accelerometer measured 264 mm/s$^2$ at 231 Hz, 213 mm/s2 at 462 Hz, 146 mm/s2 at 693 Hz, and 87.8 mm/s$^2$ at 924 Hz.

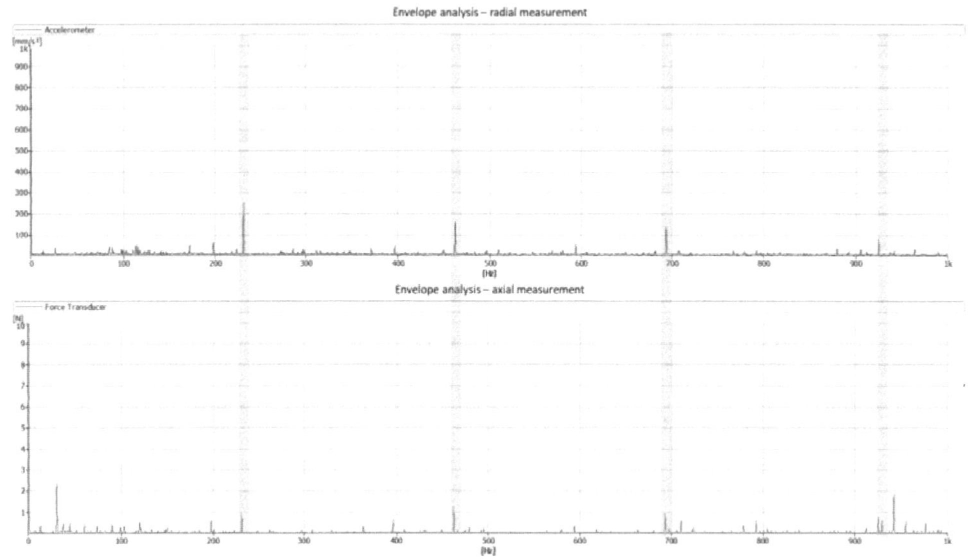


**Figure 11.** The outer ring defect level B envelope analysis.

The results for the outer ring defect with magnitude A are presented in figure 12. From the envelope spectrum, we can conclude that the defect amplitude is not of sufficient magnitude to be detected by the vibration measurement.

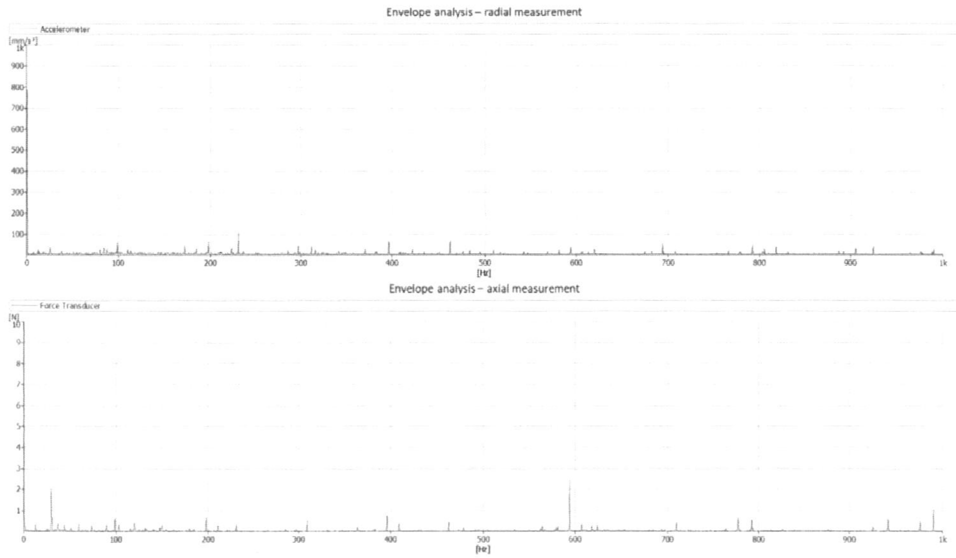


**Figure 12.** Outer ring defect level A envelope analysis.

### *3.3. The roller vibration measurement*

The next phase of this study was the identification of the roller dent defect. In figure 13 are presented the two envelope spectra from the studied measurement points. The roller defect was created on the roller body; thus, it will be translated to roller defect harmonics or BSF multiples of two. They are identified as the second, fourth, sixth, and eighth roller defect harmonics. The radial acceleration measurement more precisely identifies this defect.

The values recorded by the accelerometer are 818 mm/s$^2$ at 198 Hz, 444.9 mm/s$^2$ at 396 Hz, 346.2 mm/s$^2$ at 594 Hz, and 302.25 mm/s$^2$ at 792 Hz. The force transducer recorded amplitudes for the same frequencies with the following values: 2.57 N – 198 Hz, 1.31 N – 396 Hz, 1.1 N – 594 Hz, and 1.24 N – 792 Hz.

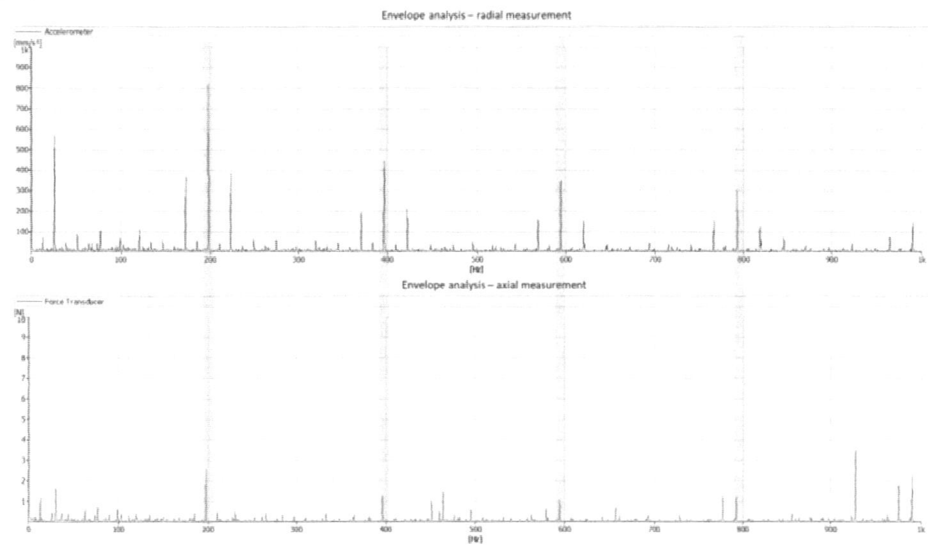


**Figure 13.** Roller defect level C envelope analysis.

Figure 14 presents the results obtained for a defect with magnitude B. Conclusions similar to the defect with magnitude C can be drawn. The radial accelerometer is more suitable for detecting this type of small imperfection.

The values recorded by the accelerometer are 589 mm/s$^2$ at 198 Hz, 325.6 mm/s$^2$ at 396 Hz, 248.1 mm/s$^2$ at 594 Hz, and 182.3 mm/s$^2$ at 792 Hz. The force transducer recorded amplitudes for the same frequencies with the following values: 1.38 N – 198 Hz, 1.53 N – 396 Hz, 1.23 N – 594 Hz, and 1.1 N – 792 Hz.

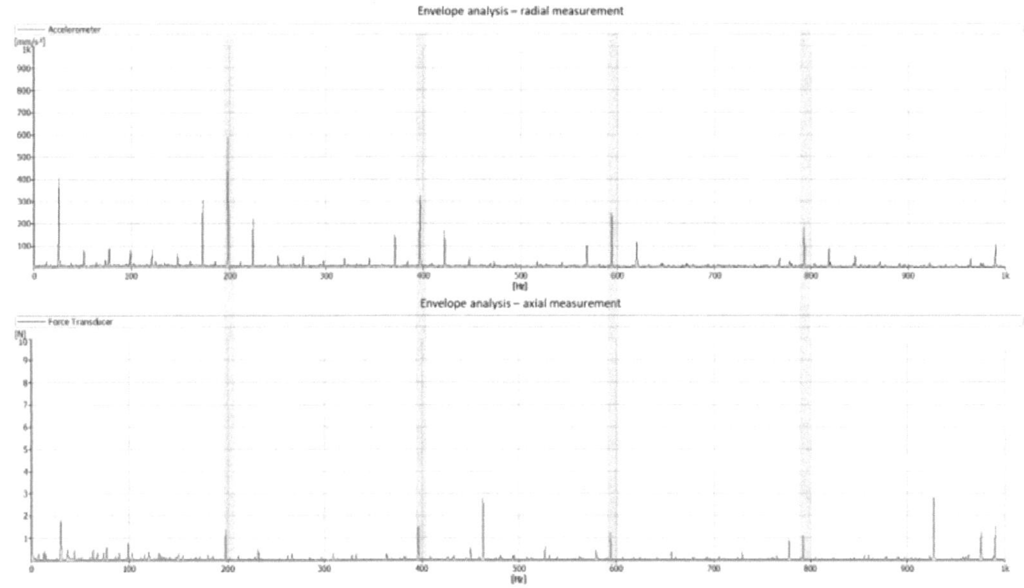


**Figure 14.** Roller defect level B envelope analysis

As for the magnitude A defect, on one roller body of the assembly its amplitude is too small to be identified either by the accelerometer or the force transducer.

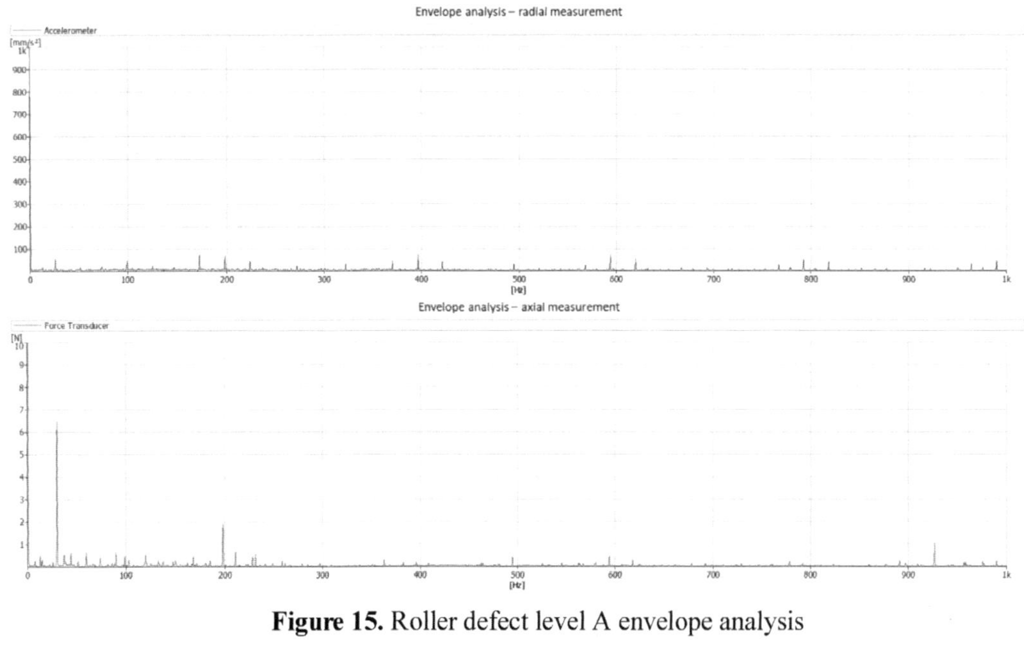


**Figure 15.** Roller defect level A envelope analysis

## 4. Conclusions

To ensure high quality, every single bearing must be inspected before sending it to the customer.

This study focused on the influence of localized surface variations on tapered roller bearing components. It achieves its main objective of finding a more suitable testing methodology to identify tiny imperfections on tapered roller bearing components.

Three defect magnitudes were created using a Rockwell indenter to simulate defects that can occur during the manufacturing process or bearing installation. The envelope analysis proved to be suitable for identifying dents on the ring raceways or the roller body having 12.5 microns or more of raised material. Compared to the fast Fourier transform method, the envelope method is more accurate in those situations.

When the recorded data from the axial force transducer and the radial accelerometer is combined, it is fast and easy to observe possible defects on the components.

One added value of this study is that most of the available research was performed on ball bearings in general, and this paper is specifically about tapered roller bearing localized defects.

**Acknowledgments**

The authors wish to thank Dr. William Hannon for the paper review and approval and Dr. Desheng Li for his good collaboration on the project development. Not least, the authors give thanks to The Timken Company policy that facilitates publishing research papers in Timken's activity domain.

**Author's ORCID ID**

Ș Voloacă: https://orcid.org/0000-0002-4726-6354
B A Sabo: https://orcid.org/0009-0001-3219-5506

**References**

[1] Hochman D and Bechhoefer E 2005 *IEEE Aerospace Conference Proceedings*, Envelope Bearing Analysis: Theory and Practice, pp 1-9

[2] Kim S, An D and Choi J H 2020 *MDPI Applied Sciences 10(20):7302*, Diagnostics 101: A Tutorial for Fault Diagnostics of Rolling Element Bearing Using Envelope Analysis in MATLAB, pp 1-23

[3] Konstantin-Hansen H and Herlufsen H *Denmark: Bruel & Kjaer*, Envelope Analysis for Diagnostics of Local Faults in Rolling Element Bearings, pp 1-8

[4] Li D and Kang YS 2014 *ASME 2014 International Design Engineering Technical Conferences and Computers and Information in Engineering Conference*, Simulation and Experimental Validation of Tapered Roller Bearing Vibration Induced by Geometrical Imperfection on Cup Raceway, pp 1-11

[5] Doguer T and Strackeljan J *The Sixth International Conference on Condition Monitoring and Machinery Failure Prevention Technologies*, Using a Dynamic Roller Bearing Model Under Varying Fault Parameters, pp 907-918

[6] Patel VN, Pandey RK and Tandon N 2010 *Journal of Tribology vol. 132*, A Dynamic Model for Vibration Studies of Deep Groove Ball Bearings Considering Single and Multiple Defects in Races, pp 1-11

[7] Stoica R M et al. 2022 *IOP Conference Series: Materials Science and Engineering, Volume 1220,* Bearing fault detection by means of frequency analysis

[8] Adamczak S, Domagalski R, Wrzochal M, Piotrowicz G and Wnuk S 2021 New Parameter Proposed for Industrial Acceptance Measurements of Rolling Bearings (CC BY 4.0)

[9] Kamiel B 2020 *JMPM: Jurnal Material dan Proses Manufaktur Vol. 4, No. 2,* Demodulation of Vibration Signal Based on Envelope-Kurtogram for Ball Bearing Fault Detection, pp 115-123

[10] Liu J, Wang L and Shi Z 2022 *Elsevier: Mechanical Systems and Signal Processing*, Dynamic modelling of the defect extension and appearance in a cylindrical roller bearing, pp. 1-26

[11] Arslan H and Aktürk N 2008 *Journal of Tribology vol 130*, An Investigation of Rolling Element Vibrations Caused by Local Defects, pp 1-13
[12] Ahmadi A, Howard C and Petersen D 2016, *Journal of Sound and Vibration 366*, The path of rolling elements in defective bearings: Observations, analysis and methods to estimate spall size, pp 277–292.

IOP Conf. Series: Materials Science and Engineering 1303 (2024) 012023 doi:10.1088/1757-899X/1303/1/012023

# Analytical analysis of the thermal comfort of the occupants inside a truck cabin

**A M Manea[1], M D Iozsa[1] and E M Vieru[2]**

[1] Automotive Engineering Department. University Politehnica of Bucharest. 313 Spl. Independentei, Bucharest, Romania
[2] Faculty of Transports, University Politehnica of Bucharest, 313 Spl. Independentei, Bucharest, Romania

E-mail: mannea_annamaria@yahoo.com

**Abstract.** Thermal comfort inside a truck cabin is a parameter that influences the well-being of the driver and passengers. From the very first steps of designing the driver's seat of a vehicle, the driver's well-being is taken into account through ergonomic parameters. To analyze the ultimate comfort inside a truck cabin, a simplified sketch of the cabin will be proposed. The three-dimensional analytical calculation model is solved using the Ansys program in the Fluid Flow (CFX) module. Through modeling and numerical simulations, the temperature inside a truck cabin, the mode of movement of the air flow and the speed of the air flow are determined. Following these tests, it is checked whether the thermal comfort conditions for the driver and passengers in a truck cabin are respected and fulfilled. These term comfort conditions are evaluated based on some comfort indices available in specialized literature.

## 1. Introduction

The ergonomic quality of a vehicle is related to the quality of the relationship between the user and the vehicle during it is use. Thermal comfort is among the main ergonomic parameters.

Thermal comfort represents the quality of the microclimate inside the passenger compartment and the thermal sensation felt by the driver when in contact with the interior surfaces of the vehicle. This is a parameter inside the body that can psychologically influence the pleasure of using a vehicle.

Occupants must feel comfortable even in adverse environmental conditions (high or low temperatures, solar radiation, heavy rain, and snow) [2].

Also, the driver's vigilance and his ability to concentrate in traffic depend to a large extent on his comfort level.

The internal temperature of the human body is naturally controlled and is maintained within a narrow range around 37°C. Thus, thermal discomfort and dangerous situations of thermal stress are avoided [1].

If the temperature of the human body increases by a few tenths of a degree, this increase is sufficient to stimulate the sweating mechanism which is capable of multiplying the heat dissipated by the human body four times [1].

During the hot season, if a vehicle's cooling system is not working properly, the bodies of the occupants of the vehicle heat up. If the internal temperature of the human body increases by more than 2°C compared to the optimal figure of 37°C, the subject's behavior changes for the worse, a feeling of weakness appears. The consequences of the increase in human body temperature are vasodilatation (increased blood flow through the skin which leads to an increase in heat conduction to the outside, and the temperature of the skin becomes similar to that of the internal tissues) and sweating (the cooling mechanism for the body, the energy required to evaporate sweat is directly subtracted from the skin) [1].

During the cold season, the heating of the passenger compartment must be taken into account. If the measures put in place to prevent cooling are inadequate, the human body gets cold. Also, it must be taken into account when evaluating thermal comfort that the hands and feet, being far from the center of the body, are the first prone to cooling. And in this case, if the internal temperature drops by more than 2°C below the optimal value, the normal behavior of the occupants changes, a feeling of numbness or stiffness appears.

In studies of the physiology of the human body, it has been shown that the body's internal temperature does not change (no cooling, heating or evaporation occurs) if the body is exposed to an ambient temperature in the range of 23°C ÷27°C (subject is clothed and in sedentary activity). Also, skin temperature ($T_{sk}$) and internal temperature ($T_{cr}$) are in neutral state at the following values: $T_{sk}$ =33.7°C and $T_{cr}$= 36.8°C [2].

The main parameters that directly affect the comfort state and microclimate of the occupants are the following: temperature, air speed, humidity, solar radiation, radiation from components, particles in the air [2].

The interior temperature of the passenger compartment depends a lot on the ambient temperature. Because of this, a constant temperature of around 22°C cannot be established. During the cold season the temperature inside the cabin must be higher to compensate for this low temperature level. During the summer, the human body adapts more easily to high temperatures. Thus, the temperature inside the cabin can exceed the value of 22°C and be evaluated as a comfortable temperature. Some vehicles are equipped with powerful air conditioning systems that cool the passenger compartment by up to 20°C at an ambient temperature of 45°C, solar radiation of 800 W/m2 and humidity of 30%. But after a period on board the vehicle this temperature is too low. Also, the human body must cope with the impact of the sudden change in temperatures when getting out of the vehicle. For this reason, studies have been done that recommend certain temperatures in the passenger compartment [2].

Also, to increase the comfort of the occupants, in addition to the optimal temperature, they can be exposed to air flow. According to [3], one can see what air speed is suitable depending on the temperatures. In the specialized literature, there are many data from different authors regarding the air flow. In conclusion, the higher the ambient temperature, the higher the air flow speed should be to reach the same level of comfort. However, the effect of an air flow is reversed if the air temperature is too high. The effect of an air flow becomes negative at temperatures above 35°C (temperature close to that of the human body) the heat is transferred to the human body.

The paper presents a model made for the study of the air flow introduced by the HVAC system in the cabin of a truck, to control the temperature level inside it.

## 2. The model

The geometric model was made based on the shape and dimensions of the passenger compartment of a truck cabin with two seats and equipped with two beds (fig.1.a). The mannequins were modeled considering the main dimensions of a 90% mannequin. Geometric modeling was performed in the geometric modeling module of ANSYS software. Discretization with elements (fig.1.b) was done only on the interior space of the passenger compartment that contains air. The discretization density is higher on the interior geometric elements and in the area of the air inlet and outlet.

The setting of the CFD analysis took into account the establishment of the main parameters that define its boundary conditions, considering that the analysis will only be performed for heating the cabin in winter conditions.

The total air flow given by the HVAC system was considered to be 0.05 kg/s [4], [5], [6], this being distributed, depending on the setting, on the ventilation panels from the windshield, dashboard and feet. The exit of air from the cabin is done at normal atmospheric pressure, through two openings, located at the bottom of the rear wall.

The boundary walls were considered no slip walls with different thermal conductivity properties [3]: windshield and side windows (glass) - 1.172 $W/m^2K$; body and dashboard (plastic) - 2.7 $W/m^2K$; floor (carpet) – 0.294 $W/m^2K$, seats and beds (foam) – 0.05 $W/m^2K$. The temperature of the outer walls was assumed to be equal to the outside temperature, except for the tunnel wall, whose temperature is influenced by the temperature of the engine compartment.

a. b.

**Figure 1.** Geometric model (a) and finite element model (b) of the passenger compartment.

Six cases of passenger compartment heating were studied, for three modes of regulation of the ventilation system and for two values of the outside temperature, 0 °C and -10 °C (table 1).

**Tabel 1.** The analysed cases.

| C1 | C2 | C3 | C4 | C5 | C6 |
|---|---|---|---|---|---|
| **Outside temperature, 0 °C** | | | **Outside temperature, 10 °C** | | |
|  |  |  |  |  |  |

## 3. Results

To study the variation of temperature inside the cabin over time, the six cases were studied using a transient CFD analysis using ANSYS software. It was analyzed how the air inside the cabin is heated for one hour, from the initial time when warm air was introduced inside the cabin through the ventilation panels.

The analysis of the temperature variation over time was done especially at eight points near the driver and passenger. Points 1 and 2 are next to the two dummies, Point 3 being positioned at the same level, in the median vertical plane of the cabin. Points 4 and 5 are located at an almost equal distance from the face and palms of the mannequins, Point 6 being positioned at the same level, in the median vertical plane of the cabin. Points 7 and 8 were considered in the leg area of the two mannequins (fig.2).

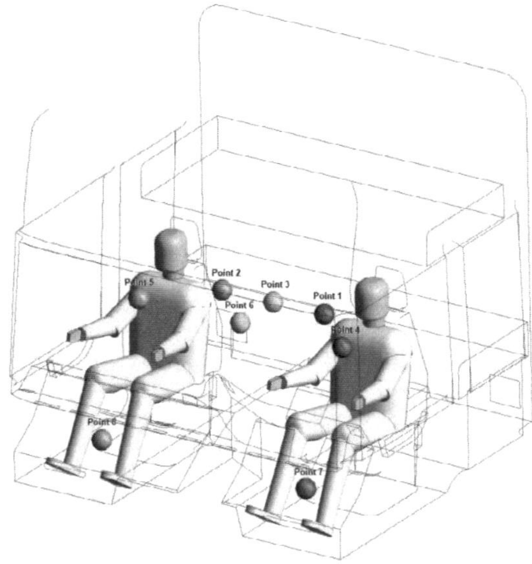


**Figure 2.** The points where the temperature variation was analyzed.

From the analysis of the temperature variation in the six cases (figure 3 - figure 8), it can be seen that in the first 5-6 minutes there is a rather rapid increase in temperature, followed by several oscillations (decreases followed by increases). After about 15-20 minutes. these oscillations subside and there follows a slow increase in temperature, which after about an hour remains approximately constant.

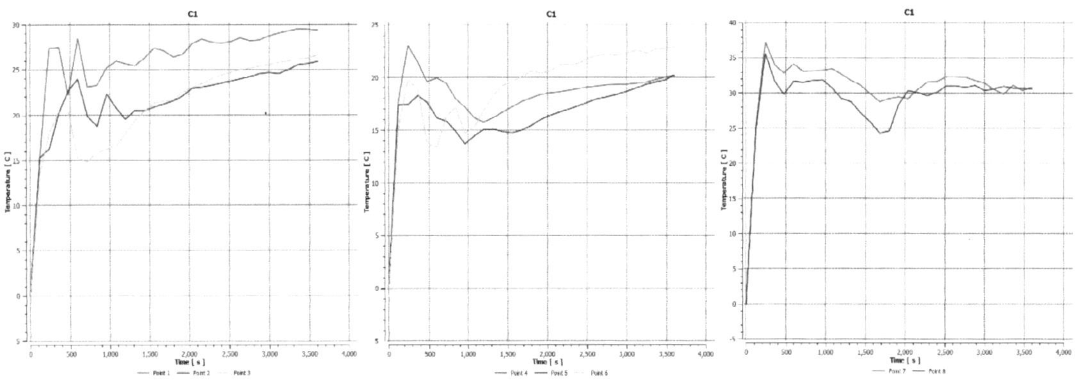


**Figure 3.** Case C1.

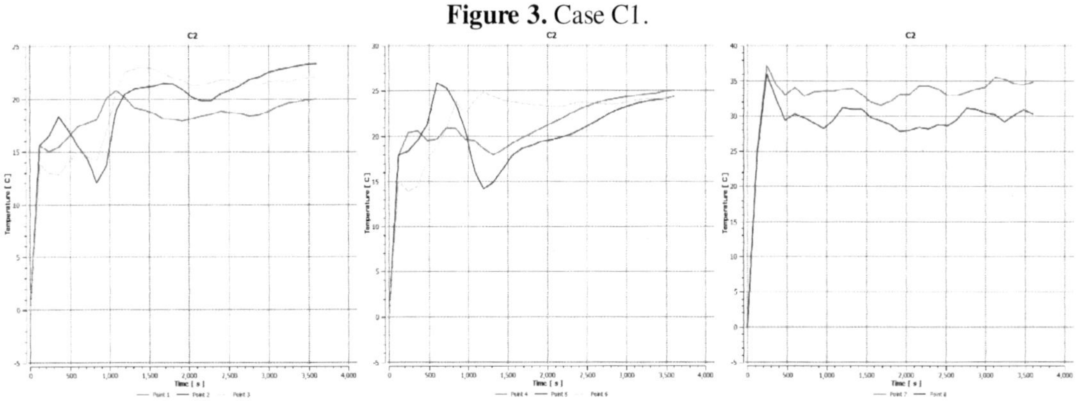


**Figure 4.** Case C2.

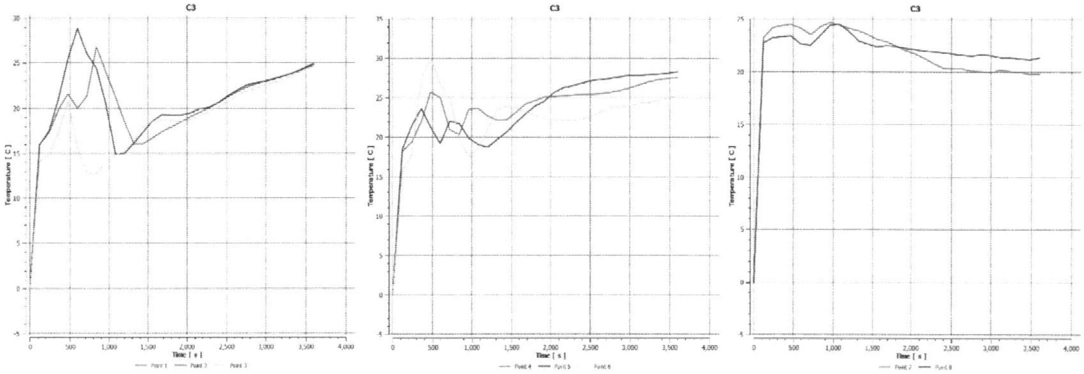


**Figure 5.** Case C3.

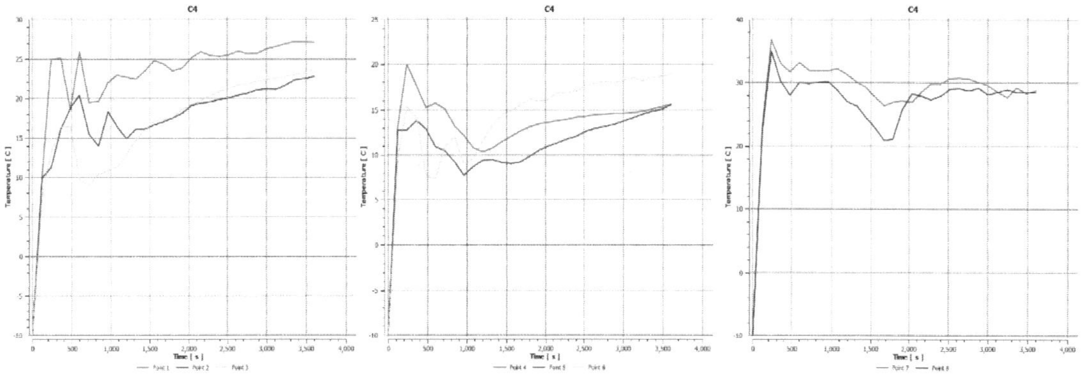


**Figure 6.** Case C4.

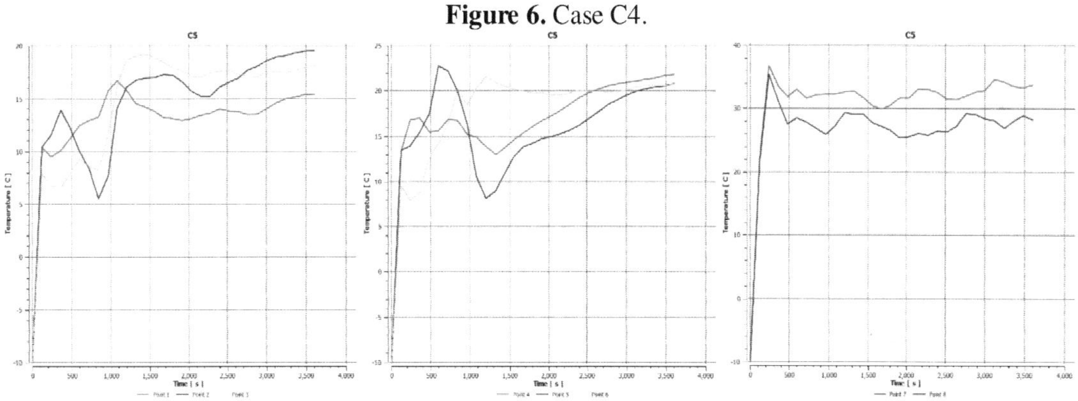


**Figure 7.** Case C5.

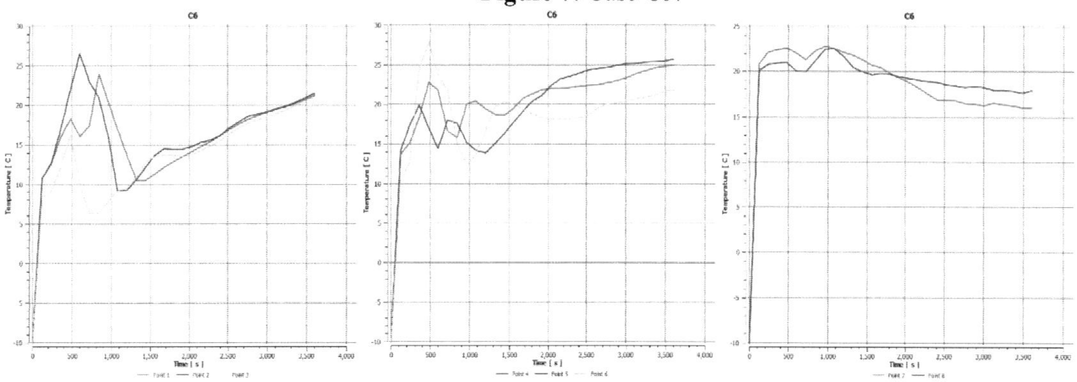


**Figure 8**. Case C6.

The oscillations that occur after the first heating can be explained by the return of the cooled air stream when it passes through the rear area near the beds, towards the front of the passenger compartment. After this stage, the air behind the passenger compartment also starts to heat up and cooling no longer occurs in the analyzed points, even if the air returns to the front (figure 9).

From the analysis of the graphs, it is observed that the shape of the curves is similar for the cases C1 and C4, C2 and C5, C3 and C6, differing in the initial temperature and the temperature up to which the first heating is done, before the temperature oscillations, this temperature being slightly higher in cases with initial temperature 0 °C. In contrast, the temperature reached after one hour is almost identical.

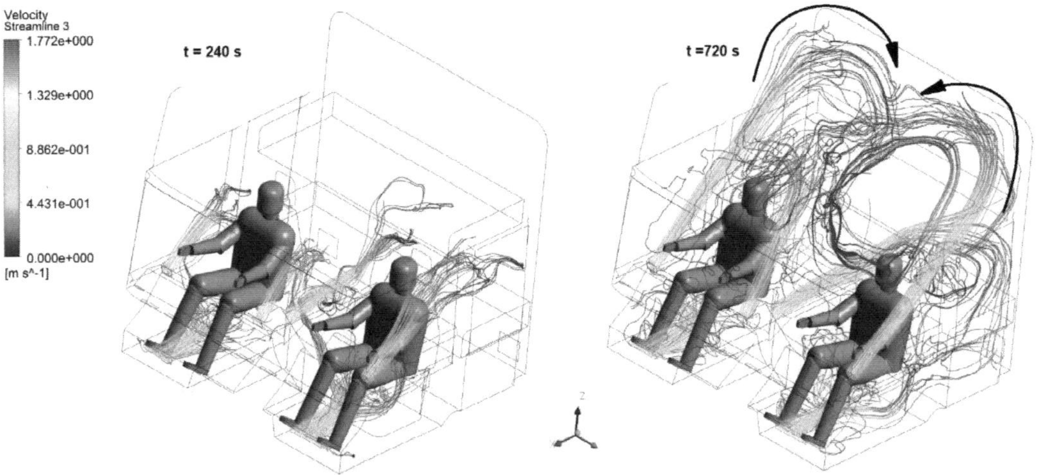


**Figure 9.** The appearance of temperature oscillations in the first 15 min in the case of C1.

In points 7 and 8, the heating is stronger already in the first 5 minutes, probably due to the fact that the air returns more difficult from the colder areas. Towards the end, the temperature remains constant but lower by about 5 °C (figure 10). Of course, in cases C3 and C6, the temperature at the feet is almost 10 °C lower than the other cases, but it is still at the level of 18-20 °C.

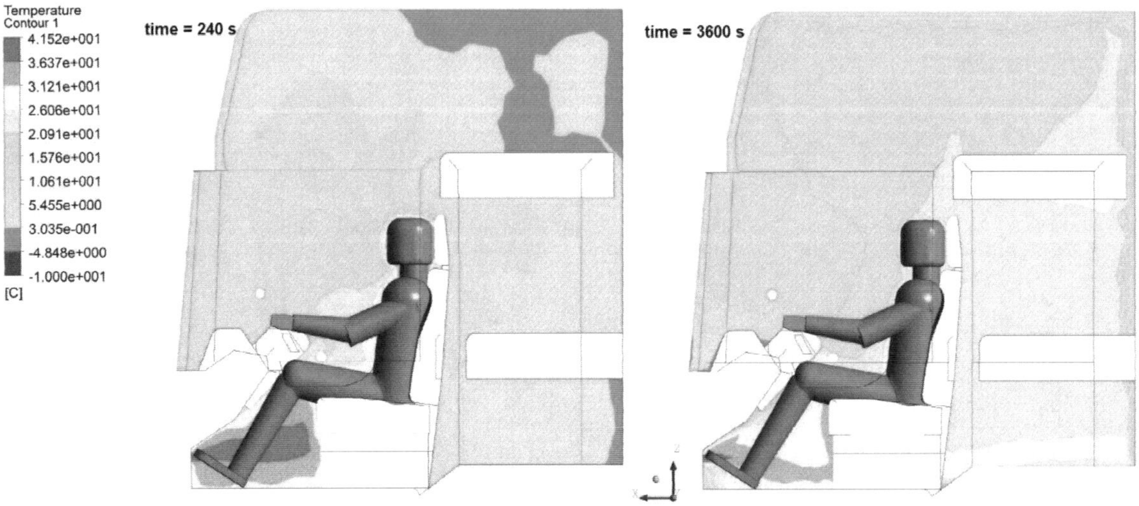


**Figure 10.** The temperature in the median plane of the driver, case C1.

In cases with an outside temperature of -10 °C, the windshield heating is done at temperatures above 15 degrees only in cases C4 and C5, in case 6, even after an hour, there may be areas that are not defrosted.

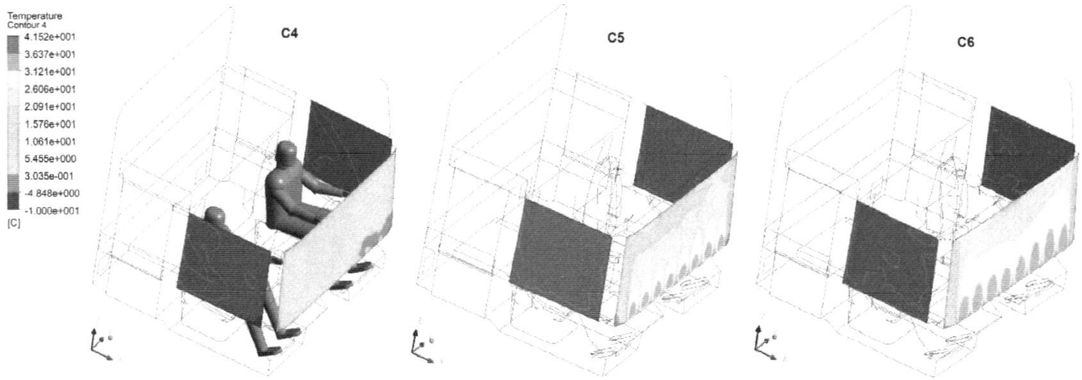


**Figure 11.** Temperature on the surface of the windshield and side windows.

## 4. Conclusions

This model represents a first step in the study of the main parameters that characterize the flow of air inside the cabin, to ensure an optimal climate for the driver and passenger. The transient CFD analysis can show the evolution of these parameters over time, being possible to determine the time when favorable temperatures are obtained in the seat area, but also in the bed area, in the case of this cabin.

It is desired to continue the study with the performance of some experimental tests, in order to be able to validate the created model. Of course, the improvement of this model will also be pursued, so that the calculation errors are reduced as much as possible. We believe that the thermal conductivity properties of the walls, as well as the air flow rates through the cabin ventilation panels, should be determined as accurately as possible, through experimental studies.

## Author's ORCID ID

M D Iozsa: https://orcid.org/0000-0002-6547-8633
A M Manea: https://orcid.org/0000-0002-3976-8848

## References

[1] Morello L et al. 2011 The Automotive Body, Volume I: Components Design, Volume II: System Design, Springer;
[2] Hucho W H 1998 *Aerodynamics of Road Vehicles*. Fourth Edition, SAE International, USA;
[3] ElDegwy A and Khalil Essam E 2018 Passengers' Thermal Comfort in Private Car Cabin in Hot Climate, *AIAA Propulsion and Energy Forum, Joint Propulsion Conference*, Cincinnati, Ohio;
[4] Paulke S 2019 Comparative comfort simulations for winter load cases using global and local comfort models HdT: 3. Tagung Fahrzeugklimatisierung Kühlen, Heizen und Komfort – Zukunftsorientierte Konzepte, *ARRK, P+Z Engineering GmbH*;
[5] Christ P *Modeling of Automotive HVAC Units Using Proper Orthogonal Decomposition*, Technische Universität München, Institut für Energietechnik Lehrstuhl für Thermodynamik;
[6] Fojtlín M, Planka M, Fišer J, Pokorný J and Jícha M Airflow Measurement of the Car HVAC Unit Using Hot-wire Anemometry;
[7] Dragan R and Ferenc Č 2012 Thermal Interaction Between a Human Body and a Vehicle Cabin, Heat Transfer Phenomena and Applications, Edited by Md Salim Newaz Kazi;
[8] Dragan R and Stjepan G 2017 Thermal radiation between the driver and the vehicle cabin interior, *Mobility & Vehicle Mechanics*, Volume 43, Number 1;
[9] Ivanescu M, Neacsu C A and Tabacu I 2010 Studies of the thermal comfort inside of the passenger compartment using the numerical simulation, *International Congress Motor Vehicles & Motors*;
[10] Iozsa M D 2016 *Bodies for road vehicles*, Ed. Politehnica Press.

IOP Conf. Series: Materials Science and Engineering 1303 (2024) 012024 doi:10.1088/1757-899X/1303/1/012024

# Finite element analysis and experimental study of strain for a vehicle rear suspension

**I D Geonea**

Applied Mechanics and Civil Buildings Department, University of Craiova, Faculty of Mechanics, Craiova, 200512, Romania.

E-mail:ionut.geonea@edu.ucv.ro

**Abstract**. In this paper, aspects of the study of the durability of rear suspension components of a vehicle are discussed. These aspects are addressed in two ways, namely the study by CAD modelling and finite element method simulation, and secondly by experimental analysis on a durability test bench. In order to develop these research problems, an experimental stand is virtually modelled and studied by simulation using ADAMS software. Results are obtained that characterize the operation of the test bench system kinematically and dynamically. The studies are then completed by finite element analysis of the dependent rear axle and helical springs in the suspension structure. These results are corroborated with those obtained experimentally, i.e. stresses and strains recorded on the stand. The usefulness of these studies is that they allow the validation of different components of the rear suspension in terms of fatigue strength and different types of rear axles from cars can be tested, as the stand has a modular construction and the test parameters can be adjusted.

## 1. Introduction

In the automotive industry, durability and fatigue resistance testing is imperative for a wide range of automotive components such as bearings, anti-roll bars, springs, shock absorbers, bushings, ball joints, as well as more complex elements such as rear axle or suspension arms.

Thus, within our team at the Faculty of Mechanics-University of Craiova, there has been a concern for the study of the durability of automotive components. Thus, in the works [1, 2], the fatigue behavior of the stabilizer bar of the rear and front axle of a car is studied. The stiffness characteristic of a torsion bar spring is calculated experimentally, as well as the fatigue strength by the finite element method (FEM) using ANSYS software. The construction of a stand for testing the stabilizer bar at the front and rear axle of a passenger car are presented in the paper [2]. The kinematics of the connecting rod-crank-balancer mechanism through which the alternating rotational motion of the stabilizer bar is achieved is presented in [3]. Another concern was to study the frictional moment in radial ball bearings. Research addressing this topic is published at the CONAT 2017 conference [4]. The 3D modeling of a stand allowing fatigue testing of the rear axle is presented in the paper [5], published at the AITS 2021-Chisinau conference. Here, the translational displacements of the stabilizer bar and the stresses and strains in the rear axle assembly are presented. Nationally and internationally, we find concerns in these directions. Thus, in the paper [6] the housing of a truck rear axle is optimized. The optimized structure was verified on a test stand and in operation on the truck by experimental measurement of deformations. Fatigue failure is a common occurrence in parts subjected to variable stresses. An example is that of a planetary shaft under torsional stress that failed [7]. A very important role in the correct operation of a mechanical system is played by vibrations, which must be eliminated from the operation of mechanical transmissions. An example of such a study, is presented in the paper [8], where the vibration behaviour of a hypoid bevel gear central transmission is simulated. One of the most commonly used tools for fatigue studies is finite element analysis in conjunction with experimental analysis, usually on test stands. Thus, in the paper [9], finite element simulation of the fatigue bending behaviour of a rigid truck rear axle is presented. The influence of motors installed in passenger car wheels on the torsion beam of the rear axle suspension is presented in paper [10]. Also, in [11] the simulation and optimization of truck cab suspension system based on ADAMS is studied. In [12] the functional testing of both vehicle suspension components and characteristics, namely the actuators and the vehicle dynamics logic control, there are performed on an MTS 6DOF bench test. These tests are made prior to physical track testing on a prototype vehicle using Ferrari facility which exists in Maranello, Italy. An important component of the suspension structure is the stabiliser bar. This component must have high elasticity and resistance to fatigue, so from the design phase of such a component the aim is to optimise the design, reducing stress concentrators [13, 14, 15]. Obviously for validation of the automotive component, tests performed on test stands are required [16, 17].

The present paper is a continuation of the paper presented at the AITS 2021 conference [5]. The paper is structured in 3 parts. After the "Introduction" the 3D modeling of the test stand is presented. In the third part a finite element analysis (FEM), of the tested rear axle is presented, corroborated with experimental analysis of deformations and stresses in the axle structure.

## 2. Design of a bench for fatigue test of the rear axle of a passenger car

Fatigue testing methodology is approached in two ways, modelling and simulation and testing on an experimental bench. The kinematic scheme of the rear axle fatigue test bench is shown in Fig. 1. The components of the stand are: 1- crank; 2- adjustable length rod; 3- rocker arm; 4- bushing; 5- assembled rear axle. According to the kinematic scheme, the virtual model is also made (Fig. 2), in order to simulate the kinematics and dynamics of the stand, as well as the finite element modeling of the rear axle.

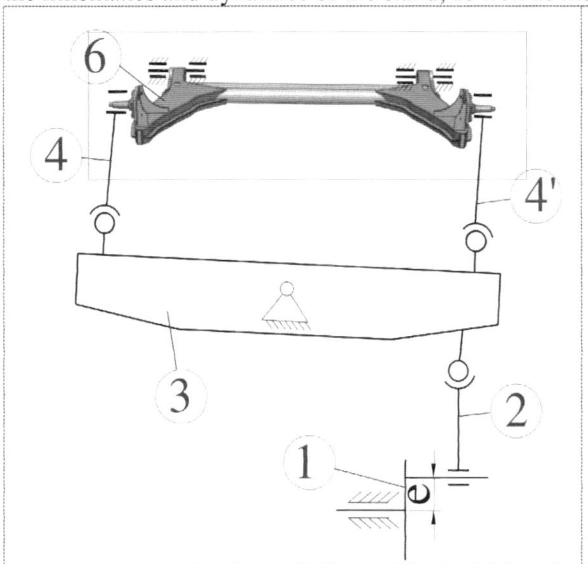


**Figure 1.** Kinematic scheme of the stand.

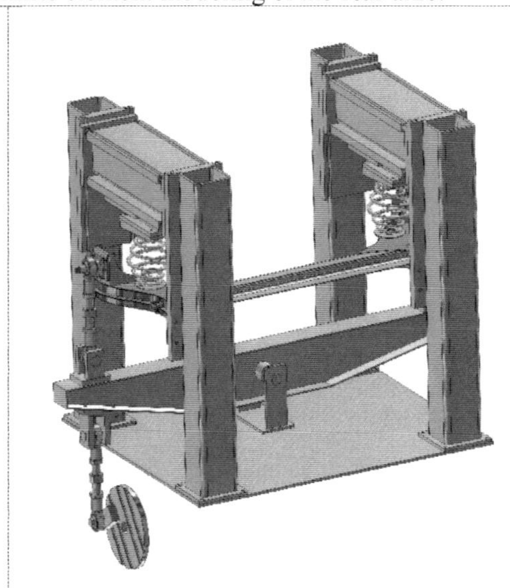

**Figure 2.** CAD model of the stand in SW.

As results of the kinematic and dynamic modelling of the stand in ADAMS, we present the kinematic parameters of the stabilizing bar considered as a deformable solid, modelled as shown in Fig. 3.

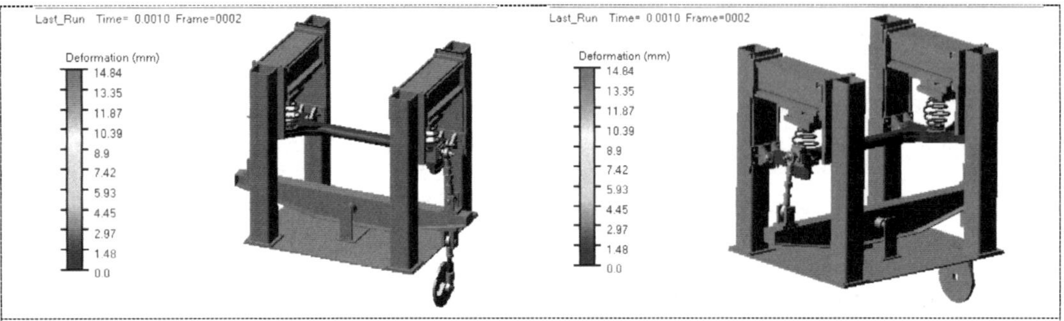


**Figure 3.** Rear axle and stabilizer bar modeled as deformable solids in ADAMS.

Figures 4 and 5 show the graphs obtained for the following parameters: displacements of the marker attached to the centre of mass of the stabiliser bar, the components along the X and Y axes (longitudinal and transversal axis). The plots shown are obtained with respect to the global reference system.

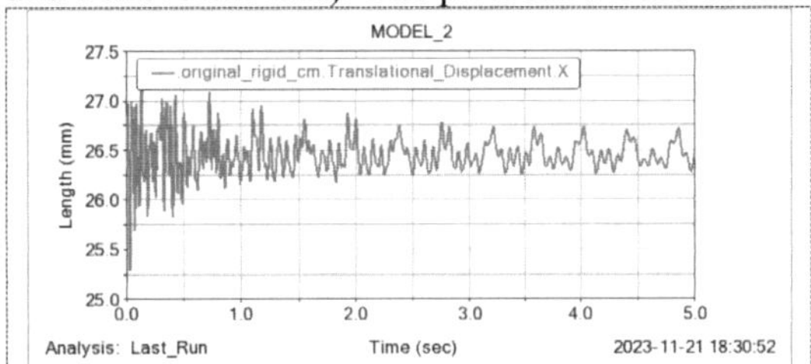


**Figure 4.** Obtained displacement of the marker from centre of mass of the stabiliser bar along the X-axis.

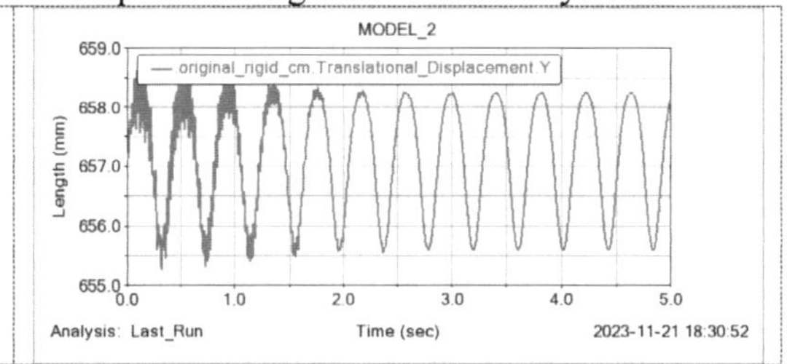


**Figure 5.** Obtained displacement of the marker from centre of mass of the stabiliser bar along the Y-axis.

Figures 6 and 7 show the plots as obtained in ADAMS for the deformations of the marker attached to the centre of mass of the stabilizer bar along the X and Y axes. The plots shown are computed in reference to the model assembly global reference system.

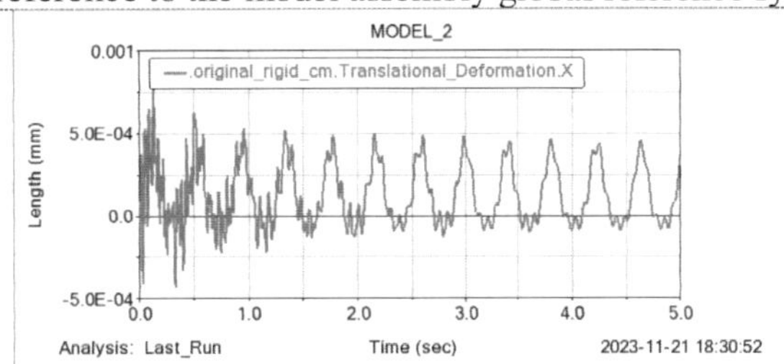


**Figure 6.** Computed deformation of the centre of mass of the stabiliser bar along the X-axis

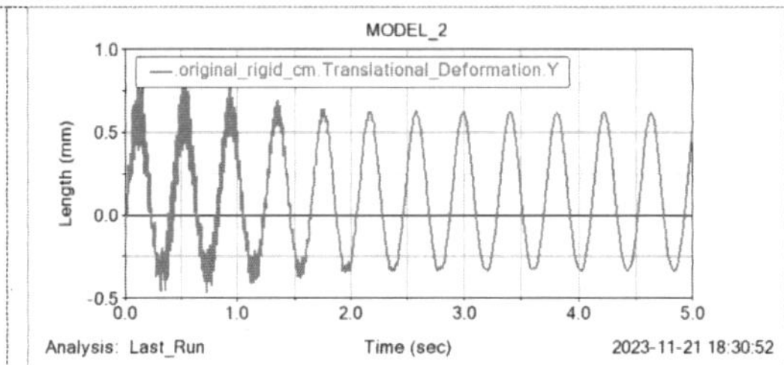


**Figure 7.** Computed deformation of the centre of mass of the stabiliser bar along the Y-axis.

The model constructed by means the ADAMS program allows us the dynamic modal analysis for every kinematic element of the mobile mechanical system and also for the entire assembly when the kinematic elements are considered as deformable solids. In a first stage of the research, we have modelled the experimental fatigue test stand of the rear axle using the 3D modelling program Solid Works. The 3D model we have produced is useful for the optimization of the experimental stand but also for the simulation in ADAMS of the dynamic modal behaviour of the axle assembly. We performed the dynamic modal modelling in ADAMS of the test bench entire whole assembly, and presented the results obtained in simulation, namely displacements, velocities, and translational accelerations of the marker attached to the centre of mass of the axle stabilizer bar mounted on the test stand.

## 3. Dynamic finite element analysis of the rear axle

In this paragraph, we propose to determine the state of stresses and strains in the experimentally tested rear axle assembly mounted on the test bench.

This is necessary for the following two reasons: we will determine where is the location of the critical areas with stress concentrators where the maximum equivalent stresses occur; the mechanical stresses and specific strains obtained by FEA simulation will be compared with the values obtained with the experimental method.

For the finite element analysis, we will use a licensed software, namely AnsysWorkbench. This FEA software has the possibility to perform several types of FEA analysis: static structural, transient structural or explicit dynamics. Since the rear axle assembly model is particularly complex, it has elements in its structure with non-linear behaviour (such as rubber flexiblocks, as shock absorbers) and the Ansys Explicit Dynamics module is suitable for use in this FEA simulation case. This module allows us to perform explicit dynamics analysis, it allows us to utilize from the database materials with non-linear characteristics (steel with non-linear behaviour, rubber).

The results that are obtained for the mechanical stresses, we will compare with the results obtained experimentally. This comparison will be carried out considering the location of the specific strain and stress measurement points as shown in Figure 8.

For the central part of the axle beam, a stress-strain rosette consisting of three strain gages was used to measure the specific strains at -45, 0 and 45 degrees to the principal direction. This rosette was set with the middle grid at 90 from the axle longitudinal axis, as shown in Figure 8.

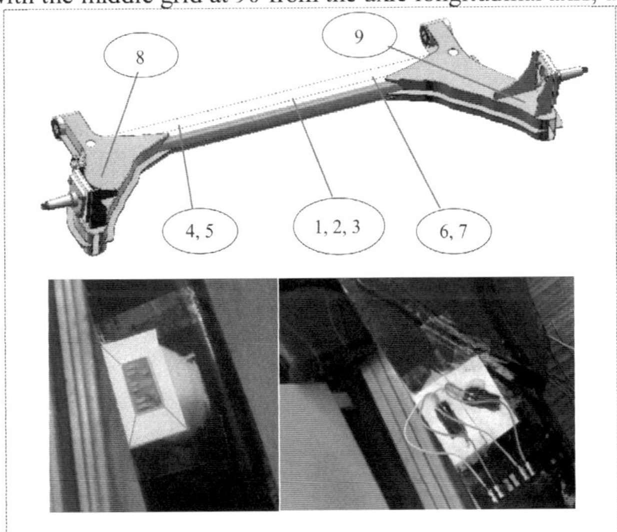


**Figure 8.** Deformation recording points.

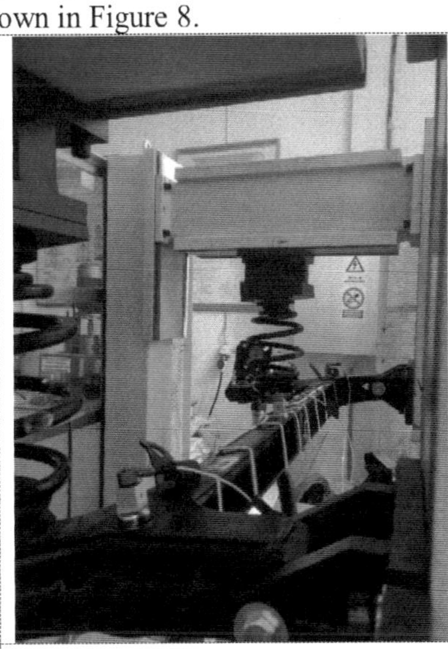

**Figure 9.** Rear axle mounted on the test stand.

The rear axle assembly model developed in Solid Works was transferred to the AnsysWorkbench software database. In order to check the correctness of the file import, it is opened and verified with the Ansys Design Modeler preprocessor. Here geometry corrections can be made, which prevent erroneous analysis results. For example, it is possible to make radii, to join several bodies in a kinematic link or structure component's, to divide the surfaces of parts into two or several surfaces. The last described utility is needed to define joints between bodies, to place loads. The constructed model of the rear axle contains 26 deformable bodies (parts). Some of the structure components bodies we have joined together by means of Boolean operations, to form a single element or part, namely: the U-shaped cross member, the arms made of the bent sheet metal. The end elements made of the bent sheet metal have been joined together to form the rear axle assembly element. As data to define the simulation model, we defined the kinematic torques corresponding to the rear axle mounting mode on the vehicle and on the test stand. We simulated the situation when a disturbing force acts on the right wheel axle. For this purpose, we considered the maximum value of the force occurring on this axle, measured in the dynamic simulation studied in ADAMS. The value of the force is 15000 N and is acting on the rear wheel right axle.

With these considerations, the construction of the finite element analysis model is completed, and we can perform a run of the analysis. Since the model of the axle is particularly complex and large, because it contains a number of over 5 million elements, the simulation takes quite a long time, namely 1400 hours. We performed the simulation on a graphics workstation with 2 Intel OCTA Core Xeon E5-2650 v2 2.60 GHz processors, and 32GB DDR3 ECC memory, a 960GB SSD drive, and an nVidiaQuadro K2000 graphics card. After the analysis there are obtained results concerning strain and stress maps and we selected some important results that are shown in Figs. 10-14.

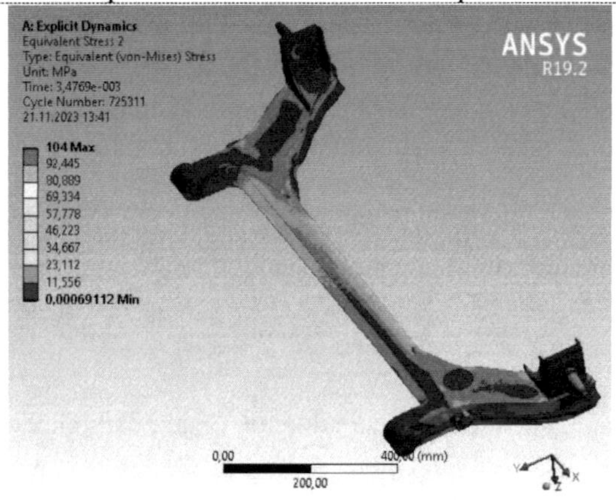


**Figure 10.** Map with the distribution of von Mises equivalent stresses (Top view).

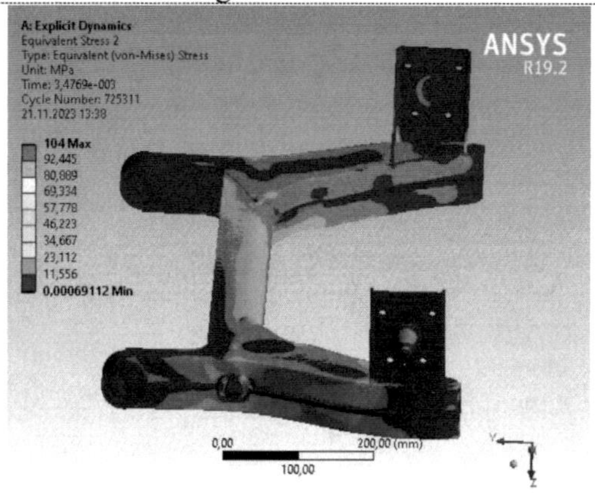


**Figure 11.** Map with the distribution of von Mises equivalent stresses (Side view).

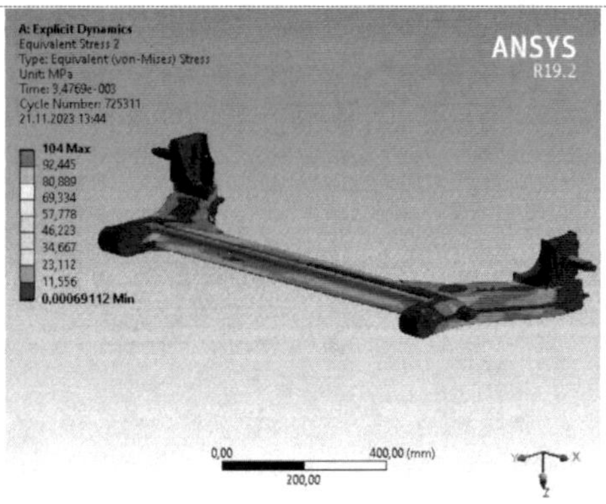


**Figure 12.** Map with the distribution of von Mises equivalent stresses (Back view).

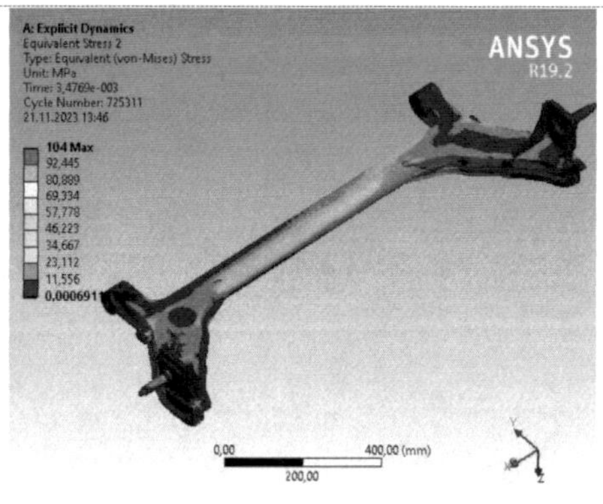


**Figure 13.** Map with the distribution of von Mises equivalent stresses (Front view).

MGCplus and CanheadDataAcquisitionSystem were used for the experimental recording of deformations and parameters:

-Longitudinal force in the rod: rod_force (N);

- Sigma 1 (MPa): mechanical stress mid beam, orientation -90 deg with respect to OX direction;
- Sigma 2 (MPa): Mechanical stress in the middle of the beam, orientation: -135 deg from OX direction;
- Sigma 3 (MPa): Mechanical stress mid beam, orientation -45 deg from OX direction;
- Sigma 4 (MPa): Mechanical stress left beam head, orientation 0 deg from OX direction;
- Sigma 5 (MPa): Mechanical stress left beam head, orientation -135 deg from OX direction;

- Sigma 6 (MPa): Right beam head mechanical stress, orientation 0 deg from OX direction;
- Sigma 7 (MPa): Mechanical stress right beam head, orientation -135 deg from OX direction;
- Sigma 8 (MPa): Mechanical stress left axle arm, orientation -45 deg from OX direction;
- Sigma 9 (MPa): Mechanical stress right deck arm, orientation -135 deg from OX direction.

The X and Y axes are oriented as shown in Fig. 10 as follows: X-axis is the longitudinal axis of the deck and y-axis the transverse axis. These results, as graphical variations laws, are shown in Fig. 14.

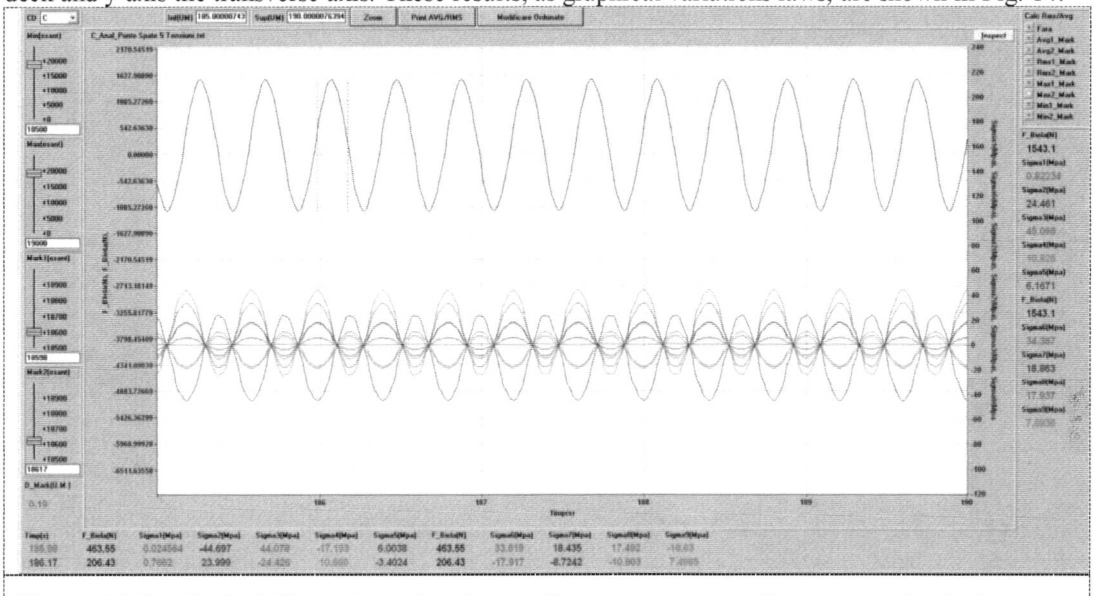

**Figure 14.** Detail of middle cycle number 4, recording parameters: rod force and mechanical stresses.

For the last three start cycles 3, 4 and 5, a stabilized area was selected for which the mean value, RMS (rms) value, maximum value and minimum value were calculated. For these values the mean value and standard deviation were calculated for the three records and are shown in Table 1. On the last line is calculated the peak-to-peak amplitude of the recorded parameters, force in the rod and mechanical stresses.

**Table 1.** Recorded values of connecting rod force and mechanical stresses.

| | F_Rod | Sigma-1 | Sigma-2 | Sigma-3 | Sigma-4 | Sigma-5 | Sigma-6 | Sigma-7 | Sigma-8 | Sigma-9 |
|---|---|---|---|---|---|---|---|---|---|---|
| | (N) | (MPa) | (MPa) | (MPa) | (MPa) | (MPa) | (MPa) | (MPa) | (MPa) | (MPa) |
| | Medium value | | | | | | | | | |
| Cycle 3 | 230.66 | 0.25 | -16.21 | 15.70 | -6.24 | 2.11 | 12.11 | 7.53 | 5.48 | -7.12 |
| Cycle 4 | 232.41 | 0.25 | -16.24 | 15.74 | -6.25 | 2.12 | 12.14 | 7.54 | 5.49 | -7.14 |
| Cycle 5 | 230.65 | 0.25 | -16.01 | 15.68 | -6.21 | 2.11 | 12.10 | 7.52 | 5.48 | -7.09 |
| Average | 231.24 | 0.25 | -16.15 | 15.71 | -6.24 | 2.11 | 12.12 | 7.53 | 5.48 | -7.12 |
| St.Dev | 0.83 | 0.00 | 0.10 | 0.02 | 0.02 | 0.00 | 0.02 | 0.01 | 0.01 | 0.02 |
| | Value (RMS efficacy) | | | | | | | | | |
| Cycle 3 | 976.07 | 0.37 | 29.26 | 28.92 | 11.69 | 3.94 | 21.92 | 12.24 | 11.46 | 11.67 |
| Cycle 4 | 976.54 | 0.37 | 29.28 | 28.94 | 11.69 | 3.94 | 21.93 | 12.25 | 11.47 | 11.68 |
| Cycle 5 | 974.90 | 0.37 | 29.12 | 28.88 | 11.66 | 3.94 | 21.89 | 12.22 | 11.45 | 11.65 |
| Average | 975.84 | 0.37 | 29.22 | 28.91 | 11.68 | 3.94 | 21.91 | 12.24 | 11.46 | 11.67 |
| St.Dev | 0.69 | 0.00 | 0.07 | 0.03 | 0.02 | 0.00 | 0.02 | 0.01 | 0.01 | 0.01 |

| | F_Rod | Sigma-1 | Sigma-2 | Sigma-3 | Sigma-4 | Sigma-5 | Sigma-6 | Sigma-7 | Sigma-8 | Sigma-9 |
|---|---|---|---|---|---|---|---|---|---|---|
| | (N) | (MPa) | (MPa) | (MPa) | (MPa) | (MPa) | (MPa) | (MPa) | (MPa) | (MPa) |
| | Maximum value | | | | | | | | | |
| Cycle 3 | 1591.95 | 0.67 | 18.94 | 50.74 | 8.01 | 6.92 | 38.45 | 21.46 | 20.00 | 6.23 |
| Cycle 4 | 1591.72 | 0.67 | 18.93 | 50.74 | 8.01 | 6.92 | 38.44 | 21.46 | 20.00 | 6.22 |
| Cycle 5 | 1591.07 | 0.68 | 19.13 | 50.65 | 8.03 | 6.93 | 38.41 | 21.43 | 20.01 | 6.26 |
| Average | 1591.58 | 0.68 | 19.00 | 50.71 | 8.02 | 6.93 | 38.43 | 21.45 | 20.00 | 6.24 |
| St.Dev | 0.37 | 0.01 | 0.09 | 0.04 | 0.01 | 0.00 | 0.02 | 0.02 | 0.00 | 0.02 |
| | Minimum value | | | | | | | | | |
| Cycle 3 | -1131.29 | -0.18 | -51.44 | -19.32 | -20.56 | -2.74 | -14.25 | -6.39 | -9.10 | -20.50 |
| Cycle 4 | -1131.29 | -0.18 | -51.40 | -19.31 | -20.55 | -2.71 | -14.23 | -6.39 | -9.07 | -20.48 |
| Cycle 5 | -1131.59 | -0.17 | -51.15 | -19.33 | -20.46 | -2.72 | -14.25 | -6.42 | -9.06 | -20.43 |
| Average | -1131.39 | -0.18 | -51.33 | -19.32 | -20.52 | -2.72 | -14.25 | -6.40 | -9.08 | -20.47 |
| St.Dev | 0.14 | 0.01 | 0.13 | 0.01 | 0.04 | 0.01 | 0.01 | 0.01 | 0.02 | 0.03 |
| | Amplitude | | | | | | | | | |
| | 2722.97 | 0.86 | 70.33 | 70.03 | 28.54 | 9.65 | 52.68 | 27.85 | 29.08 | 26.71 |

- The maximum stress in structure is 104 MPa, as shown in Figs. 10-13; It occurs in the corner area of the joint of the bent side plate with the bent side arm. This area is the most stressed in this stress case;

- Values of 46 MPa for the equivalent stresses occur on the central cross member middle, which decrease to values by 34 MPa in the area nearby to this central area;

- The values for obtained experimentally stresses in the central portion, measured by strain gauge sensors in points 1, 2 and 3 reaches values of 50.65 MPa; These values are similar to those obtained by simulation in ANSYS; - Location of points 4, 5 and 6, 7 is symmetrically placed; The values obtained for the equivalent stresses in these points are different in the finite element analysis. This obtained result is correct, because the stress force is applied only on the right spindle, and points 4, 5 are placed nearby this spindle. Thus, in by FEA simulation in points 4, 5 are obtained stresses with 34 MPa value, and in location of points 6, 7 the equivalent stresses have values of 46 MPa. We also observe an area with values of 57 MPa for the stresses, shown in figure 13. So there is a correspondence between the results obtained by simulation and experimental analysis.

## 4. Conclusions

Fatigue testing methodology is approached in two ways. First, based on the virtual model a dynamic simulation model is built in ADAMS. Through this dynamic simulation we obtain results on the kinematic and dynamic parameters of the drive mechanism in the stand structure, but also data on the stresses of the rear axle elements (springs, rear axle and stabilizer bar). In order to study the distribution of mechanical stresses and specific deformations in the rear axle components we used finite element analysis in ANSYS. Finally, the simulations are validated by experimental analysis on the test stand. Strain gauge transducers have been mounted on the rear axle to measure the deformations. These are compared with those obtained by simulation. The results obtained by simulation and experiment are similar. The result of the research is that by experimental testing each component of the rear axle can be validated in terms of fatigue strength. Furthermore, through finite element analysis, areas with stress concentrators can be found which can be eliminated by design optimization. We performed dynamic finite element analysis for the tested rear axle assembly. This is important because we can compare the stresses obtained by finite element analysis with those determined experimentally. It should be noted that experimentally the stresses can be measured at a limited number of points, in our case nine points.

Instead, by finite element analysis we obtain a map of the distribution of specific stresses, displacements and deformations at all points in the assembly.

**Author's ORCID ID**
I D Geonea: https://orcid.org/0000-0001-8844-0497

**References**

[1] Geonea I, Dumitru N, Rosca A, Craciunoiu N, Racila L and Copilusi C 2019 Computation of a torsion spring stabilizer bar rigidity and fatigue resistance. *In IOP Conference Series: Materials Science and Eng.* (Vol. 568, No. 1, p. 012013). IOP Publishing.
[2] Rosca A S, Craciunoiu N, Geonea I D and Ploscaru C 2018 Design of a test rig for vehicle stabilizer bar fatigue study, *Applied Mechanics and Materials,* 880, 226-231.
[3] Dumitru N, Copilusi C, Geonea I and Margine A 2017 Kinematic and dynamic study of a mechanism for a vehicle front and rear stabilizer bars, *In CONAT 2016 International Congress of Automotive and Transport Engineering 12* (pp. 161-169). Springer International Publishing.
[4] Geonea I, Dumitru N and Dumitru I 2017 Experimental and theoretical study of friction torque from radial ball bearings, *In IOP Conference Series: Materials Science and Engineering* (Vol. 252, No. 1, p. 012048). IOP Publishing.
[5] Ionica V, Dumitru I, Bogdan M and Geonea I 2022 Study on a test bench of a vehicle rear axle fatigue behavior. *In IOP Conference Series: Materials Science and Engineering* (Vol. 1220, No. 1, p. 012040). IOP Publishing.
*[6]* Kumar G S and Kumaraswamidhas L A 2021 Design optimization focused on failures during developmental testing of the fabricated rear-axle housing. *Engineering Failure Analysis, 120, 104999.*
[7] Chaudhary S K, Rajak A K and Ashish K 2021 Failure analysis of rear axle shaft of a heavy vehicle, *Materials Today: Proceedings, 38, 2235-2240.*
[8] Han S J, Li X and Liu K 2011 Based on ADAMS of the analysis of vibration and noise testing of micro-vehicle rear axle, I*n 2011 International Conference on Electric Information and Control Engineering* (pp. 5740-5743). IEEE.
[9] Firat M 2011 A computer simulation of four-point bending fatigue of a rear axle assembly. *Engineering Failure Analysis*, 18(8), 2137-2148.
[10] Dukalski P, Będkowski B, Parczewski K, Wnęk H, Urbaś A and Augustynek K 2021 *Analysis of the influence of motors installed in passenger car wheels on the torsion beam of the rear axle suspension.* Energies, 15(1), 222.
[11] Jiang C, Zhu Q and Li H 2019 Simulation and Optimization of Truck Cab Suspension System Based on ADAMS. *In IOP Conference Series: Earth and Environmental Science* (Vol. 242, No. 3, p. 032039). IOP Publishing.
[12] Ruiz I R, Fricke D, Stachel T and Garcia J 2012 6 DOF Bench Test on a New Active Kinematics Rear Suspension for Functional Development, *SAE International Journal of Passenger Cars-Electronic and Electrical Systems*, 5(2012-01-0550), 238-245.
[13] Zhan N, Zhang X, Jin X and Cao H 2019 Fatigue analysis of weld region in torsion beam rear suspension system, *International Journal of Automotive Technology*, 20, 247-253.
[14] Lukoševičius V, Makaras R, Rutka A, Keršys R, Dargužis A and Skvireckas R 2021 *Investigation of vehicle stability with consideration of suspension performance.* Applied sciences, 11(20), 9778.
[15] Li S, Liu X, Wang X and Wang Y 2020 Fatigue life prediction for automobile stabilizer bar *International Journal of Structural Integrity*, 11(2), 303-323.
[16] Kim S I, Kang S W, Yi Y S, Park J and Kim Y Y 2018 Topology optimization of vehicle rear suspension mechanisms, *International Journal for Numerical Methods in Engineering*, 113(8), 1412-1433.
[17] Shabolin M L 2020 Optimization of the design and experimental study of the stress-strain state

of the rear suspension balancer of an all-terrain vehicle, *Izvestiya MGTU MAMI*, 14(4), 101-104.

# On the influence of different alcohol-type biofuels on performance and engine emissions of an SI engine

**R Conțiu** [1,2] **and R Chiriac**[1]

[1] University Politehnica of Bucharest, Faculty of Mechanical Engineering and Mechatronics, Bucharest, Romania

[2] National Research and Development Institute for Gas Turbines COMOTI, Bucharest, Romania

E-mail: contiurares@yahoo.com

**Abstract.** Greenhouse gas emissions (G.H.G.) from vehicles are the main source of pollution. In this transition period from fossil fuels to the use of synthetic fuels, the diversification of alternative fuels used to fuel internal combustion engines is seen as one of the best alternatives for reducing G.H.G. In Europe, conforming to the stringent emission reduction targets for 2030, as well as trying to fulfil the regulations of the new Euro VII standard will force engine manufacturers to adopt alternative fuel solutions with a low environmental impact. However, fossil fuels will continue to be used but alternative fuels will substantially decrease our dependence on petroleum-derived fuels. Modern simulation software tools make it easy to produce a fairly accurate analysis of how an internal combustion engine works without the need for prototyping. Packages such as Ricardo WAVE or AVL Boost, are relatively cheap and represent accessible tools for developing, designing, and testing modern internal combustion engines. AVL Boost is a widely used engine simulation tool a 1D (one-dimensional) simulation software that allows engineers and researchers to model, simulate, and optimize various internal combustion engines, including spark ignition (SI) engines, diesel engines, and hybrid powertrains. The software also offers a high degree of flexibility in terms of alternative fuel blends. This paper evaluates the influences of using alcohol-based biofuels on performance metrics and pollutant emissions, such as brake power, brake thermal efficiency, and emissions, such as CO, CO2, and NOx in a spark ignition engine. The effects of varying alcohol fractions in the gasoline-alcohol blends on engine performance and emissions are analysed and explained. In this study, several types of ethanol in gasoline blends were simulated (E25, E50, E85). The one-dimensional model of the tested engine was developed based on the design dimensions of the 1.2L TCe H5FT engine produced by Renault. In the context of maintaining constant engine power output, the findings from the simulation results indicate that the utilization of alcohol-based blends with a high volumetric percentage of alcohol (ranging from 70-90%) can result in a substantial increase in fuel consumption, particularly in the case of methanol blends. Consequently, this phenomenon is associated with an elevated emission of carbon dioxide (CO2). However, it should be noted that despite this drawback, there is evidence of an inclination towards reduced emissions of other pollutants due to the enhanced combustion processes facilitated by the higher ratio of oxygenated compounds and a lower peak temperature. Methanol, one of the two alcohols investigated in this study, is not recommended for usage in fuel blends for several reasons. Firstly, the consumption rate of methanol is higher compared to ethanol, which can result in increased fuel usage. Secondly, methanol poses health risks due to its toxicity at certain levels, posing potential hazards in handling and utilization. Moreover, high concentrations of methanol are not easily miscible with gasoline without the addition of co-solvents, further limiting its feasibility as a viable fuel component.

IOP Conf. Series: Materials Science and Engineering 1303 (2024) 012025 doi:10.1088/1757-899X/1303/1/012025

## 1. Introduction

During the transition from petroleum-based fuels to electrification or the use of synthetic fuels, the exploration of clean alternative fuels to power internal combustion engines emerges as a paramount approach. Fossil fuels currently fulfil the majority of global energy demands; however, using them as fuels in a combustion process contributes significantly to greenhouse gas emissions released into the atmosphere. To address this issue, biomass has gained recognition as a crucial resource for the production of alternative fuels. Given the stringent regulations on exhaust gas emissions produced by internal combustion engines worldwide, the transportation sector is compelled to seek alternatives to fossil fuels. Notably, ethanol has gained momentum in recent years as from an additive blended into gasoline to a viable substitute for gasoline. Blending ethanol with petrol can not only extend its availability but also improve its properties like knock characteristics, Reid vapor pressure and combustion characteristics while reducing "well-to-wheel" emissions generated by internal combustion engines. Moreover, if ethanol is derived from renewable sources, it boasts a neutral carbon footprint. Enzymes and microorganisms play a vital role in extracting bio alcohols through the process of alcoholic fermentation, which involves the conversion of cellulose, glucose, starch, carbohydrates, and other sugars into ethanol. These bio alcohols are categorized into distinct types, namely bioethanol, bio propanol, and biobutanol [1].

Currently, the production methods for bio alcohols hold significant importance. The European Union has committed to including a certain proportion of bio-components in all internal combustion engine fuels distributed within its market. Among the various fuels utilized in the European Union, gasoline stands as the second most commonly used, following diesel [2]. Consequently, the consumption of these biocomponents within the fuel aligns with this ratio. The European standard EN 228 outlines the quality requirements and parameters for gasoline sold in the European market [3]. Directive 2009/30/EC mandates that Member States necessitates suppliers to ensure that gasoline with a maximum oxygen content of 2.7%vol. is made available in the market, effectively translating to 7.3% volume of ethanol [4].

Approximately half of the emissions generated can be attributed to private car transport and public transportation. Recognizing the urgent necessity to tackle climate change, governments worldwide are exerting pressure on the transport sector to curtail greenhouse gas (GHG) emissions resulting from fuel combustion. To this end, the majority of major vehicle markets have established limits on the amount of $CO_2$ they are allowed to produce [5].

The road transport sector accounts for approximately 15% of global carbon dioxide emissions. Europe has taken the lead in establishing the most rigorous targets for tailpipe $CO_2$ emissions, aiming for a 37.5% reduction within the next decade, ultimately reaching below 59 g/km by 2030. Ongoing efforts are being made to further refine this target to align with the European green initiative [6].

In recent years, the utilization of alternative fuels in engines has gained significant importance due to growing concerns regarding climate change and the imperative to reduce greenhouse gas emissions. However, despite this growing awareness, numerous governments have set targets for the gradual phasing out of internal combustion engines within the timeframe 2025 to 2040. Consequently, current research endeavours are concentrated on developing alternative solutions that not only meet tank-to-wheel carbon dioxide ($CO_2$) emission targets but also address urban air quality challenges. To accomplish these objectives, the adoption of electrification is paramount, beginning with the widespread implementation of hybrid electric vehicles (HEVs) and progressing towards the vision of zero tailpipe emission vehicles (ZEVs). ZEVs encompass diverse technologies such as battery electric vehicles (BEVs) and fuel cell electric vehicles (FCEVs) [7].

The introduction of a new emissions standard, Euro 7, is currently in progress and is scheduled to be implemented in 2025. Euro 7 will introduce more stringent regulations for internal combustion engines, with a particular focus on diesel vehicles. Notably, it will also establish the first-ever guidelines for brake and tire emissions, which will apply to electric cars as well. According to Euro 7 standards, both gasoline and diesel vehicles must limit their emissions to no more than 60 milligrams [8].

Internal combustion engines are known to emit several harmful exhaust gases, including unburned hydrocarbons (HC), nitrogen oxides (NOx), carbon monoxide (CO), and particulate pollutants. These emissions have significant environmental implications and must be effectively addressed to mitigate their impact.

Alcohol-based fuels offer a promising solution for significantly reducing carbon emissions. This research study aims to present and analyze the outcomes of an engine simulation conducted using various blends of alcohol and petrol. The objective is to gain valuable insights into the performance of different fuel blends and their impact on engine efficiency and emissions. Furthermore, the study examines the potential advantages and disadvantages of employing alcohol as a fuel source, while also comparing it to conventional petrol. By delving into these aspects, a comprehensive understanding of the viability of alcohol-based fuels can be attained.

The polarity of methanol and ethanol plays a significant role as it enables both compounds to form hydrogen bonds and achieve complete miscibility with water. In contrast, gasoline primarily consists of hydrocarbon molecules, rendering it a non-polar substance with a low dipole moment. Consequently, gasoline does not mix with water and exhibits phase segregation when combined with alcohols. This phenomenon results in the formation of distinct layers of gasoline and alcohol-water mixtures [9].

Fortunately, spark ignition engines can tolerate a certain percentage of water-contaminated mixtures without adverse effects. This alleviates the necessity to obtain stable gasoline-ethanol-water blends using co-solvents and emulsifiers [10].

## 2. Methodology

The utilization of virtual modeling through mathematical models implemented in specialized software, such as AVL Boost, Ricardo WAVE, or GT-Power, has emerged as a prominent approach, surpassing traditional prototyping methods. This shift offers significant advantages in the development of internal combustion engines, including time efficiency, cost-effectiveness, and improved overall efficiency. Simulation models provide insights that are often challenging to obtain experimentally, enhancing the understanding of complex engine processes.

Moreover, simulations that incorporate fluid flow modeling have become important tools for designing engine intake and exhaust systems. While most engine processes occur in three dimensions (3D), simulations introduce additional complexity to the model and increase compilation time. As a result, simplified versions of the typically one-dimensional models are commonly employed, striking a balance between accuracy and computational efficiency.

Recently, there has been a rise in articles and publications centring on simulating the functionality of an internal combustion engine. This trend reflects the increasing interest in employing such software for research purposes. While many of these published works primarily involve comparing experimental findings to simulation outcomes, there exist certain publications that exclusively rely on calculations [10]. This study endeavours to construct a one-dimensional (1D) model that simulates the response of a four-stroke spark ignition engine with direct in-cylinder fuel injection when the fuel type is changed from regular petrol to a mixture of various percentages of ethanol and gasoline (E25, E50, E85).

To achieve this objective, the simulation was fine-tuned by utilizing data acquired from the reference engine while operating on gasoline and establishing correlations for simulations conducted with alternative fuel types. The methodology revolves around constructing a one-dimensional model of a spark ignition engine, intending to illustrate the engine's behaviour when fuelled with various blends of alcohol and petrol. The model was created using a software solution provided by AVL namely Boost version 2019.1 and subsequently calibrated using data obtained from the selected reference model, specifically the 1.2L TCe H5FT (Gasoline Direct Injection) engine produced by Renault. Table 1 showcases some significant design attributes of the reference engine.

**Table 1.** Characteristics of the 1.2L TCe H5FT engine [11]

| Engine spec | Model: 1.2L TCe 120(**H5Ht**) | Measuring unit |
|---|---|---|
| Bore | 72.2 | mm |
| Stroke | 73.1 | mm |
| Number of cylinders | 4L | – |
| Peak power | 88 | kW |
| Peak rpm | 5500 | rpm |
| Displacement | 1198 | $cm^3$ |
| Compression ratio | 10:1 | – |

## 3. Simulation study

Figure 1 depicts the simulation program interface with the constructed virtual engine model. Each icon featured in the interface represents a mathematical model. The primary components encompassed in the model are as follows: cylinders, ducts, metering points, boundary conditions, transfer elements (such as throttle valve, injector, and restrictor), volume elements (such as plenum), assembled elements (including air cleaner, air filter, intercooler, catalytic converter, and particulate filter), and supercharger elements (such as a turbocharger or a mechanical compressor). Furthermore, the user has the flexibility to incorporate external elements (linked to AVL Cruise or Fire), control elements (such as the engine control unit, temperature, or pressure sensors) or acoustic elements (like a microphone).

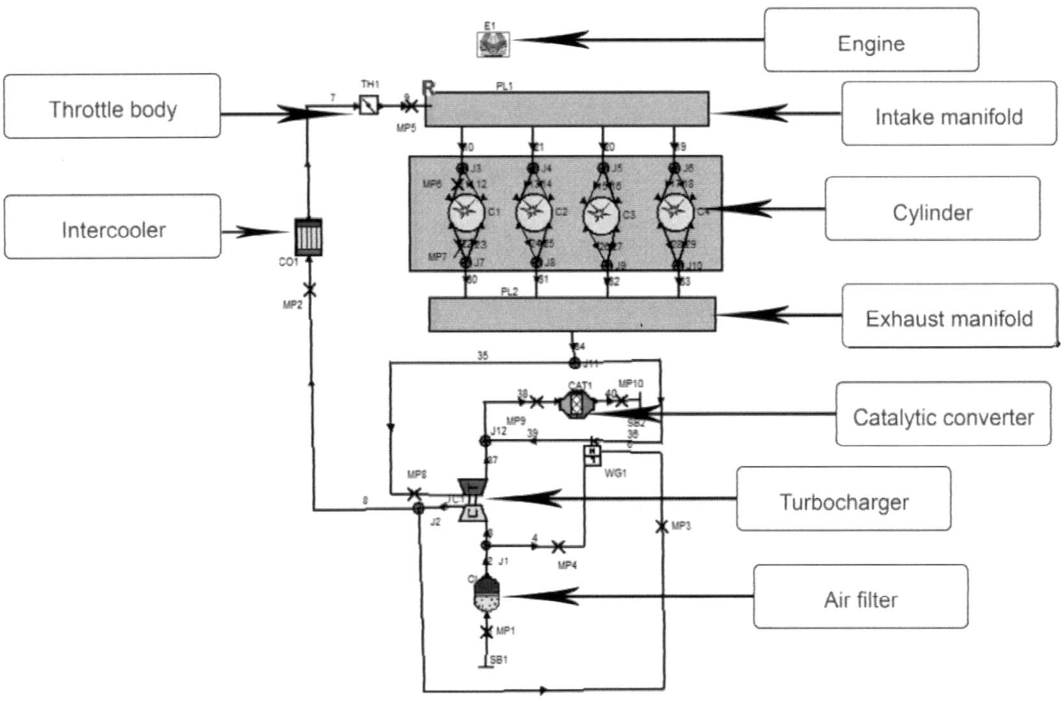


**Figure 1.** Virtual engine model in AVL Boost program

To assess the real-world applicability of the virtual model, an evaluation of its real counterpart is necessary. The simulation model must incorporate the geometric characteristics specific to the actual engine. Subsequently, the model needs to be calibrated by comparing its performance to that of the real engine when fuelled solely with regular petrol. Following the calibration process, the performance and pollutant emissions of the engine will be examined under different fuel scenarios, including gasoline, and ethanol-gasoline blends. To streamline the model calibration procedure, a map for the simulated engine has been developed using statistical data and publicly available results provided by the manufacturer of the selected reference model. This map was tailored to accommodate the stoichiometric air-fuel ratio for each fuel utilized. Table 2 displays the stoichiometric ratios for gasoline and each blend employed in the study.

**Table 2.** The stoichiometric air-fuel ratio and lower heating value for different blends of alcohols and gasoline [12]

| Fuel | A/F Ratio (-) | Lower Heating Value (MJ/kg) |
|---|---|---|
| Gasoline | 14.7: 1 | 43.5 |
| E25 | 13.2:1 | 39.56 |
| E50 | 11.8:1 | 35.61 |
| E85 | 9.85:1 | 29.93 |

The development of an air-fuel ratio map holds significant importance in the optimization of engine performance, ensuring that the engine receives the ideal mixture of air and fuel for various operating conditions. Numerous engine parameters, such as throttle position, engine speed, and load, influence the required air-fuel ratio. The process of "mapping" remains a crucial element in optimizing engine performance. By calibrating the air-fuel ratio according to the specific operating conditions, engineers can achieve the desired balance between power, efficiency, and emission control. With ongoing advancements in engine technologies, further enhancements in the techniques for developing lambda maps are anticipated, resulting in more efficient and environmentally friendly engines. Figure 2 illustrates the generated map for the simulated model. This map was maintained for all fuel mixtures utilized, adapting the ratio based on the stoichiometric air-fuel ratio specific to each mixture type.

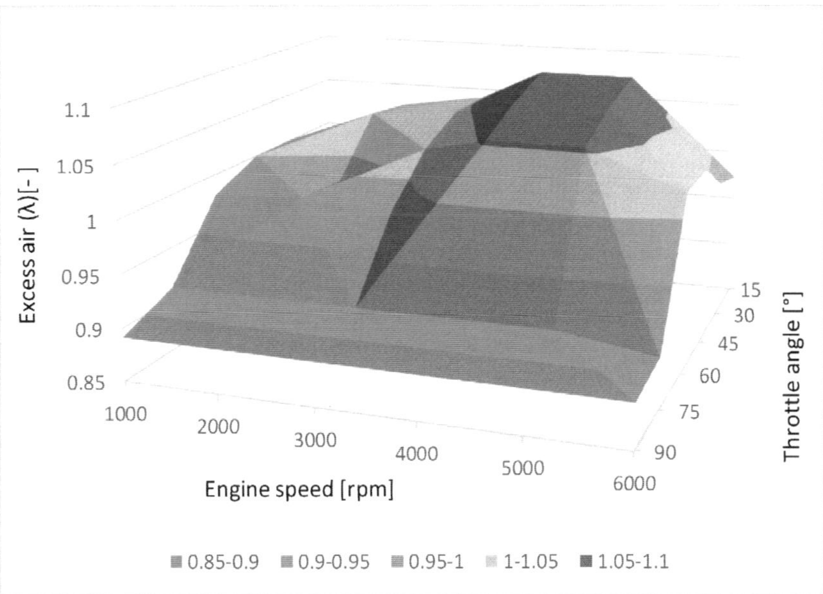


**Figure 2.** Excess air map (λ) used in the model

ESFA-2023 IOP Publishing
IOP Conf. Series: Materials Science and Engineering 1303 (2024) 012025 doi:10.1088/1757-899X/1303/1/012025

When employing fuel blends such as gasoline-ethanol or gasoline-methanol, it was necessary to define their composition using an arbitrary number of species, which is determined directly by the user. The minimum required number of species is seven, namely: fuel, $O_2$, $N_2$, $CO_2$, $H_2O$, CO, and $H_2$. For each species, a conservation equation in terms of mass fractions is solved for every model element.

### *3.1. Fuel Model*

In this simulation, the blending of gasoline and ethanol was modelled using predefined chemistry models within the software. The software offers two approaches for defining the fuel used: classical and general. In the general model, the gas composition can be described using an arbitrary number of chemical species directly specified by the user [13][14]. The minimum required number of chemical species for the general model to operate is seven: fuel, $O_2$, $N_2$, $CO_2$, $H_2O$, CO, and $H_2$. Each species is solved for conservation equations in terms of mass fractions across the various model elements.

The input chemical species for the fuel include gasoline, ethanol, $O_2$, $N_2$, $CO_2$, $H_2O$, CO, $H_2$, H, O, N, and OH. The fuel species are set based on fractions. For this study, the concentrations of the ethanol-gasoline blend were categorized as E0 (pure gasoline), E25, E50, and E85.

The specification of the chemical species, chemical patterns, and fuel composition is depicted in Figure 3.

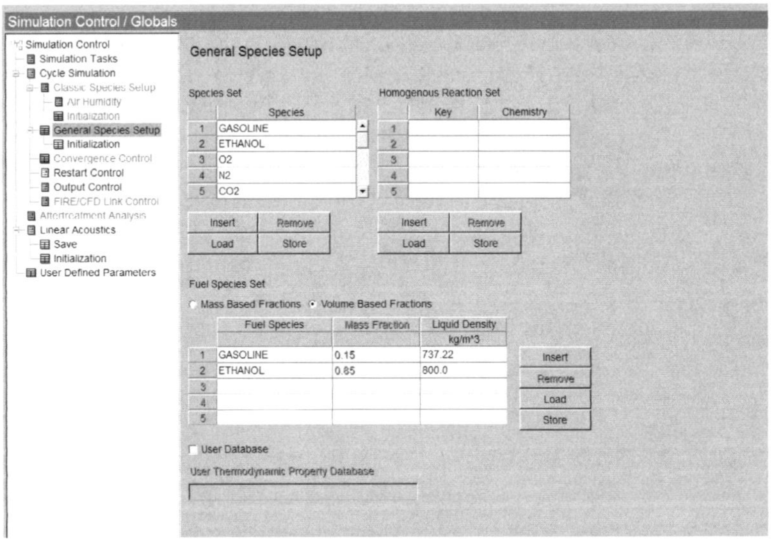


**Figure 3.** Fuel model implemented for this simulation

For this study, the AVL Boost v.2019.1 program was applied as the simulation tool. This program employs mathematical models to calculate regulated pollutant emissions. The combustion process within an internal combustion engine involves a complex series of exothermic chemical reactions. The chemical energy stored in the molecular bonds of the fuel undergoes conversion, first into thermal energy and ultimately into mechanical energy. During this high-temperature oxidation process, a multitude of chemical intermediates are generated and rapidly depleted within fractions of a second. However, certain compounds persist in very low concentrations and are released as pollutant emissions in the engine exhaust, including carbon monoxide (CO), unburned hydrocarbons (HC), and nitrogen oxides (NOx).

### *3.2.* NOx Model

The simulation program employs a kinetic model, developed by Pattas and Häfner, to estimate the generation of nitrogen oxides (NOx) within internal combustion engines. This model is derived from the widely recognized Zeldovich mechanism. It considers various factors including engine speed, fuel properties, and several other parameters such as pressure, temperature, equivalence ratio, volume, and mass of gas within the combustion zone. The incorporation of these parameters is vital as the theory of NOx formation is rooted in the dissociation of $N_2$ and $O_2$ molecules at high temperatures ahead of the flame front [14].

Among the mathematical models employed to calculate exhaust emissions, the NOx model stands out as the most intricate. It comprises six equations that operate in a state of equilibrium. The equations are derived from the following equilibrium reactions (1)-(6).

$$N2 + O \leftrightarrow NO+N \tag{1}$$
$$O2 + N \leftrightarrow NO+O \tag{2}$$
$$N + OH \leftrightarrow NO + H \tag{3}$$
$$N2O + O \leftrightarrow NO + NO \tag{4}$$
$$O2 + N2 \leftrightarrow N2O + O \tag{5}$$
$$OH + N2 \leftrightarrow NO2 + H \tag{6}$$

### *3.3. CO Model*

The model utilized in Boost draws from the research conducted by Onorati. It enables the calculation of the quantity of carbon monoxide (CO) generated during combustion by solving a differential equation derived from the following equilibrium reactions (7)-(8) [14].

$$CO + OH \leftrightarrow CO2 + H \tag{7}$$
$$CO2 + O \leftrightarrow CO + O2 \tag{8}$$

### *3.4. HC Model*

The formation of unburned hydrocarbons in spark ignition engines is a complex phenomenon influenced by multiple factors. Although a comprehensive understanding and a definitive predictive model are still under development, the software uses a phenomenological model to evaluate trends and behaviours regarding unburned hydrocarbons.

In a spark ignition engine, unburned hydrocarbons originate from various sources. The model identifies the following primary sources of unburned hydrocarbons:

- A portion of the air-fuel mixture enters the interstitial space without being burned, as the flame is extinguished upon entering.
- Fuel vapor is absorbed into the oil layer and subsequently deposited on the cylinder wall during the intake and compression strokes, respectively. Desorption occurs when the cylinder pressure drops during the expansion stroke, as complete combustion is no longer feasible.
- The flame is prematurely extinguished before reaching the walls.
- Incomplete combustion of the fresh charge or the occurrence of a misfire [14].

These identified sources are significant contributors to the presence of unburned hydrocarbons in the engine system.

### *3.5. Model calibration*

The calibration of the virtual model is a crucial step in ensuring the accuracy and reliability of the simulation outcomes. It involves adjusting the model's parameters and input data to closely mimic the actual behaviour of the simulated engine.

The calibration process commences by gathering experimental data from the real engine, including measurements of pressure, temperature readings, fuel consumption, and emissions. This data serves as a reference for comparing and validating the simulation results.

The initial stage of calibration entails fine-tuning the physical parameters of the model, such as engine geometry, valve lift laws, and combustion characteristics. These parameters undergo iterative adjustments until the simulated engine's performance closely aligns with the measured data.

In cases where experimental or manufacturer-provided data is unavailable, the model is calibrated by comparing the power and torque curves obtained from simulations at full load (wide open throttle or WOT) with an external characteristic provided by the manufacturer. Consequently, the primary simulation parameters are established within the simulation control window. To assess the parameters of interest, the simulation time is set to cover 15 engine duty cycles and the average finite element dimension is set at 25mm in order to keep the computational time to a minimum. The results are then plotted on a shared graph for comparison. The power and torque curves demonstrate a close resemblance between the simulated model and the actual engine. The maximum calculation error for actual power and torque is approximately 4.8%, with an average not exceeding 5%. The overall shape of the curves also exhibits similarity. The calibration results of the model are depicted in Figures 4,5,6.

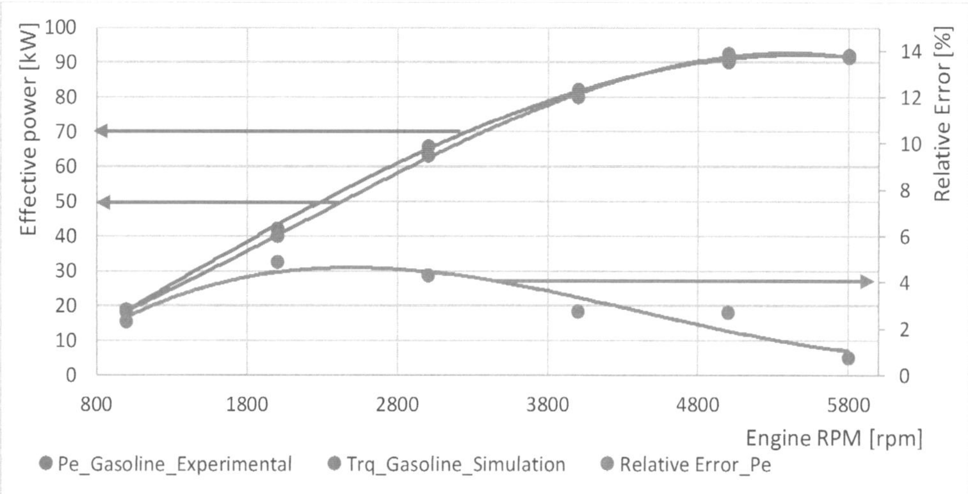


**Figure 4.** Calibrating the power developed by the engine

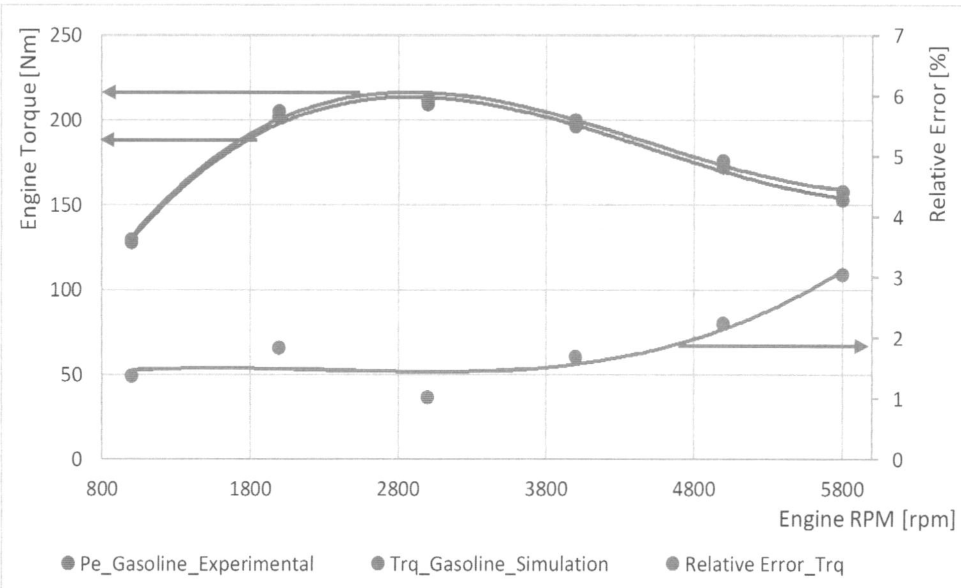


**Figure 5.** Calibration of engine torque

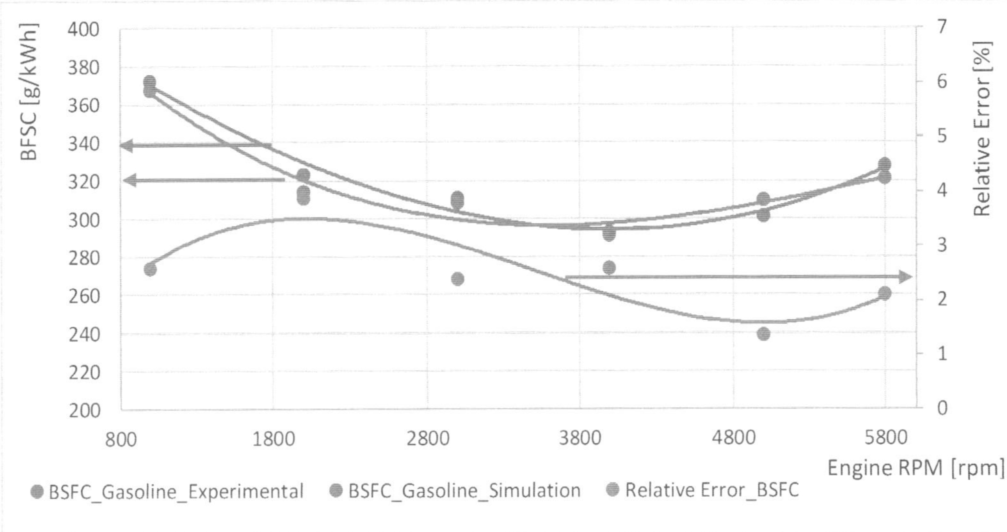


**Figure 6.** Fuel consumption obtained for calibration curves

Following these results obtained for the operation of the petrol engine, it can be considered that the model is ready for the change of the fuel type.

## 4. Results and discussions

Once the model was adjusted and its accuracy validated, the concentration of fuel alcohol was increased. The engine simulation was run at full load across a wide range of engine speeds, from 1000 to 5800 rpm with an increment of 1000 rpm. The fuels used for these simulations were E25, E50, and E85, corresponding to ethanol blended with conventional gasoline at volumetric percentages of 25%, 50%, and 85%, respectively. Simultaneously, the engine operation on standard gasoline was also maintained for comparison.

During the simulations, it was imperative to ensure that the engine's performance closely resembled that of operating on gasoline in order to not disturbed the end user experience. By utilizing the data acquired from each case ran on the engine model external characteristic, power and torque curves were derived along with other curves regarding the principal exhaust emissions (CO, HC, NOx).

### *4.1. Engine performance characteristics*

Effective power is a key parameter for evaluating the performance of an internal combustion engine. To analyse its variation with engine speed, full load conditions were considered for different fuel compositions: E25, E50, E85, and pure petrol (E0). As the ethanol content in the fuel increased, the power exhibited a small decrease across the entire range of engine RPMs even if it was tried to maintain them closely similar. This behaviour was highlighted specifically when compared to high ethanol blends like E50-E85 fuels. Running the engine on regular gasoline resulted in higher power output at all engine speeds.

The higher evaporation heat of ethanol, relative to gasoline, contributes to the cooling of the air-fuel charge and increases its density. This can potentially improve combustion by bringing the equivalence ratio of the fuel blend closer to the stoichiometric state. However, the calorific value of ethanol is lower than that of gasoline, which can offset the positive effects mentioned above. Consequently, reduced power output is observed when using blended fuels.

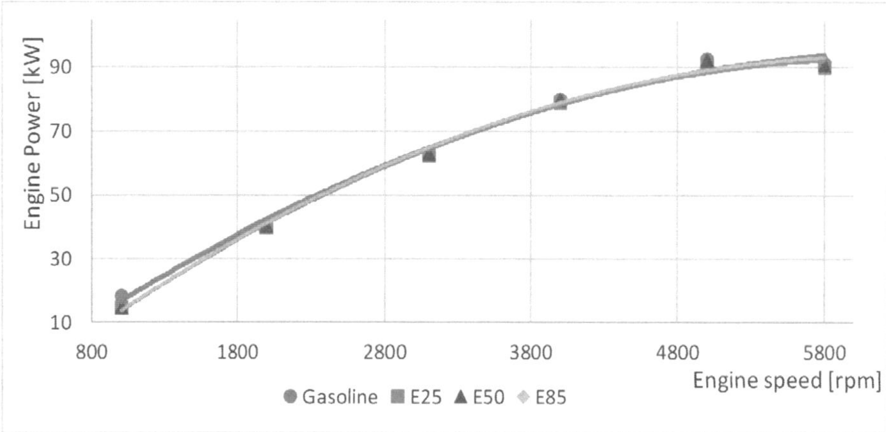


**Figure 7.** The effective power developed by the engine

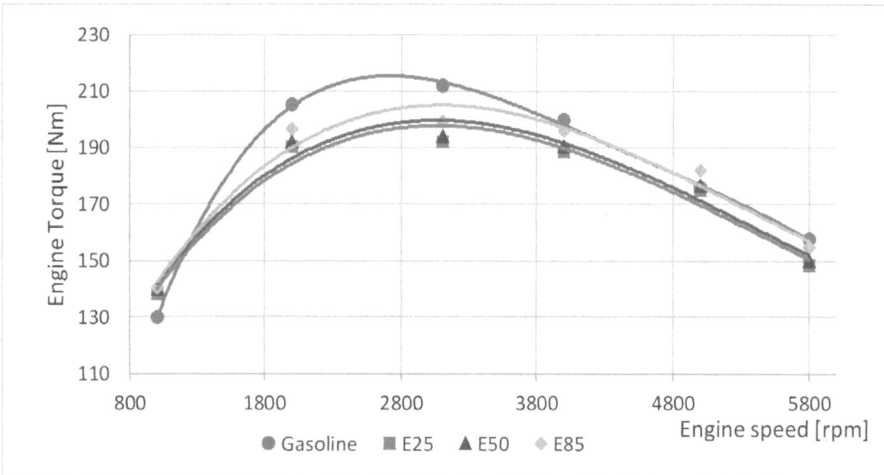


**Figure 8.** The effective torque developed by the engine

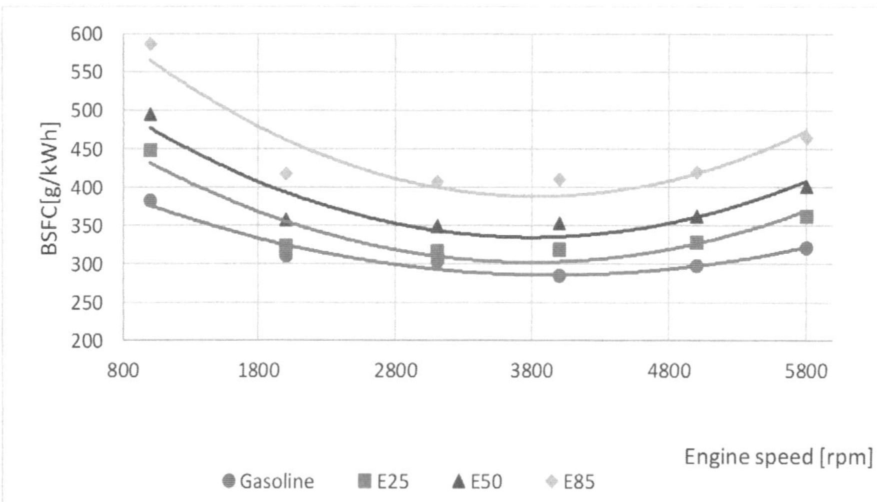


**Figure 9.** The break specify fuel consumption (BSFC) for the used fuels

The break specify fuel consumption (BSFC) serves as an important index that measures the efficiency of a thermal engine. It is a crucial parameter in the automotive industry, utilized for evaluating the effectiveness of internal combustion engines.

In Figure 9, the relationship between break specify fuel consumption and engine speed is depicted for regular petrol and ethanol-gasoline blends. The graph illustrates a notable increase in fuel consumption as the percentage of ethanol in the fuel rises. This can be attributed to the lower heating value and stoichiometric air-fuel ratio of ethanol. Essentially, when aiming to achieve the same energy output through combustion, a greater quantity of fuel is required due to these lower values. The E85 blend demonstrates the highest specific fuel consumption among the tested blends.

Despite the rise in fuel consumption associated with higher ethanol percentage blends, the benefits outweigh this drawback. These advantages include improved combustion properties, resulting in decreased engine deposits and potentially cleaner combustion, as well as the utilization of a diverse range of renewable resources as raw materials for alcohol production.

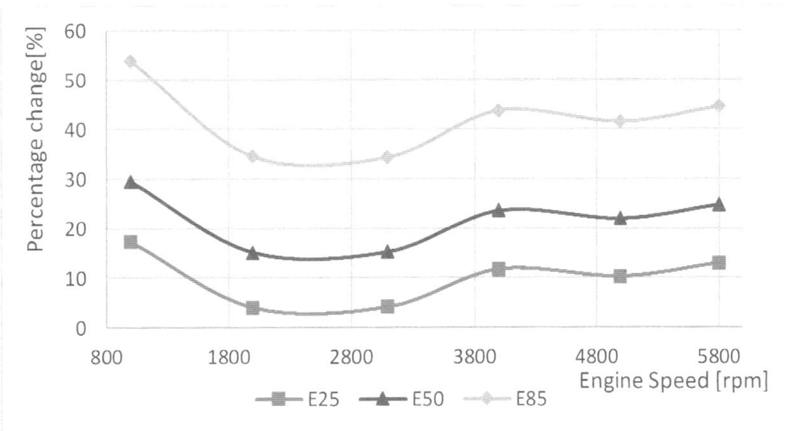


**Figure 10.** Percentage change in the break specify fuel consumption

In Figure 10, the graph demonstrates the percentage increase in brake specific fuel consumption. It clearly indicates that as the ethanol percentage rises, fuel consumption also increases, resulting in a corresponding drop in engine efficiency. As expected, the decrease in effective efficiency, as depicted in Figure 11, aligns with the increasing alcohol percentage.

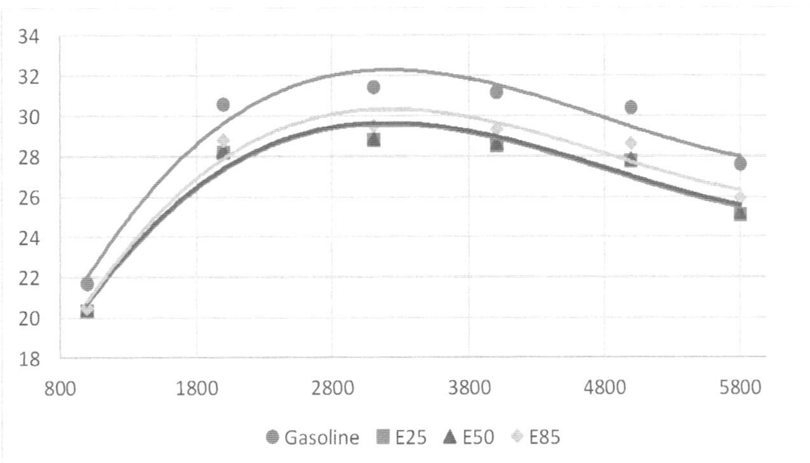


**Figure 11.** Engine effective efficiency determined for different ethanol-gasoline blends

IOP Conf. Series: Materials Science and Engineering 1303 (2024) 012025 doi:10.1088/1757-899X/1303/1/012025

*4.2. Enhancing engine performance through increasing the compression ratio.*

One significant advantage of utilizing alcohols as fuels in internal combustion engines is their higher-octane number in comparison to petrol. This characteristic enables engines to operate with higher compression ratios, effectively preventing detonation. As a result, engine performance is enhanced, leading to improved thermal efficiency.

To maximize the engine's thermal efficiency, one approach is to raise the engine's geometric compression ratio to take advantage of the alcohol's higher-octane rating. By maintaining the other parameters constant, the compression ratio was incrementally increased through iterative adjustments. The increase continued until the calculated required octane number, as determined by the model for the alcohol blends in petrol, matched the octane number calculated for regular gasoline itself. The simulation was performed for two ethanol blends, specifically E50 and E85.

Table 3 provides the research octane numbers (RON) sourced from literature for the most commonly used alcohols in spark ignition engines.

**Table 3.** Research/Motor octane number for different fuels [15]

| Fuel | RON (-) | MON (-) |
|---|---|---|
| Gasoline | 95 (min) | 85 (min) |
| Ethanol | 120 | 99 |
| Methanol | 126 | 96 |

By employing the simulation software, it is possible to compute the minimum octane number needed for the engine to function without experiencing detonation. This calculation is performed using a mathematical approach [14]. The calculated octane number is illustrated in Figure 12.

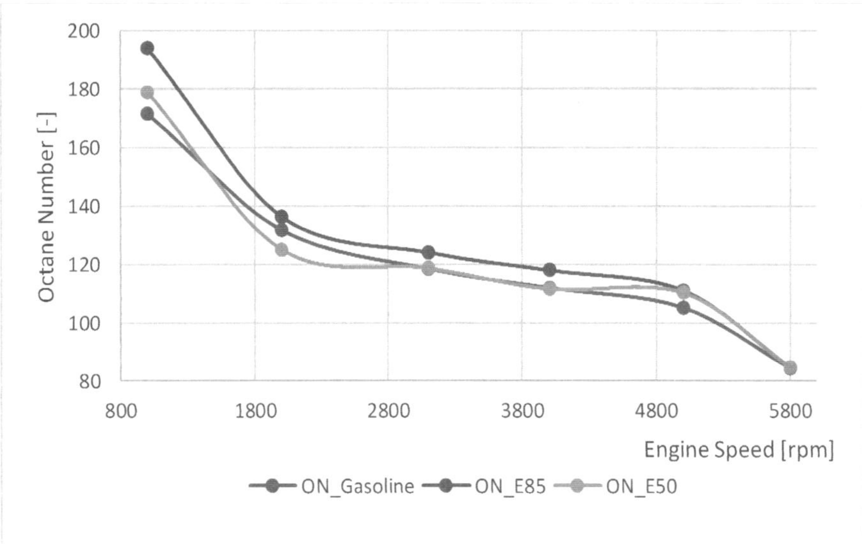


**Figure 12.** The octane number calculated from the simulation

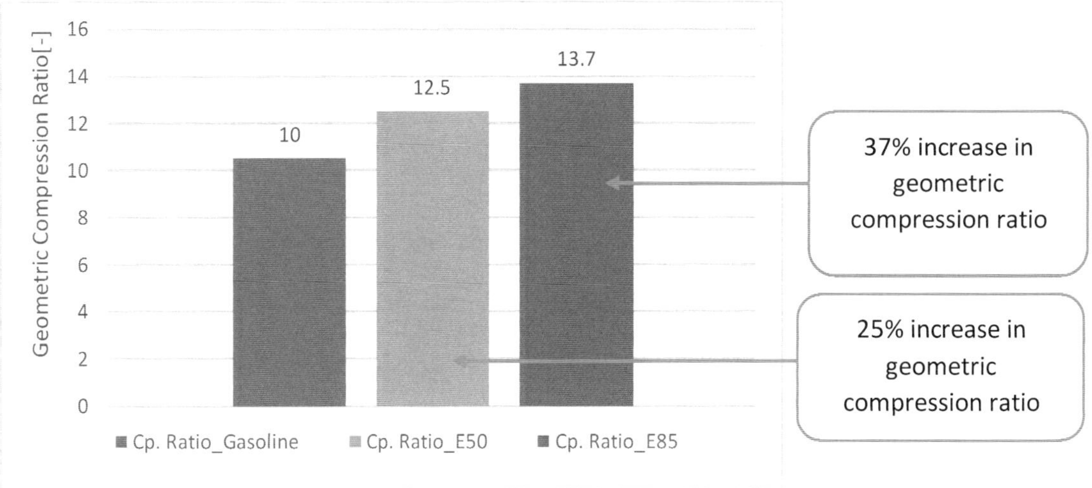


**Figure 13.** Compression ratio for obtaining an octane number similar to regular gasoline

As depicted in Figure 13, to achieve an octane number similar to that of petrol, the compression ratio is raised by 25% for E50 and 37% for E85. This elevation in the compression ratio has a direct impact on the power and torque generated by the engine.

Figure 14 illustrates the actual efficiency of the engine and it's demonstrating that increasing the compression ratio leads to enhanced engine efficiency.

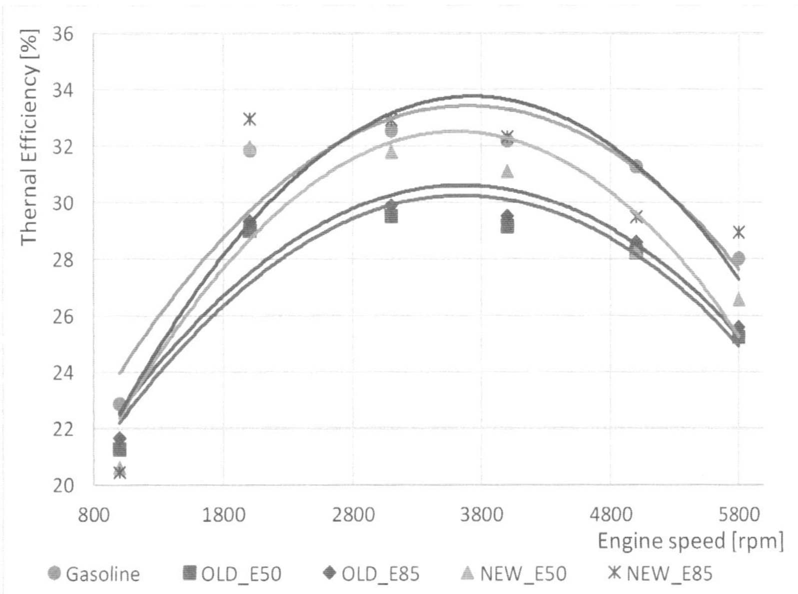


**Figure 14.** The actual engine efficiency before and after increasing the compression ratio

*4.3. Exhaust emissions*

The impact of exhaust emissions on the environment is of great importance when evaluating vehicle's engine performance. To mitigate the environmental effects, the utilization of gasoline-alcohol blends has emerged as a potential solution. These blends combine traditional petrol with a certain proportion of ethanol, typically sourced from renewable resources like corn or sugar cane. Ethanol exhibits a higher oxygen content than petrol, leading to more complete combustion and consequently lower carbon monoxide (CO) emissions. Moreover, ethanol is considered a renewable fuel with an almost neutral carbon footprint since it is primarily derived from plant materials. During the growth phase, plants absorb $CO_2$ from the atmosphere, which is subsequently released upon ethanol combustion. When considering the entire life cycle, the carbon footprint of ethanol may be lower than that of gasoline due to the ability of plants used for ethanol production to offset some of the emissions.

It is important to acknowledge that the emission levels resulting from the combustion of petrol-ethanol blends can differ based on various factors, including ethanol concentration, engine technology, and maintenance practices. Blends with higher ethanol content, such as E85 (comprising 85% ethanol), have the potential for greater reductions in pollutant emissions but may necessitate specific engine adaptations. Additionally, it is worth noting that the emission values obtained from the simulation represent the quantities generated after the combustion process, before undergoing any treatment in an after-treatment system.

Nitrogen oxides (NOx) are a group of chemical compounds composed of oxygen and nitrogen, which is formed through a reaction between these elements at high temperatures. It is commonly generated during the combustion of fuels such as diesel, natural gas, and organic matter. NOx, which encompasses nitrogen oxides like NO and $NO_2$, contributes to the formation of smog in urban areas, leading to poor air quality. These emissions are also associated with acid rain as they react with ammonium, water vapour, and other substances to produce nitric acid ($HNO_3$) and small particles.

Figure 15 illustrates the impact of fuel mixtures on NOx emissions. The graph demonstrates that NOx emissions tend to increase with higher engine speeds, reaching a peak around 4000 rpm and subsequently decreasing at maximum engine speed. Moreover, increasing the ethanol percentage in the fuel mixtures leads to a reduction in NOx emissions.

This phenomenon can be attributed to the higher ethanol content in the blends, which has the property of lowering the temperature in the combustion chamber. The lower temperature, as shown in the Figure 16, is a result of two significant characteristics of alcohol. Firstly, the latent heat of vaporization contributes to cooling the combustion chamber during fuel injection. Secondly, the production of more triatomic molecules in the combustion process increases the heat capacity of the gas, resulting in a lower combustion gas temperature.

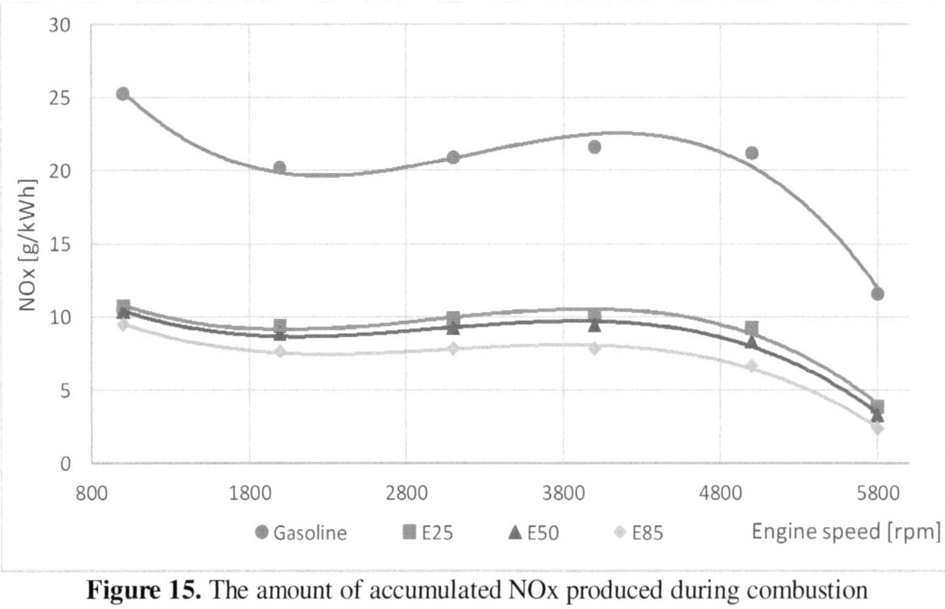


**Figure 15.** The amount of accumulated NOx produced during combustion

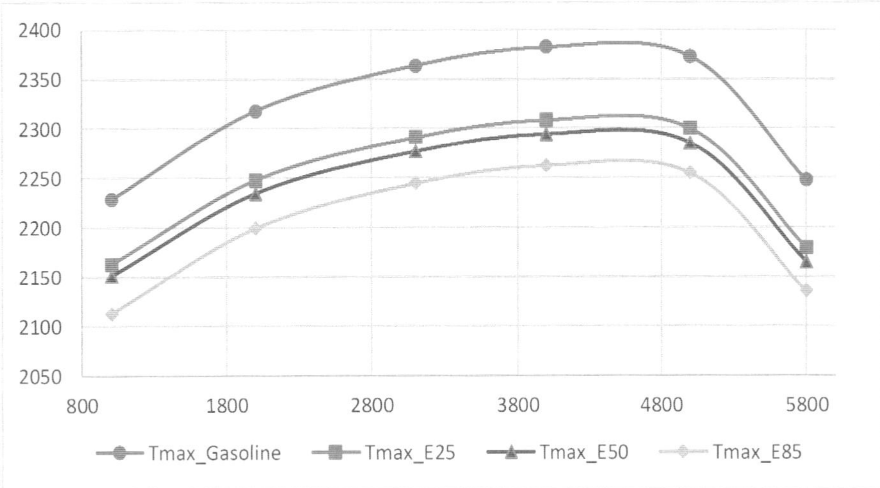


**Figure 16.** The peak temperature during the combustion process

Carbon monoxide (CO) is a hazardous by-product that results from the incomplete combustion of hydrocarbon fuels. It is commonly observed in spark ignition engines, particularly when they operate at specific loads with rich fuel mixtures. However, carbon monoxide can also be produced under conditions of excess air, although the concentrations formed in such cases are minimal. These minimal concentrations usually arise from brief periods of spark ignition engine operation with rich mixtures or from regions within the combustion chamber where inhomogeneous mixtures contain localized pockets of rich mixtures.

By examining Figure 17, it can be inferred that an increase in ethanol content leads to a decrease in the amount of carbon monoxide produced. One possible explanation for this phenomenon lies in the presence of additional oxygen within the ethanol molecule, which promotes further oxidation during the combustion process. As a result, the formation of carbon monoxide is reduced.

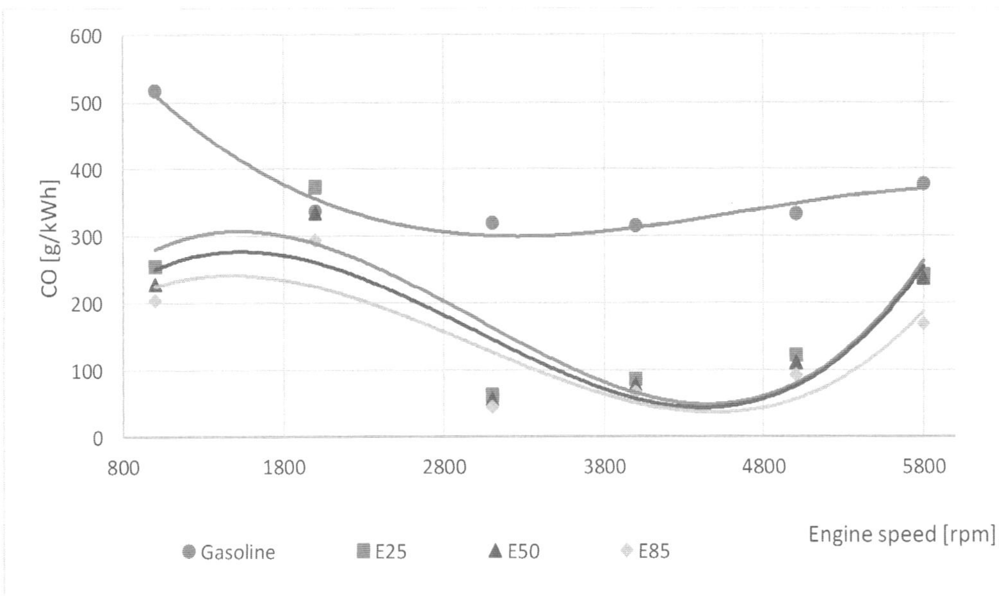


**Figure 17.** The amount of accumulated CO produced during combustion

Unburned hydrocarbons (HC) comprise a diverse group of chemical compounds consisting of hydrogen and carbon. The release of unburned hydrocarbons occurs when there is an insufficient amount of oxygen to support complete combustion. During the combustion process, hydrocarbon compounds are generated through the fragmentation of extended molecular chains and subsequent transformations of the original fuel, resulting in the formation of hydrocarbon compounds that were not initially present in the fuel. However, it is worth noting that the lower temperature in the combustion chamber may also lead to an increase in the fraction of unburned hydrocarbons, which can have environmental implications *[16]*.

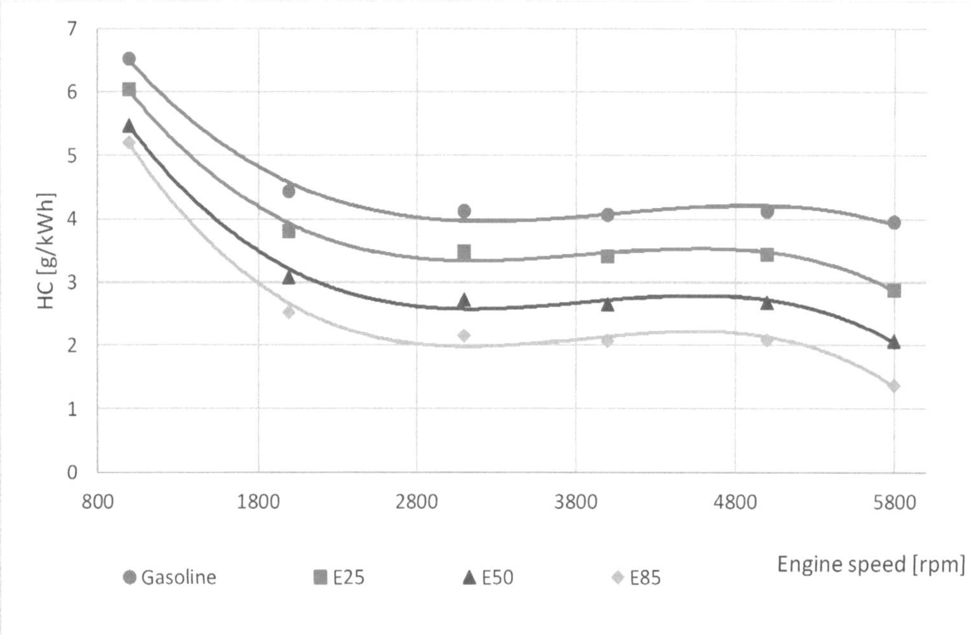


**Figure 18.** The amount of accumulated HC produced during combustion

Figure 18 provides a visual representation of the impact of ethanol-gasoline blends on HC emissions. The graph demonstrates that as the proportion of ethanol increases, the concentration of HC decreases. The reduction in HC emissions can be attributed to similar factors as the decrease in CO concentration, as both are influenced by the combustion process. Ethanol promotes more complete combustion due to its higher oxygen content, leading to a shift towards stoichiometric combustion and a decrease in the emission of unburned hydrocarbons.

During the complete combustion of fuel, the carbon present in the fuel structure is converted into carbon dioxide ($CO_2$), which accounts for approximately 13.5% of the volume of the exhaust gas composition. The quantity of carbon dioxide produced serves as a direct indicator of fuel consumption. While $CO_2$ is a natural component of the Earth's atmosphere and not classified as a pollutant, it is a significant contributor to the greenhouse effect, which contributes to global warming and climate change.

Figure 19 depicts the relationship between engine speed, as shown on an external engine characteristic, and the emissions of carbon dioxide resulting from combustion.

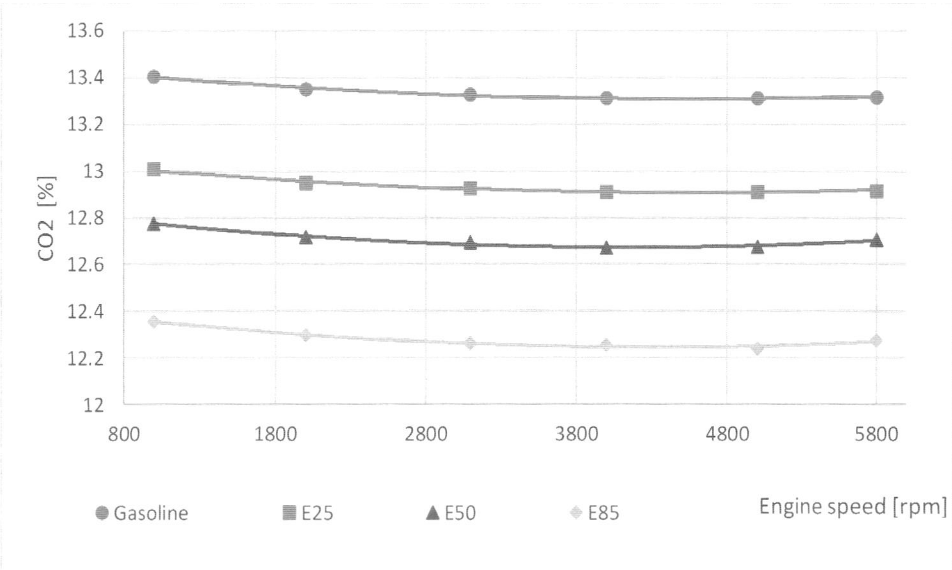


**Figure 19.** The amount of accumulated $CO_2$ produced during combustion

## 5. Conclusion

The findings of this study highlight the importance of continued research and development in the field of ethanol-blended fuels to maximize their performance and emission advantages.

When using high-volume percentage blends of alcohols in gasoline (70-90% vol.), fuel consumption can rise with about 50%, as shown in Figure 10. However, this is not a concern for countries like Brazil, which can leverage the abundant biomass from sugar production to produce ethanol. Simulations indicate that an optimal range of 25 to 50% vol. ethanol in gasoline can improve fuel quality without significantly impacting consumption.

In general, the production of CO emissions is heavily influenced by the excess air ratio and the C/H ratio of the fuel. As the percentage of ethanol increases, the concentration of both CO and HC decreases. Blends containing 85% vol. ethanol (E85) exhibit the lowest CO and HC emissions.

NOx emissions exhibit an upward trend with increasing speed, followed by a sharp decrease around the nominal speed. The use of E85 fuel yields the lowest NOx emissions.

Ethanol, derived from renewable sources, offers the potential to reduce global greenhouse gas emissions when blended with petrol. Although burning ethanol releases CO2, the carbon footprint can be offset by the plants used for its production, making it a more environmentally friendly option.

Alcohols demonstrate good resistance to detonation. By adjusting the compression ratio when utilizing ethanol blends in petrol, it is possible to enhance effective engine efficiency while avoiding detonation.

Blends of ethanol and gasoline, particularly those with higher ethanol concentrations like E85, may require specific engine modifications to optimize performance. These modifications could involve adjustments to the fuel system, materials, and engine control strategies. Ensuring engine compatibility and implementing proper maintenance practices are crucial for maximizing the benefits of ethanol and gasoline blends.

In summary, ethanol-gasoline blends offer advantages such as reduced emissions, decreased greenhouse gas emissions, and economic sustainability, albeit at the expense of slightly lower engine performance. However, achieving optimal results necessitates engine compatibility, appropriate maintenance, and continuous research efforts to further enhance the performance and emission benefits of ethanol-blended fuels.

**Acknowledgments**
This work was carried out through the "NUCLEU" Program of the National Research, Development and Innovation Plan 2022-2027, carried out with the support of Ministry of Research, Innovation and Digitization, project no. PN23.12.07.01.

The paper was realized also with the support of the National Research and Development Institute for Gas Turbines COMOTI.

The authors of this paper acknowledge the AVL-AST team from AVL List GmbH for the special support in performing the simulation activity.

**Author's ORCID ID**
R Chiriac: https://orcid.org/0000-0002-0568-4881

## 6. References

[1] Obergruber M, Hönig V, Procházka P, Kucerová V, Kotek M, Boucek J and Marík J 2021 Physicochemical properties of biobutanol as an advanced biofuel *Materials* **14**(4) 914
[2] Statistical Report 2018, Fuels Europe, Brussels, Belgium, 2018 (https://www.fuelseurope.eu/publications/publications/statistical-report-2018)
[3] European Committee for Standardization 2017 DIN EN 228:2017 Automotive fuels - Unleaded petrol - Requirements and test methods Brussels Belgium
[4] The European Parliament 2009 Directive 2009/30/EC of the European Parliament and of the Council of 23 April 2009 *Off. J. Eur. Union*
[5] Our World in Data, "Emissions by sector", (https://ourworldindata.org/emissions-by-sector#annual-co2-emissions-by-sector)
[6] (https://ec.europa.eu/info/strategy/priorities-2019-2024/european-green-deal_en)
[7] Conway G, Joshi A and Leach F 2021 A review of current and future powertrain technologies and trends in 2020 *Transportation Engineering* **5** 100080
[8] (https://ec.europa.eu/commission/presscorner/detail/en/ip_22_6495)
[9] Aravind A, Saravanan S, Bhavin K. B, Arul V and Selvan M 2021 Investigation on methanol–gasoline–aluminum oxyhydroxide nanoparticle blends on the emission characteristics of an SI engine *Enviromental Quality Management* **31**(2) 66-75
[10] Wang X, Gao J, Chen Z, Chen H, Zhao Y and Huang Y 2022 Evaluation of hydrous ethanol as a fuel for internal combustion engines: A review *Renewable Energy* **194** 504-525
[11] (https://casse-moteur-renault.weebly.com/uploads/1/1/8/9/11892129/brochure_1200_turbo_gdi_v1_maj_ 2014__1_.pdf)
[12] Beller G, Arpad I, Kiss JT, Kocsis D 2021 AVL Boost: a powerful tool for research and education *Journal of Physics: Conference Series* **1935**(1) 012015
[13] AVL Boost 2019 AVL User Guide
[14] AVL List Gmbh. 2019 AVL Boost—Theory
[15] AlRamadan A, Mani Sarathy S and Badra J 2021 Unraveling the octane response of gasoline/ethanol blends: Paving the way to formulating gasoline surrogates *Fuel* **299** 100882
[16] Simeon Iliev 2019 Comparison of Ethanol and Methanol Blending with Gasoline Using Engine Simulation *Biofuels - Challenges and opportunities*

IOP Conf. Series: Materials Science and Engineering 1303 (2024) 012026 doi:10.1088/1757-899X/1303/1/012026

# Experimental research on vehicles equipped with a GMP Hybrid for the new Euro 7 emissions standard

**Gh Bancă[1], F Ivan[2], M Toma[1], V Nişulescu[1], D A Micu[1]**

[1]University Politehnica of Bucharest, Romania,
[2]University of Piteşti, Romania

E-mail: gheorghe.banca@yahoo.com

**Abstract**. With the increase the number of vehicles on European roads, the road transport has become the biggest source of air pollution in the cities of the old continent. According to European Commission estimates in 2018, more than 39% of NOx (nitrogen oxides) emissions and 10% of PM2.5 and PM10 emissions (mechanical particles with diameters of 2.5 and 10 micrometers) were attributed to road transport. It is known that, in order to limit to some extent, the uncontrolled increase in harmful emissions from road vehicles, the European Commission has gradually introduced new pollution standards to be met by manufacturers of new vehicles. In the second half of 2022, the Commission presented a new proposal to reduce pollutant emissions from new vehicles sold in the European Union (EU), defining the new Euro 7 pollution standards. Compared to Euro 6 by 2035, according to European Commission estimates, these new Euro 7 standards, together with a progressive increase in the number of electric vehicles sold, aim to reduce pollutant emissions from private vehicles and vans by around 35% for NOx emissions and around 13% for particulate emissions. Therefore, both the Euro 7 and $CO_2$ emissions standard (CAFE) for vehicles have been designed to improve air quality in cities and reduce greenhouse gas emissions, including by increasing the use of electric vehicles. Starting from this hypothesis, the first part of the paper presents the experimental model developed in the AMESim simulation platform of a mid-size vehicle equipped with a hybrid GMP, tested on a roller bench, which simulates pollutant emissions in this new Euro 7 standard. In the second part of the paper, the simulation results and possible causes influencing hybrid GMP behavior leading to the evolution of pollutant emissions are analyzed. Finally, the paper formulates conclusions on the evolution of pollutant emissions.

## 1. Introduction

It is known that, in order to limit the negative environmental impact of the continuous increase in the number of cars in Europe, the European Commission has adopted numerous pollution standards, that we know today as Euro 1, Euro 2,..., Euro 6.

The next major step in terms of emissions regulations for cars is the transition to Euro 7 from July 2025. Although the level of accepted pollutant emissions and test conditions are currently at the proposal stage, this standard represents a significant challenge for the car and commercial vehicle industry. The European Commission's proposal to introduce the Euro 7 standard aims to reduce air pollution in major cities, caused by new cars sold in the European Union, and represents an intermediate step to meet the European Green Deal's ambitious goal of reducing to "zero pollution" or "zero $CO_2$ emissions" in 2035 [1].

As road transport is the main source of air pollution in major cities, the new pollution standards align pollutant emissions of CO (carbon monoxide) and NOx (nitrogen oxides) for vehicles equipped with SI engine (spark ignition engines) and CI engine (compression ignition engines) to their lowest value at Euro 6. The standards also impose for the first time a limit for $NH_3$ (ammonia) emissions for vehicles equipped with SCR (Selective Catalytic Reduction) and a limit for particles resulting from the braking system of the vehicle, figure 1.

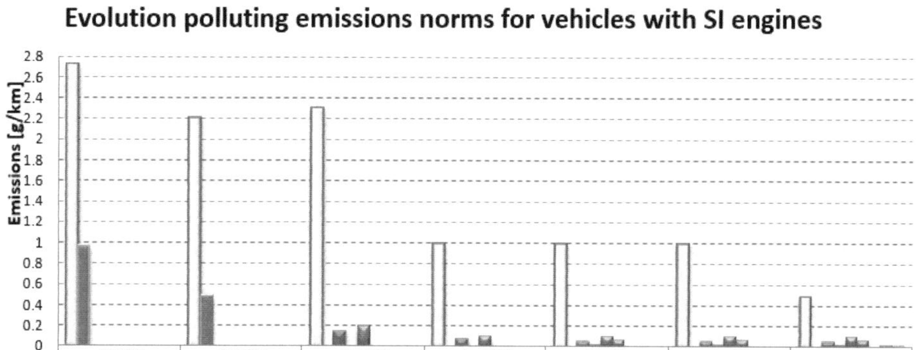


| | Iul-1992 EURO 1 | Ian-1996 EURO 2 | Ian-2000 EURO 3 | Ian-2005 EURO 4 | Sep-09 EURO 5 | Sep-14 EURO 6 | Jul-25 EURO 7 (proposed) |
|---|---|---|---|---|---|---|---|
| CO [g/km] | 2.72 | 2.2 | 2.3 | 1 | 1 | 1 | 0.5 |
| HC+NOx [g/km] | 0.97 | 0.5 | | | | | |
| NOx [g/km] | | | 0.15 | 0.08 | 0.06 | 0.06 | 0.06 |
| Particule [g/km] | | | | | 0.005 | 0.0045 | 0.0045 |
| THC [g/km] | | | 0.2 | 0.1 | 0.1 | 0.1 | 0.1 |
| NMHC [g/km] | | | | | 0.068 | 0.068 | 0.068 |
| PN [#/km] | | | | | | 6E+11 | 6E+11 |
| NH3 [g/km] | | | | | | | 0.02 |
| PM10 [Bk] [mg/km] | | | | | | | 0.007 |

**Figure 1.** Evolution of SI engine pollutant emission norms.

In order to study the impact brought by the severity of the Euro 7 standard on a mid-size car, the author proposed a mathematical model made with the help of the HOT interface (Hybrid Optimization Tool), integrated into the Simcenter AMESim simulation platform. This interface allows engineers involved in the development of new vehicle models to make an estimate at an early stage of vehicle design, in order to know the energy potential of a vehicle architecture and a chosen test cycle.

## 2. Presentation of the Model

The proposed model is specific to a car, from the Volkswagen Golf GTE class, equipped with a rechargeable parallel electric hybrid propulsion system (PHEV - Plug-in Hybrid Electric Vehicles).

This vehicle has been approved for the Euro 6 emissions standard and is equipped with a powertrain combining a 1.4l combustion engine and an electric car, coupled to a 6-speed DSG (Direct Shift Gearbox) dual clutch gearbox. This rechargeable hybrid powertrain provides the vehicle with outstanding dynamic, economical and environmentally friendly performance, such as reaching a top speed of 222km/h, acceleration from 0 to 100km/h in 7.6s and an announced fuel consumption of 1.5l/100km or 35g/km $CO_2$ emissions. The internal combustion engine is a turbocharged gasoline engine with direct injection TSI (Turbo Stratified Injection), and the main characteristics of the combustion engine are summarized in table 1 [2].

**Table 1.** Main characteristics of the 1.4TSI [2]

| Engine code | CUKB |
|---|---|
| No. of cylinders | 4 in line |
| Displacement | 1395 cm3 |
| Bore | 74.5 mm |
| Stroke | 80 mm |
| Compression ratio | 10 : 1 |
| Maximum power | 110 kW from 5000 to 6000 rpm |
| Maximum torque | 250 Nm from 1600 to 3500 rpm |
| ECU Engine Control | Bosch MED 17.1.21 |
| Fuel | Unleaded gasoline 95 RON |
| Pollution standard | UE6 |

The electric machine is of the synchronous type with permanent magnets that can be used as a motor to provide the torque necessary to move the vehicle up to 130km/h in electric mode, but also in generator mode, for brake energy recovery (deceleration). Its mass is 34kg and the torque and power characteristic are shown in figure 2 [2].

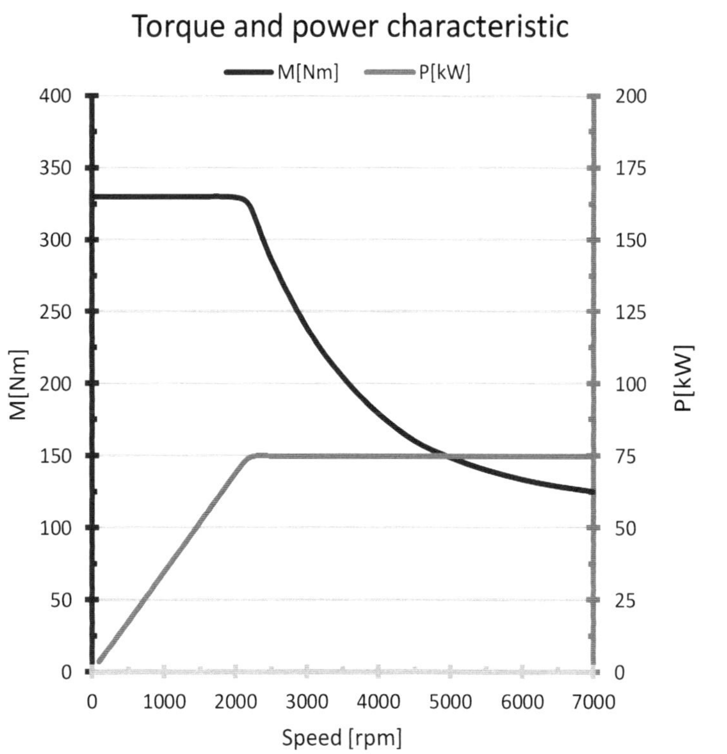


**Figure 2.** Torque and power characteristics [2]

The electric car is powered by the traction battery that stores the electrical energy needed to propel the car and provides electricity to various electrical components: the 12V battery, heating the passenger compartment, powering the air conditioning compressor, etc. The traction battery is of the lithium-ion type, cooled by water and is composed of 96 cells linked in series, divided into eight modules of 12 cells each. The main characteristics of this battery, as well as the standards or indices of protection are summarized in Table 2 [2].

**Table 2.** Main characteristics of the traction battery [2]

| Technology | Lithium-ion |
|---|---|
| Weight | 120 kg |
| Number of cells | 96 in series |
| Modules and number of cells per module | 8x12 |
| Capacity | 25Ah |
| Voltage | 345 V |
| Rated energy | 8.8 kWh |
| Voltage on the cell | 3.6 V |
| Cooling system | Liquid |
| Temperature beaches | - 28 °C à 60 °C |
| Protection | IP66, IP67, IP6K9 |

In order to reproduce as closely as possible, the behavior of the studied vehicle, when drawing up the model sketch in AMESim, using the HOT interface, a parallel rechargeable hybrid architecture vehicle was chosen, shown in figure 3 [3], [4], [5], [6].

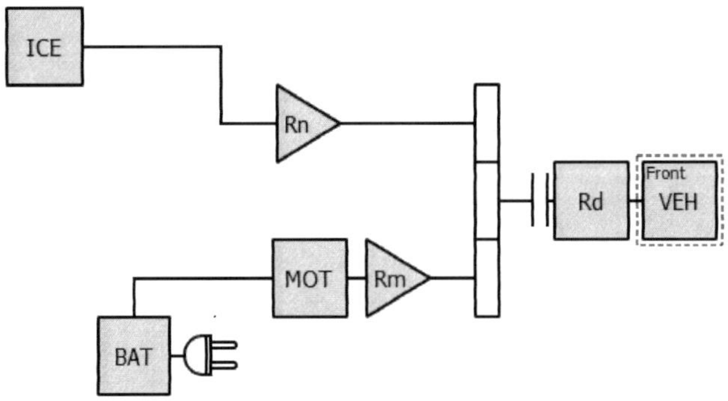


**Figure 3.** Parallel rechargeable hybrid vehicle architecture (PHEV)

The next step proposed by the interface, the parameterization of the model was performed with the data known from the previous characteristics presented in table 1, figure 2 and table 2, for each of the 7 main components presented by the vehicle architecture:

- ICE (Internal Combustion Engine);
- Rn (fixed ratios – the name is just to differentiate components in function of configuration);

- BAT (Battery);
- MOT (Electric Motor);
- Rm (fixed ratios – the name is just to differentiate components in function of configuration);
- Rd (fixed ratios – the name is just to differentiate components. Gearbox in this model);
- VEH (Vehicle).

The following two stages consist of:

- definition of the strategy for battery management, transmission modes calculated based on previously defined parameters at parameterization together with the choice of 100% electric mode;
- definition of test cases where to choose test cycle, gears selected during its cycle, altitude at which the test is performed, electrical loads activated as well as other parameters such as initial SOC (State of Charge), final SOC target and zero pollutant emission zone (ZEV) – if applicable;

Finally, after the calculation has been completed, the final post-processing step provides an overview of the vehicle's economic and dynamic performance, as well as a distribution of lost energy on each component. Figure 5 shows the sketch of the model with the desired architecture obtained from the launched calculation, which can be run in AMESim independently of the HOT interface, modified or optimized.

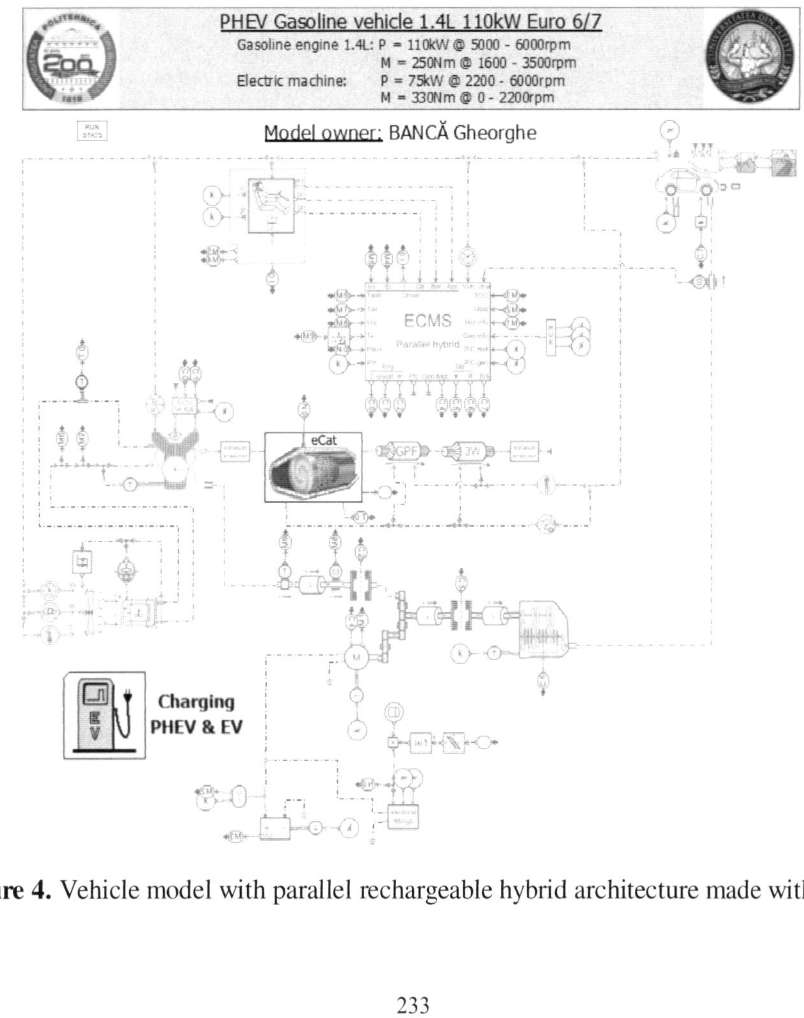


**Figure 4.** Vehicle model with parallel rechargeable hybrid architecture made with HOT

The key component of the model shown in the figure 4, responsible for the economic and ecological performance of the modelled vehicle is the ECMS (Equivalent Consumption Minimization Strategy) submodel. This component, similar to an ECU (Electronic Control Units), is dedicated to hybrid vehicles and is not found in the IFP Drive library but is available when developing the model architecture with the HOT interface. The submodel monitors vehicle behavior by optimizing fuel consumption over the selected driving cycle [3], acting on the main components:

- heat engine;
- electric motor/generator;
- gearbox clutch;
- engine clutch.

Optimizing fuel consumption is done by minimizing the cost of energy $H$, using $s$ as the price of electricity, in relation 1 [3], [4], [5], [6]:

$$H = \dot{m}_f \cdot H_{LHV} + s \cdot P_{ech} \quad (1)$$

where: $H_{LHV}$ is the lower calorific value of the fuel, $P_{ech}$ is the electrochemical consumption of the battery and $m_f$ is the fuel consumption. The strategic parameter for optimal control is the equivalence factor $s$, the theoretical value of which is given by HOT for the same vehicle configuration and driving cycle. To avoid calculation problems, a random $s$ factor cannot be used, although it has a limited effect on fuel consumption.

In the case of a parallel hybrid vehicle, the following parameters are called:

- the minimum time between two gear changes, required only if there is a gearbox, but it should be borne in mind that in the case of electric cars with large inertia (over $0{,}05 kg \cdot m^2$) problems may arise, because the system is not yet adapted;
- clutch position such that acceleration is allowed;
- factor $G_c$ during clutch slip, used to spin the clutch when the combustion engine has to be started at a reduced vehicle speed so as to ensure engine idling speed. The clutch position value $Cl$ is calculated by the formula 2 [3], [4], [5], [6]:

  $$Cl = (V_{tar} - V_{veh}) \cdot 10 \cdot G_c \quad (2)$$

  where: $V_{tar}$ is the desired vehicle speed and $V_{veh}$ is the actual vehicle speed;
- the factor for controlling the rotational speed $G_e$ during clutch slip and $M_{eng}$ are calculated with the relationship 3 [3], [4], [5], [6]:

  $$M_{eng} = (W_{idle} - W_{eng}) \cdot G_e \quad (3)$$

  where: $W_{idle}$ is the rotational speed of the engine at idle and $W_{eng}$ is the actual engine speed.

Finally, the equivalence factor $s$ per cycle is calculated using formula 4 [3]:

$$s(t) = s_0 - c1 \cdot (SOC(t) - SOC_{ref}) - c2 \cdot \int_0^t (SOC(t) - SOC_{ref}) \cdot dt \quad (4)$$

where: $s_0$ is the initial value of $s$ and the $SOC_{ref}$ is the SOC endpoint.

For the model made using the HOT interface, originally intended to estimate only the energy potential of various types of architectures, it was necessary to map the main pollutant emissions of SI engine based on pollution tests carried out at the pollution bench. The next step to the emission mapping was to model the SI engine cooling system to replicate the coolant temperature increase and the 'artisanal' creation of an electrically heated catalyst component, knowing the severity of the CO emission depollution standard and knowing that a vehicle equipped with a hybrid GMP (powertrain) has a low catalyst efficiency, because of frequent deactivations of the combustion engine on the depollution cycle. At the time of simulations, this component is not currently available in the IFP Drive library.

## 3. Results and discussion

The ultimate goal of the model used was to carry out a theoretical case study using the newly created eCat component (electric catalyst), simulating on the WLTC cycle the ecological and economic behavior of the vehicle in relation to Euro 7 standard, for a 3wCat (or TWC - three-way catalyst [7]) and four electrically heated catalysts with resistances of different powers: eCat 0,5kW, eCat 1kW, eCat 2kW and eCat 3kW.

Looking at the overall results of engine THC (total hydrocarbons) emissions shown in figure 5, an approximately linear increase is observed for the simulations launched with the 4 electrical catalysts, relative to reference, where a 3wCat was used. The increase of the THC emission is about 1,0-1,7% for every 1kW electrical power losses. This is explained by more frequent and sometimes longer activation of the internal combustion engine to compensate for the energy consumed from the battery, needed to heat up the electrical resistance of the catalyst. The overall exhaust THC emissions highlight the expected behavior of increasing the efficiency for the 4 eCat's in relation to 3wCat due to the increase in internal temperature in the 4 electric catalysts. Although we note that the Euro 7 norm is respected in all cases reporting to 3wCat (reference), eCat 0,5kW is 37,3% more efficient, eCat 1kW is 54,8% more efficient, eCat 2kW is 57,2% more efficient, and eCat 3kW is with 60,1% more efficient.

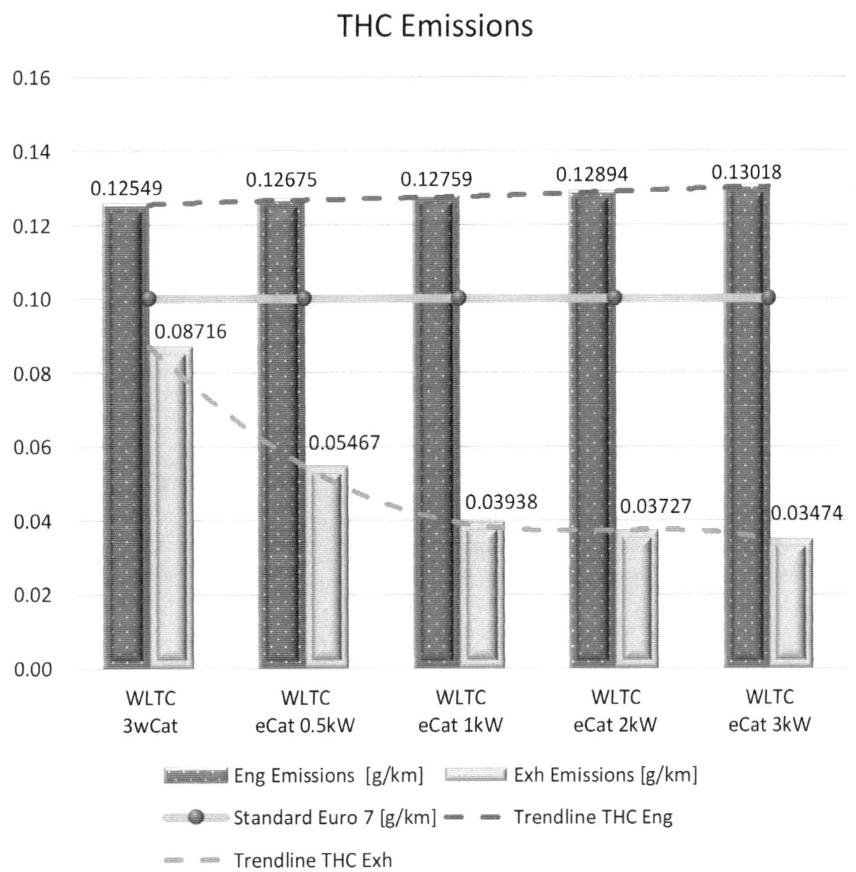


**Figure 5.** Evolution of engine and exhaust THC emissions

Figure 6 allows analysis of overall CO emissions, where a normal upward trend is observed for engine emissions compared to the WLTC 3wCat reference simulation. Motor CO global emissions increased by around 0,7 - 2,2 %, for each kW of electricity power losses additionally add when heating the eCat. In terms of exhaust emissions, we note that, in the reference situation (WLTC 3wCat), the proposed Euro 7 standard is not met, the efficiency of this catalyst being relatively low ~84,28%. At the moment, with the Euro 6 standard in force, this is not a problem, as the CO threshold is set at 1[g/km]. In the other 4 simulations (WLTC eCat 0,5kW, WLTC eCat 1kW, WLTC eCat 2kW and WLTC eCat 3kW) the proposed Euro 7 standard is respected with sufficient safety margin. In these 4 simulations, catalyst efficiency at CO aftertreatment is approximately 3,66 – 5,09% higher than in the reference situation, reaching a maximum of 89,38% for the WLTC eCat 3kW simulation.

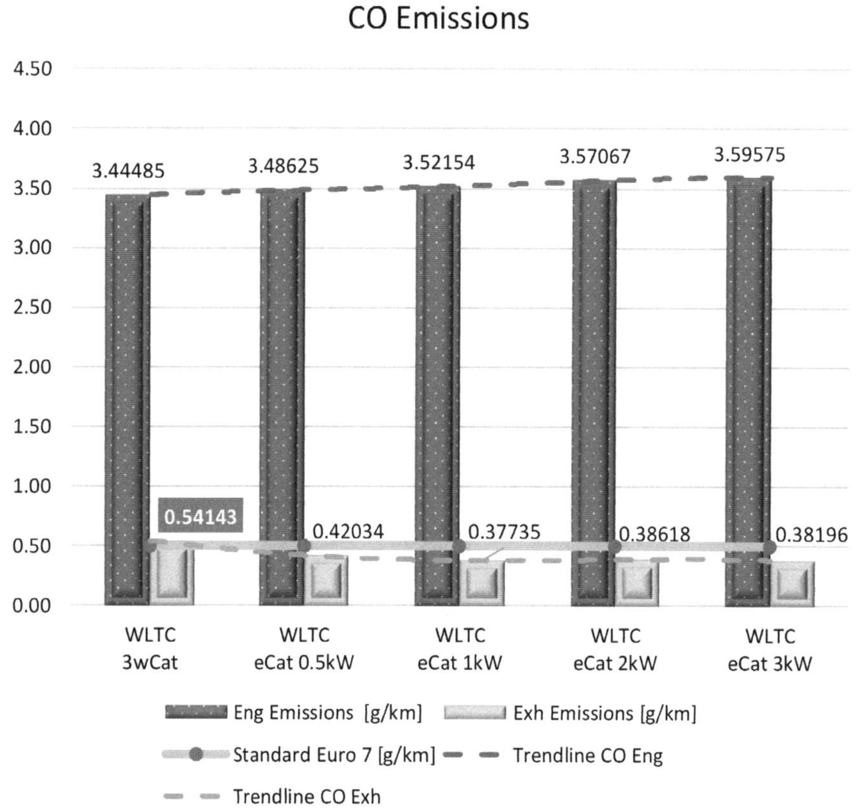


**Figure 6.** Evolution of engine and exhaust CO emissions

In figure 7 it is presented the overall engine and exhaust NOx emissions. Global NOx emissions follow the same linear upward trend depending on the electrical power consumed by eCat. Comparing the results of the simulations of the 4 eCat catalysts to the WLTC 3wCat reference simulation, an increase of 1,14% for eCat 0,5kW, 2,19% for eCat 1kW, 3,70% for eCat 2kW and 4,13% for eCat 3kW is observed in the case of engine NOx emissions. After analyzing the differences between these increases in engine NOx emissions, an average increase of around 1% was observed for each 1kW of electricity power losses additionally add when heating the eCat. NOx exhaust emissions are below the Euro 7 threshold for all 5 cases in which the simulations were launched. There is a significant decrease in exhaust emissions in the 4 eCat configurations with a maximum efficacity for the eCat 3kW. The overall NOx efficiency calculated for the 3wCat benchmark simulation is 57,59%, while for the other 4 configurations efficiencies of 73,48% were achieved for the eCat 0,5kW, 78,29% for the eCat 1kW, 80,18% for the eCat 2kW and 81,64% for the eCat 3kW.

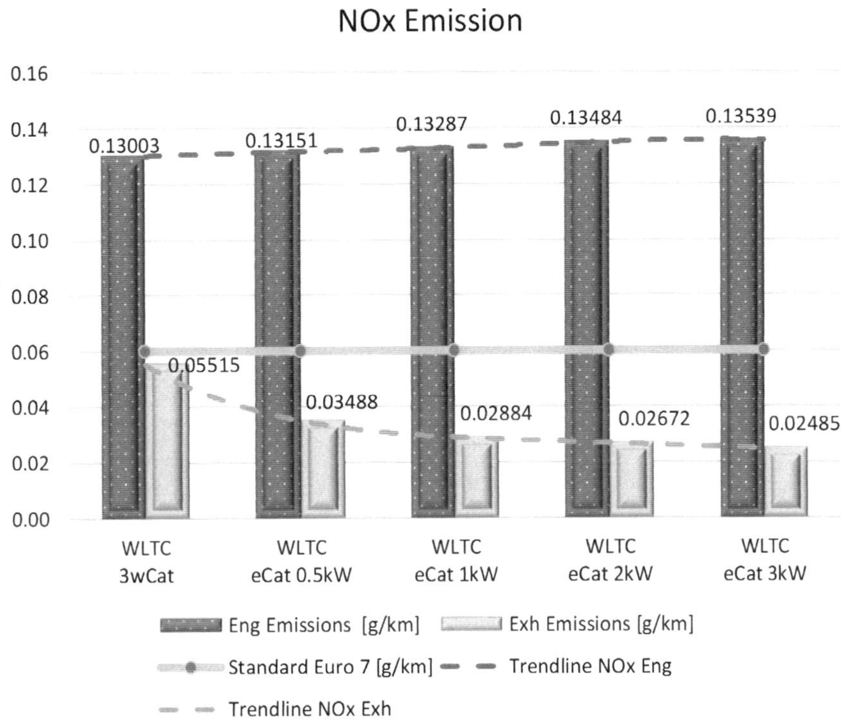


**Figure 7.** Evolution of engine and exhaust NOx emissions

The economic performance of the simulated vehicle in the 5 configurations is shown using graphs in figure 8 which showing $CO_2$ emissions and fuel consumption.

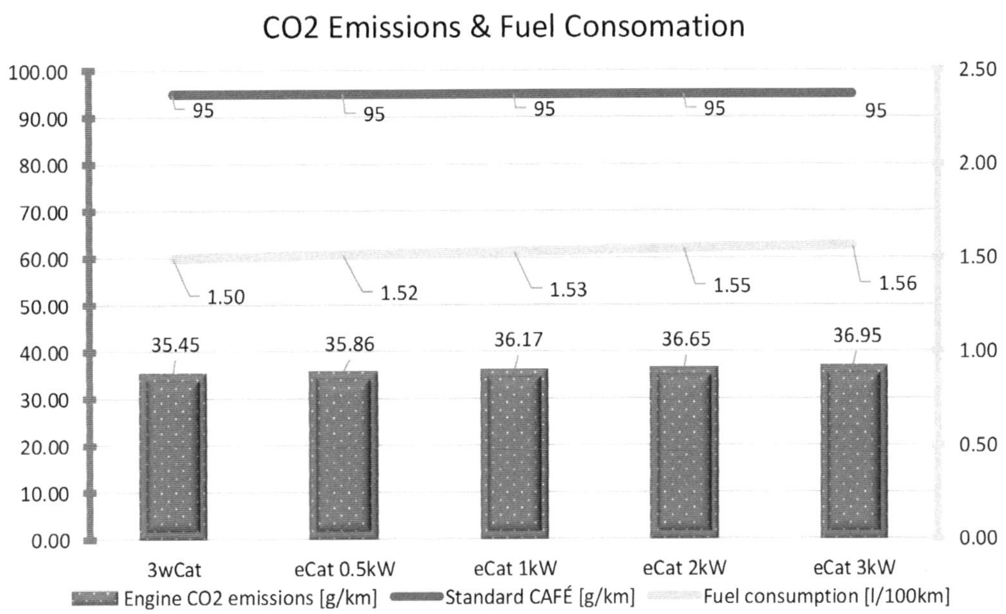


**Figure 8.** Evolution of $CO_2$ emissions and fuel consumption

It is known that to determine the fuel consumption of a vehicle it is sufficient to know the $CO_2$ emissions released by that vehicle into the atmosphere. Following the theoretical case study made using the 4 simulations, a linear increase in fuel consumption of approximately 1,4% was observed for every 1kW additional electrical power losses for heating the electrical catalyst, as a result of more frequent activation for the spark ignition engine.

## 4. Conclusions

The proposed Euro 7 standard eliminates differences in pollutant emissions between vehicles equipped with SI engine or CI engine and introduces limit thresholds for $NH_3$ and particulate matter from the braking system.

In terms of THC emissions, the behavior observed for the simulated vehicle in the 4 configurations is expected for engine and exhaust emissions.

Following the analysis of the results obtained after the launched simulations, CO emissions exceed the limit threshold for the proposed Euro 7 standard in one case (3wCat), but when is using an electrically heated catalyst on a PHEV, the standard is respected.

Following the theoretical case study made on the studied vehicle, for the 4 catalysts eCat 0,5kW, eCat 1kW, eCat 2kW and eCat 3kW an efficiency increases of up to 89,38% was observed, but this remains relatively small compared to the efficiency of aftertreatment systems that equips vehicles with 100% thermal engines, where for carbon monoxide the efficiency of 3wCat is 90...99%.

For $NO_x$ emissions as for THC emissions, the behavior observed in the 4 configurations is expected for engine emissions and exhaust emissions. The Euro 7 limit threshold for $NO_x$ emissions is at a level that allows us, according to the results obtained from the simulations, a sufficient margin of safety in case of light drifts.

According to the study, the increase in fuel consumption is directly proportional to the increase in power losses that simulate the heating resistance of the electric catalyst.

As last one conclusion, based on the case study carried out, it was concluded that the exit of the engine from the thermal operating regime has direct negative consequences on pollutants emissions and indirectly on the fuel consumption.

## References

[1] https://ec.europa.eu/commission/presscorner/detail/en/ip_22_6495 - data accesării 14.07.2023;

[2] https://www.vag-technique.fr/ext_files/ssp/vw/SSP_537_La_Golf_GTE.pdf - data accesarii 14.0.2023;

[3] Help AMESim - DRVECMS09 - *ECMS vehicle optimal control unit*;

[4] Dabadie J, Sciarretta A, Font G, and Le Berr F, *Automatic Generation of Online Optimal Energy Management Strategies for Hybrid Powertrain Simulation*, SAE Technical Paper 2017-24-0173, 2017, DOI:10.4271/2017-24-0173

[5] Guzzella L and Sciarretta A 2013, *Vehicle Propulsion Systems. Introduction to Modeling and Optimization*, 3rd edition, Springer-Verlag, Berlin Heidelberg, ISBN 978-3-642-35912-5 (Print), 978-642-35913-2 (Online), DOI:10.1007/978-3-642-35913-2.;

[6] Sciarretta A 2013, *Supervision des véhicules hybrides*, in: Badin F (ed.), Les véhicules hybrides. Des composants au système, Editions Technip, Paris, p. 381-420, ISBN 978-2-7108-0986-9. Also in English translation: Control of hybrid vehicles, in Badin (ed.), Hybrid Vehicles, p. 369-411, ISBN:978-2-7108-0994-4.;

[7] Ivan F, Liță D and Bușoi A, 2014, *Methods and means of depollution of the engines for vehicles*, ISBN 978-606-25-0038-2.

IOP Conf. Series: Materials Science and Engineering 1303 (2024) 012027 doi:10.1088/1757-899X/1303/1/012027

# On the impact of using nanoparticles type additives on the altering of diesel fuel jet characteristics

**A L Niculae**[1], **R Chiriac**[1,2] **and A Racovitza**[1]

[1]Faculty of Mechanical Engineering and Mechatronics, University Politehnica of Bucharest, Romania
[2]EA7341 CMGPCE of Conservatoire National des Arts et Metiers, Paris, France

E-mail: andrei.niculae@upb.ro

**Abstract**. Even though diesel-powered passenger cars and light-duty vehicles may not be sold in the European Union by 2035, possibly in most other countries will be. Truck diesel engines will remain in use after 2035 because electric trucks are not a viable alternative for goods transportation. This involves continuing research and development on diesel engines to reduce their pollutant emissions. Alternative (renewable) fuels with combustion improvers can diminish emissions at the source. Nanoparticle-based additives can enhance the efficiency and emissions of diesel engines through their intense catalytic activity and by improving the fuel injection process. However, their effect on the injection process has not been well tested and published in the dedicated literature. The use of nanoparticles mixed in small fractions with original diesel fuel influences the physicochemical properties of the fuel as well as the formation of fuel jets inside the combustion chamber. This research focuses on the opportunity of using different diesel fuel blends mixed with combustion improver additives such as MWCNT and $CeO_2$ nanoparticles. In this sense, a simulation investigation was conducted using the AVL BOOST Hydsim software. The physicochemical properties of fuels were used to assess the macroscopic characteristics of the fuel spray, such as penetration, spray cone angle, and Sauter mean diameter. When increasing the nanoparticle dosage, the penetration and spray cone angle values are decreasing, while the Sauter mean diameter values are increasing. This tendency is present for both nanoparticle types considered.

## 1. Introduction

Since the way of managing the injection process inside the diesel engine cylinder has become increasingly important in terms of preserving power and limiting fuel consumption and emissions, attention has been drawn to the possibility of correlating specific aspects of combustion preparation and development to the physicochemical properties of the injected fuel. Thus, different studies investigated the relevance of using several types of nanoparticles as additives for different diesel fuel blends with varying effects on ignition characteristics, rate and duration of combustion, rate of heat release and transfer, and so on. Because of their surface and thermal characteristics, modest amounts of nanoparticles with metallic or organic structures might provide encouraging outcomes in terms of combustion process operation, lowering emissions concentrations across a broad range of engine testing regimes.

Various carbon-based nanoparticles are used to cause fuel droplet micro-explosions, resulting in a more efficient air-fuel mixture and managing the combustion flame temperature through their high latent absorbing heat [1]. Consequently, the oxidation chemical reactions are accelerating, reducing the amounts of unburned hydrocarbons (UHC) and carbon monoxide (CO) [2]. Ardabili et al. [1] studied the characteristics of a natural gas diesel engine powered by water-emulsified (WE) diesel/biodiesel (D/B) blends (5% vol. biodiesel and 3% wt. water) containing 30 and 60 ppm aqueous carbon nanoparticles. At a low addition level, the tested fuel minimized hydrocarbon emissions by 8 to 40% under various engine operating loads. During an experimental one-droplet investigation, Ooi et al. [3] added 1000 mg/L graphite oxide (GO) to diesel and biofuel (B100). According to the data, adding GO to both fuels resulted in a 38% shorter ignition delay, a nearly 8% lower peak temperature, and a 12% shorter combustion duration. A similar experiment [2] was carried out on isolated droplet combustion for the MF15 fuel blend comprising 15% vol. 2-methylfuran and 85% vol. neat diesel fuel with 25, 50 and 100 ppm multi-walled carbon nanotubes (MWCNT) amounts. Higher nanoparticle content reduced the occurrence of micro-explosions while increasing their intensity, as explained by the higher viscosity and surface tension of nanofuels compared to basic fuel. The greater thermal conductivity of the MWCNT provides a shorter autoignition delay, as well as a decrease in heat release and peak temperature, resulting in improvements in terms of combustion efficiency and emissions [1-3].

Metal nanoparticles form a solid layer structure in fuel. This structure improves the thermal conductivity of the mixture, reducing the autoignition delay, increasing the cetane number of the fuel, and therefore improving the ignition characteristics. For several reasons, cerium oxide ($CeO_2$) is an excellent additive to utilize since it has a superior surface-volume ratio and a high oxygen content, delivering the necessary for the oxidation reactions of hydrocarbons as well as soot. It is also capable of reversing the chemical reaction by converting to cerous oxide ($Ce_2O_3$), allowing nitrogen oxide reduction [4]. Bao et al. [5] performed an experimental investigation to examine the spray characteristics of DE20 fuel blend (20% vol. ethanol, 80% vol. diesel), DE20Ce25 (containing 25 ppm $CeO_2$), DE20Ce50 (containing 50 ppm $CeO_2$), and DE20Ce100 (containing 100 ppm $CeO_2$) in a common rail direct fuel injection system. When compared to pure diesel, the spray penetration of DE20, DE20Ce25, DE20Ce50, and DE20Ce100 modified by -3%, -1.8%, -0.3%, and 9.5%, the spray cone angle changed by -3%, 4%, 2.8%, and 12.5%, and the spray area varied by -3%, -1.3%, 0.6%, and 31%. However, the benefits of each type of nano-additives stated above might be combined by using the metal-organic framework (MOF) crystalline nanoparticles, containing metal clusters with organic ligands [6].

The purpose of this study was to determine how the use of additives such as $CeO_2$ and MWCNT nanoparticles affects the physicochemical properties of certain fuels, as well as the effects on certain macroscopic characteristics of the fuel jet, such as penetration, spray cone angle, and Sauter mean diameter.

## 2. Simulation Details

Following a literature review, we decided to compare the effects on the macroscopic characteristics of the fuel jet produced by adding $CeO_2$ nanoparticles to the DE20 fuel blend (80% diesel and 20% ethanol) and MWCNT nanoparticles to the MF15 fuel blend (85% vol. diesel fuel and 15% vol. 2-methylfuran).

The AVL Hydsim tool is designed to simulate fuel injection. Some fuel jet characteristics, such as penetration, spray cone angle, and Sauter mean diameter, may be computed using this tool, which are key factors related to the evolution of the air-fuel mixture formation [8]. The software provides the option of selecting a set of relations to be used for calculating these parameters. Varde-Popa formulas were employed for penetration and Sauter mean diameter, whereas the Sitkei formula was utilized for spray cone angle. In work [9], the authors of this paper have provided a thorough explanation of them.

Using the AVL Hydsim tool it was successfully implemented a simulation model for the Delphi injection system of the UTB 2404055 tractor diesel engine as in prior research [9–11]. The model was developed to examine the impacts of DE20 and MF15 fuels containing $CeO_2$ and MWCNT nanoparticles, respectively.

The AVL Hydsim simulation tool necessitates the physicochemical properties of the fuel, such as bulk modulus, density, kinematic viscosity, and surface tension. Their values may be determined by utilizing the software's database or by introducing these properties as variable or constant values. The authors of this study have considered constant fluid properties for this simulation investigation. The information presented by the authors of the studies [2, 5] on the physicochemical properties of pure DE20 and MF15 fuels or those containing 25, 50, and 100 ppm of $CeO_2$ and MWCNT nanoparticles was used in the simulation model. In comparison to the reference fuels (DE20 and MF15), six additional nanofuels with varying $CeO_2$ or MWCNT fractions were considered and labelled as DE20CeXX and MF15CXX where XX denotes the $CeO_2$ or MWCNT nanoparticle dosage in ppm. These data are presented in Table 1.

**Table 1.** Physicochemical properties of the fuels [2, 5].

| | DE20 | DE20Ce25 | DE20Ce50 | De20Ce100 | MF15 | MF15C25 | MF15C50 | MF15C100 |
|---|---|---|---|---|---|---|---|---|
| Density [kg/m$^3$] | 815 | 815 | 815 | 816 | 860.9 | 860.1 | 859.8 | 858.4 |
| Surface Tension [N/m] | 23.21 | 24.03 | 24.38 | 24.73 | 28.18 | 28.43 | 28.61 | 28.77 |
| Viscosity [mm$^2$/s] | 1.95 | 2.03 | 2.11 | 2.24 | 4.031 | 4.372 | 4.652 | 4.916 |

The pressure and temperature values obtained for the engine's maximum power operating condition, characterized by 2400 rpm speed, are shown in Figure 1. These values are required by the AVL Hydsim software to establish the in-cylinder conditions during fuel injection. The in-cylinder pressure was measured experimentally for DE20 fuel, and the associated global cylinder charge temperature was computed by simulation using an engine model created in the AVL Boost v2021.1 program, considering the same DE20 fuel.

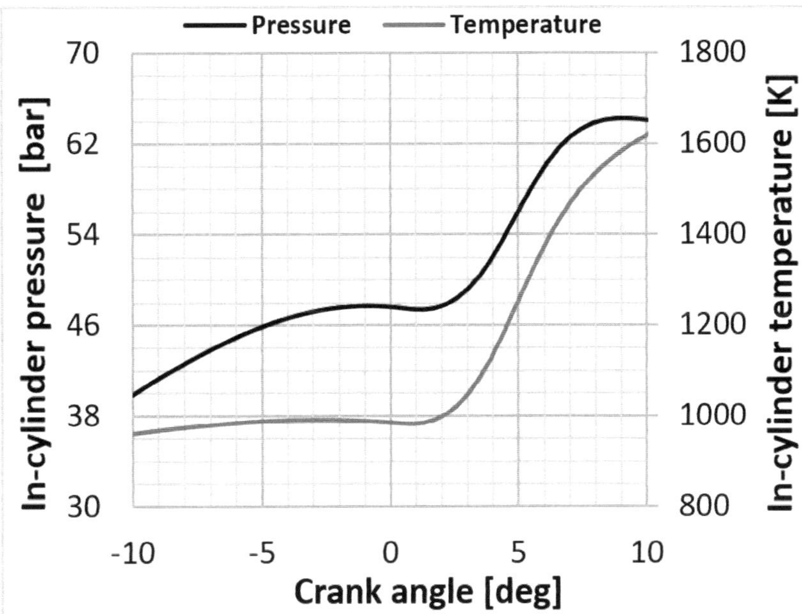


**Figure 1.** The cylinder pressure and temperature variations relative to the crankshaft position for DE20 fuel.

First, simulation tests were performed for DE20 fuel to assure model calibration for the engine's maximum power operating condition, which is characterized by a 2400 rpm speed. Figures 3 and 4 show a proper match between the simulated and experimental shapes of the injection characteristic of DE20 in terms of needle lift and line pressure, using a single injection pulse starting at 5.5 CA° before TDC. Both injection-defining parameter variations, maximum needle lift, and maximum rail-injection pressure are quite reproducible with our experimental data, with 0.35 mm needle lift and nearly 700 bar injection pressure. Because MF15 is a denser fuel than DE20, with a greater kinematic viscosity, the start of injection occurs earlier by 0.5 degrees. When utilizing biodiesel instead of diesel, a similar pattern is observed. The maximum needle lift and line pressure values are unchanged when using the MF15 fuel.

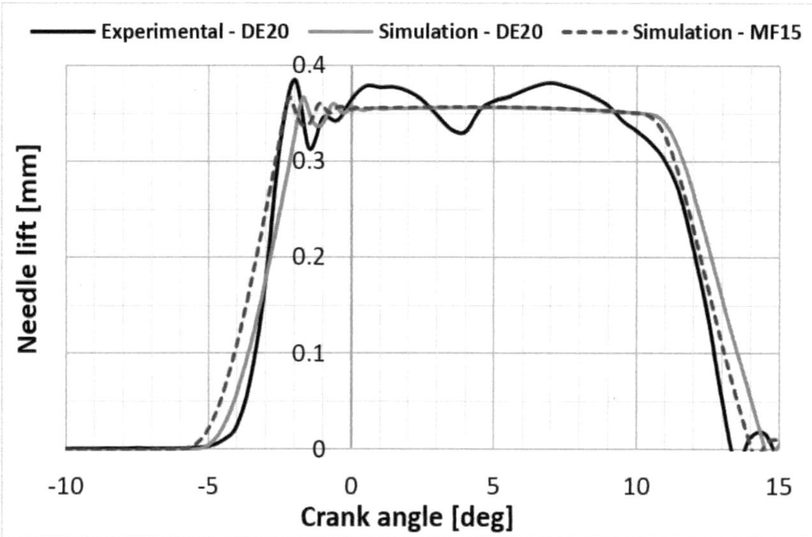


**Figure 2.** The variation of needle lift vs crankshaft position for DE20 fuel (experimental and simulation) and MF15 fuel (simulation).

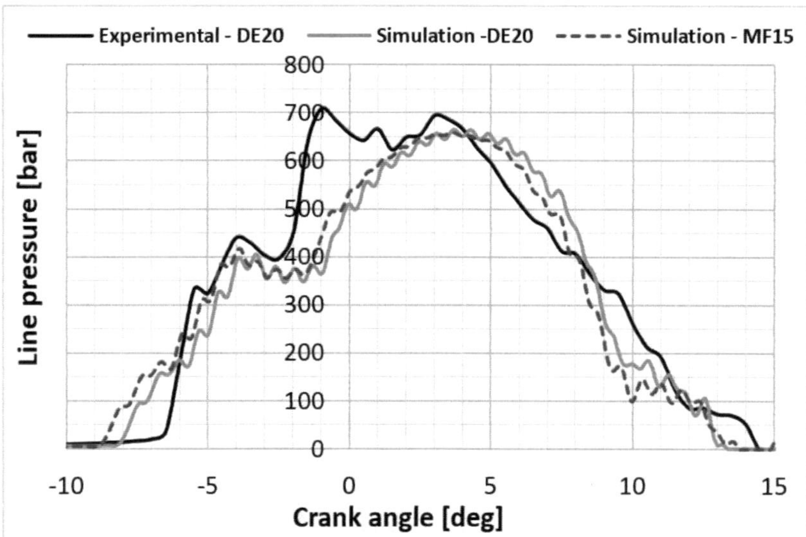


**Figure 3.** The variation of line pressure vs crankshaft position for DE20 fuel (experimental and simulation) and MF15 fuel (simulation).

## 3. Simulation Results

Figure 4 depicts the penetration values for DE20, MF15, and six nanofuels with varying doses of $CeO_2$ or MWCNT type nanoparticles depending on the crankshaft position between the start of injection (SOI) and start of combustion (SOC) moments. Penetration values increase with time between the SOI and SOC moments, thus the maximum values are reached at SOC. As the dosage of nanoparticles rises, there is a trend for maximum penetration to decrease. Thus, the minimum values were obtained for DE20Ce100 and MF15C100 fuels with decreases of 8.1% and 11.5%, respectively, compared to the reference fuels (DE20 and MF15). At -5 CA°, the penetration values for all fuels investigated are similar; for DE20, the value is 0.5% lower than the value obtained for MF15, while the value obtained for DE20Ce100 is 2.7% greater than the value obtained for MF15C100. The maximum penetration values obtained for DE20 and DE20Ce100 are 41.7% and 47.2% higher, respectively, than the values obtained for MF15 and MF15C100.

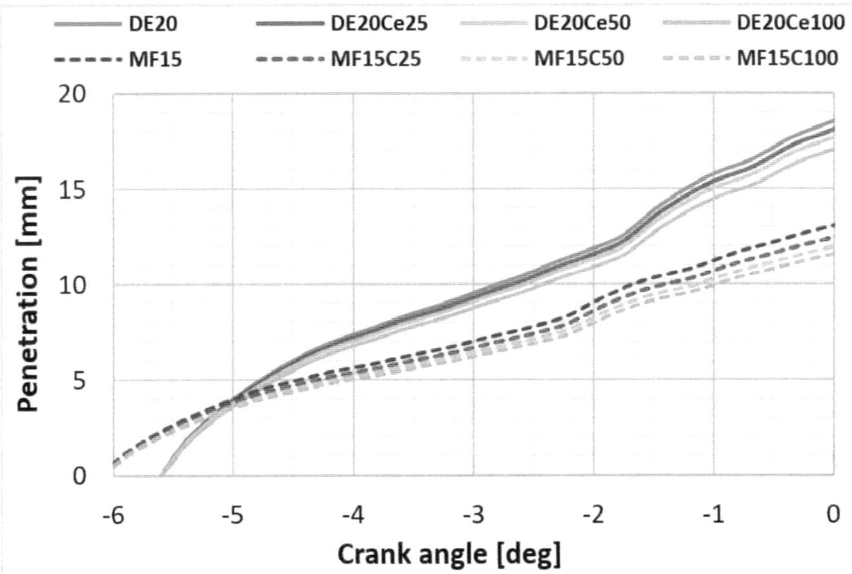


**Figure 4.** Spray penetration variation vs the crankshaft position.

Figure 5 illustrates the spray cone angle values for DE20, MF15, and six nanofuels with varying concentrations of $CeO_2$ or MWCNT type nanoparticles depending on the crankshaft position between the SOI and SOC moments. Spray angle values increase with time between the SOI and SOC moments, thus the maximum values are reached at SOC. The spray cone angle values decrease as the nanoparticle concentration increases. Thus, the minimum values were reached for DE20Ce100 and MF15C100 fuels, with average reductions of 9.5% and 13%, respectively, as compared to the reference fuels (DE20 and MF15). The spray cone angle values found for DE20 are on average 67.2% greater than the values obtained for MF15. At -5 CA°, the values obtained for DE20 and DE20Ce100 are 58.9% and 64.8% higher than those obtained for MF15 and MF15C100, while at SOC moment, the differences are greater, more precisely 67% and 74.2%.

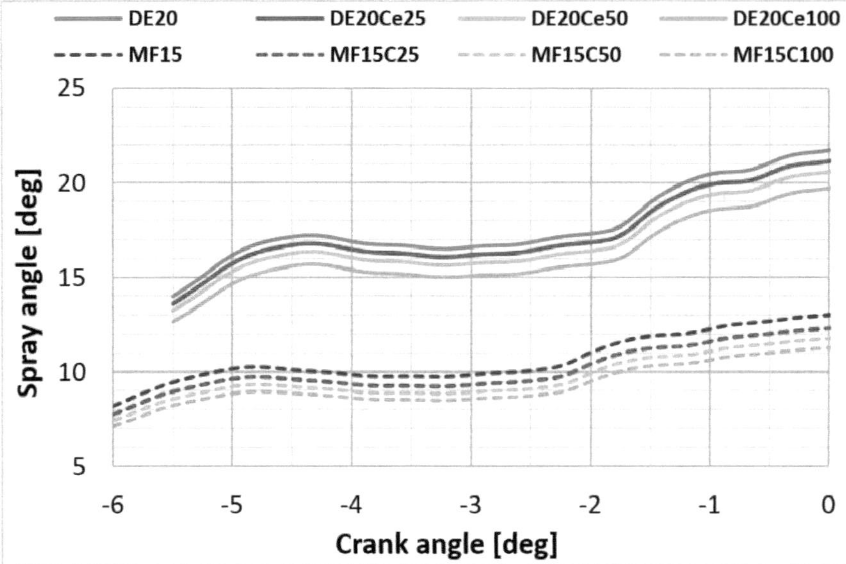


**Figure 5.** Spray cone angle variation vs the crankshaft position.

Figure 6 presents the Sauter mean diameter values for DE20, MF15, and six nanofuels with different doses of $CeO_2$ or MWCNT type nanoparticles depending on the crankshaft position between the SOI and SOC moments. Sauter mean diameter values decrease with time between the SOI and SOC moments, thus the minimum values are reached at SOC. The Sauter mean diameter values rise as the nanoparticle dose increases. Thus, the highest values were achieved for DE20Ce100 and MF15C100 fuels, with average increases of 6.1% and 6.4%, respectively, above the reference fuels (DE20 and MF15). The Sauter mean diameter values obtained for MF15C100 are on average 30.2% greater than those achieved for DE20Ce100. At -5 CA°, the values obtained for DE20 and DE20Ce100 are 17.9% and 17.8% higher than those obtained for MF15 and MF15C100, while at SOC moment, the differences are greater, more precisely 22.7% and 23.1%.

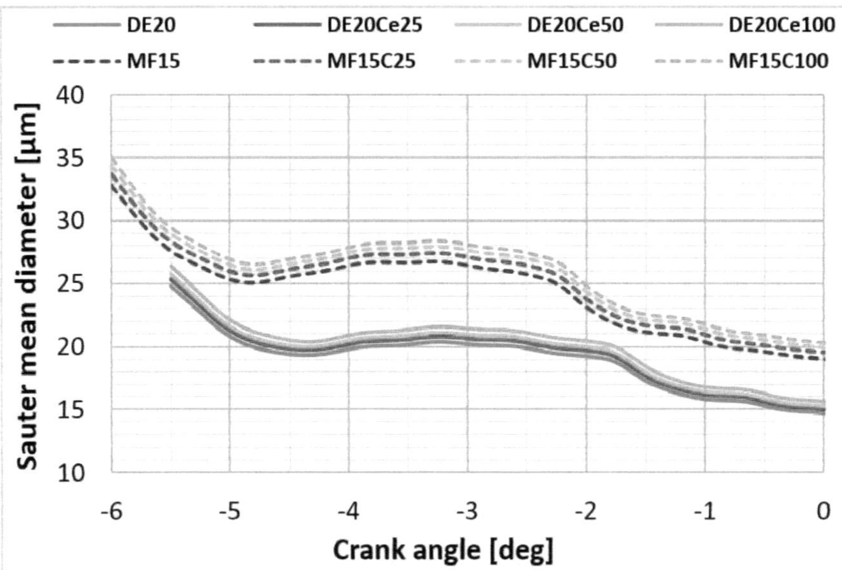


**Figure 6.** Sauter mean diameter variation vs the crankshaft position.

## 4. Conclusions

The findings of this study, which describes the evolution in time of the macroscopic spray characteristics of two different types of fuels, namely diesel-ethanol and its blends with nano-$CeO_2$, and diesel-methylfuran and its blends with multi-walled carbon nanotubes at ppm levels used in a high-pressure injection system equipped with a multi-holes injector, are as follows:

- The addition of nanoparticles to the original reference fuels in small fractions has no significant effect on the hydraulic behaviour of the injection system.
- Essential for the formation of the air-fuel mixture, the macroscopic spray characteristics, such as penetration, spray cone angle, and Sauter mean diameter, reveal almost similar time-dependent variations regardless of the original fuels and the presence of nanoparticles.
- The presence of nanoparticles at a maximum concentration of 100 ppm in both original reference fuels reduces penetration and spray cone angle while increasing Sauter mean diameter.
- The Sauter mean diameter, which is essential for the vaporization process, decreases in time in a quasi-linear trend, with the greatest values recorded for the fuels dopped with the nanoparticles chosen in the highest fractions.

Although nanoparticles influence fuel's physiochemical properties, they have a higher impact on the fuel jet because of the secondary atomization effect and the heat sink effect, which stimulate accelerated vaporization. Even though the study only considered the impact of constant fuel physicochemical properties into account, some interesting findings were obtained.

This work emphasized that, if constant fuel physicochemical properties are taken into account, the development of the fuel jet within the combustion chamber during the auto-ignition delay may be reasonably determined using fuel jet parameters.

An experimental investigation will be conducted in the future to better understand how the inclusion of nanoparticles in fuel affects spray development.

**Acknowledgement**

The authors of this paper acknowledge the AVL-AST team from AVL List GmbH for the special support offered under the University Partnership Program on performing the simulation activity.

**Author's ORCID ID**

A L Niculae: https://orcid.org/0000-0002-1174-289X
R Chiriac: https://orcid.org/0000-0002-0568-4881
A Racovitza: https://orcid.org/0000-0003-4650-5867

**References**

[1] Faizollahzadeh Ardabili S, Najafi B, Aghbashlo M, Khounani Z and Tabatabaei M 2021 Performance and emission analysis of a dual-fuel engine operating on high natural gas substitution rates ignited by aqueous carbon nanoparticles-laden diesel/biodiesel emulsions *Fuel* **294**

[2] Ittoo B, Ooi J B, Tran M V, Jaliliantabar F, Hasan Najafi G and Swamy V 2022 Effects of dispersed multiwalled carbon nanotubes on the micro-explosion and combustion characteristics of 2-methylfuran – diesel mixture droplets *Fuel* **316**

[3] Ooi J B, Rajanren J R, Ismail H M, Swamy V and Wang X 2017 Improving combustion characteristics of diesel and biodiesel droplets by graphite oxide addition for diesel engine applications *Int J Energy Res* **41** 2258–67

[4] Balamurugan S and Sajith V 2013 Stabilization studies of zirconium - Cerium oxide nanoparticle-diesel suspension *Advanced Materials Research* vol **685** pp 128–33

[5] Bao L, Fu W, Li F, Liu Y, Bi H, Jiang C and Lin Q 2020 Experiments on macroscopic spray characteristics of diesel-ethanol with dispersed cerium nanoparticles *Journal of the Energy Institute* **93** 2186–96

[6] Hosseini S H, Rastegari H, Aghbashlo M, Hajiahmad A, Hosseinzadeh-Bandbafha H,

Mohammadi P, Jamal Sisi A, Khalife E, Lam S S, Pan J, Khataee A and Tabatabaei M 2022 Effects of metal-organic framework nanoparticles on the combustion, performance, and emission characteristics of a diesel engine *Energy* **260**
[7] Mei D, Fang Y, Zhang Z, Guo D, Chen Z and Sun C 2021 Analysis of surface tension for nano-fuels containing disparate types of suspended nanoparticles *Powder Technol* **388** 526–36
[8] AVL BOOST Hydsim Users Guide 2022
[9] Niculae A L, Racovitza A and Chiriac R 2023 in-press article Some effects of nanoparticles addition on diesel fuel jet characteristics *IEEE 11th International Conference on Thermal Equipment, Renewable Energy and Rural Development (TE-RE-RD), Bucharest, Romania*
[10] Niculae A L, Chiriac R and Racovitza A 2022 The effect of using different Biodiesel fuels on jet development in a Diesel engine *IOP Conference Series: Earth and Environmental Science* vol **960** *(IOP Publishing Ltd)*
[11] Niculae A L, Chiriac R, Racovitza A and Ungureanu V 2022 Effect of injection system characteristics on performance, efficiency and emissions of agricultural tractor Diesel engines *IOP Conf Ser Mater Sci Eng* **1220** 012002

# Aerodynamic means of reducing the influence of warning accessories on drag and emissions, for intervention vehicles

**L Ilea, C Stan and M Vieru**

University Politehnica of Bucharest, Autovehicule Rutiere Department, 313 Splaiul Independentei St., Romania

E-mail: laurentiu.ilea@yahoo.com

**Abstract.** Aerodynamic performance of passenger vehicles is one of the most efficient methods to further reduce the $CO_2$ emissions and comply with the increasing constraints in the legislation. Previously performed studies on exterior warning accessories (systems with visual alerting signals that may be found on intervention vehicles e.g. ambulances), indicate a significant degradation of aerodynamic coefficient (Cd). Even if this category of vehicles does not have to comply with the standard emission rules, it is important to study and discover the means to reduce aerodynamic drag for this type of vehicles considering the general aim to reduce air pollution. The target would be to have a Cd level equal to that found in the base vehicles, before transformation (i.e. often, intervention vehicles are obtained by converting a vehicle available to general customers – multistage vehicle transformation).

The current paper's purpose is to continue the investigation on methods to improve the aerodynamic performance of intervention vehicles. This will be done by compensating the shape and dimensions of the warning devices with the addition of supplementary body elements or by adapting existing parts. The Cd value of the different configurations is assessed using computational fluid dynamics (CFD) software based on Lattice Boltzmann mathematical model. 3D models of different generic body type vehicles (sedan, hatch back, SUV) were created and immersed in a virtual wind tunnel. To simulate the real-world measurement method specific boundary conditions on walls, inlets, outlets were set while the fluid environment was defined with a network from fine (2mm length, near the studied model), to coarse (20mm length, near exterior walls).

The results indicate that there is great potential in reducing the aerodynamic drag and, by extension, fuel consumption and $CO_2$ emissions for all body types studied. Deviations could exist due to different make designs, or diversity of warning devices for example so the results obtained have a general nature. Nevertheless, the conclusion remains the same: if one's aim is to improve air quality or prepare for new legislation (that will also apply to intervention vehicles), the aerodynamic optimization should be further scrutinized as it can prove to be a robust method to obtain the targeted results.

## 1. Introduction

Depollution of automotive manufacturing industry is a challenge that is a daily reality for any role in this domain, as it was implemented 3 decades ago to improve air quality and, by extension, health benefits [1]. The $CO_2$ emission can be linked with almost all branches of automotive industry, from cradle to grave, starting with conception and manufacturing of each part, logistics, vehicle assembly process, customer use and finally recycling [2].

The current paper is part of a study concerning the reduction of aerodynamic resistance assessed by product between aerodynamic coefficient and frontal vehicle area (CdA), with the purpose of improving the $CO_2$ emission level for intervention vehicles. The focus of the research is on vehicles destinated for general use, that are further converted to be used as intervention vehicles (transport of patients, auxiliary vehicles for various tasks like fire extinguish, disaster control, etc.) and for this purpose must be equipped with warning lights, as indicated in Figure 1. This extra equipment is most often mounted on the roof of the vehicles and has a direct impact in aerodynamic resistance and $CO_2$ emissions. While current legislation exempts this category of vehicles from meeting specific $CO_2$ requirements, the aim of the paper is to identify some means to reduce the aerodynamic resistance of the warning lights [3].

**Figure 1.** Intervention vehicles (sedan, hatch back, SUV) with warning lights [4].

The influence of a generic warning light equipment on arbitrary chosen vehicles designs, corresponding to sedan, hatch back and SUV body type, was studied as a first part of the research [5], with the help of Computational Fluid Dynamics (CFD) software. A brief synthesis is shown in Figure 2.

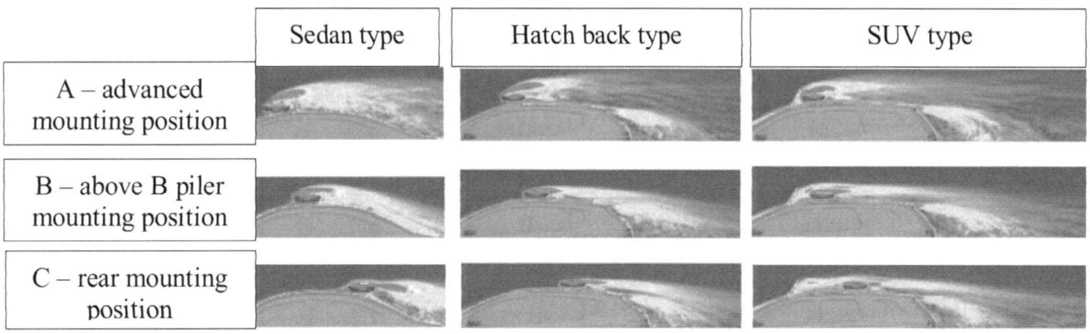


**Figure 2.** CFD analysis for air flow behaviour near warning lights to determine best mounting position.

Based on this previous study, it was concluded that the warning lights should be mounted in B pillar area and avoid extreme advanced or rear position to obtain the smallest degradation in aerodynamic resistance and $CO_2$. Numerical results are presented in Table 1.

**Table 1.** CdA results for base vehicle and warning lights in advanced, above pilar B and rear position.

| **Studied case** | **Sedan** | **Sedan - A** | **Sedan - B** | **Sedan - C** |
|---|---|---|---|---|
| **SCx [m²]** | 0.659 | 0.905 | 0.753 | 0.796 |
| **Δ SCx [m²]** | - | + 0.246 | + 0.094 | +0.137 |

| **Studied case** | **Hatch back** | **Hatch back - A** | **Hatch back - B** | **Hatch back - C** |
|---|---|---|---|---|
| **SCx [m²]** | 0.836 | 0.996 | 0.910 | 0.905 |
| **Δ SCx [m²]** | - | +0.160 | +0.074 | +0.069 |

| **Studied case** | **SUV** | **SUV - A** | **SUV - B** | **SUV - C** |
|---|---|---|---|---|
| **SCx [m²]** | 1.000 | 1.127 | 1.059 | 1.059 |
| **Δ SCx [m²]** | - | +0.127 | +0.059 | +0.059 |

## 2. Means of reducing warning lights' influence on CdA

The two main factors that led to the present study were (i) anticipation of potential legislative changes that would also require intervention vehicles to meet specific $CO_2$ emissions standards and (ii) the significant benefits brought by a cleaner, less polluted air in urban areas.

The target set at the beginning of the work was to achieve the same aerodynamic performance for the intervention vehicle as for the base vehicle, before transformation. This step represents the second phase of the work and, for traceability, will be continued with the same CFD tools and coordinates. The mounting position over pillar B was chosen as reference for the intervention vehicle considering that, in practice, this is frequently used but also it was proved to have the most reduced impact on aerodynamic drag. Considering the mentioned premises, the fallowing CdA values must be compensated to achieve the CdA target: 0.094m² for Sedan type, 0.074m² for Hatch back type, 0.059m² for SUV type.

To manage the CdA improvement the work hypothesis was to overcome the warning lights (equipment characteristics used for current CFD study are shown in Figure 3) degradation with auxiliary parts that can be easily mounted. Heavy modification of the vehicle was avoided for the purpose of the paper to ensure the application of the research also in real-life use – implementation of heavy modifications would be more difficult and thus there will be no significant overall $CO_2$ level improvement if solutions found would not be applied.

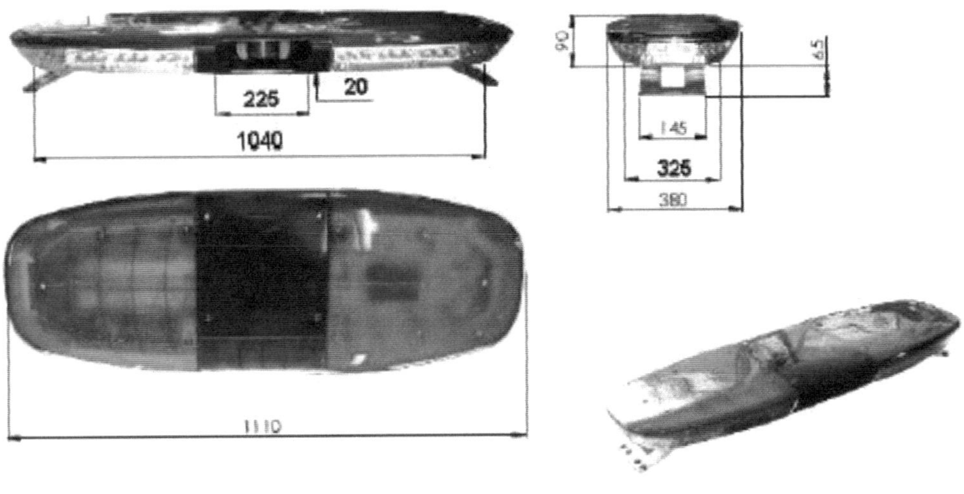


**Figure 3.** 'Warning light' shape and dimension, used for current study.

The parts that were decided to be modified/added in current study are:

a. Replacing the underbody fairings with parts extended in surface and with improved flatness

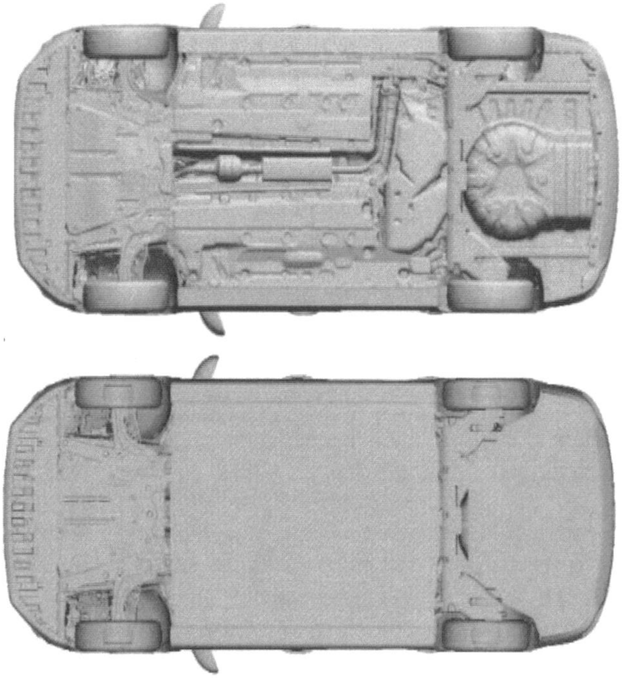

**Figure 4**. Standard underbody fairings (1st image) versus proposed underbody fairings (2nd image).

b. Adding an extra mask on wheel rim to reduce porosity, parameter directly linked to CdA reduction (brake cooling is not considered to be an issue, as the emergency braking is validated at full load and with wheels that have similar characteristics as the proposed model [6])

**Figure 5.** Standard rims (left side) versus proposed rims (right).

Performed CFD simulations on each body type, with above detailed technical evolutions indicate a various influence of added parts on the CdA overall performance, as shown in Table 2.

**Table 2**. CdA results for optimised technical definition, compared to base and intervention vehicles.

| **Studied case** | **Sedan 'base vehicle'** | **Sedan 'intervention vehicle'** | **Sedan 'optimised vehicle'** |
|---|---|---|---|
| **SCx [m²]** | 0.659 | 0.753 | 0.708 |
| **Δ SCx [m²]** | - | + 0.094 | + 0.049 |
| **Δ SCx [%]** | - | + 14 | +7.44 |
| **Studied case** | Hatch back 'base vehicle' | Hatch back 'intervention vehicle' | Hatch back 'optimised vehicle' |
| **SCx [m²]** | 0.836 | 0.910 | 0.899 |
| **Δ SCx [m²]** | - | +0.074 | + 0.063 |
| **Δ SCx [%]** | - | +8.8 | +7.54 |
| **Studied case** | SUV 'base vehicle' | SUV 'intervention vehicle' | SUV 'optimised vehicle' |
| **SCx [m²]** | 1.000 | 1.059 | 0.998 |
| **Δ SCx [m²]** | - | +0.059 | -0.002 |
| **Δ SCx [%]** | - | +5.8 | - 0.02 |

If for Sedan body type the modifications made reduce the influence of the warning lights with 52.13% and for SUV with 102.2%, for the hatch back body type the gain is almost inexistent with just 1.2%. The CdA values obtained are a direct consequence of the rear wake of the three studied cases, as indicated in Figure 6. To achieve the purpose of reducing the warning lights influence, rear wake balance must be seeked. A balanced rear wake can be visualized with the help of streamlines and must be horizontal and reduced in dimension on X axis.

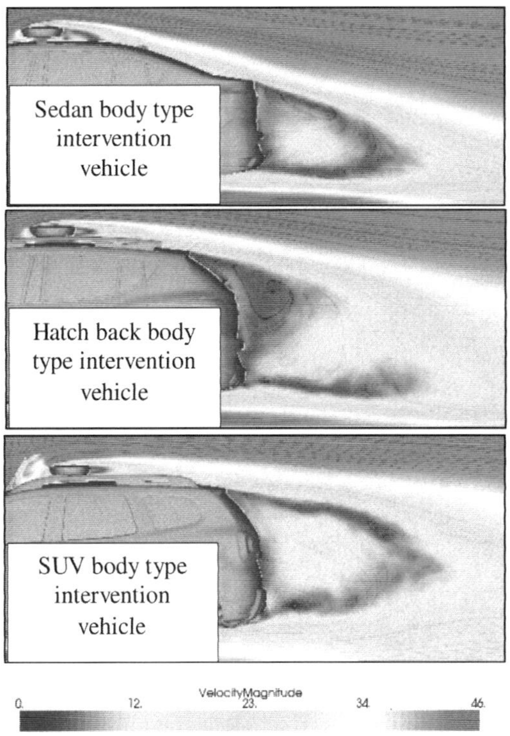


**Figure 6.** Rear wake balance for Sedan, Hatch back and SUV studied cases.

By analyzing the results obtained, several conclusions may be drawn. Firstly, as expected, a newly introduced element will not have the same effect on different body types. This is due to a different shape of the vehicle and air flow speed above and under the body (giving a specific rear wake shape and dimensions). Consequently, for obtaining better results, specific customization per body is required. Secondly, the aerodynamic resistance of warning light can be fully compensated (SUV case – balanced rear wake) or reduced to half (Sedan case – rear wake slightly unbalanced in downward direction) by using a relatively simple body kit for intervention vehicles.

When it comes to hatch back body type, further studies will need to be performed to identify a proper kit to match the wake of the base vehicle, as the current setup has an unbalanced rear wake, with big dimensions. Some options that will be tested to see if they improve the wake for the hatch back body type intervention vehicles are: partial coverage of underbody, addition of deflectors on underbody fairings or extension of rear upper diffuser.

## 3. Conclusion

The specific warning lights on intervention vehicles have a direct influence on aerodynamic resistance and $CO_2$ emissions. While current legislation grants a special status to this category of vehicles, their impact on overall $CO_2$ measured levels, especially in urban areas, remains a reality. This paper demonstrates, by using CFD simulations, that the impact of warning lights on aerodynamic resistance can be compensated to a certain extent with the modification of accessories such as underbody fairings and wheel rim porosity.

The results obtained are generic and show that $CO_2$ emissions may be reduced only by adding specific accessories – which should not be that difficult also from a practical perspective. The specific parts to be modified, their shape and position should be studied for each particular case to have accurate results. It must also be considered that CFD simulations incorporates a calculation error. The final $CO_2$ levels can be validated with physical measurements. CFD method, as used in the current paper, can be applied in the analysis phase to identify best technical solutions, followed by the validation phase that is focused on final fine-tunning iterations before the official homologation process.

## References

[1] https://climate.ec.europa.eu/eu-action/transport-emissions/road-transport-reducing-CO2-emissions-vehicles/CO2-emission-performance-standards-cars-and-vans_en

[2] Bothe D and Steinfort T 2020 *Cradle to Grave Life - Cycle Assessment in the Mobility Sector* (Frankfurt/ Main: FVV)

[3] Howett G L and Pierce E T 1978 *Emergency vehicle warning lights: State of the Art* (Washington DC - National Institute of Law Enforcement and Criminal Justice Law Enforcement Assistance Administration)

[4] https://sketchfab.com/3d-models/daciarenault-logan-politia-high-poly

[5] Ilea L, Iozsa D and Fratila G 2021 *An aerodynamic theoretical study on exterior accessories of intervention vehicles* (Iasi, The 25th Edition of IManEE 2021 International Conference)

[6] Vdovin A, Gustafsson M and Sebben S 2018 *A coupled approach for vehicle brake cooling performance simulations* (International Journal of Thermal Sciences, Volume 132)

# The impact of real driving on vehicle $CO_2$ emissions

**C Constantin[1], M Bățăuș[2], C Rențea[2] and G Frățilă[2]**
[1]Faculty of Transports, University Politehnica of Bucharest, 313 Spl. Independentei, Bucharest, Romania
[2] Automotive Engineering Department. University Politehnica of Bucharest. 313 Spl. Independentei, Bucharest, Romania

E-mail: rentea.cristian@yahoo.com

**Abstract.** On-road (real) driving of a conventional vehicle having free gear-shift strategy and driving style, non-zero positive altitude gain, and uncontrolled ambient conditions is expected to produce different emissions values, including $CO_2$, in respect to those produced in a controlled laboratory experiment following a standardized velocity profile.
This paper aims to investigate the impact of driving style characterized by driving dynamics on $CO_2$ emissions of a passenger car equipped with an internal combustion engine. The study is based on recorded on-road trips of a passenger car that are postprocessed using the EU real driving calculation procedures.
Selected segment trips of urban, rural and motorway driving are used as input for a detailed vehicle model developed in LMS Imagine.Lab AMESim in order to assess the impact of real driving on vehicle $CO_2$ emissions.

## 1. Introduction

The transport sector and in particular the emissions from passenger car internal combustion engines represent a large fraction of pollutants and $CO_2$ emissions in the European Union (EU). The latest EU regulations are imposing limits for fleets average $CO_2$ emissions and define a Real Driving Emission (RDE) test procedure in order to limit tailpipe emissions under all normal conditions of use. For now, the focus of the RDE regulation is on air pollutant emissions but RDE testing is also measuring the $CO_2$ emissions.

It became obvious that with more available data regarding the real-world performance of new vehicles in terms of $CO_2$ emissions, the vehicles manufacturers must focus on reducing them under all conditions. On-road (real) driving is different from roller bench testing due to varying rolling resistance (different road surfaces and cornering) and aerodynamic drag (wind), free gear-shift strategy and driving style, non-zero positive altitude gain, and uncontrolled ambient conditions.

The driving style is practically defined by gear-shift strategy and driving dynamics. The driving dynamics during the trip can be evaluated using specific global parameters such as: VSF (Vehicle Speed Fluctuation), average positive acceleration, RMS acceleration, RPA (Relative Positive Acceleration), a specific percentile of the product of vehicle speed per positive acceleration etc. In [1] the influence of different standardize cycles (US 06, Japanese 10-15, HWFET, NEDC and WLTC) was considered based on VSF parameter. Average positive acceleration and RMS acceleration are considered in [2] for correlation of real-world vehicle emissions. The RDE procedure uses a different method to assess the overall excess or absence of dynamics during trips [3].

This paper aims to investigate the driving dynamics influence on $CO_2$ emissions based on recorded trips on a passenger car based on RDE driving dynamics indicators. The recordings were done on a passenger car for 22 days and a total of 920 km of combined urban, rural and motorway driving were recorded. The data are postprocessed and analyzed with the aim of selecting the most relevant trips segments to be evaluated using a detailed model developed in LMS Imagine.Lab AMESim and previously validated [4].

## 2. Trip recordings and postprocessing

The recordings were made using a DAWN OBD Mini Logger on a C-class passenger car. There were recorded 920 km of trips during 22 days of combine urban, rural and motorway driving. The postprocessing of the trips in terms of driving dynamics and classification was done using RDE procedure [3].

As shown in [3], the dynamic parameters are determined from the speed signal with an accuracy of 0.1 % above 3 km/h and a sampling frequency of 1 Hz. This accuracy requirement can be usually attained by wheel rotational speed signals.

The following calculations were performed over the whole time-based speed trace: distance increment per data sample, vehicle acceleration, the product of vehicle speed per acceleration.

$$d_i = v_i/3.6 \quad (1)$$

$$a_i = (v_{i+1} - v_{i-1})/(2 \cdot 3.6) \quad (2)$$

$$(v \cdot a)_i = (v_{i+1} - v_{i-1})/(2 \cdot 3.6) \quad (3)$$

where: $d_i$ is the distance covered in time step $i$; $v_i$ is the actual vehicle speed in time step $i$ [km/h]; $N_t$ is the total number of samples.

The values $v_i$ , $d_i$ , $a_i$ and $(v \cdot a)_i$ are ranked in ascending order of the vehicle speed.

To show the high variability of the recorded data, the $(v \cdot a_{pos})$ values for each point of the trips with minimum and maximum average $(v \cdot a_{pos})$ are plotted against vehicle speed and shown in figure 1. Moreover, the RPA values for each point of the trips with minimum and maximum average RPA are plotted against vehicle speed in figure 2.

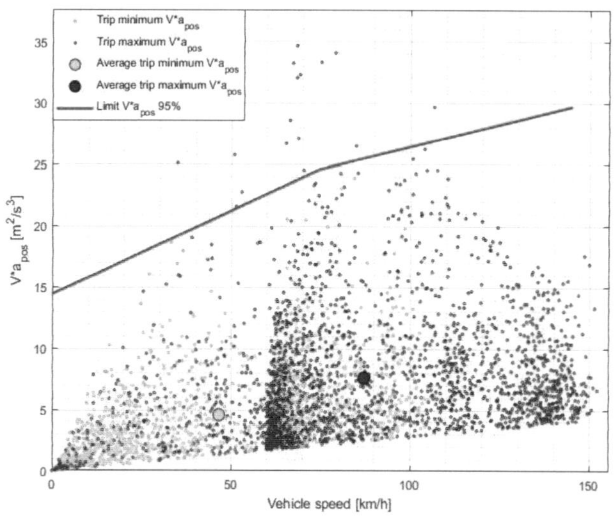


**Figure 1.** $(v \cdot a_{pos})$ values for each point of the extreme trips.

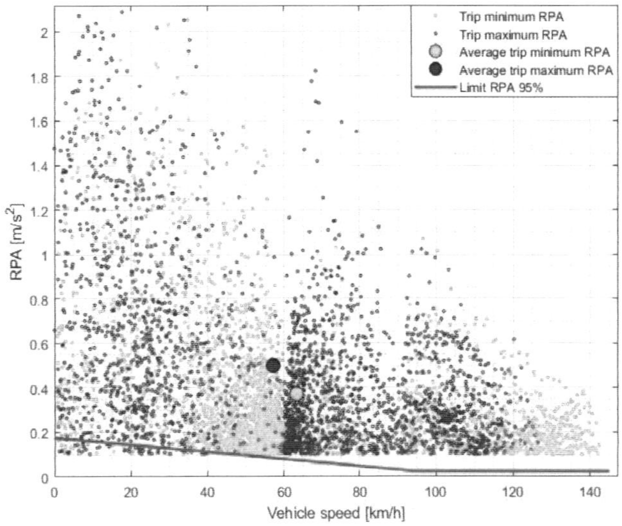


**Figure 2.** RPA values for each point of the extreme trips.

The datasets are sorted, those with $v_i \leq 60$ km/h belong to the 'urban' speed bin, for 60 km/h < $v_i \leq$ 90 km/h belong to the 'rural' speed bin and if $v_i > 90$ km/h belong to the 'motorway' speed bin. The average vehicle speed is calculated for each speed bin.

$$\bar{v}_k = \sum_i v_{i,k} / N_k \text{ for } i=1 \text{ to } N_k \tag{4}$$

where, for urban (k=u), rural (k=r) and motorway (k=m) shares: $N_k$ is the total sample number.

A number of 22 trips are recorded, but it is useful to separate into trips segments that are classified as urban, rural and motorway. Only the trips segments with a number of datasets with acceleration values $a_i > 0.1$ m/s² and bigger or equal to 150 seconds are considered. Therefore, after applying those conditions, a number of 15 urban trips segment, 6 rural trips segment and 7 motorway trips segments were left.

For each trip the 95th percentile of $v \cdot a_{pos}$ is calculated by ranking the $(v \cdot a)_i$ per trip (for $a_i \geq 0.1$ $m/s^2$) and the total number of these *Mk* is determined. Afterwards, percentile values are allocated to the $(v \cdot a_{pos})_{j,k}$ values by assigning for the lowest value the percentile *1/Mk*, for the second lowest *2/Mk*, and so on until the highest value for which the *Mk/Mk =100 %* value is assigned.
$(v \cdot a_{pos})_k$ *_[95]* is the $(v \cdot apos)_{j,k}$ value, with *j/Mk = 95 %*, or is calculated by linear interpolation between consecutive samples *j* (with *j/Mk < 95 %*) and *j+1* (with *(j+1)/Mk > 95 %*).
The relative positive acceleration is calculated for each speed bin.

$$RPA_k = \sum_j \left(\Delta t \cdot (v \cdot a_{pos})_{j,k}\right) / \sum_i d_{j,k} \text{ for } j=1 \text{ to } M_k \text{ and } i=1 \text{ to } N_k \tag{5}$$

where, for urban (*k=u*), rural (*k=r*) and motorway (*k=m*) shares: $RPA_k$ is the relative positive acceleration; $M_k$ is the sample number with positive acceleration; $N_k$ is the total sample number and $\Delta t$ is the time difference (1 s).

The average velocity, 95th percentile of the $v \cdot a_{pos}$ and *RPA* of each trip segment are calculated and shown in figure 3 and 4 together with an average of each trip segment class.
The limits considered in [3] for high dynamic and shown with red line in figures 1 and 3 are:

$$(v \cdot a_{pos})_k\text{\_}[95] > (0.136 \cdot \bar{v}_k + 14.44) \text{ for } \bar{v}_k \leq 74.6\ km/h \tag{6}$$

$$(v \cdot a_{pos})_k\text{\_}[95] > (0.0742 \cdot \bar{v}_k + 18.966) \text{ for } \bar{v}_k > 74.6\ km/h \tag{7}$$

The limits considered in [3] for low dynamic and shown with red line in figures 2 and 4 are:

$$RPA_k < (-0.0016 \cdot \bar{v}_k + 0.1755) \text{ for } \bar{v}_k \leq 94.05\ km/h \quad (8)$$

$$RPA_k < 0.025 \text{ for } \bar{v}_k > 94.05\ km/h \quad (9)$$

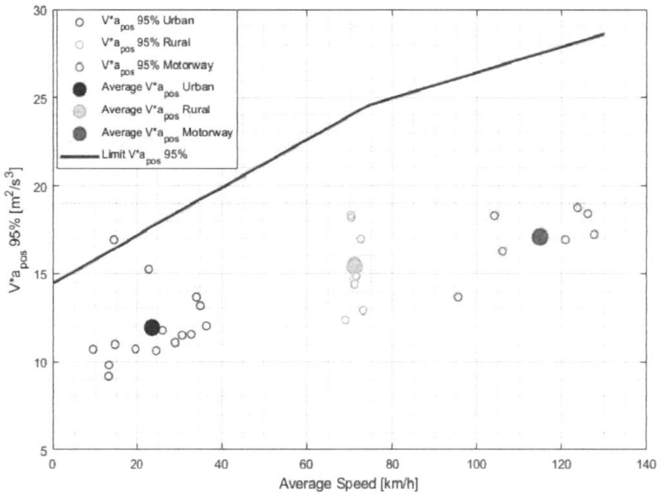


**Figure 3.** The 95th percentile of the $v \cdot a_{pos}$ and average velocity of each trip segment.

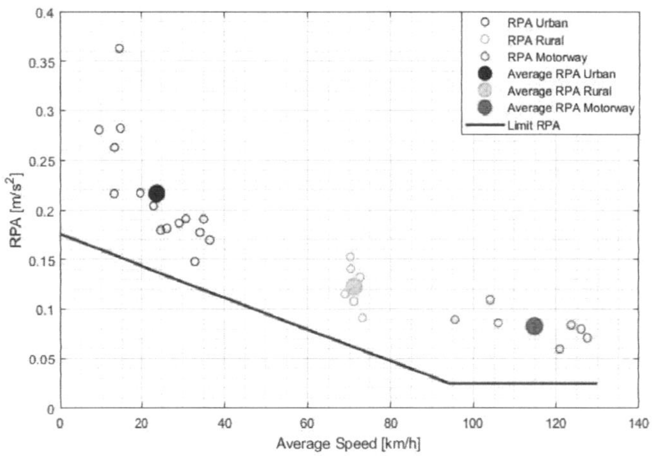


**Figure. 4.** The RPA and average velocity of each trip segment.

## 3. Results

The speed profiles for the relevant trip segments are used as test cycle input in a complex C-class SUV model developed using LMS Imagine.Lab AMESim. The model presented in [4] is based on the sub-models from the IFP Drive library and can compute the $CO_2$ emissions for a variable speed test cycle.

The $CO_2$ emissions were determined for low and high dynamics cycles in every category (urban, road and motorway) in order to assess the influence of using the proposed classification criteria. The results are shown in table 1 and figure 5.

**Table 1.** $CO_2$ emissions for low and high dynamics cycles.

| Segment type | Dynamics | RPA [m/s$^2$] | (v $a_{pos}$)_[95] [m$^2$/s$^3$] | $V_{med}$ [km/h] | $CO_2$ emisson [g/km] | | $CO_2$ emission ratio [-] |
|---|---|---|---|---|---|---|---|
| | | | | | For segment | At Vmed | |
| **Urban** | High | 0.363 | 16.94 | 14.87 | 194.1 | 151.2 | 1.284 |
| | Low | 0.147 | 11.55 | 33.37 | 110.9 | 78.4 | 1.414 |
| **Rural** | High | 0.152 | 18.33 | 70.2 | 95.9 | 91.4 | 1.049 |
| | Low | 0.091 | 12.9 | 73.08 | 96.1 | 92 | 1.045 |
| **Motorway** | High | 0.109 | 18.3 | 104.1 | 145.6 | 138 | 1.055 |
| | Low | 0.059 | 16.95 | 120.8 | 185.1 | 176.6 | 1.048 |

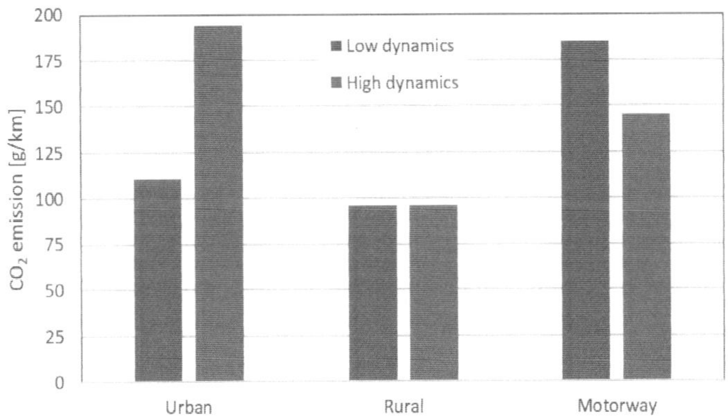


**Figure 5.** $CO_2$ emissions for low and high dynamics in every category (urban, road and motorway).

To compare the influence on fuel consumption of different test cycles we define a parameter called $CO_2$ emission ratio ($CO_2$_ER) in a similar way with FCV (Fuel Consumption Ratio) defined in [5]:

$$CO_2\_ER = \frac{CO_2\_cycle}{CO_2\_ct.speed} \cdot 100\ [\%] \quad (10)$$

where: *$CO_2$_cycle* is the $CO_2$ emission obtained in the test cycle; *$CO_2$_ct.speed* is the $CO_2$ emission obtained at constant speed movement equal to the average speed of cycle.

From the Table 1 data, we can observe that there is not a simple corelation between chosen cycle dynamics parameters and $CO_2$ emission. In urban driving, the high dynamic corresponds to a cycle with low average velocity. The engine will be operated in a disadvantageous point and the fuel consumption (thus the $CO_2$ emissions will be high), in this particular case 75% higher than low dynamic driving. For motorway driving, the low dynamics corresponds to a cycle with high average velocity and grater aerodynamic drag that will increase the $CO_2$ emissions. For the simulated examples is 27% higher when the driving dynamic is low. In rural driving the two tendencies are compensating and the differences between high and low dynamic driving are under 0.2%.

## 4. Conclusions

It was shown that in real-world the driving dynamics can have a big influence especially in urban and motorway driving. The test cycles were obtained as trip segments from 920 km of combine urban, rural and motorway driving recorded for 22 days using RDE procedure.

One limitation of the resulted trip segments is the low specific power of the test vehicle that can influence the overall driving dynamics. Therefore, the recordings must be used for studies on the vehicles with similar specific power. However, for every driving segment type substantial difference between the driving dynamics characterisation parameters was obtained: from 67% to 145% for RPA and from 8% to 47% for the 95$^{th}$ percentile of the $v \cdot a_{pos}$.

If the influence of average vehicle speed is eliminated using the $CO_2$ emission ratio it is observed that the influence of cycle dynamics is reduced from 75% to 9 % in urban driving and become insignificant (less than 0.7 %) in rural and motorway driving.

**Author's ORCID ID**
C Renţea: https://orcid.org/0009-0007-8536-8638

**References**

[1] Renţea C, Tucă A, Oprean M and Băţăuş M 2017 *The study of the energetic performance in test cycles*, IOP Conference Series: Materials Science and Engineering (Vol. 252, No. 1, p. 012087)
[2] Pathak S K, Sood V, Singh Y and Channiwala S A 2016 *Real world vehicle emissions: Their correlation with driving parameters*. Transportation Research Part D: Transport and Environment, 44, 157-176
[3] Commission Regulation (EU) 2017/1151 of 1 June 2017 supplementing Regulation (EC) No 715/2007 of the European Parliament and of the Council on type-approval of motor vehicles with respect to emissions from light passenger and commercial vehicles (Euro 5 and Euro 6) and on access to vehicle repair and maintenance information, amending Directive 2007/46/EC of the European Parliament and of the Council, Commission Regulation (EC) No 692/2008 and Commission Regulation (EU) No 1230/2012 and repealing Commission Regulation (EC) No 692/2008.
[4] Renţea C and Frăţilă G 2021 The modelling and simulation of propulsion systems to estimate $CO_2$ emissions. *Romanian Journal of Automotive Engineering*, **vol. 27**, no. 1
[5] Oota Y 2013 *The advantage of CVT in the real word*. In 12th International CTI Symposium (Berlin: Car Training Institute).

# Biogas from wastewater's sludge as potential resource for transportation. Experimental approach.

**T B Bacos[1], A E Cioabla[1], F Popescu[1], T Vintilă[2] and A T Borborean[1]**

[1]Department of Mechanical Machines, Equipment and Transportation, University Politehnica Timisoara, Bd. Mihai Viteazul no.1, Timisoara 300222, Romania
[2]Faculty of Bioengineering and Animal Resources, University of Life Science King Mihai I from Timişoara, Romania

E-mail: andrei.borborean@upt.ro

**Abstract**. Considering growing energy requirements and the quest for self-sufficiency in energy sources, the utilization of waste-derived materials has surged in various fields of research in recent years. Within this framework, waste waters, sludge, or slurries, biodegradable substances, second-generation effluents, and the resulting by-products have emerged as crucial substrates of interest to produce biogas through anaerobic digestion (AD). The objective is to employ these materials as the foundation for generating renewable energy, ultimately establishing them as the primary energy source for households or industries. This paper endeavours to assess, both physically and chemically, the waste industrial and urban wastewaters sourced from west Banat region as a foundational substrate material for co-fermentation alongside residual biomass from cereals. The results obtained are presented.

## 1. Introduction

The continuing demand of energy, year by year and rising prices of fuel grows interest in green energy such as solar power, wind turbines but also biofuel. The versatility of biofuel gives a large scale to work with, from wastewater to animal manure and to even food waste. Biofuels are an interesting topic now because of their potential energy uses.

The fundamental concept behind a circular economy involves minimizing waste and resource utilization by preserving the value of materials, products, and resources for an extended period. When a product nears the end of its useful life, it can be reclaimed to extract value. Anaerobic digestion, a method that generates biogas from diverse sources such as agricultural residues, energy crops, organic-rich wastewater, the organic fraction of municipal solid waste, and industrial organic waste, is a key process in achieving this objective. [1]

Biomass refers to a biodegradable material derived from plants or animals, encompassing sources such as municipal waste management, forestry residues, garbage, agricultural production, and the agri-food industry. These materials can be utilized to generate heat or electricity. Various biomass sources include cereal straw, legumes, oilseeds, hay, wood (waste from forestry and wood processing), used wood packaging, energy crops, dehydrated sewage sludge, pellets, briquettes, bio-carbon, processing industry waste, agricultural waste (such as slurry, manure, and plant biomass), municipal landfill waste, and agro-food industry waste (such as beet pulp, bagasse, molasses, wine residues, and waste from dairy and oil products). Rotten and expired fruits and vegetables also serve as biomass sources for energy production. [2-5]

An example of a widely employed renewable energy source today is biogas. Biogas is generated through anaerobic decomposition, primarily consisting of carbon dioxide and methane. The methane content within biogas typically falls between 40% and 70%. The predominant source of biogas production is from dairy farms, leading to the establishment of a substantial industry. Biogas can serve as an alternative to propane and liquid petroleum gas for heat energy applications.

Wastewater stands out as a potential feedstock for anaerobic digestion, and the importance of sustainability is on the rise across various industries, including wastewater treatment. This heightened focus is driven by more frequent energy crises and the escalating global costs of energy [6, 7]. Conventional wastewater treatment technologies, particularly the sludge activation process with its substantial energy consumption (approximately 1 kWh/m3, constituting up to 60% of the system's energy budget), have spurred an increasing emphasis on developing sustainable wastewater treatment approaches tailored for situations with limited resources, especially energy. [9, 10].

Preserving national public health hinges on the critical aspects of wastewater treatment and management. Stringent regulations have been implemented to guarantee the proper treatment of wastewater and the appropriate disposal of treated effluent [11]. The annual production of livestock manure has exhibited fluctuations in recent years, ranging from 80 to 100 million tons. Slurries constitute approximately 20% of natural fertilizer, with the remaining portion being comprised of farmyard manure. Among all manures produced, dairy cow excrement holds the largest mass share [12].

There are three leading technologies for managing cow manure. The basic, but the most burdensome for the environment and neighborhood is storage manure in piles, and periodic (3-4 times a year) spreading manure on the fields for fertilizing purposes. In this technology, the energy contained in fresh manure goes strictly to the soil and is used by soil microorganisms [13]

Manure composting is a more advanced technology using various machines (tractor aerator, manure spreader). Throughout the composting process, conducive aerobic conditions are established to facilitate the proliferation of bacteria. This prompts the rapid decomposition of organic matter in manure, leading to the substantial release of energy in the form of heat [14, 15]. During the thermophilic composting phase, the temperature of the manure pile can elevate to nearly 80 °C and persist for several weeks. While various concepts and technologies exist to utilize the heat generated in composting, practical challenges make it difficult to effectively harness the heat from compost heaps.

The primary advantage of the composting process lies in producing compost with a high fertilizer value and ecological significance [16].

Methane fermentation represents the third method currently employed for utilizing manure. In this process, manure undergoes anaerobic microbiological breakdown [17]. This results in the liberation of energy from the organic content of the manure in the form of methane, serving as a chemical fuel. Additionally, digestate, a high-grade fertilizer suitable for agriculture, is produced as a byproduct [18, 19]. The biogas generated through anaerobic digestion can be directly utilized in cogeneration engines (CHP) to generate heat and electricity, or in trigeneration, to produce cold as well. Alternatively, the biogas can be refined to produce biomethane, which can be injected into the gas network for future use [20, 21].

In the past, initial endeavors led to the production of first-generation biofuels, primarily derived from various feedstocks such as rapeseed [24], corn [23], and soybean [22]. While economically viable and technically accessible, these biofuels come with notable drawbacks, including competition for water resources between food and fuel production, biodiversity loss, and constraints on feedstock availability [25]. The expensive and intricate processes involved in producing these biofuels have considerably restricted their economic feasibility and profitability, despite their numerous advantages [26].

## 2. Material and Methods

Agricultural residues from barley, wheat, corn, and rye were subjected to drying in an oven and subsequently ground into particles smaller than 1 mm using a hammer mill. This procedure adhered to the necessary material documentation protocols. Following this mechanical processing, representative samples were meticulously collected and transported to the laboratory for further analysis.

To ensure the accuracy of laboratory analysis procedures, thorough consultations of equipment user manuals and adherence to biofuels standards were undertaken before initiating the laboratory determinations. Specific analytical standards designed for solid biofuels were employed to identify the primary parameters of interest for the agricultural waste samples. These parameters included the total content of carbon, hydrogen, nitrogen, sulfur, and chlorine (ISO 16948 & 16994), moisture content (ISO 1834), ash content (ISO 18122), calorific value (ISO 14918), and volatile matter (ISO 18123).

To assess the ash's melting behavior, we followed the CEN/TS 15370 requirements. All analyses were conducted using the ISO technique, and the measurements were performed on a dry basis. For solid substrates, this entailed determining the dry mass of the sample remaining after heating it to a temperature between 105 and 110 °C for an hour. This heating process was iterated until only the solid portion of the sample remained.

In the final analysis, we examined the concentrations of sulfur, nitrogen, hydrogen, carbon, and chlorine in randomly selected dry substrates. Figure 1 provides an overview of the experimental twin 5-liter anaerobic digestion reactor system.

**Figure 1.** Front image of the test rig for 5-liter volume.

After substrate analysis (on dry basis), the materials (3 liters of biogas factory substrate, 1 liter of wastewater from beer factory and 150 grams of degraded milled corn) were tested inside the 5-liter reactors (Figure 1) to determine the potential and the quality of biogas production.

## 3. Results and Discussions

Table 1 outlines the key characteristics of the degraded barley and degraded corn employed in anaerobic digestion. Both Tables 1 and 2 demonstrate that all substrates exhibit a favorable energetic potential, as evidenced by their net calorific values.

**Table 1.** Parameters of biomass used in anaerobic digestion. Part 1.

| Substrate | Moisture content (db) (%) | Ash content (db) (%) | Gross calorific value (db)(J/g) | Net calorific value (db)(J/g) | Volatile matter content (db) (%) | Sulphur content (db) (%) | Chlorine content (db) (%) |
|---|---|---|---|---|---|---|---|
| Corn | 10.2 | 1.58 | 18460 | 16887 | 85.5 | 1.03 | 0.34 |
| Biogas plant Gataia substrate | 8.46 | 18.7 | 1669 | 15444 | 60.2 | 0.43 | 4.4 |
| Wastewater beer factory | 5.1 | 26.7 | 17800 | 17200 | 42.3 | 0.64 | 2.7 |

**Table 2.** Parameters of biomass used in anaerobic digestion. Part 2.

| Substrate | Carbon content (%) | Hydrogen content (%) | Nitrogen content (%) |
|---|---|---|---|
| Corn | 40.35 | 6.5 | 1.29 |
| Biogas plant Gataia substrate | 38.3 | 5.2 | 2.59 |
| Wastewater beer factory | 36.4 | 4.2 | 3.7 |

The experimental substrate formed from a mixture of all 3 materials were subject of continuous anaerobic digestion for 20 days. The process parameters and results obtained are presented in the next figures.

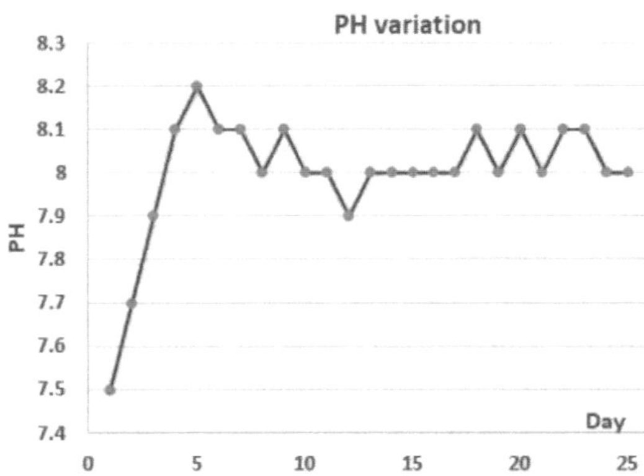


**Figure 2.** Substrate mixture pH evolution

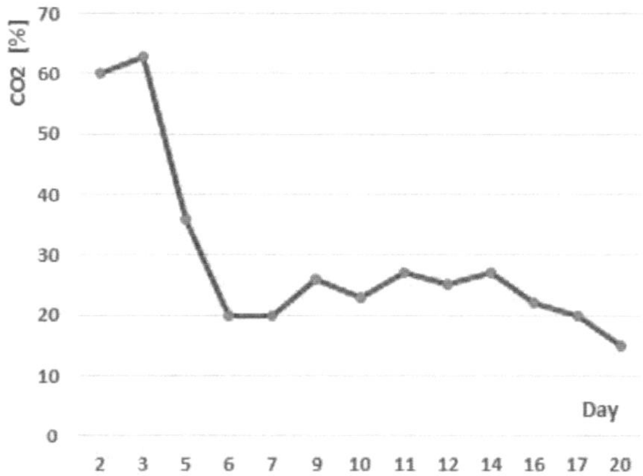


**Figure 3.** Substrate mixture $CO_2$ evolution

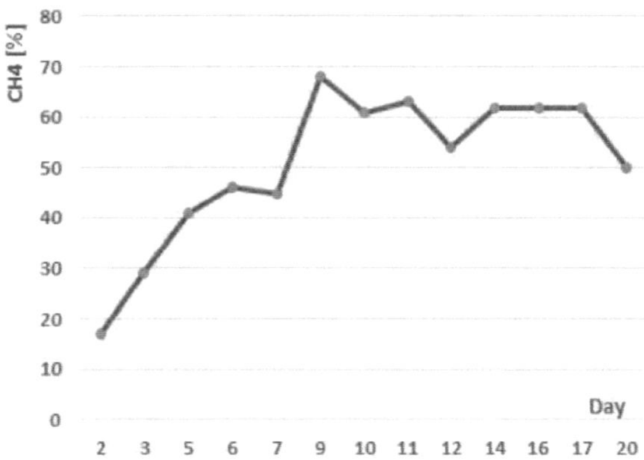


**Figure 4.** Substrate mixture $CH_4$ evolution.

The results of the experiment indicate a relatively high potential for biogas production from corn wastes, particularly when combined with wastewater and biogas plant substrate.

There is substantial potential for producing biogas from organic waste and energy crops. Nonetheless, utilizing energy crops as the source material for biogas may pose competition with arable land and water resources used for food production, which depends on regional availability and the demand for these resources. Consequently, European Union member states enforce sustainability criteria for the application of biogas as a vehicle fuel.

Production of biogas from urban and/or industrial wastewaters has an enormous and sustainable potential in Romania, an energy resource that is, at this time, vastly ignored by national decision factors. Technologies to convert biogas to biomethane is well established and the biogas potential should be used for transportation fuel, in the beginning at least in urban public transport fleets.

## 4. Conclusions

An investigation was conducted to assess the anaerobic digestion of agricultural resources under mesophilic temperature conditions in a batch operation. The findings revealed that combining agricultural waste barley with wastewater from a beer factory proves to be an excellent substrate for anaerobic digestion due to its significant capacity for producing biogas enriched with methane.

Overall, the research findings indicate that the combination of agricultural waste and beer factory wastewater is an excellent resource for anaerobic digestion because it yields high-quality biogas, containing roughly 70% methane, suitable to be used for energy production.

The presented research, which focuses on the production of biogas from various substrate combinations, including beer industry effluent and degraded agricultural materials, and presents anaerobic digestion procedures without the requirement for additional inoculums, may be of interest to the scientific community. This technology is useful for lowering greenhouse gas emissions in addition to being energy-efficient.

Growth and advancement are possible because of the benefits of biogas, which include its contributions to waste management and nutrient recycling. Therefore, local biogas producers should think about combining a variety of feedstocks such agricultural leftovers, pulp and paper mill residues, and animal manure in addition to sewage sludge and food waste. [27]

Methane can be used as a fuel source for cars that are intended to run on compressed natural gas (CNG) by enriching the gas and compressing it. When comparing biogas to natural gas and diesel, its lower emission levels make it a more appealing and eco-friendly option.

Biogas can power natural gas vehicles (NGVs) and dual-fuel vehicles in place of fossil fuel-based gas if it is cleaned to remove CO2 and H2S.

As a fact, in 2016, the leading countries in terms of biogas production for use as a vehicle fuel were Germany, Sweden, Switzerland, the United Kingdom, and the United States. [28] Globally, approximately 500 facilities generate biogas and enhance its quality to match that of natural gas, amounting to roughly 50 Petajoules (PJ) per year.

**Author's ORCID ID**
A Cioabla: https://orcid.org/0000-0001-6261-0064
F Popescu: https://orcid.org/0000-0001-7736-0758
A Borborean: https://orcid.org/0000-0001-6818-9549

## References

[1] Cucchiella F, D'Adamo I and Gastaldi M 2017 Biomethane: A Renewable Resource as Vehicle Fuel. *Resources,* Vol 6(4), p.58
[2] Wiater J and Horysz M 2017 Organic waste as a substrat in biogas production. *Journal of Ecological Engineering*, Vol 18(5), pp. 226-234
[3] Kuziemska B, Pieniak-Lendzion K, Trębicka J, Wieremej W and Klej P 2015 Alternative sources of energy, *Scientific Letters of the University of Natural Sciences and Humanities in Siedlce, Administration and Management*, Vol. 103, pp. 99–113
[4] Herbert G M J and Krishnan A. U. 2016 Quantifying environmental performance of biomass energy. *Renewable and Sustainable Energy Reviews*, Vol. 59, pp. 292–308
[5] Wrzosek J and Gworek B 2010 Biomass in renewable energy, *Environmental Protection and Natural Resources*, Vol. 43, pp.104–116
[6] Ahmad T and Zhang D 2020 A critical review of comparative global historical energy consumption and future demand: The story told so far. *Energy Reports*, Vol. 6, pp.1973–1991
[7] Milewska B and Milewski D 2022 Implications of Increasing Fuel Costs for Supply Chain Strategy. *Energies* Vol. 15, p.6934
[8] Li W W, Yu, H Q and He Z 2013 Towards sustainable wastewater treatment by using microbial fuel cells-centered technologies. *Energy & Environmental Science* Vol. 7, pp. 911–924

[9] Metcalf E 2004 *Wastewater Engineering. Treatment and Reuse*, 4th ed.; McGraw-Hill: New York, NY, USA
[10] Chachuat B, Roche N and Latifi M A 2005 Long-term optimal aeration strategies for small-size alternating activated sludge treatment plants. *Chemical Engineering and Processing - Process Intensification* Vol. 44, pp. 591–604
[11] Chua K H, Cheah W L, Tan C F and Leong Y P 2013 Harvesting biogas from wastewater sludge and food waste, *IOP Conf. Series: Earth and Environmental Science* Vol. 16, p. 012118
[12] Kozlowski K, Dach J, Lewicki A, Malinska K, Do Carmo I E P and Czekala W 2019 Potential of biogas production from animal manure in Poland, *Archives of Environmental Protection* Vol. vol. 45(5), pp. 99-108
[13] Zhang Y L, Sun C X, Chen Z H, Zhang G N, Chen L J and Wu Z J 2019 Stoichiometric analyses of soil nutrients and enzymes in a Cambisol soil treated with inorganic fertilizers or manures for 26 years, *Geoderma* Vol. 353, pp. 382-390
[14] Li Y, Manandhar A, Li G and Shah A 2018 Life cycle assessment of integrated solid state anaerobic digestion and composting for on-farm organic residues treatment, *Waste Management* Vol. 76, pp. 294-305
[15] Bernal M P, Alburquerque J A and Moral R 2009 Composting of animal manures and chemical criteria for compost maturity assessment. A review, *Bioresources Technology*, vol. 100(22), pp. 5444-5453
[16] Pergola M, Piccolo A, Palese A M, Ingrao C, Di Meo V and Celano G 2018 A combined assessment of the energy, economic and environmental issues associated with on-farm manure composting processes: Two case studies in South of Italy, *Journal of Cleaner Production*, Vol. 172, pp. 3969-3981
[17] Esteves E M M, Herrera A M N, Esteves V P P and Morgado C R V 2019 Life cycle assessment of manure biogas production: A review, *Journal of Cleaner Production*, Vol. 219. pp. 411-423
[18] Czekala W, Lewicki A, Pochwatka P, Czekala A, Wojcieszak D, J6zwiakowski K and Waliszewska H 2020 Digestate management in polish farms as an element of the nutrient cycle, *Journal of Cleaner Production*, Vol. 242, p. 118454
[19] Czekala W, Dach J, Dong R, Janczak D, Malinska K, J6zwiakowski K, Smurzynska A and Cieslik M 2017 Composting potential of the solid fraction of digested pulp produced by a biogas plant, *Biosystems Engineering*, Vol. 160, pp. 25-29
[20] Wandera S M, Qiao W, Algapani D E, Bi S, Yin D, Qi X, Liu Y, Dach J and Dong R 2018 Searching for possibilities to improve the performance of full scale agricultural biogas plants, *Renewable Energy*, Vol. 116, pp. 720-727
[21] Bidart C, Frohling M and Schultmann F 2014 Livestock manure and crop residue for energy generation: Macro-assessment at a national scale, Renewable and Sustainable Energy Reviews, Vol. 38, pp. 537-550
[22] Zhou D, Qi L, Qiao B Q, Xu Q Q and Yin J Z 2017 Continuous production of biodiesel from soybean flakes by extraction coupling with transesterification under supercritical conditions: *Journal of Supercritical Fluids*, Vol. 120, pp. 395-402
[23] Rajaeifar M A, Ghobadian B, Safa M and Heidari M D 2014 Energy life-cycle assessment and CO2 emissions analysis of soybean-based biodiesel: a case study, *Journal of Cleaner Production*, 2014, 66, pp. 233-241
[24] Gu J, Xin Z, Meng X, Sun S, Qiao Q and Deng H 2015 Studies on biodiesel production from DDGS-extracted corn oil at the catalysis of Novozym 435/super absorbent polymer, *Fuel*, Vol.146, pp.33-40
[25] Halleux H, Lassaux S, Renzoni R and Germain A 2008 Comparative life cycle assessment of two biofuels ethanol from sugar beet and rapeseed methyl ester, *The International Journal of Life Cycle Assessment*, Vol. 13, pp. 184

[26] Naik N, Goud V V, Rout P K and Dalai A K 2010 Production of first and second generation biofuels: a comprehensive review, *Renewable and Sustainable Energy Reviews*, Vol. 14, pp. 578-597

[27] Arfan M, Wang Z, Soam S and Eriksson O 2021 Biogas as a Transport Fuel. A System Analysis of Value Chain Development in a Swedish Context. *Sustainability*, Vol. 13, 4560

[28] IRENA, Biogas for road vehicles: Technology brief, *International Renewable Energy Agency*, 2018, Abu Dhabi, ISBN: 978-92-9260-060-0

IOP Conf. Series: Materials Science and Engineering 1303 (2024) 012031 doi:10.1088/1757-899X/1303/1/012031

# Determining the mechanical characteristics of some tensile specimens, depending on the material and the printing position

**C Babiş[1], A Dimitrescu[1], S A Fica[2], C Stochioiu[1], D Vlăsceanu[1], F Baciu[1], O R Chivu[1] and IC Enache[1]**

[1] National University of Science and Technology Politehnica Bucharest, Faculty of Industrial Engineering and Robotics, Splaiul Independentei no. 313, sect 6, Bucharest, Romania
[2] National Research and Development Institute for Electrical Engineering Bucharest, Splaiul Unirii no 313, sect 3, Bucharest, Romania

E-mail: claudiubab8@gmail.com

**Abstract.** 3D printing is finding more and more applications in the industrial field and represents a modern additive manufacturing process based on a digital model. There are a number of advantages of additive manufacturing through 3D printing compared to classic manufacturing processes. Additive manufacturing through 3D printing allows material savings compared to classic manufacturing processes. Material consumption is punctual according to the project, without unnecessary losses and excess material. Through additive manufacturing, the design can be optimized in the sense that rapid changes can be made to the prototype in the CAD file. Also, through additive manufacturing through 3D printing, the principle of sustainability is promoted. We can say that additive manufacturing through 3D printing ensures rapid prototyping, which leads to the development of new products, shortens the design and manufacturing cycle, improves the quality and precision of models, eliminates costly mistakes, and optimizes the way of collaboration between engineers, marketing departments and sales and management team. In addition to these advantages, it must be seen if the 3D printed elements can replace the elements manufactured by classical methods in terms of mechanical resistance. In this sense, an investigation of some mechanical properties of some samples manufactured by 3D printing from various types of materials is required. The samples will be subjected to tension and bending. In this paper, only the tensile tests of some specimens, obtained by 3D printing in three positions: horizontal, vertical and in height, using three categories of printing filaments of the type: ABS+, PETG, PLA+, are presented. They have been made 5 samples for each material and printing direction, resulting in a total of 45 samples. The samples made from the three categories of material and in the three printing positions were subjected to tension until breaking. The formulas that were the basis of the tests are presented and graphs are drawn that represent the average characteristic curves of the samples for each category of material and printing guidelines.

## 1. Introduction

The 3D printing is finding more and more applications in the industrial field and represents a modern additive manufacturing process based on a digital model [1]. There are a number of advantages of additive manufacturing through 3D printing compared to classic manufacturing processes.

Cost reduction is a first advantage [2]. Costs of thousands of euros can be reduced by eliminating some operations during the classic production process such as casting, processing by plastic deformation, processing by cutting and welding, allowing at the same time to make the required changes much faster and cheaper.

Another important advantage of additive manufacturing through 3D printing is related to design optimization in the sense that rapid changes can be made to the prototype in the CAD file [3].

Rendering 3D printed parts to a high degree of complexity is another important advantage [4]. Each layer of material that the printer deposits on the printing surface is made sequentially, which allows the creation of complex internal structures. In the additive manufacturing process, the 3D printer can create partial interior voids, filled with a honeycomb structure, resulting in particularly light and at the same time rigid parts. If the parts to be manufactured exceed in size the surfaces intended for printing, they can be printed separately in pieces, later being glued to form a whole.

Reducing manufacturing time is another very important advantage of additive manufacturing through 3D printing [5]. Depending on the complexity of the model and the printing time of the printer, the parts can be manufactured in a few hours or days.

Additive manufacturing through 3D printing allows material savings compared to classic manufacturing processes. Material consumption is punctual according to the project, without unnecessary losses and excess material.

Through additive manufacturing through 3D printing, the principle of sustainability is promoted. There is no need for a planned production. The products can be made to order according to needs and easy to recycle thanks to the versatile materials. In addition, some of the materials are ecological, compatible with the environment and resources are not consumed in excess.

In this work, the breaking tests of some specimens subjected to tension, obtained by 3D printing in three positions: horizontal, vertical and in height, using three categories of printing filaments of the type: ABS+, PETG, PLA+ [6], are presented. [7]. They have been made 5 samples for each material and printing direction, resulting in a total of 45 samples. The samples made from the three categories of material and in the three printing positions were subjected to tension until breaking. The formulas that were the basis of the tests are presented and graphs are drawn that represent the average characteristic curves of the samples for each category of material and printing direction.

## 2. 3D printed samples for tensile tests

Figure 1 shows the 3D modeling in the Autodesk Fusion program of the specimen for the tensile tests, which results in the file in STL format for slicing before its 3D printing.

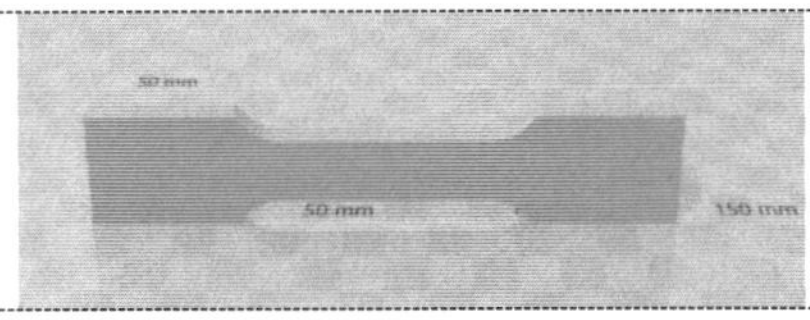


**Figure 1.** Modeling the tensile test specimen in the Autodesk Fusion program

The samples were manufactured by 3D printing from three types of filaments: ABS+, PETG and PLA+. The physical and mechanical characteristics of the filaments used to print the samples are presented in table 1.

**Table 1** The mechanical and physical characteristics of the filaments

| No | TF | TDCP (°C) | TPI (°C) | TS (Mpa) | EB (%) | FS (Mpa) | FM (Mpa) | IS (Mpa) |
|---|---|---|---|---|---|---|---|---|
| 1 | PETG | 250 | 75 | 49 | 225 | 68 | 1800 | 7,6 |
| 2 | ABS+ | 260 | 110 | 40 | 30 | 68 | 2443 | 42 |
| 3 | PLA+ | 220 | 60 | 60 | 29 | 87 | 3642 | 7 |

The following notations were used in the table: TF-Type of fillament; TDCP-Print head nozzle temperature; TPI-Printer bed temperature; TS-Tensile Strentgh; EB-Elongation at break; FS-Flexural Strentgh; FM-Flexural Modulus; IS-Impact Strentgh.

The tensile samples were made from three categories of material (ABS+, PETG and PLA+) and in three printing directions (A-horizontal direction, B-vertical direction and C-lateral direction) resulting in 45 pieces, according to the experimental plan presented in table 2.

**Table 2** Experimental plan for tensile tests

| Material | ABS+ | | | PETG | | | PLA+ | | |
|---|---|---|---|---|---|---|---|---|---|
| Position | A | B | C | A | B | C | A | B | C |
| No pieces | 5 | 5 | 5 | 5 | 5 | 5 | 5 | 5 | 5 |

The realization of the tensile specimens taking into account the three printing directions A, B and C is presented in figure 2.

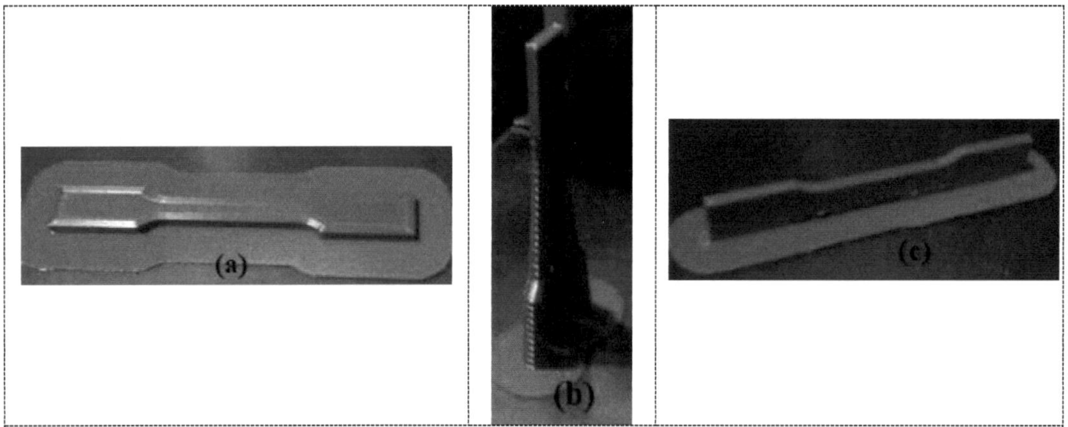


**Figure 2.** Printing the tensile samples in the three printing directions

a-printing in horizontal direction A; b-printing in vertical direction B; c-printing in the lateral direction C

## 3. Tensile tests

The tests were carried out with an Instron 8875 electrohydraulic testing machine. The force was read using the machine's force cell.

The deformation was recorded using the digital correlation method, on a surface in the calibrated area, painted with a white background and black spots.

The tension in the specimen was calculated using the relationship below where F is the force measured with the force cell and A is the cross-sectional area of the specimen, which was considered to be 40 $mm^2$.

$$\sigma = \frac{F}{A} \ (MPa) \quad (1)$$

The yield stress is the apparent yield stress, calculated as the stress value at a permissible plastic deformation of 0.2%.

The breaking limit is considered as the highest value of the stress recorded.

The results are not presented for some samples, for which the digital image correlation equipment did not record, as a result of a software error.

## 4. Experimental results

The results of tensile tests for the samples made of ABS+ material; PETG and PLA+, for the three printing directions are presented in the tables below.

The results of the ABS material tensile tests, printing position A, are presented in table 3.

**Table 3** ABS+ test results, position A

| Epruveta | E (MPa) | $\gamma$ (-) | $\sigma_C$ (MPa) | $\sigma_R$ (MPa) |
|---|---|---|---|---|
| ABS-A01 | 2016 | 0.348 | 28.440 | 29.228 |
| ABS-A02 | 2099 | 0.390 | 31.542 | 32.228 |
| ABS-A03 | 2161 | 0.387 | 31.782 | 32.553 |
| ABS-A04 | 2137 | 0.368 | 27.755 | 29.280 |
| ABS-A05 | 2174 | 0.380 | 27.909 | 29.468 |
| Medie | 2117 | 0.375 | 29.486 | 30.551 |
| StDev | 63.400 | 0.017 | 2.005 | 1.685 |
| CV | 2.994 | 4.547 | 6.798 | 5.515 |

The results of the ABS material tensile tests, printing position B, are presented in table 4.

**Table 4** ABS+ test results, position B

| Epruveta | E (MPa) | $\gamma$ (-) | $\sigma_C$ (MPa) | $\sigma_R$ (MPa) |
|---|---|---|---|---|
| ABS -B01 | - | - | - | - |
| ABS -B02 | 1471.707 | 0.286 | 12.932 | 12.932 |
| ABS -B03 | - | - | - | - |
| ABS -B04 | 1505.443 | 0.298 | 12.572 | 12.761 |
| ABS -B05 | 1442.028 | 0.275 | 12.572 | 13.052 |
| Medie | 1473 | 0.287 | 12.692 | 12.915 |
| StDev | 31.729 | 0.011 | 0.208 | 0.146 |
| CV | 2.154 | 3.922 | 1.638 | 1.132 |

The results of samples from ABS material, printing position C, are presented in table 5.

**Table 5** ABS+ test results, position C

| Epruveta | E (MPa) | $\gamma$ (-) | $\sigma_C$ (MPa) | $\sigma_R$ (MPa) |
|---|---|---|---|---|
| ABS -C01 | 2083 | 0.377 | 30.582 | 31.062 |
| ABS -C02 | 2106 | 0.376 | 31.096 | 32.038 |
| ABS -C03 | 2099 | 0.390 | 31.542 | 32.228 |
| ABS -C04 | 2161 | 0.387 | 31.782 | 32.553 |
| ABS -C05 | 2104 | 0.377 | 32.467 | 32.604 |
| Medie | 2111 | 0.381 | 31.494 | 32.097 |
| StDev | 29.636 | 0.007 | 0.711 | 0.624 |
| CV | 1.404 | 1.771 | 2.258 | 1.944 |

The following notations were made in the tables: E-modulus of elasticity; ɣ-Poisson's ratio; $\sigma_C$-flow stress; $\sigma_R$-breaking stress.

The results of the tensile specimens made of PETG material position A are presented in table 6.

**Table 6** PETG test results, position A

| Epr | E | ɣ | $\sigma_C$ | $\sigma_R$ |
|---|---|---|---|---|
| | (MPa) | (-) | (MPa) | (MPa) |
| PETG-A01 | 2087 | 0.399 | 32.741 | 42.938 |
| PETG-A02 | 2049 | 0.392 | 32.193 | 41.686 |
| PETG-A03 | 2061 | 0.402 | 33.341 | 43.160 |
| PETG-A04 | 2055 | 0.403 | 31.456 | 42.012 |
| PETG-A05 | 1994 | 0.386 | 33.358 | 42.183 |
| Medie | 2049 | 0.396 | 32.618 | 42.396 |
| StDev | 34.196 | 0.007 | 0.809 | 0.627 |
| CV | 1.669 | 1.790 | 2.480 | 1.480 |

The results of the tensile specimens made of PETG material position B are presented in table 7.

**Table 7** PETG test results, position B

| Epr | E | ɣ | $\sigma_C$ | $\sigma_R$ |
|---|---|---|---|---|
| | (MPa) | (-) | (MPa) | (MPa) |
| PETG-B01 | 1399 | 0.289 | 15.743 | 16.650 |
| PETG-B02 | 1413 | 0.284 | 16.359 | 18.484 |
| PETG-B03 | 1453 | 0.296 | 16.548 | 18.158 |
| PETG-B04 | - | - | - | - |
| PETG-B05 | 1307 | 0.285 | 15.09083 | 16.873 |
| Medie | 1393 | 0.289 | 15.935 | 17.541 |
| StDev | 61.820 | 0.006 | 0.659 | 0.915 |
| CV | 4.438 | 1.925 | 4.139 | 5.215 |

The results of the tensile specimens made of PETG material position C are presented in table 8.

**Table 8** PETG test results, position C

| Epr | E | ɣ | $\sigma_C$ | $\sigma_R$ |
|---|---|---|---|---|
| | (MPa) | (-) | (MPa) | (MPa) |
| PETG-C01 | 2087 | 0.389 | 36.888 | 47.993 |
| PETG-C02 | 2128 | 0.418 | 35.534 | 48.901 |
| PETG-C03 | 2091 | 0.421 | 36.323 | 47.667 |
| PETG-C04 | 2051 | 0.396 | 37.265 | 48.078 |
| PETG-C05 | 2118 | 0.429 | 36.271 | 48.249 |
| Medie | 2095 | 0.411 | 36.456 | 48.178 |
| StDev | 30.168 | 0.017 | 0.660 | 0.456 |
| CV | 1.440 | 4.153 | 1.811 | 0.947 |

The results of the tensile specimens made of PLA+ material position A are presented in table 9.

**Table 9** PLA+ test results, position A

| Epr | E | ɣ | $\sigma_C$ | $\sigma_R$ |
|---|---|---|---|---|
| | (MPa) | (-) | (MPa) | (MPa) |
| ABS-A01 | 2320 | 0.336 | 32.789 | 37.041 |
| ABS-A02 | 2518 | 0.351 | 30.479 | 33.716 |
| ABS-A03 | 2510 | 0.329 | 26.653 | 26.653 |
| ABS-A04 | 2525 | 0.327 | 31.379 | 33.882 |
| ABS-A05 | 2266 | 0.327 | 29.999 | 33.636 |
| Medie | 2428 | 0.334 | 30.260 | 32.986 |
| StDev | 124.404 | 0.010 | 2.279 | 3.818 |
| CV | 5.124 | 3.028 | 7.532 | 11.575 |

The results of the tensile specimens made of PLA+ material position B are presented in Table 10.

**Table 10** PLA+ sample results, position B

| Epr | E | ɣ | $\sigma_C$ | $\sigma_R$ |
|---|---|---|---|---|
| | (MPa) | (-) | (MPa) | (MPa) |
| PLA-B01 | 1957 | 0.223 | 17.250 | 17.250 |
| PLA-B02 | - | - | - | - |
| PLA-B03 | 1951 | 0.215 | 14.296 | 14.296 |
| PLA-B04 | 1981 | 0.234 | 14.642 | 14.642 |
| PLA-B05 | 2213 | 0.236 | 18.635 | 18.635 |
| Medie | 2026 | 0.227 | 16.206 | 16.206 |
| StDev | 125.865 | 0.010 | 2.088 | 2.088 |
| CV | 6.214 | 4.380 | 12.886 | 12.886 |

The results of the tensile specimens made of PLA+ material position C are presented in table 11.

**Table 11** PLA+ sample results, position C

| Epr | E | ɣ | $\sigma_C$ | $\sigma_R$ |
|---|---|---|---|---|
| | (MPa) | (-) | (MPa) | (MPa) |
| PLA-C01 | 2954 | 0.349 | 45.307 | 48.297 |
| PLA-C02 | 2886 | 0.340 | 44.846 | 46.839 |
| PLA-C03 | 2970 | 0.543 | 45.806 | 47.451 |
| PLA-C04 | 2817 | 0.191 | 46.806 | 47.309 |
| PLA-C05 | 2940 | 0.323 | 44.094 | 47.184 |
| Medie | 2913 | 0.349 | 45.372 | 47.416 |
| StDev | 62.638 | 0.126 | 1.020 | 0.542 |
| CV | 2.150 | 36.043 | 2.248 | 1.144 |

## 5. Interpretation of the results

The centralization of data from tables 3-11 is done graphically by drawing the average characteristic curves for the tensile samples, depending on the material and the three printing positions.

For the tensile specimens made of ABS+ material, the average curves in figure 3 result.

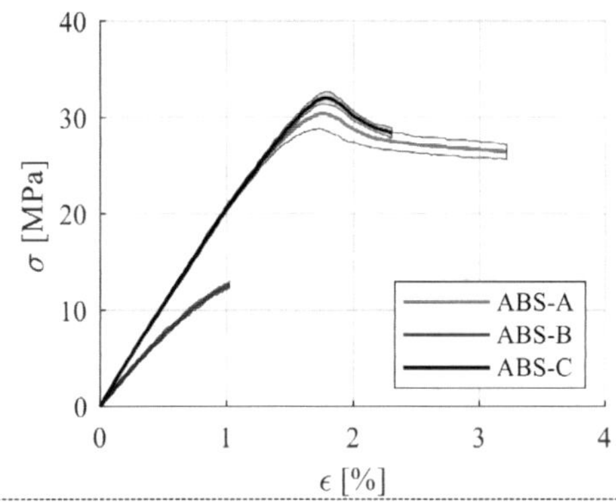


**Figure 3.** Average curves for tensile specimens printed from ABS+

For the tensile specimens made of PETG material, result the average curves in figure 4.

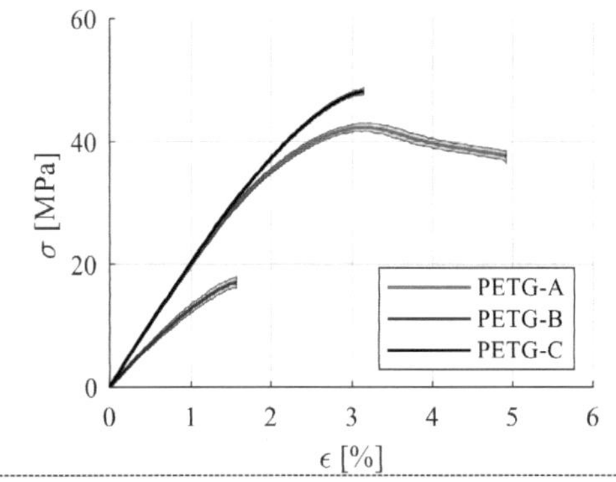


**Figure 4.** Average curves for the tensile specimens printed from PETG

For the tensile samples from the PLA+ material, t result he average curves from figure 5.

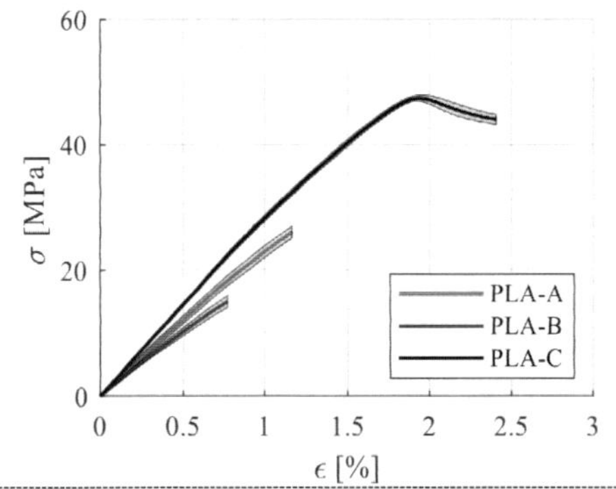


**Figure 5.** Average curves for tensile specimens printed from PLA+

## 6. Conclusions

From the average characteristic curves in figures 3; 4 and 5, it can be seen that regardless of the type of material used to manufacture the tensile test specimens, the specimens printed in the lateral position C have the highest breaking resistance, the specimens printed in the horizontal position A have an intermediate breaking resistance and the minimum resistance is the samples printed in the vertical B position.

For the samples printed in the horizontal position A, it is found that the highest resistance was obtained by the samples from PLA+ materials, i.e. over 40 MPa, an intermediate resistance by the samples from PETG material, i.e. around 40 MPa, and the lowest resistance around 30 MPa the samples from the ABS+ material obtained it.

For the samples printed in position B, the highest resistances are obtained by the samples from the PETG and PLA materials, i.e. a little under 20 MPa, and the lowest resistance is obtained by the ABS+ material, i.e. a little over 10 MPa.

For the samples printed in the lateral C position, the highest but close resistances were obtained by the samples printed from the PETG and PLA materials, i.e. over 40 MPa, and the lowest resistance was obtained by the ABS material, i.e. a little over 30 MPa.

We can conclude that the samples printed in lateral C position have the highest tensile strength and the most resistant material is PLA+, followed by PETG and ABS+.

**Author's ORCID ID**
C Babis: https://orcid.org/0009-0009-4174-056X
A Dimitrescu: https://orcid.org/0009-0008-9650-804X
S A Fica: https://orcid.org/0009-0002-4797-3216
C Stochioiu: https://orcid.org/0000-0002-7413-6133
D Vlasceanu: https://orcid.org/0000-0002-6609-7496
F Baciu: https://orcid.org/0000-0002-0574-6149
O Chivu: https://orcid.org/0009-0009-9586-148X
I C Enache: https://orcid.org/0000-0003-2876-4128

## References

[1] Toiba T and Ajaz AM 2023 *Materialstoday* A review of 3d printing technology-the future of sustainable contructiopn **1**
[2] Minguella-Canela J, Morales Planas S, Gomà Ayats JR and Santos López MA 2019 *Procedia Manufacturing* Study and comparison of the different costs' schema associated to geometry, material and processing between 3D printing, injection molding and machining manufacturing technologies **41** 280-7
[3] Zhai C, Wang J, Tu Y, Chang G, Xiaolei R and Ding C 2023 *Additive Manufacturing* Robust optimization of 3D printing process parameters considering process stability and production efficiency **71**
[4] Yan Z, Rahimizadeh A, Zhang Y, Zhou Y and Lessard L 2023 *Composite Structures* A finite element model for 3D printed recycled parts from end-of-life wind turbine blades **320**
[5] Qamar S, Xujiang W, Wenlong W, Yi W, Guolin L, Changzai R and Yanpeng M 2020 *Construction and Building Materials* Coordinated adjustment and optimization of setting time, flowability, and mechanical strength for construction 3D printing material derived from solid waste **259**
[6] Abhay M, Mukul K, Rajeev K and Anuj KS 2023 *Materialstoday* An investigation on tensile strength of 3d printed parts of acrylonitrile butadiene styrene material **1**
[7] Imran K, Muhammad T, Muhammad A, Muhammad S, Fatima H, Ans AR and Muammer K 2023 *Composites* Parametric investigation and optimisation of mechanical properties of thick tri-material based composite of pla-petg-abs 3d printed using fused filament fabrication **1** C

IOP Conf. Series: Materials Science and Engineering 1303 (2024) 012032 doi:10.1088/1757-899X/1303/1/012032

# Comparison between gas flame welding and shielded gas welding in case of aluminium alloys

**C Babiş[1], A Dimitrescu[1], A M Rugescu[2], O R Chivu[1], CI Enache[2]**

[1]University Politehnica of Bucharest, Faculty of Industrial Engineering and Robotics, Splaiul Unirii, No. 313, sector 6, Bucharest, Romania
[2] University Politehnica of Bucharest, Faculty of Aerospace Engineering, Splaiul Unirii, No. 313, sector 6, Bucharest, Romania

E-mail: claudiubab8@gmail.com

**Abstract.** Although the assembly by welding of aluminum alloys is a common assembly process, it is done in quite difficult conditions, being forced to take into account a number of shortcomings, such as: high thermal conductivity which translates into losses of high heat in the welding area; formation of aluminum oxide films that melt at high temperatures, well above the melting temperature of the alloy; large expansion coefficients that result in the appearance of large deformations, the appearance of cracks in the joint at high temperatures due to loss of mechanical strength of the alloy, but also the appearance of pores, inclusions of aluminum oxide and corrosion in various aggressive environments. Despite all these shortcomings that occur in welding, aluminum alloys can be joined indestructibly by welding.
The present paper aims, is to make a comparative study, based on experimental data, of two welding assembly processes, one with gas flame and another with shielded inert gas environments-WIG, for aluminum alloys of group 6000. The base material will be presented, the preparation of the basic material for welding through the two mentioned welding procedures and the steps taken to make the welded joints. The comparative study will include the parameters of welding regimes, non-destructive defectoscopy methods with which certain discontinuities can be identified and the destructive mechanical tests. The experimental results will be able to lead to pertinent conclusions on the possibilities of aluminum alloys welding assembly, using those two welding processes mentioned above, the results being easily generalized in the industrial fields.

## 1. Introduction

The production of non-ferrous metals and alloys is growing worldwide and occupies an important place in the construction of welded structures. Of these non-ferrous alloys, aluminum alloys are widespread.

Depending on the alloying elements and heat treatment, aluminum grades can exhibit a wide variety of properties, from good appearance, ease of manufacture, corrosion resistance, high power-to-weight ratio, good weldability and high crack resistance. The appropriate selection of aluminum grade depends on the required application and working conditions [1].

Non-ferrous metals and alloys differ from steel by their composition and by their characteristic properties. The welding behavior is also appreciated depending on the melting and boiling temperature, the thermal and electrical conductivity, the mechanical characteristics at high and low temperatures, the gas dissolution capacity, etc. Welding these aluminum alloys requires highly concentrated heat sources to compensate for conduction losses [2].

There are a multitude of aluminum alloys that can be assembled in non-removable assemblies by electric arc welding and gas flame welding. In both cases there are advantages and disadvantages. In this sense, in this paper we will analyze for a certain aluminum alloy from the 6000 series, the possibility of welding with electric arc, in a shielded gas environment (Argon) - the WIG process, with as well as the possibility of welding with gas flame. In this sense, will be presented the basic material, the non-removable assembly procedures with their specific particularities, the filler materials used, as well as the mechanical fracture tests. The purpose of the research is to determine which of the two non-removable assembly processes is more suitable for use. The topic is of interest because both assembly processes referred to above are widely used in the industrial environment.

## 2. Base material

Aluminum is characterized by a series of specific properties such as: specific weight of 2.7g / $cm^3$, melting point of 658℃, boiling point of 2500℃, has a high thermal conductivity, low electrical resistivity, good plasticity, high resistance to low temperatures and high corrosion resistance. Instead, aluminum has low mechanical strength $\sigma_r$ <100N / $mm^2$, high coefficient of expansion and low modulus of elasticity.

Aluminum alloys are divided into two major categories: deformable and cast. The deformable ones are used in untreated state or hardened by dispersed precipitation. Most of the aluminum constructions are non-heat treated, but lately, joints from hardened alloys have also appeared through precipitation treatments.

Aluminum alloys in the 6xxx series (6061, 6063) contain silicon and magnesium in approximate proportions necessary for the formation of magnesium silicide ($Mg_2Si$), thus making them heat-treatable. Although not as strong as most 2xxx and 7xxx series alloys, 6xxx series aluminum alloys have good toughness, weldability, machinability and relatively good corrosion resistance, with medium strength.

Two plates of dimensions 2x150x150 mm were used for the butt joint, for each process-WIG and oxy gas welding, as can be seen in figures 1 and 2.

In this paper was used as a base material for the butt assembled samples, material type AlMgSi0.5 3.3206 EN AW-6060 [AlMgSi] 2 mm thick [3].

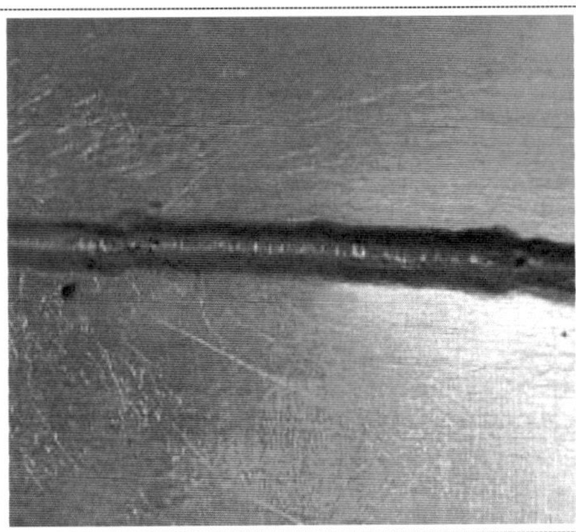

**Figure 1. WIG** Welded samples.

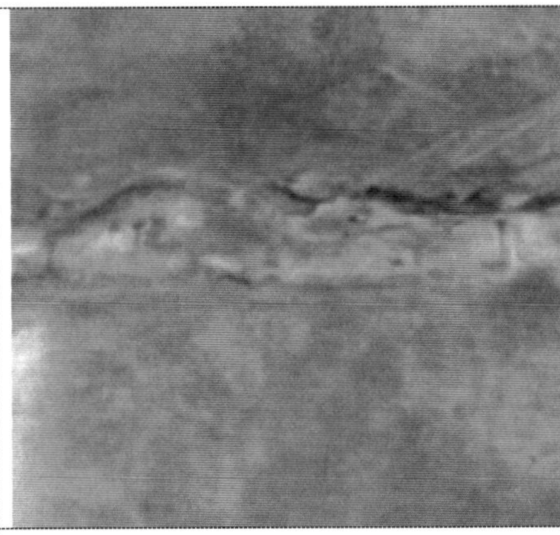

**Figure 2.** Oxy-gas welded samples.

## 3. Characteristics of welding and brazing of aluminium alloys

The characteristic element that appears when welding aluminum is the $Al_2O_3$ aluminum oxide film, which is formed due to the high affinity of aluminum for oxygen. This film has a high melting point of 2050°C and stop the base metal from melting, often remaining in the form of non-metallic inclusions

IOP Conf. Series: Materials Science and Engineering 1303 (2024) 012032 doi:10.1088/1757-899X/1303/1/012032

in the weld seam. This aluminum oxide must be removed either due to the bombardment effects in the electrical arc welding, with positive ions when fed in DC with reverse polarity or by dissolution with the help of fluxes in case of the oxy gas welding.

The rupture of the oxide film is generally obtained for small thicknesses, for large thicknesses being necessary mechanical or chemical removal measures.

Another problem appeared when welding aluminum is the appearance of gases in the weld seam, near the melting line of the base metal. The main cause of the appearance of pores in welded aluminum seams is considered to be hydrogen. Removing the pores from the weld seam is one of the most important concerns in establishing welding technology. For this it is necessary to perform a good cleaning of oxides and greases on the surfaces of the parts subjected to welding [4].

An important difficulty that occurs when welding aluminum alloys is the phenomenon of hot cracking. Cracking is generally caused by the silicon content. Quantities up to 0.6% Si can cause cracking. The presence of iron has an inverse influence. A content of 0.7% Fe leads to increased crack resistance. In the same sense, the additions of 5-6% Mg also influence.

Welding of aluminum alloys is much more difficult due to the occurrence of cracking phenomena, especially those that contain alloying elements. This can be combated due to aging treatments and precipitation of intermetallic phases [5].

The filler metal used to the welding of aluminum and its alloys is chosen from the condition of not forming in the metal bath fragile products such as $Mg_2Si$, $CuAl_2$, $MnAl_6$ and Al-Fe-Si. They can penetrate the weld seam and substantially reduce plasticity.

The wire used for welding must have a mechanical strength at least equal to that of the annealed base material.

An important condition when choosing the wire for welding aluminum is the solidification interval. The lower it is, the lower is the risk of cracking. The solidification interval is recommended to be below 50 °C.

Welding wires are chosen with the same composition as the base metal if the heat treatment is done after welding, or with a higher mechanical strength if no subsequent heat treatment is applied. These wires contain alloying elements based on Zr, Cr and Ti. It is recommended that the wires be clean for welding and the coatings and fluxes well dried.

In this work, two non-removable assembly processes were used, one by WIG welding and another by welding with a gas flame.

## 4. WIG welding

The welding process in shilded gas environments has many advantages: easy handling of the gun used for welding, good arc control, narrow molten welding bath, spray-free spring electrical arc welding and clean welding seam.

The AC current was used for WIG welding process, to cope with the high melting temperature of the oxide layer on the surface of the molten metal bath. By using the AC current, the aim is to break the oxide layer and remove it by means of ions that pass from the electrode to the part. AC welding is associated with a continuous change in the polarity of the infusible tungsten electrode. There are two phases (half waves), a positive phase and a negative phase. The positive phase causes the aluminum oxide layer on the surface of the material to break (the so-called cleaning effect). At the same time a cap is formed on the tip of the tungsten electrode. The size of this cap depends on the duration of the positive phase. It will be taken into account that too large a cap leads to the formation of a diffuse electric arc welding with low penetration of the weld. The negative phase cools the tungsten electrode on the one hand and reaches the required penetration on the other hand. It is important to choose correctly the ratio (balance) between the positive phase (cleaning effect and the size of the cap) and the negative phase (depth of penetration to the weld). In this sense, it is necessary to adjust the AC balance. The default (zero) balance adjustment is 65% and this ratio refers to the negative half-wave part. Rupture of the oxide film can generally be done for small thicknesses. For large thicknesses, additional measures are required to remove the oxide film mechanically or chemically.

The shielding gas welding machine is a Stahlwerk TIG 200 AC / DC inverter. The device has a high frequency and voltage generator that allows the ignition of the arc without electrical contact at an output current up to 200 A. The IGBT technology used by the machine is based on powerful IGBT transistors, which have the advantages of high blocking voltage and robustness. . The flow of the shielding gas when the button is pressed on the gun, before closing the welding circuit, protects the infusible electrode from excessive wear and protects the welding seam from oxidation. The shielded gas flow released after welding protects the infusible electrode from excessive wear and protects the welding molten bath from oxidation. The operating modes of the control can be determined via the 2T / 4T switch. This mode provides an extended control over the current flow. The smooth slope of the arc extinguishing prevents the appearance of end craters at the end of the welding. After the button on the gun has been released, the device switches to an automatic lowering phase. Downslope only works in 4T mode.

The filler material chosen is the Al-4043-AlSi5 rod according to EN ISO 18273 (2004) or ER 4043 according to AWS-A-5.10. The filler material was chosen with a chemical composition similar to the base material, possibly with a surplus of 1-2% Mg, to compensate for combustion losses. The shielded gas used for welding was high purity Argon. The gas flow is 15 l / min.

The parameters of the welding regime are presented in table 1

**Table 1** WIG welding parameters

| Characteristics | Value |
|---|---|
| Material thikness (mm) | 2 |
| Filler material diameter | 1.2 |
| Wolfram electrode diameter (mm) | 2 |
| Amperage (A) | 50 |
| Voltage (V) | 25 |

After welding the sample, 6 specimens with a width of 25 mm were cut from it, as seen in figure 3. Subsequently, these specimens will be tested for fracture, by static traction

**Figure 3.** WIG welded samples

It should be mentioned that, before cutting the test pieces, in order to detect possible discontinuities, a visual and penetranting liquid tests was done for the welding seam obtained by the WIG welding process. Following the non-destructive investigations, no notable defects were highlighted, as can be seen in figure 4.

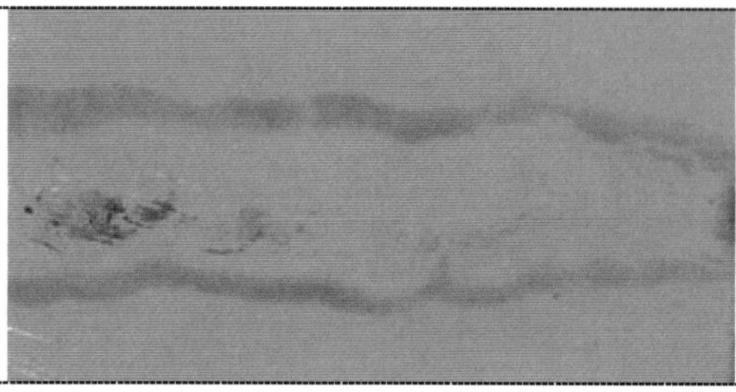

**Figure 4.** Non destructive tests with penetrant liquids for the WIG welding

## 5. Oxy-gas flame welding

In the gas flame welding of aluminum and its alloys, the parameters from table 2 were used and a series of technological recommendations were taken into account.

**Table 2** Oxy-gas welding parameters

| Characteristics | Value |
|---|---|
| Material thikness (mm) | 2 |
| Type of filler material: wire from base material | AlMgSi0.5 3.3206 EN AW-6060[AlMgSi] |
| Dimention of the filler material (mm) | 2x2 |
| Burner number: | 2 |
| Oxygen pressure (N/mm2) | 2.5 |
| Acetylene pressure (N/mm2) | 0.3 |
| Type of flame | Fuel flame |
| Pre heating temperature (̊C) | 350 |
| Distance burner-plate (mm) | 21 |
| Welding speed (m/h) | 10 |

The preparation of the sheets to be welded is done by cleaning, degreasing and pickling the sheets on a width of 30 to 40 mm, on both sides of the joint. Degreasing is done with a solution of 3 ... 5% sodium hydroxide and 1 ... 3% sodium silicate, followed by a wash with hot water and a pickling with a solution of 10% nitric acid, again following a wash with water. .

It will then be preheated to 300 ... 350 ˚C which helps to transmit the heat well, during welding. To prevent the parts from cooling during welding, they are placed on thermally insulating material (asbestos or refractory bricks).

For a thickness of 2 mm, an I-joint will be used, with a very small gap between the sheets.

Regarding the welding regime, several elements were respected.

Due to the lack of finding a suitable filler material, thin fillets cut from the base material were used as filler material. The main difficulty that appears when welding, consists in the formation on the surface of the molten metal bath of an aluminum oxide film, very difficult to fuse. To remove this drawback, a stripping flux with the following composition is used: 28% sodium chloride, 50% potassium chloride, 14% lithium chloride and 8% sodium fluoride. The flux will be applied both on the edges to be welded and on the filler material.

When welding, a light fuel flame with a volumetric ratio between oxygen and acetylene from 0.95 to 1 will be used, which helps to prevent the formation of $Al_2O_3$ aluminum oxide.

Regarding the power of the flame, bottled acetylene and a clean, soot-free burner were used. The burner was chosen with a larger number than the one used for welding steel of the same thickness, because aluminum alloys have a high thermal conductivity.

Taking into account the fact that our application involves butt welding with thicknesses below 5 mm, we choose the welding process to the left. At the beginning, until the molten metal bath is formed, the burner is kept almost in a vertical position, after which it is tilted at an angle of 45… .60°. The filler material is held with the same inclination.

After welding the sample, 6 specimens with a width of 25 mm were drawn and cut, as seen in figure 5, which will then be tested for breaking by static traction.

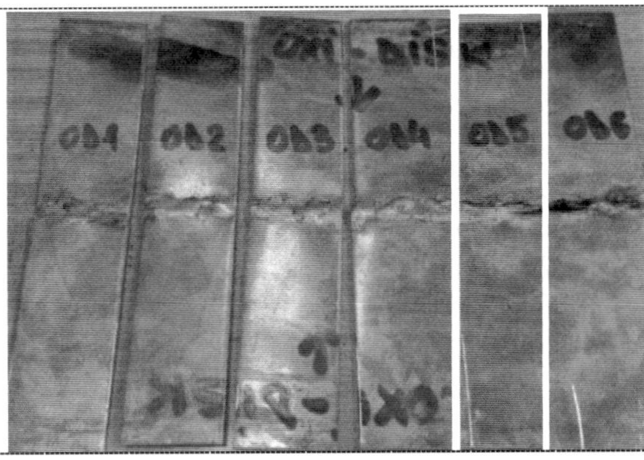

**Figure 5.** Oxy gas samples

It should be mentioned that, before cutting the test pieces, in order to detect possible discontinuities, a visual and a penetrating liquids control was made of the welding seam. Following the non-destructive investigations, an irregular geometric shape of the welding bead was highlighted, some isolated pores and some short longitudinal cracks, as can be seen from figure 6.

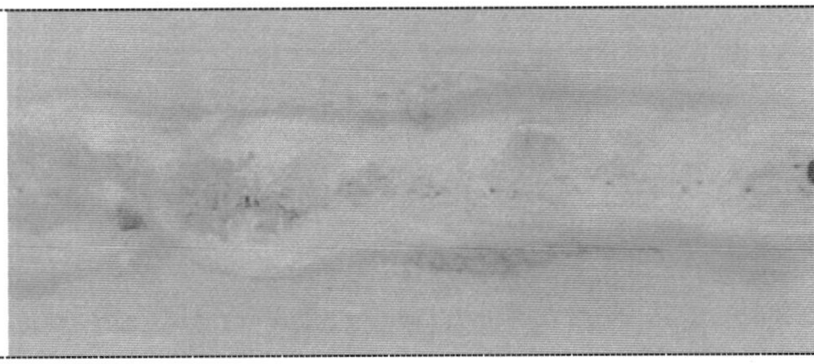

**Figure 6.** Non destructive tests for the Oxy-gas welding

## 6. Mechanical tests

As shown above, after cutting, there were 6 specimens marked with WS1-WS6 for WIG welding and with OD1-OD6 for gas flame welding.

All specimens were tested for tensile tests traction, on the LVF 1000 test machine. The load was executed until the specimens fractured.

The fracturing force was recorded for each test piece. Some fractured specimens are shown in figures 7 and 8.

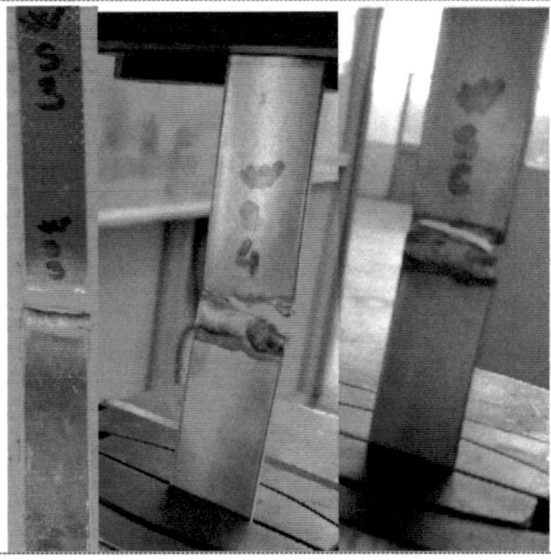

**Figure 7.** WIG Welded samples.

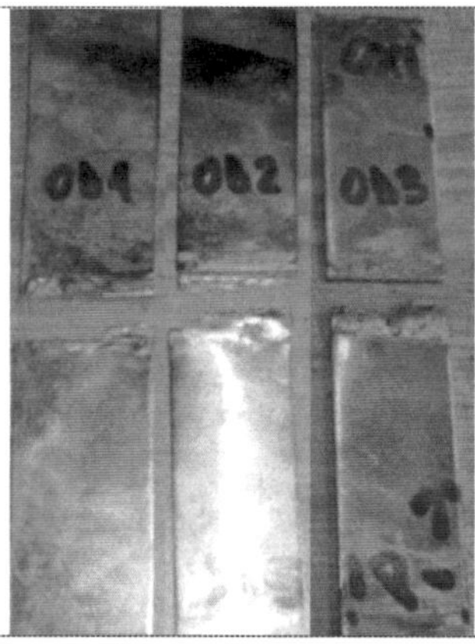

**Figure 8.** Oxy-gas welded samples.

## 7. Results and interpretation

The recorded values of the fracturing forces for the specimens obtained from the WIG welded sample are presented in table 3.

**Table 3** Fracturing forces for WIG welded samples

| Sample indicative | Fracture forces (daN) |
|---|---|
| WS1 | 900 |
| WS2 | 1000 |
| WS3 | 1050 |
| WS4 | 960 |
| WS5 | 1260 |
| WS6 | 940 |

The recorded values of the fracturing forces for the specimens obtained from the welded sample with gas flame are presented in table 4.

**Table 4** Fracturing forces for oxy-gas welded samples

| Sample indicative | Fracture forces (daN) |
|---|---|
| OD 1 | 650 |
| OD 2 | 950 |
| OD 3 | 920 |
| OD 4 | 960 |
| OD 5 | 800 |
| OD 6 | 750 |

The graphical representation of the values of the fracturing forces depending on the type of test piece is presented in figure 9.

Analyzing the samples examined by the optical-visual and penetrating liquids described in points 4 and 5, we find that in the case of WIG welding, the welding seam has a normal appearance, has a regular geometric shape and no discontinuities that can be highlighted. with that methods.

On the other hand, also examining by optical-visual methods and non-destructive defectoscopy with penetrating liquids, we find that the welding seam made by the oxy-gas flame welding process showed an irregular geometric shape, some isolated pores and some longitudinal short cracks.

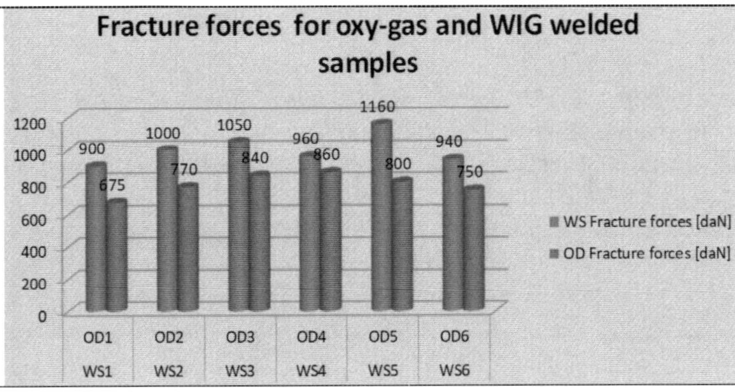


**Figure 9.** Fracturing forces

Regarding the results of the static tensile tests, as can be seen in Figure 9, the forces at which the WS1; WS2; WS3; WS4; WS5 and WS6 specimens fractured, welded by the WIG process, are about 25% higher than the forces at which the OD1; OD2; OD3; OD4; OD5 and OD6 specimens fractured welded by gas flame welding process.

## 8. Conclusions

The comparative study was performed between the WIG welding process and the oxy-acetylene flame welding process using different filler materials, but close in chemical composition to the base material. It is mentioned that in the case of welding with oxyacetylene flame, rods of filler material made of the basic material were used.

It was observed that in the case of WIG welding, the specimens investigated by the optical-visual method and with penetrating liquids did not show visible defects that would influence the mechanical fracture tensile tests.

On the other hand, in the case of specimens obtained by gas flame welding, both in the optical visual control and in the one with penetrating liquids, the irregular shape of the welding seam and some surface pores and cracks along the welding bead were observed. These highlighted discontinuities act as local stress concentrators and cause a decrease in breaking forces at static tensile tests. It is found that the fracture tensile tests forces of WIG welded specimens, are approximately 25% higher than the fracture forces of oxyacetylene welded specimens.

The explanation for the appearance of porosity defects and longitudinal cracks in samples welded with oxyacetylene flame could be given by the use of an inappropriate preheating temperature or a stripping flux that does not properly remove oxides from the bath. As a result of not removing the oxides from the bath, problems occur because their melting temperature is very high, ie above 2200°C and cannot be removed, because the base material melts at about 650C.

**Author's ORCID ID**
C Babis: https://orcid.org/0009-0009-4174-056X
A Dimitrescu: https://orcid.org/0009-0008-9650-804X
A M Rugescu: https://orcid.org/0000-0003-0591-3901
O Chivu: https://orcid.org/0009-0009-9586-148X
I C Enache: https://orcid.org/0000-0003-2876-4128

## References

[1] Banoth S, Babu VS, Raghavendra G and Shankar PH 2023 *Materialstoday* Study of various types of aluminium alloy grade series and their effect on mechanical performance **1**
[2] Schneider A, Avilov V, Gumenyuk A and Rethmeier M 2013 *Physics Procedia* Laser beam welding aluminium alloys under the influence of an electromagnetic field **41** 4-11

[3] Lesniak D and Gromek P 2020 *Procedia Manufacturing* Estimation of extrusion welding conditions for 6xxx aluminium alloys **47** 253-60
[4] Brijesh KY, Shailendra SB and Varun S 2023 *Materialstoday* A review on fracture and fatigue behavior of FSW joints of Al alloys **1**
[5] Rui DF, Jian FZ, YimJL, Ju K, Hui JL and Fu CZ 2013 *Materials Science and Engineering* Effect of welding heat input and post-welding natural aging on hardness of stir zone for friction stir-welded 2024-T3 aluminium alloy thin sheet **559** 319-24

# Comparisons between Assembly Technologies by Welding and Brazing of Aluminium Alloys from the Sight of Assembly Quality

**A Dimitrescu, C Babiş and A M Rugescu**

National University of Science and Technology Politehnica Bucharest, Faculty of Industrial Engineering and Robotics, Splaiul Independentei no. 313, sect 6, Bucharest, Romania

E-mail: andrei_dimitrescu@yahoo.com

**Abstract.** Aluminum and its alloys are finding use in more and more industrial fields. Since the early 1970s, when many companies in the automotive, pharmaceutical and aeronautical industries replaced the classic components of iron alloys with those of aluminum alloys, they faced various difficulties in the non-removable assembly of designed components.
The first assembly technologies used were welding, followed in the 1980s by brazing. Due to its economic advantages, in many cases brazing technologies are preferred by manufacturers. The present paper aims to make a comparative study, based on experimental data, between the non-removable assembly technology by welding and brazing for aluminum alloys of group 6000. Due to the high costs of materials and labor, it is preferable to use relevant non-destructive control methods that can indicate any discontinuities that may occur in welded or brazed structures. These defects can provide valuable information and may indicate either non-compliance with the required technological conditions or the wrong choice of assembly method. Non-destructive testing methods that can highlight interior defects, such as radiographic control with penetrating radiation and ultrasound will be used, as well as less expensive equipment and labor methods that may highlight surface defects or which communicates with the surface such as the method of control with penetrating liquids. Thus, the joint areas of the base and filler materials will be analyzed, as well as the areas in the immediate vicinity that have been thermally influenced. The experimental results will be able to lead to pertinent conclusions in the way of choosing the non-removable assembly technology, results that can be easily generalized in the industrial fields. Thus, each company will be able to choose according to the conditions imposed by the product specifications a certain technology that will satisfy its requirements.

## 1. Introduction

Due to its very good mechanical properties, aluminum alloys have revolutionized many technological fields.

Although it is lighter than other metallic materials, it has a very good mechanical power, which is why it is often used in fields such as aeronautical engineering, machine building and due to its resistance to many forms of corrosion that do not require high maintenance costs in the pharmaceutical and food industry.

Another advantage of using aluminum alloys is the relatively easy recycling and reshaping of this material. It is known that the energy required for recycling is only 5% of the total consumed for its initial production. It is therefore not surprising that 85% of aluminum used in construction at European level comes from recycling [1], as can be seen in Figure 1 [3].

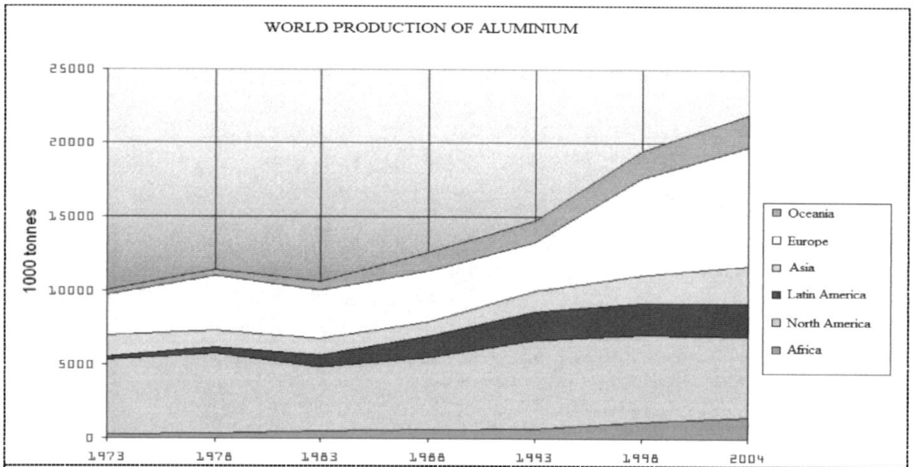


**Figure 1.** World production of aluminium.[3]

We can say that aluminum is an "environmentally friendly" material that can be recycled an unlimited number of times, while maintaining its characteristics, without degrading over time. [2]

## 2. Used materials

The basic material to be used for the experimental stage is an aluminum alloy of class 6000, a general purpose alloy 6061 with a thickness of 2 mm. This aluminum alloy was chosen because it behaves very well in non-removable joints by welding and brazing. [4]

The generic chemical composition of Aluminum Alloy 6061 is:

- Silicon - at least 0.4%, not more than 0.8% by weight,
- Iron - no minimum, maximum 0.7%,
- Copper - minimum 0.15%, maximum 0.4%,
- Manganese - no minimum, maximum 0.15%,
- Magnesium - minimum 0.8%, maximum 1.2%,
- Chromium - at least 0.04%, maximum 0.35%,
- Zinc - no minimum, maximum 0.25%,
- Titan - no minimum, maximum 0.15%,
- Other elements not more than 0.05% each, 0.15% total,
- The remaining Aluminum - minimum 95.85%, maximum 98.56%.

The chemical element that generates the biggest problems is Silicon, which if it exceeds 0.6% makes it difficult to apply the brazing technology.

The filler material must be chosen so that it can be applied by both the welding process and the brazing process. This choice will be made because it is desired to eliminate any discontinuities from the formed assemblies introduced by the deposition material.

For brazing, avoid materials with a Silver content that have a very good brazability, as well as tubular rods with a built-in flux content, because their melting temperatures are too low. [4] Precision rod-type additives shall also not be used Durafix [5] and A103.

Thus, an AlSi5 or SAl4043 rod type filler material will be chosen. It is used for welding aluminum and aluminum alloys with a maximum silicon content of 7% in forged and cast. Recommended for

6000 series Al-Mg-Si alloys and for heterogeneous joining, for example 6000/1000 or 6000/3000. [6] The chemical composition is shown in Table 1 [7] according to EN ISO 18273 or AWS A5.10

**Table 1.** Chemical Composition in %

| Alloy symbol | | Si | Fe | Cu | Mn | Mg | Cr | Zn | Ga, V | Ti | Zr | Al | Be | Other each | Other total |
|---|---|---|---|---|---|---|---|---|---|---|---|---|---|---|---|
| Numerical Al4043 | Chemical AlSi5 | 4.5 – 6.0 | 0.8 | 0.3 | 0.05 | 0.05 | - | 0.1 | - | 0.2 | - | Rem | 0.0003 | 0.05 | 0.15 |

## 3. Technologies used

The WIG welding process will be used for the welding operation because the assembly quality is very good using the parameters presented in the table 2:

**Table 2.** WIG welding regime parameters

| Characteristics | Value |
|---|---|
| Alternative current | |
| Tension | 25 [V] |
| Starting intensity | 50 [A] |
| Welding intensity | 75 [A] |
| Frequency | 55 [Hz] |

The brazing process with oxy-acetylene flame is used for the brazing operation, with the characteristics presented in the table 3.

**Table 3.** Brazing parameters

| Characteristics | Value |
|---|---|
| Flame | Oxy-acetylene |
| The type of acetylene used | High purity acetylene 2.6. |
| The type of flame used | Carbide flame |
| Submitting the additive material | Both surfaces |

One of the most important operations for both welding and brazing aluminum alloys is surface preparation. In the experimental model that will be presented below, a simple preparation was chosen which consists of the following steps:

- Washing with warm water and detergents that do not contain more than 5% soda;
- Natural drying;
- Mechanical cleaning with stainless steel wire brush;
- Degreasing with acetone of the surfaces that will come into contact.

These operations must be performed no later than 30 minutes before assembly.

## 4. Collectoin of Experimental Data

The WIG welding and brazing operations with filler material resulted in the sample sets shown in Figures 2 and 3.

**Figure 2.** Welded samples.

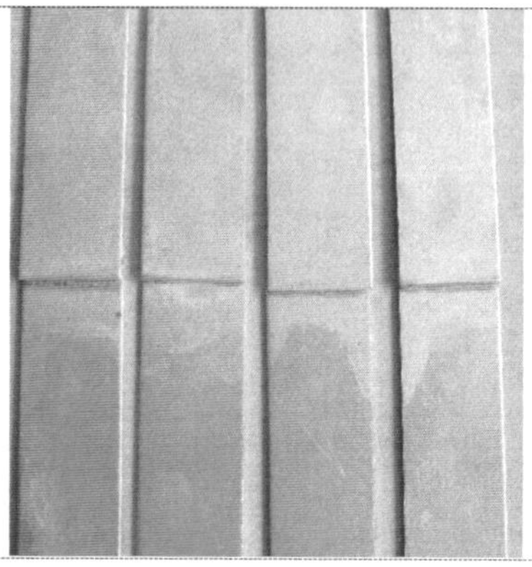

**Figure 3.** Brazed samples.

The non-destructive testing methods that have been analyzed are:

*4.1. Examination method with penetrating liquids*

This method highlights discontinuities in the surface of the material or that communicates with the surface. In the case of samples obtained by welding, this method was performed before cutting, and for brazed samples after the cutting operation.

Following the control technology described in detail in the paper results the following data sets presented in Figures 4 and 5

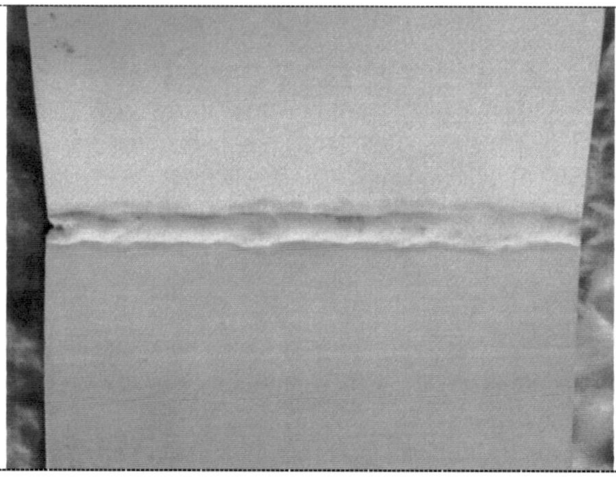

**Figure 4.** Welded samples examined with penetrating liquids.

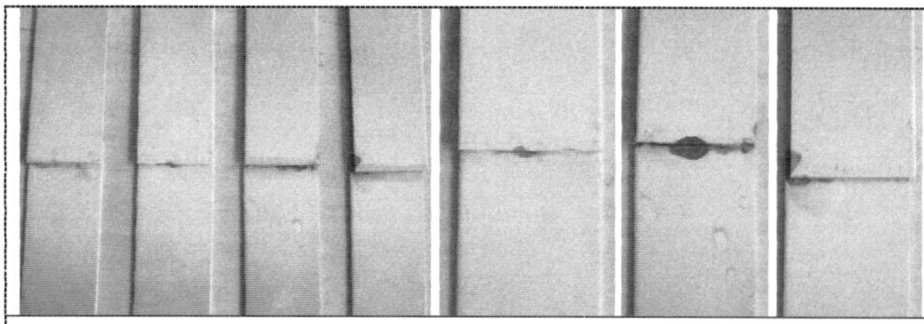

**Figure 5.** Brazed samples examined with penetrating liquids.

### *4.2. Ultrasound examination method*

This method was used to check the area of thermal influence of the base material. There is a possibility that in case of overheating, especially after the WIG welding process, there will be major changes in the structure of the base material.

Also by this method, changes in the thickness of the base material in the immediate vicinity of the joint cord can be highlighted.

Several points have been established for performing the examination shown in Figures 6 and 7.

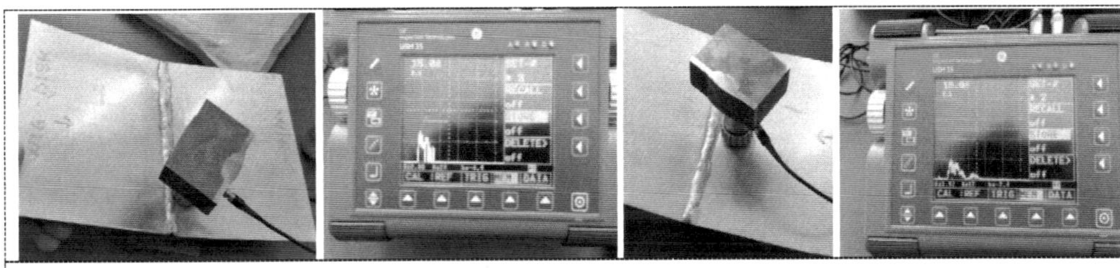

**Figure 6.** Welded samples examined with ultrasound.

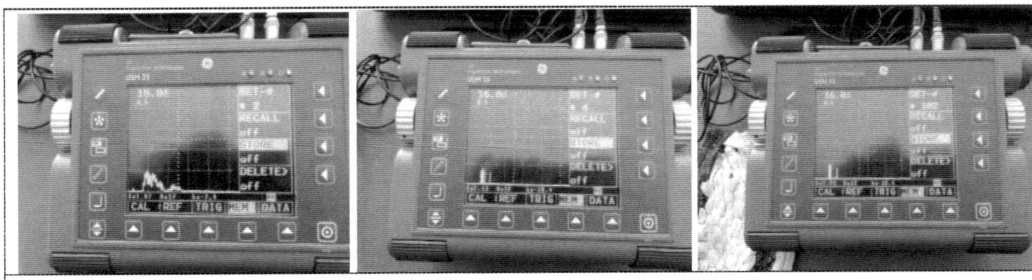

**Figure 7.** Brazed samples examined with ultrasound.

### *4.3. Ultrasonic immersion examination method*

Due to the thickness of the base material which is very small (2mm), and the irregularity of the surfaces resulting from the joining processes, this method of examination is especially recommended for checking the weld bead by the WIG process. In this case the entire weld bead was scanned by the "C-scann" mode and the areas where the indications were inconclusive were checked by the "A-scann" mode.

**Figure 8.** Welded samples examined by ultrasonic immersion.

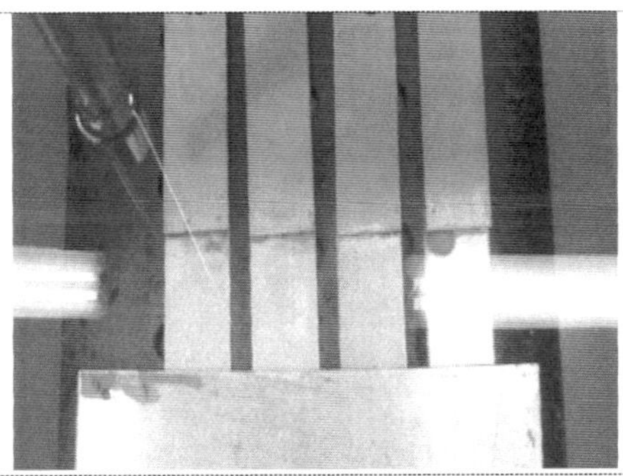

**Figure 9.** Brazed samples examined by ultrasonic immersion.

For the WIG welded samples, several interpretations shown in the figure 10 were performed.

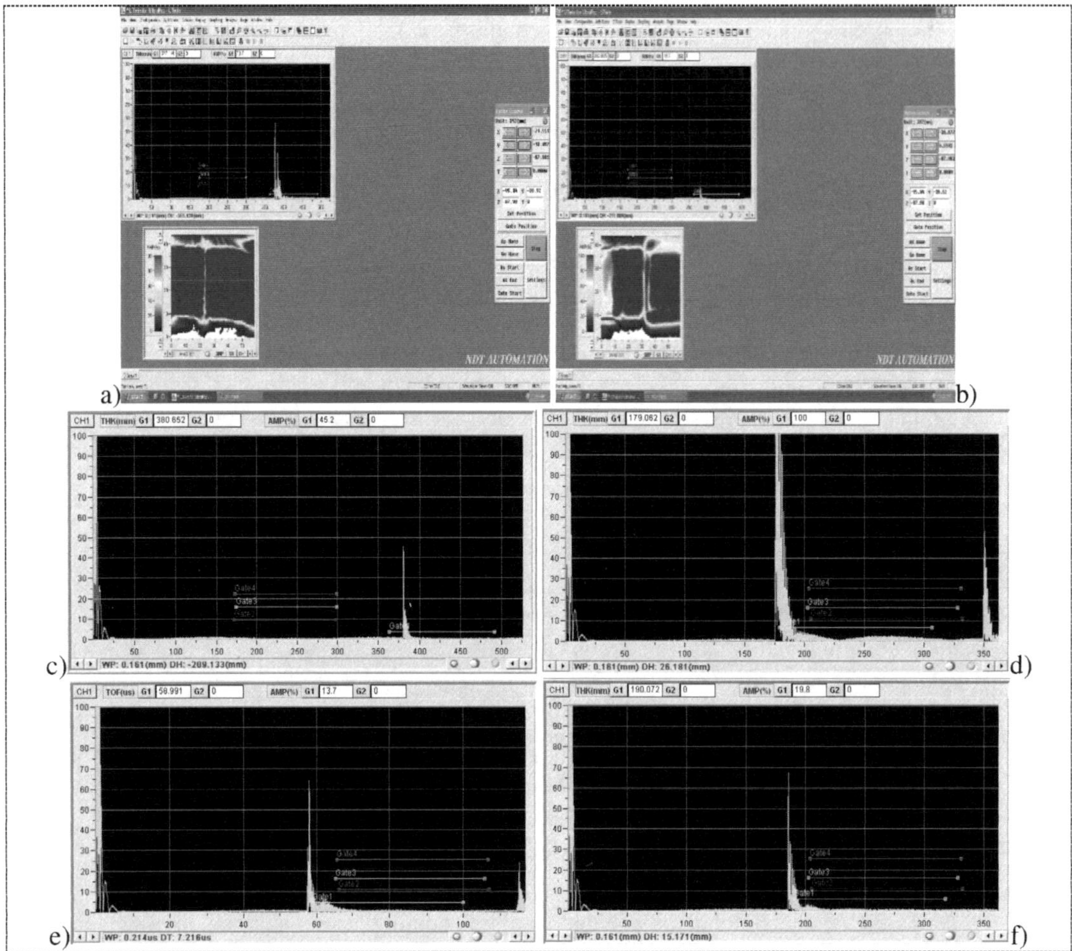


**Figure 10.** Scan welded samples: a) A+C-scann, b) A+C-scann, c) A-scann defect-free area, d) A-scann defective area, e) A-scann defective area, f) A-scann defective area.

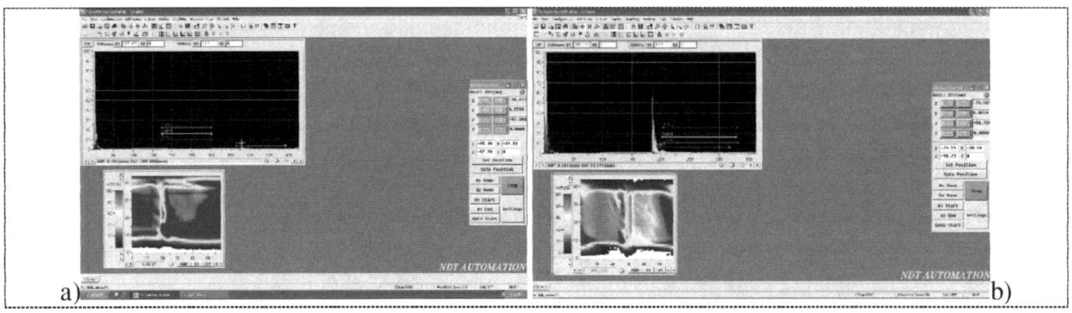

c) 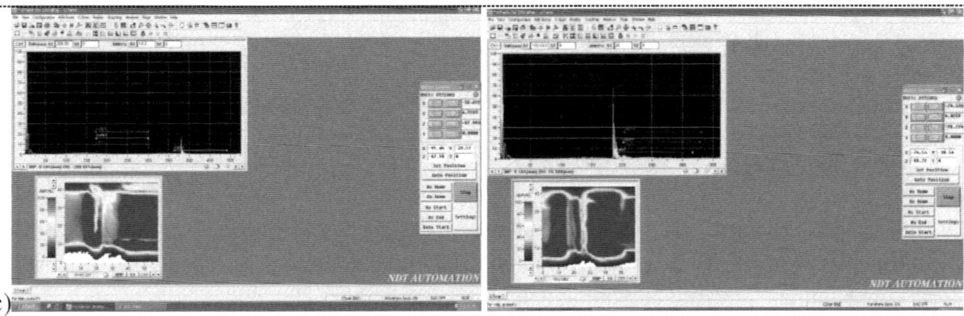 d)

**Figure 11.** Scan brazed samples: a) defect-free area, b) defective area, c) defective area, d) defective area.

## 5. Interpretation of results

Analyzing the samples examined by the method with penetrating liquids described in Chapter 4.1, we notice that we have no indications of discontinuities in the case of welding by the WIG process, and in the case of brazing in a single sample there is an indication of high color intensity which confirms a defect. In other brazed samples, color stains may be the result of poor cleaning of excess penetrating liquid.

In the case of the ultrasonic examination method, it can be seen from Figure 6 that the base material did not change its structure significantly, and no oscillations of defect appear on the oscillograms and neither did the end echo significantly decrease its amplitude.

Performing a comparative analysis shown in Figure 12, a thinning of the base material in the thermally influenced area.

**Figure 12.** Change in the thickness of the base material in welded samples.

After measuring the thickness of the base material in areas further away from the weld cord, it was found that its thickness is 2.13 mm, while in the immediate vicinity of the weld the thickness has values of 1.94 mm, 1.97 mm and 2 mm.

For brazed samples the use of this method was difficult due to the small thickness of the materials. In many cases the echo was distorted as seen in the first image in Figure 7 as it reached the edge of the material. However, no discontinuities in the base or filler material were observed after the examination.

From the analysis performed by the immersion ultrasound method, many discontinuities can be observed in the welding bead: some minor ones as seen in figure 10 e) and f), but also major defects as in figure 10 d). Many areas are flawless and this is also highlighted in the first image.

In the case of brazed samples, few minor defects were identified. The end echo decreased its intensity very little as can be seen in Figure 11 a) and c).

IOP Conf. Series: Materials Science and Engineering 1303 (2024) 012033 doi:10.1088/1757-899X/1303/1/012033

## 6. Conclusions

The comparative study was performed between the WIG welding process and the oxy-acetylene flame brazing process using the same filler material and the same base material. The experimental difficulty lies in the melting temperature of the filler material which is close to the melting temperature of the base material.

In the case of both assembly technologies, both WIG welding and brazing do not produce surface defects or those that communicate with the surface, defects that can be highlighted by the method of non-destructive examination with penetrating liquids.

Using the ultrasonic examination method, the change in the thickness of the base material in the immediate vicinity of the WIG weld bead was determined, which can have a very large influence by reducing the quality of the assembly. This thinning is mainly due to the way the molten bath is formed. A detailed study can provide additional information on the correlation of thickness and mechanical properties.

In the case of WIG welding, minor defects appear, following the analysis by the ultrasonic immersion method. The main cause of their occurrence may be the uniform misalignment between the base material and the filler material. The major defects found are of the type of oxide inclusions, which in the case of aluminum have higher melting temperatures.

Few minor defects were identified for the brazing operation, which proves that the surface preparation technology, the choice of flow and the choice of brazing technology were well chosen and led to the achievement of flawless joints.

**Author's ORCID ID**

A Dimitrescu: https://orcid.org/0009-0008-9650-804X
C Babis: https://orcid.org/0009-0009-4174-056X
A M Rugescu: https://orcid.org/0000-0003-0591-3901

## References

[1] Source: https://www.alumil.com/extrusion/ro/info/knowledgebase-in-aluminium accessible 11.02.2023
[2] Ionescu O C 2008 *University Politechnica of Bucharest* Studies regarding the characterization of secondary aluminium alloys obtained by waste recycling Bucharest p 26
[3] Source: https://upload.wikimedia.org/wikipedia/commons/4/49/Production_alu_primaire.png accessible 11.02.2021
[4] Dimitrescu A 2016 *Ed. Printech* Researches on reconditioning technologies Bucureşti p168
[5] Source: https://www.saf-fro.com/ro/altig-alsi5 accessible 12.02.2023
[6] Source: http://www.weldwire.net/weld_products/product-2/ accessible 12.02.2023
[7] Thornton M, Han L, Shergolg M, 2012 *Elsevier NDT & E International* Progress in NDT of resistance spot welding of aluminium using ultrasonic C-scann **48** pp30-38

# Theoretical and experimental research on the mechanical tests of 3D printed samples from various materials

**A Dimitrescu[1], C Babiş[1], S A Fica[2], C Stochioiu[1], D Vlăsceanu[1] and F Baciu[1]**
[1] National University of Science and Technology Politehnica Bucharest, Faculty of Industrial Engineering and Robotics, Splaiul Independentei no. 313, sect 6, Bucharest, Romania
[2] National Research and Development Institute for Electrical Engineering Bucharest, Splaiul Unirii no 313, sect 3, Bucharest, Romania

E-mail: andrei_dimitrescu@yahoo.com

**Abstract.** At this moment, a real industrial, scientific and technological revolution is being discussed worldwide, consisting in the development of Additive Manufacturing AM technologies through the use of 3D printing. The FDM (Fused Deposition Modelling) 3D printing technology consists in the deposition of fusible material layer by layer in the XOY plane, the thickness of the 3D printed layer is $g = 0.12 \div 0.24$ mm, the resolution of 3D printing is $r = 0.1$mm, while the print head advances vertically on the OZ axis. The purpose of the work is to determine the breaking moment through destructive mechanical bending tests of some samples made by 3D printing from the following materials: ABS+, PETG and PLA. 5 samples were made in 3 printing directions: horizontal, vertical and lateral, resulting in 15 samples for each material. The total number of samples used is 45. Knowing these values, depending on the requirements of the specifications, a certain material and a certain printing position can be used.

## 1.Introduction

The need to obtain components with the highest possible material utilization efficiency, with the lowest possible material consumption and with the lowest possible number of machines represents the future of technology.

FDM (Fused Deposition Modeling) 3D printing technology is considered the most feasible and with the best price / quality ratio 3D printing technology [1,2]. This was first developed for desktop 3D printers by the American company Maker Bot, which use PLA+, ABS+ or PETG filament. These filaments have a maximum yield strength of $\sigma C = 60$ MPa, which is a big disadvantage when 3D printing high strength parts such as shafts, axles, gears, sprockets, couplings, etc. but which is highly valued for most other types of 3D printed landmarks [3].

The multitude of advantages offered by 3D printing manufacturing technology consists in the way of automation from the level of the CAD design system where there are libraries of designed products to the production system using robotic cells recommends the expansion of use in all industrial fields and non-industrial [4,5].

Originating in the early 1980s, this technology has moved from the family of non-conventional technologies to that of conventional technologies, at the moment these printers can be accessed at reasonable prices.

## 2.Used materials

The 3D printer Creality Ender 3 V2 was used for this printing. The characteristics of the materials used are presented in table 1.

**Table 1**. The Characteristics of the materials

| No | Type of filament material for FDM 3D printing | Temp. Nozzle 3D Print Head°C | Temperature bed printer°C | Tensile Strentgh [MPa] | Elongation at Break [%] | Flexural Strentgh [MPa] | Flexural Modulus [MPa] | Strentgh [MPa] |
|---|---|---|---|---|---|---|---|---|
| 1 | ABS+ | 260 | 110 | 40 | 30 | 68 | 2443 | 42 |
| 2 | PLA | 210 | 70 | 60 | 29 | 87 | 3642 | 7 |
| 3 | PETG | 250 | 75 | 49 | 225 | 68 | 1800 | 7,6 |

The three print direction of the samples are: horizontal, vertical and lateral.

## 3.Construction of experimental samples

The first material to be printed using the 3 directions is ABS +. The tensile samples are shown in figure1.

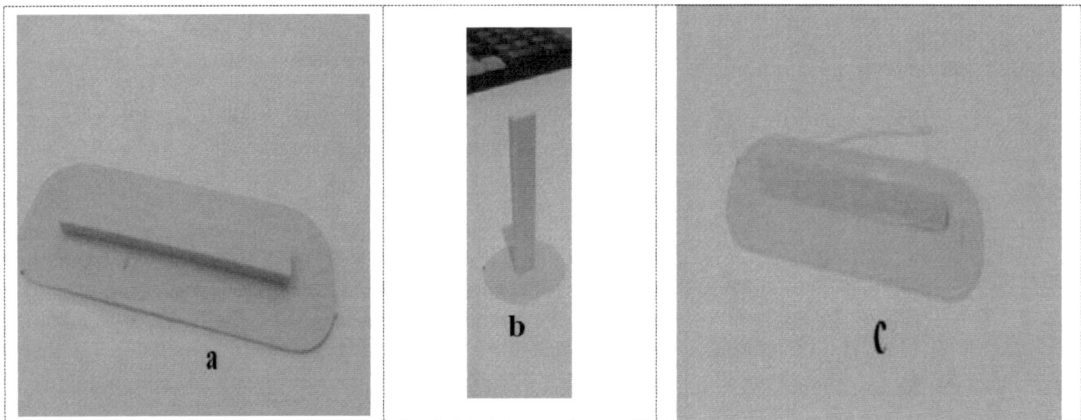


**Figure 1.** Printing the tensile samples from ABS+ in horizontal-a, vertical-b and lateral-c direction

The second material printed in the three positions is PETG and is presented in figure 2.

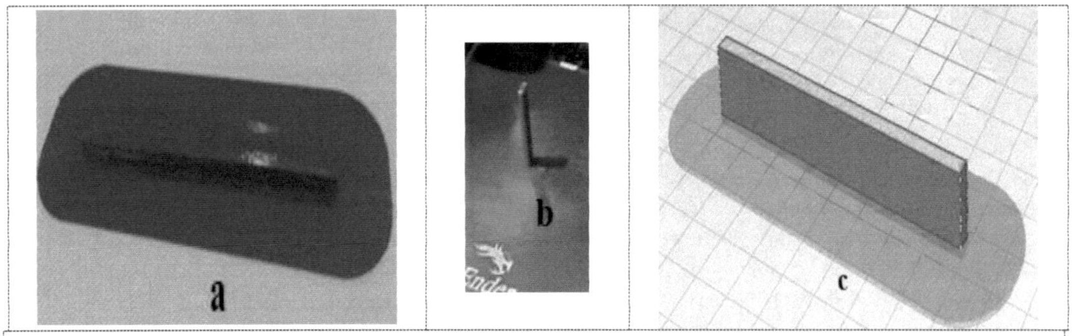


**Figure 2.** Printing the tensile samples from PETG in horizontal-a, vertical-b and lateral-c direction

The third material used for printing in the three positions is PLA, presented in figure 3.

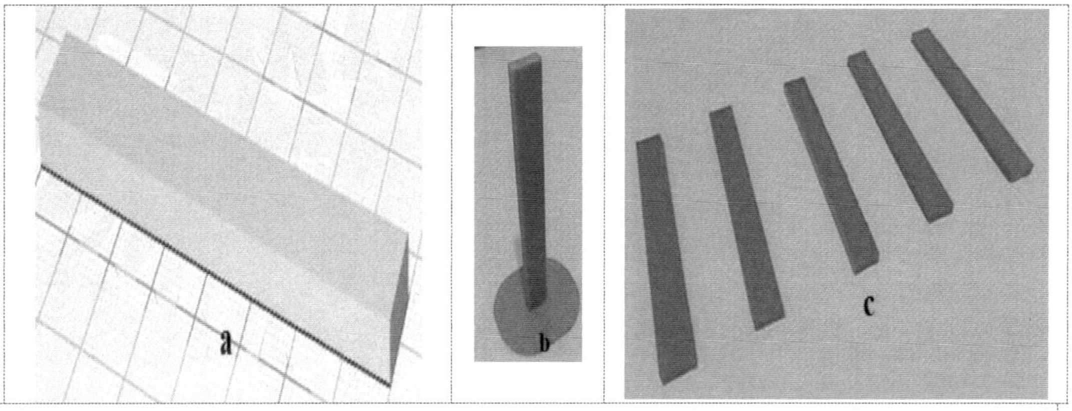


**Figure 3.** Printing the tensile samples from PLA in horizontal-a, vertical-b and lateral-c direction

## 4.Mechanicalbending tests

The stress was calculated as the maximum value of the bending stress, at the extremity of the section where the force is applied, with the below relation 1.

$$\sigma = \frac{3Fl}{2bh^2}\ [MPa] \tag{1}$$

In the relation 1, F is the force read using the force cell, L is distance between supports (60mm), b is section width measured and h is section height, measured. The specific strain was calculated as the maximum value of the longitudinal specific strain, at the extremity of the section where the force is applied, with the relation 2.

$$\varepsilon = \frac{6wh}{L^2} * 100\ [\%] \tag{2}$$

In the relation 2, w is the read arrow, with the help of the position transducer of the machine`s crossbar. For samples that did not break during the test, the test was stopped at 12 mm deflection or approximately 8 % specific strain.

## 5.Experimental data

The results for the ABS+ material in the horizontal printing Ais presented in table 2.

**Table 2**. Values for ABS+, horizontal printing A

| Sample | Ɣ | σC | σR |
|---|---|---|---|
| ABS_A01 | 1500 | 38.817 | 45.509 |
| ABS_A02 | 1524 | 36.198 | 44.204 |
| ABS_A03 | 1528 | 35.682 | 43.875 |
| ABS_A04 | 1555 | 36.682 | 44.548 |
| ABS_A05 | 1564 | 38.162 | 45.339 |
| Average | 1534 | 37 | 45 |
| StDev | 25.720 | 1.330 | 0.709 |
| CV [%] | 1.676 | 3.584 | 1.587 |

The results for the ABS+ material in the vertical printing B, is presented in table 3.

**Table 3**. Values for ABS+, vertical printing B

| Sample | Ɣ | σC | σR |
|---|---|---|---|
| ABS_B01 | 1196 | 15.572 | 15.572 |
| ABS_B02 | 1178 | 21.043 | 21.629 |
| ABS_B03 | 1255 | 14.447 | 14.447 |
| ABS_B04 | 1265 | 17.571 | 17.571 |
| ABS_B05 | 1197 | 20.049 | 20.071 |
| Average | 1218 | 18 | 18 |
| StDev | 39.235 | 2.820 | 3.003 |
| CV [%] | 3.221 | 15.900 | 16.814 |

The results for the ABS+ material in the lateral printing C is presented in table 4.

**Table 4**. Values for ABS+, lateral printing C

| Sample | Ɣ | σC | σR |
|---|---|---|---|
| ABS_C01 | 1653 | 41.979 | 50.590 |
| ABS_C02 | 1675 | 42.263 | 50.142 |
| ABS_C03 | 1638 | 44.187 | 50.518 |
| ABS_C04 | 1620 | 40.265 | 49.162 |
| ABS_C05 | 1699 | 42.600 | 50.907 |
| Average | 1657 | 42 | 50 |
| StDev | 31.039 | 1.404 | 0.673 |
| CV [%] | 1.873 | 3.322 | 1.340 |

The results for material PETG material in the horizontal printing A, is presented in table 5.

**Table 5**. Values for PETG, horizontal printing A

| Sample | Ɣ | σC | σR |
|---|---|---|---|
| PETG.A01 | 1518 | 53.882 | 60.977 |
| PETG.A02 | 1576 | 51.126 | 60.291 |
| PETG.A03 | 1549 | 53.608 | 59.879 |
| PETG.A04 | 1621 | 53.097 | 60.895 |
| PETG.A05 | 1472 | 51.336 | 57.510 |
| Average | 1547 | 53 | 60 |
| StDev | 56.63 | 1.292 | 1.416 |
| CV [%] | 3.661 | 2.456 | 2.363 |

The results for material PETG in the vertical printing B, is presented in table 6.

**Table 6**. Values for PETG, vertical printing B

| Sample | Ɣ | σC | σR |
|---|---|---|---|
| PETG.B01 | 1357 | 25.008 | 25.008 |
| PETG.B02 | 1113 | 21.857 | 21.857 |
| PETG.B03 | 1169 | 24.811 | 24.811 |
| PETG.B04 | 1121 | 18.543 | 18.543 |
| PETG.B05 | 1211 | 24.599 | 26.062 |
| Average | 1194 | 23 | 23 |
| StDev | 99.32 | 2.785 | 3.063 |
| CV [%] | 8.316 | 12.128 | 13.170 |

The results for material PETG in the lateral printing C, is presented in table 7.

**Table 7**. Values for PETG, lateral printing C

| Sample | Ɣ | σC | σR |
|---|---|---|---|
| PETG.C01 | 1814 | 59.782 | 68.079 |
| PETG.C02 | 1640 | 61.811 | 64.719 |
| PETG.C03 | 1718 | 62.605 | 66.460 |
| PETG.C04 | 1705 | 60.115 | 65.312 |
| PETG.C05 | 1590 | 64.128 | 66.063 |
| Average | 1694 | 62 | 66 |
| StDev | 85.03 | 1.797 | 1.28 |
| CV [%] | 5.021 | 2.913 | 1.938 |

The results for PLA material in the horizontal printing A, is presented in table 8.

**Table 8**. Values for PLA, horizontal printing A

| Sample | Ɣ | σC | σR |
|---|---|---|---|
| PLA.A01 | 1889 | 44.005 | 58.732 |
| PLA.A02 | 1878 | 44.733 | 57.439 |
| PLA.A03 | 1791 | 40.530 | 55.084 |
| PLA.A04 | 1864 | 41.795 | 55.258 |
| PLA.A05 | 2001 | 48.920 | 60.713 |
| Average | 1885 | 44 | 57 |
| StDev | 75.589 | 3.227 | 2.382 |
| CV [%] | 4.011 | 7.334 | 4.146 |

The results for PLA material in the vertical printing B, is presented in table 9.

**Table 9**. Values for PLA, vertical printing B

| Sample | Ɣ | σC | σR |
|---|---|---|---|
| PLA.B01 | 1689 | 36.177 | 40.105 |
| PLA.B02 | 1697 | 29.277 | 29.277 |
| PLA.B03 | 1752 | 26.685 | 26.685 |
| PLA.B04 | 1724 | 29.351 | 29.351 |
| PLA.B05 | 1820 | 30.707 | 30.719 |
| Average | 1737 | 30 | 31 |
| StDev | 52.842 | 3.522 | 5.173 |
| CV [%] | 3.043 | 11.571 | 16.565 |

The results for PLA material in the lateral printing C, is presented in table 10.

**Table 10**. Values for PLA, lateral printing C

| Sample | Ɣ | σC | σR |
|---|---|---|---|
| PLA.C01 | 2350 | 66.911 | 74.285 |
| PLA.C02 | 2355 | 59.197 | 69.176 |
| PLA.C03 | 2165 | 59.787 | 67.214 |
| PLA.C04 | 2497 | 62.379 | 71.700 |
| PLA.C05 | 2248 | 62.565 | 69.879 |
| Average | 2323 | 62 | 70 |
| StDev | 125.238 | 3.049 | 2.678 |
| CV [%] | 5.391 | 4.905 | 3.801 |

IOP Conf. Series: Materials Science and Engineering 1303 (2024) 012034 doi:10.1088/1757-899X/1303/1/012034

## 6.Interpretation of results

The tests were performed on an INSTRON 8875 universal electrohydraulic testing machine.

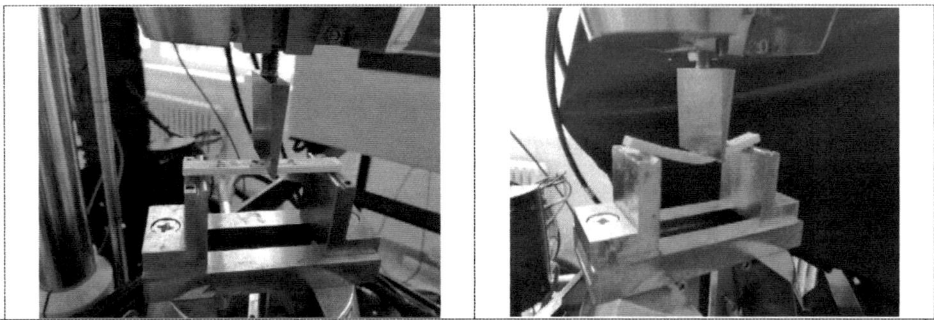

**Figure 4.** Samples subjected to bending on the universal machine INSTRON 8875

After processing the results presented in tables 2, 3 and 4, the following curves were generated. For ABS+ material in the 3 printing positions figure 5.

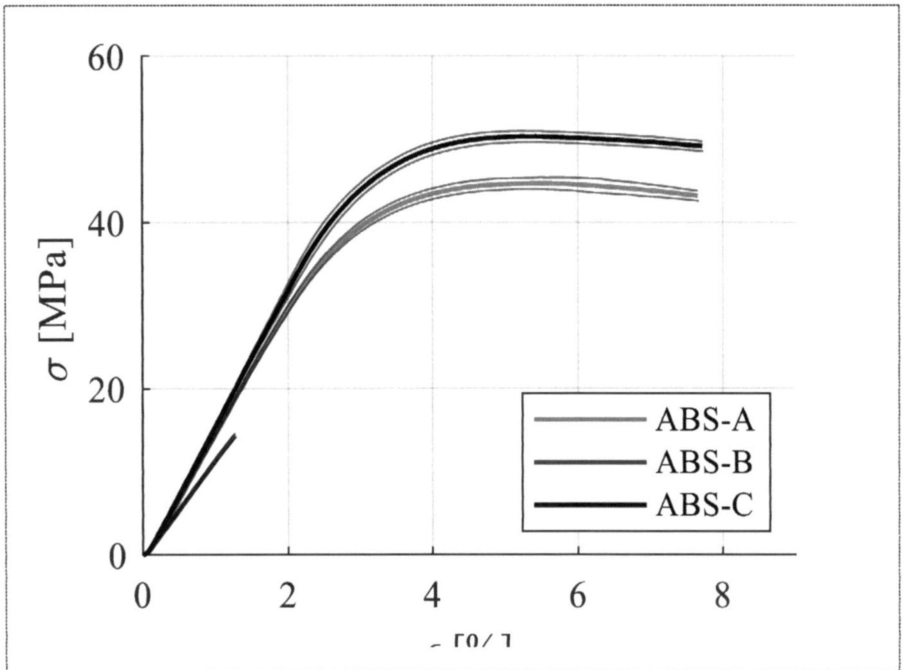


**Figure 5.**Average characteristic curves for bending tests, material ABS+

For PETG in the three printed directions figure 6.

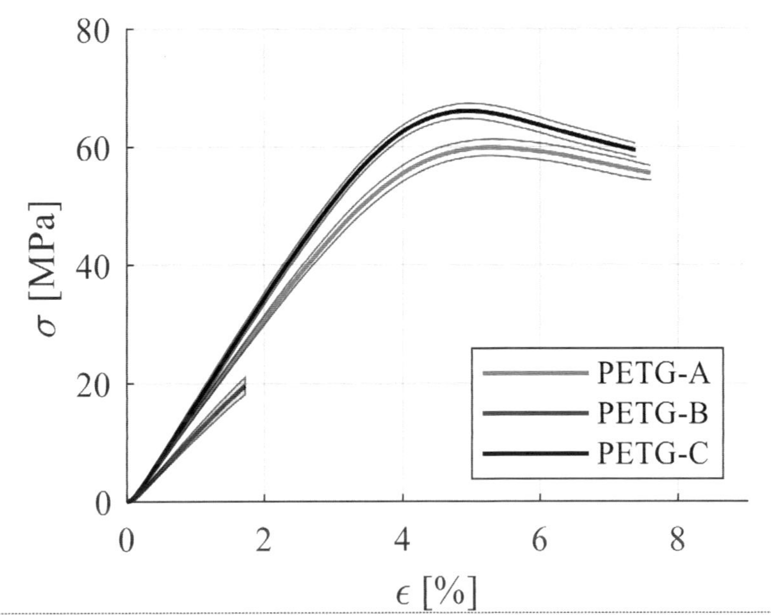


**Figure 6.** Average characteristic curves for bending tests, material PETG

For PLA in the three printed directions figure 7.

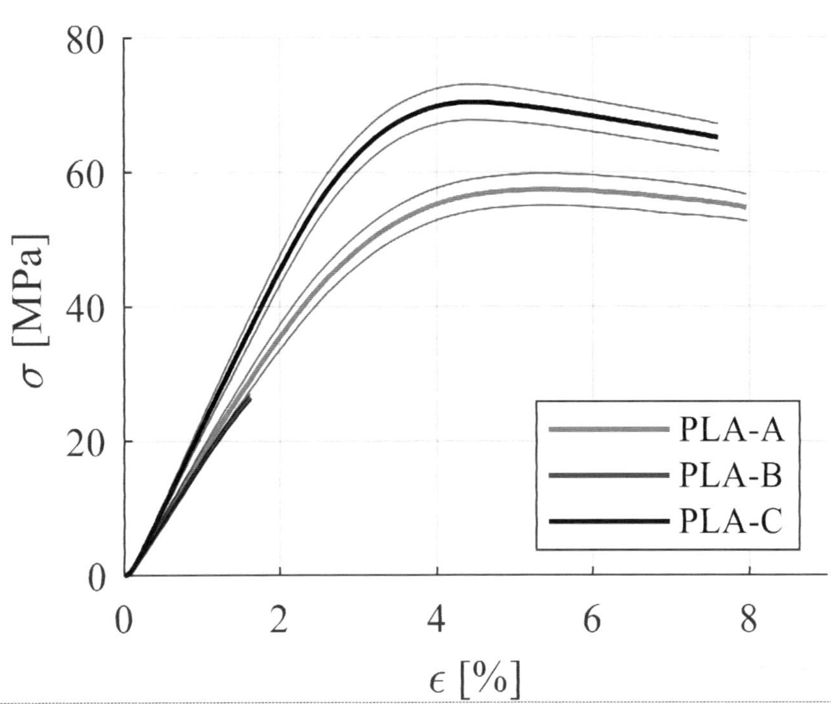


**Figure 7.** Average characteristic curves for bending tests, material PLA

## 7. Conclusions

Analyzing the graphs in Figures 4, 5 and 6, it can be seen that the PLA material from which the samples are made has a better bending strength than the PETG material and much better than the ABS+ material.

Average value of bending stress as also shown in table 2 to table 10 is 2323 (MPa) for PLA, 1694 (MPa) for PETG and 1657 (MPa) for ABS+.

From the point of view of the 3D printing direction for all 3 materials, the maximum value in the case of bending tests is obtained when the specimen is positioned in variant C, lateral direction.

From the point of view of the PLA material, having a melting temperature between 125-185$^0$C, higher than the other materials used, the bonds formed between the deposition layers have greater strength.

Through the 3D printing mode in the C variant - lateral direction - it forms continuous field lines as in the case of hot lamination of metallic materials because between the deposition layers there is no total cooling of the previously deposited row and thus a more stable connection takes place. This phenomenon could be followed in detail through a study at the microcrystalline level, which will be the subject of future research.

## Author's ORCID ID

A Dimitrescu: https://orcid.org/0009-0008-9650-804X
C Babis: https://orcid.org/0009-0009-4174-056X
S A Fica: https://orcid.org/0009-0002-4797-3216
C Stochioiu: https://orcid.org/0000-0002-7413-6133
D Vlasceanu: https://orcid.org/0000-0002-6609-7496
F Baciu: https://orcid.org/0000-0002-0574-6149

## References

[1] Zhengshu Yan, Amirmohammad Rahimizadeh, Yixue Zhang, Yuheng Zhou, Larry Lessard 2023 *Composite Structures* A finite element model for 3D printed recycled parts from end-of-life wind turbine blades **320/2023** p167

[2] Abhay Mishra, Mukul Kumar, Rajeev Kumar, Anuj Kumar Sharma 2023 *Materialstoday* An investigation on tensile strength of 3D printed parts of Acrylonitrile butadiene styrene (ABS) material **1/2023** pp112-118

[3] Qamar Shahzad, Xujiang Wang, Wenlong Wang, Yi Wan, Guolin Li, Changzai Ren, Yanpeng Mao 2020 *Construction and Building Materials* Coordinated adjustment and optimization of setting time, flowability, and mechanical strength for construction 3D printing material derived from solid waste **259/2020** p267

[4] J. Minguella-Canela, S. Morales Planas, J.R. Gomà Ayats, M.A. de los Santos López 2019 *Procedia Manufacturing* Study and comparison of the different costs' schema associated to geometry, material and processing between 3D printing, injection molding and machining manufacturing technologies **41/2019** p201

[5] Imran Khan, Muhammad Tariq, Muhammad Abas, Muhammad Shakeel, Fatima Hira, Ans Al Rashid, Muammer Koç 2023 *Composites Part C* Parametric investigation and optimisation of mechanical properties of thick tri-material based composite of PLA-PETG-ABS 3D-printed using fused filament fabrication **1/2023** pp106-110

# Dummy kinematics in frontal impact, comparison of simulations in Madymo with sled tests

**C Nedelescu, A Chiru, P Vrabie and D Trusca**

Transilvania University of Brasov, Department of Automotive and Transport Engineering, Brasov, Romania

E-mail: claudiu.nedelescu@unitbv.ro

**Abstract**. The paper aims to respond to the current requirements regarding the usage of modern virtual analysis tools in accidentology research, presenting the methods of performing the virtual and experimental studies. The main objective of this paper is to determine the movement of the dummy by comparing the kinematic parameters provided by the simulations performed in Madymo with the experimental tests (sled tests). A prototype test device was used to analyse the movement of the dummy in a vehicle cabin at different impact situations of a frontal collision. The experiment was performed for two different cases: with seatbelt and without seatbelt. For the virtual study in Madymo two different setups were prepared in order to obtain the results for both testcases based on a parameterized model. The analysis of the results is performed by comparison, between experimental results and simulation results. Three different phases are taken into consideration in order to make a study of a frontal collision: pre-collision, collision and post-collision.

## 1. Introduction

In the past years, the number of road accidents has increased significantly, and the development of prevention techniques has become imminent. Whether we are talking about active safety systems, road accident reconstruction programs or virtual analyzes through computer-aided simulation tools, their continuous improvement must lead to a reduction of the number of road accidents and to have an important contribution in decreasing the severity of injuries 0.

To study the responses of occupants and injury criteria in vehicle collisions two approaches are used in this work. One approach consists of performing experimental tests of real collisions in polygon or using a sled test device and crash test dummies. This aspect has been highlighted in recent studies where a prototype dummy was developed and compared to a Hybrid III 50th percentile male dummy through sled test 11]. Other recent studies present various dummies used to determine head injury criteria. For example, a prototype dummy head is used on a test-rig, recording the kinematic parameters through which the head injury criterion (HIC) will be determined 0, or when head injury index is assessed for standard Hybrid III 5th, 50th and 95th dummies using the same restraint system 0.

Other approach is to perform virtual studies conducting computerized simulations in which the occupant kinematics is predicted and assessed by software packages such as Madymo or LS Dyna. In a relevant study 0, published in 2019, the reconstruction of a frontal accident is carried out, the movement of the vehicle being simulated with the help of PC-Crash and to obtain the kinematic parameters and implicitly the response of an occupant Hybrid III 50th male, Madymo is used. With the kinematic parameters obtained, an assessment of the severity of head injuries was also made.

The two methods of conducting the experimental and virtual study were compared in the study 0, where a THOR dummy restrained by a standard 3-point shoulder subjected to two frontal crash sled tests was evaluated. By comparison, numerical simulations were carried out following the same procedures as in the case of the experimental test. The dummy kinematics and the interaction forces of dummy with the sled and the belt were recorded from the experimental test and compared with the simulations. In 2021, the study 0 was published, in which a comparison is made between the capability of the Madymo to predict the response of the occupant and human volunteers in the event of a frontal collision.

The main objective of this study is to determine the dummy movement by comparing the kinematic parameters provided by the simulations performed in Madymo with the experimental tests (sled tests) 0. The aim of conducting this comparison is to calculate the probability of injury correlated with the kinematics of the dummy, using the head injury criterion in a time interval of 36 ms.

## 2. Methodology

### *2.1. Procedures and techniques for the organization of the experimental research*

#### *2.1.1. Testing procedure of the experiment*

For the test environment, the parking lot from the Research and Development Institute of the Transilvania University of Brasov was used. In this location was placed a prototype test device, *Figure 1,* used to *analyse* the movement of an 84.12 kg prototype dummy in a vehicle cabin within different impact scenarios in a frontal collision. The experiment was performed for two different cases: with seatbelt and without seatbelt.

**Figure 1.** The prototype test device.

#### *2.1.2. Equipment features and performance*

A high speed camera which is capable of recording at 600 fps, was used to record the actual impact. To ensure the correct light regardless of the test conditions, led projectors were used, *Figure 2.*

### *2.2. 3D modelling in the Madymo*

The 3D modelling was performed in Madymo using a parameterized model of a frontal impact. The occupant was a Hybrid III 50% male dummy of approximately 78 kg. Two different setups were prepared in order to get the results for both test cases: with seatbelt and without seatbelt.

The pre-processor XMADgic was used for editing the .xml content. For second test case (without seatbelt) the corresponding node was disabled (Figure 3).

**Figure 2.** LED Projectors.

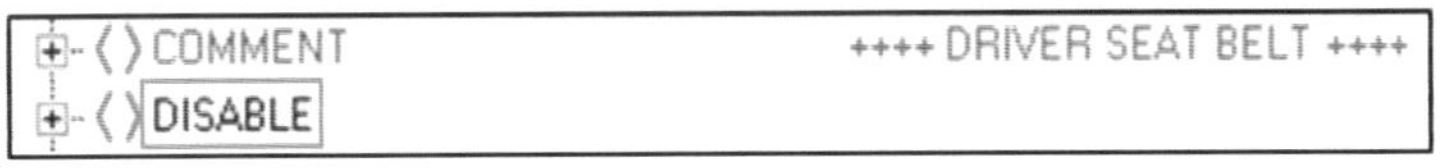


**Figure 3.** Seat belt removing.

Furthermore, for the two load cases, the simulation parameters start and end time were modified (Figure 4), but the time step was used by default. The period of the simulation time, measured in seconds, was correlated with the values of the time obtained from the experimental data.

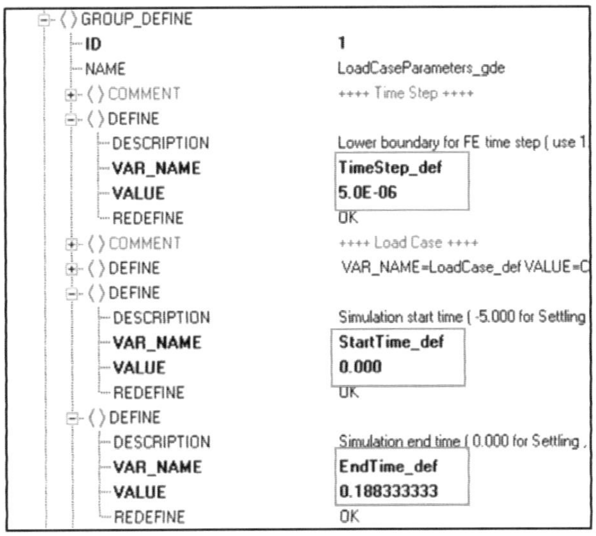


**Figure 4.** Load Case Parameters.

To include the motion of the sled in the simulation, its velocity and the acceleration on X component were added in the .xml content (Figure 5). For this type of crash, the acceleration on X component is the most significant.

IOP Conf. Series: Materials Science and Engineering 1303 (2024) 012035 doi:10.1088/1757-899X/1303/1/012035

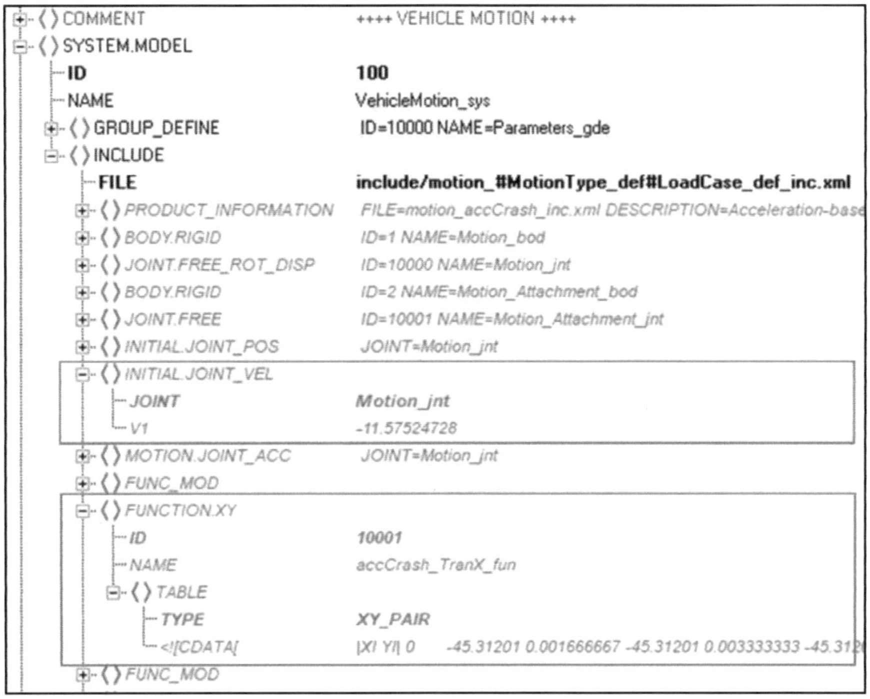


**Figure 5.** Motion parameters of the sled.

As output request the .CSV format was chosen to get the output signals of the dummy 0. The next step was to validate the document (Figure 6) and to ensure that the solving process can start.

**Figure 6.** Document validation.

Following the indications from 0, two setups were created, as presented in Figure 7.

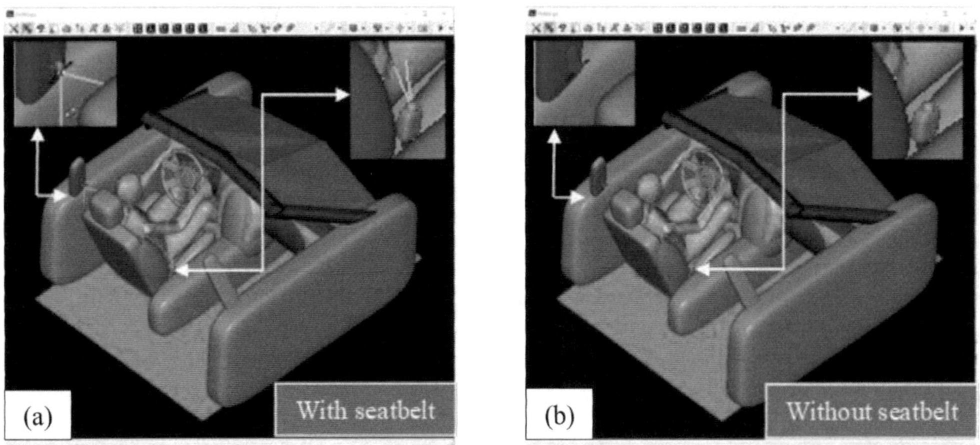


**Figure 7.** Setups creation.

For the processing part was used the default solver and the job was submitted (Figure 8).

**Figure 8.** Processing part - Solve.

2.3. *Data acquisition and processing*

No accelerometers were used for measurements, instead the acquisition of the experimental data was performed using the Tracker software which is a video analysis tool performing both, measurements and calculations, on X & Y axis in a relatively short period of time for different landmarks in the videos recorded with the high speed camera.

In Madymo some CFC (Channel Frequency Class) filters were automatically introduced in post-processing as a result of processing 0. For instance CFC 1000 for head accelerations, CFC 180 for thorax accelerations and CFC 180 for velocities. All the values were exported in .csv format and processed in the Microsoft Excel.

For a better visualization of the velocity graphs, the values corresponding to the rebound movement for both, Madymo and sled test, were multiplied by -1. The software Origin was used to obtain smooth curves on the graphs from Microsoft Excel, this being done with the help of FFT filtering.

A measure that can be used to study and analyse the causes of a road accident, regarding the mechanism of brain injury, is the HIC (Head Injury Criterion, defined by NHTSA), which is determined with the formula:

$$HIC = \max\left\{(t_2-t_1)\cdot\left[\frac{1}{t_2-t_1}\int_{t_1}^{t_2} a_R dt\right]^{2.5}\right\} \quad (1)$$

Where:

$a_R = \sqrt{a_x^2 + a_y^2 + a_z^2}$

$a_R$ - the resultant acceleration of the components ($a_x$, $a_y$, $a_z$) [g]

$t_1$,$t_2$ - the initial moment, respectively the final moment of the time interval during which the evolution of the resultant acceleration is considered, in which the HIC reaches its maximum value for $0 < t_1 < t_2 < t$ [s]

t - impact duration [s]

The risk of injury is evaluated by the probability of producing injuries of a certain severity 0:

$$p = [1 + \exp(5.02 - 0.00351 \cdot HIC36)]^{-1} \quad (2)$$

The severity of injuries is assessed according to the AIS scale and Table 1 shows the most common types of injuries that occur in road accidents 0.

**Table 1.** AIS scale.

| AIS code | Severity of injury |
|---|---|
| 1 | Low |
| 2 | Moderate |
| 3 | Serious (no danger of death) |
| 4 | Severe (with danger of death) |
| 5 | Critical (uncertain survival) |
| 6 | Unsurvivable |

Figure 9 shows an equivalence between the HIC criterion for head injury and the AIS scale, which assesses the severity of injuries. The limit value of human tolerance is considered 1000 for the HIC criterion, a value above which human life is in danger. The limitation of HIC is that it does not take into account the angular acceleration of the head, as it is calculated based on the resulting translational acceleration on the 3 axes 0.

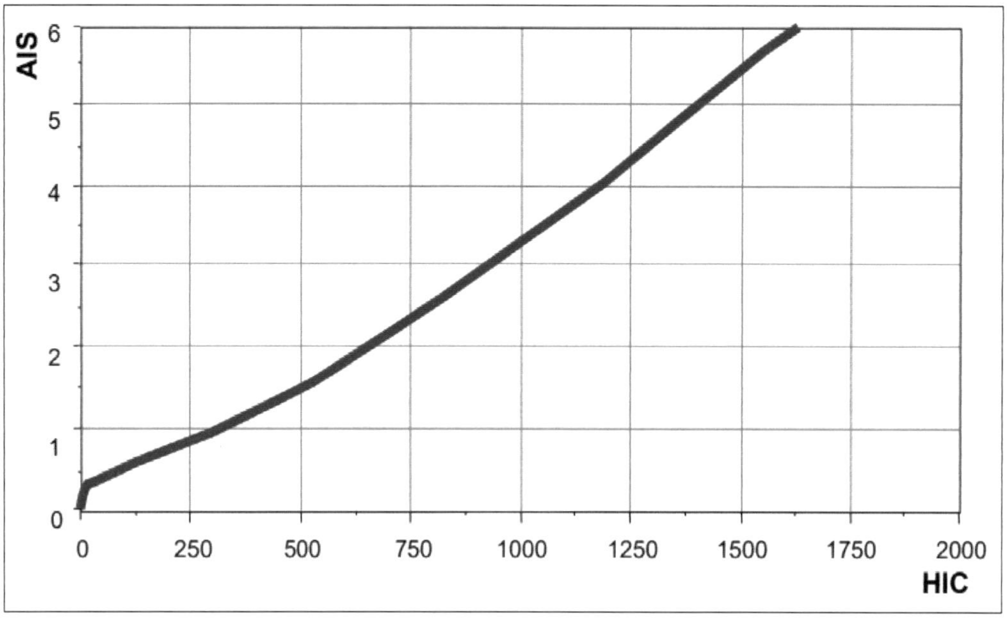


**Figure 9.** Equivalence between HIC and AIS (Abbreviated Injury Scale) 0.

## 3. Results analysis

The kinematic parameters for the head and thorax of an 84.12 kg humanoid anthropometric device (dummy) were recorded through two experimental frontal impact tests conducted at a velocity of approximately 42 km/h. Two simulations were performed in Madymo with a Hybrid III 50% dummy of approximately 78kg, and their results were used for the comparison with the measurements results of the experimental frontal impact tests. The analysis of the results is performed for two cases when the occupant is equipped with seatbelt and without seatbelt. In order to study a frontal collision, three different phases are considered: pre-collision, collision and post-collision.

Both movements of the belted dummy during the experimental test, respectively the Madymo simulation, are illustrated in the Figure 10 describing the collision phases.

The dummy movement is analyzed in 3 different moments at 0 ms, 50 / 70 ms and 183 ms. The collision phase of the experimental test has a delay of approximately 20 ms, the aspect being visible on the plots in Figure 11 which presents a comparison of the kinematic parameters obtained to the time interval [0, 0.1883 s]. It can be observed on the graphs that the velocity was approximately 42 km/h (11.7 m/s) before the collision.

For the case when the occupant is equipped with seatbelt, the peaks of the kinematic parameters are reached when the head hits the airbag, being protected from significant damage with the interior parts inside the dummy compartment. The thorax is held on the seat with the help of the safety belt, without significantly rotating 0, 0. Some regulations (ECE-R 17 in Europe and FMVSS 201 in the USA) provide as a performance criterion for the head impact the limit of deceleration in the antero-posterior direction. The head injury tolerance limit is 80 g over a maximum period of 3 ms 0.

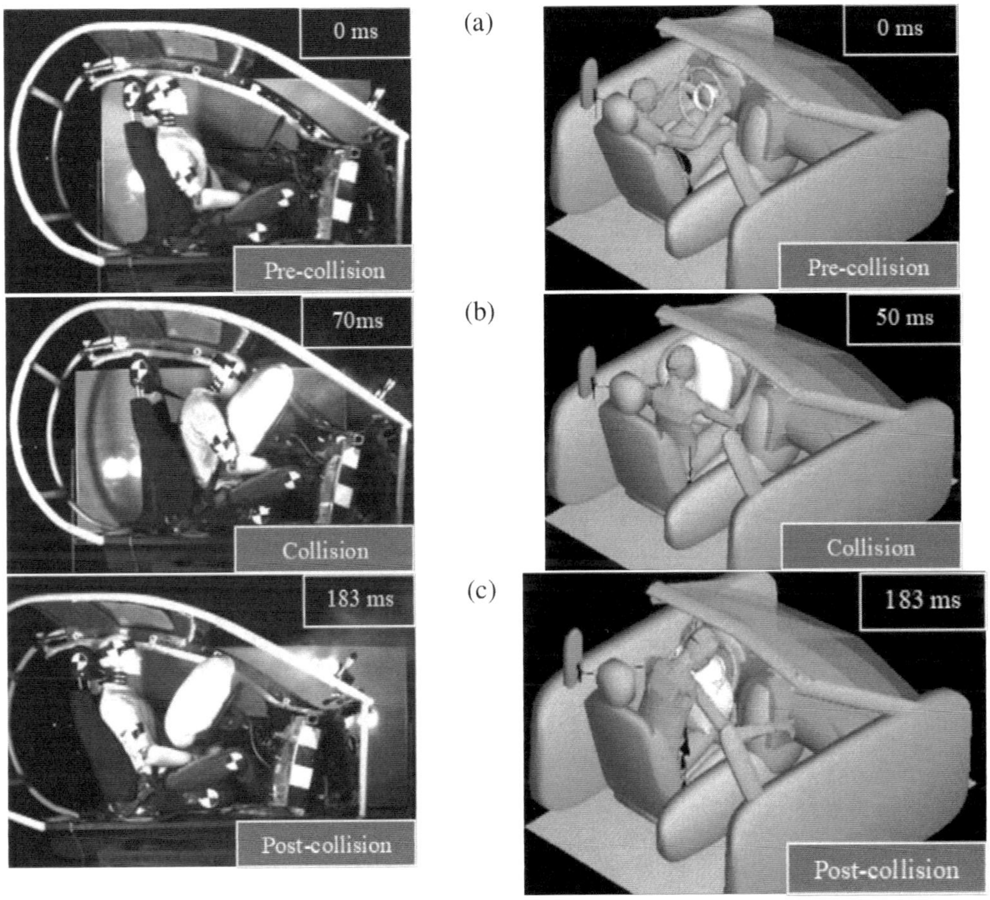


**Figure 10.** Collision phases when the dummy is belted.

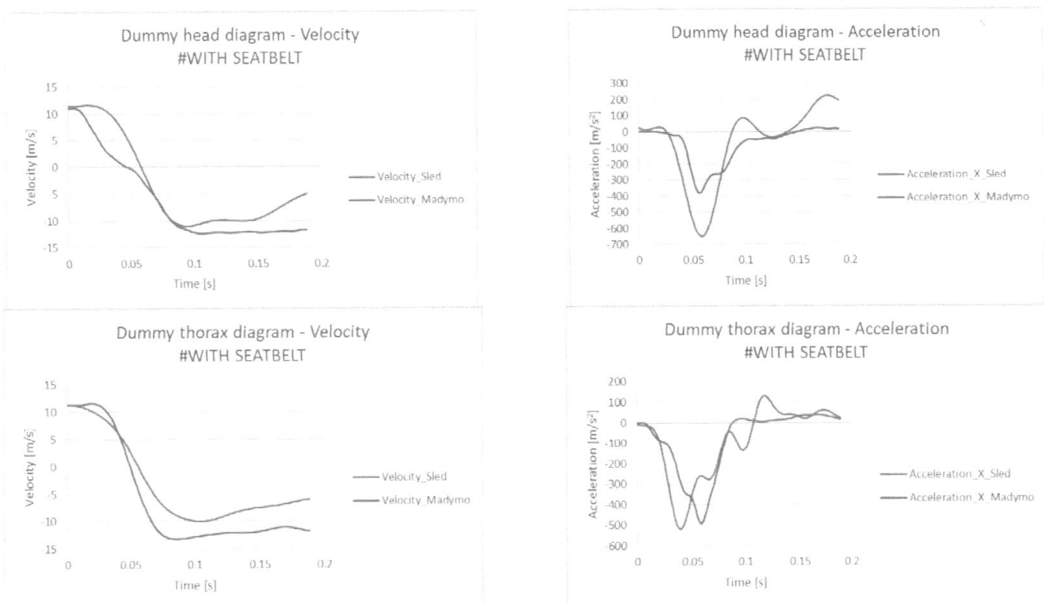


**Figure 11.** Kinematics diagrams when the dummy is belted.

Regarding the biomechanical limit of the thorax, the FMVSS 208 regulation provides that the limit is the acceleration value of 60 g for a maximum duration of 3 ms when the occupant is belted 0. In this case the greatest acceleration values on X axis of the head and thorax when the occupant is belted don't exceed the limits. The seatbelt can cause injuries only at relatively higher velocities or if it is not positioned correctly, which consists of breaking the clavicle on the side of the shoulder belt, tearing the liver, damage to the small intestine etc. 0.

Collision phases for the unbelted dummy are presented in Figure 12.

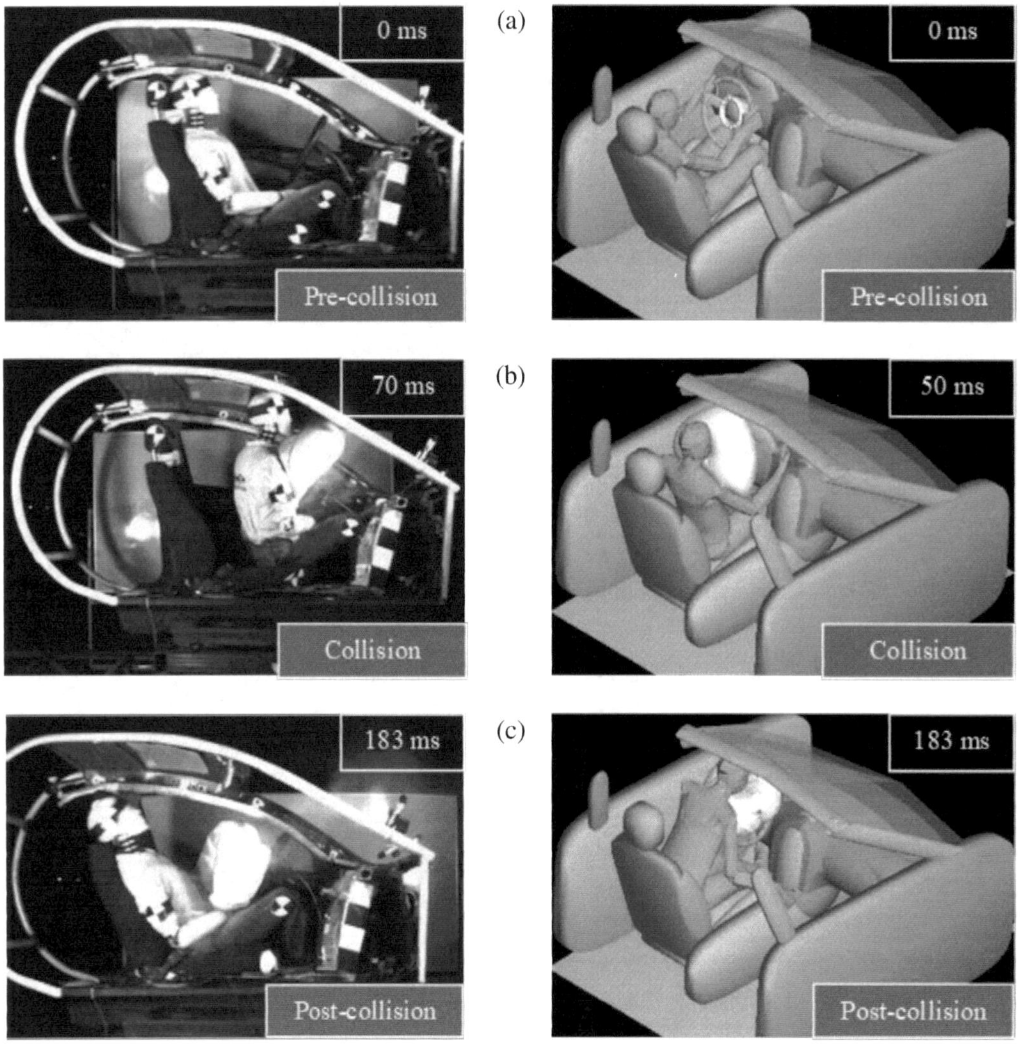


**Figure 12.** Collision phases when the dummy is unbelted.

The values obtained from the experimental test as well as from the simulation in Madymo are illustrated comparatively by representative graphs (Figure 13), observing that all tests were conducted at similar velocities before the collision:

For the experimental test, the unbelted occupant hits the ceiling with his head during the impact representing the peak of the head acceleration value displayed on the graph at 0.07 s approximately. In the Madymo simulation, this aspect results from the similar acceleration values obtained, being close with the ones from the experiment 2, 0. Even if the limits according to the regulations ECE-R 17 and FMVSS 201 are exceeded, the occupant is not wearing a seatbelt as tested by these standards.

**Figure 13.** Kinematics diagrams when the dummy is unbelted.

According to some specialized literatures, the tolerance limits of the head are described more detailed and more classifications are available, for example for the forehead and chin 80g, respectively 60g and for the crown of the head the limit is 200 g. It is observed that the greatest acceleration values of the occupant's head on X axis when not wearing the seatbelt are between 112 g and 128 g and even if the biomechanical limit for the crown of the head is not exceeded, the impact occurs in the area of the dummy's forehead exceeding the injury tolerance limit.

The occupant can suffer external injuries such as skull fractures and internal injuries such as displacement of the brain in the skull or other injuries to the face or brain, with the risk that some of these may be irreversible, even fatal 0. The thorax injury tolerance limit for an occupant who is not wearing a belt, according to the FMVSS 208 is 60 g for a maximum duration of 3 ms. The thorax acceleration values don't exceed the tolerance limit value using both analysis methods. However, a risk can be associated if the unbelted occupant is close to the airbag at the time of its inflation, fatal injuries can occur through a strong impact to the neck and chest like an explosion, which can cause ventricular fibrillation, tearing of liver or lung injuries. Injuries can also occur in the upper cervical area of the spine because the airbag reaches under the chin 0.

It is important to mention that the study has some limitations, and the accuracy of the measurements was influenced by factors that were not taken into account, such as the position of the seat, the height of the safety belt or the position of the anchor points, the differences in the design of the dummies, the position of the markers, as well as in the generation of collision forces, amplifying the rebound movement, the test device being equipped with air cushions mounted in the frontal side. For technical reasons of visibility, the thorax markers were mounted on the upper part of the dummy's arm, the difference in weight mainly leading to different measured kinematic parameters compared to the values provided by Madymo. The values obtained for the acceleration of the dummy's head on X axis were close in the experimental test compared to Madymo when the occupant is unbelted. However, higher head acceleration values on X axis are observed in the results provided by the experimental test compared to Madymo, the difference being more noticeable when the occupant is belted. The differences are mainly caused by a different design of the dummy in the experimental test, more specifically a neck joint with a different stiffness 0.

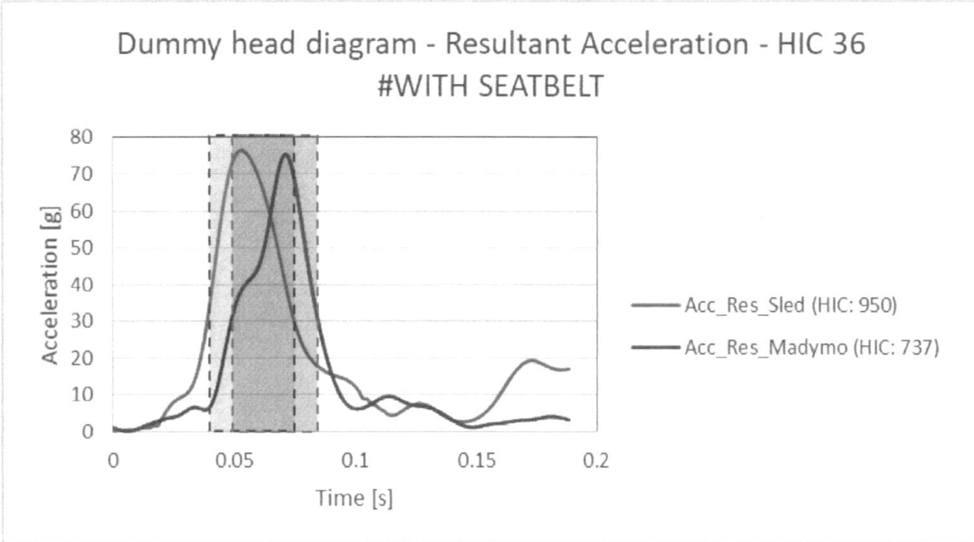


**Figure 14.** Resultant acceleration for belted dummy on 36 ms time interval (HIC 36)

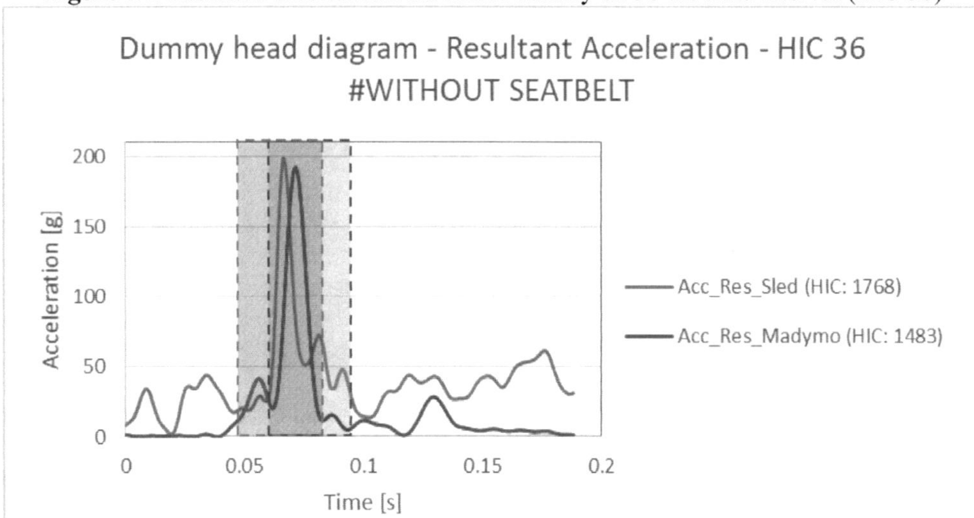


**Figure 15.** Resultant acceleration for unbelted dummy on 36 ms time interval (HIC 36)

In the current study, the HIC criterion was calculated in a range of 36 ms, as can be seen in Figure 14 and Figure 15, the range in which the head injury criterion reaches the maximum value being marked according to the origin of the resultant acceleration values, namely sled tests or Madymo.

In the case of tests where the occupant is equipped with a seat belt, the calculated HIC value is 950 for the sled and for the Madymo 737. The values do not exceed the human tolerance threshold of 1000, but correlated with the AIS scale, for the sled corresponds to code 3, and for Madymo the code 2, which represents a moderate or even serious severity, but without the risk of fatal injuries. The probability of injury is 8.1% for Madymo and 15.6% for sled tests.

When the occupant is unbelted, the calculated HIC value is 1768 for the sled and for the Madymo 1483. The human tolerance limit is exceeded and the severity of the injuries is fatal for the occupant in the case of the sled test, reaching index 6 on the AIS scale and critical with uncertain survival in the case of the Madymo simulation with index 5 on the AIS scale. The probability of injury is 54.6% for Madymo and 76.6% for sled tests.

During a real collision, the vehicle absorbs through its deformation the kinetic energy and protects the occupant, while for the sled test device, the kinetic energy produced during the impact is transferred to the dummy 0. The obtained results highlight the limitation of a sled device, as it is not able to reproduce the current response of a real vehicle during a collision and both the HIC values and the probability of producing injuries are higher in the case of tests performed on the sled compared to the results provided by Madymo, where the values and probabilities of injuries are lower, simulating a real impact.

## 4. Conclusions

The occupant movement in case of a crash is seriously analyzed by the car manufacturers so they can develop new and more performant systems that reduce the effects of the collision on the passengers. The in-car movements experienced by the occupants depend on multiple aspects such as the type of impact, vehicle type, adjustment of the interior parts and vehicle occupancy.

During the impact, the injuries of the occupants are mainly caused by the movements of the occupants and different parts of the vehicle as a result of the forces developed in the impact, also known as collision forces.

The experimental research was carried out by creating simulations of the situations in which the road accidents occur, from the point of view of the occupant's behavior in case of an impact. This work represents a research of the kinematics of the occupant, in situations where the occupant is equipped or not with a safety belt. This type of impact is very common in real-life cases of giving way at intersections, roundabouts and more. The situations presented can occur when the occupant neglects to wear the seatbelt or when the restraint device breaks down.

The novelty element of this work presents the usage of a prototype test device specially designed and executed for carrying out impact tests. A prototype dummy was also used. The obtained results are compared with the data provided by Madymo software. The advantage of using the research test devices versus testing by using the vehicles on a scale of 1:1 in the polygon is primarily the possibility of controlling the test conditions and ensuring the repeatability of the tests, and then the difference in logistics costs.

The methodology used and the data obtained through the experimental tests can be used to improve the test device, during the study being identified the main technical limits that led to the differences in the kinematic behavior of the occupant.

Once the identified limitations of this study are removed, an improved correlation of the experimental test with the model of Madymo software can be achieved. This aspect will lead to a reference Madymo simulation set-up which can be used for further simulations, resulting into a cut of the financial and time resources for the experimental set-up. On top of that, the numerical simulation method used in this research can also be applied for the verification and validation of other test devices or prototype dummies used in crash simulations 0.

## References

[1] Bosma F, Van Hooijdonk P A, Kietlinski K, Tijssens M G and Unger M F 2019 Simcenter Madymo *DHM and posturography* Academic Press pp 155-168
[2] Fung K, Xu R, Jung S and Sobanjo J 2019 Development and testing of a simplified dummy for frontal crash *Experimental Techniques* **43** 7-14
[3] Alhaddad A Y, Cabibihan J J, Hayek A and Bonarini A 2019 A low-cost test rig for impact experiments on a dummy head *HardwareX Elsevier* **6** *e00068*
[4] Gu H, Lou L and Jiang X 2020 Injury Characteristics of Different Dummies Based on Frontal Impact Test *Big Data Analytics for Cyber-Physical System in Smart City: BDCPS 2019* Springer Singapore **1117** pp 1347-54
[5] Jie W and Yutong H 2019 Research on frontal collision reconstruction model based on coupling of PC-CRASH and MADYMO *IEEE 2019 4th International Conference on Control and Robotics Engineering (ICCRE)* 54-9

[6] Untaroiu C D, Shin J and Lu Y C 2013 Assessment of a dummy model in crash simulations using rating methods *International Journal of Automotive* Technology **14** 395-405
[7] Valdano M, Jiménez Octavio J R, Vives Torres C M, López Valdés F J and Pipkorn B 2021 Assessment of madymo active human body model kinematics and dynamics by means of human volunteer response at low-speed frontal impacts *IRCOBI Conference*
[8] Deng G, Gan S, Chen W, Zhou Q, Peng Y, Cui T and Nie B 2021 Effect of simulated muscle activation level on driver neck response in frontal motor vehicle crashes *SAE Technical Paper 2021-01-5008*
[9] Wismans J and Hermans J H A 1988 MADYMO 3D simulations of Hybrid III dummy sled tests *SAE Technical Paper 880645*
[10] Post A, Clark J M, Robertson D G E, Hoshizaki T B and Gilchrist M D 2017 The effect of acceleration signal processing for head impact numeric simulations *Sports Engineering* **20** 111-9
[11] Todoruț I A, Barabas I and Burnete N 2012 *Siguranța autovehiculelor și securitatea în transporturi rutiere* U.T. Press Cluj-Napoca chapter 2 pp 292-301
[12] Gaiginschi R 2009 *Reconstrucția și expertiza accidentelor rutiere* Editura tehnică București chapter 2 pp 66-89
[13] Shojaati M 2003 Correlation between injury risk and impact severity index ASI *Proceedings of the 3rd Swiss Transport Research Conference* 19-21
[14] Xiao S, Qie Y, Huang J, Yang J and Crandall J 2020 Influence of restraint load on injury biomechanics in frontal impact based on Dummy Test *Iranian Journal of Science and Technology Transactions of Mechanical Engineering* **44** 1065-75
[15] Ilie S 2012 *Modelarea fenomenelor de impact ale autovehiculelor* Editura Agir București chapter 2 pp 66-74
[16] Trotta A, Clark J M, McGoldrick A, Gilchrist M D and Annaidh A N 2020 Biofidelic finite element modelling of brain trauma: Importance of the scalp in simulating head impact *International Journal of Mechanical Sciences* **173**
[17] Yoganandan N, Baisden J L, Maiman D J, Gennarelli T A, Guan Y, Pintar F A, Laud P and Ridella S A 2010 Severe-to-fatal head injuries in motor vehicle impacts *Accident Analysis & Prevention* **42** 1370-8
[18] Thorbole C 2017 Small occupant neck injury biomechanics in frontal crash: a study to address the variation in restraint performance with a conventional 3-point single loop belt system *SAE Technical Paper 2017-26-0003*

IOP Conf. Series: Materials Science and Engineering 1303 (2024) 012036 doi:10.1088/1757-899X/1303/1/012036

# Numerical analysis and evaluation of a semi-heavy guardrail system

**M I Gheres, N Cordos and A Todorut**

Department of Automotive Engineering and Transports, Technical University of Cluj-Napoca, RO

E-mail: Marius.Gheres@auto.utcluj.ro

**Abstract**. The continuous increase in traffic accidents, following the rise in the number of vehicles in traffic and their speed, requires the development of research on the design of active and passive safety systems. This research mainly aims to compare the performance of commonly used passive safety systems (semi-heavy W-beam guardrails) mounted on roadside and pedestrian bridges to protect vehicles and pedestrians. It is intended to reduce the deformations of the vehicle upon impact with the protective device mounted between the roadside and the pedestrian area for the protection of the vehicle occupants; reduce the parapet working width for the pedestrian's safety; and maintain the vehicle's trajectory from the moment of impact to the final movement, between the guardrail and the median axis of the road, to ensure the protection of other road users. A CAD model of the vehicle-guardrail scenario was developed to achieve these objectives. Load simulation cases are performed using the finite element method according to the European standards SR EN 1317 for impact tests of vehicles - TB 42, recommended for semi-heavy guardrails.

## 1. Introduction

Guardrails represent the most widely used protective device mounted on roads and highways. They are designed to absorb the energy of a crash and prevent vehicles from leaving the roadway, reducing the risk of passenger injury. The European standard EN 1317 establishes the general requirements of the European road restraint system. It provides the specification for determining the level of vehicle containment, impact test acceptance methodology, and terminology.

Teng et al. [1] evaluated the crashworthiness of the W-beam guardrail, which is studied for four post sections: C-shaped, I-shaped, U-shaped, and Sigma-shaped. The vehicle was chosen according to the TB11 test regulation for EN 1317-2. For the simulation of natural work conditions, boundary conditions were chosen, such as the guardrail continuation connected to the elastic springs added at both ends of the guardrail. The interaction between the posts and the soil was modelled using spring elements after simulations demonstrated that the Sigma-post represents the best performance to maintain the vehicles on the road.

Wolny et al. [2] studied the W-beam system crash behaviour when influenced by the lighting column, affecting the working width of the guardrail. A modified TB51 full-scale crash test is presented according to EN 1317 standard, considering a 13-tonnes bus with a 70 km/h velocity and a 20° impact angle for two test combinations: with lighting column and without. The actual experiments of the crash were validated by the numerical model developed in Ls-Dyna software.

The performance of the guardrail system is recommended to be tested before use on public roads to minimize the in-situ testing costs and to find the optimal shape of the system during the impact. A single impact case can be made in a crash test because, after the impact, both vehicle and guardrail are damaged.

To evaluate the W-beam system, actual researches, developed according to EN 1317 standard and validated by the finite elements model dedicated software [3-12], are studied.

This study evaluates a semi-heavy bridge traffic W-beam guardrail system designed in four supporting post configurations: U-shaped, C-shaped, Sigma-shaped, and I-shaped. The main purpose is to determine the working width (Wm) and mechanical behaviour of the guardrail system when this is in collision with a 10000 kg Heavy Goods Vehicle (HGV), according to the TB42 crash test descriptions. The geometric model of the guardrail system is designed using the SolidWorks CAD environment and the preprocessing of the finite element model is performed using HyperMesh software. To calculate the numerical impact between HGV and guardrail, the explicit Radioss solver will be used.

## 2. Material and method

In this chapter are presented the necessary steps to evaluate the performance of the W-beam guardrail made up of supporting post with different cross sections.

### *2.1. W-beam model*

The W-beam guardrail structure is designed to mitigate the vehicle's impact and redirect it on the road in the crash event. In this study, a detail of the guardrail barrier is presented in Figure 1, which can be observed in all parts of the assembly. The connection between the guardrail parts is made through welding and bolted assembly. The plate (7) is fixed on the ground surface by the four screws, generically called connection parts (4). The plate welds the lower side of the supporting U-post, the plate welds the gusset (8) and the supporting U-post, and the reinforcement part (2) is welded by the plate and supporting U-post to reinforce the structure. The other parts of the guardrail are assembled with bolts. The supporting U-post connects the W-beams (5) through the spacer element (3).

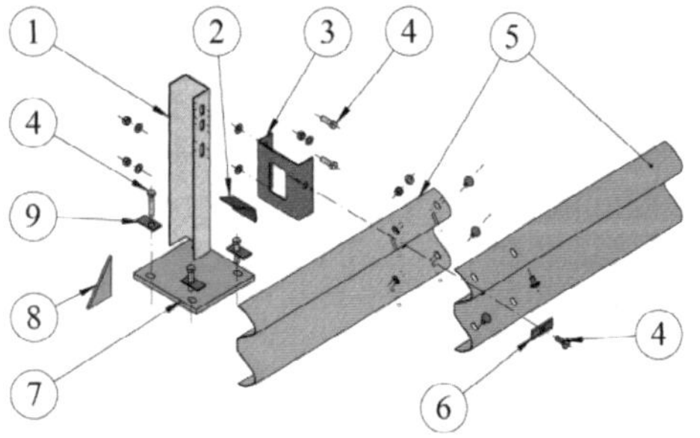


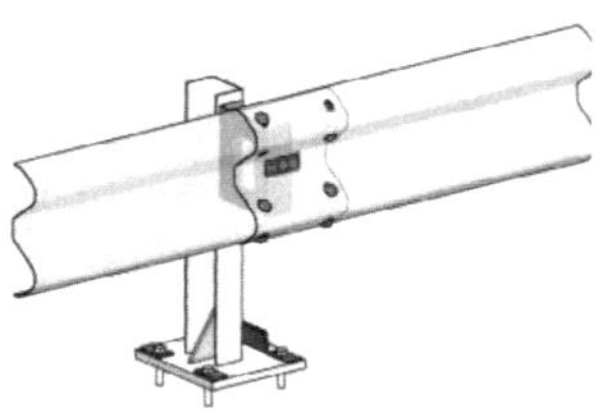

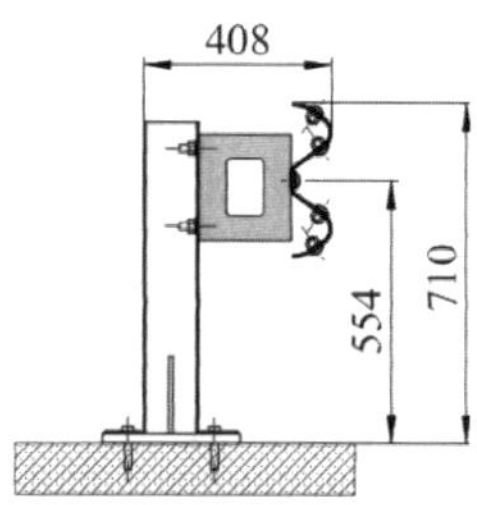


a). Exploded view of W-beam assembly b). W-beam assembly c). Overall dimensions

**Legend**: 1. Supporting U-post; 2. Reinforcement part; 3. Spacer element; 4. Connection parts; 5. W-beam; 6. Plate; 7. Base plate; 8. Gusset; 9. Washer plate.

**Figure 1.** Semi-heavy bridge W-beam guardrail parts.

Considering the guardrail system is mounted on a bridge, the distance between the supporting posts is 1.5 m, and the total length is 18 m.

According to the W-beam guardrail model with a supporting U-post, the other configurations are made, replacing the supporting post. The reinforcement part (2) is adapted to the post geometry in the guardrail configuration with supporting Sigma and I-post. The supporting post cross section used for all guardrail configurations is presented in Figure 2.

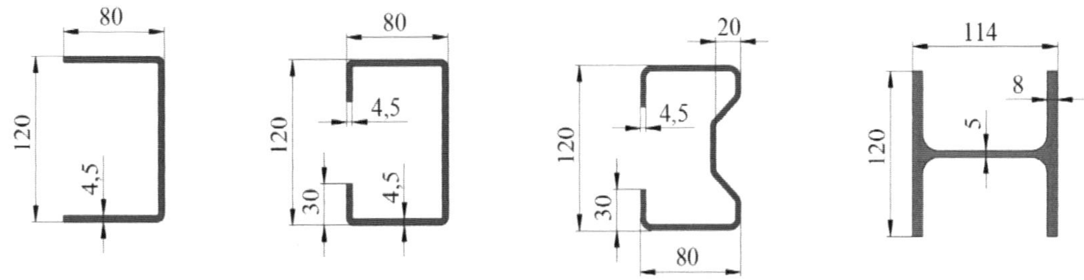


a). U-shaped b). C-shaped c).Sigma-shaped d). I-shaped (HEA120)

**Figure 2.** Guardrail post cross sections.

### *2.2. European standards (EN) 1317*
The European Standard EN 1317-2 refers to the test conditions for parapets and various vehicles' impact crash test criteria, the severity of the passenger impact and levels of working width. In this research, the guardrails are simulated respecting the TB 42 impact crash test criteria. Accordingly, the Table 1 presents the impact vehicle's requirements.

**Table 1.** Vehicle impact crash test criteria [2].

| Test | Impact speed (km/h) | Impact angle (°) | Total mass (kg) | Type of vehicle |
|---|---|---|---|---|
| **TB 42** | 70 | 15 | 10000 | Rigid HGV |

**Table 2.** Levels of working width – accordingly with EN 1317 standard [2].

| Working width classes | W1 | W2 | W3 | W4 | W5 | W6 | W7 | W8 |
|---|---|---|---|---|---|---|---|---|
| Working width $W_m$ (m) | ≤ 0.6 | ≤ 0.8 | ≤ 1.0 | ≤ 1.3 | ≤ 1.7 | ≤ 2.1 | ≤ 2.5 | ≤ 3.5 |

The working width divides the guardrails into eight classes, as shown in Table 2. The working width represents the maximum lateral distance between any part of the guardrails on the undeformed face before collision oriented to the traffic side and the maximum dynamic position of any part of the safety device after the crash. The studied W-beam guardrails are classified at the W2 level with a working width of up to 0.8 m.

### *2.3. FEM - setup*
The FEM (Finite Element Model) was created for each part of the guardrail assembly using HyperMesh dedicated software. This part includes the preprocessing and processing of the simulation stages. The preprocessing set consists of geometry cleanup, the meshing model, materials assignments, and load case creation. Processing steps represent the calculation of the process using a numerical solver. The analysis process ends with the visualization and interpretation of the obtained results in HyperView.

*2.3.1. W-beam model.* The models of W-beam guardrails have meshed into quad and tria shell elements with 10 mm target length and different thicknesses for each part, presented in Figure 3. All guardrail parts present in the connection area of the W-beam assembly are fine-meshed to increase the result accuracy. The back view of the assembly shows the gusset position, welded in the middle of the C-post section. The welding between the supporting C-post, gusset, reinforced part, and plate is represented using rigid elements connected to the parts nodes.

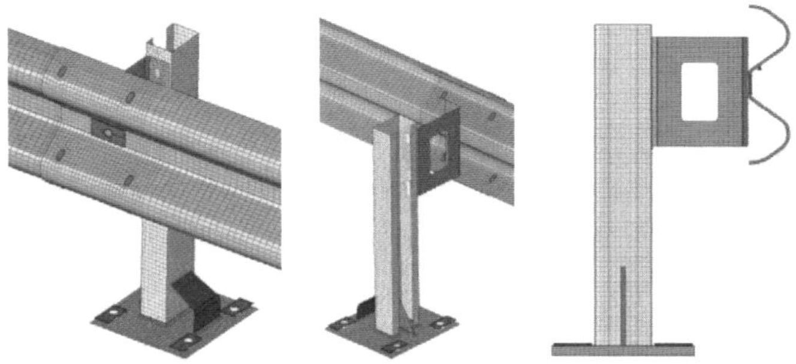

| Parts | **Thick** (mm) |
|---|---|
| Supporting C-post | 4.5 |
| W-beam | 3 |
| Base plate | 20 |
| Plate | 5 |
| Washer plate | 5 |
| Spacer element | 5 |
| Reinforcement part | 6 |
| Gusset | 10 |

a). Front view b). Back view c). 2D detailed view

**Figure 3.** Finite element model of W-beam (C-post configuration).

All the bolted assemblies of the W-beam assembly are represented by the RB2 elements linked to the multiple rigid nodes. The guardrail assembly with supporting C-post configuration has meshed into 113991 shell elements, the model whit U-post into 108921 elements, the model with Sigma-post configurations into 115525 elements and the model with HEA-post configurations into 117215 elements. Figures 4.a and b present the finite element model of the W-beam guardrail with Sigma-post configurations. This assembly model's reinforcement part geometry is adapted to Sigma and HEA -post shapes. The reinforcement part on the Sigma-post configuration is adapted to the post shape as can be observed in Figure 4.b. The details on the HEA-post area assembly are presented in the front view in Figure 4c and the back view in Figure 4.d. In the case of the HEA-post configuration, all other parts remain unchanged regarding the U-post assembly.

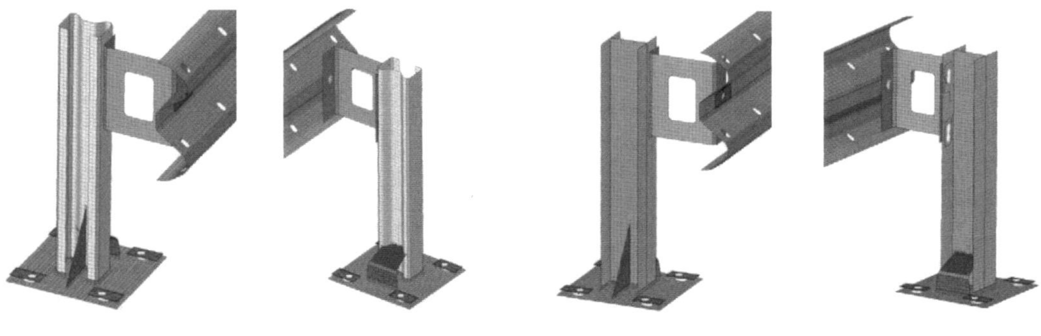

a). Sigma-shaped (front view) b). Sigma-shaped (back view) c). HEA-shaped (front view) d). HEA-shaped (back view)

**Figure 4.** Finite element model of W-beam guardrail (Sigma and HEA-post configuration).

The guardrail continuations are rigidly constrained because the safety system is mounted on the bridge with an 18 m length. The washer plate hole is connected to the base plate using the center of the rigid element constrained for translational displacement.

*2.3.2. Impactor model.* To imitate the real Rigid HGV, the impactor model, presented in Figure 5, is created respecting the real overall dimensions and shape of the vehicle. The P1 part is built from aluminium, and the others are steel. The finite elements model meshes into 13232 shell elements with a 100 mm target length. The P4 impactor parts are created from steel and have the role of chassis. The thickness of the components is chosen to allow the deformation of the part in case of a crash event. According to TB 42 impact crash test criteria, the impactor's mass is 10000 kg.

The thickness of the components determines the total mass of the impactor. It has been chosen so as not to attach an additional mass, which can destabilize the impactor moving, considering that it is without wheels and implicitly without contact with the road surface.

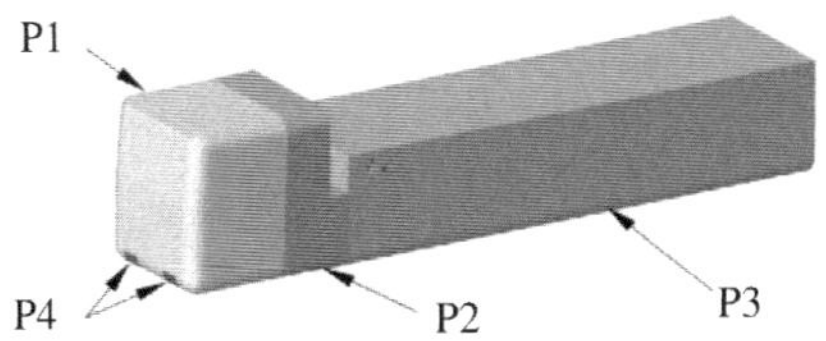


| Vehicle parts | Thickness (mm) |
|---|---|
| P1 | 5 |
| P2 | 5.5 |
| P3 | 9.8 |
| P4 | 16 |

**Figure 5.** Finite elements model of an impactor.

*2.3.3. Assigned materials model.* In this study two isotropic elastoplastic materials are used: steel and aluminium. These added materials have nonlinear behaviour that uses the Zerilli-Amstrong plasticity model. The mechanical properties of steel are presented in Table 3, and in Table 4 are presented aluminium mechanical properties.

**Table 3.** Mechanical properties of the assigned steel [4].

| | |
|---|---|
| Initial density | $7.85e^{-6}$ kg/mm$^3$ |
| Poisson ratio | 0.3 |
| Hardening exponent | 0.5 |
| Hardening parameter | 0.450 |
| Maximum Stress | 0.425 GPa |
| Failure plastic strain | 0.3 |
| Young's modulus | 210 GPa |

**Table 4.** Mechanical properties of the assigned aluminum alloy [5].

| | |
|---|---|
| Initial density | $2.7e^{-6}$ kg/mm$^3$ |
| Poisson ratio | 0.33 |
| Ultimate tensile stress | 0.175 |
| Yield stress | 0.09026 GPa |
| Failure plastic strain | 0.75 |
| Young's modulus | 60.04 GPa |

The load case is created according to TB 42 vehicle crash test description; the impactor hits the W-beam guardrail at an initial velocity of 70 km/h at a 15° impact angle, as seen in Figure 6. This case has been chosen because the 15° impact angle on impact is larger to reach, except in the case of a crash with another vehicle. After all, the parapet is placed on the right portion of a bridge.

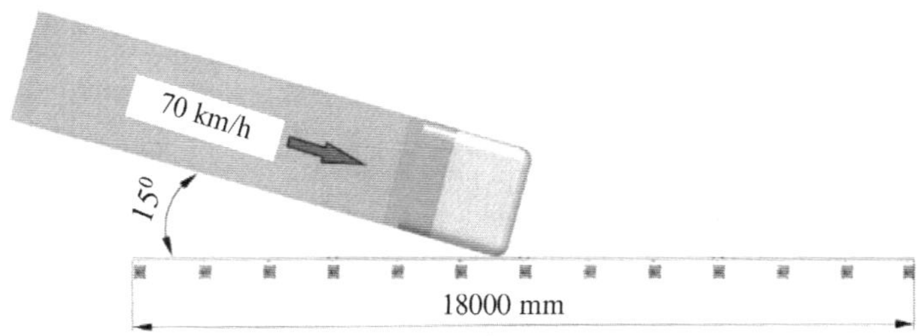


**Figure 6.** Load case setup.

The simulation of the crash is provided using the explicit Radioss solver. The contact between the parts is assured by the general TYPE 7 interface, with a 0.2 Columb friction coefficient.

## 3. Results

*3.1. Results – C-shaped post*

The simulation results of impact with C-supporting post W-beam guardrail configuration are presented in Figure 7, where the von Mises stress concentration plot and sequential figures show the impactor position and guardrail deformation during impact. Until 100 ms, the impactor hit and deform the guardrail parts in the supporting post area. After 250 ms, the impactor is redirected to the roadside, resulting in the working width of the guardrail barrier being 590 mm. At 400 ms, the impactor's speed decreases, and its geometry aligns with the initial state of the parapet, which determines that the impactor is redirected on the roadside. In sequential figures, the impactor's initial position and the guardrail's undeformed shape have the contour represented wireframe. It has been found that there are six affected pillars, spacers, and three W-beam profiles.

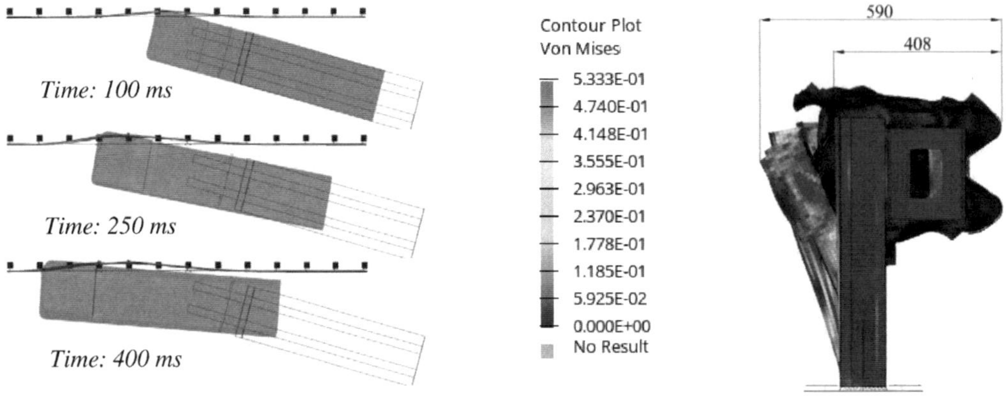


a). Sequential figures from guardrail with C-post b). Guardrail deformations (mm)

**Figure 7.** C-shaped post assembly simulation results according to TB 42 crash criteria.

### *3.2. Results – U-shaped post*

The result of the impact of U-supporting post configuration is presented in Figure 8. The position of the impactor at 100, 250, and 400 ms can be observed in sequential figures. The plot of the Von Mises stress concentration is also presented.

In the comparison between this case and the C-post configuration case, it is found that the impactor's trajectory is approximately the same as the number of affected elements. The large deformation in the case of the C-post guardrail highlights the difference. The measured working width of the guardrail after impact is 667 mm.

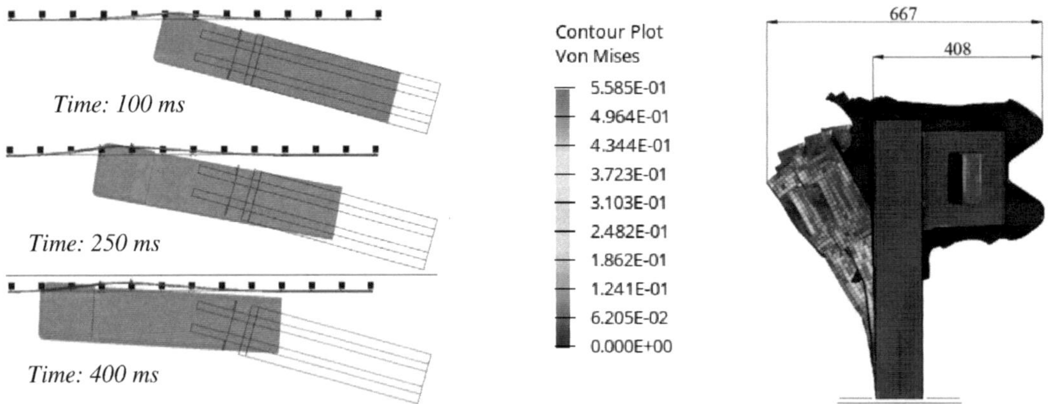


a). Sequential figures from guardrail with U-post b). Guardrail deformations (mm)

**Figure 8.** U-shaped post assembly simulation results according to TB 42 crash criteria.

### *3.3. Results – Sigma-shaped post*

Figure 9 presents the results of the SIGMA-supporting post configuration; the impactor is redirected on the roadside. After impact, the resulting working width is 584. The shape of the Sigma-post section offers a higher degree of rigidity in cases with C and U-post configurations. The deformation of the affected pillars is placed in connection with the base plate.

### *3.4. Results – HEA-shaped post*

The results of the impact of HEA-supporting post-guardrail configuration are presented in Figure 10. In this configuration, the guardrail's working width is 487 mm. This study evaluated the configuration with HEA-posts configuration as the most rigid in case of an accident.

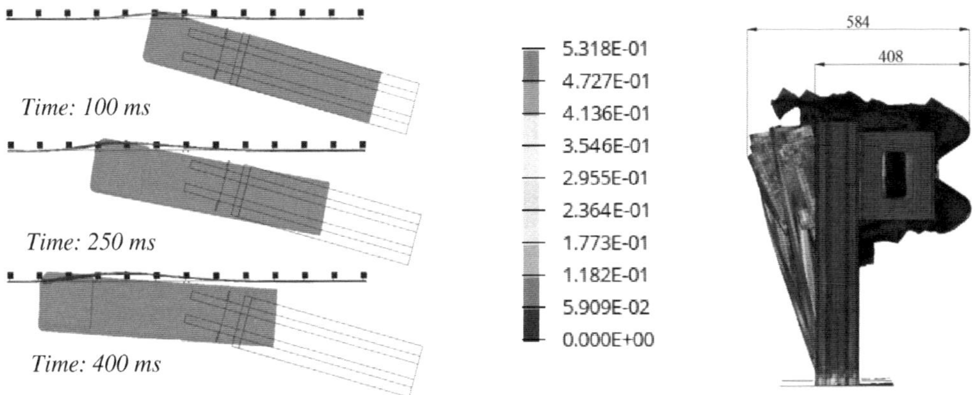


a). Sequential figures from guardrail with Sigma-post b). Guardrail deformations (mm)

**Figure 9.** Sigma-shaped post assembly simulation results according to TB 42 crash criteria.

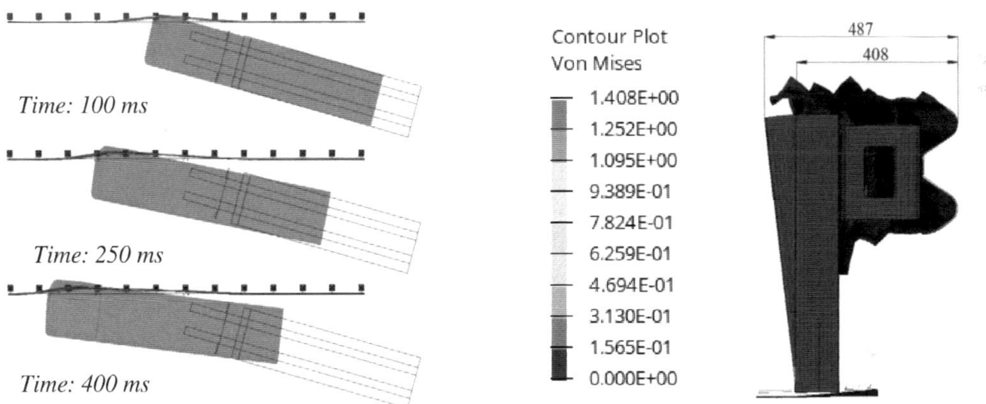


a). Sequential figures from guardrail with HEA-posts b). Guardrail deformations (mm)

**Figure 10.** HEA-shaped post assembly simulation results according to TB 42 crash criteria.

Table 5 presents the simulation's comparative results for all shapes of supporting posts. It can be observed that the HEA post configuration shows a low deformation in comparison to other configurations.

**Table 5.** Resulted working width.

| | U-shaped | C-shaped | Sigma-shaped | HEA -shaped |
|---|---|---|---|---|
| Initial width (m) | 0.408 | 0.408 | 0.408 | 0.408 |
| Deformed width (m) | 0.667 | 0.590 | 0.584 | 0.487 |

## 4. Conclusions

This research presents a comparative study of four different types of post configurations used in semi-heavy bridge W-beam guardrail systems and crash simulation using a numerical solver. The simulation cases are provided following EN 1317 standards, using TB 42 crash criteria.

This study is performed for economic reasons to determine the optimal type of supporting post used in the W-beam configuration. It is important to highlight that differences can occur in the validation of the virtual model by performing the physical experiment, which can be different by the lack of wheels and implicitly by the contact with the road, which influences the dynamics of the vehicle in the event of an impact. The supporting post area was the focus of the crash in this study, but the results may differ if the effect is obtained between the supporting posts. Also, the angle of impact or how the mass is placed in the vehicle can vary from reality. If the impact angle is smaller, the damage will be lower, and the vehicle will be redirected back onto the road more easily.

The results show that the U-shaped post configuration has a large working width compared to other structures, so using this system in areas without sidewalks is recommended. If it is necessary to have a sidewalk behind the guardrail, it is recommended to use the HEA-shaped configuration, which is more rigid. The Sigma and C-shaped structures present similar deformations.

**Author's ORCID ID**
M I Gheres: https://orcid.org/0000-0001-8194-5126
N Cordos: https://orcid.org/0000-0002-9877-6530
I A Todorut: https://orcid.org/0000-0003-3368-3701

**References**

[1] Teng T L, Liang C C, Hsu Y C, Shih C J and Tran T T 2016 *Impact Performance of W-beam Guardrail Supported by Different Shaped Posts,* International Journal of Mechanical Engineering and Applications. Vol. 4, No. 2, 2016, pp. 59-64, doi: 10.11648/j.ijmea.20160402.14
[2] Wolny R, Bruski D, Budzyński M, Pachocki L and Wilde K 2022 *Influence of a lighting column in the working width of a W-beam barrier on TB51 crash test*, Materials, 15(14), 4926, https://doi.org/10.3390/ma15144926
[3] Soltani M, Moghaddam T B, Karim M R and Sulong N R 2013 *The safety performance of guardrail systems: review and analysis of crash tests data*, International Journal of Crashworthiness, 18(5), 530-543, DOI: 10.1080/13588265.2013.815020
[4] Gheres M I and Scurtu I L 2022 *Crash testing and evaluation of W-beam guardrail using finite elements method,* In IOP Conference Series: Materials Science and Engineering (Vol. 1220, No. 1, p. 012049). IOP Publishing, doi:10.1088/1757-899X/1220/1/012049
[5] Scurtu L, Szabo I, Mariasiu F, Moldovanu D, Mihali L and Jurco A 2019 *Numerical analysis of the influence of mechanical stress on the battery pack's housing of an electric vehicle,* In IOP Conference Series: Materials Science and Engineering (Vol. 568, No. 1, p. 012054). IOP Publishing, doi:10.1088/1757-899X/568/1/012054
[6] Scurtu IL and Lupea I 2014 *Frontal Crash Simulation of a Chassis Frame*, vol 57-III, Acta Technica Napocensis-Series: Spplied Mathematics, Mechanics, and Engineering, pp 411-414
[7] Teng T L, Liang C C and Tran T T 2015 *Effect of various W-beam guardrail post spacings and rail heights on safety performance,* Advances in mechanical engineering, 7(11), 1687814015615544, https://doi.org/10.1177/1687814015615544
[8] Atahan A O 2018 *Development of a heavy containment level bridge rail for Istanbul*, Latin American Journal of Solids and Structures 15 https://doi.org/10.1590/1679-78254684
[9] de Albuquerque F D B and Awadalla D M 2020, *Roadside Fixed-Object Collisions, Barrier Performance, and Fatal Injuries in Single-Vehicle, Run-Off-Road Crashes*, Safety 6, no. 2: 27. https://doi.org/10.3390/safety6020027
[10] Li Z, Fang H, Fatoki J, Gutowski M and Wang Q 2021 *A numerical study of strong-post double-faced W-beam and Thrie-beam guardrails under impacts of vehicles of multiple size classes, Accident Analysis & Prevention*, 159, 106286. doi:10.1016/j.aap.2021.106286
[11] Atahan A O, Yücel A Ö and Erdem M M 2014 *Crash testing and evaluation of a new generation L1 containment level guardrail,* Engineering Failure Analysis, 38, 25–37. doi:10.1016/j.engfailanal.2014.01.00310.1016/j.engfailanal.2014.01.003
[12] Tomasch E, Sinz W, Hoschopf H, Gobald M, Steffan H, Nadler B and Schneider F 2011 Required length of guardrails before hazards. Accident Analysis & Prevention, 43(6), 2112–2120. doi:10.1016/j.aap.2011.05.034
[13] Radioss Theory Manual, Large Displacement Finite Element Analysis, 2021

[14] European Committee for Standardization, European Standard EN 1317-1, EN 1317-2, *Road Restraint Systems*, 2010

# Dynamic model for an independent suspension

**D Popa, C M Popa and S Marcu**

Automotive Department, University of Pitesti, Romania

E-mail: dinel_popa@yahoo.com

**Abstract**. The paper presents the dynamic model with 4 degrees of freedom of an independent suspension with MacPherson type coil springs, taking into account the anti-roll bar. Starting from the constructive scheme, a kinematic analysis necessary to obtain the dynamic model is carried out. The differential displacement equations are obtained starting from the expressions of kinetic and potential energy, by applying Lagrange's equations. Next, the expressions of the inertia, stiffness and damping matrices and, finally, the matrix expression of pitch-roll vibrations are presented. At the end of the paper, in the numerical application carried out with a calculation program in MATLAB, the influence of the values of the damping constants on the dynamic forces at the wheels is studied. The numerical results are presented in the form of diagrams.

## 1. Introduction

The dynamic and passing performance of a vehicle moving on unimproved roads or in an advanced state of wear is characterized, first of all, by the qualities of the suspension and, secondly, by the energy performance of the engine [1], [2].

An incorrect operation of the suspension together with the wheel guidance system will influence the handling, maneuverability and stability of the vehicle [3], [4].

The suspension analyzed in this paper is that of an independent suspension with wheels guided by a crank mechanism - oscillating slide, also known as the MacPherson suspension. This type of suspension, in which the shock absorber also plays the role of a pivot, has the advantage of an insignificant change in the gauge and the inclination of the wheel during cornering, the reduction of the unsuspended mass and a simple design [1], [4]. Such a suspension system, when heavily damped, increases vehicle handling but decreases comfort. When the suspension system is slightly damped, passenger comfort is high, but during sharp turns the vehicle may lose stability [5], [6].

A semi-active suspension system can achieve a compromise between comfort and stability, choosing appropriate values for the elasticity constant of the spring and the damping constant of the shock absorber [6], [7].

To establish the limits between which the values of the elastic and damping constants can vary, a dynamic model with 4 degrees of freedom will be presented. This model can be used for a control strategy of semi-active suspensions [8], [9].

## 2. The kinematic scheme of the suspension and the model used for the study of vibrations

Figure 1 shows the model of a front drive axle with suspension with helical springs, type MacPherson, in a certain position.

The used notations were: 1 - suspended mass, 2 - suspension spring, 3 - shock absorber, 4 - rim, 5 - knuckle, 6 - side shafts, 7 - lower control arm, 8 - anti-roll bar.

To simplify the calculations, the projection of the suspension mechanism in the plane $ZOY$ is considered. The adopted model is a model with 4 degrees of freedom (left wheel displacement, right wheel displacement, bounce and roll angle of suspended mass).

The used notations are: OZ – the symmetry axis of the suspension in the resting position; $C$ - the center of gravity of the suspended mass; O - the point of intersection between the axis $OZ$ and the line that joins the centers of the wheels, with the vehicle being in the resting position; $O_0$ - the point of intersection between the axis $OZ$ and the plane tangent to the inside of the rim, with the suspension being in the resting position; $O*$ - the point of intersection between the axis of symmetry $OZ$ and the ground; $A$, $A'$ - centers of the joints of the arms to the body; $A*$ - the point of intersection between the axis of symmetry $OZ$ and the line $AA'$; $B$, $B'$ - centers of the joints of the arms to the knuckles; $F$, $F'$ - the centers of the wheels; $D$, $D'$ - the attachment points of shock absorbers and springs to the knuckles ; $\widetilde{D}$ - the projection of the point $D$ on the line $OF$ ; $M$ , $M'$ - the points of attachment of the anti-roll bar on the suspension arms; $MNPRR'P'N'M'$ - points on the anti-roll bar; $G$ , $G'$ - the points of articulation of the anti-roll bar on the body; $\gamma$ - specific constructive angle.

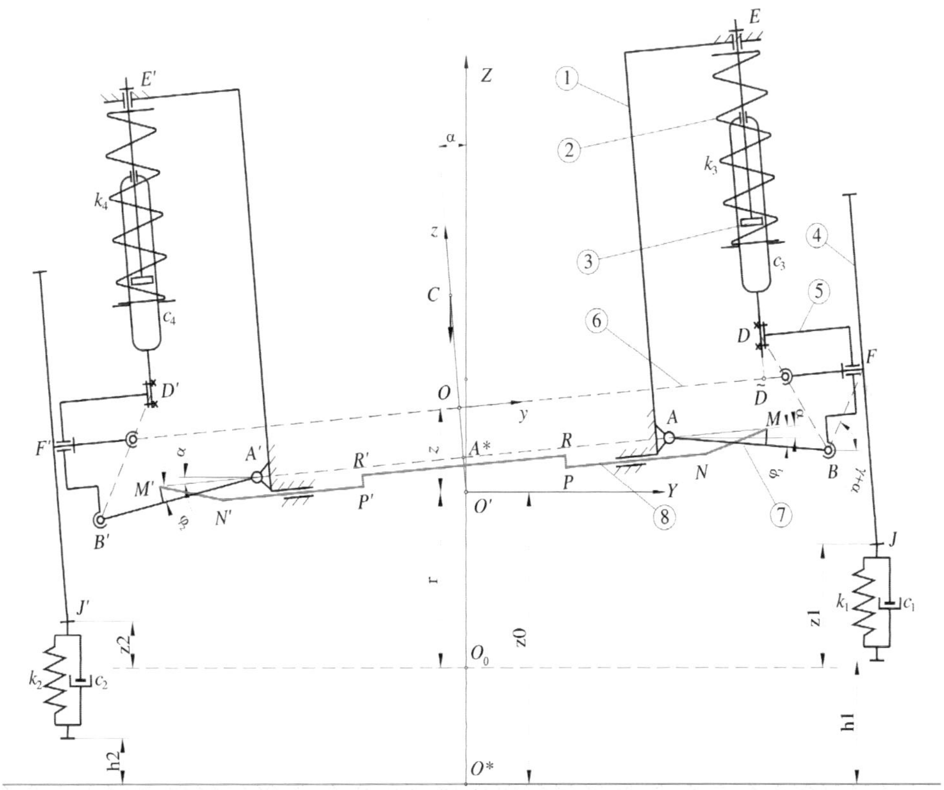


**Figure 1.** Suspension model in a certain position.

The following notations are used for the constant geometric parameters:

$$z_0 = OO^*,\ l_0 = O_0C,\ l_1 = OA^*,\ l_2 = A^*A = A^*A',\ l_3 = AB = A'B',\ l_4 = BF = B'F',$$
$$r = FJ = F'J' = O_0O',\ l_5 = D\widetilde{D} = D'\widetilde{D}',\ l_6 = AM = AM'.$$

## 3. The dynamic model

For the dynamic model, represented in figure 1, the next notations where used: $z$ - the displacement of the body relative to the equilibrium position on an horizontal road, $\alpha$ - the roll angle, $\varphi_1$, $\varphi_2$ - the rotation angles of the arms $AB$ and $A'B'$, relative to the body; $z_1$, $z_2$ - the displacements perpendicular to the ground of the points $J$, $J'$; $h_1$, $h_2$ - the unevenness of the rolling path, $k_1$, $k_2$ - the elastic constants of the tires, $c_1$, $c_2$ - the damping constants of the tires, $k_3$, $k_4$ - the elastic constants of the springs, $c_3$, $c_4$ - the damping constants of the suspension dampers, $k_5$ - the elastic constant of the anti-roll bar.

To obtain the expression of the angle $\varphi_1$, the projection on the axis $OZ$ of the vector equation is used:

$$\overline{O^*O'} + \overline{O'AA^*} + \overline{A^*A} + \overline{AB} + \overline{BF} + \overline{FJ} = \overline{O^*J} \tag{1}$$

To simplify the expressions, it is taken into account that the displacements $\alpha$, $z$ and $\varphi_1$ are small.

The expression is thus obtained:

$$z_0 + (z - l_1) + l_2\alpha - l_3(\phi_1 - \alpha) + l_4 \sin(\gamma + \alpha) - r = z_0 - r + z_1 \tag{2}$$

from which it results:

$$\phi_1 = \frac{z - z_1 + l_7\alpha + l_8}{l_3}, \tag{3}$$

where:

$$l_7 = l_2 + l_3 + l_4 \cos\gamma,\ l_8 = -l_1 + l_4 \sin\gamma. \tag{4}$$

To obtain the expression of the angle $\varphi_2$, the projection on the $OZ$ axis of the vector equation is used:

$$\overline{O^*O'} + \overline{O'A^*} + \overline{A^*A'} + \overline{A'B'} + \overline{B'F'} + \overline{F'J'} = \overline{O^*J'} \tag{5}$$

It is also taken into account that the displacements $\alpha$, $z$ and $\varphi_2$ are small and the following expression is obtained:

$$z_0 + (z - l_1) - l_2\alpha - l_3(\varphi_2 + \alpha) + l_4 \sin(\gamma - \alpha) - r = z_0 - r + z_2, \tag{6}$$

from which results:

$$\varphi_2 = \frac{z - z_2 - l_7\alpha + l_8}{l_3}. \tag{7}$$

## 4. Differential equations

The following notations are used: $m_0$ - the suspended mass of the car, $J_C$ - the moment of inertia of the suspended mass in relation to the longitudinal axis passing through the point $C$, $J_B$ - the moment of inertia of the suspension arm in relation to the axis of rotation of the body, $m_1$ - the equivalent mass for a wheel, hub and 2/3 of the side shaft [6], $m_2$ - two thirds of the mass of the suspension arm [6], $s_{01}$, $s_{02}$ - the static deformations of the tires, $s_{03}$, $s_{04}$ - the static deformations of the springs.

Lagrange's equations are used to establish the differential equations of displacement. We will first need to perform a kinematic analysis. The reference system $O'XYZ$ is chosen.

The coordinates of the points $C$, $F$, $F'$, $D$ and $D'$ are given by the expressions:

$$Y_C = -l_0 \sin\alpha,\ Z_C = r + z + l_0 \cos\alpha, \tag{8}$$

$$Y_F = OF \cos\alpha,\ Z_F = z_1 + r \cos\alpha, \tag{9}$$

$$Y_{F'} = -OF' \cos\alpha,\ Z_{F'} = z_2 + r \cos\alpha, \tag{10}$$

$$Y_D = OD^* \cos\alpha + DD^* \sin\alpha,\ Z_D = r + z + OD^* \sin\alpha + D^*D \cos\alpha, \tag{11}$$

$$Y_{D'} = -OD^* \cos\alpha + DD^* \sin\alpha,\ Z_{D'} = r + z - OD^* \sin\alpha + D^*D\cos\alpha. \quad (12)$$

The expression for kinetic energy will be rewritten as:

$$E_C = \frac{1}{2}m_0 v_C^2 + \frac{1}{2}J_C\dot{\alpha}^2 + \frac{1}{2}m_1(v_F^2 + v_{F'}^2) + \frac{1}{2}m_2(v_G^2 + v_{G'}^2) + J_B\dot{\alpha}^2. \quad (13)$$

To obtain the speeds, relations (8) - (12) are derived in relation to time.
Neglecting the non-linear terms, the next expressions are obtained:

$$v_C^2 = \dot{z}^2 + l_O^2\dot{\alpha}^2;\ v_F^2 = \dot{z}_1^2;\ v_F^2 = \dot{z}_1^2,\ v_{F'}^2 = \dot{z}_2^2,\ v_G^2 = (\dot{z}_1 - l_6\dot{\alpha})^2,\ v_{G'}^2 = (\dot{z}_2 - l_6\dot{\alpha})^2. \quad (14)$$

which, replaced in the kinetic energy expression, lead to the expression:

$$E_C = \frac{1}{2}m_0(\dot{z}^2 + l_0^2\dot{\alpha}^2) + \frac{1}{2}(J_C + 2J_B)\dot{\alpha}^2 + \frac{1}{2}m_1(\dot{z}_1^2 + \dot{z}_2^2) + \frac{1}{2}m_2[(\dot{z}_1 - l_6\dot{\alpha})^2 + (\dot{z}_2 + l_6\dot{\alpha})^2]. \quad (15)$$

The potential energy has the expression:

$$V = m_0 g(r + z + l_0\cos\alpha) + (m_1 + m_2)(z_1 + 2r\cos\alpha + z_2)g + \frac{1}{2}\left[k_1(z_1 - h_1 - s_{01})^2 + k_2(z_2 - h_2 - s_{02})^2\right] +$$
$$+ \frac{1}{2}\left[k_3(z - z_1 + l_7\alpha + l_8 - s_{03})^2 + k_4(z - z_2 - l_7\alpha + l_8 - s_{04})^2\right] + \frac{1}{2}k_5\left(\frac{l_6}{l_3}\right)^2(z_2 - z_1 + 2l_7\alpha)^2. \quad (16)$$

In the previous relationship $s_{01}$, $s_{02}$, $s_{03}$ and $s_{04}$ indicate the static deflections.
The static deflections check the system of equations

$$\frac{\partial V}{\partial z} = 0;\ \frac{\partial V}{\partial \alpha} = 0;\ \frac{\partial V}{\partial z_1} = 0;\ \frac{\partial V}{\partial z_2} = 0, \quad (17)$$

in the case where:

$$z = 0;\ \alpha = 0;\ z_1 = 0;\ z_2 = 0. \quad (18)$$

To obtain the system of differential equations, we apply Lagrange's equations:

$$\begin{cases} \dfrac{\mathrm{d}}{\mathrm{d}t}\left(\dfrac{\partial E_C}{\partial \dot{q}_k}\right) - \dfrac{\partial E_C}{\partial q_k} + \dfrac{\partial V}{\partial q_k} = 0 \\ k = 1, 4 \end{cases} \quad (19)$$

where the generalized coordinates $q_k$ are:

$$q_1 = z_1,\ q_2 = z_2;\ q_3 = z\ q_4 = \alpha. \quad (20)$$

The static deflections resulting from equations (17) are:

$$s_{01} = \left(m_1 + m_2 + \frac{m_0}{2}\right)\frac{g}{k_1},\ s_{02} = \left(m_1 + m_2 + \frac{m_0}{2}\right)\frac{g}{k_2},\ s_{03} = \frac{m_0 g}{2k_3} + l_8,\ s_{04} = \frac{m_0 g}{2k_4} + l_8. \quad (21)$$

After performing the calculations, the expression of the inertia matrix is obtained:

$$[M] = \begin{bmatrix} (m_1 + m_2) & 0 & 0 & -m_2 l_6 \\ 0 & (m_1 + m_2) & 0 & m_2 l_6 \\ 0 & 0 & m_0 & 0 \\ -m_2 l_6 & m_2 l_6 & 0 & (m_0 l_0^2 + J_C + 2J_B) \end{bmatrix} \quad (22)$$

and the expression of the stiffness matrix:

$$[K]=\begin{bmatrix} k_1+k_3+k_5\frac{l_6^2}{l_3^2} & -k_5\frac{l_6^2}{l_3^2} & -k_3 & -\left(k_3+2k_5\frac{l_6^2}{l_3^2}\right)l_7 \\ -k_5\frac{l_6^2}{l_3^2} & k_2+k_4+k_5\frac{l_6^2}{l_3^2} & -k_4 & \left(k_4+2k_5\frac{l_6^2}{l_3^2}\right)l_7 \\ -k_3 & -k_4 & k_3+k_4 & (k_3-k_4)l_7 \\ -\left(k_3+2k_5\frac{l_6^2}{l_3^2}\right)l_7 & \left(k_4+2k_5\frac{l_6^2}{l_3^2}\right)l_7 & (k_3-k_4)l_7 & k_{44} \end{bmatrix}, \quad (23)$$

where the term $k_{44}$ has the expression:

$$k_{44}=-g[m_0l_0+2r(m_1+m_2)]+\left(k_3+k_4+4k_5\frac{l_6^2}{l_3^2}\right)l_7^2. \quad (24)$$

To obtain the damping matrix, the constants $k_5$ and $g$ are replaced by zero in the stiffness matrix and $k_1\to c_1$, $k_2\to c_2$, $k_3\to c_3$ and $k_4\to c_4$ . And so is obtained:

$$[C]=\begin{bmatrix} c_1+c_3 & 0 & -c_3 & -c_3l_7 \\ 0 & c_2+c_4 & -c_4 & c_4l_7 \\ -c_3 & -c_4 & c_3+c_4 & (c_3-c_4)l_7 \\ -c_3l_7 & c_4l_7 & (c_3-c_4)l_7 & (c_3+c_4)l_7^2 \end{bmatrix}. \quad (25)$$

The excitation vector $\{F\}$ is deduced from the equations (19):

$$\{F\}=(k_1h_1+c_1\dot{h}_1 \quad k_1h_2+c_1\dot{h}_2 \quad 0 \quad 0)^T. \quad (26)$$

According to relations (20), the displacement vector $\{Z\}$ has the expression:

$$\{Z\}=(z_1 \quad z_2 \quad z \quad \alpha)^T. \quad (27)$$

The system of differential equations of displacement, written in matrix form is:

$$[M]\{\ddot{Z}\}+[C]\{\dot{Z}\}+{+}[K]\{Z\}=\{F\} \quad (28)$$

Equation (28) obtained is the matrix differential equation of bounce-roll vibrations.

## 5. Numerical application

The numerical application is made on the data of a hypothetical medium class vehicle for which it was adopted: $m_0=515$ kg suspended mass (unsuspended mass is subtracted from the mass of the front axle), $m_1=31$ kg - unsuspended mass (11.25 kg - wheel, 6.2 kg hub and knuckle, 2.85 kg - spring, 3.57 kg – shock absorber, 3.69 kg – break disk, 2/3×5.21 kg - 2/3 of the side shaft mass), $m_2=1.36$ kg - 2/3 from the mass of the (lower) arm (2/3×2.05 kg), $J_C=287\text{kg}\cdot\text{m}^2$ - the moment of inertia of the suspended part, $J_B=0.065\text{kg}\cdot\text{m}^2$ - the moment of inertia of the lower arm, $z_0=0.311$ m, $l_0=0.35$ m, $r=0.190$ m, $l_1=0.088$ m, $l_2=0.397$ m, $l_3=0.311$ m, $l_4=0.159$ m, $l_5=0.077$ mm, $l_6=0.188$ m, $l_7=0.787$ m, $l_8=0{,}050$ m, $\gamma=60°$, $k_1=120000$ N/m , $k_2=120000$ N/m - the elastic constants of the tires, $k_3=22000$ N/m , $k_4=22000$ N/m - the elastic constants of the springs, $k_5=10400$ N/m - the elastic constant of the anti-roll bar, $c_1=180$ Ns/m , $c_2=180$ Ns/m - the damping constants of the tires, $c_3=1000\div2000$ Ns/m , $c_4=1000\div2000$ Ns/m - the average damping constants of the shock absorbers.

To obtain numerical results, a calculation program is drawn up in MATLAB based on the previous relations. The following numerical results are obtained:

- the static deflections for tires: $s_{01}=0.0237$ m, $s_{02}=0.0237$ m;
- the static deflections for springs: $s_{03}=0.1648$ m, $s_{04}=0.1648$ m;

- the static forces at the wheels: $F_{s1} = 2843.5\text{N}$, $F_{s2} = 2843.5$ N;
- the static forces at the springs: $F_{s3} = 3626.1$ N, $F_{s4} = 3626.1$ N;
- the natural frequencies: $f_1 = 1.3503$ Hz, $f_2 = 1.4090$ Hz, $f_3 = 10.5590$ Hz, $f_4 = 10.8454$ Hz.

Next, the behavior of the structure, in the case of harmonic and random excitations, is studied [6].

In the case of harmonic excitations, sinusoidal unevenness of the roadway was considered. The car is moving at a constant speed of $v = 30$ m/s. 3 values of the damping constants are considered:

$$c_3 = c_4 = 1000\,\text{Ns/m},\; c_3 = c_4 = 1500\,\text{Ns/m},\; c_3 = c_4 = 2000\,\text{Ns/m}.$$

The parameter $\omega$ is given values in the range $[0 \div 155]\text{s}^{-1}$, with constant step $\Delta\omega = 1$ [6].

The displacement amplification factors are:

$$\xi_1 = \frac{z_1^0}{h_0},\; \xi_2 = \frac{z_2^0}{h_0},\; \xi_3 = \frac{z^0}{h_0},\; \xi_4 = \frac{\alpha^0}{h_0}. \tag{29}$$

Due to the symmetry of the curves of the amplification factors, $\xi_1$, $\xi_2$ are identical, and the amplification factor $\xi_4$ is practically zero. In figure 2 and figure 3 the curves of the amplification factors $\xi_1$, $\xi_2$ and respectively $\xi_4$ are represented, for the 3 values of the damping constants

The maxima of the amplification factors $\xi_1$ and $\xi_2$ are around their natural frequencies. If the first maximum is not influenced by the value of the damping constant of the damper, the second maximum decreases with the increase of the value of the damping constant of the damper: from 1.81 at the frequency of 9.55 Hz for $c = 1000$ Ns/m to the value of 1.01 at the frequency of 7.95 Hz for $c = 2000$ Ns/m.

Also, the damping factor $\xi_3$ is slightly influenced by the values of the damping constants.

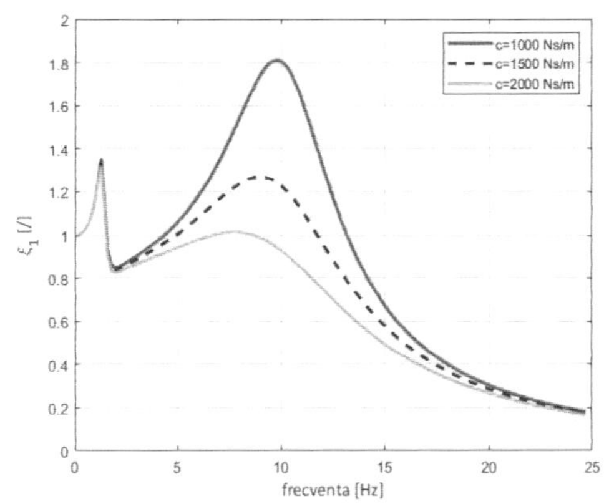


**Figure 2.** The amplification factor $\xi_1$, $\xi_2$.

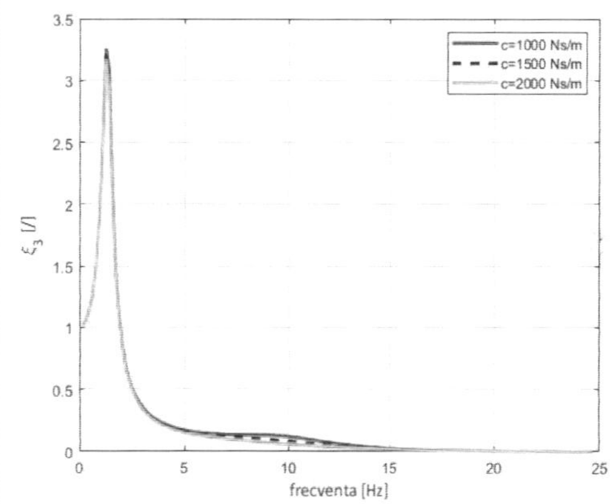


**Figure 3.** The amplification factor $\xi_3$.

Figure 4 and figure 5 present the amplification factors of the dynamic normal forces at the wheels, given by the relations:

$$F_{1D} = -k_1(z_1 - h_1) - c_1(\dot{z}_1 - \dot{h}_1),\; F_{2D} = -k_2(z_2 - h_2) - c_2(\dot{z}_2 - \dot{h}_2). \tag{30}$$

The graphics have the same allure due to symmetry.

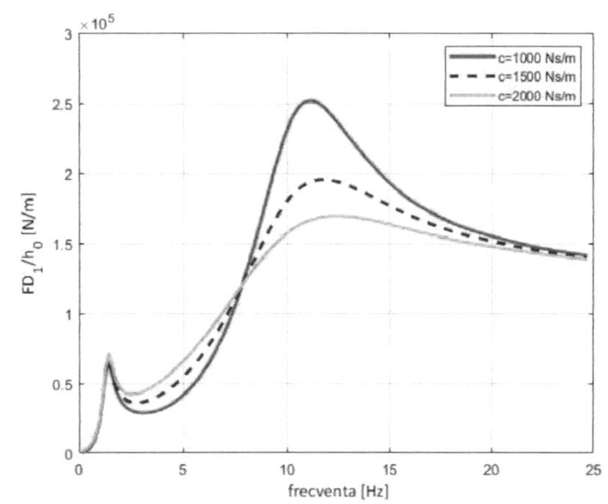


**Figure 4.** The dynamic force amplification factor at the right wheel.

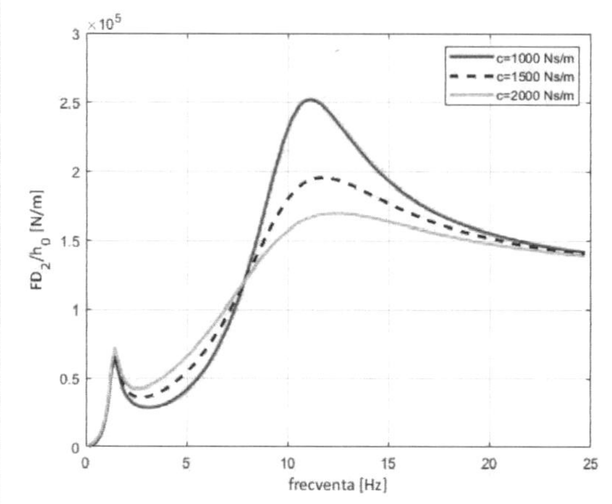


**Figure 5.** The dynamic force amplification factor at the left wheel.

The allure of the amplification factors of the dynamic normal forces at the wheels is influenced by the value of the damping constant of the damper, along with the increase of the frequency value.

A first maximum is at the frequency $f_1 \cong f_2$, a maximum that is very little influenced by the value of the damping constant of the damper.

The amplification factor curve continues to have increasing values towards the frequencies $f_3$ and $f_4$, for $c_3 = c_4 = 1000$ Ns/m the maximum is $F_{1D} = F_{2D} = 251667$ N/m (at $f = 11.14$ Hz frequency), and for $c_3 = c_4 = 2000$ Ns/m the maximum is $F_{1D} = F_{2D} = 168021$ N/m (at $f = 11.46$ Hz frequency).

As the static force has the value of $F_{s1} = F_{s2} = 2843.5$ N and therefore for values greater than $\frac{2843.5}{251667} = 0.0113$ m of the oscillation amplitude of the ground level, i.e. approximately 11 mm, separations from the ground occur at the wheels for $c_3 = c_4 = 1000$ Ns/m, and for $c_3 = c_4 = 2000$ Ns/m for $h_0$ greater than $\frac{2843.5}{168021} = 0.0169$ m, i.e. approximately 17 mm, separations occur at the wheels.

Next, the effective values of the displacements $z_1$, $z_2$ and the effective force at the wheels are determined, depending on the 3 values of the damping constant $c_3 = c_4$. The obtained values are those in table 1.

**Table 1.** The effective values of the displacements $z_1$, $z_2$ and of the effective forces at the wheels.

| | $z_{1ef}$ [mm] | $z_{2ef}$ [mm] | $F_{1Def}$ [N] | $F_{2Def}$ [N] |
|---|---|---|---|---|
| $c =$ **1000 Ns/m** | 25.1 | 25.9 | 1879.1 | 1902.0 |
| $c =$ **1500 Ns/m** | 23.6 | 24.3 | 1740.2 | 1763.9 |
| $c =$ **2000 Ns/m** | 22.7 | 23.2 | 1705.6 | 1734.9 |

There is a small difference between the displacements of the two wheels, a difference due to the anti-roll bar. All these values were obtained for a vehicle speed of $v = 30$ m/s.

Comparing the dynamic values obtained, with the static ones ( $F_{s1} = F_{s2} = 2843.5$ N), it is observed that, on this type of paved road and at this speed of movement ( $v = 30$ m/s), the wheels do not come off the roadway.

## 6. Conclusions

The presented model has 4 degrees of freedom and is closer to the case of a real vehicle because it takes into account the anti-roll bar. Compared to the models in which the differential equations of motion are obtained by applying the momentum theorem and the theorem of kinetic moment in relation to the center of gravity, in this paper the differential equations of motion were obtained using Lagrange's equations.

Considering the 4 generalized coordinates (left wheel displacement, right wheel displacement, bounce and roll angle of suspended mass) and applying Lagrange's equations, the 4 differential equations of motion were obtained. To solve the system of differential equations of motion, the differential equations were put in matrix form.

In the presented example, equal and in-phase excitations at the two wheels were considered. Thus, the bouncing vibrations of the center of gravity were highlighted. In this way it was possible to check the accuracy of the presented model in comparison with the model with 4 degrees of freedom obtained with the theorem of impulse and kinetic moment in relation to the center of gravity

The model can be extended for 7 degrees of freedom, so taking also into account the movements of the rear axle. In this way, the roll movement and the influence of the anti-roll bar can be correctly appreciated.

Analyzing the form of the matrices of inertia, stiffness and damping obtained for this model, a strong similarity is observed with the forms of the matrices obtained for the model with 4 degrees of freedom with independent suspensions, model without anti-roll bar [6].

The dynamic model obtained can be used for a control of a semi-active suspension and is useful when the excitations of the wheels on the same axle are not equal as in the analyzed example.

## References

[1] Crolla D 2009 *Automotive Engineering: Powertrain, Chassis System and Vehicle Body* (Elsevier)

[2] Johanson C and Stockel M 2021 *Automotive Suspension and Steering Systems* 5th Edition Online Textbook (The Goodheart-Willcox Company)

[3] Phalke T and Mitra A C Design And Analysis Of Vehicle Suspension System *International Engineering Research Journal* p 165-172

[4] Untaru M, Poțincu G, Tabacu I, Stoicescu A and Peres G 1981 *Dinamica autovehiculelor pe roți* (Editura Didactică și Pedagogică, București)

[5] Gervasi G 2010 *Variable Stiffness Mechanical Leaf Spring Suspension and A Method For Adjusting Such Stiffness* EP1541392-B1 Patent

[6] Pandrea N, Popa D and Parlac S 2001 *Modele Pentru Studiul Vibratiilor Automobilelor* (Tiparg, Pitești)

[7] Balamurugan L and Jancirani J 2012 An Investigation On Semi-Active Suspension Damper And Control Strategies For Vehicle Ride Comfort And Road Holding *Proc. Inst. Mech. Eng. Part J. Syst. Control Eng*

[8] Guglielmino E, Sireteanu T, Stammers C, Ghita G, Giuclea M and Casey T 2008 *Semi-active Suspension Control* (Springer-Verlag London Limited)

[9] Giua A, Seatzu C and Usai G 2011 Semiactice Suspension Design with an Optimal Gain Switching Target *Vehicle System Dynamics - International Journal of Vehicle Mechanics and Mobility*

# Dynamic system for the study of bouncing vibrations of a car with active suspension model for an independent suspension

**D Popa, C M Popa and S Marcu**
Automotive Department, University of Pitesti, Romania
E-mail: dinel_popa@yahoo.com

**Abstract**. At the beginning of the paper, the quantities that characterize a dynamic system are presented. Based on them, the state equations of the dynamic system are obtained, a system that models an active suspension. The types of used controllers, their block diagrams and the influence of the adjustment parameters on the process are reviewed. Next, numerical applications are made for the study of vibrations on models with 2 degrees of freedom, for an active suspension equipped with PID and LQR controllers. These controllers, as well as the simulation of the two systems, passive and active, were modelled and simulated using MATLAB/Simulink software. The numerical results obtained after the simulation are presented in the form of diagrams. In the end of the paper, it is concluded that the designed active suspension system can improve the ride quality by minimizing the displacement of the suspended mass.

## 1. Introduction

The active suspension is developed from the passive suspension which, in addition to the components of the passive suspension, has a force actuator, respectively a device that controls the elastic and damping characteristics [1], [2], [8], [9]. It, through a controlled supply of energy, quickly adjusts the elastic and damping characteristics to the specified suspension mode: normal, comfortable or sporty [2], [3].

To do this, the necessary information about the current driving/use conditions of the vehicle is first collected, using various sensors that collect data such as the type and condition of the road surface, driving parameters, driving style and other data [4].

The data provided by the sensors goes to the vehicle's electronic control unit, where the processed information - depending on the control algorithms - is transmitted to the active components.

The complexity of the system implies a wide range of control algorithms, and most of the control methods are based on the exact knowledge of the suspension system, thus, the requirement of sensors increases and also increases the cost of the suspension system [5].

The calculation of the command of an active suspension requires the knowledge of dynamic systems theory notions. Therefore, in the following, the theoretical notions are presented and applications are made on systems with two degrees of freedom.

The dynamic system consists of a set of elements that communicate with each other and the physical quantities that help to transmit information in technical systems are called signals.

The control force is one of the defining characteristics of semi-active and active suspensions. It is responsible for controlling the elastic and damping characteristics of the suspension, which means that it has a significant impact on the behavior of the vehicle while driving. The purpose of this force is to minimize oscillations and displacements of the sprung mass, which contributes to improve the comfort and stability of the vehicle. In the case of semi-active suspensions, the controlled force is generated by adjusting the damping characteristics of the dampers. These can be achieved by changing the viscosity of the fluid or by adjusting the size of the holes in the damper. Modern semi-active suspensions use technologies such as magnetorheological dampers, electrorheological dampers and hydraulic dampers to achieve rapid adjustment of damping characteristics.

Active suspensions use an actuator to provide external compensating force. These systems are able to adapt at road conditions changing and can satisfy the requirements of comfort and stability while driving. The controlled force in active suspensions is generated by actuators acting directly on or in parallel with the suspension components. These actuators can be of various types, such as pneumatic actuators, hydraulic actuators, and magnetic or electromagnetic actuators. The control force, managed through appropriate controllers, is essential to achieve optimal performance of the semi-active and active suspension system [10], [11].

## 2. The characteristics of a dynamic system

A dynamic system is characterized by the fact that the output $y(t)$ depends on the entire internal evolution of the system under the influence of the input $u(t)$ (figure 1).

A linear continuous system has the following form [2]:

$$\begin{cases} \dot{\boldsymbol{x}} = \boldsymbol{A}\boldsymbol{x} + \boldsymbol{B}\boldsymbol{u} \\ \boldsymbol{y} = \boldsymbol{C}\boldsymbol{x} + \boldsymbol{D}\boldsymbol{u} \end{cases} \tag{1}$$

where, the first equation is the state equation and the second is the output vector equation.

The used notations are: $\boldsymbol{A} = (a_{ij})_{\substack{1\le i\le n \\ 1\le j\le n}}$ is the state parameter matrix, $\boldsymbol{x} = (x_1 \quad x_2 \quad \cdots \quad x_n)$ is the state vector, $\boldsymbol{B} = (b_{ij})_{\substack{1\le i\le n \\ 1\le j\le m}}$ is the matrix of input parameters, $\boldsymbol{u} = (u_1 \quad u_2 \quad \cdots \quad u_m)$ is the input vector, $\boldsymbol{y} = (y_1 \quad y_2 \quad \cdots \quad y_p)$ is the output vector, $\boldsymbol{C} = (c_{ij})_{\substack{1\le i\le p \\ 1\le j\le n}}$ is the matrix of output parameters, $\boldsymbol{D} = (d_{ij})_{\substack{1\le i\le n \\ 1\le j\le k}}$ is the matrix of transmission parameters [2].

The structı
can be represe

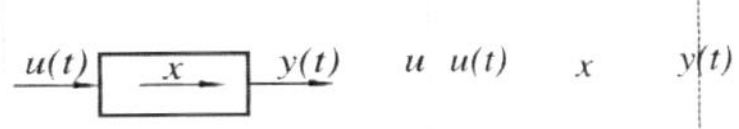


Figure 1. Representation of a dynamic system.

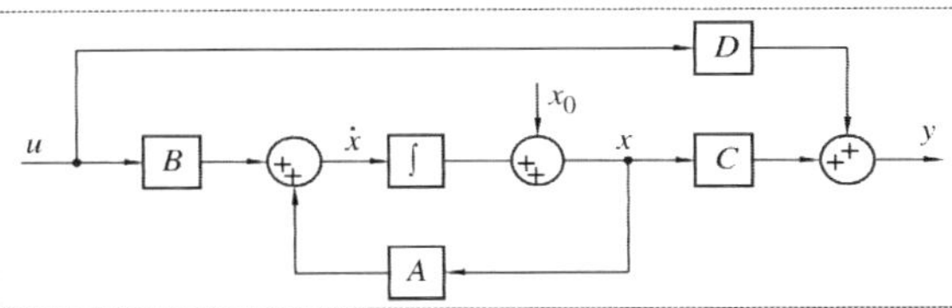


**Figure 2.** Block diagram of a dynamic system.

Stability is an important characteristic of dynamic systems, being one of the conditions of existence and functionality. A dynamic system is unstable if it cannot achieve the purpose for which it was designed (that of being stable).

In the field of system stability, two notions are used: the notion of internal stability (referring to the state of the system) and the notion of external stability (referring to the output of the system).

A system is internally strictly stable if the state of the system evolves in a free regime towards the origin:

$$\lim_{t\to\infty} \boldsymbol{x} = 0 \tag{2}$$

State controllability is the property of a system in which the defined input $u(t)$ transfers the state of the system $x(t)$ from the initial state $x_{initial}$ to the final state $x_{final}$. The controllability study is based on the controllability matrix:

$$\boldsymbol{\Omega}_C = [\boldsymbol{B} \quad \boldsymbol{AB} \quad \boldsymbol{A}^2\boldsymbol{B} \quad \dots \quad \boldsymbol{A}^{n-1}\boldsymbol{B}] \tag{3}$$

Observability is studied using the observability matrix $\boldsymbol{\Omega}_O$:

$$\boldsymbol{\Omega}_O = [\boldsymbol{C} \quad \boldsymbol{CA} \quad \boldsymbol{CA}^2 \quad \dots \quad \boldsymbol{CA}^{n-1}]^T \tag{4}$$

## 3. Controllers

The Linear Quadratic Regulator (LQR) is a variant of regulation in which the state equation of the model is linear, the performance index is quadratic and the initial conditions are known [2]. LQR is an optimal control problem for linear systems. For a dynamic system of the form:

$$\begin{cases} \dot{\boldsymbol{x}}(t) = \boldsymbol{Ax}(t) + \boldsymbol{Bu}(t) \\ \boldsymbol{x}(0) = \boldsymbol{x}_0 \end{cases} \tag{5}$$

a controller $\boldsymbol{u}(t)$ must be found in order to minimize the performance index:

$$J_{LQR} = \int_0^{\infty} \boldsymbol{x}(t)^T \boldsymbol{Qx}(t) + \boldsymbol{u}(t)^T R\boldsymbol{u}(t)dt \tag{6}$$

The first term on the right side of relation (6) represents the error, and the second term represents the energy consumption.

The block diagram of an LQR controller is the one in figure 3, where the matrix K has the role of minimizing the performance index, which describes the cost function of the system. Thus, to find an optimal control law for the dynamic system, the reference signal will have to be $r(t) = 0$

rgy
ight

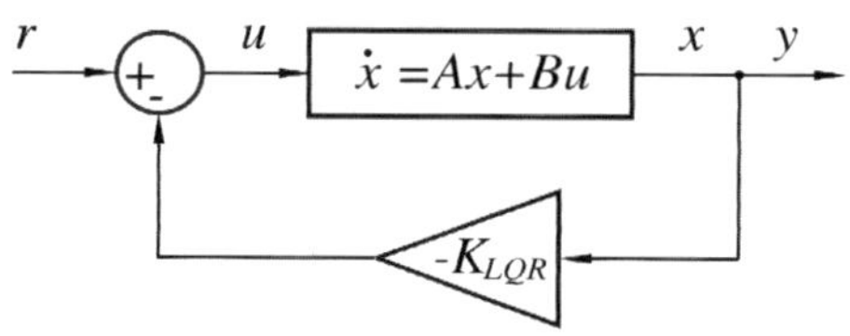


**Figure 3**. Block diagram of an LQR controller.

The PID (Proportional Integral Derivative) control method, a method proposed since the 1940s, is still used today in the control of industrial processes in most industrial procesess.

PID control is combined with sequential functions, logic and functional blocks for the construction of automated systems [7], [2].

The optimal performance of a controller can only be achieved if its parameters are chosen correctly.

The PID controller has in its composition three elements P, I and D which are summed to calculate the control signal. The controller is characterized by the equation:

$$u(t) = k_p e(t) + k_i \int_0^t e(\alpha)d\alpha + k_d \frac{d}{dt}e(t) \tag{7}$$

where: $k_p$ - is the tuning parameter, **P**roportional, $k_i$ - is the tuning parameter, **I**ntegrated, $k_d$ - is the tuning parameter, **D**erived and $e$ - is the error (the process variable). The block diagram of a PID controller is shown in figure 4.

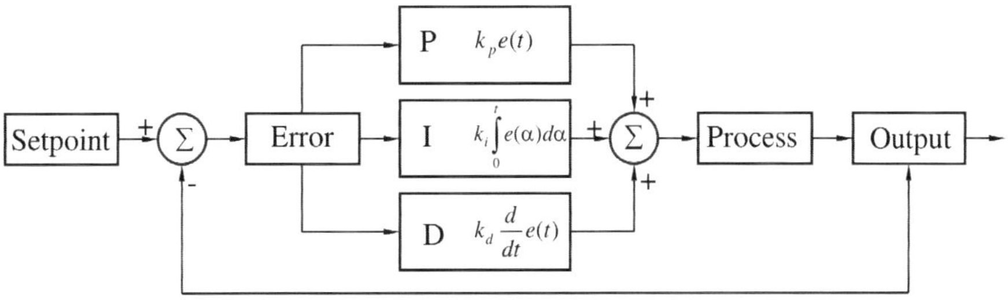


**Figure 4.** Block diagram of a PID controller.

To obtain optimal process performance, determining the PID control parameters is obviously a crucial part of closed-loop control systems.

Various methods have been developed to adjust these parameters. Regardless of the method used, it is necessary to monitor the key elements of the controller: load disturbances, sensor sensitivity and reference signals. Methods used to adjust a PID controller are: Ziegler-Nichols, Cohen-Coon and Software tools. The best known methods are those developed by Ziegler and Nichols. The two methods proposed by Ziegler and Nichols help to adjust P, PI and PID controllers [5].

These were presented in 1942, but are still used today as a starting point for adjusting controllers [6]. Next, based on these notions, a numerical application on a model with 2 degrees of freedom will be created.

## 4. Numerical results on models with 2 degrees of freedom

### *4.1. The mechanical model for the study of bounce vibrations*

In figure 5, the active suspension example is presented for the study of vibrations on a model with two degrees of freedom, also called a quarter car. The following notations were used: $m_1$ - the unsprung mass, $m_2$ - the suspended mass, $k_1$ - the elastic constant of the tire, $k_2$ - the elastic constant of the spring, $c_1$ - the damping constant of the tire, $c_2$ - the damping constant of the shock absorber, $h$ - the unevenness of the running path, $z_1$ - the displacement of the mass $m_1$, $z_2$ - the displacement of the mass $m_2$, $F_C$ - the force controlled by the controller.

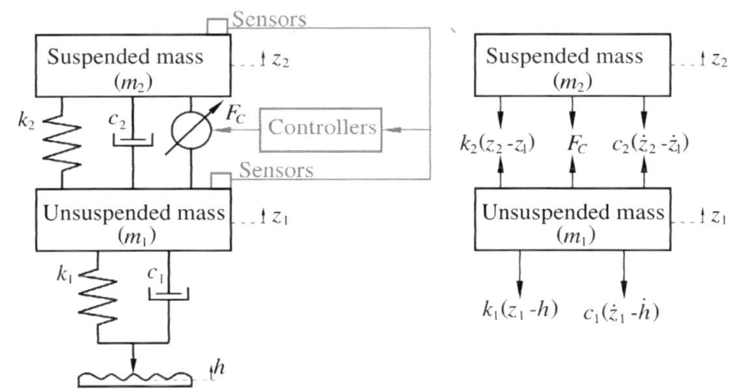


**Figure 5.** Mechanical model for the study of bouncing vibrations.

From the dynamic balance equations, the differential equations of motion are obtained:

$$\begin{gathered} m_1\ddot{z}_1 + (c_1 + c_2)\dot{z}_1 + (k_1 + k_2)z_1 - c_2\dot{z}_2 - k_2 z_2 = k_1 h + c_1\dot{h} - F_C \\ m_2\ddot{z}_2 - c_2\dot{z}_1 - k_2 z_1 + c_2\dot{z}_2 + k_2 z_2 = F_C. \end{gathered} \tag{8}$$

The numerical application will be made for a car for which the following data is adopted: $m_1 = 32{,}5\text{kg}$, $m_2 = 345\text{kg}$, $k_1 = 120000\, N/m$, $k_2 = 27900\, N/m$, $c_1 = 6000\,\text{Ns}/m$, $c_2 = 1500\,\text{Ns}/m$.

The purpose of the application is to compare the performances of an active system, by reducing the oscillations of the suspended mass, in which the damping force is controlled by two controllers: PID and LQR.

The main stages in determining the performance of an active system are the following: the representation of the state-space model (figure 2), the model based on equations (8), the determination of the properties of the linear system, the design of the two PID and LQR controllers.

*4.2. Presenting the model space-state*

To implement a controller, the state-space model for representing differential equations can be used. The general form of the model is given by equation (1).

Using the notations in figure 5, we can write the state vector $\boldsymbol{x}$, the output vector $\boldsymbol{y}$ and the input vector $\boldsymbol{u}$ in the following form:

$$\boldsymbol{x} = \begin{bmatrix} z_2 - z_1 \\ \dot{z}_2 \\ z_1 - h \\ \dot{z}_1 \end{bmatrix}, \boldsymbol{y} = \begin{bmatrix} z_2 - z_1 \\ \ddot{z}_2 \end{bmatrix}, \boldsymbol{u} = \begin{bmatrix} \dot{h} \\ F_C \end{bmatrix}. \tag{9}$$

The components of the state vector $\boldsymbol{x}$ are: $z_2 - z_1$ the suspension stroke, $\dot{z}_2$ the speed of the suspended mass, $z_1 - h$ the deflection of the tire, $\dot{z}_1$ the speed of the unsprung mass.

The components of the input vector $\boldsymbol{u}$ are: $\dot{h}$ the derivative of the road height, $F_C$ the force of the actuator, and the components of the output vector $\boldsymbol{y}$ are: $z_2 - z_1$ the stroke of the suspension, $\ddot{z}_2$ the acceleration of the suspended mass. We assume that the tire and the shock absorber have linear characteristics and based on the differential equations of displacement we can calculate the matrices $\boldsymbol{A}$, $\boldsymbol{B}$, $\boldsymbol{C}$ and $\boldsymbol{D}$:

$$\boldsymbol{A} = \begin{bmatrix} 0 & 1 & 0 & -1 \\ -\frac{k_2}{m_2} & -\frac{c_2}{m_2} & 0 & \frac{c_2}{m_2} \\ 0 & 0 & 1 & 0 \\ \frac{k_2}{m_1} & \frac{c_2}{m_1} & -\frac{k_1}{m_1} & -\frac{c_2+c_1}{m_1} \end{bmatrix}, \boldsymbol{B} = \begin{bmatrix} 0 & 0 \\ 0 & \frac{1}{m_2} \\ -1 & 0 \\ \frac{c_1}{m_1} & -\frac{1}{m_1} \end{bmatrix}, \tag{10}$$

$$\boldsymbol{C} = \begin{bmatrix} 1 & 0 & 0 & 0 \\ -\frac{k_2}{m_2} & -\frac{c_2}{m_2} & 0 & \frac{c_2}{m_2} \end{bmatrix}, \boldsymbol{D} = \begin{bmatrix} 0 & 0 \\ 0 & \frac{1}{m_2} \end{bmatrix} \tag{11}$$

*4.3. The properties of the linear system*

*The stability***.** A linear system is stable if all poles have the real part located in the negative complex half-plane. The poles represent the roots of the characteristic equation, which is of the form:

$$\boldsymbol{P}(s) = det(s\boldsymbol{I} - \boldsymbol{A}) \tag{12}$$

where **I** is the unit matrix.

After substitution we determine the poles of the system:

$$\boldsymbol{P} = 1 \cdot 10^2 \begin{bmatrix} -2.099 + 0.00i \\ -0.217 + 0.00i \\ -0.017 + 0.79i \\ -0.017 - 0.79i \end{bmatrix} \tag{13}$$

It is observed that all poles of the system have negative real part and we deduce that the system is stable.

*The controllability.* The state of the system is controllable if the rank of the controllability matrix is equal to the size of the matrix **A**.

Matrix of the controllability of the state of the linear system is built based on the matrices **A** and **B**.

The general form of the controllability matrix is:

$$\boldsymbol{\Omega}_C = [\boldsymbol{B} \quad \boldsymbol{AB} \quad \boldsymbol{A}^2\boldsymbol{B} \quad \dots \quad \boldsymbol{A}^{n-1}\boldsymbol{B}] \tag{14}$$

With the help of functions in the MATLAB software, *ctrb* and *rank*, the controllability matrix $\boldsymbol{\Omega}_C$ and its rank are determined:

$$\boldsymbol{\Omega}_C = 1 \cdot 10^9 \begin{bmatrix} 0 & 0 & 0 & 0 & 0 & 0 & -0.0083 & 0 \\ 0 & 0 & 0 & 0 & -0.002 & 0 & 0.0330 & 0 \\ 0 & 0 & 0 & 0 & 0 & 0 & 0.0082 & 0 \\ 0 & 0 & 0 & 0 & 0.0082 & 0 & -1.7164 & 0.0003 \end{bmatrix} \tag{15}$$

$$rang(\boldsymbol{\Omega}_C) = 4$$

It is observed that the rank of the controllability matrix is equal to the size of the matrix **A**.

*The observability.* A linear dynamical system is observable if the rank of the observability matrix is equal to the matrix size **A**.

The observability matrix of the state of the linear system is built based on the matrices **A** and **C**.

$$\boldsymbol{\Omega}_O = [\boldsymbol{C} \quad \boldsymbol{CA} \quad \boldsymbol{CA}^2 \quad \dots \quad \boldsymbol{CA}^{n-1}]^T \tag{16}$$

With the help of the MATLAB functions *obsv* and *rank* the observability matrix $\boldsymbol{\Omega}_O$is determined and it's rank.

$$\boldsymbol{\Omega}_O = 1 \cdot 10^8 \begin{bmatrix} 0 & 0 & 0 & 0 \\ 0 & 0 & 0 & 0 \\ 0 & 0 & 0 & 0 \\ 0 & 0 & 0.0002 & 0 \\ 0 & 0 & 0 & 0 \\ -0.0082 & -0.0004 & 0.0348 & 0.0020 \\ 0.0021 & 0.0001 & -0.0087 & -0.005 \\ 1.7296 & 0.0848 & -7.299 & -0.4150 \end{bmatrix} \tag{17}$$

$$rang(\boldsymbol{\Omega}_O) = 4$$

It can be noted that the rank of the observability matrix is equal to the size of the matrix **A**.

### 4.4. The controller LQR

If the two matrices, system matrix **A** and input matrix **B**, are controllable we can use the LQR optimization method.

A function must be found, $F_C$, so as to minimize the performance criterion.

$$J = \int_0^\infty \boldsymbol{x}(t)'\boldsymbol{Q}\boldsymbol{x}(t) + RF_C(t)^2 dt \tag{18}$$

The matrices **Q** and **R** are weight matrices.

Since it is possible to intervene only on the force of the actuator $F_C$, not on the disturbance of the path $h$, the new matrices **A** and **B** are the following:

$$\boldsymbol{A} = \begin{bmatrix} 0 & 1 & 0 & -1 \\ -\frac{k_2}{m_2} & -\frac{c_2}{m_2} & 0 & \frac{c_2}{m_2} \\ 0 & 0 & 1 & 0 \\ \frac{k_2}{m_1} & \frac{c_2}{m_1} & -\frac{k_1}{m_1} & -\frac{c_2+c_1}{m_1} \end{bmatrix}, \quad \boldsymbol{B} = \begin{bmatrix} 0 \\ \frac{1}{m_2} \\ 0 \\ -\frac{1}{m_1} \end{bmatrix}. \tag{19}$$

The solution to the problem is to find a vector **K** such that $F_C = -\boldsymbol{Kx}$.
So the new matrix of the system is of the form:

$$\dot{\boldsymbol{x}} = \boldsymbol{Ax} + \boldsymbol{B}(-\boldsymbol{Kx}) = (\boldsymbol{A} - \boldsymbol{BK})\boldsymbol{x} \tag{20}$$

The calculation of the **Q** matrix is done with the following relation:

$$\boldsymbol{Q} = \boldsymbol{C}^T\boldsymbol{C} \tag{21}$$

$$\boldsymbol{Q} = 1 \cdot 10^3 \begin{bmatrix} 6.5409 & 0.3516 & 0 & -0.3516 \\ 0.3516 & 0.0189 & 0 & -0.0189 \\ 0 & 0 & 0 & 0 \\ -0.3516 & -0.0189 & 0 & 0.0189 \end{bmatrix} \tag{22}$$

With the help of the calculation program made in MATLAB for a value of R = 1, the gain vector is obtained:

$$\boldsymbol{K} = [0{,}1172 \quad 0{,}0282 \quad 0{,}0948 \quad -0{,}0011].$$

### *4.5. The PID controller*

The PID controller, like the LQR, continuously monitors the operation of the system and calculates the error. The general form of a PID controller is:

$$C_{PID} = k_p + \frac{k_i}{s} + k_d s = \frac{k_d s^2 + k_p s + k_i}{s} \tag{23}$$

Determining the coefficients and adjusting them is done with the PID Tuner function in the MATLAB/Simulink software. These coefficients have the following values: $k_p = 1{,}05$, $k_i = 6{,}86$, $k_d = 0{,}04$.

### *4.6. Numerical results*

Based on the equations of motion and the previous numerical results, the following graphs will show the influence of the two controllers on the passive system with two degrees of freedom. The red color (the curve with the highest amplitude) represented the response of the passive system, with a solid green line the response of the system controlled by the LQR controller and with a broken blue line the response of the system controlled by the PID controller.

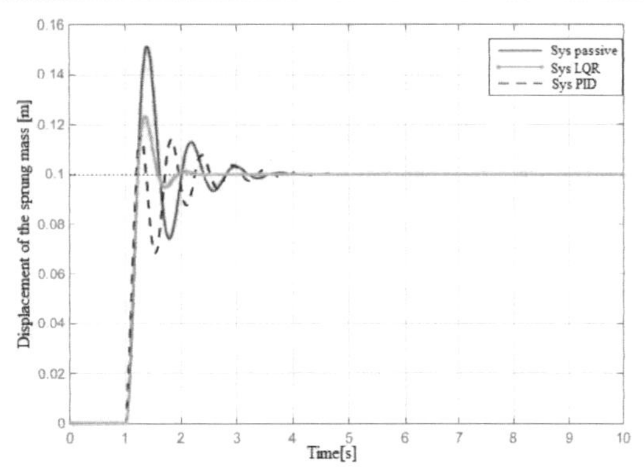


**Figure 6.** Displacement of the suspended mass.

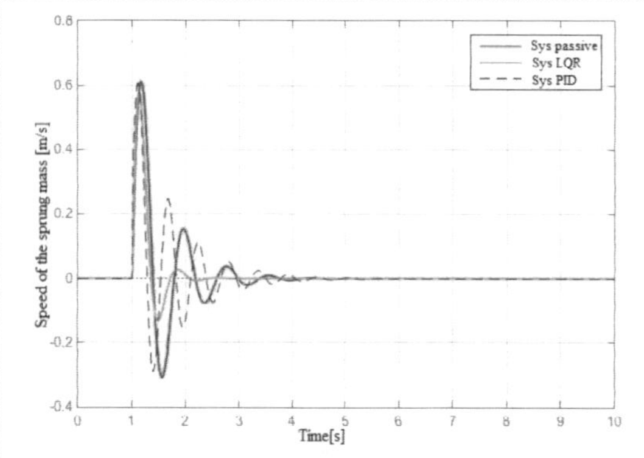


**Figure 7.** The speed of the suspended mass.

In figure 6 a significant reduction, approximately 50%, of the displacement of the suspended mass is observed for the system controlled by one of the PID or LQR controllers.

In the case of the system controlled by the LQR controller, the oscillations of the suspended mass are eliminated after approximately 2 seconds.

In figure 7 the speed of the suspended mass for the system controlled by the LQR controller has the lowest value among the three compared systems, and its value becomes zero after about 2 seconds.

To obtain the two graphs, a unity step type signal having the value of 0.1 m was used to observe the influence of the two controllers on the passive system.

The value of the amplitudes for the suspended mass decreased, but also the period of oscillation of the suspended mass was reduced.

## 5. Conclusions

The active suspension offers high comfort compared to a passive suspension, but this consistent improvement in comfort requires energy consumption.

Active suspensions are used especially in luxury class vehicles and sports cars.

In this work, an active suspension for a car was designed by designing a controller, which has the role of improving the performance of the system, in terms of the design objectives compared to the passive suspension system.

The mathematical modelling was carried out using an example with two degrees of freedom on the quarter car model for the passive and active suspension system.

The state of the road surface was considered to be a unit step type excitation. In order to increase passenger comfort and vehicle handling, two LQR and PID controllers were developed. These controllers and the simulation of the passive and two active systems were modelled and simulated using the MATLAB/Simulink software.

The paper presents a way of adjusting the parameters of the PID and LQR controllers with the help of specialized functions of the MATLAB software. The steps used in designing the controllers represent a model that can be used for an easy understanding of how the controllers work. Thus, it is shown that the presented active suspension system offers a reduction of the displacements of the sprung mass.

## References

[1] Tseng H E and Hrovat D 2015 *State Of The Art Survey: Active And Semi-Active Suspension Control. Vehicle System Dynamics*

[2] Pandrea N, Popa D and Parlac S 2001 *Modele Pentru Studiul Vibratiilor Automobilelor* (Tiparg)

[3] Guido P. A. Koch 2011 *Adaptive Control Of Mechatronic Vehicle Suspension Systems* (Sprimger)

[4] Martins I, Esteves J, Marques G D and Pina Da Silva F 2006 *Permanent-Magnets Linear Actuators Applicability In Automobile Active Suspensions. Vehicular Technology* (IEEE Transactions)

[5] Bourmistrova A, Storey I and Subic A 2005 *Multiobjective Optimisation Of Active And Semi-Active Suspension Systems With Application Of Evolutionary Algorithm, In Modsim* (05 16th Congress Of The Modelling And Simulation Society Of Australia And New Zealand)

[6] Voicu M 2008 *Teoria Sistemelor* (Editura Academiei Române)

[7] Hudha K, Ahmad F, Kadir Z and Abd Jamaluddin H 2011 *Pid Controller With Roll Moment Rejection Or Pneumatically Actuated Active Roll Control (Arc) Suspension System - Ch.: Pid Control, Implementation And Tuning* (Tamer Mansour. Rijeka, Croatia: Intech) pp 41

[8] Rijumon K, Murtaza M A and Ajith Krishnan 2013 *A Comparison Between Passive & Semi Active Suspension Systems*

[9] Sharad S Jadhav, S S Kulkarni and Amod P Shrotri 2016 *Suspension System With Broad Classification And Various Models: A Review*

[10] Martins I, Esteves J, da Silva F P and Verdelho P 1999 *Electromagnetic Hybrid Active-Passive Vehicle Suspension System.* IEEE 49th Vehicular Technology Conference (Cat. No. 99CH36363)

[11] Aboud W, Mohamed Haris S and Yaacob Y 2014 *Advances in the Control of Mechatronic Suspension Systems* Journal of Zhejiang University SCIENCE C volume 15, pages 848–860

IOP Conf. Series: Materials Science and Engineering 1303 (2024) 012039 doi:10.1088/1757-899X/1303/1/012039

# Seat belt modelling and testing through virtual prototyping

**O V Stănciuc-Oțăt, I. Dumitru, D Tutunea and B Florescu**

Faculty of Mechanical Engineering, University of Craiova, Romania

E-mail: otatoana@yahoo.com

**Abstract**. Road safety requires optimal collaboration and coordination between several factors directly or indirectly involved in order to ensure safety for all public road users within a national transport system. Within this system the tasks of each component must be very clearly defined and regulated so that the whole system functions properly. The paper presents a case study on the operation of the driver restraint system in the driving position. For this purpose, the three-point seat belt was modeled using the LS-Dyna software package. For the proposed analysis, the seat belt with grip on the central pillar of the vehicle was considered. This type of seat belt is the most common passive safety system in series vehicles.

## 1. Introduction

Passive safety of vehicles aims to reduce the effects of accidents, even during their occurrence. In the construction of motor vehicles, especially passenger cars, the aim is to ensure minimum safety standards aimed at avoiding serious or fatal injuries to the occupants.

One of the most important roles a vehicle must perform is to ensure the highest possible degree of safety. In recent years, driving safety has taken on a new dimension, with increasingly complex systems designed to decrease the severity of injuries in the event of an accident. Most common road safety engineering deficiencies were identified by the authors in [1], for the South Eastern Europe, and have been presented in the paper as a part of road safety unbreakably connected to the safe System approach.

The group of specialists the study and improvement of security measures is also seen through the numerous published studies. In [2] are treated both research regarding the Active Safety Systems from a theoretical point of view, but the most common road safety engineering deficiencies are also highlighted. Starting from this principle of the safety position, in [3], it is presented a research study regarding the occupant kinematics in the vehicle at the front and rear impact.

## 2. Restraint system of occupants of a vehicle-seat belt

The seat belt is so effective that it has retained its features and purpose since 1949, when Volvo presented the first vehicle with such equipment, and to nowadays. The safety belt as standard on all vehicles is the one with three-point grip. The retractable three-point belt operates at pressure and locks when some tension is felt blocking the retractor drum.

The restraint system is actually a safety harness designed to insure the passenger against being accelerated due to a sudden collision or stop. As part of the passive safety system, the belts are designed to reduce the degree of injury, stopping the impact between the passenger and blunt objects in vehicles, or by other passengers. They absorb energy, being designed to stretch slowly at the moment of collision, reducing the speed of inertia of the passenger, for increased protection especially for passengers with special needs - eg. pregnant women. In [4] it is presented a research facility whose purpose is to analyze the stresses for a pregnant woman's abdomen. It was developed a complex measuring system that registers and processes the movement parameters of the pregnant driver, both independently, but also in relation to the vehicle movement.

### 2.1. *Seat belt functioning*

During an impact, at a speed of 50 km/h, on an adult of 80 kg acts a force of approx. 2 tons. An impact at 40 km/h subjects the human body to pressures similar to a fall from the third floor of a building.

Seat belts are the most effective means of protection against injury in the event of an accident, being the basic component of the passive safety system of any motor vehicle. The use of the belt reduces the risk of fatal injuries by up to 50% for front seat occupants and up to 75% for rear seat passengers. In the event of an accident, a passenger not wearing a seatbelt is randomly projected, increasing the risk of other occupants being injured or killed by 40%.

To be effective, the belts must be used correctly and adjusted according to the height of the occupant, well-adjusted on the body, without being twisted or passed over blunt objects in pockets, such as pens or glasses.

The Airbag, which is an additional retention system, is effective only under the condition of using the seat belt, and can generate serious injuries in its absence. According to [5], the injuries can be severe in specific passenger positions called out of position or when the airbag receives the inflation command with delay. A preoccupation in this direction is presented in [6], where the study was directed at two types of airbag miss-deployments: late deployment and non-deployment.

The deployment process of the seat belt pretensioners coincides with that of the airbag. In the case of the device in figure 1, the pyrotechnic device (detonator) placed in body 1 forces the fluid from the tube 2 onto the blades on the rotor of a turbine 4. The rotational motion of the rotor with turbine blades causes the axle of the seat belt roller to rotate 3, so that the seat belt is stretched over the body of the occupant.

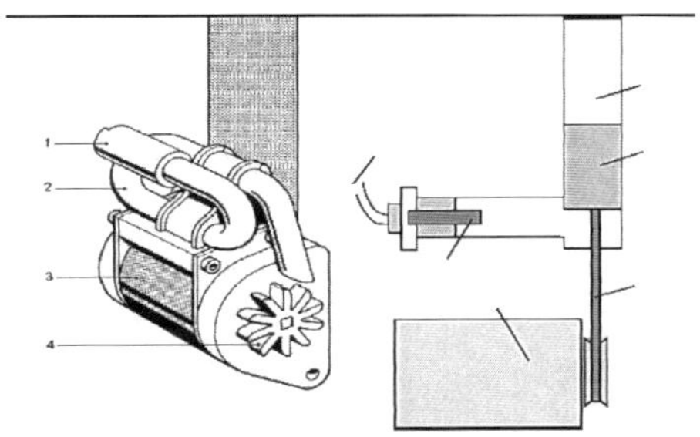


**Figure 1.** The process of triggering pretensioners: 1 - body with detonating staple, combustion chamber and piston; 2-tube; 3-seat belt roller; 4 - turbine blade rotor.

The detonating staple is activated by means of an electrical impulse. The high pressure that occurs in this way forces the piston to move through the tube that is filled with a liquid. This displacement causes the rupture of a membrane that closes the other end of the tube. The liquid, a mixture of water and glycerin (so as not to freeze at the low temperatures that are commonly encountered in winter), is directed at high speed to the turbine rotor blades through the opening that appeared at the end of the tube after the membrane ruptures. The rotation of the turbine drives the roller shaft with the seat belt in the direction of winding, which causes the belt to be tensioned.

### 2.2. *The future of restraint systems-seat belt*

Large car companies have ongoing concerns to improve the safety features of cars, and the belt plays a particularly important role in this direction and can be radically improved.

In this regard, several partnerships have been initiated between major companies and research institutes to bring to market a new seat belt that will provide increased protection for drivers.

The new seat belt that will equip the cars of the future will also have a special cover for the driver's seat, which will know all the time data about his condition. If the driver has health problems that can affect traffic decisions and reactions, he will be alerted by the new technology, which has sensors that provide real-time information on heart rate and breathing.

## 3. Seat belt model with three-point grip

For the proposed analysis, the seat belt with grip on the central pillar of the vehicle was considered. This type of seat belt is the most common passive safety system in series vehicles.

The three-point seat belt elements (figure 2) are as follows: lower grip (anchorage), upper grip (anchorage) (on the central pillar of the vehicle), grip (buckle) at the level of the pelvis and retractor.

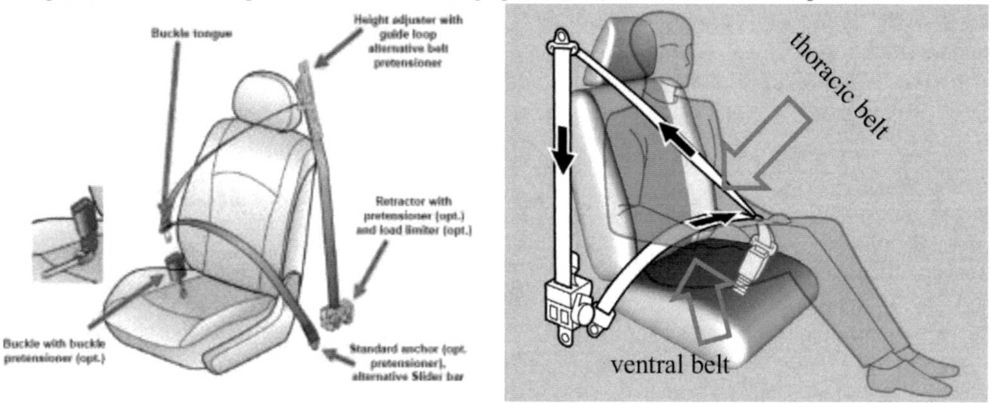


**Figure 2**. Safety belt elements with three-point fastening.

**Figure 3**. Seat belt attachments.

**Figure 4**. Ventral belt and thoracic belt.

The virtual model used in the simulation includes all the elements of a three-point restraint system (seat belt) [7]. The component elements are shown in the following pictures:

In the first stage, the three seat belt attachment points are defined, respectively: the lower grip, the upper grip and the occupant's pelvis grip. Subsequently, it is necessary to define the connection between the three grips, as seen in figure 4.

## 4. Initial data and definition of the motion law of the assembly

The complete model consists of ten parts, comprising in addition to the seat belt and seat, and a part called vehicle. This part is built so that a law of motion can be imposed on the whole assembly. In figure 5 is represented the law of motion of the vehicle, and implicitly of the assembly, being imposed restrictions of connection between the vehicle and the assembly. A vehicle impact speed of 8.2 mm/ms (about 30 km/h) was considered. The simulation was performed for a time interval of 110 ms.

The properties for the materials used to define the parts of the model are shown in the following figures:

- material properties for belt sections-seatbelt material
- belt load curve (tension vs. force)

**Figure 5**. Material properties for belt sections.

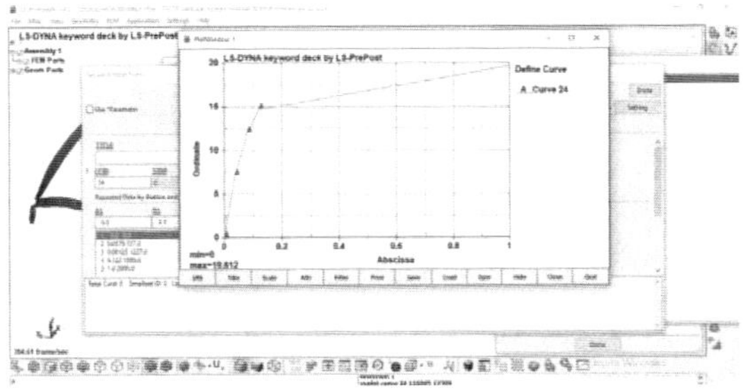

**Figure 6**. Belt load curve (tension vs. force).

## 5. Model processing and presentation of the results obtained

Following the processing of the virtual model, results can be extracted for the analysis of the behavior of the seat belt. The interface with LS-Dyna software allows monitoring the behavior of the system both in video format and in the form of numerical results of the parameters of interest.

Further, the visibility of the passenger seat has been disabled for the graphical representation of the contour of the displacements, speeds, stresses, deformations and accelerations of the seat belt, in order to be able to observe the results only at the level of the belt system:

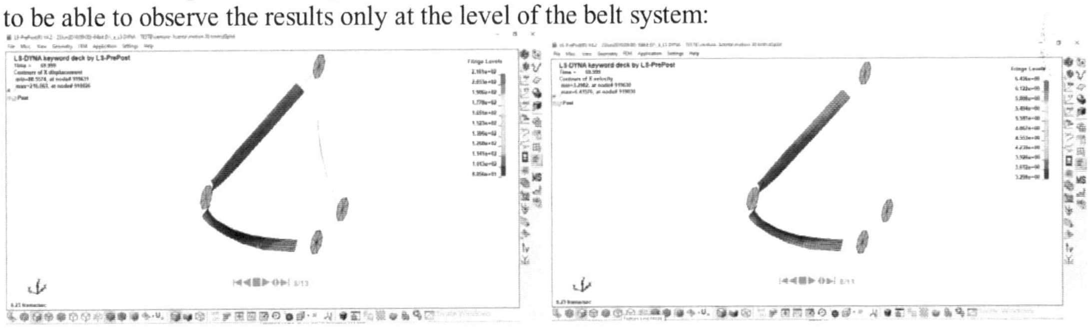

**Figure 7**. Contour of displacements [mm] and velocities [mm/ms] by longitudinal direction-x-axis.

**Figure 8**. Contour of deformations and stresses [kN/mm$^2$] after the longitudinal direction-x axis.

**Figure 9**. Contour of the resulting acceleration [mm/ms$^2$].

The kinematic parameters of both the vehicle and the occupant's seat were also tracked. The two parts are defined as rigid bodies, being possible graphical representation of the variation of displacement, speed and acceleration, after the three directions, with the post processing module for *Rigid body (rigid body).*

For monitoring the action of seat belts were first defined shell-type elements. Thus, three elements were chosen for the ventral section and three other elements for the thoracic section. These elements were arranged towards the ends of the section, and one was placed in the middle area of the sections – figure 10. The elements defined on the thoracic belt are: 166959, 166981 and 167012, and those defined on the ventral belt are: 167116, 167145 and 167160.

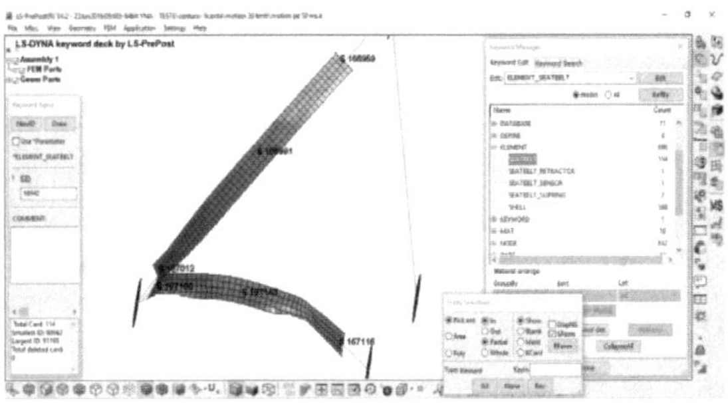


**Figure 10**. Definition of shell elements.

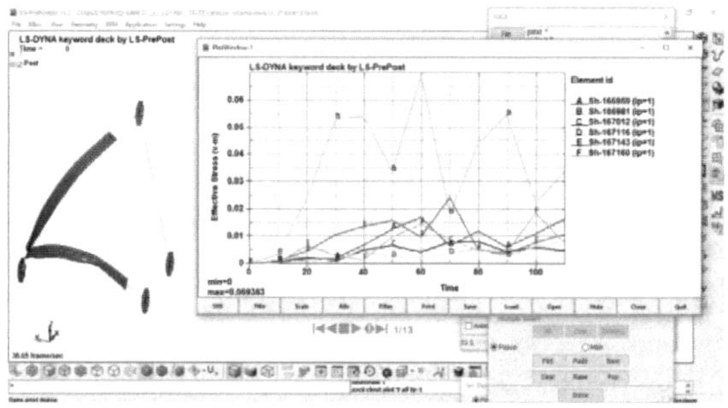


**Figure 11**. Effective strain [kN/mm$^2$] by segment.

For the belt elements defined above, both the strain developed by longitudinal direction and the effective strain were plotted. From figure 11 it is noted that the maximum strain value is reached for the element 166981 defined in the midline area of the thoracic belt.

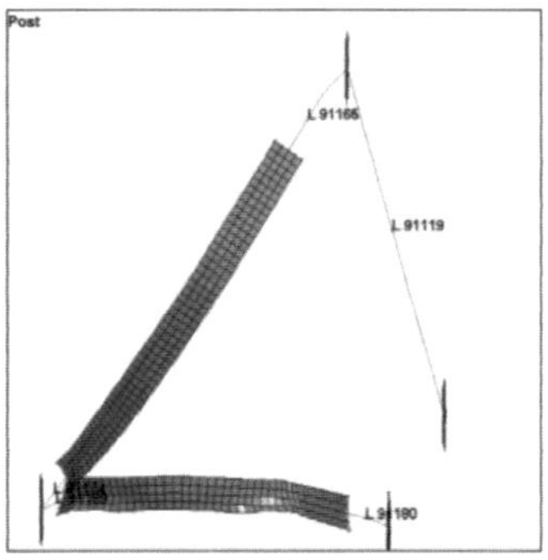


**Figure 12**. Defining seatbelt elements on connecting areas.

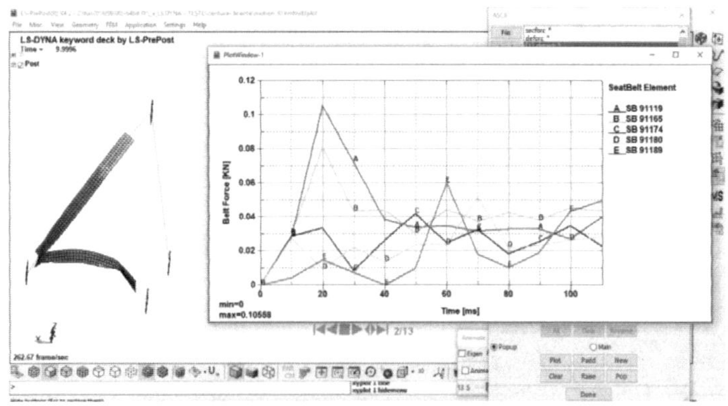


**Figure 13**. Force variation [kN] in belt connecting areas

The analysis of the belt behavior also involves monitoring the connection zones between the grip and the ventral and thoracic belts, respectively. For this purpose, five seatbelt type elements have been defined, each element being centrally located on the connecting areas. The elements defined and monitored are: 91119, 91165, 91174, 91180 and 91189 – figure 12.

From the graph shown in figure 13 it is noted that the maximum force value is recorded for the element 91119, which was defined on the area connecting the lower grip and the upper grip (on the central pole). According to figure 14, it is found that the element for which the maximum force value was recorded has the smallest length compared to the other four seatbelt elements analyzed.

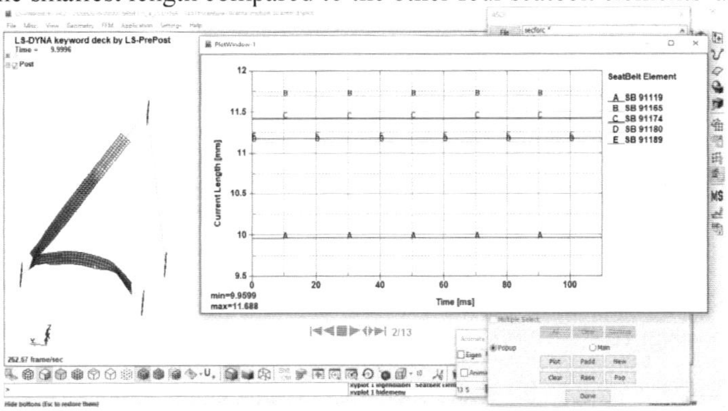


**Figure 14**. Length [mm] of the elements in the connecting areas of the belt.

The seat belt assembly also includes two elements that allow the belt to slide, both for tightening and for relaxation.

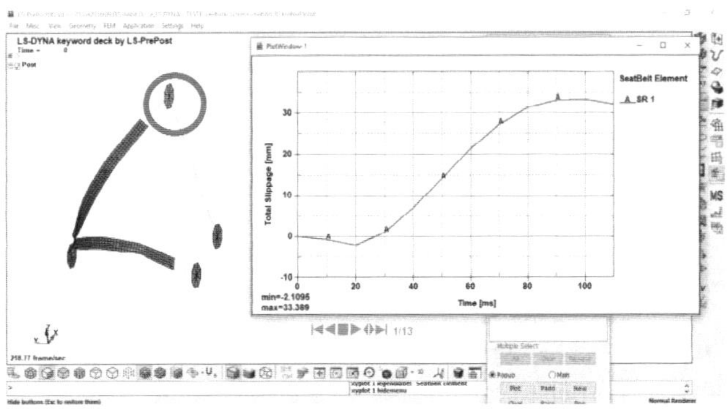


**Figure 15**. Total belt slip for the first point.

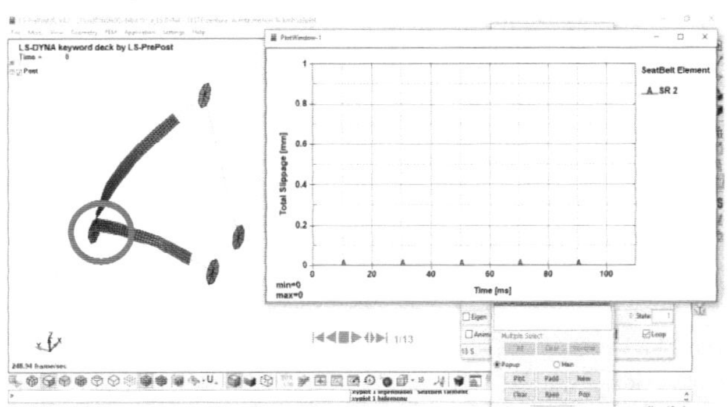


**Figure 16**. Total belt slip for the second point.

The total slip of the belt at the first point, defined at the level of the upper grip on the central pillar of the vehicle, is plotted in figure 15. The maximum value of the belt slip recorded at this point is 33 mm. The total slip of the belt at the second point, defined at the level of the middle grip of the belt, is plotted in figure 16. It is noted that at this point the slipping of the belt is not allowed, being the area containing the seat belt buckle.

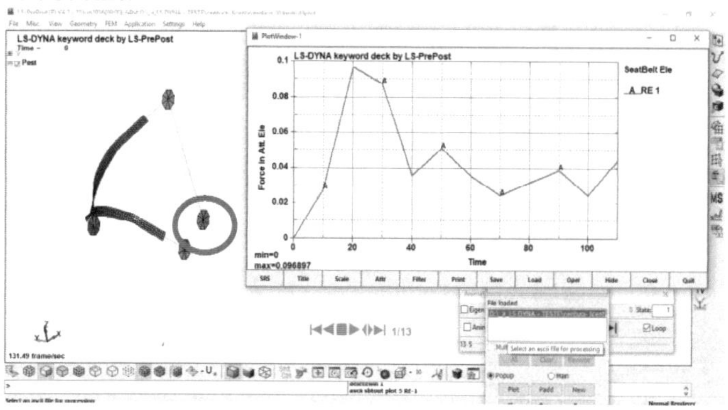


**Figure 17**. The force developed by the seat belt retractor.

The seat belt retractor is positioned in the lower area of the center pillar of the vehicle. The forces acting it are presented in figure 17.

## 6. Conclusions

The present study proposes a virtual prototype and testing method of the seatbelt system functioning in order to protect the driver in the event of a frontal collision. For this purpose, the three-point seat belt was modeled using the LS-Dyna software package. LS-Dyna software is a specialized software for the analysis and reconstruction of traffic accidents, using the finite element method.

For the proposed analysis, the seat belt with grip on the central pillar of the vehicle was considered. This type of seat belt is the most common passive safety system in series vehicles. A vehicle impact speed of 8.2 mm/ms (about 30 km/h) was considered. The simulation was performed for a time interval of 110 ms.

Following the processing of the virtual model, results were extracted for the analysis of the behavior of the seat belt. The interface with LS-Dyna software allows monitoring the behavior of the system both in video format and in the form of numerical results of the parameters of interest. They were plotted the contours of the displacements, speeds, stresses, deformations and accelerations of the seat belt. The kinematic parameters of both the vehicle and the occupant's seat were also tracked.

**Author's ORCID ID**
O V Otat: https://orcid.org/0000-0002-1939-7533

## References

[1] Jovanov D, Vollpracht H J, Beles H, Popa V and Tolea B A 2017 *Most common road safety engineering deficiencies in South Eastern Europe as a part of safe System approach*, International Congress of Automotive and Transport Engineering - Mobility Engineering and Environment (CAR2017), IOP Conference Series-Materials Science and Engineering, Volume: 252, Article Number: UNSP 012021, DOI: 10.1088/1757-899x/252/1/012021, Pages: 12-21, Published: 2017, Publisher: IOP Publishing LTD, England, ISSN: 1757-8981

[2] Beles H, Preda I, Dragomir G, Blaga V and Ardelean F 2008 *Theoretical Research Regarding the Active Safety Systems*, Annals of DAAAM for 2008 & Proceedings of the 19th International DAAAM Symposium, Pages: 87-88, Published: 2008, ISBN: 978-3-901509-68-1, ISSN: 1726-9679

[3] Popa V, Beles H, Tolea B, Dragomir G and Mitran T 2016 *Studies Regarding Occupant*

*Kinematics in the Vehicle at the front and Rear Impact*, Applied Mechanics and Materials, Vol. 822, Pages: 512-522, Jan. 2016, Publisher: Trans Tech Publications, ISSN print: 1660-9336, ISSN web: 1662-7482, ISBN: 978-3-03835-703-2

[4] Crauciuc D, Sifft M, Thierheimer W, Alexandru C and Thierheimer A 2018 *Forensic science Experimental Investigation over the Victim's behavior in Road Events*, International Journal of Environmental Science, http://www.iaras.org/iaras/journals/ijes, Volume 3

[5] Toganel G. R. and Soica A. 2017 *A late and failure of airbag deployment case study for drivers of passenger cars in rear-end collisions*, IOP Conference Series: Materials Science and Engineering, Volume 252, CAR2017 International Congress of Automotive and Transport Engineering - Mobility Engineering and Environment 8-10 November 2017, Pitesti, Romania

[6] Soica A. 2016 *Aspects of the passenger airbag E.C.U. location*, IOP Conference Series: Materials Science and Engineering, Volume 147, 7th International Conference on Advanced Concepts in Mechanical Engineering 9-10 June 2016, Iasi, Romania

[7] http://www.lstc.com/models

IOP Conf. Series: Materials Science and Engineering 1303 (2024) 012040 doi:10.1088/1757-899X/1303/1/012040

# Possibilities to evaluate the vehicle's accelerations through MATLAB Simulink simulations

**I Duma, A Molea, N Cordoş, H Cărăuşan, N Burnete, I Barabás, A Todoruț and B M Jurchiș**

Department of Automotive Engineering and Transports, Technical University of Cluj-Napoca, RO

E-mail: nicolae.cordos@auto.utcluj.ro

**Abstract**. This scientific present paper explores the methods for assessing the longitudinal and lateral accelerations of a vehicle through the combination of experimental validation and MATLAB Simulink simulations. Comprehending these accelerations is essential for a number of uses, such as stability, overall performance analysis, and vehicle safety.The first step of the study is building a detailed Simulink model of the car. With an emphasis on lateral and longitudinal accelerations, this model makes it easier to simulate a range of driving scenarios and explore how a vehicle handles and behaves under different circumstances. A key element of the study is the experimental validation of the Matlab Simulink model, which involves gathering real-world data from vehicle testing and contrasting it with simulated outcomes. Simulink model iterative refinement is made possible by the integration of simulation and experimentation, which ultimately improves the model's predictive accuracy. The study's conclusions demonstrate the integrated approach's potential to advance vehicle dynamics analysis and make cars safer and more effective. Furthermore, it offers insights into the optimization of vehicle designs and the development of innovative control strategies to enhance overall road safety and vehicle performance.

## 1. Introduction

### *1.1. Aim and objectives of the present paper*
Accurate measurement of the lateral and longitudinal accelerations of vehicles is critical in modern automotive research to maximize safety, efficiency, and control systems. Present-day techniques typically depend on isolated experimental testing or theoretical models, which frequently lack the comprehensive insights needed for in-depth analysis.

This research paper closes this gap by presenting a novel method that combines experimental validation with MATLAB Simulink simulations in a seamless manner, providing a comprehensive understanding of vehicle dynamics. The combination of these methods represents a significant breakthrough in automotive engineering and has the potential to improve control schemes, safety features, and vehicle design.

### *1.2. State of the art*
Several studies have been conducted using this tool for vehicle dynamics simulations, such as [1], in which the advantages of using Simulink for modeling and simulating vehicle dynamics is highlighted,

including its ease and flexibility. Another relevant study is [2], where authors use Simulink for modeling and analyzing the dynamics of a vehicle's suspension system. It also highlights the importance of accurate models for the suspension system, to better understand how performance and safety can be improved.

Moreover, the paper [3] presents a mathematical model for the kinematics and dynamics of a four-wheeled vehicle using MATLAB Simulink. The authors demonstrate the effectiveness of this simulation tool in accurately modeling vehicle dynamics and for evaluating the influence of different parameters on vehicle's performance and safety, together with providing insights into the use of Simulink.

The paper [4] presents a model for evaluating tire rolling resistance using MATLAB Simulink by modeling the interaction between tires and deformable pavements. The authors highlight the effectiveness of the model in accurately simulating tire and road surface interaction and evaluating the impact of pavement texture and tire characteristics on rolling resistance. The study provides insights into improving tire design for the reduction of fuel consumption.

Simulink can also be used to model and evaluate the dynamics of electric vehicles. Thus, the study [5] proposes an acceleration slip regulation (ASR) strategy for four-wheel drive electric vehicles based on sliding mode control (SMC). The authors show the effectiveness of the proposed ASR-SMC strategy in improving vehicle stability and maneuverability under different road conditions and different dynamic loads on vehicles' axles. The study provides information regarding the use of SMC for developing advanced ASR systems in electric vehicles.

In terms of experimental tests for studying vehicle dynamics, the scientific paper [6] proposes a method for estimating a vehicle's lateral load transfer and normal forces using experimental data from accelerometers and gyroscopes. The method is based on a dynamic model of the vehicle and can estimate these critical quantities accurately even under dynamic conditions. This information is critical for developing vehicle control systems, such as anti-lock braking systems, traction control systems and electronic stability control systems, which are meant to improve vehicle handling and passenger safety.

## 2. Working methodology

### *2.1. Experimental test*

For the experimental test, an M1 category automobile was considered, for which the main functional and dimensional parameters are presented in Table 1.

**Table 1.** Main functional and dimensional parameters of the considered vehicle.

| | |
|---|---|
| Engine capacity | 1.997 $cm^3$ |
| Maximum power | 110 kW |
| Total length | 4.365 m |
| Total width (without mirrors) | 1.837 m |
| Frontal trackwidth | 1.532 m |
| Rear trackwidth | 1.527 m |
| Wheelbase | 2.613 m |
| Turning circle | 11 m |
| Ground clearance | 150÷180 mm |
| Gross mass | 1507 kg |
| Total mass (considered in tests) | 1639 kg |
| Tire model | 215/60 R16 95H |
| Nominal tire pressure | 2.4 bars |

The technical condition of the automobile was verified prior to the tests, i.e. tire pressure, uniformity of tires wear and suspension system through EUSAMA test. The experimental dynamic tests must be conducted on a road section with dry asphalt. Its longitudinal and transverse inclination should be as small as possible. The device for measuring the stability parameters contains an MPU6050 module and Arduino Uno board.

IOP Conf. Series: Materials Science and Engineering 1303 (2024) 012040 doi:10.1088/1757-899X/1303/1/012040

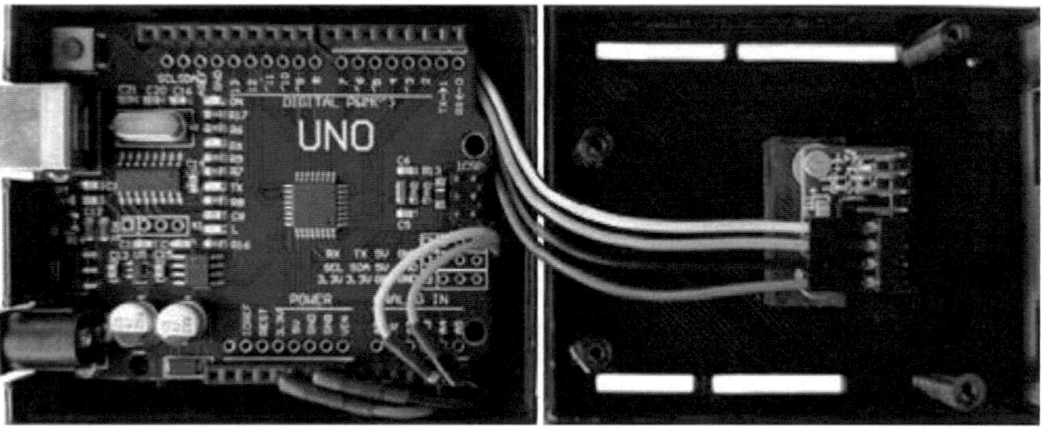


**Figure 1.** The device used for measuring the stability parameters.

The MPU6050 is a module composed of a motion measurement sensor (figure 1), which integrates a 3-axis accelerometer (on X, Y and Z directions), a 3-axis gyroscope, a digital motion processor and additional circuits that serve for the communication of the module with the microcontroller. It allows the measurement of angular velocity around all three axes, static acceleration due to gravity and dynamic acceleration due to motion, shock or vibration. The module is positioned so that its three axes (X, Y, Z) correspond to the axes of the vehicle's center of gravity. The collected and processed data are transmitted to the computer, thus displaying the results to the user in the Serial Monitor.

**Figure 2.** The universal support and acquisition device for the stability parameters.

The device was placed on a universal support (figure 2), that can be used for any vehicle model, and located in the center of gravity in the passenger compartment. The monitored parameters were the accelerations of the vehicle on the Y axis, the rotational speed of the vehicle around the X axis in order to determine the roll, and the rotational speed of the vehicle around the Z axis in order to be able to determine the route traveled. The tests were carried out at 40 km/h and nominal tire pressure at 2.4 bars.

*2.2. Computational simulation*

The validation of obtained results was conducted through the Driving Scenario Designer tool of MATLAB. The first stage of the simulation was represented by the digitization process of the used trajectory during the experimental tests. The considered track was extracted from the Open Street Maps (OMS) platform, its format allowing for direct selection of roads. The OSM file was uploaded to the Driving Scenario Designer, thus digitizing the track automatically.

In the second stage, the vehicle used in the experimental test was defined, together with the considered trajectory. When defining the vehicle, its dimensional parameters were considered, such as total length, trackwidth and total height. The defined trajectory is also highlighted in figure 3.

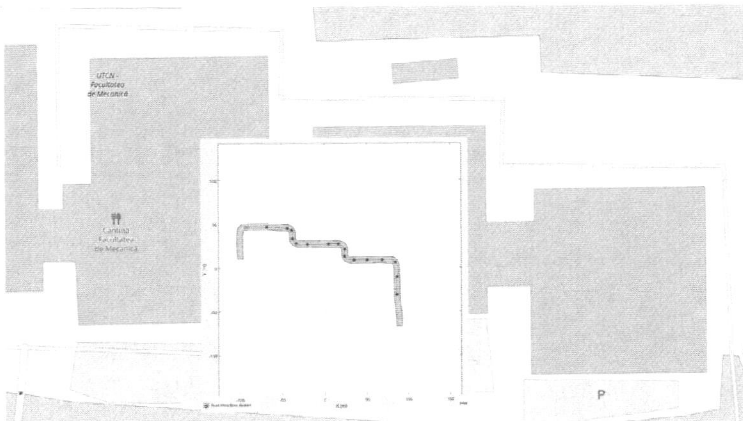

**Figure 3.** The digitized track using the OMS format file, with the defined trajectory in Simulink.

In order to extract data from the computational simulation, the vehicle was equipped with an Inertial Navigation System (INS) module. This Simulink module is responsible for the acquisition of data regarding the accelerations that occur during the vehicle's movement, being positioned in the vehicle's center of mass, similarly to the acquisition of data from the experimental test. After defining the model, all considered movement scenarios were simulated, identical to the ones considered in the testing phase. The model implemented in Simulink is depicted in figure 4.

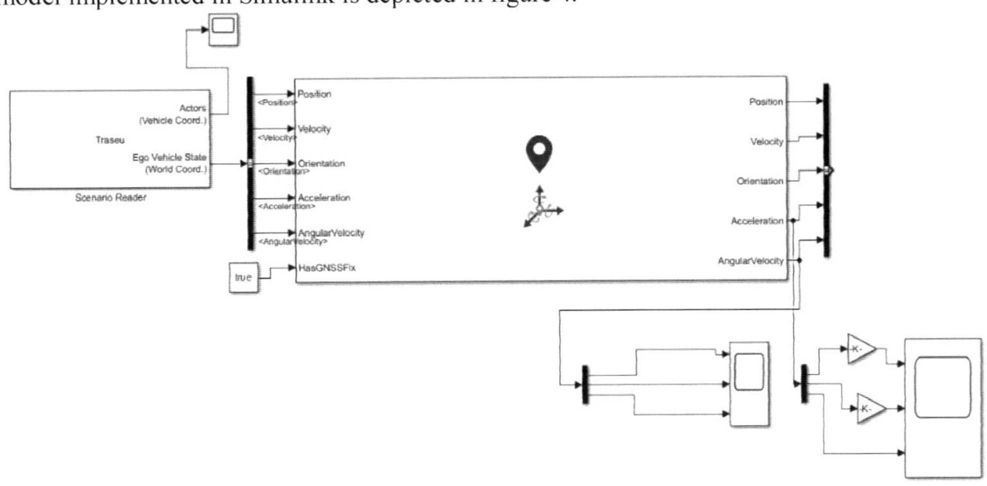


**Figure 4.** The Simulink model.

## 3. Obtained results

The experimental test was conducted with a quasi-constant travel velocity of the vehicle, hence the longitudinal acceleration (on the X axis) has values around zero. The increase in the longitudinal acceleration from the first 15 s of the experimental test is due to the time needed for the vehicle to stabilize its velocity. Also, in cornering conditions, the driving velocity was slightly variable due to the test conditions. The lateral acceleration (on the Y axis) has peaks in cornering, with positive values for left cornering and with negative values for right cornering. The vertical acceleration is also variable, depending on the surface of the road and the gravitational acceleration, as observed in figure 5.

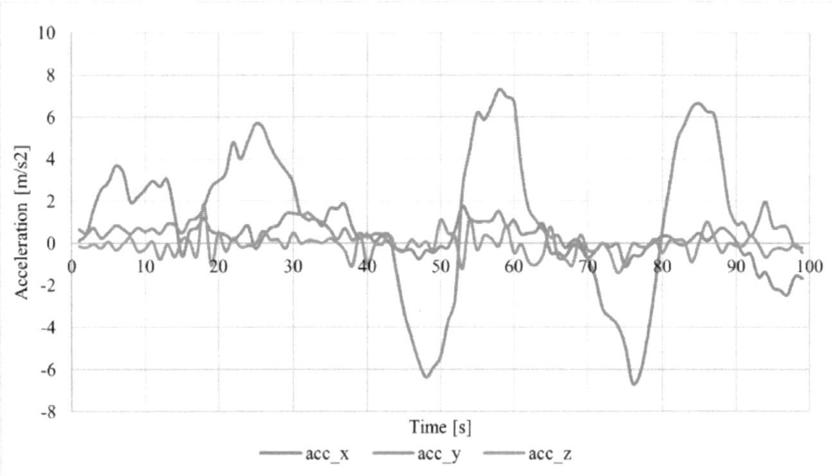


**Figure 5.** Accelerations on the three axes obtained from the experimental test.

According to the data obtained from the computational simulations in MATLAB Simulink, there is a quasi-null longitudinal acceleration (on the X axis of the vehicle), given by the constant travel velocity of the considered vehicle. On the Y axis, there are the peaks in acceleration with positive values when the vehicle is cornering. At the same time, the vertical acceleration (on the Z axis) has negative values for the same cornering conditions, as shown in figure 6.

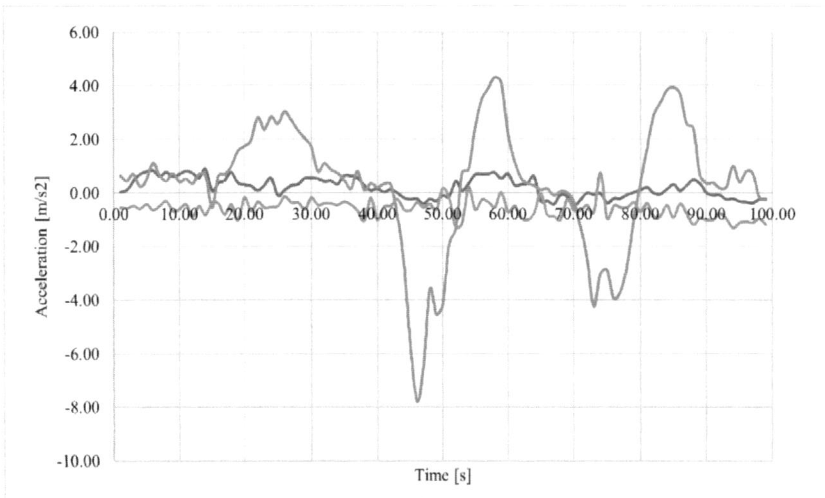


**Figure 6.** Accelerations on the three axes obtained from the computational simulation.

In terms of accelerations obtained from the experimental testing (see figure 5) and computational simulation (see figure 6), it is observed that the shape of the graphs is similar, while the values may differ because of the test and simulation conditions. The relative error (figure 7) for each of the directions of acceleration was calculated with the following relation:

$$\varepsilon_{acc} = \frac{a_{exp} - a_{sim}}{a_{exp}} \cdot 100\ [\%], \tag{1}$$

where $\varepsilon_{acc}$ is the relative error, $a_{exp}$ – the acceleration obtained experimentally and $a_{sim}$ – the acceleration obtained from computational simulation.

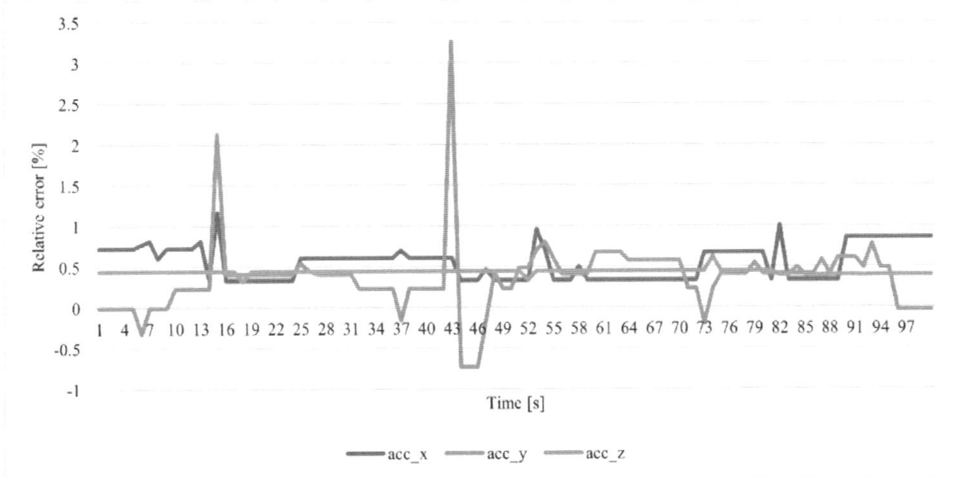


**Figure 7.** Variation of the relative error for accelerations in time.

The maximum absolute values of the relative errors appear in cornering conditions, being higher for the lateral acceleration. The average values of the relative error between the accelerations measured during road test and those obtained from the computational simulations, are included in Table 2.

**Table 2.** Mean values for the relative error between the accelerations determined experimentally and from the Simulink model.

| | $a_x$ | $a_y$ | $a_z$ |
|---|---|---|---|
| $\varepsilon$ [%] | 0.54 | 0.35 | 0.43 |

## 4. Conclusion

In this study, an integrated approach that combines MATLAB Simulink simulations with experimental validation to assess both longitudinal and lateral accelerations in a vehicle was successfully explored. Understanding and accurately evaluating these accelerations is critical for improving vehicle stability, performance analysis, and safety.

The development of a detailed Simulink model with a focus on lateral and longitudinal accelerations aided in the simulation of various driving scenarios, allowing for a better understanding of the vehicle's behavior under various conditions.

Collecting real-world data and comparing it to simulated results was a critical component of experimental validation. This integration enabled iterative refinement of the Simulink model, which improved its predictive accuracy.

We discovered a strong qualitative similarity in the shape of acceleration graphs when comparing experimental and simulation data, despite quantitative differences due to test and simulation conditions. We also computed relative errors for each acceleration direction, which provided useful information about the model's accuracy and reliability.

In the end, this integrated approach could lead to safer and more efficient vehicles on the road by making significant advancements in vehicle dynamics analysis. It also provides opportunities for improving road safety and vehicle performance through creative control strategy development and vehicle design optimization.

**Author's ORCID ID**

I Duma: https://orcid.org/0000-0003-2303-4636
A Molea: https://orcid.org/0000-0003-3848-2722
N Cordos: https://orcid.org/0000-0002-9877-6530
C Horatiu: https://orcid.org/0000-0002-0952-8157
N Burnete: https://orcid.org/0000-0001-6647-459X
A Todorut: https://orcid.org/0000-0003-3368-3701

**References**

[1] Abdussalam A A, Johnson S and Aldbea F W 2020 *Vehicle Dynamics Modeling and Simulation with Control Using Single Track Model*, 2020 IEEE International Women in Engineering (WIE) Conference on Electrical and Computer Engineering (WIECON-ECE), Bhubaneswar, India, p. 1–4 DOI: 10.1109/WIECON-ECE52138.2020.9397983.

[2] Rao K D 2014 *Modeling, simulation and control of semi active suspension system for automobiles under MATLAB Simulink using PID controller,* IFAC Proceedings Volumes p. 827–31 DOI: 10.3182/20140313-3-IN-3024.00094.

[3] Manuel C J T, Santos M M and Tusset A M 2021 *Mathematical modeling attributed to kinematics and dynamics of a vehicle with 4-wheels*, The European Physical Journal Special Topics, DOI: 10.1140/epjs/s11734-021-00238-2.

[4] Ding Y, Wang H, Qian J and Zhou H 2021 *Evaluation of tire rolling resistance from tire-deformable pavement interaction modeling*, Journal of Transportation Engineering Part B Pavements, DOI: 10.1061/JPEODX.0000295.

[5] He H, Peng J, Xiong R and Fan H 2014 *An acceleration slip regulation strategy for four-wheel drive electric vehicles based on sliding mode control,* Energies, DOI: 10.3390/en7063748.

[6] Doumiati A, Victorino A, Charara A and Lechner D 2009 *Lateral load transfer and normal forces estimation for vehicle safety: experimental test,* Vehicle System Dynamics, Vol. 47, p. 1511–33, DOI: 10.1080/00423110802673091.

# Kinematic analysis of the windshield wiper mechanism with two parallel rocker blades

**O Antonescu[1], D Antonescu[2] and A Ionita[3]**

[1] Dept. of Mechanism and Robot Theory, Politehnica University of Bucharest, Romania,
[2] "Iuliu Maniu" High School, Bucharest, Romania
[3] Dept. of Road Vehicles, Politehnica University of Bucharest, Romania

e-mail: oval33@hotmail.com

**Abstract.** The paper deals with a windshield wiper mechanism with two parallel rocker blades having a complex topological structure. In the first part of the paper, the mobility of the mechanism is analyzed by means of the simplest formula for planar linkages. Also, the topological structure consisting of adyad and a tetrad has been outlined. In the second part of the paper, a kinema-tic analysis of the multi-contour mechanism is performed, the angular displacements of the two rocker blades being obtained. In this regard the independent contour method has been used, the angular displacements of all kinematicelements of the mechanism being calculated. Moreover, the displacement systems have been derived in order to obtain the velocities and late-on the accelerations of the mechanism elements. Finally, the paper ends with the displaying of the kinematic charts of the mechanism elements of which those with themovement of the two parallel rocker blades are the most wanted. This is due to the requirement of maximization of the wiping surface which is a very important safety issue.

## 1. Introduction

The importance of the windshield wiper mechanisms has been emphasized by many authors in their researches [1]. For instance, S. Ciunel in his book “The driving system and modern devices for insuring the vehicle safety” (in Romanian) outlines that, among other mechanisms and devices, the windshield wiper mechanisms are directly linked to the way in which the driver receives information on traffic situation.

At the same time, the windshield wiper mechanism has an important role in the accident prevention by placing it according to the driving post design (accomplishing the anatomic, physiologic and psychologic proper conditions), according to the report “The evaluation on traffic safety in Romania” (in Romanian) by C. Uta, Romanian Automotive Register.

There are several types of windshield wiper mechanisms of which two types, i.e. those with “parallel” blades and with opposite blades, are the most used on passenger motor vehicles [9].

From the very beginning a simplifying hypothesis has been imposed: the windshield wiper mechanism with two parallel rocker blades studied in the paper to be a planar linkage. This is due to no more than 5° angle between the two fixed rotate axes of the “parallel” rocker blades that exists in reality in usual passenger motor vehicles. That means that apart from the three rotate joints (the two pairs between the rocker blades and the car body, and the articulation between the crank and the car body) the real -mechanism has spherical joints. Therefore, the analyzed mechanism has only revolute joints (figure 1).

## 2. The topological structure and geometry of the windshield wiper mechanism

The kinematic scheme of a windshield wiper mechanism with two parallel rocker blades consisting of articulated bars is presented in figure 1 [1, 2, 3, 6-10].

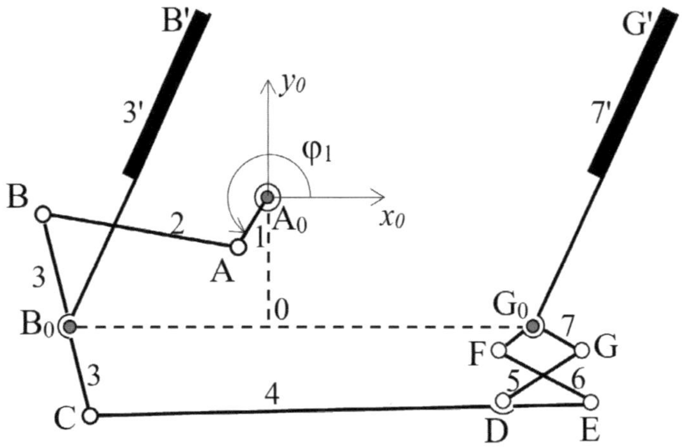


**Figure 1**. Kinematic scheme of the windshield wiper mechanism [1]

This special type of wiper mechanism has a complex structure having seven articulated elements of which the crank 1 is the driving element, and the rockers 3 and 7 are the driven elements on which the wiper blades 3' and 7' are attached. The mechanism is mounted in the vehicle body denoted with 0 as fixed element.

The mobility of the planar mechanism is calculated by means of the formula [4, 5]:

$$M = 3n - 2C_5 - C_4 \quad (1)$$

where we denoted:

- $n = 7$ (kinematic elements);
- $C_5 = 10$ (mono-mobile kinematic joints);
- $C_4 = 0$ (bi-mobile kinematic joints).

By substituting these numerical values in formula (1) the mobility is obtained:

$$M = 3 \times 7 - 2 \times 10 - 0 = 1$$

In the topological structure of the mono-mobile mechanism, with the "actuator mechanism" $MA(0,1)$, we remark one simple kinematic chain LD(2,3) type dyad and one complex kinematic chain LTt(4,5,6,7) type tetrad.

Therefore, the structural formula of the „motor mechanism" (MM) is:

$$MM = MA(0,1) + LD(2,3) + LTt(4,5,6,7) \quad (2)$$

The geometry of the mechanism is defined by the following parameters:

1) Fixed Cartesian coordinates:

$x_{B_0} = -240mm;\ y_{B_0} = y_{G_0} = -35mm;\ B_0G_0 = 500mm;$

2) Constant lengths of the kinematic bars:

$A_0A = 50mm;\ AB = 235mm; B_0B = 60mm; B_0C = 35mm; CD = 450mm;$

$DE = 65mm; EF = 67mm; DG = 67mm; FG_0 = 23mm; GG_0 = 23mm; FG = 43mm.$ 3) Constant angles:

$\angle BB_0B' = \alpha_3; \angle BB_0C = 180^0; \angle CDE = 180^0; \angle FG_0G = \alpha_7; \angle GG_0G' = \alpha_7'.$

## 3. The analytic calculation of the mechanism bar displacements

Due to the fact that the mobility of the wiper mechanism is equal to 1, there is a single independent parameter ($\varphi_1$) for the calculation of the driven elements ($\varphi_2, \varphi_3$) [1].

The movement of the crank 1 is given by the angular displacement $\varphi_1$, angular velocity $\dot{\varphi}_1$ and angular acceleration $\ddot{\varphi}_1$.

The instantaneous position of each kinematic element (articulated bar) is deduced by means of the independent kinematic contour method (figure 2).

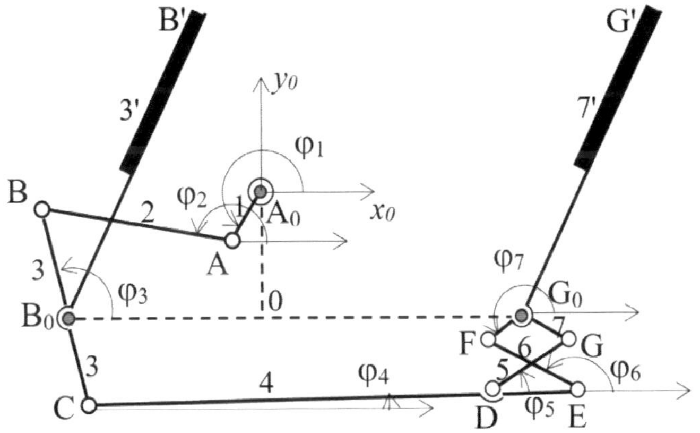


**Figure 2**. The angular displacements of the kinematic elements

The first closed kinematic contour is $A_0ABB_0A_0$ on which we choose conveniently the orientation of each vector (figure 3).

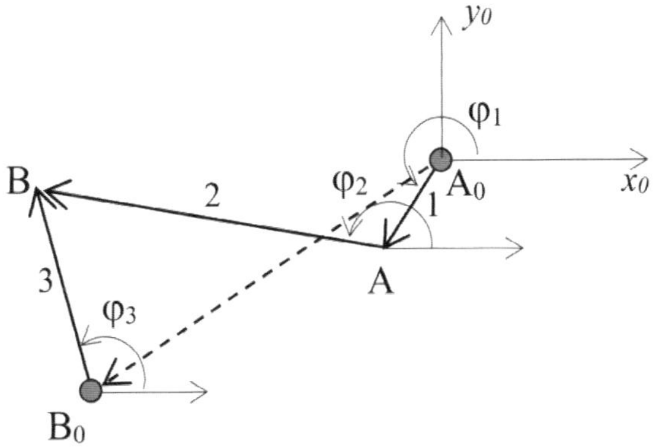


**Figure 3**. The vector contour of the kinematic chain $A_0ABB_0A_0$

The vector equation of the closed contour $A_0ABB_0A_0$ is:

$$\overrightarrow{A_0A} + \overrightarrow{AB} = \overrightarrow{A_0B_0} + \overrightarrow{B_0B} \tag{3}$$

Then we deduce the scalar equations equivalent to the vector equation (3):

$$l_2 \cos\varphi_2 - l_3 \cos\varphi_3 = x_{B_0} - l_1 \cos\varphi_1; \tag{4}$$

$$l_2 \sin\varphi_2 - l_3 \sin\varphi_3 = y_{B_0} - l_1 \sin\varphi_1 \tag{5}$$

Using the system consisting of the scalar equations (4) and (5), we calculate the angles $\varphi_2$ and $\varphi_3$ which are the positions of the coupler 2 and the rocker 3.
Having $\varphi_3$ , we obtain the position of the point C written by Cartesian coordinates:

$$x_C = x_{B_0} + l'_3 \cos(\pi + \varphi_3);\ y_C = y_{B_0} + l'_3 \sin(\pi + \varphi_3) \tag{6}$$

On the other hand, in the tetradic kinematic chain LTt(4,5,6,7) we identify two independent closed contours (figure 4).

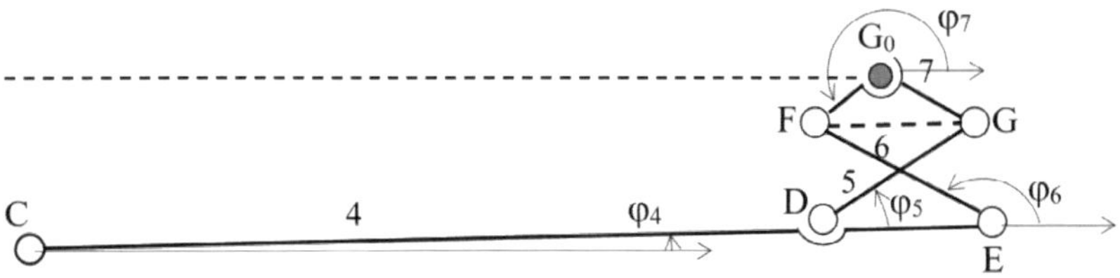


**Figure 4**. The tetradic chain with the two independent contours

The two independent kinematic contours are: $CDGG_0C$ and $CEFG_0C$.
For each of these contours we write the vector equation of closing:

$$\overrightarrow{CD} + \overrightarrow{DG} = \overrightarrow{CG_0} + \overrightarrow{G_0G} \tag{7}$$

$$\overrightarrow{CE} + \overrightarrow{EF} = \overrightarrow{CG_0} + \overrightarrow{G_0F} \tag{8}$$

From the two vector equations we deduce four scalar equations with four unknowns:

$$l_4 \cos\varphi_4 + l_5 \cos\varphi_5 - l'_7 \cos(\alpha_7 + \varphi_7) = x_{G_0} - x_C \tag{9}$$

$$l_4 \sin\varphi_4 + l_5 \sin\varphi_5 - l'_7 \sin(\alpha_7 + \varphi_7) = y_{G_0} - y_C \tag{10}$$

$$l'_4 \cos\varphi_4 + l_6 \cos\varphi_6 - l_7 \cos\varphi_7 = x_{G_0} - x_C \tag{11}$$

$$l'_4 \sin\varphi_4 + l_6 \sin\varphi_6 - l_7 \sin\varphi_7 = y_{G_0} - y_C \tag{12}$$

The four unknowns are the angles $\varphi_4, \varphi_5, \varphi_6$ and $\varphi_7$ which are calculated from the system formed by the nonlinear scalar equations (9), (10), (11) and (12).

## 4. The calculation of the velocities and accelerations of the mechanism elements

In this stage of kinematic analysis, we impose the angle $\varphi_1$, the constant angular velocity $\omega_1 = \dot{\varphi}_1 = \pi \cdot n/30$ and the angular acceleration $\varepsilon_1 = \ddot{\varphi}_1 = 0$ [1]. We mention that the position angles $\varphi_2$ and $\varphi_3$ have been calculated at section 3.

For the velocity calculation, we derive with respect to time the two scalar equations of positions and we deduce the scalar equations of velocities:

$$l_3\dot{\varphi}_3 \sin\varphi_3 - l_2\dot{\varphi}_2 \sin\varphi_2 = -l_1\dot{\varphi}_1 \sin\varphi_1 \tag{13}$$

$$l_3\dot{\varphi}_3 \cos\varphi_3 - l_2\dot{\varphi}_2 \cos\varphi_2 = -l_1\dot{\varphi}_1 \cos\varphi_1 \tag{14}$$

This system has two linear scalar equations with the unknowns $\dot{\varphi}_2$ and $\dot{\varphi}_3$ for which we deduce the formulas:

$$\dot{\varphi}_2 = \dot{\varphi}_1 \begin{vmatrix} l_3 \sin\varphi_3 & l_1 \sin\varphi_1 \\ l_3 \cos\varphi_3 & l_1 \cos\varphi_1 \end{vmatrix} \Big/ \begin{vmatrix} l_3 \sin\varphi_3 & -l_2 \sin\varphi_2 \\ l_3 \cos\varphi_3 & -l_2 \cos\varphi_2 \end{vmatrix} \tag{15}$$

$$\dot{\varphi}_3 = \dot{\varphi}_1 \begin{vmatrix} l_1 \sin\varphi_1 & -l_2 \sin\varphi_2 \\ l_1 \cos\varphi_1 & -l_2 \cos\varphi_2 \end{vmatrix} \Big/ \begin{vmatrix} l_3 \sin\varphi_3 & -l_2 \sin\varphi_2 \\ l_3 \cos\varphi_3 & -l_2 \cos\varphi_2 \end{vmatrix} \tag{16}$$

For the open contours (4570) and (4670), we derive with respect to time the four scalar equations of positions (9, 10, 11, 12) and we deduce the scalar equations of velocities:

$$-l_4\dot{\varphi}_4\sin\varphi_4 - l_5\dot{\varphi}_5\sin\varphi_5 + l_7'\dot{\varphi}_7\sin(\alpha_7+\varphi_7) = -l_3'\dot{\varphi}_3\sin\varphi_3 \tag{17}$$

$$l_4\dot{\varphi}_4\cos\varphi_4 + l_5\dot{\varphi}\cos\varphi_5 - l_7'\dot{\varphi}_7\cos(\alpha_7+\varphi_7) = -l_3'\dot{\varphi}_3\cos\varphi_3 \tag{18}$$

$$-l_4'\dot{\varphi}_4\sin\varphi_4 - l_6\dot{\varphi}_6\sin\varphi_6 + l_7\dot{\varphi}_7\sin\varphi_7 = -l_3'\dot{\varphi}_3\sin\varphi_3 \tag{19}$$

$$l_4'\dot{\varphi}_4\cos\varphi_4 + l_6\dot{\varphi}_6\cos\varphi_6 - l_7\dot{\varphi}_7\cos\varphi_7 = l_3'\dot{\varphi}_3\sin\varphi_3 \tag{20}$$

We notice that the terms from the right are known because they contain only the angular velocity $\dot{\varphi}_3$ which has been previously calculated. Therefore, using this system we calculate the four angular velocities $\dot{\varphi}_4, \dot{\varphi}_5, \dot{\varphi}_6, \dot{\varphi}_7$.

The acceleration calculation is performed by the derivation of the scalar equations of velocities (13) and (14) in which we consider $\ddot{\varphi}_1 = 0$. For the contour (01230), the acceleration equations are:

$$l_3\ddot{\varphi}_3\sin\varphi_3 - l_2\ddot{\varphi}_2\sin\varphi_2 = -l_1\dot{\varphi}_1^2\cos\varphi_1 - l_3\dot{\varphi}_3^2\cos\varphi_3 + l_2\dot{\varphi}_2^2\cos\varphi_2 \tag{21}$$

$$l_3\ddot{\varphi}_3\cos\varphi_3 - l_2\ddot{\varphi}_2\cos\varphi_2 + l_3\dot{\varphi}_3^2\sin\varphi_3 = l_1\dot{\varphi}_1^2\sin\varphi_1 + l_2\dot{\varphi}_2^2\sin\varphi_2 - l_3\dot{\varphi}_3^2\sin\varphi_3 \tag{22}$$

From this system consisting of the scalar equations (21) and (22), we calculate the angular accelerations $\ddot{\varphi}_2$ and $\ddot{\varphi}_3$. For the contours (04570) and (04670), we derive with respect to time the four scalar equations of velocities (17, 18, 19, 20) and we deduce the scalar equations of accelerations. Then, using these equations we calculate the angular accelerations $\ddot{\varphi}_4, \ddot{\varphi}_5, \ddot{\varphi}_6, \ddot{\varphi}_7$.

## 5. The numerical calculation of the displacements, velocities and accelerations

For the numerical calculation of the nonlinear equations of displacements and for the linear equations of velocities and accelerations, we use the Mathcad® software [11].

In this regard, we run the programming code file and we obtain the numerical values of the kinematic elements' lengths for the first chain LD(2,3) type dyad [1].

Then, for the second chain LTt(4,5,6,7) type tetrad, we obtain the charts for the angular displacements (figure 5).

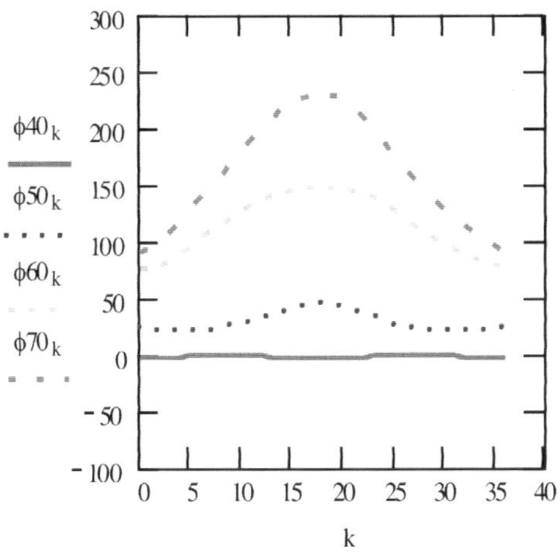


**Figure 5**. Angular displacement chart for the bars 4, 5, 6, and 7

Using MS-Excel® software we plot the charts for the displacements, velocities and accelerations of the kinematic elements by means of the numerical values obtained by the running Mathcad code file (figures 6, 7 and 8) [12].

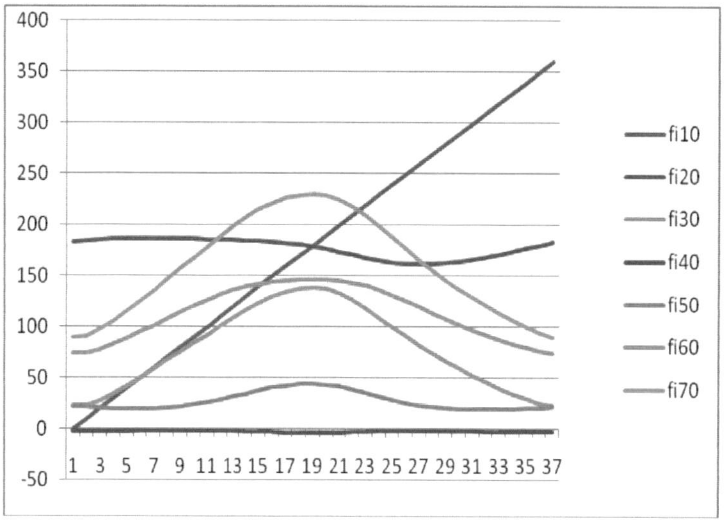


**Figure 6**. The chart of the angular displacements for all kinematic elements

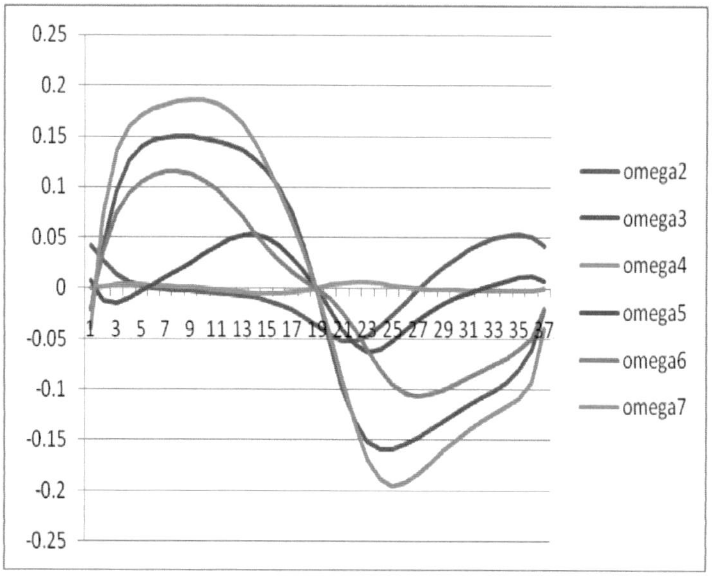


**Figure 7**. The chart of the angular velocities for the kinematic elements

The kinematic charts allow us to perform the dynamic calculation of the mechanism which help us to predict the operation behavior. This can be done by means of calculation of all forces that influence the linkage movement such as inertial forces and moments, and reaction forces.

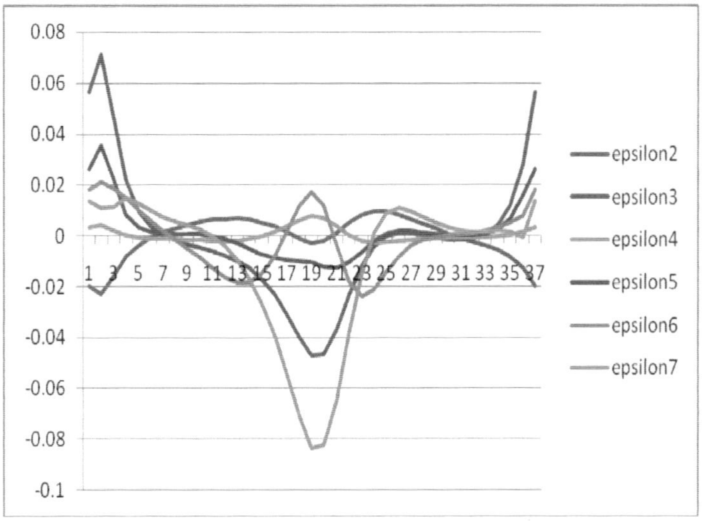


**Figure 8**. The chart of the angular accelerations for the kinematic elements

Finally, we obtain the driving torque at the shaft of the input element of the mechanism (crank 1) to know what type of actuator we choose in order to overcome all forces implied.

## 6. The modeling and kinematic simulation of the windshield wiper mechanism

In order to perform the modeling of all mechanism components, to assemble them and then to kinematically simulate the linkage we use the Inventor® software which allows us to see the mechanism behavior during a kinematic cycle [13].

The virtual modeling begins with a planar prototyping of each component of the mechanism. After planar designing of the element layout, we create the third dimension (pad) of it and obtain the 3D component [1].

The assembly starts with the inserting of the fixed element (vehicle body) on which we mount one at a time the rest of the components. Thus, we obtain the mechanism assembly by means of the concentricity and off-set constraints imposed on each rotate joint (figure 9).

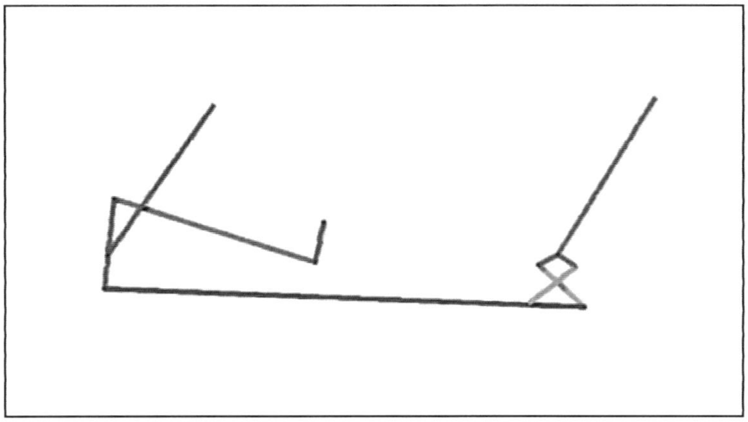

**Figure 9**. Shot of the windshield wiper mechanism

The kinematic simulation has been accomplished by defining of each joint, similar to the assembly, through imposing the controlled motion at the driving element.

Further, we can implement virtual sensors on each component in order to obtain the kinematic charts of displacements, velocities and accelerations. Thus, we can define the formulas with respect to time for the kinematic calculation by means of the digital mock-up.

## 7. Conclusions

The authors have been presented the structural-topological analysis and the kinematic analysis of a windshield wiper mechanism with two parallel rocker blades.

Even if the wiper linkages with opposite blades are more and more used on passenger motor vehicles, those with parallel blades are still the most used. When we choose the wiper mechanism type, we have to take account to the windshield design. Thus, the classical windscreens (more wide than tall) demand the use of the parallel type while the windshields having more or less a square shape are fitted to use the opposite type.

**Author's ORCID ID**
O Antonescu: https://orcid.org/0009-0009-3595-2806

## 8. References

[1] Antonescu D 2019 *Research regarding the analysis and synthesis of the complex planar mechanisms applied on passenger motor vehicles* (in Romanian), PhD Thesis (Bucharest: University Politehnica of Bucharest)
[2] Alexandru C and Buta A C 2006 *Windshield wiper mechanisms - Modeling and virtual prototyping* (in Romanian), (Brasov: Transilvania University Press)
[3] Authors group: 1969 *Getriebetechnik Lehrbuch.*, Veb Verlag Technik, Berlin
[4] Antonescu O Antonescu P 2016 *Mechanism and Machine Science* - coursebook, (Bucharest: Politehnica Press)
[5] Antonescu O and Antonescu P 2006 *Mechanisms and manipulators* (in Romanian), (Bucharest: Printech Press)
[6] Buta A C Alexandru C 2005 *Kinematic optimization of a windshield wiper mechanism*, In: SYROM 2005, vol I, ed Antonescu P (Bucharest: Printech Press) pp 17–20
[7] Antonescu P, Cocosila M and Antonescu D 1992 *Geometric and constructive structure of the windshield wiper mechanisms* (in Romanian), In: ESFA 1992, vol II (Bucharest) pp 375-386
[8] Antonescu P, Cocosila M and Antonescu O 1993 *Geometric synthesis of the windshield wiper mechanisms* (in Romanian), In: SYROM 1993, vol IV, ed Manolescu N I (Bucharest: UPB Press) pp 125-132
[9] Antonescu D, Popescu I and Strimbeanu M 2016 *Types of mechanisms used as windshield Wipers*, In: Proceedings of the 26$^{th}$ IFToMM Working Meeting on Standardization of Terminology - WMST 2016, ed Antonescu P (Bucharest: Politehnica Press) pp 37-42
[10] Antonescu P and Tempea I 1993 *The synthesis of the driving mechanisms of the windshield wipers used on passenger motor vehicles* (in Romanian), In: SYROM 1993, vol IV, ed Manolescu N I (Bucharest: UPB Press)
[11] *** Mathcad software documentation
[12] *** MS-Excel software documentation
[13] *** Inventor software documentation

# Contributions to the synthesis of the windshield wiper mechanism with one rocker-slider blade

**O Antonescu[1], D Antonescu[2] and A Ionita[3]**

[1] Dept. of Mechanism and Robot Theory, Politehnica University of Bucharest, Romania,
[2] "Iuliu Maniu" High School, Bucharest, Romania
[3] Dept. of Road Vehicles, Politehnica University of Bucharest, Romania

e-mail: oval33@hotmail.com

**Abstract.** The paper deals with synthesis of the windshield wiper mechanism with one rocker-slider blade. Starting from the kinematic scheme of the constructive model designed by Mercedes-Benz motor vehicle company, a structure analysis on this complex mechanism consisting of three kinematic chains, of which two are articulated bar mechanisms and one is planetary mechanism, has been accomplished. The novelty of this paper consists in the presentation of an optimum synthesis method of the planar bar mechanism in two variants for the wiper blade driving. The first variant is a planar 4-bar mechanism type crank-rocker where the cosines theorem has been applied in order to obtain the lengths of the coupler and the crank. The second analyzed variant is a bi-contour linkage consisting of two 4-bar mechanisms linked in series. Further, the calculation of the planetary gear angular velocity on the complex mechanism with dyadic kinematic chains has been achieved. In order to do this, the immobilization of the planetary carrier method has been applied. Finally, the geometry of the driven linkage of the wiper mechanism type crank-slider which has two mobilities has been studied. A vector equation has been written in order to obtain the length of the variable blade.

## 1. Introduction

The windshield wiper mechanism special type analyzed in the paper had been developed by Mercedes-Benz® in the 1980's [1]. It consists of several bars, including a tetradic chain, and an annulus planetary gear which drive a rocker-slider blade (figure 1).

**Figure 1**. The windshield wiper mechanism with one rocker-slider blade

Two types of windshield wiper mechanisms are the most used on present automobiles:

- with two “parallel” blades, which still the most used;
- with two opposite blades, which are increasingly used.

The mechanism with one rocker-slider blade is special due to the fact that motion of the blade is composed of both rotation and translation which makes it more efficient than the classical mentioned types, from the covered area point of view.

## 2. The topological structure and geometry of the windshield wiper mechanism

The kinematic scheme of the windshield wiper mechanism type Mercedes-Benz® with complex structure consisting of articulated bars and an annulus planetary gear is presented (figure 2) [2, 4].

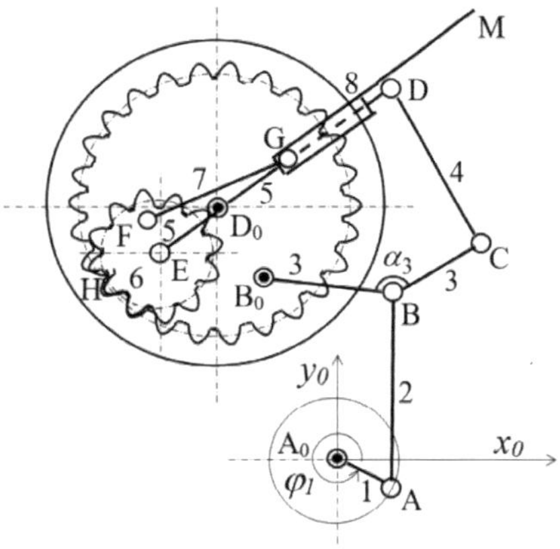


The geometry of the kinematic scheme of the mechanism is defined by the following linear and angular parameters:

$x_{B_0} = -9mm;\ x_{D_0} = -10mm;$

$y_{B_0} = 28mm;\ y_{D_0} = 35mm;$

$A_0A = 4{,}5mm;\ B_0B = 7;\ \alpha_3 = \angle(B_0BC) = 145^0;$

$AB = 30mm;\ BC = 8;$

$CD = 14;\ D_0E = 16;\ FG = 16{,}5;\ R_6 = HE = 5;$

$DD_0 = 7;\ EF = 7;\ GM = 84;\ R_0 = HD_0 = 22.$

**Figure 2**. Kinematic scheme of the wiper mechanism [2]

The mobility of the mechanism is calculated by the formula [5]:

$$M_b = \sum_{m=1}^{5} mC_m - \sum_{r=2}^{6} rN_r \tag{1}$$

where $m$ is the mobility of the kinematic joint, $C_m$ is the number of the kinematic joints of mobility $m$, $r$ is the rank of the associated space, and $N_r$ is the number of the closed contours of rank $r$.

By analyzing the kinematic scheme (figure 2) the following numerical values are identified:

- $m = 1; C_1 = 11;$ mono-mobile joints (planar articulations and sliding joints);
- $m = 2; C_2 = 1;$ bi-mobile joint (spur gear: planet gear 6 + fixed annulus gear 0);
- $r = 3; N_3 = 4;$ independent kinematic contours of the rank 3 (planar).

By substituting these numerical values in formula (1) the mobility is obtained:

$$M_b = (1 \times 11 + 2 \times 1) - 3 \times 4 = 1$$

The driving element 1 defines the “actuator mechanism” $MA(0,1)$ having a complete rotation (360°) and being positioned by the angle $\varphi_1 = 0,...,2\pi$ (counter clockwise measured).

If instead the superior joint H(6,0) which is an annulus gear, we represent the equivalent bar element $e_{60}$, then the structural formula of the equivalent „motor mechanism” can be written:

$$MM = MA(0,1) + LD(2,3) + LD(4,5) + LD(6,e_{60}) + LD(7,8) \tag{2}$$

We notice that the first dyadic chain LD (2,3) defines the first closed kinematic chain Lc(0,1,2,3) of type crank-rocker, and then the dyadic chain LC(4,5) defines Lc(0,3,4,5) of type rocker-rocker.

The third dyadic chain LD(6,$e_{60}$) defines Lc(0,5,$e_{60}$,6) where the bar $e_{60}$ has the direction on the shared normal direction to the two tangent profiles in point H (figure 2). The fourth dyadic chain LD(7,8) defines Lc(5,6,7,8) in which there are four planar articulations $(D_0, E, F, G)$ and a sliding joint (8,5). The point M belongs to the slider 8 which translates with respect to the bar 5 that executes an oscillating rotation with respect to the fixed point $D_0$.

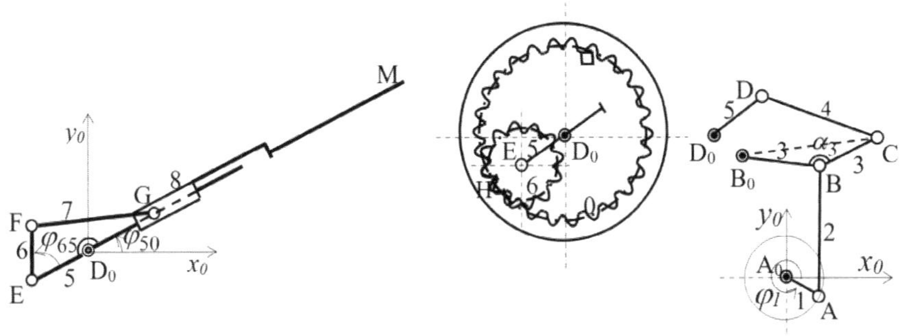


a b c

**Figure 3**. The three sub-mechanisms (independent kinematic chains): crank-slider with oscillating guide (a), annulus planetary gear (b) and two 4-bar chained linkages (crank-rocker and rocker-rocker) (c)

For each sub-mechanism (figure 3) the mobility is calculated by the formula (1):

a) $M_{b1} = 1\times 5 - 3\times 1 = 2$ (angular displacements $\varphi_{50}, \varphi_{65}$);

b) $M_{b2} = (1\times 2 + 2\times 1) - 3\times 1 = 1$ (angular displacement $\varphi_{50}$);

c) $M_{b3} = 1\times 7 - 3\times 2 = 1$ (angular displacement $\varphi_{10}$).

The linking mechanism is consisting of the joint kinematic elements of the two neighbor mechanisms. The mobility of such linking mechanism is:

$M_{bl1} = 2$ (open kinematic chain 0, 5, 6); $M_{bl2} = 1$ (kinematic chain 0, 5).

We verify the mobility of the complex mechanism by means of the linking formula of the mechanisms:

$$M_b = M_{b1} + M_{b2} + M_{b3} - (M_{bl1} + M_{bl2}) = 2 + 1 + 1 - (2 + 1) = 1 \quad (3)$$

By means of the two 4-bar sub-mechanism (figure 3c) we can obtain an oscillating rotation of the planetary carrier 5 by an obtuse angle $\psi$ equal to circa 145°. This angle results from the condition that the wiper blade (fixed together to the slider 8) to cover 80% from the entire windshield surface.

## 3. The geometric synthesis of the blade driving linkage

### *3.1. The 4-bar planar mechanism type crank-rocker*

The oscillating rotation movement by the obtuse angle of 145° can be obtained by means of the first analyzed variant of the planar articulated mechanism type crank-rocker (figure 4b) [2].

For the geometric synthesis of this mechanism we consider as known parameters: the base length $A_0D_0 = l_0$, the rocker length $D_0D = l_3$ and the angles $\angle A_0D_0D_1 = \varphi_3$, $\angle D_1D_0D_5 = \psi$.

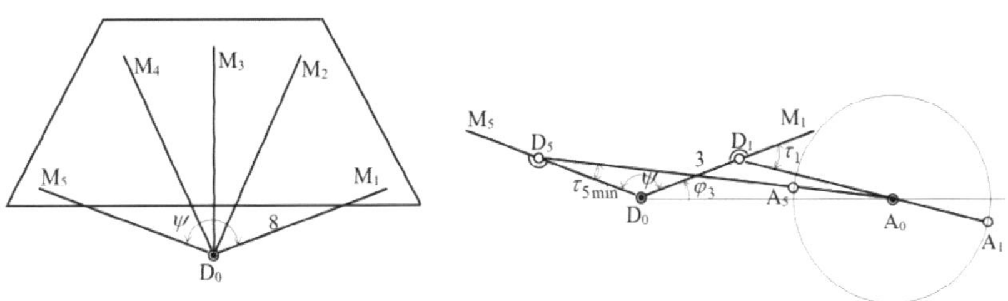


a b

**Figure 4**. a) The discrete positions of the wiper blade on the windshield; b) The 4-bar mechanism type crank-rocker for the angle $\psi$ displacement achieving

We denote by $A_0A = l_1$ (as crank) and by $AD = l_2$ (as coupler), these lengths being unknown.

We notice that in the case of the crank-rocker mechanism (figure 4b), in the extreme positions of the rocker ($D_1$ and $D_5$), the articulations $A_0, A$ and $D_1$, $D_5$ respectively, are colinear which means that the crank and the coupler are overlapped ($D_1A_0A$) or in prolongation ($D_5AA_0$).

In order to calculate the lengths $l_1$ and $l_2$, the cosine theorem in the triangles $(A_0D_0D_1)$ and $(A_0D_0D_5)$ is written:

$$l_2 - l_1 = \sqrt{l_0^2 + l_3^2 - 2l_0l_3\cos\varphi_3}; \quad l_2 + l_1 = \sqrt{l_0^2 + l_3^2 - 2l_0l_3\cos(\varphi_3 + \psi)} \tag{4}$$

From the two linear equations (4), the lengths for the coupler and the crank are resulting:

$$l_2 = \frac{1}{2}[\sqrt{l_0^2 + l_3^2 - 2l_0l_3\cos\varphi_3} + \sqrt{l_0^2 + l_3^2 - 2l_0l_3\cos(\varphi_3 + \psi)}] \tag{5}$$

$$l_1 = \frac{1}{2}[-\sqrt{l_0^2 + l_3^2 - 2l_0l_3\cos\varphi_3} + \sqrt{l_0^2 + l_3^2 - 2l_0l_3\cos(\varphi_3 + \psi)}] \tag{6}$$

The disadvantage of this solution (figure 4b) lies in the fact that the length of the crank $l_1$ results too big, and the transmitting angle in the extreme positions $D_1$ and $D_5$, $\tau_{5\min} = \angle(A_0D_5D_0)$, is too small, under 20°. This is due to the geometry of the solution and it could cause the stall of the mechanism.

### *3.2. The 4-bar bi-contour planar mechanism*

The second analyzed variant for the blade driving linkage is consisting of two 4-bar planar mechanisms linked in series (figure 5), of which the first is a crank-rocker mechanism (0, 1, 2, 3) and the second is a rocker-rocker mechanism (0, 3, 4, 5), that allows the multiplication of the oscillating rotation angle (from the angular displacement $\psi_3$ to the imposed angular displacement $\psi_5$) [2]. This solution has been studied at section 2 from the mobility point of view.

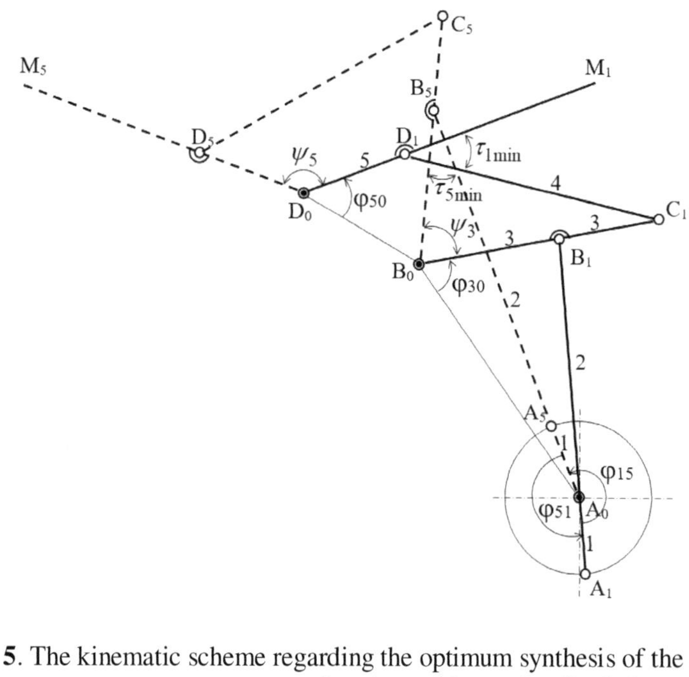


**Figure 5**. The kinematic scheme regarding the optimum synthesis of the articulated mechanism in the second variant with two dyadic chains

The geometric synthesis of the mechanism with two kinematic contours in series (figure 5) requires two stages:

a) The design of the 4-bar mechanism type rocker-rocker (0, 3, 4, 5) on which, in order to obtain the rocker 5 rotation by the imposed angle $\psi_5$, the angle $\tau_{1\min} \geq 30^0$ is chosen and the oscillating angle $\psi_3 \leq 60^0$ for the rocker 3 $(B_0C_1)$ is deduced.

b) The design of the 4-bar mechanism type crank-rocker (0, 1, 2, 3) through choosing the length $l_3$ for the rocker 3 $(B_0B_1)$ on the same direction with $B_0C_1$. Conveniently we choose the position of the fixed articulation $A_0$ taking account the condition that the transmitting angle to be at least the minimum value $\tau_{5\min} \geq 30^0$.

We use the procedure applied in the case of the first variant of driving linkage (figure 4b), using the formulas (4), by imposing the angle $\varphi_{30}$. We denote $A_0A = l_1$ and $AB = l_2$, and from the triangles $A_0B_0B_1$ and $A_0B_0B_5$ the following relations are deduced:

$$l_2 - l_1 = \sqrt{l_0^2 + l_3^2 - 2l_0l_3\cos\varphi_{30}};$$
$$l_2 + l_1 = \sqrt{l_0^2 + l_3^2 - 2l_0l_3\cos(\varphi_{30} + \psi_3)} \quad (7)$$

From the two linear equations (7) the following solutions are obtained:

$$l_1 = \frac{1}{2}[-\sqrt{l_0^2 + l_3^2 - 2l_0l_3\cos\varphi_{30}} + \sqrt{l_0^2 + l_3^2 - 2l_0l_3\cos(\varphi_{30} + \psi_3)}] \quad (8)$$

$$l_2 = \frac{1}{2}[\sqrt{l_0^2 + l_3^2 - 2l_0l_3\cos\varphi_{30}} + \sqrt{l_0^2 + l_3^2 - 2l_0l_3\cos(\varphi_{30} + \psi_3)}] \quad (9)$$

In this case (figure 5) we notice that the length of the crank is smaller and the minimum transmitting angle $\tau_{5\min}$ is maintained over the limit of 30° [5].

## 4. The angular velocity calculus of the planetary gear for the complex mechanism with dyadic kinematic chains

For the windshield wiper blade driving we consider the second variant with two dyadic chains (figure 6) on which the bar 5 is fixed together to the planetary carrier p(5) on which the planetary gear 6 is articulated. This gear 6 is geared with the annulus fixed gear 0 [2].

We notice that in the initial position $D_0D_1$, the rocker 5 is positioned through the angle $\varphi_{50}$ with respect to the direction of the fixed articulations $B_0$ and $D_0$.

Together with the rocker 5 (having the angular velocity $\omega_5$) the planetary carrier p(5) geometrically defined by the segment $D_0E$ articulated in E by the planetary gear 6 that has the angular velocity $\omega_6$.

With respect to the point $F_1$ (rotation instantaneous center), the planetary gear 6 is rolling inside on the dividing circle of the fixed annulus gear 0.

For the calculation of the angular velocity $\omega_6$ of the planetary gear 6, we apply the immobilization method of the planetary carrier p(5), resulting the following ratio:

$$i_{60}^p = \frac{\omega_6 - \omega_p}{\omega_0 - \omega_p} \quad (10)$$

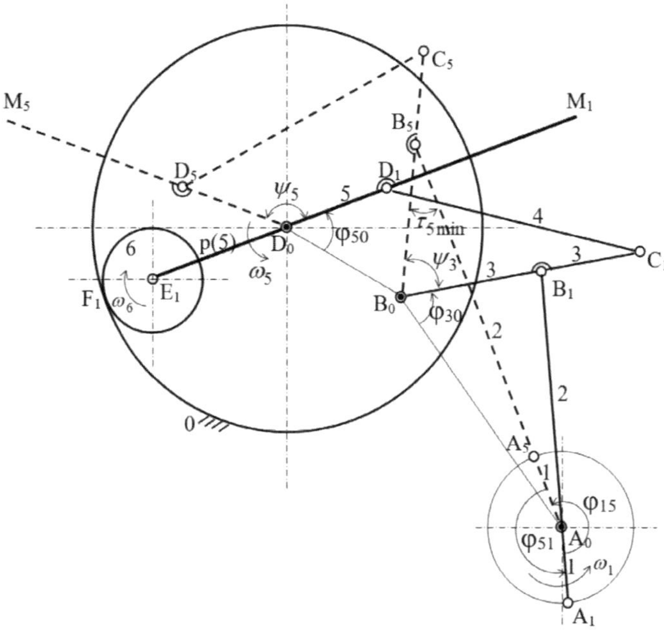


**Figure 6**. Kinematic scheme of the windshield wiper mechanism in two positions

Due to the fact that the annulus gear is fixed ($\omega_0 = 0$) the formula (10) becomes:

$$i_{60}^{p} = \frac{\omega_6 - \omega_p}{-\omega_p} = 1 - \frac{\omega_6}{\omega_p} \tag{11}$$

The transmitting ratio of the annulus gearing F(6, 0) is positive and it is function of the teeth numbers $z_6$ and $z_0$ [3]:

$$i_{60}^{p(5)} = \frac{z_0}{z_6} > 1 \tag{12}$$

By means of formula (11) we deduce:

$$\frac{\omega_6}{\omega_{p(5)}} = 1 - \frac{z_0}{z_6} \tag{13}$$

Finally, from the equation (13) we obtain the angular velocity of the planetary gear:

$$\omega_6 = \omega_{p(5)}(1 - \frac{z_0}{z_6}) \tag{14}$$

## 5. The geometry of the driven mechanism type crank-slider

The driven mechanism is type crank-slider with wiper blade (figure 7) and it represents the third kinematic chain in the configuration of the complex windshield wiper mechanism with bars and gears (figure 2) [2].

Point M (the end of the wiper blade) executes a rotation with respect to the fixed articulation $D_0$ (figure 7) simultaneously with a sliding movement in respect of bar 5 through the rotation of the crank 6 with respect to the mobile articulation E.

Basically, the segment GM represents the wiper blade which is fixed together with the slider 8, performing a rotate-sliding movement on the windshield surface.

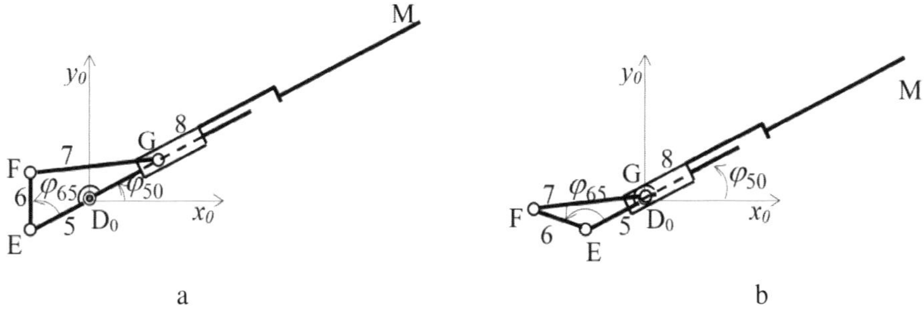


**Figure 7**. Kinematic scheme of the bi-mobile driven mechanism

The bar 5 is a rocker with the fixed articulation $D_0$ (figure 7a) being positioned by the angle $\varphi_{50}$.

The crank-slider mechanism geometry is defined by constant and variable parameters of the vector contour EFG (figure 8b).

The rotation angle of the crank 6 (fixed together with the planetary gear 6) is obtained by means of the formula (14):

$$\varphi_{65} = \varphi_{50}(1 - \frac{z_0}{z_6}) \tag{15}$$

We choose the mobile coordinate system $x_5Ey_5$ with the origin in articulation E, and we write the vector equation of the kinematic contour closing EFG, according to the chosen directions:

$$\overrightarrow{EF} - \overrightarrow{GF} - \overrightarrow{EG} = \vec{0} \tag{16}$$

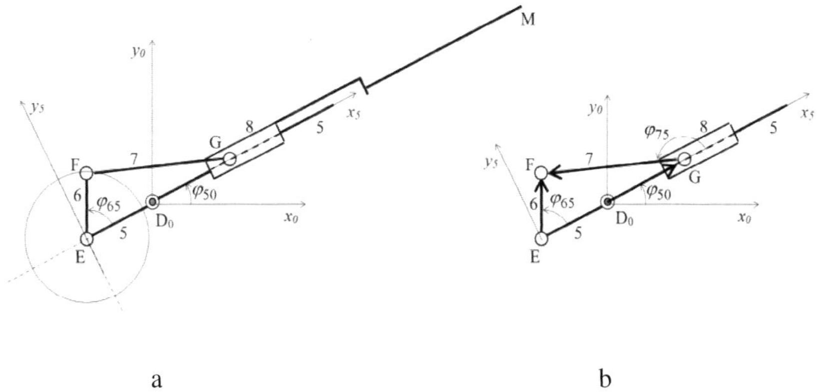


**Figure 8**. Kinematic scheme of the crank-slider mechanism (a) with the vector contour (b)

We denote $\overrightarrow{EF} = \vec{l}_6$; $\overrightarrow{GF} = \vec{l}_7$; $\overrightarrow{EG} = \vec{s}_8$ and we write the equation (16) by means of two equivalent scalar equations (figure 8b):

$$\sin\varphi_{75} = \frac{l_6}{l_7}\sin\varphi_{65}; \; s_8 = l_6\cos\varphi_{65} + l_7\cos\varphi_{75} \tag{17}$$

Finally, we denote $\overrightarrow{GM} = \vec{l}_8$ and we deduce the vector radius $\rho_M = D_0M$, of the point M, by means of the formula:

$$\rho_M = s_8 + GM = l_6\cos\varphi_{65} + l_7\cos\varphi_{75} + l_8 \tag{18}$$

IOP Conf. Series: Materials Science and Engineering 1303 (2024) 012042 doi:10.1088/1757-899X/1303/1/012042

## 6. Conclusions

The windshield wiper mechanisms are safety and comfort systems on any vehicle. These spatial mechanisms are assimilated as planar linkages with articulated bars situated in parallel planes. The advantage of the windshield wiper mechanism with one rocker-slider blade is that it covers the wiping surface with 20% higher than the classical mechanism with two parallel blades. However, the disadvantage of it is the decreased speed due to use only one blade.

Only two of four studied solutions for the driving linkages of this wiper mechanism have been highlighted: one 4-bar mechanism type crank-rocker and two 4-bar mechanisms linked in series type crank-rocker and rocker-rocker. The other two solutions that have been studied in another paper are: mechanism with a triadic chain and mechanism with a tetradic chain.

**Author's ORCID ID**
O Antonescu: https://orcid.org/0009-0009-3595-2806

## 7. References

[1] Mercedes-Benz Public Archive Homepage, https://mercedes-benz-publicarchive.com, last accessed 2023/04/28

[2] Antonescu D 2019 *Research regarding the analysis and synthesis of the complex planar mechanisms applied on passenger motor vehicles* (in Romanian), PhD Thesis (Bucharest: University Politehnica of Bucharest)

[3] Antonescu O and Antonescu P 2016 *Mechanism and Machine Science* – coursebook (Bucharest: Politehnica Press)

[4] Mekanizmalar Homepage, https://www.mekanizmalar.com/mercedes-benz-mono-wiper-animation.html, last accessed 2023/04/25

[5] Antonescu O and Antonescu P 2006 *Mechanisms and manipulators* (in Romanian), (Bucharest: Printech Press)

IOP Conf. Series: Materials Science and Engineering 1303 (2024) 012043 doi:10.1088/1757-899X/1303/1/012043

# Analysis of different dimensions of a virtual wind tunnel and different wall conditions

**E M Vieru, D M Iozsa, A M Manea and L Ilea**

Automotive Engineering Dept., Faculty of Transports, POLITEHNICA University of Bucharest, Romania

E-mail: vieru_mihai_e@yahoo.com

**Abstract**. Growing interest in improving the performance of vehicles is a subject that allows for continuous research. One such branch of these researchers is energy and consumption performance. These performances describe, on a global level, the need for lower fuel consumption but also a higher energy and power performance. Therefore, to create this balance, direct control of polluting emissions, especially C02 emissions, is necessary. Creating the best mix between low fuel consumption, maximum power of internal combustion engine and low pollutant emissions is influenced by many factors. One such factor, of prime importance, is represented by the aerodynamics of vehicle, in particular the aerodynamic performance, which needs to be tested to be confirmed, so numerical values are needed. Therefore, different ways of testing in the physical or virtual environment are necessary. To facilitate the optimization of a vehicle from an aerodynamic point of view, the most practical way is the virtual wind tunnel, given the possibility to test, at the same time, more configurations. But the CFD (Computational Fluid Dynamics) simulations, sometimes, have the disadvantage of very long compilation times. So, in the presented paper the construction of an optimal virtual wind tunnel for trucks will be presented, focusing on the test area (chamber) and several configurations will be presented such as: different proximity of the tunnel walls to the tested object so a different geometry of the test chamber and condition for the type of floor with different roughness than base case to simulate a more realistic road surface. Following the CFD simulations, it was observed the influence that increasing the dimensions of the test chamber has on the results obtained, so that moving the walls away from the test model generates much more realistic results. In addition, another thing that has an impact is the roughness of the floor, which, in this particular case, generate a 10% difference between the values obtained. Therefore, it will be possible to optimize the tunnel as best as possible to have results as close as possible to reality and in a reduced time.

## 1. Introduction

Nowadays, the demand for the transportation of goods from a place to another is constantly growing. As a result of this demand, there is also a significant diversity in the ranks of vehicles that transport goods and valuables. Transport must be done quickly, easily and safely, in one word: efficient. This efficiency is given, among other things, by the dynamic capacity of the vehicle; in this paper the emphasis is placed on the aerodynamic performance. This performance is dictated by the aerodynamic coefficient (Cd) of the vehicles which can also be translated through a reduced fuel consumption. Cd is a dimensionless numerical value that is difficult to estimate but can be determined through various physical tests and/or in the virtual environment. The physical tests take place in large wind tunnels, tests used both for optimizing the bodywork and/or its various auxiliary elements or used for homologation. A physical test requires the presence of a model of the vehicle and can be at scale 2:5, 1:1 or any other scale. This model must be in very good condition and be built mainly from body parts or clay or 3D printed elements. In addition, it must have a very solid construction to withstand the high wind speed as well as possible. A physical test in a wind tunnel has advantage, especially in the case of 1:1 scale model, of precise results and very close to the real exploitation of the vehicle, but given the relatively high dimensions and costs, they are produced in a reduced number. Most of the time, different car manufacturers test their products in the same wind tunnel. This involves rigorous and though organisation to meet the all-testing requirements. This describes a time limitation in terms of testing multiple configurations of the same model to optimize it. That's why, to streamline and speed up the development process of the vehicles it switched in parallel, to testing in the virtual environment, of the CFD type, with wind tunnel. This type of testing has the advantage that it is not necessary to model the entire wind tunnel, but only the test chamber (the area of interaction between the model and the air flow). Besides the gain from point of view of taking space (because it takes place in the virtual environment), it has the advantage that several tests (simulations) can be started at the same time; in other words, a bigger image on several models in a much shorter time. However, there is the disadvantage of not having results close to reality; these being dictated by discretization method of the models, which will be discussed in the following lines. Next, different types of wind tunnels will be presented. These are used in automotive industry, aviation, construction and others[1], [2].

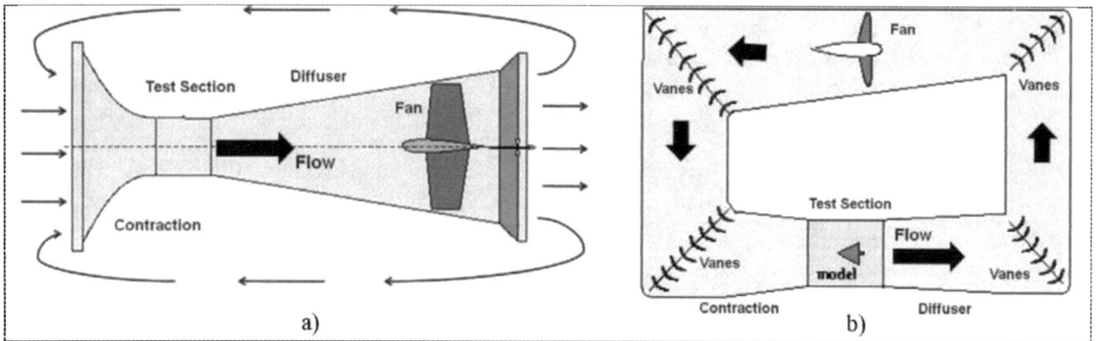


**Figure 1.** Types of wind tunnels: a) Open-return wind tunnel [4], b) Close-return wind tunnel [5]

Based on Figure 1 can be distinguishing two different wind tunnels: open-return in where the air that reach in the test section is gathered from the room where the wind tunnel it is placed and close-return where the air is conducted from the exit of the test section back to the fan by a series of turning vanes or blades. The close-return wind tunnel has a superior quality of the flow in the test section due to the fact that it is relatively uniform in comparison with the open-return where it has a poor flow quality. [3], [4], [5]. For this paper, the idea is to have a close-return type wind tunnel from which only the test chamber (test section) will be modelled. This will be configured in different dimensions to be able to present which of them have the best potential to be used for tests in order to determine an aerodynamic performance. Based on this paper, the purpose of the importance of the good definition an selection of appropriate dimensions for the virtual test chamber will be concluded and the necessity of adding different parameters, such as the roughness of the floor, to obtain the most conclusive results will be presented.

## 2. Virtual wind tunnel

For this paper, the construction of an aerodynamic tunnel for trucks will be presented and the emphasis will be placed only on the test section. In the CFD virtual environment (we have the possibility to simulate only an area of the entire tunnel, the rest can be entered as parameters. For example, to simulate the fan we will give a speed of the air as an input parameter and in addition it is not necessary to specify which type of tunnel will be used (open-return or close-return). Having to do with a numerical simulation, a very important thing in the accuracy of the results is represented by discretization. The finer the mesh, the better and more realistic the results will be. In the actual paper, for truck with a scale of 1:1, it is necessary that the discretization to be as fine as possible. However, certain details that can complicate the calculation process can be removed, such as: design details of door handles, of the mirrors, elements that meet the air flow such as those inside the passenger compartment. In this way, the time to obtain the results is minimized. Therefore, in the numerical simulation program, a wind tunnel will be modelled with a rectangular test chamber and the minimal starting dimensions will have an average, approximate value, following the verification of several wind tunnels after, it will be used with various other configurations. In Figure 2, it can be observed a test chamber for a truck assembly (tractor unit and semi-trailer) and, at the same time, it can be seen the large dimensions of this type of structure can have. In this case, the open jet test section it is part of a test hall of 50m x 30m x 20 [3]. It can be said that these give the chance of having the best and most conclusive results, eliminating the possibility of measurements errors as a result of unwanted air interference with the tunnel walls.

**Figure 2.** LLF DNW *(Large Low-Speed Facility, German-Dutch Wind Tunnels)* Wind tunnel [6]

The physical aerodynamic tunnels for vehicles have rotation lanes for each wheel, thus simulating the operation in reality as best as possible. In addition, the wheels are a significant generator of vortices that influence the overall aerodynamic performance of the vehicle. In the virtual environment, this rotation band/lane can be simulated through a rotation domain, in other words: each wheel will be given the condition to rotate with a certain angular speed (the same for all wheels). For this paper, a wheel rotation domain was not considered given the simplistic model used for the test.

In the virtual environment, as was mentioned in the rows above, there is a much greater possibility of having large dimension wind tunnels, but the waiting time for solving the results will also be longer. At the same time, in addition to large dimensions, discretization is another important factor. In addition, there is the possibility that many virtual tunnels are over-dimensioned and the values on the 3 axes (X, Y, Z) are far above the maximum limit at which relevant results can be obtained. In this case, sometimes the excessive size of dimensions leads to very long waiting times with, mainly the same results as smaller tunnels. That is why it is important to analyse several configurations to observe from which point the increase in dimensions is no longer necessary [7], [8].

## 3. Set-up

To observe the possibility of having measurements influenced by the dimensions of the wind tunnel, several simulations were made at different dimensions and different conditions for the walls. Also, a simplified model was used for these simulations. By analysing the results, it will be possible to approximate, the point from which increasing the dimensions of the tunnel is not necessary, thus shortening the simulation period.

### *3.1. Geometrical model*

The geometric model was created to simulate a basic shape of a truck, so that an air flow similar to that of a normal operation or a real truck can be generated. As can be seen in Figure 3, the test model has similar dimensions to a real truck and does not have complex elements that could complicate the calculation process. In addition to this aspect, as mentioned previously, for the truck, the condition of the field rotation for the wheels was considered, therefore, these are a common (solid) body with the base geometry of the truck. Auxiliary structural elements such as: mirrors, air tanks, storage boxes, signalling elements were, also, removed to simplify the model so that the processing time of the simulation is as low as possible.

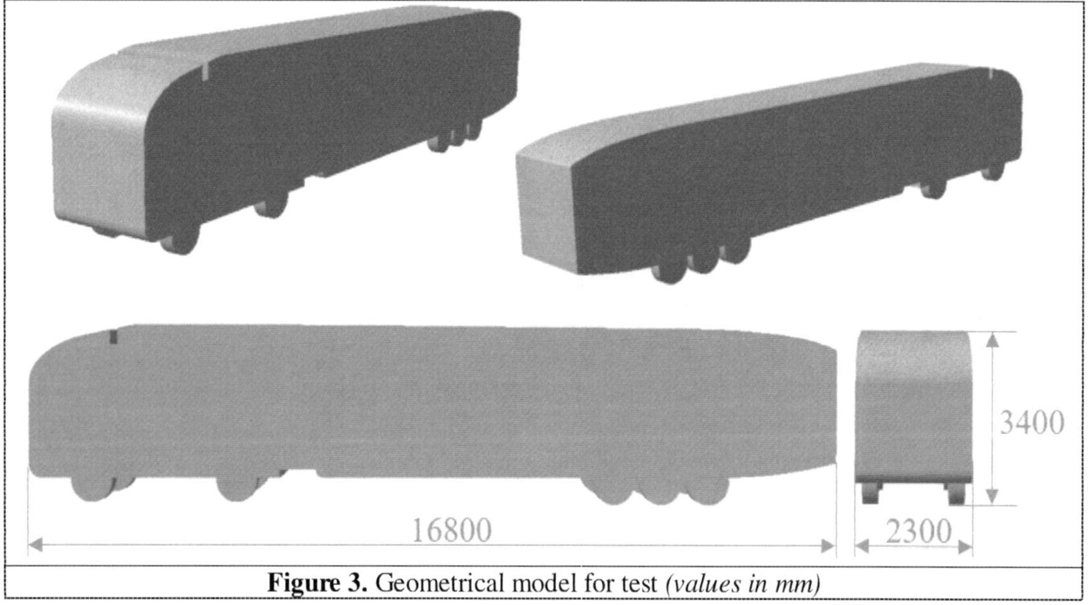


**Figure 3.** Geometrical model for test *(values in mm)*

### *3.2. Mesh*

Discretization is one of the most important factors in terms of delivering the best possible results. As a consequence of the fact that a test model with a 1:1 scale was used for this paper this lead, as a result of low computing power (12 GB RAM, 2.50 GHz CPU, 2GB GPU), to use a relatively quality criteria sufficient to satisfactorily cover all the geometry; it is good to remember that the discretization it is made both: for the geometric model and for the test chamber with the same value.

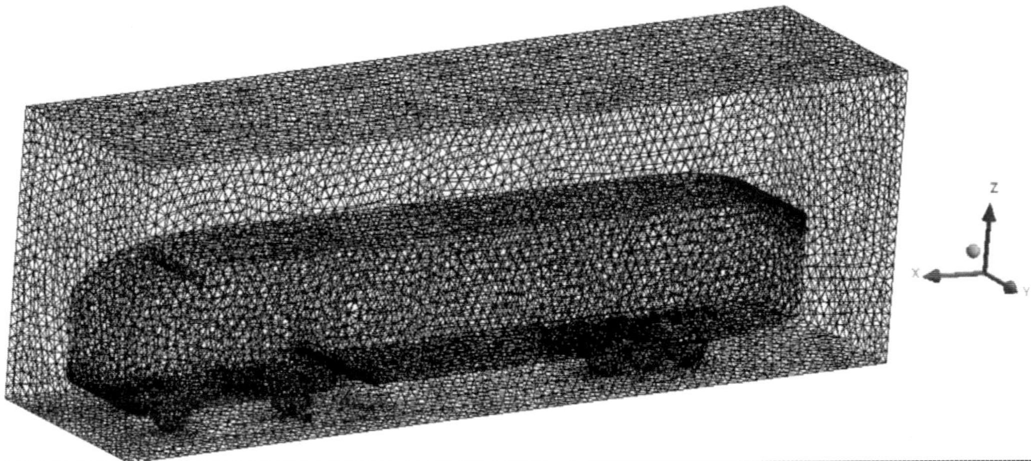


**Figure 4.** Mesh of the geometrical model and the test chamber (case when elements have a size of 260 mm)

As can be seen in the Figure 4, discretization of the wind tunnel and the geometrical model are unified as a single block due to the meshing of the solid (truck) and the area where the airflow should be (test chamber). From meshing point of view, four dimensions were used for the elements: 60 mm, 160 mm, 260 mm and 360 mm. For these, the following results of the forces on the X axis were obtained: 407297 N (for 60 mm), 397974 N (for 160 mm), 416721 N (for 260 mm) and 412081 N (for 360mm). It can be seen that there is a difference between the forces and the reason why it was chosen that the discretization of all the models in this paper should be done with elements that have size of 60 mm is given by the fact that, in general, a finer meshing determines better results. However, for the value of 160 mm, the force on X axis is lower than that at 60 mm, 260 mm and 360 mm. This particular case describes way in which the value of the elements and the type of discretization must be carefully chosen because improper values for these parameters can generate final results that are not definitively conclusive. In addition meshing with a value of 60 mm for element size sums up 2854386 nodes and 16438941 elements.

### 3.3. Preparation

Given the fact that for these CFD simulations (performed in Ansys software) only the test chamber will be used, the airflow will be given as a parameter to simulate the other areas of a real wind tunnel. Therefore, for all simulations, an air speed of 150 km/h will be used for the entrance in the chamber and a pressure of 1 atm (atmospheric pressure) will be used as a data for the exit of the air flow from the test chamber, also air temperature it is considered at 298,15 K and density at 1,185 kg/m$^3$. The value of velocity for these tests was chosen as a result of the fact that as a vehicle's forward speed increases, its forward resistance also increases. For trucks, although the speed is much lower, the purpose of this paper is to present in an informative way the influence of the dimension of the test chamber on the model. In the case of the analysis of the specific aerodynamic performance of a truck or its various specific equipment, after the correct dimensioning of the test chamber, the velocity can have values of 80 km/h or 90km/h, so specific to this kind of vehicle. In addition, at this speed the air flow can be observed more easily and more clearly. Other parameters such as wall conditions will be presented for each individual configuration. Regarding the values for the distance in X between the air inlet in the test chamber and the geometrical model for the 11 configurations they are (in this order): 100 mm, 500 mm, 1000 mm, 1500 mm, 2000 mm, 2500 mm, 3000 mm, 4000 mm, 5000 mm and 700 mm for the last two.

## 4. Configurations

As previously mentioned, for this paper, different dimensions will be checked for a virtual wind tunnel to estimate the point from which additional dimensioning is no longer necessary. For this, it was chosen to use as a starting point a base simulation in which to successively increase the dimensions on the 3 axes. The base simulation, in terms pf dimensions, was chosen to be the most constraining model. In addition, for few simulations, different conditions were set for the walls to observe the influential possibilities they could have on the results obtained. Table 1 shows the dimensions used for the test chambers and also the boundary conditions for the walls.

**Table 1.** Dimensioning of the test chamber and boundary conditions.

| | Dimensions ( L x W x H) (mm) | Boundary condition |
|---|---|---|
| Simulation 1 | 17000 x 2500 x 3500 | Free slip walls |
| Simulation 2 | 18300 x 3300 x 3900 | Free slip walls |
| Simulation 3 | 19800 x 4300 x 4400 | Free slip walls |
| Simulation 4 | 21300 x 5300 x 4900 | Free slip walls |
| Simulation 5 | 22800 x 6300 x 5400 | Free slip walls |
| Simulation 6 | 24300 x 7300 x 5900 | Free slip walls |
| Simulation 7 | 25300 x 8300 x 6400 | Free slip walls |
| Simulation 8 | 27300 x 10300 x 7400 | Free slip walls |
| Simulation 9 | 29300 x 12300 x 8400 | Free slip walls |
| Simulation 10 | 19000 x 4580 x 6050 | Free slip walls |
| Simulation 11 | 19000 x 4580 x 6050 | Rough wall of 8 mm for the floor. Rest of wall with free slip condition |

As can be seen in the Table 1, Simulation 1 will be considered as a base starting point simulation and from this one, dimensions were increased up to Simulation 9. For all this nine-simulation condition for all walls it is to be free slip so, no friction between fluid and wall. Another two iterations are Simulation 10 and Simulation 11 where was tested a condition of rough wall for the floor but it is kept the free slip condition for the other three walls. Dimensions for simulations 10 and 11 were chosen randomly. For an easy identification of the impact generated by each of the configurations in Table 1, it was chosen to present the force on the X axis for each of the 11 iterations. Through this method, having a numerical value, it is possible to determine the simulation that has the best solution, respectively finding the point from which dimensioning is no longer necessary.

**Table 2.** Forces on X axis.

| Simulation | Force (N) |
|---|---|
| 1 | 407297 |
| 2 | 34597 |
| 3 | 13702 |
| 4 | 9337 |
| 5 | 7379 |
| 6 | 6318 |

| | |
|---|---|
| 7 | 5656 |
| 8 | 4885 |
| 9 | 4484 |
| 10 | 19021 |
| 11 | 20975 |

As can be seen in Table 2 forces on X axis for the 11th simulations are presented. Focusing on the iterations from 1 to 9, it can be observed that as the dimensions of the test chamber increase from one simulation to another, the value of the force on the X axis decreases, which means that the resistance force is reduced. In parallel, for simulations 10 and 11, the impact of the rough wall condition of the floor can be observed, which significantly influences the results.

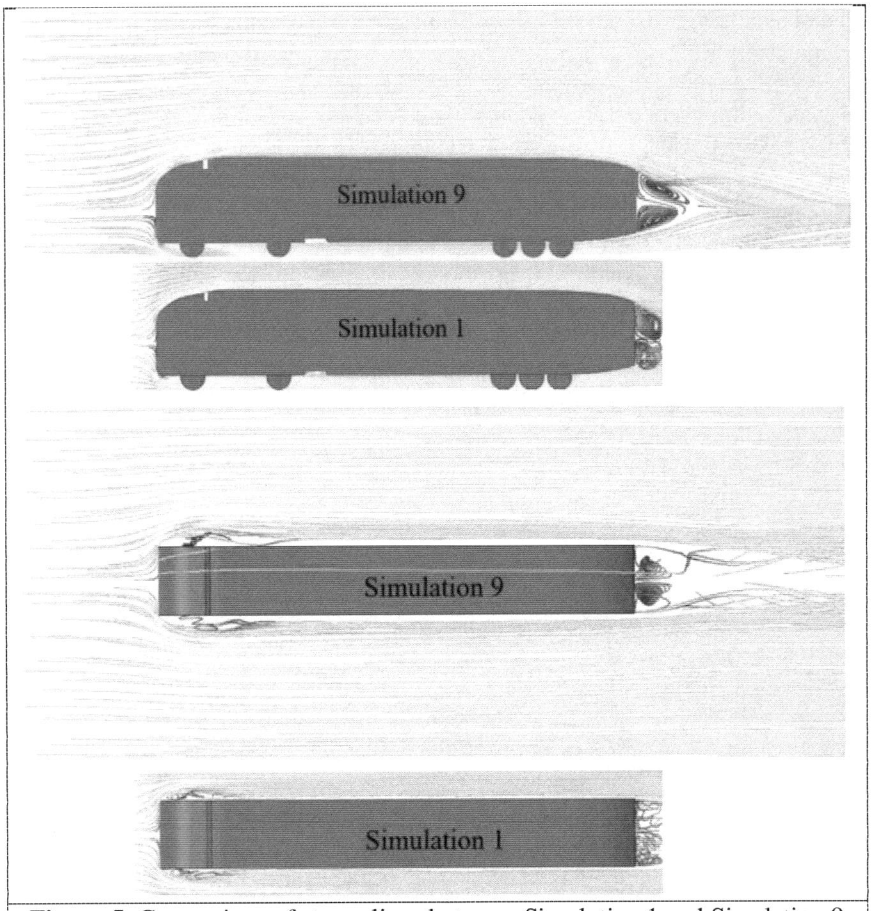

**Figure 5.** Comparison of streamlines between Simulation 1 and Simulation 9

In terms of airflow, impact can be seen in Figure 5 where is presented a comparison between Simulation 1 – one with the smallest dimensions of test chamber and Simulation 9 – one with the largest dimensions of the test chamber. It is very easily visible that for the Simulation 9, especially in the rear part of the vehicle, the airflow allows the generation of a more natural wake compared to Simulation 1 in which the wake is almost non-existent, the airflow being forced to end up suddenly. At the same time, in the front part for Simulation 9, the air has the advantage of covering the body (cabin) of the vehicle much better so that the flow is as natural as possible. It can also be seen on the sides that in Simulation 1 the air is very compressed between the semi-trailer and the walls, which does not allow the simulation of a normal, natural flow.

## 5. Conclusion

As it was presented in the lines above, by means of this paper, an attempt was made to present how the dimensions of a wind tunnel, in this case of a test chamber, influence the obtained results. Based on the numerical values achieved by measuring the force on the X axis, a decrease of this index is clearly observed, which concludes that a larger chamber test size also leads to good results, close to reality. The difference between the 1$^{st}$ simulation and the 9$^{th}$ one shows the impact that the proximity of the walls has to the tested model and describes the fact that due to an inadequate dimensioning the results obtained can be considered wrong. Although the simulation was stopped for a test chamber value of 29300 mm x 12300 mm x 8400 mm, the values show that an extra increase in the values of dimensions will lead to even better results. Based on values from Table 2 and images from Figure 5, it can be said, in a general way, that the dimensions of the tunnel for a geometric model must be large enough to avoid pressure zones between the walls and the geometry and also to have the possibility to allow the generation of good and natural wake. Regarding the boundary conditions, the influence of the floor wall with a roughness of 8 mm was observed. Compared to Simulation 10, Simulation 11 had about 10% higher value in terms of force on X axis. Therefore, this describes the need to integrate the road conditions (roughness) into the simulations. As a general conclusion, it is important to consider the dimensions of the geometry that will be tested and to choose the appropriate values so that there is no high pressure and friction between walls and model. In addition, the insertion of condition for the runway has the potential to generate better results.

**Author's ORCID ID**
M D Iozsa: https://orcid.org/0000-0002-6547-8633
A M Manea: https://orcid.org/0000-0002-3976-8848

## 6. References

[1] Tabatabaei N, Örlü R, Vinuesa R and Schlatter P 2021 *Engineering Mechanics*, KTH Royal Institute of Technology, Stockholm, Sweden, July
[2] Hermans J 2015 *Energy and transportation*, Huygens Laboratory, Leiden University, 2015
[3] Arnab M B and Ahmed Z U 2022 Dept. of Mechanical Engineering, Khulna University of Engineering & Technology, December
[4] https://www.grc.nasa.gov/www/k-12/airplane/tunoret.html, accessed in June 2023
[5] https://www.grc.nasa.gov/www/k-12/airplane/tuncret.html, accessed in June 2023
[6] https://llf.dnw.aero/test-sections, accessed in June 2023
[7] Lyu Z 2016 *Aerodynamic Wind tunnel in Passenger Car application*, Env. & Fluid Dyn, Centre, Volvo Car Corporation
[8] Watkins S 1990 *Wind-Tunnel modelling of vehicle aerodynamic: with emphasis on turbulent wind effects on commercial vehicle drag*, Department of Manufacturing and Process Engineering, Victorian University of Technology, R.M.I.T. Campus

# Analysis of the aerodynamic performance of truck semi-trailers

**E M Vieru, C Stan and L Ilea**

Automotive Engineering Dept., Faculty of Transports, POLITEHNICA University of Bucharest, Romania

E-mail: vieru_mihai_e@yahoo.com

**Abstract.** In the last time, trucks have shown a very high interest in terms of their technical and fuel consumption performance. Prioritizing the phenomenon of having trucks that transport as much, as quickly and as economically as possible, this represent a new challenge for industry and engineering, which are in continuous development. The attempt to create a bridge as stable as possible between technical and consumer performance is supported by the multitude of areas that, based on studies and experimental trials, allow optimization without interfering, substantially, with the actual transport of the goods (loading capacities and transport). The main objective of this paper is to present the possibility of improving the aerodynamic performance of trucks, focusing on semi-trailers. For this study, a truck modelled at one to one scale was used, whose aerodynamic performance was verified in the virtual environment (CFD –Computational Fluid Dynamics); this environment has the possibility to test several geometric models without requiring the layout of a vehicle in physical form. To check the performance, a basic model was created, a conventional truck (tractor unit with semi-trailer), to which various semi-trailer modifications were made, such as: total and partial coverage of the side areas, idea to create a more aerodynamic profile of the semi-trailer in the back and the addition of fairings under the platform (underbody of the semi-trailer). The results describe the impact these elements have and allow the visualization of the possible future profile that is as aerodynamic as possible for semi-trailers (therefore also for trucks) from both theoretical and technical point of view. Through the proposed improvements the capacity and the transport of the goods is not affected, in a significant way so it can be said that there is the possibility of implementation in the industry.

## 1. Introduction

Transport of goods on the roads is mostly done with truck assembly of tractor unit with semi-trailer. This is due to the fact that the transport is done between longer distances, which involves the use of highways or another's high-speed roads; therefore a relatively high speed for trucks and this implies a higher fuel consumption. To prevent, or better said to minimize this negative impact which has implications in the economic part, over the time more and more possibilities have been studied to improve the aerodynamic performance of truck semi-trailers. Given the precise shape for this type of structure, rectangular one, the areas where optimizations can be made are quite limited. Zones with the biggest potential were and remain: the wheel housing, rear, side areas and the roof. Most of the cases for the rear part, on5 the ceiling, the changes are made by bevelling the last third in order to have the possibility of generating a good wake. But this modification is constrained by the way of loading the goods, which, most of the time, in done through the rear.

a) b)

**Figure 1.** Example of aerodynamic semi-trailers [1].

In Figure 1 are presented two types of semi-trailers that mainly have the same areas covered: sides and improvements in the rear part. Image from Figure 1a has a more aerodynamic profile so a lower drag resistance; the second one, from Figure 1b, has a profile adapted to the current transport status.

**Figure 2.** Example of aerodynamic semi-trailer [2].

As was mentioned previously, Figure 2 shows a semi-trailer with aerodynamic improvements on the roof. In this case, although it seems that the rear area can reduce the access for loading goods, an alternative can be considered for the lateral side. In the actual research, the possibilities of improving the aerodynamic performance of semi-trailers will be analysed by working on the side and rear side of the roof.

## 2. Geometrical models and set-up

The geometric models used in this paper have a real scale and it was tried to be build up as close as possible to reality: were modelled with storage boxes, axles, shock bar and other elements that could influence the air flow in any way. Also the geometric model was inspired based on the global observation of European trucks, in circulation on public roads, without following a specific model. The areas where optimization was attempted are those mentioned in the rows above. The base geometry with which the simulations improvements will be compared is represented by a truck assembly consisting of a tractor unit and a semi-trailer without aerodynamic elements. In Figure 3 it is presented the base model without any aerodynamic elements on it. Based on it, several types of configurations will be tried in which different areas will be closed or optimized to obtain the best aerodynamic performance.

**Figure 3.** Geometrical base model *(values in mm)*

Based on this model a virtual wind tunnel(test chamber) was built and tested in Ansys software with the following dimensions: length of 25738mm, height of 6744mm and width of 7600 mm. The dimensions were chosen to have fastest results after the simulations; in addition, an attempt was made to avoid influence of the air flow with the test chamber walls in order not to have erroneous results. As input data for the velocity of the air, the value introduced was of 150 km/h and the walls have condition of free slip walls. The speed value of 150 km/h was chosen only to have a better and clearer picture of the airflow. Exit parameter from the virtual wind tunnel is of 1 atm and air temperature it is considered at 298,15 K and density at 1,185 kg/m$^3$. In terms of meshing, due to the fact that the simulations were performed on real scale models value for elements was of 90 mm. Regarding the power of calculation, the values for parameters are: 12 GB RAM, 2.50 GHz CPU, 2GB GPU. In Figure 4 it can be seen the test chamber where the base model was positioned and also the discretization (image from Figure 4b it is as a presentation model, the size of elements for meshing is 700 mm because the actual discretization of the model with 90 mm would be hard to be understandable).

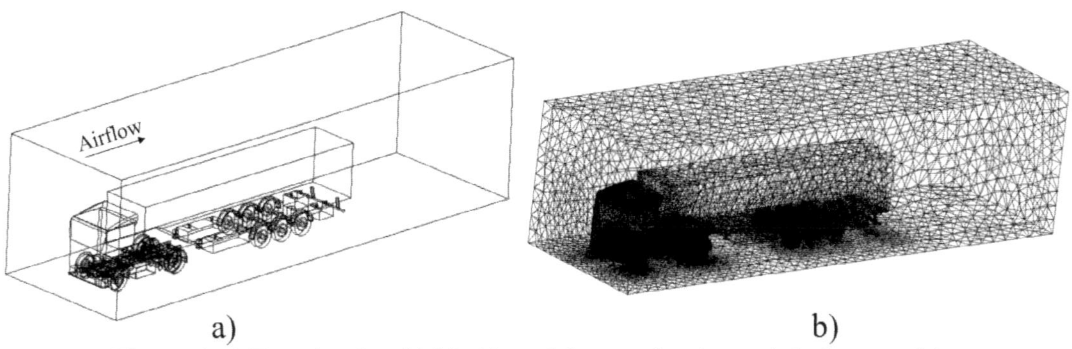


**Figure 4.** a) Test chamber, b) Meshing of the test chamber and the base model

To characterize the flow of the fluid, in this case air, the Reynolds number can be used. In this way, it can be found out if the airflow is laminar, transient or turbulent.

$$R_e = \frac{\rho \cdot v \cdot L}{\mu} \tag{1}$$

As input parameters for calculation we have: airflow velocity v=41,66 m/s, length of the assembly L=16,238 m, dynamic air viscosity $\mu=1{,}837\text{x}10^5$ kg/m·s, air density $\rho=1.185$ kg/m$^3$ correspondent to temperature of T=298,15 K. Based on formula (1) this number can be calculated and as a result it can into the following regimes: if Re<5x$10^5$ describes a laminar air flow, if Re is between 5x$10^5$<Re<$10^7$ describes a transient air flow and if Re>$10^7$ describes a turbulent air flow. So, based on the value obtained from formula (1) the Reynolds number has the value of 4,36x$10^7$ which describes a turbulent airflow. [3], [4]. In the following lines the configurations with the proposals for optimization are presented. The first aspect that was emphasized is the rear part of the roof of the semi-trailer. As can be seen in Figure 5, three configurations are presented: a - base model, b - starting from the base model, the roof was chamfered starting from the last axle of the semi-trailer, c - staring, also, from the base model, the roof was chamfered starting from the first axle of the semi-trailer. Considering these changes, the impact will be checked for configurations "b" and "c". In these cases, given the chamfered areas at the back, the two configurations have the potential to have much better aerodynamic performance than the base model. In addition, between the 3 iterations there are no changes for other areas of the semi-trailer, so the impact will be specific to the roof.

**Figure 5.** Configurations for the roof: *a) base model, b) chamfered starting from the last axle, c) chamfered starting from the first axle*

In the following lines the configurations with the proposals for optimization are presented. The first aspect that was emphasized is the rear part of the roof of the semi-trailer. As can be seen in Figure 5, three configurations are presented: a - base model, b - starting from the base model, the roof was chamfered starting from the last axle of the semi-trailer, c - staring, also, from the base model, the roof was chamfered starting from the first axle of the semitrailer. Considering these changes, the impact will be checked for configurations b and c. In these cases, given the chamfered areas at the back, the two configurations have the potential to have much better aerodynamic performance than the base model. In addition, between the 3 iterations there are no changes for other areas of the semi-trailer, so the impact will be specific to the roof. [5]

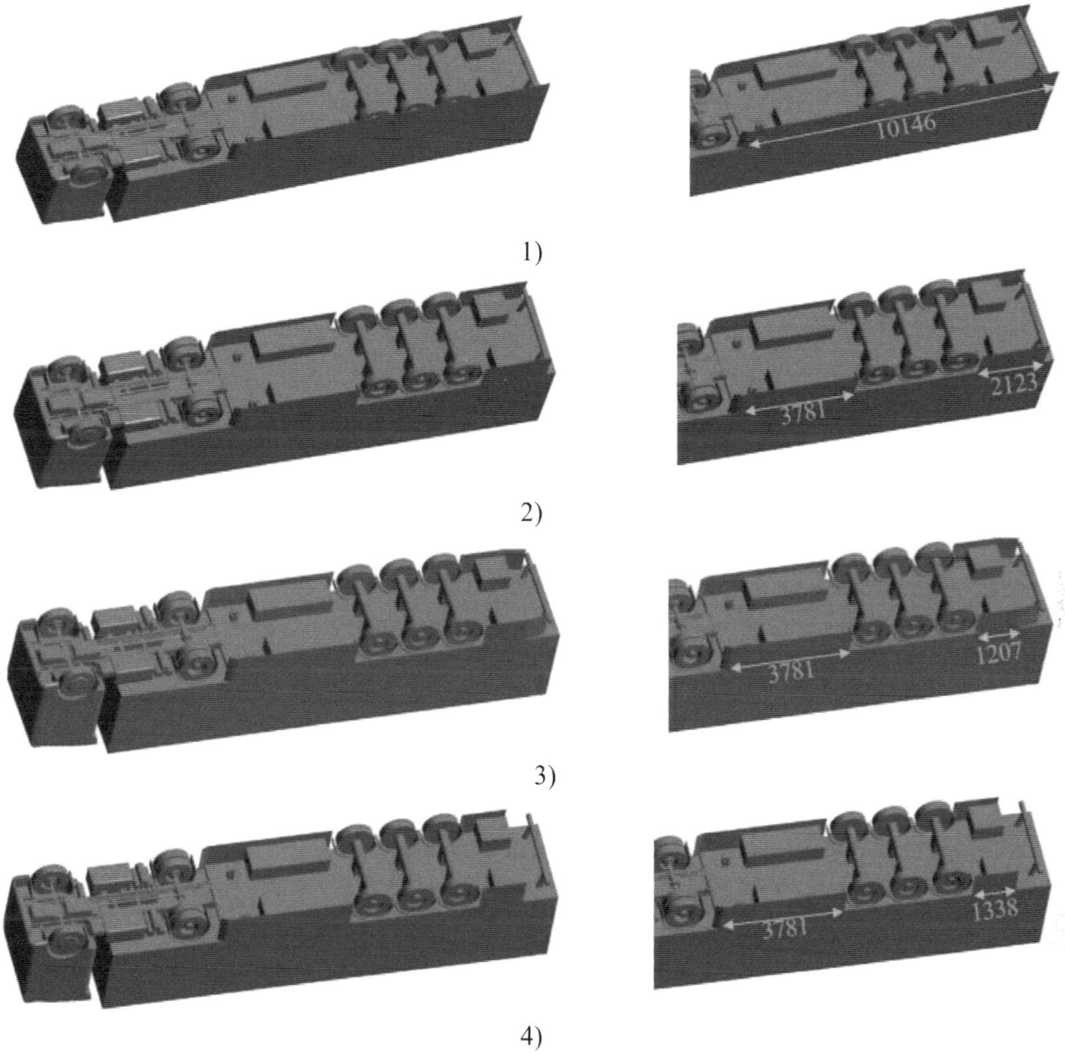


**Figure 6.** Configurations for the sides: 1) *full cover for sides, 2) partial cover for sides (open on wheel areas and straight profile in the back), 3) partial cover for sides and incline profile in the back, 4) partial cover o sides and straight partial cover in the back*

Figure 6 shows the configurations for the optimization possibilities on the side areas. As follows, a presentation of each configuration will be made: 1- starting from the base model “a”, the side areas were completely closed and for the front part of the side panels an inclination (angle of 20 deg compared to X axis) was made to allow air a smoother attachment on the surface of the panels; 2 - starting from “1” configuration, the area that cover wheels was eliminated, and the rear part of the panels stopped near the shock bar; 3- starting from “2” configuration the only change made is in the rear area of the panels where an inclination in Y axis (to the interior of the semi-trailer with an angle of 10 deg compared to X axis) was added; 4- starting from configuration “2” the rear area of the panels was limited to cover only the storage boxes. All seven configurations will be checked both from the point of view of the air flow and by comparing the force on X axis. For the four simulations in Figure 6, the back part of the roof was considered standard, so the square shape without chamfering.[6]

## 3. Results

Based on the CFD simulations for the configurations presented in the previous chapter, an analysis will be made to see which iteration offers the most possibilities for improvement. As will be presented in the following lines, the analysis of the air flow on the 7 configurations will be done in 2 stages: the analysis for "a", "b" and "c" to observe the impact for the ceiling chamfer and the analysis for "1", "2", "3" and "4" to observe the different air flows for the side areas.

**Figure 7.** Comparison of airflow between configurations *a*, *b* and *c*

Figure 7 shows the air flow in vector form on a plane at the middle of the truck on the 3 types of semi-trailers. The significant impact of this chamfer in the back of the truck is visible. It can be considered that configuration "a" has the worst wake (big in X and Z axis), therefore the worst aerodynamic performance. As for iterations "b" and "c", it can be seen that the wake it is improved and also although "c" has a larger chamfer of the roof, it can be observed that the air detaches before the end of the semi-trailer and this leads to a slight imbalance of the wake. Between iteration "b" and "c" the wake for "c" it is smaller in Z axis, concluding that this has the best aerodynamic performance. Also based on the forces on X axis values shows the same idea: a - 15046 N, b - 14545 N, c -14574 N.

IOP Conf. Series: Materials Science and Engineering 1303 (2024) 012044 doi:10.1088/1757-899X/1303/1/012044

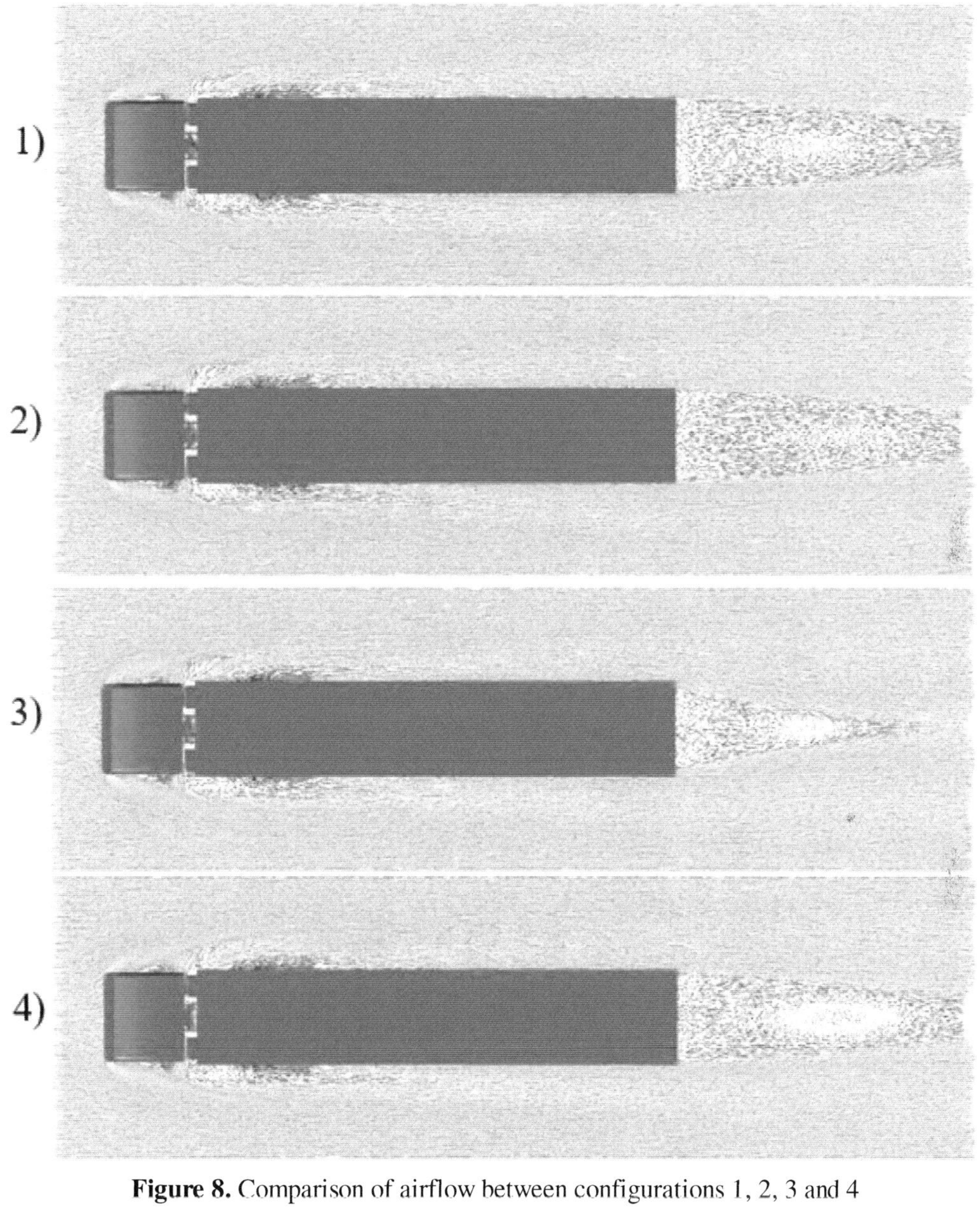


**Figure 8.** Comparison of airflow between configurations 1, 2, 3 and 4

Regarding the air flow in the sides of the truck, in Figure 8 are presented the 4 configurations with side panels in which the air flow is presented under form of vector on a plane positioned at the middle of the wheels. It can be observed that in all cases the wake is different because of airflow complexity. From a theoretical point of view, the best performance should be given by the situation in which the sides are completely closed, so the configuration "1". Of all the iterations, the smallest wake on X axis is given by the configuration "3" and this is because there is an inclined panel in the back, at the shock bumper. Although "3" can be considered the iteration with the best performance, to have even more robust image, it is necessary to consult the air flows in a plane on Y axis at the middle of the truck.

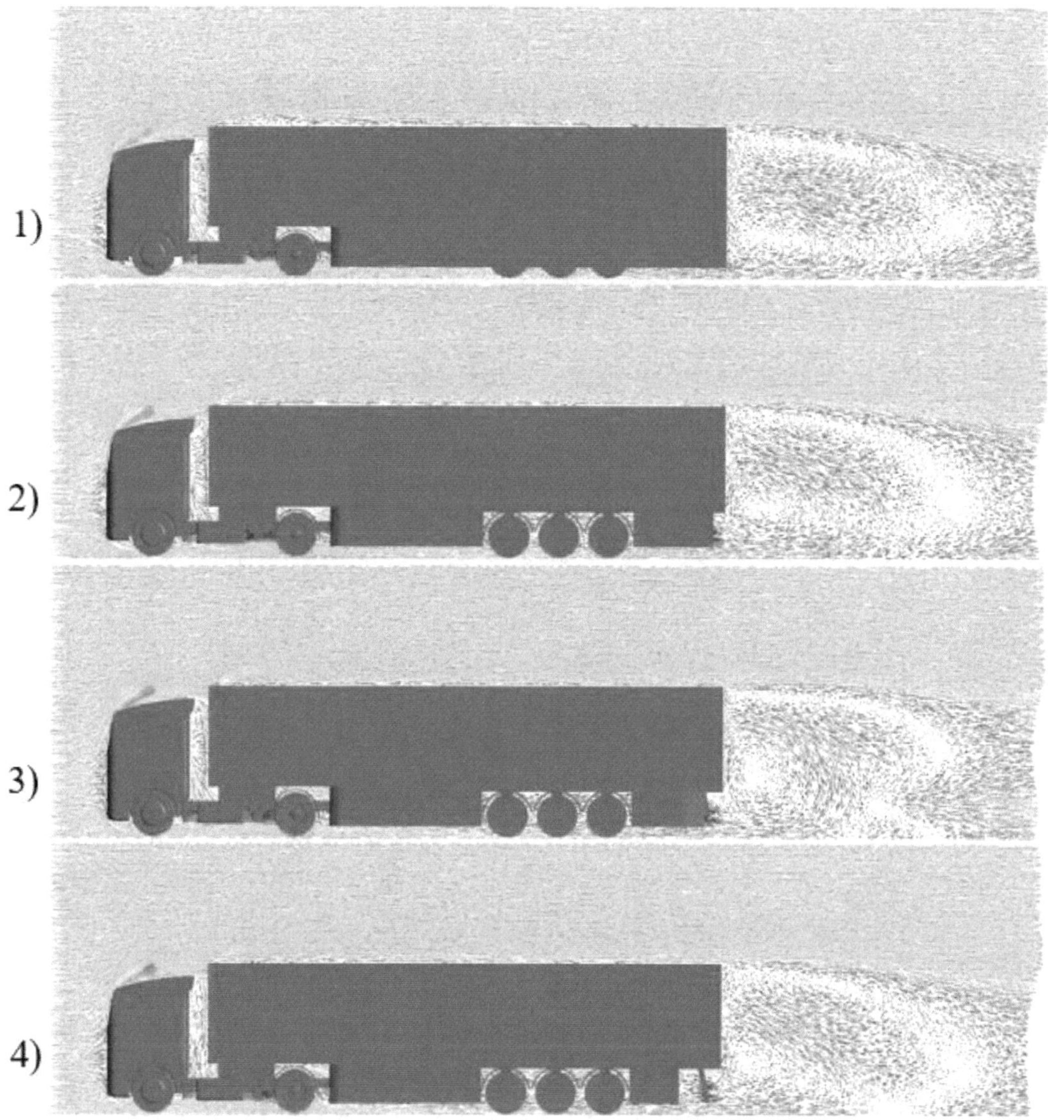


**Figure 9.** Comparison of airflow between configurations 1, 2, 3 and 4

Therefore, in Figure 9, based on the analysis of the wake, it can be observed that, indeed, the best performance is given by the configuration "1". In addition to this, the configuration "3", although it seemed to have a small wake, is quite unbalanced. Concluding from the air flows presented in Figure 8 and Figure 9, we can say that the best performance is given by configuration "1", where the wake is balanced in Y and also in Z. Also, configurations "2" and "4" are approximately at the same level of performance, but the "2" version has a more balanced wake. Configuration in which we have side panels inclined in the back ("3") describes an unbalanced wake on Y axis, which leads to the degradation of the aerodynamic performance. Regarding the forces on the X axis, the values are: d - 14143 N, e - 14274 N, f - 14400 N, g - 14287 N. Based on the numerical values, it is emphasized that the best aerodynamic performance is given by the configuration "1" and the worst one given by the configuration "3". Comparing with the truck without improvements (configuration a), configuration "1" has an aerodynamic performance with 6% better than it.

## 4. Conclusion

As was presented, current studies to improve aerodynamic performance are successfully achieving their goal. The areas with the biggest potential for improving aerodynamic performance are the rear (roof) and the side areas of the semi-trailer. Regarding the chamfering of the roof of semi-trailer, it was observed that the generation of a geometric profile that is too aerodynamic will bring a degradation in performance. That is why it is necessary to establish the air detachment area very good, in order to create the best wake and at the same time, the chamfer (height) must be correlated with the loading capabilities of the semi-trailer. In addition to this aspect, closing the side areas completely generates a very good aerodynamic performance. Considering the above, it can be concluded that in order to have a small forward resistance as possible in the case of trucks, the analysis on semi-trailers must be done very well and specifically on certain areas.

## 5. References

[1] https://www.betterflow.com/futureflow/, accessed in June 2023

[2] https://www.cargobull.com/ro/presa/2021/ecogeneration, accessed in June 2023

[3] Scurtu I L and Gheres M I 2021 *Numerical evaluation of vehicles aerodynamics in platoon using CFD simulation*, Department of Automotive Engineering and Transports, Technical University of Cluj-Napoca, Romania

[4] Harun C, Hazim M, Abdulkadir A, Iftekhar K, Firoz A, Simon W 2013 *A study on aerodynamic drag of a semi-trailer truck*, 5$^{th}$ BSME International Conference on Thermal Engineering, Melbourne

[5] Bukreev K R 2023 Shape optimization of teardrop trailers to minimize aerodynamic drag in articulated lorries, School of Mechanical Engineering Sciences, Faculty of Engineering and Physical Science, University of Surrey, Stag Hill, Guilford, UK, Vol 18, 2023

[6] Chilbule C, Upadhyay A and Mukkamala Y 2014 *Analyzing the profile modification of truck-trailer to prune the aerodynamic drag and its repercussion on fuel consumption*, Automotive Engineering, SMBS, VIT University Vellore, India

[7] Cavusoglu O F 2017 *Aerodynamic around wheels and wheel-houses*, Chalmers University of Technology, Gothenburg, Sweden

# Tyre and Rim Dimensions through Statistical Estimation

**G Anghelache[1] and T Ciobotaru[2]**

[1] University POLITECHNICA of Bucharest, Splaiul Independenţei no.313, Bucharest,
[2] Military Technical Academy Ferdinand I, Bd. G. Coşbuc, no. 81-83, Bucharest

E-mail: gabriel.anghelache@upb.ro

**Abstract**. Accurate vehicle dynamics calculations and simulations require robust estimations of the tyre and rim dimensions. This paper presents some findings on the estimation of tyre radii and rim width for passenger cars, mainly based on the statistical investigation of a large amount of data from tyre catalogues. The tyre free radius calculated with the data contained in the tyre marking code has, in principle, the same value as the tyre free radius obtained from catalogues. Thus, the classical equation is enough accurate for free radius estimation. This paper presents statistical investigations of the data in tyre catalogues and the simple equations found to estimate the static and rolling tyre radii. Moreover, numerical considerations on the proper match of tyre and rim widths are revealed. An important finding is the definition of the ratio between the rim and tyre width and the statistical calculation of this ratio vs. tyre aspect ratio. This method allows calculation of the nominal rim width.

## 1. Introduction

The early stages of the road vehicle design process require robust estimations of the tyre and rim dimensions. Regardless of the study complexity, all the road vehicle dynamics investigations generally request entry data of wheel radii (in particular rolling radius, but frequently free, static, and dynamic wheel radii as well). Some study categories need swift and more accurate input values for tyre radius, such as student theses on road vehicles, road vehicle design, automotive modelling software, investigation of road accidents etc. Inevitably, these input data have to be simplified, although the tyre dimensions are strongly influenced by its geometry, construction and exploitation conditions (speed, vertical load, inflation pressure etc.).

The width of the wheel determines the vehicle's interior wing space, some parameters for the wheel steering system, space for the spare wheel, but also an appropriate selection of tyre and rim.

Due to the continuous evolution of tyre design, there is generally insufficient information and outdated recommended values or equations in technical literature relating to these dimensions.

This paper presents some findings specified in the technical literature on tyre radii and rim width for passenger cars. The original paper's objective refers to the statistical investigation of a large amount of data, paying particular attention to the selection of contemporary types of tyres and models. The final results present more consistent equations for estimating the tyre radii and rim width.

## 2. The tyre main radii

Figure 1a and 1b show the tyre main radii in non-rolling, respectively in rolling conditions [1].

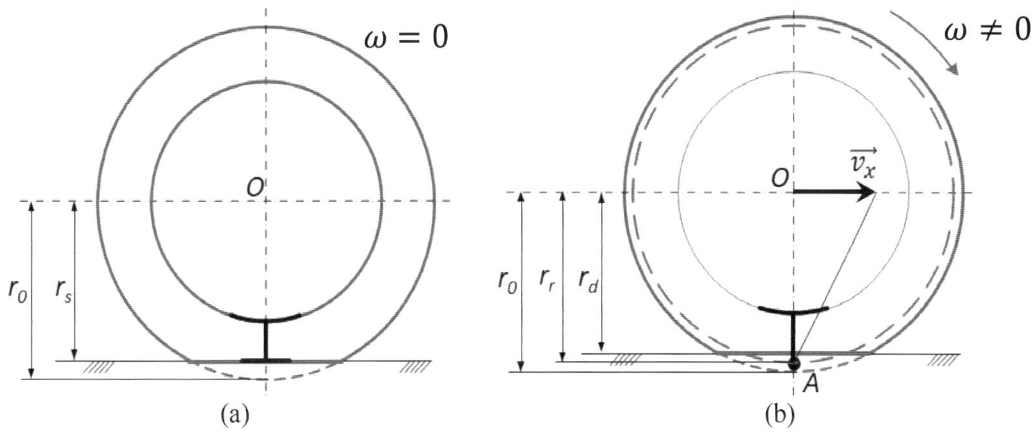


**Figure 1.** The tyre radii: non-rolling (a) and rolling (b) [1].

In Figure 1 the following notations were used:
$r_0$ = free radius of the non-rolling tyre and without contact with the road;
$r_s$ = static radius of the non-rolling tyre in contact with the road;
$r_d$ = dynamic radius of the tyre rolling on the road;
$r_r$ = tyre rolling radius;
$O$ = wheel centre;
$A$ = imaginary wheel sliding point [2];
$\omega$ = wheel angular speed;
$v_x$ = velocity of wheel centre.

## 3. Free radius

The tyre free radius value ($r_0$ in Figure 1(a)) can be determined by:

- **direct measurement** – relatively simple method, but affected by fairly large errors;
- **calculation based on the measured free diameter** $D_u$ of the tyre:

$$r_0 = \frac{D_u}{2}; \tag{1}$$

- **calculation using the tyre circumference length** $C$:

$$r_0 = \frac{C}{2 \cdot \pi}; \tag{2}$$

- **calculation using the data contained in the tyre marking code** (all dimensions are in [mm]):

$$r_o = 25.4 \cdot \frac{D_{as}}{2} + \frac{\rho_{na} \cdot B_u}{100} , \tag{3}$$

where:
$D_{as}$ is the rim diameter, Figure 2;
$\rho_{na}$ is the tyre aspect ratio, which is calculated with the relation:

$$\rho_{na} = \frac{H}{B_u} \cdot 100\,[\%], \tag{4}$$

$H$ is the tyre section height, Figure 2;
$B_u$ is the (cross-section) tyre width, Figure 2;
$B_{as}$ is the rim width.

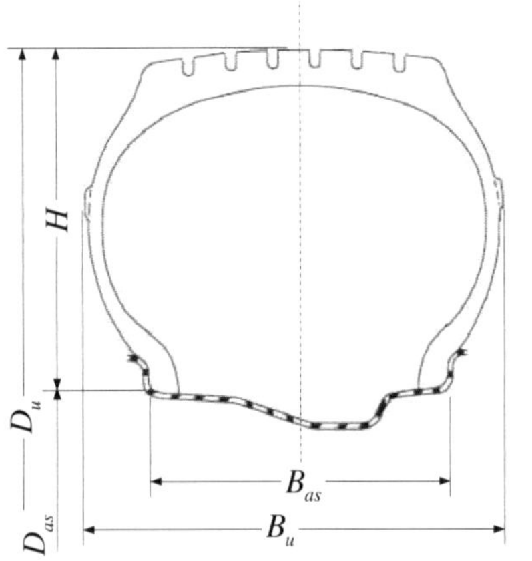


**Figure 2.** The main dimensions of the wheel.

- **using data provided in the tyre catalogues**: for 1614 different types of tyres in the catalogues [3], [4] and [5], the free radii were calculated with equation (3). For each tyre, the ratio $\gamma_{r0}$ between the free radius value obtained using equation (3) and the free radius value obtained from the catalogue was calculated:

$$\gamma_{r0} = \frac{(r_0)_{calculated}}{(r_0)_{catalogue}} [-] . \quad (5)$$

The statistical data processing for the ratio $\gamma_{r0}$ are shown in Table 1.

The very low value of standard deviation indicates a strong concentration of the data around the mean value.

**Table 1.** Statistical values of the ratio $\gamma_{r0}$.

| Average value | Maximum value | Minimum value | Standard Deviation |
|---|---|---|---|
| **0.9952** | 1.0495 | 0.8082 | 0.01582 |

Note: the tyre free radius calculated with equation (3) has, in principle, the same value as the tyre free radius obtained from catalogues.

## 4. Static radius

The tyre static radius ($r_s$ in Figure 1(a)) depends nonlinearly vs. vertical load and tyre radial stiffness, decreasing as the load increases [6]. The value of the static radius can be determined by:

- **direct measurement** – a relatively simple method, but affected by fairly large errors;
- **calculation based on a simple analytical formula**:

$$r_s = \lambda_{sr} \cdot r_0 , \quad (6)$$

where $\lambda_{sr}$ is the deformation coefficient of radial tyre. The specialized literature e.g. [7] indicates $\lambda_{sr} = 0.92$.

- **from tyre catalogues**, for the maximum load of the inflated tyre at the recommended internal pressure. Since the actual tyre load is lower than the maximum allowed value, the static radius indicated in catalogues represents its minimum value.

The deformation coefficient of the radial tyre can be calculated using the data from the catalogues with the equation:

$$(\lambda_{sr})_{catalogue} = \frac{(r_s)_{catalogue}}{(r_0)_{catalogue}} . \quad (7)$$

The statistical values calculated for 833 types of tyres and sizes are presented in Table 2.

**Table 2.** Statistical values of the tyre deformation coefficient $(\lambda_{sr})_{catalogue}$.

| Average value | Maximum value | Minimum value | Standard Deviation | Standard error |
|---|---|---|---|---|
| **0.9234** | 0.9592 | 0.8725 | 0.0147 | 0.00051 |

In this case, the static radius can be calculated using statistical data with the formula:

$$(r_s)_{catalogue} = (0.9234 \pm 0.0005) \cdot (r_0)_{catalogue} . \quad (8)$$

Comparing the values of tyre deformation coefficients mentioned in equations (6) and (8), due to the modern tyre's higher stiffness compared to the data in the previous scientific literature, a higher value "from the catalogue" is obtained.

- **computation using analytical or finite element tyre models** (e.g. [6]); this models require experimental validation which increases costs and time.

## 5. Rolling radius

Actually, the rolling radius $r_r$ represents the radius of a rigid, free-rolling virtual wheel whose centre moves a distance equal to the rolling circumference when the wheel makes a complete rotation:

$$L_c = 2 \cdot \pi \cdot r_r, \tag{9}$$

where $L_c$ is the distance travelled by the tyre in one complete rotation of the free rolling wheel.

The rolling radius is one of the fundamental quantities used in any wheeled vehicle dynamics project.

The size of the rolling radius depends on numerous factors: design (tyre construction - radial or diagonal, carcass construction etc.), functional and operating factors (internal pressure, vertical load, rolling speed, rolling regime - free / drive / braked wheel etc.). Unfortunately, a direct measurement of the rolling radius is difficult, therefore approximate calculation methods are usually used.

The main methods for determining the rolling radius are the following:

- **by calculation, measuring the distance** $L_c$ travelled by the tyre during a complete rotation of the free rolling wheel (see equation (9)):

$$r_r = \frac{L_c}{2 \cdot \pi}. \tag{10}$$

The method is influenced by several errors, for example the value of the wheel vertical load, the position of the wheel, the accuracy of the distance $L_c$ measurement etc. Accuracy is increased by making several full rotations of the wheel.

- **by calculation, simultaneously measuring the speed of the wheel centre and the wheel angular speed**:

$$r_r = \frac{v}{\omega}. \tag{11}$$

The method can be applied both in real road rolling conditions, as well as on dedicated indoor rigs, but it requires expensive and accurate measurement systems (e.q. reference [8]);

- **by estimation**, based on a simple analytical formulation:

$$r_r = \lambda_r \cdot r_0, \tag{12}$$

where $\lambda_r = 0.96 \div 0.98$ is a coefficient for radial passenger car tyres [7];

Note: in many older automotive dynamics references, the coefficient value $\lambda_r$ is 3% ÷ 5% lower, and therefore considerable errors are introduced in most automotive dynamics projects;

- **by calculation**, using $r_0$ and $r_s$ [9]:

$$r_r = \frac{2}{3} \cdot r_0 + \frac{1}{3} \cdot r_s, \tag{13}$$

where $r_0$ and $r_s$ are determined according to the previous indications;

- **by calculation**, using the distance $L_{catalogue}$ (mentioned in the catalogue) travelled by the tyre at a complete rotation of the free rolling wheel and under certain operating conditions:

$$r_{r_{catalogue}} = \frac{L_{catalogue}}{2 \cdot \pi}; \tag{14}$$

- **by calculation**, based on a statistically determined coefficient $\lambda_{rs}$:

$$r_{rs} = \lambda_{rs} \cdot r_0. \tag{15}$$

To use the equation (15), numerical values of the free radius $r_0$ and coefficient $\lambda_{rs}$ have to be available in advance.

In this paper, the value of the coefficient $\lambda_{rs}$ is statistically determined using the formulae:

$$\lambda_{rs} = \frac{r_{r\,catalogue}}{r_0} = \frac{L_{catalogue}}{2\cdot\pi\cdot r_0}, \tag{16}$$

where $L_{catalogue}$ was presented in equation (14) and $r_0$ was calculated with equation (3).

From the references [3, 4, 5], 1566 types of tyres were selected for statistical analysis. The results are presented in Table 3.

**Table 3.** Statistical values of the coefficient $\lambda_{rs}$.

| Average value | Maximum value | Minimum value | Standard Deviation | Standard error |
|---|---|---|---|---|
| **0.9702** | 0.9991 | 0.8592 | 0.0078 | 0.0002 |

The histogram of the analysed data is presented in Figure 3; a very good grouping of the data relative to the average value is found.

The data shown in Table 3 gives the coefficient $\lambda_{rs}$ the following value: $\lambda_{rs} = 0.9702 \pm 0.0002$ which is in the range mentioned for the coefficient in the equation (12).

Thus, equation (15) becomes:

$$r_{rs} = (0.9702 \pm 0.0002)\cdot r_0 .$$

Note: $L_{catalogue}$ in equation (16) is determined under certain conditions, generally different from the real operating ones: wheel vertical load, tyre internal pressure and rolling speed.

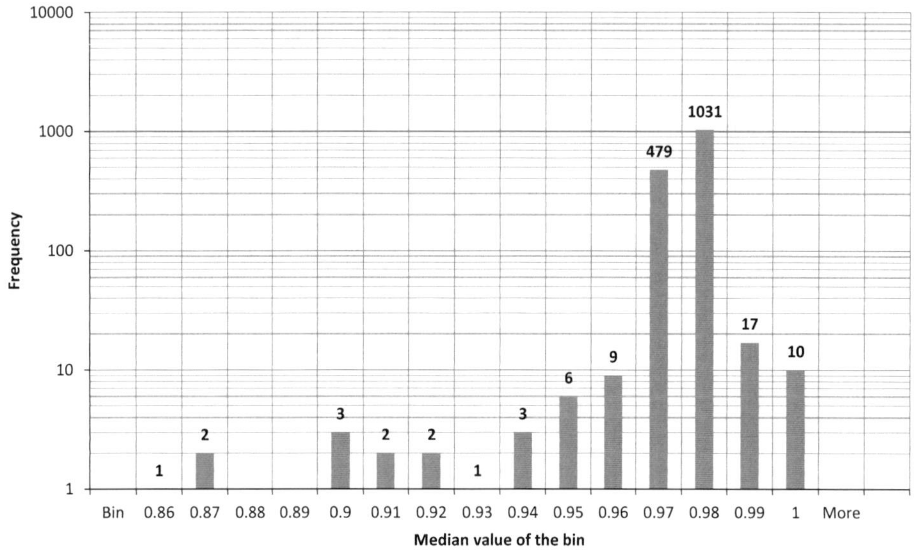


**Figure 3.** Histogram of the coefficient $\lambda_{rs}$ values.

## 6. The rim width

The main wheel cross-sectional dimensions are: tyre width ($B_u$), rim width ($B_{as}$) (Figure 2) and rim offset. The rim width ($B_{as}$) [inch] is marked on the rim. Due to the sidewall elasticity, tyres with different widths can be mounted on the same rim.

Regarding the matching of the tyre-rim assembly (according to the rim width), the following situations are frequently encountered in operation:

a) the matching of the tyre-rim assembly imposed by the car manufacturer as nominal solution;
b) for a specific tyre, a user intents to choose a different rim width (generally larger) to the nominal one;
c) for a specific rim, a user intents to choose a different tyre width.

Cases b) and c) listed above affect vehicle dynamics and endanger traffic safety; therefore restrictions are introduced by various regulations. This paper discusses case b). Thus, for each type of tyre, the manufacturing company mentions a recommended rim width ($B_{asR}$). In this case, the resulting tyre width is denoted $B_{uR}$.

A certain type of tyre can generally be mounted on rims with widths up to 1.5 inch larger / smaller compared to the recommended value. Mounting the tyre on rims with widths different to the recommended value leads to a corresponding modification in tyre width. For each type of tyre, the reference [3] recommends a set of rim widths presented in Table 4.

**Table 4.** Tyre width change when mounting on different rim widths (adapted form [1]).

| Rim width change [inch] | $B_{asR}$ – 1” | $B_{asR}$ – ½” | $B_{asR}$ | $B_{asR}$ + ½” | $B_{asR}$ + 1” | $B_{asR}$ + 1½” |
|---|---|---|---|---|---|---|
| | Tyre width change [mm] | | | | | |
| Version #1 | ╳ | $B_{uR}$ – 5 | $B_{uR}$ | $B_{uR}$ +5 | $B_{uR}$ +10 | $B_{uR}$ +15 |
| Version #2 | $B_{uR}$ – 10 | $B_{uR}$ – 5 | $B_{uR}$ | $B_{uR}$ +5 | $B_{uR}$ +10 | ╳ |

The following observations can be obtained from the data shown in Table 4:

- information provided by the tyre manufacturer shall be required for the correct selection of the rim width;
- every ½ inch change in the rim width results in a variation of 5 mm of the tyre width;
- the maximum variation range of the rim width is 2 inches (although in certain situations, this recommended range is 1 inch or 1½ inches).

## 7. Choosing the rim width without additional information provided by the tyre manufacturer

The primary input data are the following:

- the width of the cross section of the tyre ($B_u$), information found on the side wall of the tyre;
- the aspect ratio $\rho_{na}$.

The ratio $k_l$ is defined as:

$$k_l = \frac{B_{as}}{B_u}[-]. \tag{17}$$

From [3, 4, 5] a number of 1541 types of tyres was selected; for each tyre the sizes $\rho_{na}$, $B_u$ and $B_{as}$ were extracted and recorded. Moreover, for each tyre, the ratio $k_l$ was calculated using equation (17). The results obtained were grouped according to the aspect ratio values. The values included in Table 5 are the result of a statistical analysis of the data for each aspect ratio value.

**Table 5.** Statistical data of ratio $k_l$ values.

| $\rho_{na}$ [%] | **25** | **30** | **35** | **40** | **45** | **50** | **55** | **60** | **65** | **70** | **75** | **80** |
|---|---|---|---|---|---|---|---|---|---|---|---|---|
| Maximum | 0.938 | 0.919 | 0.919 | 0.929 | 0.896 | 0.847 | 0.829 | 0.824 | 0.824 | 0.798 | 0.755 | 0.737 |
| Minimum | 0.904 | 0.877 | 0.877 | 0.877 | 0.827 | 0.778 | 0.757 | 0.726 | 0.716 | 0.726 | 0.677 | 0.681 |
| Average | **0.919** | **0.898** | **0.895** | **0.897** | **0.845** | **0.801** | **0.802** | **0.757** | **0.757** | **0.755** | **0.707** | **0.710** |
| Number of tyres | 24 | 70 | 90 | 100 | 208 | 174 | 253 | 220 | 202 | 132 | 22 | 46 |

The statistical information included in Table 5 is shown in Figure 4. The statistical value appears to be very close for some adjacent values of the aspect ratio, making it possible to merge a number of columns in Table 5. The statistical values for the resulting columns are shown in Table 6.

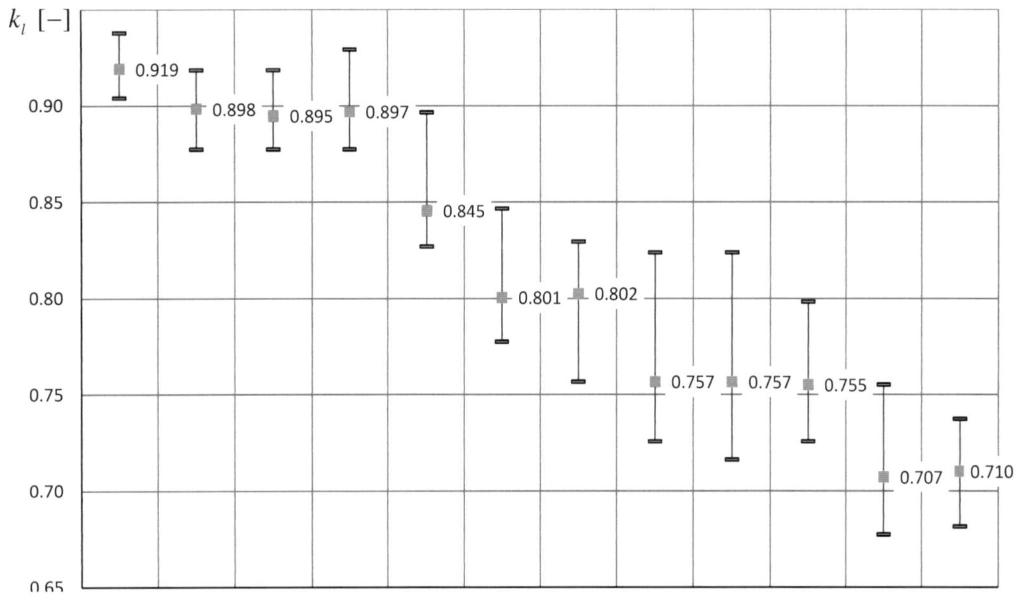


**Figure 4.** Statistical values for the ratio $k_l$.

**Table 6.** Statistical data of ratio $k_l$ values (a number of columns merged).

| $\rho_{na}$ [%] | **25** | **30; 35; 40** | **45** | **50; 55** | **60; 65; 70** | **75; 80** |
|---|---|---|---|---|---|---|
| Maximum | 0.938 | 0.929 | 0.896 | 0.847 | 0.824 | 0.755 |
| Minimum | 0.904 | 0.877 | 0.827 | 0.757 | 0.716 | 0.677 |
| Average | **0.919** | **0.896** | **0.845** | **0.802** | **0.756** | **0.709** |
| Number of tyres | 24 | 260 | 208 | 427 | 554 | 68 |

The data in Table 6 are shown in Figure 5.

The following observations may be made after analysing the data in Figures 4 and 5:

- for the types of tyres analysed, all statistical values are subunit (the ratio $k_l < 1$), which means that **the rim width** (where the tyre beads sit) **is smaller than the tyre width**;
- very close statistical values for the ratio $k_l$ correspond to each value of the aspect ratio;
- by decreasing the nominal aspect ratio, the $k_l$ value increases, thus the tyre cross section geometry approaches a rectangular shape (a specific of modern passenger car tyres);
- for nominal aspect ratio of 30 or higher (Figure 3), the average value of the ratio $k_l$ decreases approximately linearly with this ratio (or with the merged ratios);
- the ratio $k_l$ does not depend on the tyre or rim diameter.

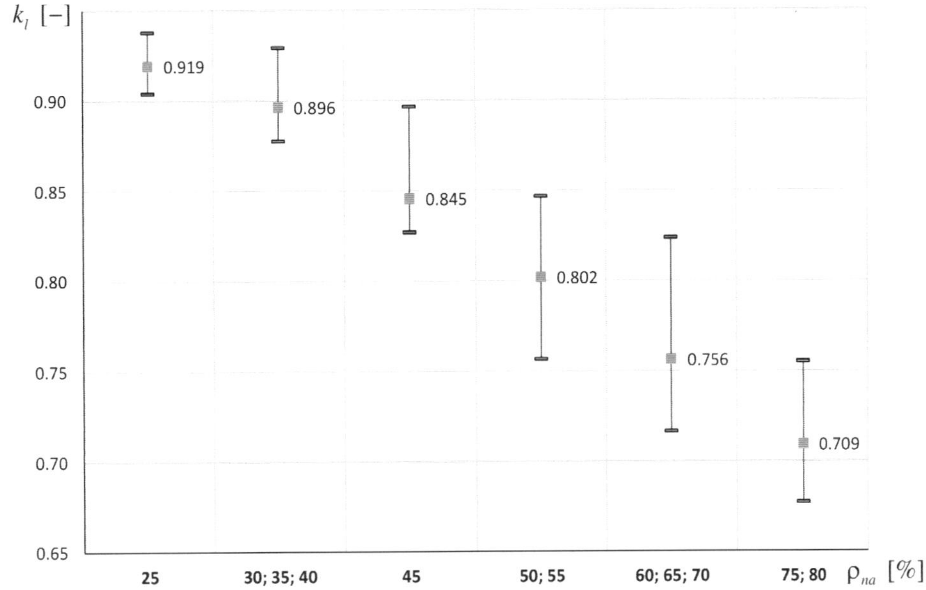


**Figure 5.** Statistical values for the ratio $k_l$ resulted by merging some of the data columns

If there is no information available from the tyre manufacturer, and the rim width needs to be chosen for a particular type of tyre, the following important steps should be followed:

- the values of tyre width and aspect ratio are found on the sidewall of the tyre;
- the average value of the ratio $k_l$ corresponding to the aspect ratio is obtained from Table 6 or from Figure 5;
- the rim width $\left(B_{as_c}\right)$ [mm] is calculated from equation (17):

$$B_{as_c} = k_l \cdot B_u [\text{mm}] \; ; \tag{18}$$

- the standardized rim width [mm] closest to the value resulting from equation (18) is selected from Table 7, second row;
- the standardized rim width ($B_{as}$) [inch] is selected from the first row of the same column in Table 7.

**Table 7.** Usual rim width values, $B_{as}$.

| | | | | | | | | | | | | |
|---|---|---|---|---|---|---|---|---|---|---|---|---|
| [inch] | **4,5** | **5** | **5,5** | **6** | **6,5** | **7** | **7,5** | **8** | **8,5** | **9** | **9,5** | **10** |
| [mm] | 114.3 | 127 | 139.7 | 152.4 | 165.1 | 177.8 | 190.5 | 203.2 | 215.9 | 228.6 | 241.3 | 254 |

Note: the above method can also be used to calculate the tyre width $\left(B_{u_c}\right)$ [mm], when the rim width ($B_{as}$) [inch] is known, writing equation (17) in the following form:

$$B_{u_c} = 25.4 \cdot \frac{B_{as}}{k_l} [\text{mm}] \; . \tag{19}$$

## 8. Conclusions

The following main conclusions can be drawn from this detailed analysis and the results obtained:

- using the data provided by the code marked on the sidewalls of the radial tyre, the free radius, the diameter of the tyre on the rim, and the aspect ratio are obtained, directly or by calculation;

- the catalogues and data sheets of the tyre manufacturers offer additional data regarding the free outer diameter, static radius, rolling circumference (rolling radius), tyre width and the width of the rims;
- this paper provides simple equations for estimating tyre/rim sizes when accurate data from the manufacturing companies are not available;
- the free radius estimated with equation (3) has in principle the same value as the free radius obtained from the catalogue data provided by the tyre manufacturer;
- the deformation coefficient of the radial tyre, that is statistically calculated using the data from the catalogue, has a slightly higher value than the data found in specialized literature; this behaviour is due to the modern tyre's higher stiffness in radial direction;
- the coefficient $\lambda_{rs}$ for the calculation of the rolling radius was statistically determined using data in catalogues for a large number of passenger car tyres; the obtained value ($0.9702 \pm 0.0002$) is in the range mentioned in the previous technical literature;
- it was found that the tyre cross-sectional width changes by 5 mm for every ½ inch change in rim width;
- for a certain tyre type, the maximum variation range of the rim width is 2 inches;
- the rim width is smaller than the width of the usual radial tyre;
- very close statistical values for the ratio $k_l$ correspond to each value of the aspect ratio;
- by decreasing the nominal aspect ratio, the $k_l$ value increases, thus the tyre cross section geometry approaches a rectangular shape;
- for nominal aspect ratio of 30 or higher, the average value of the ratio $k_l$ decreases approximately linearly with this ratio (or with the merged ratios);
- the ratio $k_l$ does not depend on the diameter of the tyre or of the rim.

Finally, it may be considered that the calculation outcomes and methodology mentioned in this paper were capable of achieving the paper objectives. The methodology described represents an important contribution for better simulation of vehicle dynamics.

**Author's ORCID ID**
G Anghelache: https://orcid.org/0000-0002-8709-6070
T Ciobotaru: https://orcid.org/0009-0003-0129-5867

## References

[1] Anghelache G and Ciobotaru T 2022 – 2023 *Motor Vehicle Dynamics*, course support (in Romanian), University POLITEHNICA of Bucharest
[2] Zegelaar P 1998 *The Dynamic Response of Tyres to Brake Torque Variations and Road Unevennesses*, PhD Thesis, Delft University of Technology, ISBN 90-370-0166-1.
[3] *Continental - Tyre Databook 2019 – 2020 Car / 4x4 / Van.*
[4] *MICHELIN® PASSENGER AND LIGHT TRUCK TIRES – 2018 Data Book*, May Edition.
[5] *Hankook Tires - 2013~2014 PATTERN DIGEST*
[6] Anghelache G 1999 *Tyre – Road Interaction at High Speed of Rolling* (in Romanian), University POLITEHNICA of Bucharest, PhD Thesis, Bucharest
[7] Anghelache G 2019 – 2020 *Motor Vehicle Dynamics*, course support (in Romanian), University POLITEHNICA of Bucharest
[8] Anghelache G and Moisescu R 2017 *The Measurement of Dynamic Radii for Passenger Car Tyre*, Proceedings of CAR2017 International Congress of Automotive and Transport Engineering - Mobility Engineering and Environment, IOP Conference Series: Materials Science and Engineering, Vol. 252, No. 1, 012014, ISSN 1757-8981, IOP Publishing
[9] Rill G 2012 Road Vehicle Dynamics: Fundamentals and Modelling, CRC Press, ISBN: 978-1-4398-9744-7.

IOP Conf. Series: Materials Science and Engineering 1303 (2024) 012046 doi:10.1088/1757-899X/1303/1/012046

# Comparative analysis of FEM tire-soft snow interaction and theoretical model based on Bekker's coefficients

**Y Surkutwar[1], A Vilsan[2], C Sandu[2] and C Untaroiu[1]**

[1] Department of Biomedical Engineering and Mechanics, Center for Injury Biomechanics (CIB), Virginia Tech, Blacksburg, 460 Turner St., VA 24060, USA
[2] Department of Mechanical Engineering, Terramechanics, Multibody, and Vehicle Systems (TMVS)Laboratory, Virginia Tech, 1600 Innovation Drive, Blacksburg, VA 24061, USA

Corresponding author's e-mail address: yogeshvs@vt.edu

**Abstract**. Tire-deformable terrain interaction is a complex phenomenon that has proven difficult to characterize accurately and comprehensively using numerical models. While researchers have attempted to develop various numerical models to estimate tire behavior on deformable terrain, the most accurate models to date rely on empirical data and are only applicable to specific rolling conditions. In this research paper, a comparison between theoretical mechanics and a finite element method (FEM) model of a rigid tire rolling on a layer of soft snow is presented. The goal is to compare the data obtained from the FEM tire-soft snow interaction analysis with the data generated by a theoretical numerical model. The FEM snow model is developed using the Drucker-Prager cap material model while the tire is assumed rigid. The assumption of the rigid tire can provide reasonably accurate results in the case of the soft snow interaction model. The theoretical pressure-sinkage coefficients of the Bekker equation ($k_c$, $k_\phi$ and $n$) for soft snow are calculated by simulating vertical penetration tests in the FEM snow model. The results of the pressure distribution and rolling resistance due to compaction obtained by the FEM and the theoretical model results are then compared.

## 1. Introduction

The interaction between tires and soil is a highly intricate phenomenon, posing significant challenges in accurately and comprehensively characterizing it through current numerical models. Numerous researchers have made attempts to develop various numerical models to assess the behavior of tires rolling on soft soil. However, to date, the most precise models rely on empirical data and are limited in their usability to specific rolling conditions.

The study on the motion resistance of two truck tires of different sizes on three distinct soil types was carried out near Buenos Aires, Argentina, at an elevation of 22 meters above sea level, and featured a slope gradient of 0.5%. The soil at the location belonged to drainage class 4 [1]. The "Shear Interface Imaging Analysis Tool," a novel technique for experimentation and analysis, was introduced by Moreland [2]. This technique is specifically designed to explore the fundamental principles of Terramechanics in great detail, enabling the visualization and analysis of soil motion both at and below the interface between the wheel and soil [2]. Pressure-sinkage experiments were performed for utilizing penetration plates of varying dimensions to characterize a model based on Bekker's formula [3]. In order to simulate the soil conditions, a mixture of bentonite and water was selected as a substitute for seabed sediments, based on experimental data collected from China's ocean poly-metallic mining contract area [3].

Our research motivation is represented by the idea of integration between the theoretical mechanics of tire-soil interaction with the experimental data obtained in the field for a set of operation conditions. The primary objective of this research is to compare data acquired from a finite element method (FEM) analysis of tire-soft snow interaction with the data generated by a theoretical numerical model. For the FEM snow model we employ the Drucker-Prager cap material model, while assuming the tire to be rigid. This assumption is expected to yield reasonably accurate results for the soft snow interaction model. The coefficients utilized in the FEM snow model will be sourced from our comprehensive literature review, primarily relying on experimental measurements [4]. Under typical conditions, snow displays intricate properties characterized by non-linear viscoelastic behavior and significant deformations, encompassing both volumetric and deviatoric strains [5],[6]. The literature presents various approaches for modeling the interaction between snow and tires, including analytical and semi-analytical methods [7], finite element methods [8], and, more recently, particle-based methods [9]. Finite element methods (FEM) employ diverse material models to represent snow, specifically chosen to accurately capture its highly compressible nature [10]. To determine the theoretical pressure-sinkage coefficients ($k_c$, $k_\phi$, and $n$) for soft snow soil, vertical penetration tests will be conducted, using our FEM model. By calculating these coefficients, our objective is to compare the results of theoretical models, which account for pressure distribution and rolling resistance due to compaction, with the outcomes obtained from the FEM model. Furthermore, we aim to evaluate the thrust capabilities as a function of wheel slip, using the shear stress distribution in the case of a rigid wheel from the FEM analysis. Through these comparisons, our aim is to incorporate relevant experimental data to further validate the accuracy and reliability of the FEM analysis and numerical modeling.

## 2. Research Methodology

This section presents the methodology for developing the FEM snow model and the Bekker-Wong coefficients. Furthermore, this section presents the analytical approach to estimate the sinkage and the motion resistance due to terrain (snow) compaction of a rigid wheel.

### *2.1. FEM snow model development*

The snow model was implemented using the explicit version of the commercial software ABAQUS, incorporating a methodology influenced by the research of Shoop [10]. The Arbitrary Lagrangian Eulerian (ALE) meshing technique was utilized to model the snow. To capture the behavior of soft snow, the Drucker-Prager cap model was employed to define its material properties. Table 1 provides the specific values utilized to comprehensively describe the properties of the snow model.

**Table 1.** Coefficients used to characterize the soft snow material properties.

| Drucker-Prager material cohesion $C$ [MPa] | Drucker-Prager material angle of friction $\beta$ [°] | The cap eccentricity parameter $R$ [-] | Initial cap yield surface position $\varepsilon$ [-] | The flow stress ratio $K$ [-] | Density $\rho$ [kg/m$^3$] |
|---|---|---|---|---|---|
| 0.005 | 22.53 | 0.0002 | 0.001 | 1.0 | 200 |

The Drucker–Prager failure criterion is a model that considers the pressure-dependent behavior of materials and determines if they have failed or undergone plastic yielding. This material model is applicable to various pressure-dependent materials such as rock, concrete, polymers, foams, and compacted snow. It provides a three-dimensional, pressure-dependent model for calculating the stress state at which a material reaches its ultimate strength [11]. An important physical parameter of snow is represented by the snow density. According to Shoop [12], the snow density can vary from 100 kg/m$^3$

(fresh snow) to 600 kg/m$^3$ (snowpack subjected to melt-freeze cycles), however the compacted snow that automobiles can experience is typically classified as having a density ranging from 370 to 560 kg/m$^3$. In order to verify the accuracy of the FEM snow model, a load-sinkage convergence test was conducted. Figure 1 presents the mesh convergence study of the soft snow model.

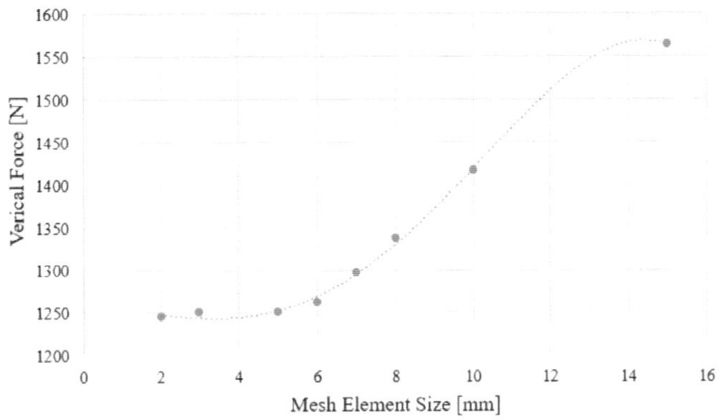


**Figure 1.** Mesh convergence study of the soft snow model.

In the mesh convergence study, the elements size varied from 15 mm (714 nodes) to 2 mm (38152 nodes). The load-sinkage tests are performed for a snow model that has a height of 500 mm and a diameter of 600 mm. A rigid disc with a diameter of 230 mm is deforming the snow in the vertical direction with a prescribed sinkage of 200 mm. The output of the tests is the resulting vertical force. It can be observed from the plot that the last 3 iterations offer similar results for the resulting vertical force, therefore the element size of 2 mm is selected for further study.

### *2.2. Bekker-Wong coefficients calculations based on the FEM pressure-sinkage results.*

This section introduces a methodology for calculating the Bekker-Wong coefficients. Equation (1) provides a suitable description of the pressure sinkage relationship for a homogeneous terrain [13]:

$$p = \left(\frac{k_c}{b} + k_\phi\right) * z^n \tag{1}$$

Where $p$ is the normal pressure, $b$ is the radius of a circular plate, $z$ is the sinkage, and $n$, $k_c$ and $k_\phi$ are the coefficients which need to be calculated for every type of soil. Some researchers consider that the coefficient $k_c$ is related to the cohesive property of the soil while $k_\phi$ is related to the frictional properties of the soil. However, later research proves that these parameters don't necessarily have a physical meaning and their values are obtained through fitting equation (1) to the measured data. The values of $k_c$, $k_\phi$, and $n$ can be calculated from the results of a minimum of two tests with two sizes of plates having different radii (or widths of rectangular plates). The tests produce two pressure equations ($p_1$ and $p_2$), for each one of the plate dimensions ($b_1$ and $b_2$):

$$p_{1,2} = \left(\frac{k_c}{b_{1,2}} + k_\phi\right) * z^n \tag{2}$$

If natural logarithm function is applied, then equation (2) becomes:

$$\ln p_{1,2} = ln\left(\frac{k_c}{b_{1,2}} + k_\phi\right) + n * \ln z \tag{3}$$

For sinkage $z = 1$, the natural logarithm will have the value of zero, therefore the expected shape of the equation (3) is presented in Figure 2 [13].

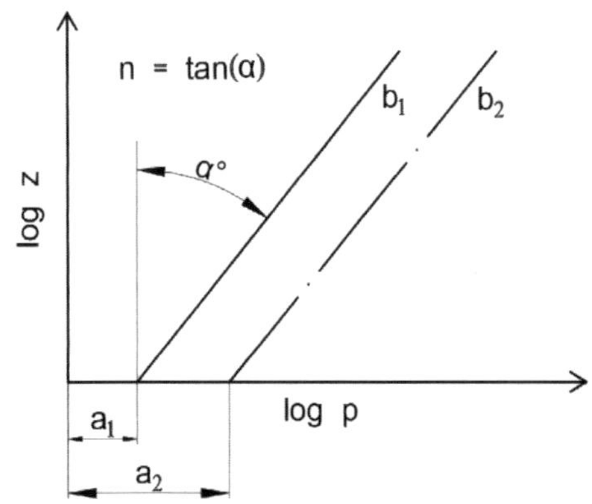


**Figure 2.** Graphical method for determining sinkage moduli and exponent (adapted from [13]).

$$k_{\phi} = \frac{a_2 * b_2 - a_1 * b_1}{b_2 - b_1} \tag{4}$$

$$k_c = \frac{(a_1 - a_2) * b_1 * b_2}{b_2 - b_1} \tag{5}$$

To find the values for the Bekker-Wong's coefficients, two different axial compression tests have been performed on the soft snow modelled in ABAQUS. Figure 3 presents the test conditions.

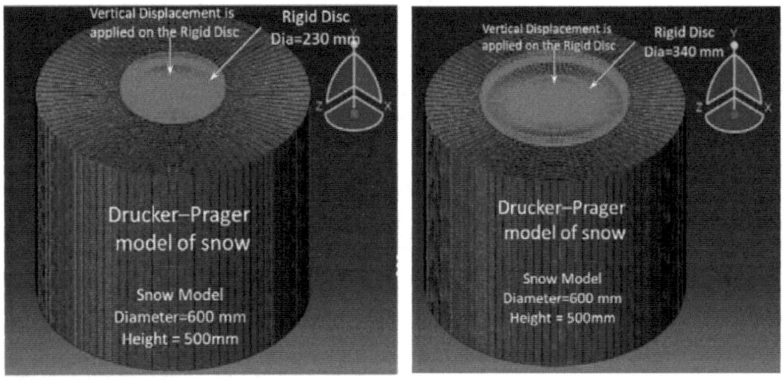


**Figure 3.** Axial compression tests conditions performed on the soft snow model.

In the first test, a circular rigid plate with a diameter of 230 mm was utilized. For the second test, a circular rigid plate with a diameter of 340 mm was employed. Both tests involved a quasit-static loading condition, which was applied by imposing a vertical displacement on the rigid disc. The sidewall of the model was subjected to roller boundary conditions, while the bottom face of the model was constrained in all directions. The snow mesh was created using C3D8R solid elements with reduced integration. The rigid plates were represented as analytical rigid surfaces and modeled using a planar analytical surface combined with a rigid body constraint. The density of the snow was assumed to be constant throughout the simulation, with a value of 200 kg/m³. The analysis has been carried out with Abaqus Explicit with double precision.

### 2.3. *Estimation of Motion Resistance due to snow compaction using Bekker-Wong's coefficients*

The total motion resistance encountered by a tire rolling on soft terrain comprises three components: motion resistance due to terrain compaction, motion resistance due to tire deflection, and motion resistance caused by the bulldozing effect. However, in our project, we focus solely on estimating the motion resistance attributed to terrain compaction. This is because we assume that the snow possesses low density, and we consider the tire to be rigid.

To estimate the motion resistance due to terrain compaction, we intend to utilize the Bekker-Wong coefficients, as suggested by Wong [13]. Based on these coefficients, the estimation involves assuming that the terrain reaction at all points within the contact patch is purely radial. It is further assumed to be equivalent to the normal pressure experienced beneath a horizontal plate positioned at the same depth during a pressure-sinkage test. A simplified wheel-soil interaction model is illustrated in Figure 4, which depicts this estimation approach.

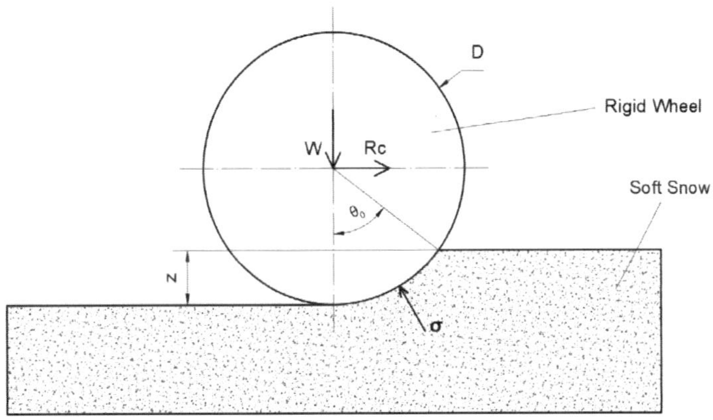


**Figure 4.** Simplified wheel-soil interaction model (adapted from [13])

The equilibrium equations for this model are:

$$R_c = b * \int_0^{\theta_0} \sigma * r * sin\theta \, d\theta \tag{6}$$

$$W = b * \int_0^{\theta_0} \sigma * r * cos\theta \, d\theta \tag{7}$$

If it is assumed that the normal pressure $\sigma$ acting on the wheel rim is equal to the normal pressure $p$ beneath a plate at the same depth $z$, the equation to calculate $R_c$ becomes:

$$R_c = b * \left[\left(\frac{k_c}{b} + k_\phi\right) * \frac{z^{n+1}}{n+1}\right] \tag{8}$$

$$z = \left[\frac{3*W}{b*(3-n)*\left(\frac{k_c}{b}+k_\phi\right)*\sqrt{D}}\right]^{\left[\frac{2}{2*n+1}\right]} \tag{9}$$

Where $b$ is the width of the tire, $\sigma$ is the normal pressure, $W$ is the vertical load, $r$ is the radius of the wheel, and $k_c$, $k_\phi$ and $n$ are the Bekker -Wong coefficients.

## 3. Results and discussion

This section presents the results obtained for the wheel sinkage, motion resistance and thrust capabilities of a rigid wheel rolling on soft snow. To determine the Bekker-Wong's coefficients for various snow depths, load-sinkage tests were carried out on the FEM soft snow model. Based on these coefficients, analytical equations were employed to predict the wheel sinkage and motion resistance. Subsequently, the results obtained from the analytical study were compared with those obtained from an FEM analysis of a rigid wheel rolling on the soft snow model.

### *3.1. Estimating Wheel Sinkage and Motion Resistance utilizing Bekker-Wong's coefficients*

As presented in section 2.2, vertical load-sinkage tests were performed on the FEM snow model using two rigid discs with diameters of 230 mm and 340 mm for different snow depths. The diameters of the disks were chosen to have the same value as those used by Shoop [10] in her laboratory compression tests. Figure 5 presents the FEM compression tests results for different snow depths.

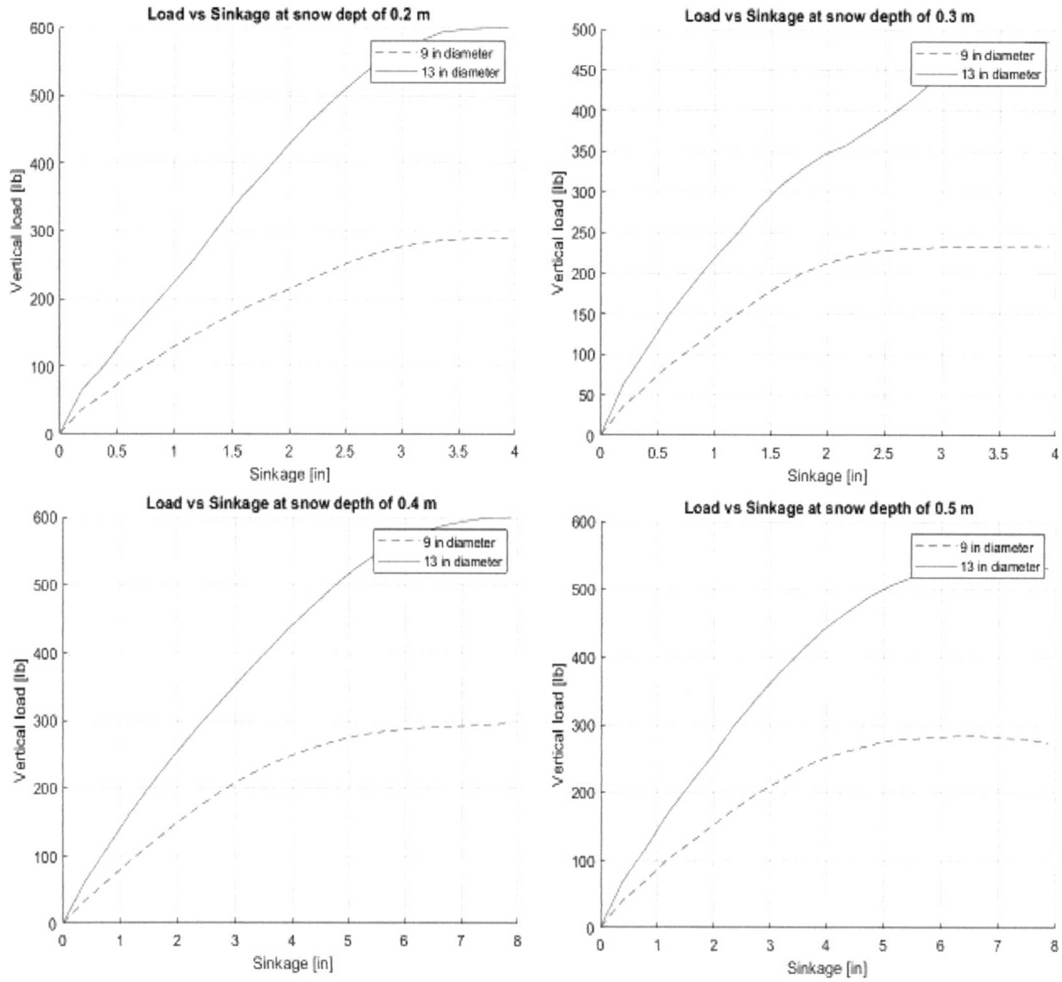


**Figure 5.** FEM compression tests results for different snow depths

The output of the FEM analysis is the resulting vertical force for a prescribed sinkage of the circular disc. The vertical load data is converted into pressure by dividing the force value with the area of the circular plates. The curves presented in Figure 5 are resulted after post-processing the raw data from the FEM analysis using the smoothing algorithm offered in MATLAB. As presented in Figure 2, in order to find the required curves to determine Bekker-Wong's coefficients, the data needs to be further processed by applying a natural logarithm function. From Figure 5 it can be observed that the load-sinkage curves are different for each snow depth, therefore the Bekker-Wong coefficients will have different values. Table 2 presents the values for the Bekker-Wong coefficients for the different snow depth values.

**Table 2.** Bekker-Wong coefficients for the different snow depth values.

| Snow depth [m] | $k_c$ [lb/in$^{n+1}$] | $k_\phi$ [lb/in$^{n+2}$] | $n$ [-] |
|---|---|---|---|
| 0.2 | 6.22 | 0.93 | 0.83 |
| 0.3 | 6.81 | 0.82 | 0.83 |
| 0.4 | 5.11 | 0.68 | 0.87 |
| 0.5 | 6.69 | 0.49 | 0.81 |

The coefficients found by our algorithm are different than the ones presented in the literature for snow [13]. However, the values of the coefficients might not have a physical meaning and are used only to curve-fit the experimental data with the Bekker-Wong pressure-sinkage equation. The coefficients from the literature are calculated from experimental data obtained in the field where snow has different compaction levels and different densities. Our FEM snow model assumes that the snow is soft (low compaction level), homogeneous, and has a constant density. As presented in section 2.3, the Bekker's coefficients obtained from the FEM snow model will further be used to estimate the sinkage and motion resistance, considering the rigid wheel case.

To analyze the accuracy of our analytical model, we conducted a comparison between our model's output and the empirical data collected in the field by Shoop [10]. In order to do that, we employed identical wheel dimensions, $D$ = 29.1" (0.74 m) and $b$ = 10.7" (0.27 m). The vertical load is considered to be $W$ = 1402 lbs (6236 N). Figure 6 presents a comparison between the theoretical and the empirical data for wheel sinkage and motion resistance coefficient. The motion resistance coefficient was calculated as a ratio between the longitudinal motion resistance force and the vertical load.

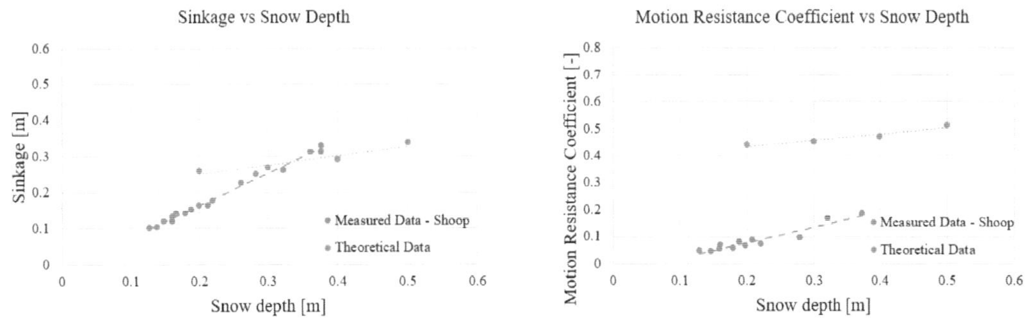


**Figure 6.** Wheel sinkage and motion resistance coefficient comparison

As it can be observed from the empirical data [10], at a snow depth of 0.4 m, the measured sinkage was 0.3 m (11.8") while the motion resistance coefficient was 0.2. At the same snow depth, our theoretical model output is given a sinkage of 0.29 m and a motion resistance coefficient of 0.47. Therefore, we can conclude that the accuracy of the theoretical model regarding the sinkage results is fairly good with respect to the measured values; however it can be observed that the motion resistance coefficients obtained using the theoretical model are significantly higher than the ones measured in the field.

*3.2. Estimating Wheel Sinkage and Motion Resistance utilizing FEM snow model*

Using the snow model described in section 2.1 (ALE method with C3D8R elements), we conducted an ABAQUS Explicit Analysis to examine the behavior of a rigid wheel rolling on soft snow. The purpose of this analysis was to investigate the changes in sinkage and motion resistance at different snow depths.

For this FEM analysis, the same wheel dimensions and properties have been used as for theoretical analysis in section 3.1. The ABAQUS Explicit software does not offer an input for the desired slip ratio, therefore the input longitudinal velocity and angular speed must be carefully chosen in order to simulate a desired slip ratio. For the analysis of wheel sinkage and motion resistance, a longitudinal speed of 2 m/s and an angular speed of 5.41 rad/s was applied to the rigid tire, therefore the simulated slip ratio was 0%. The boundary conditions for the snow model was selected as follow: the bottom face of the model was constrained in all directions and for the sidewalls roller boundary conditions were used to allow displacement only in the vertical direction. The simulated snow depth values were 0.2, 0.3, 0.4, and 0.5 m. Figure 7 presents the raw FEM output data for the longitudinal resistance force.

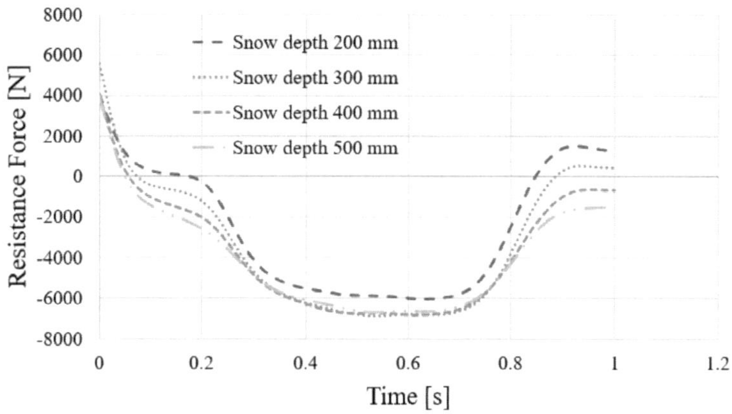


**Figure 7.** Effect of Snow depth on Longitudinal resistance force

Due to computational resources, for this type of explicit analysis, the mesh is coarse, therefore the output reaction force in the longitudinal direction varies significantly with respect to the simulation time. The resistance force data is post-processed by averaging the raw output data from the FEM model. After that, the motion resistance is calculated for every snow depth by dividing the resistance force to the vertical load. The raw output data regarding the wheel sinkage is relatively constant and it is processed in a similar manner. Figure 8 presents a comparison between the output of our own FEM model, the FEM model and the experimental results presented by Shoop [10], and the results from the analytical model of sinkage and motion resistance based on Bekker-Wong's coefficients.

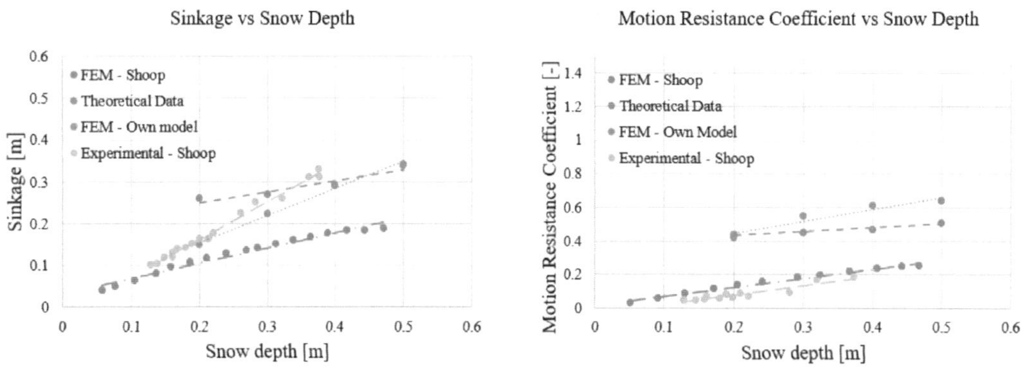


**Figure 8.** Output of FEM Model, Analytical Model and Experimental data

It can be observed from Figure 8 that the sinkage output of our FEM model and the analytical model is fairly similar to the experimental data recorded in the field. This indicates that the rigid wheel approximation does not impact in a significant manner the accuracy of the predicated sinkage. The motion resistance coefficient of our FEM model is different from the FEM model and the experimental data presented by Shoop; however, it is similar to the analytical model. As presented before, the theoretical data is obtained by using the Bekker-Wong's coefficients, which were calculated based on the FEM snow model. Therefore, it is expected that the output of our FEM model and the output of the analytical model to be similar. In order to improve the output of our FEM model, further refinement of the mesh should be considered for the 3D explicit analysis.

### 3.3. *Estimating Thrust capabilities utilizing FEM snow model*

According to Wong [13], by integrating the horizontal component of tangential stress over the entire contact area, the total tractive effort F can be determined using equation (10). Also, the total motion resistance can be determined by integrating the pressure values over the entire contact area. If the total motion resistance is subtracted from the total thrust, the total drawbar pull force can be estimated using equation (11).

$$F = r * b * \int_0^{\theta 0} \tau(\theta) * \cos(\theta)\, d\theta \tag{10}$$

$$F_d = r * b * [\int_0^{\theta 0} \tau(\theta) * \cos(\theta)\, d\theta - \int_0^{\theta 0} p(\theta) * \sin(\theta)\, d\theta] \tag{11}$$

The data for the pressure and shear stress distribution over the contact area is subtracted from the FEM Explicit analysis and then it is further post-processed using MATLAB. For this analysis the snow depth was 0.5 m. Figure 9 presents the evolution of pressure and shear stress distribution over the contact length of the rigid wheel for different slip ratios.

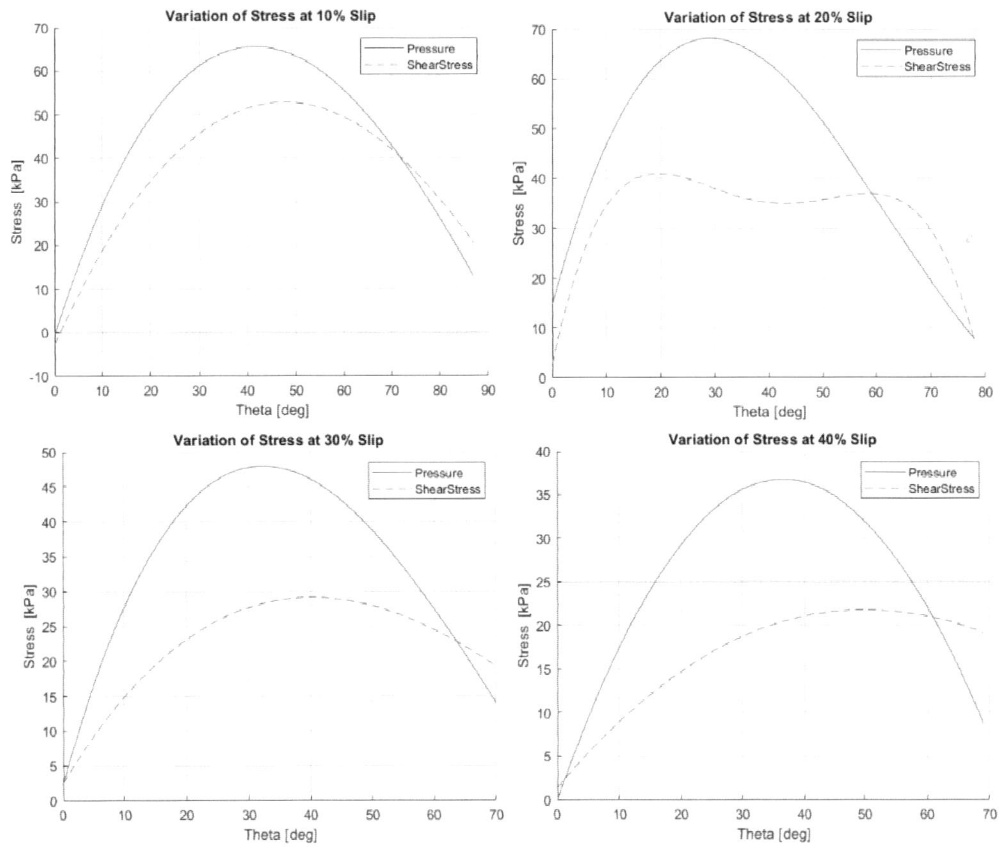


**Figure 9.** Pressure and Shear Stress distribution over the contact length of the rigid wheel

It can be observed from Figure 9, that the variation of pressure and shear stress is similar to the one presented in the literature [13]. By applying equations (10) and (11) to the data presented in Figure 9, the total thrust capabilities and the total resistive force can be predicted. Figure 10 presents the evolution of total thrust and resistive force with respect to slip ratio.

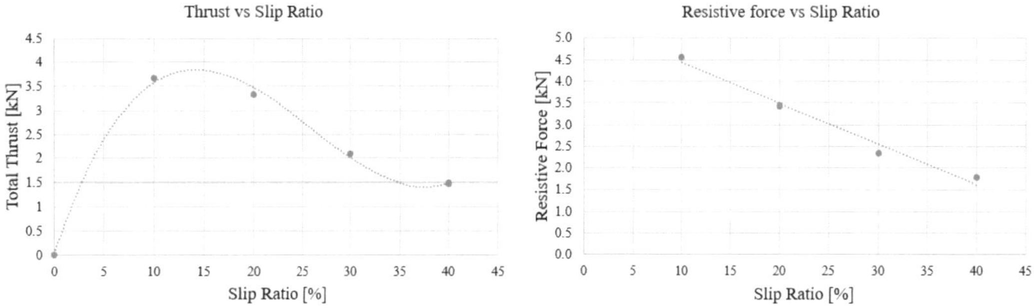


**Figure 10.** Variation of Total Thrust and Resistive Force with Slip Ratio

From Figure 10 it can be observed that the maximum thrust has a value in the close range of 4 kN for a slip ratio in the range of 10-20%, while the resistive force is estimated to have the same value for the same slip ratio range. This means that the drawbar pull has a small value, close to the limit where the wheel cannot develop enough thrust to overcome the motion resistance force. As presented in section 3.2, our FEM output produces considerable higher values for the motion resistance coefficient than those presented in literature, therefore the drawbar pull might have a higher value than the one presented in this section. Also, for this analysis, the wheel was estimated to be rigid. It is known that for snow rolling conditions, the pneumatic tire generates most of the thrust from the hysteresis component of the total longitudinal force, not from the adhesion component. Also, the rigid wheel has no tread attached. Therefore, the true values of the total thrust and the drawbar pull will be higher for a real-world scenario.

## 4. Conclusions

The implementation of the snow model using the explicit version of the commercial software ABAQUS followed a methodology similar to that of Shoop [10]. To ensure accurate results, a mesh convergence study was conducted by varying the element sizes from 15 mm (714 nodes) to 2 mm (38,152 nodes).

The mesh convergence study ensures that subsequent analyses are performed with an appropriate element size, leading to reliable and precise simulations of load-sinkage behaviour in soft snow conditions. A comprehensive approach was employed to determine the values for Bekker-Wong's coefficients using the load-sinkage results from a soft snow FEM model. Two axial penetration tests were conducted, each utilizing a circular rigid plate with different diameters: 230 mm for the first test and 340 mm for the second test. The purpose of these tests was to evaluate the behaviour of the snow under quasistatic loading conditions. These coefficients play a crucial role in modelling the mechanical behaviour of the soft snow and are essential for accurate simulations and predictions.

The analysis aimed to investigate the behaviour of a rigid wheel rolling on soft snow with different depths, using both an FEM model and an analytical model. The primary focus was on examining the variations in sinkage and motion resistance at different snow depths. The results obtained serve as a basis for further refinement and validation of the analytical model, enhancing our understanding of wheel-snow interactions in practical scenarios. By comparing the empirical data [10] with theoretical model's result, it is observed that at a snow depth of 0.4 m, the measured sinkage was 0.3 m (11.8"), while the motion resistance coefficient was 0.2. In contrast, our theoretical model predicted a sinkage of 0.29 m (11.4") and a motion resistance coefficient of 0.47 at the same snow depth. Consequently, based on the findings of the study, it can be concluded that the theoretical model demonstrates a good level of accuracy in predicting the sinkage results when compared to the corresponding measured values.

However, it is noteworthy that the motion resistance coefficients obtained using the theoretical model are slightly higher than those measured in the field. On the one hand, the analysis of the sinkage output revealed that all models, including both FEM models and the analytical model, produced fairly similar results. This suggests that the approximation of a rigid wheel does not significantly impact the accuracy of the predicted sinkage. On the other hand, the output for the motion resistance coefficient exhibited variations between the two FEM models, although it was similar to the analytical model. As previously mentioned, the analytical model's data was obtained using Bekker-Wong's coefficients, which were calculated based on the FEM snow model. Consequently, it is expected that the output of our FEM model and the analytical model would be relatively similar.

To improve the performance of our finite element method (FEM) model, we plan to enhance the mesh refinement for the 3D explicit analysis. By incorporating this refinement, a higher level of detail regarding the behavior of the snow can be captured, resulting in enhanced accuracy in predicting the motion resistance coefficient. The optimization of the mesh size allows for a more accurate representation of the snow behavior, thus increasing the fidelity of the results obtained from the FEM model.

**Acknowledgements**
This work has been partially supported by the TMVS Lab and the CIB at Virginia Tech.

**Author's ORCID ID**
Y Surkutwar: https://orcid.org/0009-0008-0153-3251
A Vilsan: https://orcid.org/0009-0003-4991-7944
C Sandu: https://orcid.org/0000-0002-3907-5080
C Untaroiu: https://orcid.org/0000-0002-1813-669X

**References**

[1] Botta G F, Tolon-Becerra A, Tourn M, Lastra-Bravo X and Rivero D 2012 Agricultural traffic: Motion resistance and soil compaction in relation to tractor design and different soil conditions *Soil and Tillage Research* **120** 92–8.
[2] Moreland S, Skonieczny K, Inotsume H and Wettergreen D 2012 Soil behavior of wheels with Grousers for Planetary Rovers *2012 IEEE Aerospace Conference*
[3] Zeng Y, Zhou Y, Liu D and Zuo Q 2013 A smart calibration model on track's pressure-sinkage characteristic of a tracked vehicle moving on soft seabed sediments *Journal of Central South University* **20** 911–7
[4] Surkutwar Y, Sandu C and Untaroiu C 2023 Review of modeling methods of compressed snow-tire interaction *Journal of Terramechanics* **105** 27–40
[5] Mellor M 1974 *A review of Basic snow mechanics* (Hanover, NH: U.S. Army Cold Regions Research and Engineering Laboratory)
[6] Shenvi M N, Sandu C and Untaroiu C 2022 Review of compressed snow mechanics: Testing methods *Journal of Terramechanics* **100** 25–37.
[7] Nakajima Y 2003 Analytical model of longitudinal tire traction in snow *Journal of Terramechanics* **40** 63–82
[8] Seta E, Nakai T, Heggli M, Jaggi M, Löwe H, Schneebeli M and Szabo D 2010. Prediction of tire traction on a compacted snow road *Physics and Chemistry of Ice* 45-49.
[9] Bui H H, Sako K and Fukagawa R 2007 Numerical simulation of soil–water interaction using smoothed particle hydrodynamics (SPH) method *Journal of Terramechanics* **44** 339–46
[10] Shoop S A 2001 *Finite element modeling of tire-terrain interaction* thesis.
[11] Drucker D C and Prager W 1952 Soil mechanics and plastic analysis or limit design *Quarterly of Applied Mathematics* **10** 157–65
[12] Shoop SA, Phetteplace GE and Wendy LW 2010 Snow Roads at McMurdo Station, Antarctica *ERDC/CRREL* **TR-10-5**
[13] Wong J Y 2022 *Theory of ground vehicles* (Hoboken, N.J, USA: John Wiley & Sons, Inc)

IOP Conf. Series: Materials Science and Engineering 1303 (2024) 012047 doi:10.1088/1757-899X/1303/1/012047

# Considerations on traffic management in isolated intersections

**V Dragu, F C Oprea and E A Roman**

University Politehnica of Bucharest, Faculty of Transport, Splaiul Independenţei nr. 313, sector 6, Bucharest, Romania, 060042

E-mail: v_dragu@yahoo.com

**Abstract**. The continuous growth in the number of motor vehicles in major cities, combined with a stagnant development of road infrastructure, has caused traffic congestion and a range of negative direct and indirect effects. Even if measures were proposed to improve the road infrastructure, such as increasing connectivity or the size of the roads, the benefits would only be temporary, as it is well known that any infrastructure improvement will attract new traffic. The aim of this study is to identify traffic management measures at conflict points in road infrastructure that can improve traffic flow parameters. The study categorizes conflict points as either saturated or unsaturated intersections in order to determine a precise mathematical model for the timing of traffic light cycles that is suitable for local traffic conditions. The theoretical assertions are supported by a case study and conclusions and recommendations are formulated for use in the practice of traffic lights at isolated intersections.

## 1. Introduction

The use of personal cars in most of the daily trips has accentuated the negative external effects and led to levels of congestion that are hard to bear for those involved (Dragu et al. 2017). Among the negative external effects induced by traffic congestion, we mention: chemical pollution, noise, additional social costs, effects on health, degradation of the landscape, increased risk of accidents and, implicitly, a decrease in the quality of life of the inhabitants (Banister, 2008; Banister, 1997).

In terms of health, traffic congestion causes excessive fatigue, cardiovascular disease, but even affects entire systems such as the respiratory system and the nervous system, resulting in a lower standard of living (Qadri et al, 2020).

In 2021, Bucharest was in eighth place in the world for traffic congestion, according to the annual index carried out by Tomtom. Road traffic congestion was estimated at 50%, up 8% from 2020 but down 2% from 2019 (https://www.tomtom.com/).

The 10 most congested cities in the world at the end of 2021 were: Istanbul (Turkey), Moscow (Russia), Kiev (Ukraine), Bogota (Colombia), Mumbai (India), Odessa (Ukraine), Saint Petersburg (Russia), Bucharest (Romania), Novosibirsk (Russia), Bengaluru (India).

Although the term congestion is used extremely frequently, it has different interpretations from stakeholders who, with often divergent interests and points of view, interfere on the transportation market. We can name here: the person in charge of the transport infrastructure development strategy, the infrastructure user, the traffic engineer, the transport beneficiary and the transport economist. For each of them the notion of congestion has different meanings depending on the degree of involvement in the transportation process (Raicu, 2007).

The plan proposed by the Bucharest Metropolitan Transport Association (an NGO structure established with the agreement of the General Council of the Municipality of Bucharest) aims at creating a complex public transport network that would solve the current congestion problems. The goal for 2030 is to increase the number of public transport users from 20% at present to at least 80%. This objective must be achieved in the area with the highest population density in the country - over 8000 inhabitants/km$^2$.

From the data supplied by the Driving Licenses and Vehicle Registration Directorate of the Ministry of Internal Affairs – DRPCIV (https://www.drpciv.ro/), information was obtained regarding the number of vehicles registered in Bucharest. Table 1 shows the situation of the number of registered vehicles between 1990 and 2021.

**Table 1.** The number of registered vehicles between 1990 and 2021

| Year | 1990 | 1992 | 1995 | 2002 | 2003 | 2013 | 2018 | 2019 | 2021 |
|---|---|---|---|---|---|---|---|---|---|
| No. ff vehicles | 336,928 | 377,206 | 468,414 | 655,042 | 695,365 | 1,125,591 | 1,381,620 | 1,457,889 | 1,529,221 |
| Occupied area (km$^2$) | 3.71 | 4.15 | 5.16 | 7.21 | 7.65 | 12.39 | 15.20 | 16.04 | 16.82 |
| % of the city occupied | 1.56 | 1.75 | 2.17 | 3.03 | 3.22 | 5.21 | 6.39 | 6.74 | 7.08 |

From table 1 it is noticeable that the number of registered vehicles and occupied area (11m$^2$/vehicle) has increased extremely fast and intensely.

The current main challenge of any urban system deals with the externalities produced by the road transportation system. It is well-known that about 50% of pollution is from road transport and, in particular, from internal combustion engines (De Luca and al, 2020). Within this context, different actions have been deployed in the last few decades to reduce pollution, such as acting on the vehicle technology or different types of fuels, and through different and sophisticated mobility/travel demand management policies (De Luca and Papola, 2001) or traffc flow control strategies.

Currently, Intelligent Transport Systems (ITS) are used at different levels (micro, meso and macro levels). The positive effect of their use is the reduction of time lost by traffic participants and the increase in the efficiency of the performance of a transport network, measured in the increase of transport capacity, but also of traffic safety (Ravish and Swamy, 2021).

As a result of the dynamic development of new technologies in the field of ITS, it became necessary to model transport processes in order to forecast traffic flow parameters. Thus, a widely used transport model concept was born to evaluate traffic flow parameters from the design phase (Darapalyuk et al, 2020).

The local control method is the most used road intersection coordination method, applied in most cases (Zhankaziev et al. 2018). This provides several modes of traffic control resulting from the traffic parameters of the intersection. This management method does not require the installation of expensive equipment, since traffic lights are the main control tool.

## 2. Determining the duration of the traffic light cycle

### *2.1 Intersection with unsaturated traffic*

From the point of view of demand, road intersections at the same level can be classified into unsaturated intersections, saturated intersections and oversaturated intersections. When all the vehicles in the queue pass through the intersection during the green period, we are dealing with an intersection with unsaturated traffic. Saturated intersections are those in which, after the end of the green phases, there are still queues of vehicles, or in other words, during the duration of the green, all the vehicles in the waiting line fail to transit the intersection. If vehicle queues tend to lengthen and block upstream intersections, then the intersection becomes oversaturated and other methods of streamlining traffic, reducing the length of queues and delay times must be sought. Traffic saturation can occur on a single phase or on several phases (Darapalyuk et al, 2020).

Regardless of the traffic light time, there is a time interval when a saturation flow ($q_s$) flows through the intersection, which represents the maximum possible flow, achieved when the distance between vehicles is the minimum safe distance and the speed is maximum and approximately constant. There is a time $t_o$ called lost time and it corresponds to the time when the first vehicle in the line enters the perimeter of the intersection after the light turns green. This is partially recovered by extending the green duration to the yellow duration for the last vehicle in the line to leave the intersection andenteron green. The duration of the yellow light, $t_g$, is also a lost time imposed by traffic safety. It represents the time required for the intersection to be cleared by the last vehicle entering the intersection at the end of the green time of a traffic phase.

The values of the saturation flows depend on the starting performance of the vehicles composing that flow and on the width of the traffic lanes ($l_b$). For forward ($q_{si}$) and turning ($q_{sv}$) directions, the saturation flux is given by relation (1) (Popa and Beloiu, 1999).

$$q_{si} = 525 \cdot l_b \left[V_e / h\right] \; ; \; q_{sv} = \frac{1800}{1 + \frac{1{,}542}{R}} \left[V_e / h\right] \tag{1}$$

For $l_b$, the minimum allowed value is 3,5 m and R is the average turning radius provided by the geometry of the intersection, but not less than 6,55 m, (5,3 + 1,25)m, 5,3m is the minimum value allowed for the innermaneuver in grange of long vehicles by EEC - UN Regulation no. 36, adopted in Romania by STAS 12139/1 – 91, and the maximum width allowed for road vehicles is 2,5m.

Determining the duration of the traffic light cycle for an intersection with unsaturated traffic involves the following steps:

1) Determination of lost time for traffic phases:

$$T_P = t_p + t_g, \tag{2}$$

2) Determining the effective green time:

$$t_{ve} = t_v - T_P \tag{3}$$

3) Determining the green time required for crossing the intersection by a number of $n_v$ vehicles:

$$t_v = (n_v - 1) h_u + T_P, \tag{4}$$

where $h_u$ represents the time interval between two successive vehicles ($h_u = 2_s$).
The number of vehicles that transit the intersection during the green period is:

$$n_v = \frac{t_v - (T_p + h_u)}{h_u} \tag{5}$$

4) Determining the critical flow ($q_{cr}$) of vehicles that can transit the intersection during the green period:

$$q_{cr} = \frac{3600}{t_v} n_v \left[V_{e/h}\right] \tag{6}$$

5) Determining the total flow that transits the intersection in the interval of one hour in the F phases of the traffic light cycle:

$$\sum_{i=1}^{F} q_{cr,i} = \frac{3600}{\sum_{i=1}^{F} t_{v,i}} \sum_{i=1}^{F} n_{v,i} = \frac{3600}{\sum_{i=1}^{F} t_{v,i}} \sum_{i=1}^{n} \left[\frac{t_v - (T_p + h_u)}{h_u}\right] \tag{7}$$

where it is obtained:

$$\sum_{i=1}^{F} q_{cr,i} = \frac{3600}{h_u \sum_{i=1}^{F} t_{v,i}} \left[\left(\sum_{i=1}^{F} t_{v,i} - F(T_p - h_u)\right)\right] \tag{8}$$

If $\sum_{i=1}^{F} t_{v,i} = C$ is considered the cycle duration, then the expression for C is:

$$C = \frac{3600\, F (T_p - h_u)}{3600 - h_u \sum_{i=1}^{F} q_{cr,i}} \tag{9}$$

The maximum allowed duration of C, resulting from comfort conditions, is 120s (to reduce waiting time).

From the analysis of relation (9) it can be seen that it is not valid for large flows.The limit value is obtained from the expression:

$$3600 - h_u \sum_{i=1}^{F} q_{cr,i} = 0, \tag{10}$$

which leads to values of the total traffic flow that transit the intersection equal to 1800 $V_e$ /h, under the conditions where $h_u = 2$ s

There are antagonistic conditions in increasing the flow. One of the measures is aiming the increasing of the number of access lanes in the intersection, but this leads to an increase of the area of the intersection and implicitly an increase in the duration of the intermediate times, with consequences on the reduction of the flow of vehicles passing through the intersection.This is because it is known that the duration of a cycle is composed of the sum of the durations of green and the durations of intermediate times.

$$C = \sum_{i=1}^{F} \left(T_{v,i} + T_{i,i}\right) \tag{11}$$

In conclusion, for a succession interval $h_u = 2$s the maximum value of the sum of the flows transiting the intersection is 1800Ve/h for the intersection to be classified as unsaturated intersections.

The maximum critical flow is assimilated to the saturation flow and is determined with the relation:

$$\sum_{i=1}^{F} q_{cr,i} = \frac{3600}{h_u}\left[\frac{C - F\left(T_p - h_u\right)}{C}\right] \left[V_e / h\right] \tag{12}$$

The critical flow values depend on the cycle duration, while the saturation flow, which is obtained when the cycle duration tends to infinity, is a constant that does not depend on any of the parameters involved.The values of critical flow and saturation flow for 2, 3 and 4 operating phases are shown in table 2.

**Table 2.** Critical flow and saturation flow

| F | $T_p$[s] | $h_u$ [s] | $\sum' q_{cr}$ [$V_e$/h] | C[s] | $q_s$[$V_e$/h] |
|---|---|---|---|---|---|
| **2** | 6 | 2 | $1800 - \frac{14400}{C}$ | $\to \infty$ | |
| **3** | 6 | 2 | $1800 - \frac{21600}{C}$ | $\to \infty$ | 1800 |
| **4** | 6 | 2 | $1800 - \frac{28800}{C}$ | $\to \infty$ | |

The variation of cycle time for $T_p = 6$s, $h_u = 2$s and number of phases equal to 2, 3 or 4 is shown in Table 3.

**Table 3.** The variation of the traffic light cycle time depending on the size of the flows

| $\sum' q_{cr}$ | 100 | 200 | 300 | 400 | 500 | 600 | 700 | 800 | 900 |
|---|---|---|---|---|---|---|---|---|---|
| **C(F=2) [s]** | 8,47 | 9,0 | 9,6 | 10,28 | 11,07 | 12 | 13,09 | 14,4 | 16 |
| **C(F=3) [s]** | 12,7 | 13,5 | 14,4 | 15,43 | 16,61 | 18 | 19,63 | 21,6 | 24 |
| **C(F=4) [s]** | 16,94 | 18 | 19,2 | 20,57 | 22,15 | 24 | 26,18 | 28,8 | 32 |
| $\sum' q_{cr}$ | 1000 | 1100 | 1200 | 1300 | 1400 | 1500 | 1600 | 1700 | 1800 |
| **C(F=2) [s]** | 18 | 20,57 | 24 | 28,8 | 36 | 48 | 72 | 144 | ∞ |
| **C(F=3) [s]** | 27 | 30,85 | 36 | 43,2 | 54 | 72 | 108 | 216 | ∞ |
| **C(F=4) [s]** | 36 | 41,14 | 48 | 57,6 | 72 | 96 | 144 | 288 | ∞ |

Experience has shown that a correctly sized traffic light cycle should have a minimum value of 35s for 2 phases and 45s for 3 or 4 phases and a maximum value of 120s, with the recommendation that for a 4-way intersection not to exceed 80s (Salter, 1989).

Plausible values are shown in Table 4.

**Table 4.** Plausible values of the traffic light cycle time

| $\sum q_{cr}$ | 400 | 600 | 1100 | 1300 | 1500 | 1600 |
|---|---|---|---|---|---|---|
| **C(F=2) [s]** | 36 | 72 | – | – | – | – |
| **C(F=3) [s]** | – | – | – | 43,2 | 72 | 108 |
| **C(F=4) [s]** | – | – | 41,14 | 57,6 | 96 | 144 |

From the analysis of the data representing the cycle duration as a function of the sum of the hourly flows, it is observed that the relation for determining the cycle time provides different values depending on the number of operating phases of the traffic light installation. Two-phase operation is suitable for small flows while three - or four-phase operation for large flows, which corresponds to practical reality and increases traffic safety in intersections through the existence of separate traffic phases for left turns.

*a. Intersection with saturated traffic*

Determining the cycle time is done with the relationship: (Salter, 1989; Boroiu and Neagu, 2015)

$$C_s = \frac{k T_p + 5}{1 - \sum_{i=1}^{n} y_i} \tag{13}$$

where k is a constant which for typical intersections has the value 1,5.

$T_p$– the total time lost for the entire traffic light cycle,

$y_i$ – ratios, for each access, between the actual traffic flow and the saturation flow.

The folowing relationship is used to determine lost time:

$$T_p = \sum_{j=1}^{F} (T_{ij} - t_g) + \sum_{j=1}^{F} i_{pj}, \tag{14}$$

where$T_i$ is the intermediate time required for the phase change,

$i_p$ the delay caused by inertia when starting and stopping the vehicle (approximatelly 2 seconds).

From the analysis of relation (13) it can be seen that this also has limits in application. The relation can only be applied under the conditions in which:

$$1 - \sum_{i=1}^{n} y_i > 0 \tag{15}$$

In this case, the application limit cannot be rigorously determined, as in the case of unsaturated intersections because it depends on the sum of the y values that are given by the ratio of the access flow to the saturation flow. The more the number of accesses or phases increases, the more the inequality tends to no longer be respected and the relation for determining the cycle can no longer be applied.

After the green color appears, a certain amount of time is lost until the first vehicle enters the intersection, but after that the movement of the line of vehicles enters a steady state at an approximately constant rate called saturation flow ($q_s$).

If at the end of the green period there are still vehicles in line, those that have crossed the stop line will continue their movement during the yellow period to clear the intersection.Thus, the flow of vehicles moves during the green period and partly during the yellow period, but the vehicle flow has a lower value than the saturation flow ($q_s$). This can be seen in fig. 1.

After determining the cycle time ($C_s$), the total effective green time is distributed among the traffic phases in proportion to their corresponding y value. Thus, for an intersection with 2 phase operation (F1 NS direction and F2 EV direction) we can write:

$$t_{ve}^{NS} = \frac{y_{NS}}{Y_{NS} + Y_{EV}} \cdot C_s, \qquad t_{ve}^{EV} = \frac{y_{EV}}{Y_{NS} + Y_{EV}} \cdot C_s \tag{16}$$

To make the transition from effective green duration to actual duration, the following relations are used:

$$t_v^{NS} = t_{ve}^{NS} + i_p - t_g \ , \ t_v^{EV} = t_{ve}^{EV} + i_p - t_g \tag{17}$$

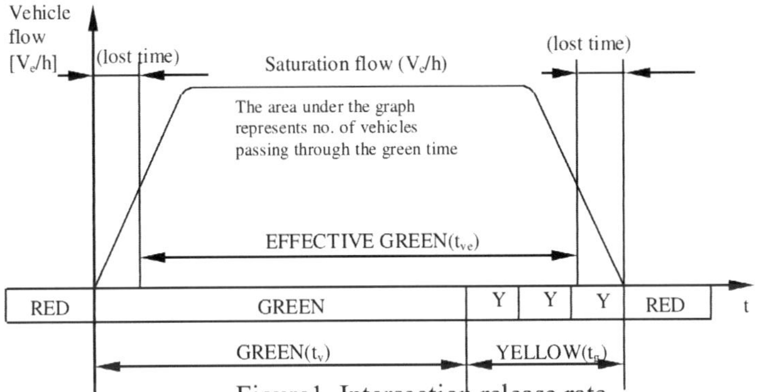


Figure1. Intersection release rate

In the case of fixed traffic lights, it is necessary to reach a compromise regarding the duration of the traffic light cycles so that the traffic variations are satisfied by a single traffic light program that does not cause very long delays.

To make some comparisons and observations on the application of relation (13), an intersection with 4 accesses operating in 2 phases is assumed, the total lost time is 10s, equal values of the green time and the value of the saturation flow 1800 Ve/h on all 4 accesses. The calculated values of the cycle are presented in table 5.

**Table 5.** Values of the duration of the traffic light cycle

| Total flow [Ve/h] | Access flow [Ve/h] | $y_i$ | $T_p$ [s] | $\sum_{i=1}^{n} y_i$ | $C_s$ [s] | $C_{smin}$ [s] | $C_{smax}$ [s] |
|---|---|---|---|---|---|---|---|
| 3000 | 750 | 0,42 | 10 | 0,84 | 125 | 94 | 188 |
| 2800 | 700 | 0,39 | 10 | 0,78 | 91 | 68 | 137 |
| 2400 | 600 | 0,33 | 10 | 0,66 | 59 | 44 | 89 |
| 1600 | 400 | 0,22 | 10 | 0,44 | 36 | 27 | 54 |

Figure 2 shows the variation of the average delay per vehicle as a function of the cycle duration,

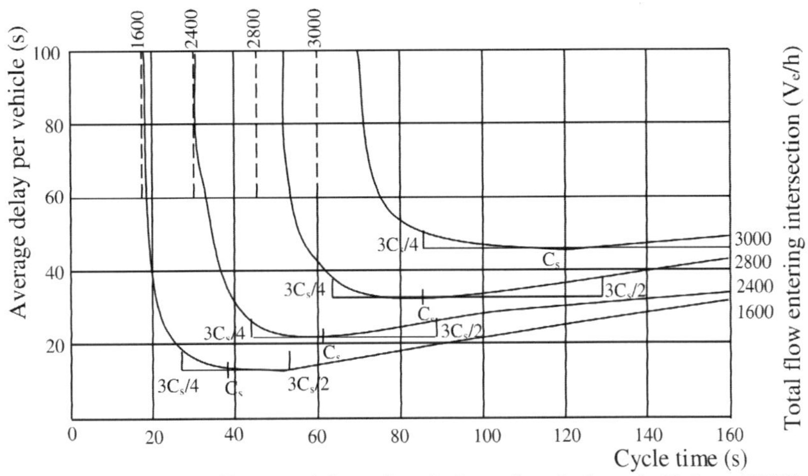


**Figure 2.** Effect on delay of variation of cycle length (Salter, 1989)

variation obtained through simulation. It is observed that as the cycle duration decreases, the value of the delay increases, eventually reaching infinity (the reis a vertical asymptote).

It is considered that a variation of the cycle time between $3C_s/4$ and $3C_s/2$ does not lead to an increase of delays by a maximum of 10%. The minimum value of the cycle is considered 25s and the maximum 120s.

Determining the duration of the cycle consists of going through certain stages. These are shown in fig. 3.

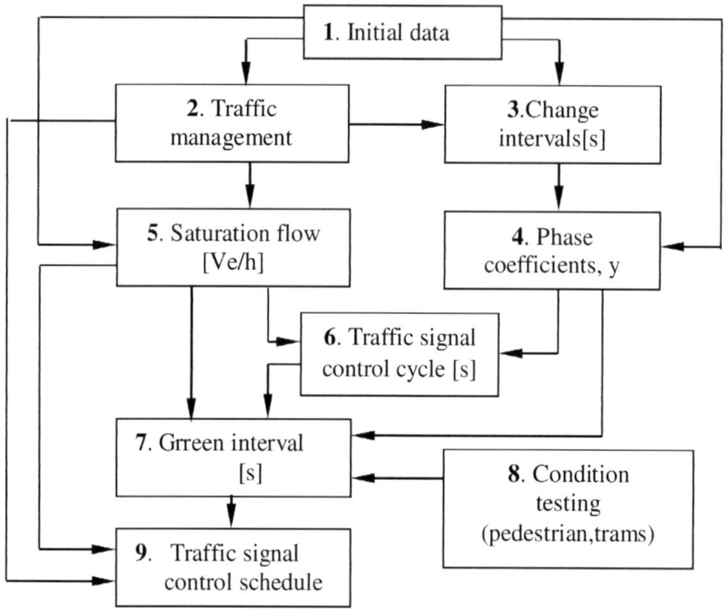


**Figure 3.** Flow chart for the calculation of the duration of the cycle (Drapalyuk et al, 2020)

## 3. Conclusions

The paper presented some considerations regarding the dimensioning and regulation of traffic in isolated intersections. These modelings are differentiated according to the characteristics of the traffic in the area and the influence of exogenous factors of the system. The way of dimensioning the traffic light cycle for intersections with unsaturated traffic was critically analyzed and it was identified that the cycle duration determination relationship has application limits. It can be applied only in cases where the sum of the flows of vehicles requesting the intersection does not exceed the value of 1800 reference vehicles/h, for the case where the tracking interval between vehicles is 2s. This value can increase if the interval between successive vehicles decreases, but this fact leads to an increase in the degree of risk because any unexpected braking of one of the vehicles in the row can lead to accidents due to the failure to ensure the necessary braking space. It is observed that this value corresponds to the value of the traffic capacity of a lane in ideal conditions, which is reflected in the practical reality.

It was concluded that the saturation flow in the intersection has a constant value regardless of the value of the critical flow and the other parameters that define the traffic. Two-phase operation is suitable for small flows while three-or four-phase operation for larger flows, which corresponds to practical reality and increases traffic safety through the existence of separate traffic phases for left turns.

From the analysis of the methodology for determining the duration of the traffic light cycle for saturated intersections, it was concluded that this also has application limits, but these are more relaxed and depend on the number of phases, the number of accesses and obviously the ratio between the arriving flows and the flow of saturation related to each access.

In conclusion, the analytical models for dimensioning the duration of traffic light cycles have limitations in application and then, in the situation of increased values of traffic flows, a possible solution would be that of simulations using specialized programs and the realization of a grouping of traffic flows on arteries with large capacity and organization of traffic according to the *green wave* system.

Another model for streamlining traffic is that of moving from the coordination of isolated intersections to the coordination of an entire congested area with an emphasis on giving priority to flows leaving the locality. This model requires specialized programs for routing and operational management of traffic that could also have an integrated function to inform users about the choice of the most appropriate moment to travel, which would lead to the flattening of traffic peaks and the rational use of the capacity of road infrastructures.

**Author's ORCID ID**
V Dragu: https://orcid.org/0000-0002-1703-101X
F C Oprea: https://orcid.org/0000-0003-2602-2534
E A Roman: https://orcid.org/0000-0002-4667-7987

**References**

[1] Dragu V, Burciu Ş and Roman E A 2017 *Dezvoltarea transportului public urban de mare capacitate – Soluţie pentru un oraş inteligent [Development of high capacity urban public transport - Solution for a smart city]*, Buletinul AGIR nr. 1/2017, Romanian. Available from: https://www.buletinulagir.agir.ro/articol.php?id=2818
[2] Banister D 2008 *The sustainable mobility paradigm*, Transp. policy, vol. **15**, p. 73-80; https://www.sciencedirect.com/science/article/abs/pii/S0967070X07000820?via%3Dihub
[3] Banister D 1997 *Reducing the need to travel*, Environ. and Planning B: Planning and Design, **24**, p. 437-449
[4] Qadri S S S M, Gökçe M A and Öner E *2020 State-of-art review of traffic signal control methods: challenges and opportunities*, Eur. Transp. Res. Rev., vol. **12**, artic. no. 55, https://doi.org/10.1186/s12544-020-00439-1
[5] Available from: https://www.tomtom.com/)
[6] Raicu Ş 2007 *Sisteme de transport* [*Transport systems*]. Publishing house AGIR, Bucharest, 2007, Romanian
[7] Available from: https://www.drpciv.ro/
[8] De Luca S, Di Pace R, Memoli S and Pariota L. 2020 *Sustainable Traffic Management in an Urban Area: An Integrated Framework for Real-Time Traffic Control and Route Guidance Design*, Sustainability, vol.**12**, 726; Available from: doi:10.3390/su12020726
[9] De Luca S and Papola A 2001 *Evaluation of travel demand management policies in the urban area of Naples*, WIT Trans. Built Environ., vol. **52**, 10.
[10] Ravish R and Swamy S R 2021 *Intelligent Traffic Management: A Review of Challenges, Solutions, and Future Perspectives*, Transp. and Telecom, vol.**22**, no. 2, p. 163–182
[11] Darapalyuk M, Dorokhin S and Artemov A 2020 *Estimation of efficiency of different traffic management methods in isolated area*, Transp. Res. Proc., **50**, p. 106–112.
[12] Zhankaziev S, Gavrilyuk M, Morozov D and Zabudsky 2018 *A Scientific and methodological approaches to the development of a feasibility study for intelligent transportation systems. Transp. Res. Proc.*, **36**, p. 841–847.
[13] Popa G and Beloiu D M 1999 *Trafic şi securitate rutieră - Îndrumar de laborator, [Traffic and road safety - Laboratory guide]* Publishing house Datavision, Bucharest
[14] Available from: STAS 12139/1 – 91
[15] Salter R J 1989 *Highway traffic analysis and design,* Macmillan Education LTD, London, 1989
[16] Boroiu A A and Neagu E 2015 *Trafic rutier şi siguranţa circulaţiei rutiere – Aplicaţii, [Road traffic and road traffic safety - Applications],* Publishing house of the University of Piteşti

IOP Conf. Series: Materials Science and Engineering 1303 (2024) 012048 doi:10.1088/1757-899X/1303/1/012048

# The influences of human factors on road accidents in Romania in the period 2013 – 2022.

**C I Dimitriu**[1, 2], **G Frățilă**[1]

[1]POLITEHNICA University of Bucharest
[2]Romanian Road Authority – ARR

E-mail: cydimitriu@yahoo.com

**Abstract**. The main objective of the study is to establish a correlation between the main causes of road traffic accidents in Romania and "human factors". The aim of the study is to show that accidents are attributed exclusively to human mistakes in a higher percentage than. The data used in the research are taken from the database of the Traffic Police-for Road traffic accidents during 2013 -2022 and from the database of the Romanian Road Authority - ARR - for road safety inspections during 2018-2021. From the perspective of human factors, the research will give answers on the nature of the causes that lead to the production / favouring of road accidents in Romania and will propose solutions to address road risks to improve road infrastructure safety. As a case study, the research will focus on the human factors that contribute to road accidents on National Road 2. Compared to other national roads, on DN 2 road accidents occur with the highest number of fatalities, approximately 8.28% of all road deaths nationwide. The analysis of findings from road safety inspection reports carried out on DN2 shows that there is a correlation between road infrastructure non-conformities and causes of road traffic accidents.

## 1. Introduction

According to EU statistics, as presented in the National Road Safety Strategy - SNSR, 2022-2030 [1], Romania consecutively registers the highest annual rate of road deaths among all Member States of the European Union. In addition, the perceived quality of road infrastructure in Romania is at the lowest level in the entire EU.

Reality shows that some aspects of the road infrastructure or road environment can influence driver behaviour. The road (including the road environment), the vehicle and the human (driver and vulnerable users) are key factors of the road transport system. Each of these factors can individually contribute to accidents. In most cases, complex combinations and interactions between factors favour the occurrence of road accidents. The Road Safety Manual published by the Permanent International Association of Road Congresses – PIARC, pp. 422-441 [2], shows that interactions between road and vehicles are determined by physical factors, interactions between vehicle and driver are determined by ergonomic needs, and interactions between man (driver, users) and the road (road environment) are determined by „human factors". Thus, unwittingly, drivers or road users can make errors in assessing road risk, leading to a significant number of road accidents because of involuntary mistakes.

The Executive Director of the European Transport Safety Council (ETSC), Antonio Avenosso, criticizes in an open letter, published by the Financial Times, the perpetuation of the „myth" related to the main cause of road accidents, namely human error [3]. Mr Antonio Avenosso states that this myth: „human error causes 90% of road accidents", „is misleading and dangerous", above all because it is repeated and perpetuated. To accurately assess the claims made and identify the causes of accidents attributable to „human factors"', it is essential to carry out detailed analyses of accident data in parallel with the results of road safety inspections.

In Romania, traffic police bodies go to the scene of serious accidents to make research, measurements, collect information and write down details for drawing up the database of road accidents [4]. One of the criteria for compiling the database is the „root cause of the accident". There are 42 main cases in the proceedings. There is a significant correlation between the establishment of „main causes" and the provisions of the Romanian Government Emergency Ordinance No 195/2002, concerning traffic on public roads [5]. As a rule, each „case in the main proceedings" has a corresponding breach of a legal provision. The analysis of the causes main led to the identification of a correlation between each cause and the key factor responsible for the accident.

During the last years (2017-2022), at national level, „speed not adapted to road conditions" was the main cause of serious accidents, contributing 16.33% of the overall total [6]. „Speed not adapted to road conditions" is responsible for 17.18% of road deaths and 18.29% of serious injuries. If we refer only to the statistical data of the national road family, the "speed not adapted to road conditions" is responsible for 21.38% of serious accidents resulting in 18.32% deaths and 24.68% serious injuries.
Starting from this finding, the research ranks the most dangerous national roads in Romania in terms of road deaths registered between 2017 and 2022. The case study is represented by the DN2 national road, on which the highest percentage of road deaths of the national road family was recorded. For DN2, the results of road safety inspections conducted by the Road Authority Romanian - ARR between Bucharest and Adjud are presented. The road segment was selected due to its unusual cross-sectional configuration, which stands out from that of other national roads. There are two traffic lanes of 3.5 m and reinforced verges of 2.5 m. This configuration favors the movement of vehicles at high speeds and the performance of overtaking maneuvers of vehicles. This segment of the DN2 road, which constitutes 62% of the total length of the road, accounted for 80% of serious accidents and 83% of road deaths between 2017-2022.

The Road Authority Romanian – ARR has the competence to carry out road safety inspections on the entire national public road network. Cumulative statistical data of periodic road safety inspections and supplementary data for accidents resulting in road deaths are available on the institution's website [7]. Between 2018 and 2022, ARR carried out regular road safety inspections covering approximately 9.5% of the total length of the national road and motorway network. Currently, the procedures for road safety inspections have been changed through the provisions of OG nr. 3/2022 [8] which, amends Law nr. 265/2008, on traffic safety management on road infrastructure [9]. Accordingly, targeted road safety inspections will primarily focus on road sections with a high level of road hazard, identified based on road infrastructure safety assessment.

The Romanian Road Safety Handbook (pages 17-21) stresses the importance of analyzing road risks generated by "human factors" in road safety procedures [10]. These factors are a significant component in the system of ensuring road traffic safety. Human errors can occur when an individual's cognitive capacity is exceeded, when coordination of basic tasks related to driving and moving is deficient, judgment is impaired, and there is a discrepancy between the individual's perception and surrounding reality. Changing the behavior of the human requires a process that unfolds over time.

The American Road Safety Handbook for Highways (Chapter 2—HUMAN FACTORS) deals with human factors and shows how human factors affect the interaction between drivers and roads [11]. About 90% of the information drivers use to perform driving tasks is visual. Visual acuity, contrast sensitivity, peripheral vision and visual search are aspects of the driver's vision that must be considered when designing roads as well as when conducting road safety inspections. The road must meet the expectations that drivers develop in both the short and long term.

The PIARC Road Safety Guidelines set out the basic requirements of road infrastructure-specific human factors [12]:

- the road should provide enough reaction time – roads with forgiving edges,

- the road must provide a field of vision allowing the driver to check that the vehicle is kept safe in the lane – ensuring visibility and readability,
- the future configuration of the road must correspond to the expectations of the driver formed by recent perceptions – predictable roads.

Research shows that accident databases do not contain enough information to determine what percentage of „human factors" caused or contributed to road accidents. The nature of information about serious accidents and the structuring of data by the Traffic Police does not allow an assessment of the level of interaction between the factors responsible for the occurrence of an accident.

## 2. Database analysis

Two databases were used during the research. One of them includes the dispositions for the execution of the intervention works on the road infrastructure for the elimination/ reduction of road risks on the road DN2-E85. The dispositions are in the inspection reports that are drawn up by road safety auditors, employees of the Romanian Road Authority – ARR. The second database of Romanian Police includes the serious road accidents resulting in dead persons and serious injuries, from 2017-2022.

### *2.1. Responsible factors in the occurrence of road accidents*

Following the analysis of each "main cause", according to the situation in Table 1, an allocation of causes was made for each of the key factors responsible for the occurrence of a road accident – man, road (road environment), motor vehicle.

**Table 1.** The distribution of the main causes of road accidents by responsible factors.

| Responsible factors | | Main causes - according to the classification of the Traffic Police | No. Causes |
|---|---|---|---|
| **ROAD** | | | |
| | 1 | other causes related to the road | |
| | 2 | animals or other objects | |
| | 3 | road damaged or under construction | |
| | 4 | unmarked obstacle on the road | |
| | 5 | incomplete/insufficient road signs | |
| | | **Total causes attributed to the responsible factor ROAD** | 5 |
| **MAN (Driver)** | | | |
| | 6 | falling asleep at the wheel | |
| | 7 | other offenses committed by drivers | |
| | 8 | other distracting concerns | |
| | 9 | oncoming traffic | |
| | 10 | aggressive driving | |
| | 11 | driving without a license | |
| | 12 | driving under the influence of alcohol | |
| | 13 | driving under the influence of drugs | |
| | 14 | irregular overtaking | |
| | 15 | infirmities or medical conditions | |
| | 16 | irregular return | |
| | 17 | not giving priority to pedestrians | |
| | 18 | not giving priority to vehicles | |
| | 19 | lack of insurance when changing the direction of travel | |
| | 20 | no insurance went back | |
| | 21 | no lane change insurance | |

| | | |
|---|---|---|
| 22 | failure to ensure load stability | |
| 23 | failure to respect the distance between vehicles | |
| 24 | non-observance of mandatory or regulatory road signs | |
| 25 | non-compliance with CF crossing rules | |
| 26 | non-compliance with traffic lights | |
| 27 | stopping, irregular idling | |
| 28 | speed not adapted to road conditions | |
| 29 | irregular speed | |
| 30 | overload | |
| 31 | incorrect use, lights & signalling devices | |
| | **Total causes attributed to the responsible factor MAN (Driver)** | 26 |
| **MAN (Vulnerable Participants)** | | |
| 32 | deviations of drivers of teams or animals | |
| 33 | deviations of machine operators | |
| 34 | cyclist violations | |
| 35 | passenger/traveller/companion violations | |
| 36 | other pedestrian violations | |
| 37 | unsupervised minors | |
| 38 | pedestrians on the roadway | |
| 39 | irregular pedestrian crossing | |
| | **Total causes attributed to the responsible factor MAN (Vulnerable Participants)** | 8 |
| **VEHICLE** | | |
| 40 | vehicle technical faults | |
| 41 | lack of traffic safety devices | |
| 42 | windshield, rear window, mirrors missing or unusable | |
| | **Total causes attributed to the responsible factor VEHICLE** | 3 |
| **TOTAL** | | 42 |

To find out the influence of the determinants in the occurrence of a road accident, the weight of each factor to which the accidents resulting in casualties and injuries are allocated during the reference period is calculated. The values calculated with the Excel - PivotTable tool is shown in Table 2.

**Table 2.** The percentage distribution of serious accidents and their consequences on each factor.

| Responsible factor | Causes % | Accidents (%) | Deaths (%) | Seriously Injured (%) | Slight Injured (%) |
|---|---|---|---|---|---|
| Road | 11.90 | 0.34 | 0.20 | 0.31 | 0.38 |
| Vehicle | 7.14 | 0.22 | 0.47 | 0.27 | 0.27 |
| Man Factors | 80.95 | 99.43 | 99.33 | 99.43 | 99.35 |
| *Driver* | *61.90* | *70.62* | *65.00* | *71.79* | *80.12* |
| *Vulnerable Participants* | *19.04* | *28.81* | *34.33* | *27.65* | *19.23* |
| Total | 100 | 100 | 100 | 100 | 100 |

The analysis of the results in Table 2 shows that the "man" factor is responsible in proportion of 99.43% for the occurrence of serious road accidents, respectively 70.6% driver responsibility and 28.8% responsibility for vulnerable participants.

The PIARC Road Safety Manual, p.46, Figure 3-1, [2], presents the results of studies showing that the interaction between the factors responsible for road accidents must be considered in the analysis of road risks, especially the interaction between the human factor and the road factor (road environment). By intervening on road elements, immediate gains can be achieved to increase road safety.

### *2.2. The main cause of serious accidents on the DN 2 national road*

While at national level the main cause of serious accidents is „speed not adapted to road conditions" [6], Table 3. Main causes of serious accidents on DN2. shows the order of main causes of serious accidents recorded on DN2.

**Table 3**. Main causes of serious accidents on DN2.

| No.crt. | Main causes | Deaths | Seriously injured | Serious accidents |
|---|---|---|---|---|
| 1 | irregular pedestrian crossing | 18.70% | 7.75% | 15.26% |
| 2 | falling asleep while driving | 16.60% | 12.47% | 10.63% |
| 3 | speed not adapted to road conditions | 11.76% | 14.89% | 11.68% |
| 4 | pedestrians on the carriageway | 6.51% | 1.94% | 4.84% |
| 5 | oncoming traffic | 6.09% | 4.72% | 4.53% |
| 6 | failure to give priority to vehicles | 5.88% | 12.83% | 10.00% |

The main causes of serious accidents on DN2 are: „irregular pedestrian crossing", which contributes significantly to deaths (18.70%) and serious injuries (7.75%), „falling asleep while driving", which has a major impact on the number of fatalities (16.60%) and serious injuries (12.47%) and „speed not adapted to road conditions", which contributes significantly to deaths (11.76%) and serious injuries (14.89%).

Compared to the situation at national level, where the main cause of road deaths is 'speed not adapted to road conditions', on the DN2 national road, irregular pedestrian crossing is the main cause of a significant number of road deaths.

These conclusions suggest that interventions to reduce serious accidents on the DN2 should focus on preventing irregular pedestrian crossings, managing driver fatigue, and promoting speed appropriate to road conditions.

### *2.3. National roads*

DN2-E85 is identified as the most dangerous national road in Romania in terms of the number of people killed in road accidents during 2017-2022. By processing the serious accident database with the Excel – PivotTable tool, Table 4 shows the ranking of the 5 most dangerous national roads.

**Table 4.** National roads according to the rate of deaths from serious road accidents.

| National road name | Deaths (%) | Seriously Injured (%) | Serious accident (%) | National road length (km) | Annual deaths density (persons / 100 km) |
|---|---|---|---|---|---|
| DN 2 | 9,01 | 6.76 | 6.53 | 482 | 16.46 |
| DN 1 | 7,10 | 7.72 | 8.40 | 642 | 9.74 |
| DN 6 | 6,98 | 5.67 | 5.33 | 639 | 9.62 |
| DN 7 | 5,33 | 5.01 | 5.19 | 597 | 7.65 |
| DN 15 | 2,82 | 3.07 | 3.09 | 338 | 7.35 |

The values in the table represent the statistics for serious accidents and their consequences in relation to the family of national roads in Romania. These statistics make it possible to compare roads from a road safety perspective. The DN2, for example, has a higher rate of fatalities, relative to road length, compared to DN1, DN6, DN7 and DN15, suggesting that DN2 needs increased attention in terms of road safety measures.

*2.4. Road safety inspection*

Between September 2018 and February 2022, the Road Authority Romanian – ARR carried out 89 periodic road safety inspections on the national roads and motorways network [7]. For the DN2 national road, in Table 5 presents the results of the quantitative analysis of the information contained in the road safety inspection reports carried out on the road segment with a consolidated verge width of 2.5 m.

**Table 5.** Distribution of provisions on DN2 segments with a reinforced verge of 2,5 m.

| Road safety inspection – no. ARR | Km – start (km + m) | Km – final (km + m) | Section length (km) | Seriously Injured (nr.) | Deaths (no.) | Provisions of inspection reports (no.) |
|---|---|---|---|---|---|---|
| 8 | 10+000 | 51+200 | 41.20 | 98 | 76 | 69 |
| 9 | 51+200 | 106+790 | 55.59 | 87 | 50 | 87 |
| 10 | 114+500 | 144+070 | 29.57 | 76 | 34 | 115 |
| 11 | 147+870 | 185+993 | 38.123 | 155 | 53 | 229 |
| 12 | 185+993 | 228+470 | 42.477 | 119 | 57 | 194 |
| 13 | 231+020 | 280+753 | 49.733 | 123 | 71 | 226 |
| 14 | 292+350 | 328+692 | 36.342 | 83 | 43 | 167 |
| Total | - | - | 293.035 | 741 | 385 | 1087 |

The analysis excluded the road segments that were under the administration of local authorities, such as the Buzau belt and the road sections that ensured the transit through the urban localities of RM. Salty, Adjud and Bacău.

Table 6 provides an overview of the results of the analysis of inspection reports and the provisions which have been grouped according to the criteria used to assess road traffic safety.

**Table 6.** The allocation of provisions on road traffic safety assessment criteria.

| The evaluation criterion | Criterion code | Number of provisions | Observations |
|---|---|---|---|
| **1**. Visibility: between users and on road elements | V | 539 | 49,6 % |
| **2**. Perception/predictability: understanding the road infrastructure and surroundings | P | 307 | 28,2 % |
| **3**. Match between infrastructure and dynamic limits | D | 21 | 1,9 % |
| **4**. Possibility of recovery/recovery | R | 220 | 20,2 % |
| **5**. Collision mitigation | A | 0 | - |
| **6**. Coherence/ consistency/ compatibility of road elements with the surrounding environment | C | 0 | - |
| **7**. Management of traffic flows from the perspective of road safety | M | 0 | - |
| **Total** | - | 1087 | 100 % |

As a result of road safety inspections, the auditors issued provisions to carry out specific road infrastructure improvement works aimed at reducing certain types of road risks. Each provision has been allocated to a specific criterion for assessing road traffic safety. The provisions are contained in Annex 5 of Order of the Minister of Transport no. 606/2017 [13]. These provisions are classified into 13 chapters and comprise 50 types of specific works.

Almost half of the provisions analysed are allocated to the visibility criterion. To this criterion, provisions have been allocated that include works for restoring/completing road markings or signs and trimming vegetation obstructing visibility over infrastructure elements or between road users. All these deficiencies and non-conformities of the infrastructure can cause errors of assessment of the driver, respectively the lack of reaction time that generates errors in driving.

From the observations on the movement of the vehicle flow on DN2 was found that the width of the road platform and long alignments allow and even favour the frequent carrying out of the overtaking manoeuvre. On the other hand, wide angles of vision are created that lead to driving behaviour to increase the speed of travel compared to driving on other roads that do not have such a wide width or that have another structure of the lateral marking of the road boundary [11].

## 3. Conclusions

Through this research, a correlation was made between the causes of serious accidents and the factors responsible. The results show that the accident database compiled by the traffic police does not help determine the percentage of accidents determined by "human factors". The results show that the human factor is responsible for serious accidents in a proportion of almost 100%. The database does not contain elements to help identify causes of accidents that are the result of the interaction of responsible factors, e.g., man – road (road environment). It can be concluded that cooperation between police and road safety auditors is necessary to identify appropriate ways to collect additional data from the scene of serious accidents. These data must enable a clear analysis of the conditions under which the accident occurred and enable the elements of the infrastructure responsible for human error to be identified by the driver or other road users.

Based on the results of road safety inspections conducted on a critical segment of Romania's most dangerous road, DN2, a quantitative analysis of the provisions issued by road safety auditors was performed. This involved establishing a correlation between the road traffic safety evaluation criterion (RTS) and the nature of the corrective measures required for each directive related to road infrastructure non-compliance. Comparing the approximately 50% allocation to the provisions related to driver visibility issues with the approximately 90% representing the portion of information volume - primarily visual - used by drivers for driving tasks, it can be concluded that the infrastructure of DN2 road is responsible for serious driving errors that lead to severe accidents.

Linking road safety assessment criteria to road risk relief provisions can be a way of analysing and addressing the structuring of road safety procedures, including the collection of serious accident scene data.

It is crucial that the ongoing professional development of specialized inspectors and the periodic training programs for road safety auditors be enriched with information and procedures to assist in identifying those elements of road infrastructure related to human factors.

## Author's ORCID ID

C I Dimitriu: https://orcid.org/0000-0001-9433-8165

## References

[1] The Government of Romania. Annex to Government Decision no. 682/2022 for the approval of the National Road Safety Strategy for the period 2022-2030. Official Gazette of Romania, Part I, no. 535 bis of May 31, 2022. 2022 May 25: p. 72.

[2] Technical Committee on Road Safety , PIARC. www.piarc.org. [Online].; 2003 [cited 2023 August 29. Available from: https://www.piarc.org/en/search.htm?q=Manual&l=current.

[3] Avenoso A. https://e-sigurantarutiera.ro. [Online].; 2022 [cited 2023 July 07. Available from: https://e-sigurantarutiera.ro/avenoso-etsc-90-din-accidentele-rutiere-provocate-de-erori-umane-este-un-mit-periculos-si-inselator.

[4] Romanian Police. www.politiaromana.ro. [Online].; 2022 [cited 2023 July 07. Available from: https://www.politiaromana.ro/ro/utile/statistici-evaluari/statistici.

[5] The Government of Romania. Emergency ordinance no. 195/2002, republished, with subsequent changes and additions, regarding traffic on public roads. The Official Monitor of Romania. 2006 August 3: p. 1-80.

[6] Dimitriu CI, Frățilă G. Top causes of road traffic accidents in Romania. In IOP Conference Series - Materials Science and Engineering; 2022; Timisoara, EAEC MVT: IOP Publishing Ltd. p. 8.

[7] Romanian Road Authority - A.R.R. www.arr.ro. [Online]. [cited 2022 July 07. Available from: https://www.arr.ro/arr_doc_230_rapoarte-si%20studii_pg_0.htm.

[8] The Government of Romania. Ordinance no. 3/2022 for the amendment and completion of Law no. 265/2008 regarding the management of traffic safety on the road infrastructure. The Official Monitor of Romania. 2022 January 28: p. 1-8.

[9] The Romanian Parliament. Law no. 265/2008, republished and updated with subsequent amendments, regarding the management of traffic safety on the road infrastructure. The Official Monitor of Romania. 2012 August 23 07: p. 6-18.

[10] Burlacu FA, Burlacu A, Răcănel C. The auditor's manual road safety Bucharest: Conspress; 2019.

[11] U.S. Department of Transportation. https://highways.dot.gov/. [Online].; 2009 [cited 2023 06 28. Available from: http://www.highwaysafetymanual.org/.

[12] International Permanent Association of Road Congresses, PIARC. Road Safety Evaluations Based on Human Factors Method. Road Safety. Paris: PIARC, Technical Committee; 2019. Report No.: ISBN: 978-2-84060-561-4.

[13] Ministry of Transport of Romania. Order no. 606/2017, with subsequent changes and additions, for the approval of some measures regarding the road safety inspection. The Official Monitor of Romania. 2017 June 19: p. 6-13.

IOP Conf. Series: Materials Science and Engineering 1303 (2024) 012049 doi:10.1088/1757-899X/1303/1/012049

# Methodology for assessing the road traffic risk in urban areas

**D Costescu, C R Bujor and A M Ifrim**

Doctoral School of Transports, Polytechnic University of Bucharest, Splaiul Independentei 313, 060042 Bucharest

E-mail: dorinela.costescu@upb.ro

**Abstract**. In the assembly of concerns for reducing the negative effects of urban road traffic, special consideration is paid - both at the local and global level - to issues related to road traffic safety. Spotting urban zones vulnerable to traffic crashes, understanding the causes of crashes, identifying some directions of action to reduce the traffic risk and their application from the urban planning stage can improve road safety. In this framework, the paper presents a methodology for estimating the intrinsic risk linked to road traffic. The main objective is identifying the relationships between the crash frequency and the intrinsic factors contributing to their occurrence. The proposed methodology is particularized for Bucharest based on the disaggregated statistical data analysis on severe accidents recorded in 2017-2022. The resulting model allows the classification of risk levels and, further, identification, planning and scheduling measures to improve urban road safety.

## 1. Introduction

In contemporary society, the risks caused by anthropic activities have been aggravated [1]. The concepts of sustainable development tend to globalize concerns for diminishing multiple risks of ecological and sociological nature (greenhouse effect, pandemic, fossil fuels resources, social inequalities, etc.). Road traffic risk is part of the risks produced by human society. The programs for road safety are undoubtedly on the political agenda of the authorities at different levels [2, 3].

Traffic rules and control mechanisms have always been established in road systems, regardless of their type of road infrastructure. Parallel, the technical conception of road vehicles has considerably evolved to reduce collision severity and determine behavioral changes. Complementary, starting from the "vision of zero victims (dead and seriously) in road traffic", adopted by the Swedish Parliament in October 1997 [4], new concepts regarding road safety responsibilities have been added. The main responsibility for low road safety performance is not assigned to drivers but to the system planners. The main goal of the new vision is to eliminate the likelihood of events causing victims in road crashes. Key actions based on scientific and tested experience must be adopted at the beginning of the decision-making process [5].

Consequently, the one who designs and implements the system is the first responsible for road safety. From this perspective, at the urban level, an objective of the Sustainable Urban Mobility Plan (SUMP) consists in enhancing road safety from the planning phase [6]. The solutions to ensure mobility demand, dependent on the urban area's size, shape and pattern, are potentially responsible for the volume and structure of the related traffic. The mobility solutions (correlated to the goals of sustainable development) must harmonize the accessibility needs (for people and freight) with the life quality (including the imperious necessity for road safety) and with the requirements of the protection of the natural and anthropic environment [7].

From a phenomenological point of view, the urban area where the road traffic risk occurs is a complex spatial system. The defining input classes for associated road traffic risk can be grouped into four sets: features (mobile entities - vehicles, pedestrians, etc. and functional entities of the traffic infrastructure), stakeholders (local authorities, managers of traffic infrastructures, etc.), spatial pattern and temporal pattern [5, 7]. In Bucharest - the study area of this research, all these classes have faced significant changes in the last two decades. Radical transformations in socio-economic life, major changes in the pattern of the commercial sector (caused by the development of large shopping centers), new dense residential and offices zones, structural and spatial changes of the interest points (for work, education, leisure time, etc.), as well as changes in the working practices and travel behaviors determined by Covid-19 lockdown have led to traffic heterogeneity increase, i.e., road traffic risk exposure. Identifying vulnerable urban areas to road crashes, understanding the causes, identifying action directions to reduce the traffic risk, and implementing them from the urban planning stage can improve road safety performance [8-10].

In this framework, this research develops a methodology for a detailed analysis of the recorded crash and spotting urban areas with intrinsic traffic risk. The purpose is to rank the zones with high traffic risk and to ensure the necessary information support for assessments in cost-effectiveness analysis (CEA) [5, 11] and, further, for planning the proposed actions to improve road safety performance.

## 2. The risk associated with urban road traffic

### *2.1. Traffic risk analyses*

Usually, at the macroeconomic level, four intercorrelated influencing factors of traffic risk are emphasized [5, 10]:

- Drivers of road vehicles (behavior with perceived traffic risk)
- Vehicles (technical characteristics, maintenance, total load, destination)
- Infrastructure (location, road geometry, type and structure of the road system, infrastructure management, traffic management)
- Socio-economic activities (dimension, pattern, temporal and spatial distribution).

The characteristic variables of the socio-economic activities determine the interactions between the first three classes of factors. The traffic, as well as the travel/transport demand (being a derived demand) depend on the trade-off between the mobility offer and demand [5]. The temporal dynamics of these interactions and the particularities of the four classes of factors generate spatial and temporal specificity for the size and nature of the traffic risk [8, 9]. In other terms, the spatial mobility supply and demand, the traffic risk exposure, and the frequency, location and severity of road crashes are influenced by the socio-economic space and, undeniably, by the implemented road safety measures [11].

Concerns for increasing road safety belong mainly to the state. Local authorities are insufficiently involved. Urban development plans usually do not consider road safety at meso- and macroscopic levels. The lack of a "culture of road safety" is reflected in the deficiency of traffic risk assessment in the large projects financed for urban areas [6]. The most common methodologies for characterizing road safety are based on crash statistics. They intend to identify deficiencies and dysfunctions of the road infrastructure that cause driver errors. The data available in Romania compile the reports made by the Road Traffic Police following serious crashes. The collected data refers to the location and time of the crash occurrence, the road category, type of the road infrastructure feature (road section or junction, their type), local characteristics for traffic management, type of road surface, environmental settings (lighting, weather conditions), type and severity of the crash (number of injured and dead, number of involved vehicles), primary causes of the crash.

The traffic risk evaluations considering all the factors involved in crash occurrence are not frequent. This state is caused mainly by the sectoral character of the data provided by different sources and the difficulty of correlating the information (such as time series datasets on traffic - volume, debit and structure of the road traffic, urban space settings - exposure, transformations of the urban zones and functions). For instance, based on data available for Bucharest, Figure 1 shows a decrease in crashes during and after the COVID-19 lockdown (2020, 2021). Obviously, the reduction is associated with diminished road and pedestrian traffic flows. However, due to the lack of traffic time series data (in terms of traffic volume and structure), it is difficult to perform complex analysis to determine the correct dependencies between the size of the road and pedestrian traffic and the crash frequency.

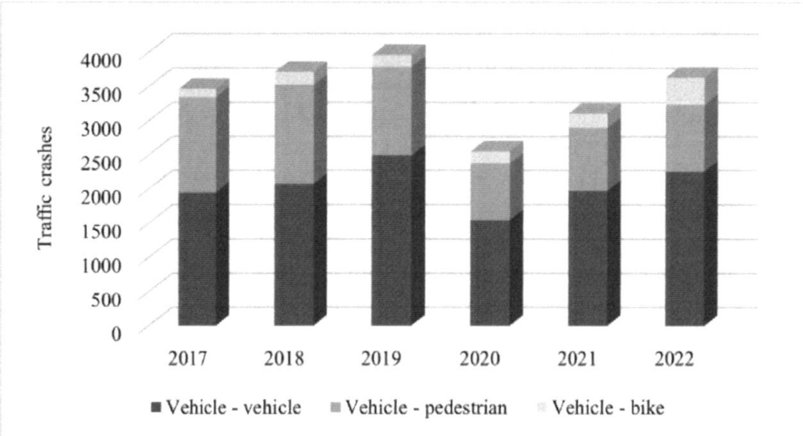


**Figure 1.** Traffic crashes in Bucharest, period 2017-2022.

Therefore, a comprehensive interdisciplinary approach to the key concepts of the binomial" road safety – traffic risk" is necessary to lead to systematic modeling of phenomena. By defining and identifying the concept of "risk situation" in a spatial model appropriate for functional analysis, criteria can be established for representing the different particular aspects of road traffic in connection with the urban environment [9]. Only in this manner can the road traffic risk assessment be included in an "ex-ante" hierarchy of project financing in urban mobility system planning.

### *2.2. Assessment of the traffic risk*

Different types of models for estimating the number of crashes are presented in the literature [12-14]. Most models are proposed to estimate the crash frequency for different types of intersections [14, 15] and homogeneous segments of road arteries [16]. There are also models developed to identify the relationships between the crash occurrence and road network pattern [17], respectively, the urban functions [8, 9].

In this research, the basic equation used to estimate the number of crash victims, $k$, is [18]:

$$k = E \cdot \mu \cdot \gamma \tag{1}$$

where $E$ represents the risk to exposure (vehicles-km), $\mu$ the risk related to crash frequency (crashes/vehicles-km), and $\gamma$ the risk related to crash severity (victims/crash).

A study of the dependences of exogenous and endogenous variables (speed, driver behavioral variables, technical-functional characteristics of vehicles, attributes of the road, lighting and weather circumstances, etc.) is required for each of the three factors ($E$, $\mu$ and $\gamma$),

The methodology developed and synthetically presented in the next section aims to create data sets that allow calibrating the dependencies between the number of crash victims and the composite risk exposure.

## 3. Data and methodology

### *3.1. The spatial model of the study area*

This research aims to develop a model that allows ranking the urban road infrastructure features with an increased risk of crashes and estimating the number of victims. The model results represent inputs in the cost-efficacy analyses necessary to prioritize financing measures to improve road traffic performance [5]. The model is developed for the particularities of the urban space and the road infrastructure network from Bucharest.

The topological, geometric and structural design of urban road networks determine the traffic safety performance [7, 9]. In order to define and calibrate the traffic risk estimation functions, the development of the macroscopic digital model of the urban network in Bucharest is necessary.

The road network models are different in terms of applied disaggregation level, depending on the specific analysis objectives. In this study, the model of the road network includes (Figure 2):

- Urban Road Intersection Dataset with attributes related to section length, number of lanes, tram lines existence (simple, double), bus public transport lines existence, street side parking existence.
- Urban Road Intersection Dataset with attributes related to traffic lights existence, configuration (3-way intersections - T, Y or skewed Y intersections, 4-way intersections - regular cross, skewed 4-legged intersections, more than 4-way intersections), type of infrastructure: only road traffic intersections and intersections including tram infrastructure, public transport stations existence, pedestrian crossing existence/type.

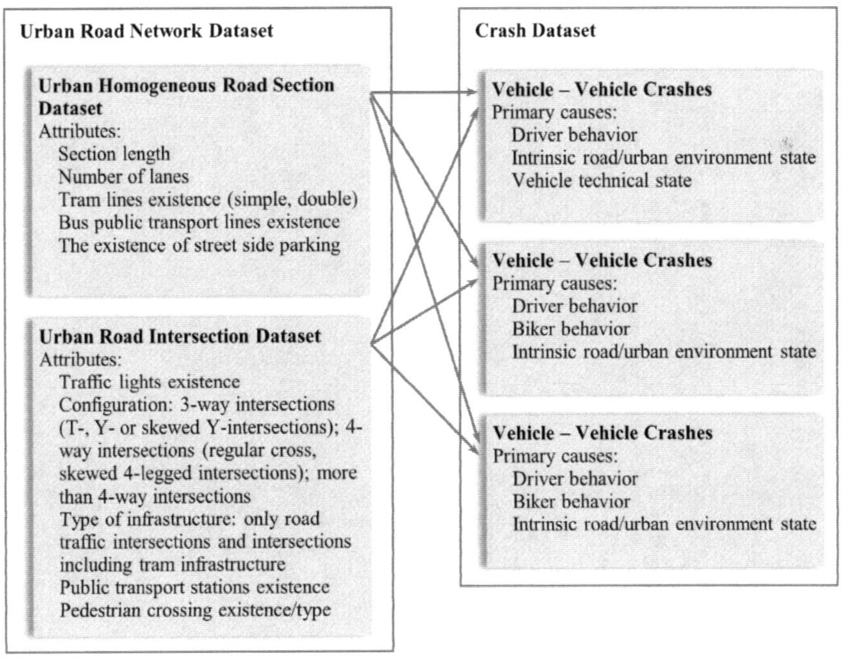


**Figure 2.** Selection criteria for disaggregate crash classes.

Several studies [8, 9, 19] indicate that different socio-demographic variables, such as the population, the active population, the degree of poverty and the economic activity, influence urban road safety. In order to be able to consider these features in traffic risk estimation, the digital model includes (Figure 2):

- The geographical datasets of urban zones defined by their functions (the available data for Bucharest include 9675 zones delimitated for 12 urban functions).

- The geographical datasets of traffic analysis zones (TAZ) defined in Bucharest – Ilfov SUMP [20]. The urban area is divided into 366 zones. Based on these data, for each TAZ, the densities of population, employees, school population, and road network are computed (as attributes that could influence the crash frequency) [7, 9].

The data available in this study for traffic risk estimation contain the records related to severe crashes (involving injuries, dead and/or significant financial damages) in Bucharest from 2017 – 2022 (summarized in Figure 1). In order to identify the properties of the most vulnerable network features, the crash data is processed in Geographic Information System (GIS) environment based on procedures presented in [9]. The data sets are used to disaggregate crashes in homogeneous classes. The crashes, grouped into three categories (those involving only motor vehicles, bicycles and motor vehicles, and pedestrians and motor vehicles), are assigned to feature classes of the urban road network (Figure 2). However, if the resulting size of certain samples for some crash causes is too small, then the level of data disaggregation will be reviewed [9, 16].

### *3.2. Data analysis*

The data sets presented above are necessary for estimating the risk associated with road traffic and then applying CEA for ranking the measures to improve road safety performance (Figure 3).

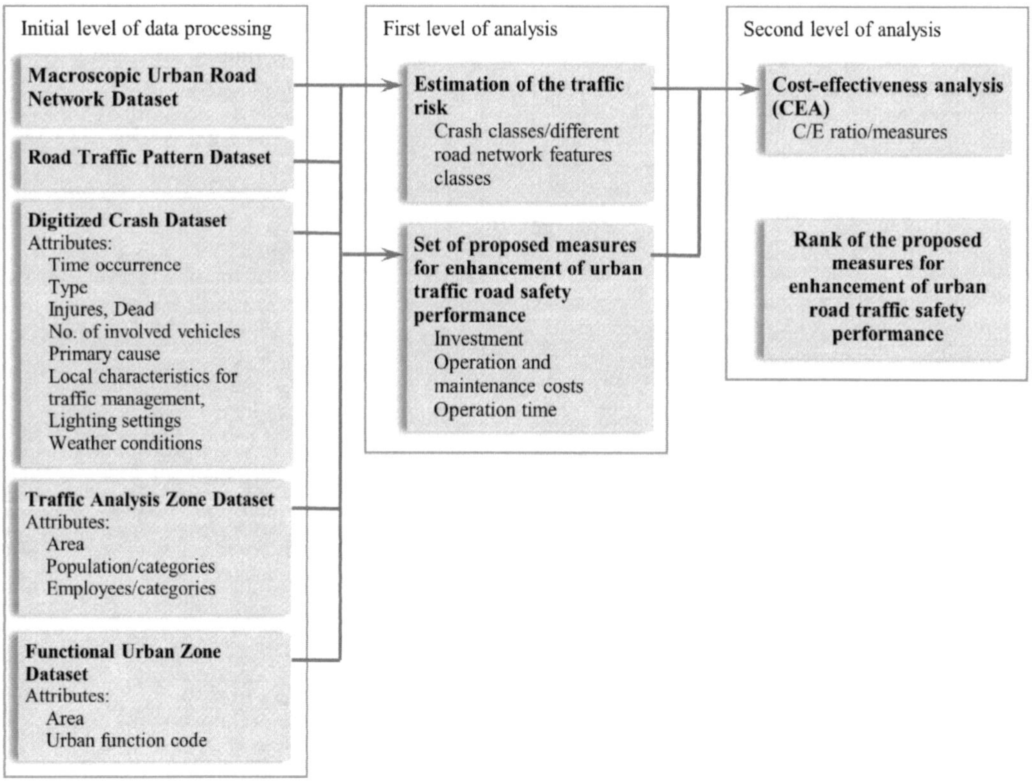


**Figure 3.** Flowchart for ranking the measures for enhancement of urban traffic road safety. performance.

The first phase involves data spatial processing in order to select potentially dependent variables for crash occurrence. It is assumed that a type of crash that repeats in the same location represents a predictable event [2, 10, 12]. Crash frequency can indicate the existence of an influence of the road system in the cause of that type of accident. If crashes are systematic in certain types of locations, it is unlikely to be caused only by driver error or vehicle dysfunctionality. They depend on technical and urban environmental identifiable factors which can be analyzed and a priori reduced. Therefore, datasets of yearly disaggregate homogeneous crash classes are compared and intersected (Figure 4).

In this research stage, the set of road features with repeated crash frequency is obtained. Further research is needed to define and calibrate the traffic risk estimation functions associated. In order to complete this step, besides selecting features characteristic of urban road network and urban zones (presented above), supplementary processing of traffic flow data is necessary.

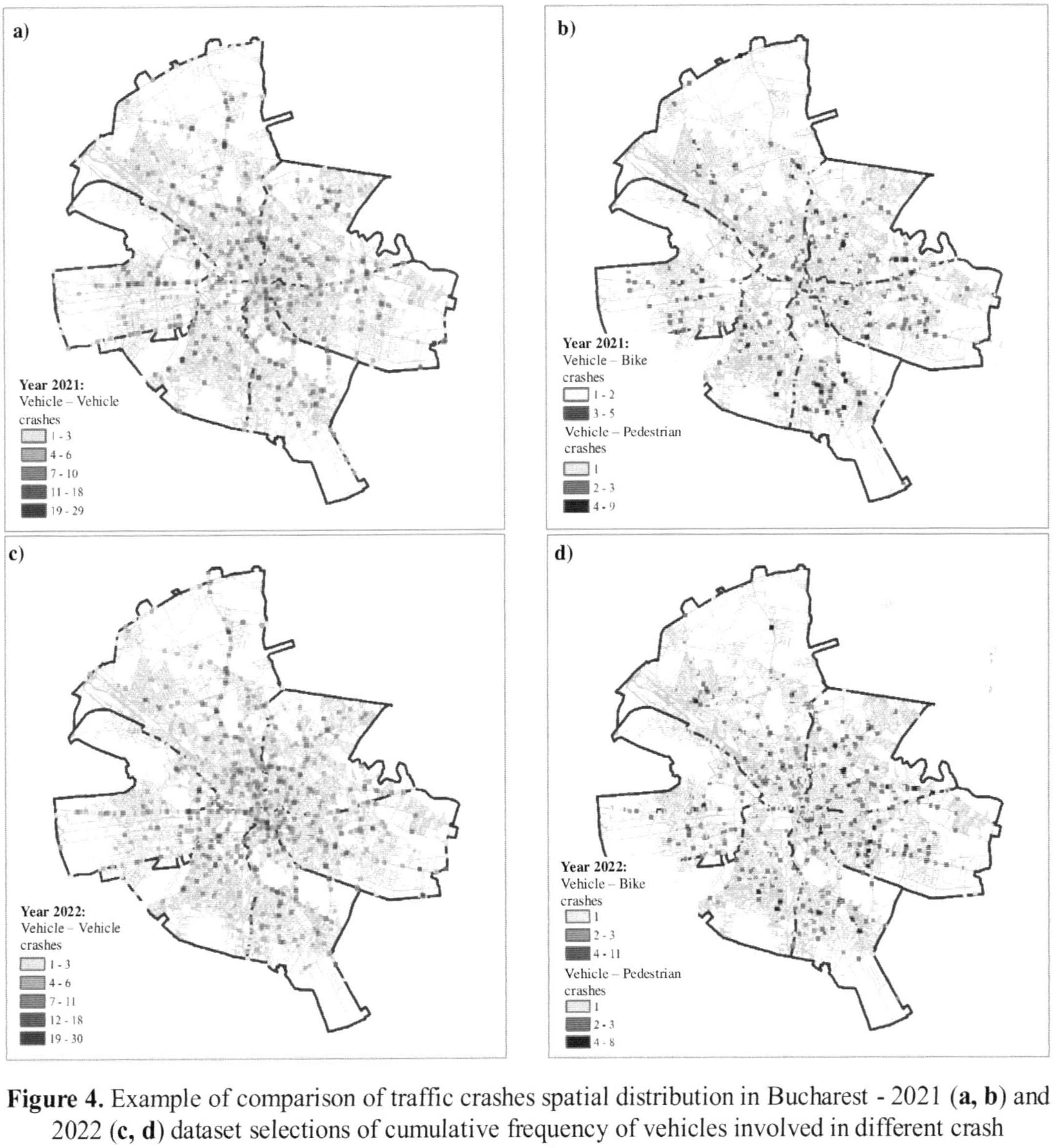


**Figure 4.** Example of comparison of traffic crashes spatial distribution in Bucharest - 2021 (**a, b**) and 2022 (**c, d**) dataset selections of cumulative frequency of vehicles involved in different crash categories.

Traffic characteristics are considered important factors in crash occurrence [2, 21]. Generally, average annual daily traffic values are used to characterize the flow in this type of estimation, even if, thus, significant effects of traffic non-uniformity are excluded [14, 21].

## 4. Conclusion

The risk associated with urban road traffic detaches from other socio-technical risks through the diversity of the influencing factors and the consequence extent. Therefore, it requires particular examinations. The dispersed responsibilities of various authorities and institutions impose a systematic approach to political actions to increase urban road traffic safety performance.

Under the new technical and ethical visions, the binomial "road traffic risk - road traffic safety" represents a topic of interdisciplinary research oriented towards the responsibility of those who design and operate the system. The transfer of the crash responsibility from the user to the one who designed and implemented the system is the fundamental characteristic of the new paradigms that will gradually restructure the current situation in road traffic.

In this line, the research aims to provide a model for assessing the risk associated with urban road traffic and for identifying measures to increase urban road traffic safety performance from the planning stage (before crash occurrence and blackspot registration drawing attention to the need for safety actions). The paper presents the first stage of the study. The developed model of the urban space allows processing data regarding the urban road network, traffic patterns, urban patterns, and crash data necessary to analyze the factors influencing traffic risk at the disaggregated level. Further research is needed to calibrate the traffic risk estimation functions and obtain appropriate results to carry out the necessary CEA in prioritizing the proposed measures to improve road safety.

## References

[1] Battistelli F and Galantino M G 2019 Dangers, risks and threats: An alternative conceptualization to the catch-all concept of risk *Currt. Soc.* **67** (1) 64–78

[2] Yannis G, Dragomanovits A, Laiou A, Richter T, Ruhl S, La Torre F, Domenichini L, Graham D, Karathodorou N and Li H. 2016 Use of accident prediction models in road safety management – An International inquiry *Transp. Res. Proc.* **14** 4257-4266

[3] Choi D and Ewing R 2021 Effect of street network design on traffic congestion and traffic safety *J. of Transp. Geogr.* **96** 103200

[4] Johansson R 2009 Vision Zero – Implementing a policy for traffic safety *Safe. Sc.* **47** 826–831

[5] Raicu S and Costescu D 2016 *"Risc rutier - Siguranță rutieră". În sprijinul unui plus de cunoaștere și responsabilizare (Traffic risk - Road safety. Support to advance knowledge and responsibility)* Bul. AGIR Supl. **2** 5-12 (In Romanian)

[6] Rupprecht Consult (editor) 2019 Guidelines for developing and implementing a sustainable urban mobility plan. 2nd Edition

[7] Raicu S, Costescu D and Burciu S 2014 Evaluation of road safety performances in urban areas *Urban Transport XX. The Built Environment* vol 38 ed Brebbia C A (WIT Press, Southampton, Boston) 447–458

[8] Pulugurtha S S, Duddu V D and Kotagiri Y 2013 Traffic analysis zone level crash estimation models based on land use characteristics *Acc. Anal. and Prev.* 678– 687

[9] Raicu S, Costescu D and Burciu S 2017 Analysis of intrinsic factors contributing to urban road crashes *Intern. J. of Safe. and Sec. Eng.* **7** (1) 1-9

[10] Colagrande S 2022 A methodology for the characterization of urban road safety through accident data analysis *Transp. Res. Proc.* **60** 504-511

[11] Elvik R 2019 How to trade safety against cost, time and other impacts of road safety measures, *Acc. Anal. and Prev.* **127** 150-155

[12] Ambros J, Jurewicz C, Turner S and Kiec M 2018 An international review of challenges and opportunities in development and use of crash prediction models *Europ. Transp. Res. Rev.* **10** 35

[13] Hughes B P, Newstead S, Anund A, Shu C C and Falkmer T 2015 A review of models relevant to road safety *Acc. Anal. and Prev.* **74** 250-270

[14] Lord D and Mannering F 2010 The statistical analysis of crash-frequency data: A review and assessment of methodological alternatives *Transp. Res. Part A* **44** 291–305

[15] Miaou S P and Lord D 2003 Modeling traffic crash-flow relationships for intersections: dispersion parameter, functional form, and Bayes versus Empirical Bayes methods *Transp. Res. Rec.* **1840** 31-40
[16] Hauer E 2001 Overdispersion in modelling accidents on road sections and in Empirical Bayes estimation *Acc. Anal. and Prev.* **33** (6) 799–808
[17] Marshall W E and Garrick N 2011 Does street network design affect traffic safety? *Acc. Anal. and Prev.* **43** (3) 769-781
[18] Gaudry M J I and Lassarre S (eds.) 2000 Structural road accident models: The International Drag Family (Pergamon Press)
[19] Kim K, Brunner IM and Yamashita E Y 2006 The influence of land use, population, employment and economic activity on accidents *Transp. Res. Rec.* **1953** 56–64
[20] Sustainable Urban Mobility Plan 2016 – 2030. Bucharest - Ilfov Region 2016 (Rom Engineering Ltd & AVENSA Consulting SRL Bucharest) (*In Romanian*)
[21] Quddus M A 2008 Modelling area-wide count outcomes with spatial correlation and heterogeneity: An analysis of London crash data *Acc. Anal. and Prev.* **40** 1486–1497

IOP Conf. Series: Materials Science and Engineering 1303 (2024) 012050 doi:10.1088/1757-899X/1303/1/012050

# Research on the influence of traffic conditions on noise level in the passenger compartment of motor vehicles

**G Neamţu[1], A M Ţîţu[2,3], A A Boroiu[4] and A Boroiu[4]**

[1]University Politehnica of Bucharest, Faculty of Industrial Engineering and Robotics, 313 Splaiul Independenţei, 6th District, Bucharest, Romania
[2]Lucian Blaga University of Sibiu, 10 Victoriei Street, Sibiu, România
[3]The Academy of Romanian Scientist, 54 Splaiul Independenţei, Bucharest, Romania
[4]University of Piteşti, 1 Târgu din Vale Street, Piteşti, Romania

E-mail: mihail.titu@yahoo.com

**Abstract.** The research investigates the noise levels measured in the passenger compartments of several automobiles. These vehicles are powered by various engines, travel on various highways, and function at differing speeds and acceleration. The results are visually processed using specialist software. The study investigates the effects of three major sources of automotive noise: engine operation, tire-road contact, and friction between the car body and the air. The immis noise in the passenger compartment is reported to be mostly impacted by the engine's operational characteristics (load and speed) at lower speeds. In contrast, as travel speed increases, the noise produced by tire-road contact becomes more substantial, and at greater speeds, the noise produced by the interaction between the car body and the air becomes more significant. As a result, the engine type, tire and road characteristics, vehicle speed, and passenger compartment acoustic insulation are all important elements in influencing noise emission in the passenger compartment.

## 1. Introduction

The motivation for investigating sounds generated in the passenger compartments of motor vehicles arises from the World Health Organization's suggestion to decrease motor vehicle speeds in urban areas from 50 km/h to 30 km/h. This advice is motivated by the increasing rates of traffic fatalities and the significant environmental degradation caused by pollutants. [1]. People frequently experience the impacts of sounds produced by the road transportation system, especially in densely populated urban areas characterized by high-intensity road traffic. Noise pollution originating from vehicle traffic stands as a persistent and significant environmental threat, impacting millions of individuals on a daily basis.

This sort of pollution is caused by frictional forces between moving components, structural elements, or organs in rotational or translational motion during motor vehicle operation. Furthermore, noise pollution occurs during motor vehicle dynamic states when air friction interacts with exterior portions of their bodies or when tires make contact with the ground (highway) [3, pp. 103-118]. At the European level, roughly 20% of the population is impacted by car-related sounds [4], revealing a gap in attaining regional noise pollution control targets. The significant noise released by road traffic has a wide range of long-term consequences on human health, causing pain, sleep disruptions, negative effects on the cardiovascular and metabolic systems, and cognitive impairments in children [5]. According to the World Health Organization (WHO), approximately 40% of the EU population is subjected to road traffic noise above 55 decibels (dB), motivating continuing study to fully understand the many health impacts of noise [6]. The EU Noise Directive of 2003, which has been in operation in all Member States since

2006, specifies a daily noise exposure limit value of 87 dB (8 hours) [4, p. 117]. The widespread continuation of high noise levels from road traffic, on the other hand, implies that the aims specified in European laws to reduce noise pollution have not been entirely fulfilled.

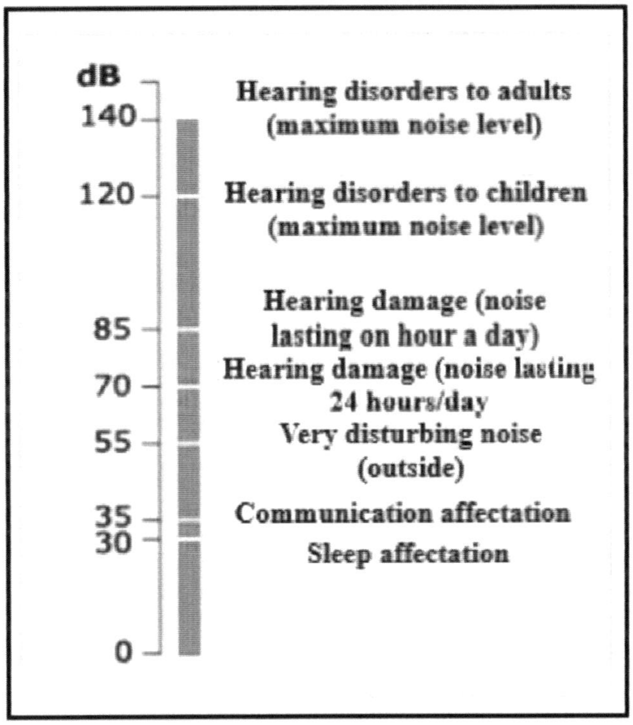


**Figure 1.** Noise's detrimental impact on human health [16].

Acoustic comfort [7] is defined by the different sounds emitted by motor vehicles, such as engine noises, the movement of rotational or translational organs, frictional forces, tire contact with the ground, and air friction as the vehicle is moving [8, p. 106]. The perception of motor vehicle noises by the human ear is dependent on their sound intensity, which can exceed 80 dB and cause health dangers. Figure 1 illustrates the adverse effects of noise on human health. The resonance of mechanical waves inside the structure of the motor vehicle [9], [10], as sensed by passengers at the body level, is referred to as vibrational comfort. The rotational or translational aspects of the car's organs, as well as abnormalities in the surfaces it travels, create these waves. The oscillations of the vehicle must not endanger its structural integrity, the safety of products, or the well-being of passengers since they might cause tiredness and pain [11]. Road traffic noise is measured in decibels (dB) [12, p. 1.025] [13], which indicate mechanical vibrations discharged into the environment and conveyed to live beings' auditory organs. For example, the human ear records sound waves with frequencies ranging from 16 to 20,000 vibrations per second and perceives their strength as 0-120 dB. Sound changes in the 30-60 dB range are normal in human conversation. The human ear is unaffected by levels between 20 and 30 decibels, with the top limit accepted by humans being roughly 80 decibels. A sound level of about 130 dB causes acoustic discomfort, whereas levels beyond 150 dB are unbearable to the human ear.

## 2. Vehicle noise statistics for reference

Directive 70/157/EEC, published February 6, 1970, encapsulates the basic legislative framework addressing pollution rules for motor vehicles. This regulation, enacted at its inception, introduced pioneering restrictions on noise emissions from motor vehicles, outlining technical specifications that control the quantifiable levels of sound released into the atmosphere during vehicular operation. Over time, this directive has undergone iterative modifications through Directives 73/350/EEC, 77/212/EEC, 81/334/EEC, 84/372/EEC, 84/424/EEC, 87/354/EEC, 89/491/EEC, 92/97/EEC, 96/20/EC, 99/101/EC, 2006/96/EC, 2007/34/EC, and in compliance with UN-ECE Equivalent Regulation No. On December 9, 2011, the European Commission proposed to the European Parliament and Council the adoption of the Regulation on the Sound Level of Motor Vehicles [14]. Importantly, this proposal annulled Directive 70/157/EEC, which had been effective since February 6, 1970. Table 1 provides a comprehensive description of the legally mandated de-noise level limit values, precisely as outlined in Appendix III to the Regulation on the Sound Level of Motor Vehicles.

**Table 1.** The legal noise level limit values given in Annex III to the Regulation on the Sound Level of Motor Vehicles [14].

| Vehicle classification | The vehicle category's description | Limit values expressed in dB(A) [decibels (A)] | | |
|---|---|---|---|---|
| | | Stage 1 is applicable to new automobiles beginning July 1, 2016. | Stage 2 is applicable to new automobiles beginning July 1, 2020, and to initial registrations beginning July 1, 2022. | Stage 3 is applicable to new automobiles beginning July 1, 2024, and to initial registrations beginning July 1, 2026. |
| Vehicles for transporting M-class passengers | | | | |
| $M_1$ | Power-to-weight ratio: 120 kW/1.000 kg | 72(*) | 70(*) | 68(*) |

**Note** (*) Specific conditions apply to M1 category cars developed from the N1 category. M1 vehicles having a point R more than 850 mm above the ground and a permitted loaded mass greater than 2,500 kg must adhere to the limit values established for the N1 category (where the mass is larger than 2,500 kg but not greater than 3,500 kg). It is important to remember that the raised limit values applicable to off-road vehicles are only applicable to M1 category vehicles when the maximum permissible mass exceeds 2 tons.

## 3.Results and discussions

### *3.1 Experiment Concept*

Experimentation [15] was carried out on automobiles fitted with compression ignition engines that were within the European pollution limit range of [Euro 3 - Euro 5]. The study included three unique routes: a 30-kilometer-per-hour urban route, a 50-kilometer-per-hour urban route, and a 130-kilometer-per-hour extra-urban highway route. The following are the processes for determining the noise level measured in the passenger compartment of a motor vehicle: Reset the dashboard mileage to zero; install a sound-level meter (Sound-level meter SL-400) on the car; start the engine and begin moving the car from a stop, incorporating acceleration and gear changes, until it reaches the set speed (30 km/h, 50 km/h, 130 km/h); maintain the set speed as consistently as feasible along the authorized route or prescribed travel itinerary, taking into account traffic values and road infrastructure conditions; during the experimental investigation, follow defensive driving guidelines to the greatest degree feasible; conduct tests in the urban area on the same route at both speeds (30 km/h and 50 km/h) and on the extra-urban route (highway) at 130 km/h, covering the same 10 km distance.

**Table 2.** Centralizer employed to concisely present the critical data related to noise pollution identified in the passenger compartments of vehicles equipped with compression ignition engines. This includes summarizing the lowest, average, and maximum values of noise levels.

| The vehicle's model and year | dB(A) values for sound measured within the vehicle's passenger compartment | | | | | | | | |
|---|---|---|---|---|---|---|---|---|---|
| | Urban V = 30 km/h | | | Urban V = 50 km/h | | | Extra-urban V = 130 km/h | | |
| | min | med | max | min | med | max | min | med | max |
| RENAULT CAPTUR 1,5 dCI, Euro 5 | 51,0 | 59,2 | 71,3 | 53,0 | 62,9 | 75,3 | 68,0 | 73,4 | 80,2 |
| VOLKSWAGEN JETTA TDI, Euro 4 | 56,2 | 63,9 | 72,0 | 56,1 | 65,6 | 75,7 | 64,3 | 72,4 | 82,0 |
| RENAULT SYMBOL 1,5 dCI, Euro 3 | 50,9 | 61,8 | 75,4 | 55,8 | 65,0 | 74,1 | 71,0 | 76,2 | 80,4 |

**The Legend:**

| The legal standard is found in Annex II of the Regulation on the Sound Level of Motor Vehicles, which was adopted by the European Commission on April 16, 2014. | **Vehicle emissions standards in Europe** | Maximum allowable sound limit value dB(A) | Values that are normal and do not exceed the permissible limit |
|---|---|---|---|
| | Vehicles for passenger transport in the M1 category with a power-to-weight ratio of 120 kW/1.000 kg | **72** | Values that exceed the legal limit |

The measurements were taken during travel on urban routes at speeds of 30 km/h and 50 km/h, as well as on extra-urban routes (highway) at 130 km/h. The centralizer offers a comprehensive overview of the noise pollution levels in these diverse driving scenarios.

*3.2 Noises made by the Renault Captur 1.5 dCI, euro 5 car's passenger compartment*

The results of the analysis of sound evolution measured in the passenger compartment of a Renault Captur 1.5 dCI Euro 5 vehicle are presented below (Figure 2):

- When going at 30 km/h on the urban route:
  - 51.00 dBA is the minimum sound value.
  - 59.20 dBA is the average sound pressure level.
  - Maximum sound pressure level: 71.30 dBA
- When going at 50 km/h on the urban route:
  - Minimum sound pressure level: 53.00 dBA
  - Average sound pressure level: 62.90 dBA
  - Maximum sound pressure level: 75.30 dBA
- When finishing the A1 - Sibiu Highway route:
  - Minimum sound pressure level: 53.00 dBA
  - Average sound pressure level: 62.90 dBA
  - Maximum sound pressure level: 75.30 dBA

IOP Conf. Series: Materials Science and Engineering 1303 (2024) 012050 doi:10.1088/1757-899X/1303/1/012050

Figure 2 illustrates the progression of sound levels (dBA) measured in the passenger compartment of a Renault Captur 1.5 dCI, Euro 5 automobile, during both urban and highway travel.

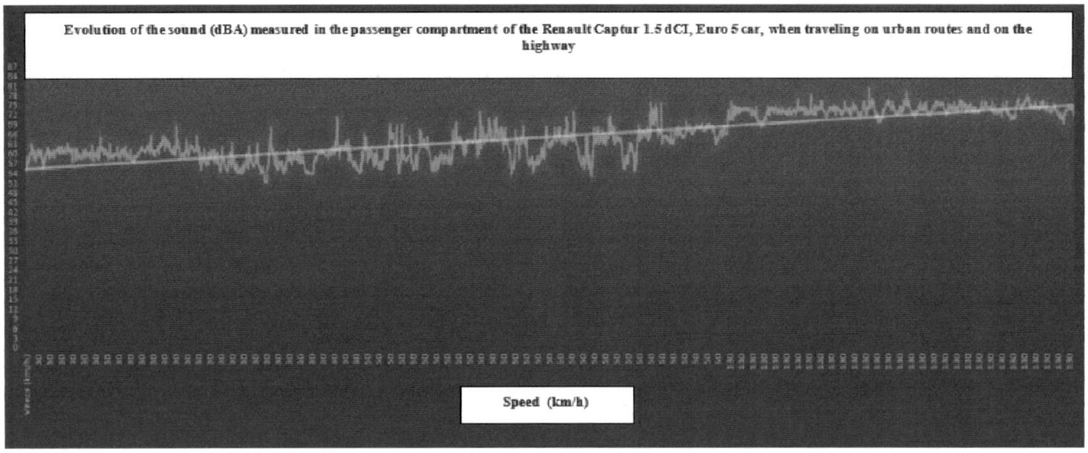


**Figure 2.** The development of sound (dBA) in the passenger compartment of a Renault Captur 1.5 dCI Euro 5 automobile when traveling on urban (30 km/h, 50 km/h) and highway (130 km/h) routes.

*3.3 Noises made by the Volkswagen Jetta 2.0 TDI, euro 4 car's passenger compartment*

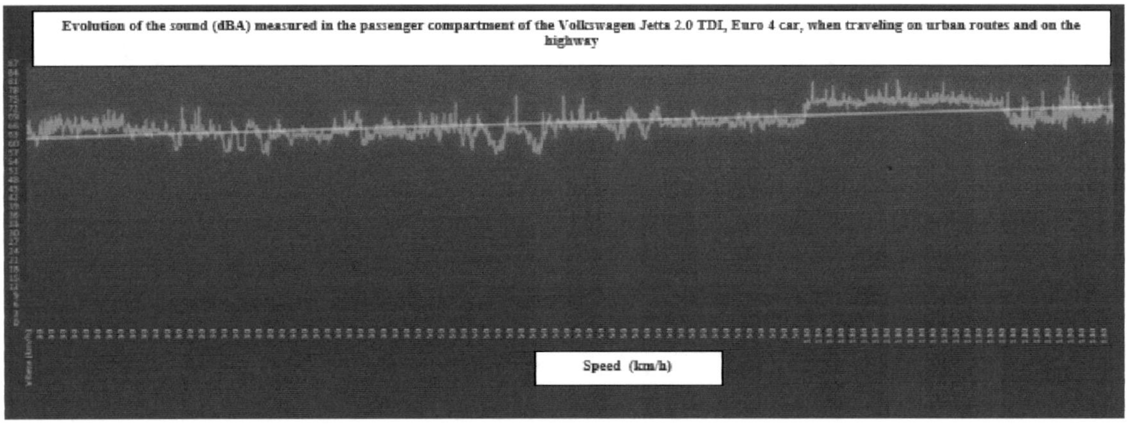


**Figure 3.** The evolution of sound (dBA) measured in the passenger compartment of the Volkswagen Jetta 2.0 TDI is depicted during urban driving at speeds of 30 km/h and 50 km/h, as well as highway driving at 130 km/h.

The results of the analysis of sound evolution measured in the passenger compartment of a Volkswagen Jetta, 2,000 cm3, TDI, Euro 4 automobile are outlined below (Figure 3):

- When going at 30 km/h on the urban route:
  - Minimum sound pressure level: 56.20 dBA
  - 63.9 dBA is the average sound pressure level.
  - Maximum sound pressure level: 72.00 dBA
- When going at 50 km/h on the urban route:
  - Minimum sound pressure level: 56.10 dBA
  - Average sound pressure level: 65.60 dBA
  - Maximum sound pressure level: 75.70 dBA

- When finishing the A1 - Sibiu Highway route:
  - Minimum sound pressure level: 64.30 dBA
  - 72.40 dBA is the average sound pressure level.
  - Maximum sound pressure level: 82.00 dBA

Figure 3 illustrates the progression of sound levels (dBA) recorded in the passenger compartment of a Volkswagen Jetta 2.0 TDI, Euro 4 automobile, during both urban and highway travel.

*3.4 Noises made by the Renault Symbol 1.5 dCI, euro 3 car's passenger compartment*

The results of the analysis of the measured sound development in the passenger compartment of the Renault Symbol 1.5 dCI, Euro 3 automobile, are presented below (Figure 4):

- When going at 30 km/h on the urban route:
  - Minimum sound pressure level: 50.90 dBA
  - 61.80 dBA is the average sound value.
  - Maximum sound pressure level: 75.40 dBA
- When going at 50 km/h on the urban route:
  - Minimum sound pressure level: 55.80 dBA
  - 65.00 dBA is the average sound pressure level.
  - Maximum sound pressure level: 74.10 dBA
- When finishing the A1 - Sibiu Highway route:
  - 71.00 dBA is the minimum sound value.
  - Average sound pressure level: 76.20 dBA
  - Maximum sound pressure level: 80.40 dBA

Figure 4 illustrates the progression of sound levels (dBA) measured in the passenger compartment of a Renault Symbol 1.5 dCI, Euro 3 automobile, during both urban and highway travel.

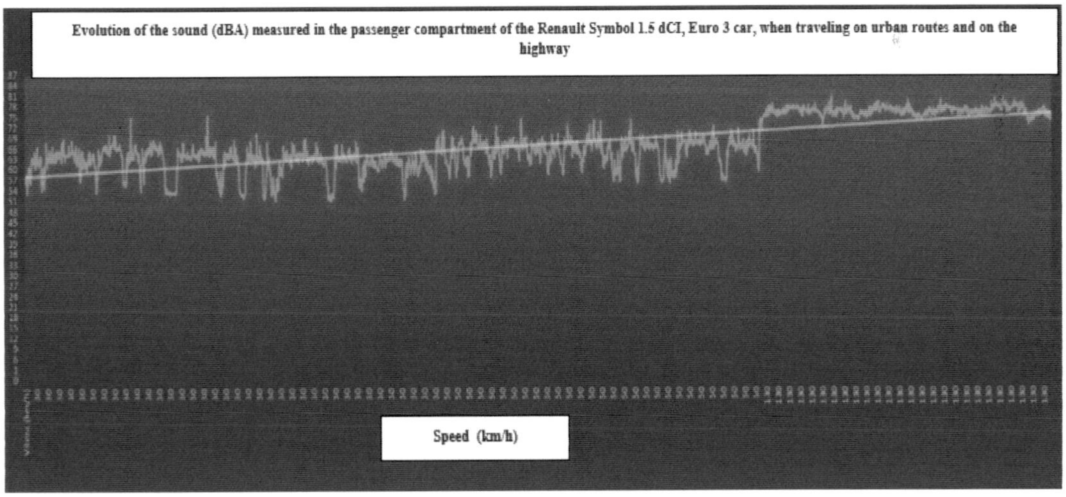


**Figure 4.** The evolution of sound (dBA) measured in the passenger compartment of the Renault Symbol 1.5 dCI Euro 3 vehicle is presented during urban driving at speeds of 30 km/h and 50 km/h, as well as highway driving at 130 km/h.

## 4. Conclusions

A. The observed trend in the evolution of sound in the passenger compartment of the Renault Captur 1.5 dCI, Euro 5 vehicle, reveals a consistent increase on two urban routes and on the highway. Noise levels escalate with the increase in speed. At 30 km/h, sound readings on urban roads stay within the regulation limit of 72 dBA. Minimum and average sound values are under the legal limit at 50 km/h, but maximum values surpass it. Minimum sound values on the A1 - Sibiu highway match the legal requirement, while maximum and average sound values exceed it. Noise levels in cities fall when traffic speeds decrease from 30 to 50 km/h. The linear sound growth along all three paths shows a rapid increase in levels.

B. The trend in the evolution of sound in the passenger compartment of the Volkswagen Jetta, 2,000 cm3, TDI, Euro 4 car, across two urban routes and on the highway, indicates a noticeable upward trajectory. Noise levels rise in direct proportion to speed. At 30 km/h, sound levels on urban highways consistently remain within the permissible limit of 72 dBA. The minimum and average sound levels are within the legal limit at 50 km/h, while the highest values surpass it. Minimum sound values on the A1 - Sibiu highway meet the legal requirement, but maximum and average sound values exceed it. Noise levels in cities drop when traffic speeds decrease from 30 to 50 km/h. On all three pathways, the linear sound progression shows a progressive increase in values.

C. The apparent trend in the evolution of sound in the passenger compartment of the Renault Symbol 1.5 dCI, Euro 3 car, indicates an ascending trajectory along two urban routes and on the highway. Noise levels rise as speed increases. The minimum and average sound levels on urban highways at 30 km/h are under the legal limit of 72 dBA, but the highest levels surpass it. The minimum and average sound levels are within the legal limit at 50 km/h, while the highest values exceed it. Minimum sound values on the A1 - Sibiu highway match the legal standard, while maximum and average sound values exceed it. Noise levels in cities drop when traffic speeds decrease from 30 to 50 km/h.
On all three paths, the linear sound growth demonstrates a quick increase in values.

D. The level of noise pollution detected in the passenger compartment is determined by several factors, such as engine speed, tire type and profile, road surface condition, vehicle maintenance status, noise generated by transmission components and organs, vehicle age, physical wear, and fuel type.

E. Decreasing the speed of traffic in urban areas from 50 km/h to 30 km/h leads to a noteworthy reduction in the measured levels of noise pollution in motor vehicle passenger compartments.

**Author's ORCID ID**
A M Titu: https://orcid.org/0000-0002-0054-6535
A Boroiu: https://orcid.org/0000-0002-6908-5546
A A Boroiu: https://orcid.org/0000-0001-7313-3398

## 5. References

[1] Kountoura E 2021 "Resolution on road safety 2021-2030," European Parliament, Bruxelles

[2] ANPM, „Environmental pollution," 13 January 2014. [Interactiv]. Available: https://web.archive.org/web/20140113020730/http://apmbt.anpm.ro/poluare_sonora. [Accesat 22 November 2022].

[3] Untaru M, Pereș G, Stoicescu A, Poțincu G and Tabacu G 1981 Dinamics of wheeled vehicles, D. a. P. P. House, Ed., Bucharest, pp. 103-118.

[4] Șchiopu N and Bardac D I 2012 Noise and its effects - References," *Acta Medica Transilvania,* vol. II, nr. 1, pp. 117-118

[5] E. A. European, Noise pollution is a major problem for human health and the environment, 11 May 2021. [Online]. Available: https://www.eea.europa.eu/ro/articles/poluarea-fonica-este-o-problema. [Accessed 22 November 2022].

[6] Bogahty Z 2000 Handbook of Work and Organizational Psychology, P. P. House, Ed., Bucharest,.

[7] Potel C and Bruneau M 2006 General acoustics, Elipse, Ed.

[8] Berkeley 1981 Berkeley Physics Course, Vol. %1 din %2I, Mechanics, D. a. P. P. House, Ed., Bucharest, Romania

[9] Wang X 2010 Vehicle noise and vibration refinement, W. P. Ltd., Ed., Canbridge, p. 416.

[10] Temple R 1969 The ABCs of Chassis Frames and Suspensions," *Popular Mechanics,* vol. 132, no. 3, pp. 128-133.

[11] I. Tabacu și M. Marinescu, „Mijloace de combatere a zgomotelor la autoturisme," Seria Autovehicule rutiere, Pitești, 1996.

[12] Mark J E 2007 Physical Properties of Polymers, Springer, Ed., Berlin

[13] Yost W Fundamentals of Hearing: An Introduction, Second edition ed., New York: New York, Holt, Rinehart and Winston., 1985, p. 269.

[14] E. Comisia, „Regulation (EU) No. 540/2014 of the European Parliament and of the councilL of April 16, 2014, on the noise level of motor vehicles and replacement silencers, amending Directive 2007/46 / EC and repealing Directive 70/157 / EEC," *Official Journal of the EU,* nr. L 158/131, p. 131, 27 May 2014.

[15] Țîțu M and Oprean C 2006 Experimental research and data processing, T. p. H. o. ". B. U. o. Sibiu, Ed., Sibiu, p. 415.

[16] Boroiu A A 2017 Studies and research regarding the reduction of noise pollution produced by motor vehicles through the organization of road traffic- Doctoral thesis, Pitesti, Romania: University of Pitesti

# Mathematical Modeling of the Normal Fuel Consumption of Certain Vehicles

**A M Țîțu[1,2], G Neamțu[3], I Țincu[4] and A Boroiu[5]**

[1]Lucian Blaga University of Sibiu, 10 Victoriei Street, Sibiu, România
[2]The Academy of Romanian Scientist, 54 Splaiul Independenței, Bucharest, Romania
[3]University Politehnica of Bucharest, Faculty of Industrial Engineering and Robotics, 313 Splaiul Independenței, 6th District, Bucharest, Romania, geluneamtu@yahoo.com
[4]Lucian Blaga University of Sibiu, 10 Victoriei Street, Sibiu, România
[5]University of Pitești, 1 Târgu din Vale Street, Pitești, România

E-mail: mihail.titu@yahoo.com

**Abstract.** The scientific paper introduces original research and unique perspectives on the technical and economic aspects of fuel consumption in motor vehicles. Numerous experts argue that conducting practical experiments, specifically through active determinations, can yield noteworthy outcomes in this domain. To substantiate this viewpoint, the paper includes practical examples of experiments and considerations related to the mathematical model used for validating objective functions. These functions encompass the average fuel consumption in liters per 100 kilometers and the quantity of $CO_2$ emitted per kilometer by the thermal engine of motor vehicles. The research also articulates the authors' stance, citing the World Health Organization's recommendation to decrease urban traffic speed from 50 km/h to 30 km/h. This reduction aims to mitigate traffic accidents and curb chemical noise and vibration pollution. The study amalgamates results from tests conducted on both urban and extra-urban routes, incorporating eco-friendly driving practices for comparative purposes. The primary focus of the mathematical model is to validate fuel consumption under real driving conditions. The paper concludes with insights and future research directions in modeling.

## 1. Introduction

The motivation for investigating motor vehicle fuel consumption stems from the World Health Organization's recommendations to decrease urban speed limits from 50 km/h to 30 km/h. This suggestion is rooted in the high number of casualties resulting from road accidents and the escalating chemical pollution in the environment. We advocate for specific life-saving measures, including lower speed limits, particularly establishing an implicit 30 km/h limit for cities, expeditious transition to alternative transportation modes, and the establishment of a new European Road Transport Agency [1].

Analyzing the gathered data reveals a persistent global increase in environmental pollution, particularly carbon dioxide and other harmful gases. Concurrently, road accidents continue to rise. In Romania, this trend is attributed to an outdated national vehicle fleet, inadequate road infrastructure for high traffic volumes (limited high-speed roads and poorly connected streets), absence of bypasses, and major traffic arteries passing through city centers (dating back to the '50s-'60s infrastructure). This situation contributes to a significant number of casualties, with children and young adults emerging as the primary victims [2].

Fuel consumption represents the maximum allowable quantity of gasoline, diesel, liquefied petroleum gas (LPG), or biofuel a heat engine vehicle can utilize over a specified distance (typically 1000 meters). This evaluation considers prevailing operating conditions, road category, and environmental factors. Stringent emission requirements for vehicle type approval align with evolving European standards. The Worldwide Harmonized Light Vehicle Test Procedure (WLTP), introduced by the EU to standardize global fuel consumption and CO2 emissions, replaced the previous New European Driving Cycle (NEDC) method from September 1, 2017 [5]. The WLTP incorporates dynamic parameters to provide more realistic information on fuel consumption and pollutant emissions [6].

Vehicle fuel consumption is influenced by technical characteristics, equipment, and environmental conditions. Factors such as operating conditions, passenger and cargo weight, driving style, ambient temperature, tire type, vehicle aerodynamics (Cx), road category, traffic conditions, and driver comfort all contribute to fuel consumption variability [7, 103-118 p.]. Fuel consumption measurement serves as a performance examination, offering insights into design decisions that contribute to economical operation and reduced operating costs associated with fuel provision [8, p. 96]. Considering these aspects, conducting experimental studies is recommended to practically demonstrate whether reducing urban traffic speed leads to a reduction in fuel consumption.

## 2. Organizational measures

The steps of the process (procedure) for determining the fuel consumption of motor vehicles are as follows:

- Check the tire pressure [9], refuel the tank of the vehicle to full and reset the mileage in the dash to zero;
- The engine starts and the car starts, accelerates, and shifts until the set speed is reached (30 km/h, 50 km/h, 130 km/h). This speed is maintained as uniform as possible along the entire route (prescribed route of travel), taking into account traffic conditions or requirements dictated by the road infrastructure.;
- During the experimental research, the rules of defensive driving of the vehicle shall be observed as much as possible;
- In urban areas, measurements are carried out both for speeds on one route (30 km/h and 50 km/h), and on a route outside the city (motorway) at a speed of 130 km/h, covering the same distance (10 km);
- After completing the routes established for carrying out the experimental research, the vehicle and the engine are stopped, the tank is filled in full and calculations on the fuel consumption will be made according to the relation (1).

The calculation relationship through wich the fuel consumptions is determinated is the following:

$$FC = \frac{AFF}{d} \cdot 100 \qquad \text{[liter/100 km]} \tag{1}$$

where,

*FC* represents fuel consumption;
*AFF* - the amount of fuel feed, in litres, completed after completion route for filling the tank, full;
*d* - the distance in kilometers of the route traveled (it is read from the dashboard of the vehicle).

## 3. Results and discussions

### *3.1 Proposed experiment*

Experimental studies [10][11] were conducted on a vehicle equipped with a compression ignition (MAC) engine compliant with the European pollution standard Euro 3. The tests followed predefined sections and were conducted on three routes: an urban highway at a speed of 30 km/h, urban routes at a speed of 50 km/h, and highways outside the city at a speed of 130 km/h. The outcomes of these tests are summarized in Table 1.

**Table 1.** Fuel consumption data, derived from experimental research, for the Renault 1.5 dCI, Euro 3 car, under urban conditions at speeds of 30 km/h and 50 km/h, as well as extra-urban conditions at 130 km/h.

| The type and make of the vehicle | Urban consumption [l/100 km] v=30 km/h | $CO_2$ emissions [Kg la 8,1 l/100 km] | Urban consumption [l/100 km] v=50 km/h | $CO_2$ emissions [Kg la 14,2 l/100 km] | Extra-urban consumption [l/100 km] v=130 km/h | $CO_2$ emissions [Kg la 3,5 l/100 km] |
|---|---|---|---|---|---|---|
| **0** | **1** | **2** | **3** | **4** | **5** | **6** |
| RENAULT SYMBOL 1,5 dCI vehicle | 8,1 | 21,4 | 14,2 | 37,5 | 3,5 | 9,3 |

Table 1 and Figure 1 display the fuel consumption in urban settings (30 km/h, 50 km/h) and extracity conditions (130 km/h) for the Renault Symbol 1.5 dCI, as determined through experimental studies. The table also highlights the $CO_2$ emmissions related to consumption.

The analysis of the evolution of fuel consumption (table 1 and figure 1) shows the following data:

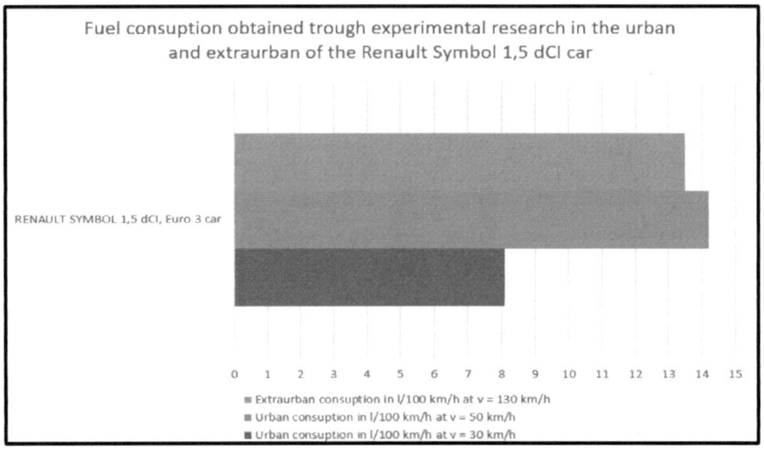


**Figure 1.** The fuel consumption obtained through experimental research in both urban and extra-urban settings for the Renault.

- When driving a city route at a speed of 30 km/h, fuel consumption is 8.1 l/100 km.
- When driving a city route at a speed of 50 km/h, fuel consumption is 14.2 l/100 km.
- On the A1 – Sibiu highway, the consumption during the route is 3.5 l/100 km.

Chemically pure hexadecane [12], with a molecular formula n = 16, shares similar properties to diesel fuel, despite diesel fuel containing additional molecules and additives. A straightforward calculation enables the determination of hexadecane's molar mass: 12 x 6 + 1 x (2 x 16 + 2) = 226 g/mol. The CO2 produced by burning one mole of hexadecane is 44 x 16 = 704 g. Consequently, the diesel/CO2 emission ratio is 704 g/226 g = 3.11 g. The density of diesel fuel is 0.850 kg/l. Therefore, the emission of carbon dioxide per gram of burned diesel fuel is 3.11 g. A simple computation reveals that a car utilizing diesel in a heat engine emits 0.85 x 3.11 = 2.64 kg CO2/liter. Utilizing this information, we calculated the corresponding CO2 amounts associated with fuel consumption in columns 2, 4, and 6 of Table 1.

Analysis of the dynamics of fuel consumption (Table 1 and Fig. 1) shows the following data::

- When driving a city route at a speed of 30 km/h, fuel consumption is 8.1 l/100 km.
- When driving a city route at a speed of 50 km/h, fuel consumption is 14.2 l/100 km.
- On the A1 – Sibiu highway, the consumption during the route is 3.5 l/100 km.

The automaker initially endorsed the Worldwide Harmonized Light Vehicle Test Procedure (WLTP) method [5], [6] with the specified fuel consumption values: urban consumption = 5.3 l/100 km; extracity consumption = 3.5 l/100 km (refer to Table 2 and Figure 2). Nevertheless, experimental studies conducted under urban conditions (50 km/h) revealed that the actual fuel consumption was higher than the values declared by the manufacturer. Table 2 and Figure 2 present the manufacturer-approved fuel consumption figures for Renault Symbol 1.5 dCI, Euro-3 vehicles, both in urban and extracity settings.

**Table 2.** Urban and extra-urban consumption approved by the manufacturer, for the Renault Symbol 1.5 dCI Euro 3, car.

| The vehicle's type and brand | Urban consumption [l/100 km] | Extra-urban consumption [l/100 km] |
|---|---|---|
| RENAULT SYMBOL 1,5 dCI Euro 3, car | 5,3 | 3,5 |

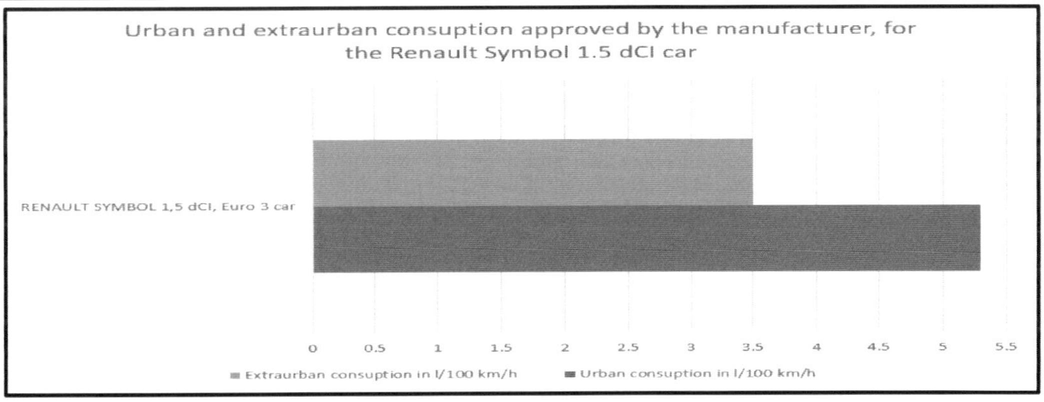


**Figure 2.** Urban and extra-urban consuption approved by the manufacturer, for the Renault Svmbol 1.5 dCI car.

*3.2 Mathematical model proposed in the framework of experimental research of active type*

The proposed mathematical model faithfully reproduces the average fuel consumption obtained in the real case through experimental research. The determination of the parameters a,b,c, so that h (v; a, b, c) solves the problem of minimum, is done by using ***the method of least squares*** [13], which consists in determining the parameters a, b, c so that the sum $S(a, b, c) = \sum_{k=1}^{3}[h(v_k; a, b, c) - f(v_k)]^2$ is minimal.

The least squares define the solution that is obtained for minimizing the sum of squares of deviations from the values of the equations. The most important application is to determine the coefficients of a mathematical function that best approximates a set of data. [14].

The result is a *Cramer* system or a *uniquely determined system*, the values of the standard fuel consumption, necessary for the validation of the mathematical model, are obtained:

$$a^* = \frac{\Delta 1}{\Delta S} = \frac{3e\gamma + f\theta\gamma + \beta g\theta - g\gamma^2 - e\theta^2 - 3\beta f}{3\alpha\gamma + 2\beta\gamma\theta - \gamma^3 - \alpha\theta^2 - 3\beta^2} \tag{2}$$

$$b^* = \frac{\Delta 2}{\Delta S} = \frac{3\alpha f + \beta\gamma\gamma + \gamma e\theta - f\gamma^2 - \alpha g\theta - 3\beta e}{3\alpha\gamma + 2\beta\gamma\theta - \gamma^3 - \alpha\theta^2 - 3\beta^2} \tag{3}$$

$$c^* = \frac{\Delta 3}{\Delta S} = \frac{\alpha\gamma g + \beta\theta e + \beta\gamma f - e\gamma^2 - \alpha\theta f - \beta^2 g}{3\alpha\gamma + 2\beta\gamma\theta - \gamma^3 - \alpha\theta^2 - 3\beta^2} \tag{4}$$

The Kramer system or custom-defined system provides a unique solution. A mathematical model for calculating values of θ, g, γ, β, α, e, and f is illustrated in Table No. 3.

**Table 3.** Mathematical calculation model for obtaining the values of θ, g, γ, β, α, e şi f.

| | v | y | $v^2$ | $v^3$ | $v^4$ | $v^2y$ | vy |
|---|---|---|---|---|---|---|---|
| | 30 | 8,1 | $30^2$ | $30^3$ | $30^4$ | $30^2 \cdot 8{,}1$ | $30 \cdot 8{,}1$ |
| | 50 | 14,2 | $50^2$ | $50^3$ | $50^4$ | $50^2 \cdot 14{,}2$ | $50 \cdot 14{,}2$ |
| | 130 | 3,5 | $130^2$ | $130^3$ | $130^4$ | $130^2 \cdot 3{,}5$ | $130 \cdot 3{,}5$ |
| $\sum_{k=1}^{3}$ | **210** | **25,8** | **20.300** | **2.349.000** | **292.670.000** | **101.940** | **1.408** |
| | **θ** | **g** | **γ** | **β** | **α** | **e** | **f** |

where, $v_1$ = 30 km/h; $v_2$ = 50 km/h; $v_3$ = 130 km/h; y – fuel consumption (8,1 liters la 30 km/h; 14,2 liters la 50 km/h şi 3,5 liters la 130 km/h).

$$\text{Thus, } y \simeq h^* = h\,(v;\, a^*, b^*, c^*) \tag{5}$$

*Note:* For v = $v_0$ (for example for the other speeds than those used in the research such as: 10 km/h, 20 km/h, 30,km/h, 40 km/h, 50 km/h, 60 km/h, 70 km/h, 80 km/h, etc, possible to obtain:

$$y \simeq h^* = h\,(v_0;\, a^*, b^*, c^*) = a^* \cdot v_0^2 + b^* \cdot v_0 + c^* \tag{6}$$

After obtaining the values of θ, g, γ, β, α, e şi f, according to the calculations in Table no. 3, the will be replaced in the mathematical relations (2), (3) and (4), the results are obtained:

$$a^* = \frac{\Delta 1}{\Delta S} = -0.0044; \qquad b^* = \frac{\Delta 2}{\Delta S} = 0.656; \qquad c^* = \frac{\Delta 3}{\Delta S} = -7.631. \tag{7}$$

The obtained mathematical model is:

(Litres/100) km FC = $a^* \cdot v^2 + b^* \cdot v + c^*$; (Litres/100 km) FC = $-0.0044 \cdot v^2 + 0.656 \cdot v + (-7.631)$ (8)

The graphical representation of the mathematical model is depicted in Figure 3.

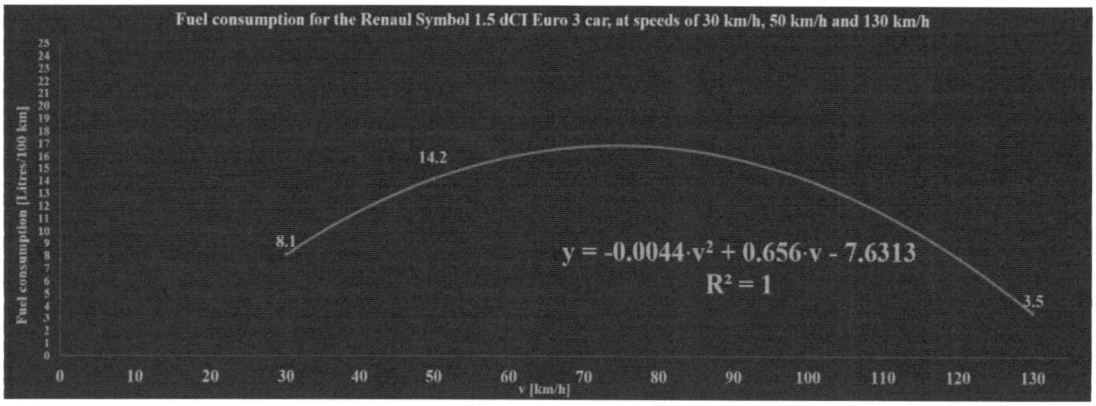


**Figure 3.** The graphical representation of the mathematical model for the fuel consumption of the Renault Symbol 1.5 dCI Euro 3 car at speeds of 30 km/h, 50 km/h, and 130 km/h.

Applying in relation (6), the results obtained for speeds in the range [0-130] km/h, fuel consumption values are returned, coresponding to each sped in the range **y = FC**$_{\textbf{Renault Symbol 1,5 dCI vehicle}}$ **= [0 – 14.2] liters/100 km**, thus:

a) For $v_0$ = 10 km/h, is obtained: $y \simeq h^* = h\,(-0.0044; 0.656; -7.631) = -0.0044 \times 10^2 + 0.656 \times 10 + (-7.631)$ = **-1.51** l/100 km;

b) For $v_0$ = 20 km/h, is obtained: y ≃ h* = h (-0.0044; 0.656; -7.631) = -0.0044x $20^2$ + 0.656 x 20 + (-7.631) = **3.73** l/100 km;
c) For $v_0$ = 30 km/h, is obtained: y ≃ h* = h (-0.0044; 0.656; -7.631) = -0.0044x $30^2$ + 0.656 x 30 + (-7.631) = **8.10** l/100 km;
d) For $v_0$ = 40 km/h, is obtained: y ≃ h* = h (-0.0044; 0.656; -7.631) = -0.0044x $40^2$ + 0.656 x 40 + (-7.631)= **11.59** l/100 km;
e) For $v_0$ = 50 km/h, is obtained: y ≃ h* = h (-0.0044; 0.656; -7.631) = -0.0044x $50^2$ + 0.656 x 50 + (-7.631) = **14.20** l/100 km;
f) For $v_0$ = 60 km/h, is obtained: y ≃ h* = h (-0.0044; 0.656; -7.631) = -0.0044x $60^2$ + 0.656 x 60 + (-7.631) = **15.93** l/100 km;
g) For $v_0$ = 70 km/h, is obtained: y ≃ h* = h (-0.0044; 0.656; -7.631) = -0.0044x $70^2$ + 0.656 x 70 + (-7.631) = **16.79** l/100 km;
h) For $v_0$ = 80 km/h, is obtained: y ≃ h* = h (-0.0044; 0.656; -7.631) = -0.0044x $80^2$ + 0.656 x 80 + (-7.631) = **16.77** l/100 km;
i) For $v_0$ = 90 km/h, is obtained: y ≃ h* = h (-0.0044; 0.656; -7.631) = -0.0044x $90^2$ + 0.656 x 90 + (-7.631) = **15.87** l/100 km;
j) For $v_0$ = 100 km/h, is obtained: y ≃ h* = h (-0.0044; 0.656; -7.631) = -0.0044x $100^2$ + 0.656 x 100 + (-7.631) = **14.09** l/100 km;
k) For $v_0$ = 110 km/h, is obtained: y ≃ h* = h (-0.0044; 0.656; -7.631) = -0.0044x $110^2$ + 0.656 x 110 + (-7.631)= **11.44** l/100 km;
l) For $v_0$ = 120 km/h, is obtained: y ≃ h* = h (-0.0044; 0.656; -7.631) = -0.0044x $120^2$ + 0.656 x 120 + (-7.631) = **7.91** l/100 km;
m) For $v_0$ = 130 km/h, is obtained: y ≃ h* = h (-0.0044; 0.656; -7.631) = -0.0044x $130^2$ + 0.656 x 130 + (-7.631) = **3.50** l/100 km.

For the speed range developed by the Renault Symbol 1.5 Dci euro 3 car, the fuel consumption values (litres/100 km) are presented with a step of 10 km/h in figure 4.

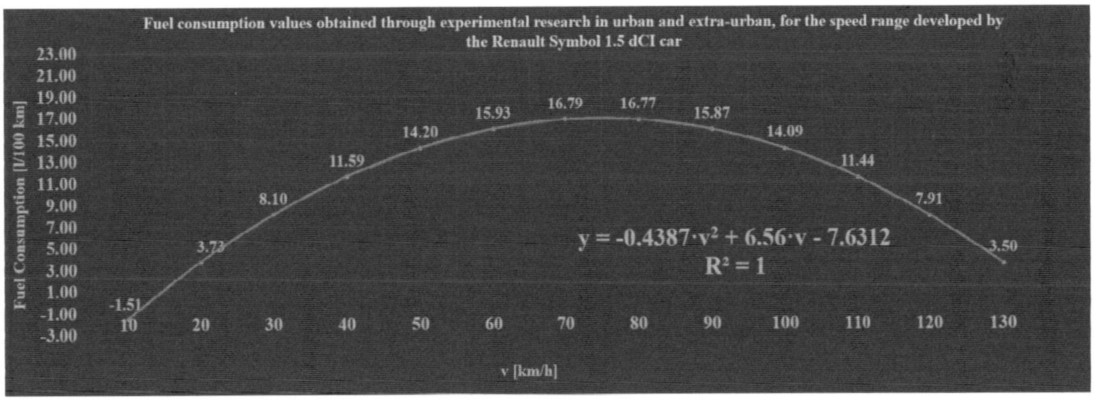


**Figure 4.** Fuel consumption values obtained through experimental research in urban and extra-urban, for the speed range developed by the Renault Symbo1,5 dCI Euro 3 car.

## 4. Conclusions

The dynamics of fuel consumption and $CO_2$ emissions in Renault Symbol 1.5 dCI, Euro 3 vehicles reveal an increasing trend based on experimental studies conducted on two urban routes. The rise in fuel consumption and $CO_2$ emissions is more pronounced. In suburban conditions on the highway, fuel consumption decreases to 3.5 l/100 km, accompanied by a corresponding reduction in CO2 emissions. Within urban settings, reducing the driving speed from 50 km/h to 30 km/h leads to a decrease in fuel consumption from 14.2 l/100 km to 8.1 l/100 km, representing a percentage reduction of 42.95%.

Initially, the car manufacturer endorsed the following fuel consumption values using the WLTP method: urban consumption - 5.3 l/100 km; extracity consumption - 3.5 l/100 km (refer to Fig. 2 and Table 2). However, experimental studies in urban conditions (50 km/h) revealed higher actual fuel consumption than the manufacturer's claims. The increase in fuel consumption in urban conditions is 62.67% compared to the manufacturer's declared consumption, with a corresponding increase in $CO_2$ emissions. In suburban conditions, the observed fuel consumption aligns with the manufacturer's approval.

Utilizing the least squares method, we determined the fuel consumption values "y" for speeds in the [0-130] km/h range: y = FCRenault Symbol 1.5dCI [0-130] km/h = [0-14.2] l/100 km. The mathematically calculated fuel consumption values at speeds of 30 km/h, 50 km/h, and 130 km/h (refer to Fig. 4) closely resemble the average fuel consumption values obtained experimentally at the same speeds (30 km/h = 8.10 l/100 km; 50 km/h = 14.2 l/100 km; 130 km/h = 3.5 l/100 km) (see Table 1 and Fig. 1).

The mathematical model demonstrates a high degree of correlation, as evidenced by a correlation coefficient - determination equal to 1 (refer to Fig. 3). Therefore, for a Renault Symbol 1.5 dCI, Euro 3 car, the applied mathematical model for average fuel consumption per liter/100 km is valid. Both mathematical formulas and calculations affirm the results obtained during experimental studies.

**Author's ORCID ID**
A M Titu: https://orcid.org/0000-0002-0054-6535
A Boroiu: https://orcid.org/0000-0002-6908-5546
A A Boroiu https://orcid.org/0000-0001-7313-3398

**References**
[1] Kountoura E 2021 Resolution on road safety 2021-2030, European Parliament, Bruxelles
[2] Neamțu G and Țîțu A 2022 The road transport system in Romania in the context of sustainable development. News, perspectives and sustainability,," in *"STRATEGICA" International Conference; The 10th edition: Sustainable Development and Strategic Growt; October 20-21, 2022,*, Bucharest, Ro
[3] Falkiner J 2003 Liquefied Petroleum Gas," in *Fuels and lubricants handbook: technology, proprieties, performance and testing,* , vol. 2nd printing. ed, P. A. I. a. S. R. S. E. West Conshohocken
[4] Unipetrol, „Propane-butane Fuel Mixture (Summer, Winter)," 14 Noiembrie 2010. [Interactiv]. Available: http://www.unipetrolrpa.cz/en/our-products/refinery_products/other-refinery-products/propane-butane.html
[5] Volkswagen.ro/WLTP, „WLTP: new standards for determining consuption," 2022. [Interactiv]. Available: https://www.volkswagen.ro/wltp. [Accesed 12 September 2022]
[6] Skoda.ro, „New standards for consumer values," 2022. [Interactiv]. Available: https://www.skoda.ro/despre-skoda/wltp. [Accesat 12 septembrie 2022]
[7] Untaru M, Pereș G, Stoicescu A, Poțincu G and Tabacu G 1981 Dinamics of wheeled vehicles, D. a. P. P. House, Ed., Bucharest, pp. 103-118
[8] Hilohi C, Untaru M, Soare I and Druță G 1982 Methods and means of testing automobiles, T. P. House, Ed., Bucharest
[9] T&A tyrepress, "The website for tyre and whell professionals.," 29 October 2015. [Online]. Available: https://www.tyrepress.com/2015/10/anglian-water-spot-on-with-pressure-test/. [Accessed 18 November 2022]
[10] Țîțu M and Oprean C 2006 Experimental research and data processing, T. p. H. o. ". B. U. o. Sibiu, Ed., Sibiu, p. 415
[11] Țîțu M, Oprean C and Boroiu A 2011 Applied experimental research in improving the quality of products and services. Experimental Data Processing Collection, P. H. AGIR, Ed., Bucharest, p. 684

[12] N. L. o. Medicine, “National Center for Biotechnology Information,” 2022. [Online]. Available: https://pubchem.ncbi.nlm.nih.gov/compound/11006#section=Top. [Accessed 14 November 2022]
[13] Bretscher O 1995 Linear Algebra With Applications, P. Hall, Ed., Upper Saddle River NJ
[14] A. RPR 1962-1966 Romanian Encyclopedic Dictionary, P. H. Politics, Ed., Bucharest

# Road infrastructure safety assessment using the integrated method - case study DN2.

**C I Dimitriu[1, 2] and G Frățilă [1]**

[1] POLITEHNICA University of Bucharest
[2] Romanian Road Authority – ARR

E-mail: cydimitriu@yahoo.com

**Abstract**. This study aims to practically test, identify vulnerabilities and analyses the limits of working tools for assessing road network safety and validate the integrated method proposed by the European Commission in the context of different technical classes of national roads in Romania. The research aims to adapt, convert, and transform the calculation values of the formulas of the proactive evaluation tool to the values used by national norms for road design and construction. 2 evaluation parameters are analysed, which represent the width of the traffic lane, respectively the widths and type of shoulders. On the other hand, the study is a tool for analysing road infrastructure safety, for efficiently allocating resources and for prioritising specific road safety inspections on the most dangerous sections of roads. The case study is the comparison of the results obtained from the safety assessment of DN 2 road infrastructure, both by the proposed test method and by the method adopted by the European Commission following Member States' comments.

## 1. Introduction

In accordance with paragraph 5, Article 5 of Directive (EU) 2019/1936 [1] amending Directive 2008/96/EC [2], The European Commission has made available to Member States a tool and a guide containing „guidelines on the methodology for conducting systematic road safety assessments and network-wide safety ratings" – pdf file [3].

Romania transposed the provisions of Directive (EU) 2019/1936 into national legislation by amending Law no. 265/2008, republished, on traffic safety management on road infrastructure [4], with the provisions of GO 3/2022 [5]. In line with the strands of action set out in Directive (EU) 2019/1936 – also called the „RISM Directive" – Member States will complete the first road infrastructure safety assessment, at motorway and national road network level, by the end of 2024.

In the first phase of the implementation process of the methodology, Romania and the other Member States finalised pilot programmes for the evaluation of a section of national road. Following Member States' comments, the European Commission amended the working tools - Excel smart files - and published on the European Union's website the adopted working version. One of the objectives of the research is the comparative analysis of the results obtained between the evaluation method used to test the evaluation methodology and the evaluation method adopted by the European Commission. The influence of removing a parameter from the proactive assessment tool, the parameter referring to „lighting", is also analysed.

Two tools are used for evaluation, one for proactive evaluation and another for reactive evaluation. With the proactive tool, NWA-p, the built-in safety of the road is determined according to its configuration and the geometric characteristics of the road infrastructure - Excel file No.1, [3]. With reactive instrument, NWA-r, the level of reactive safety of road infrastructure is determined according to the annual daily average of motor vehicles - AADT and the number of serious accidents resulting in fatalities / serious injuries - Excel file No. 2, [3].

The integrated road infrastructure safety assessment methodology consists of combining the results of the assessments obtained with the two assessment tools – NWA-r and NWA-p. Through the graphic integration scheme, by colour, the road infrastructure safety classification for each road segment is obtained according to a scale on 5 risk levels.

The research focuses on studying the influences of 2 of the 9 evaluation parameters of the proactive tool. For these parameters, solutions are identified to modify the ranges of values used to achieve some correspondence with the provisions of road construction standards in Romania [6]. These values are used to identify correlations between road cross-sectional characteristics and impact modification factors (CMFs). These correlations are then used in calculation formulas in Excel files to ensure that the results obtained are assimilable to those available on the specialized site [7].

The research centres on conducting a comparative analysis of outcomes derived from employing two road infrastructure safety assessment methods, encompassing a total length of 472.230 km along DN 2. Additionally, the study will investigate the influence of excluding the "Lighting" parameter on road safety classification results within a 132 km segment, stretching from km 10 to km 142 on the same road. This choice was decided by the fact that DN 2 ranks first in the ranking of the most dangerous roads in Romania and registered the highest number of deaths in road accidents between 2017 and 2020, being known as the „road of death". It is analyzed comparatively the result of the evaluation by adapting the two calculation parameters to the standards of Romanian roads.

The data used in the research comes from three sources. The official results of the safety assessment of the infrastructure of the national road DN 2 are obtained from the Romanian Road Authority– A.R.R. For the values of the annual daily average of vehicles (AADT) on national roads in Romania, the database of the Center for Road Technical Studies and Informatics - CESTRIN is used, which is found on the institution's website [8]. For data on serious road accidents resulting in deaths and serious injuries, the database of the Romanian Police is used [9]. **Table 1**Table 1 presents the ranking of the most dangerous national roads in Romania in terms of road safety because of the consequences of serious accidents between 2017 and 2020.

**Table 1**. The consequences of serious accidents and the classification of national roads.

| Name of national road | Serious accident | Deceased persons | Seriously injured persons | Ranking of roads according to the consequences of serious accidents: | |
|---|---|---|---|---|---|
| | (%) | (%) | (%) | Deaths | Seriously injured |
| DN 1 | 7.85 | 7.09 | 8.56 | II | I |
| DN 2 | 6.62 | 8.42 | 6.57 | I | II |
| DN 6 | 5.50 | 6.98 | 5.12 | III | III |
| DN 7 | 5.08 | 5.07 | 5.53 | IV | IV |
| DN 15 | 3.18 | 3.19 | 3.14 | V | V |

The classification of roads according to the consequences of serious accidents suggests that DN 2 has the biggest problems in terms of fatalities in accidents, while DN 1 has the biggest problems in terms of seriously injured people. The DN 2 national road has the highest percentage of serious accidents and fatalities, indicating a serious safety problem on this road. The safety assessment of the DN2 national road infrastructure is carried out on a length of 472.23 km, from the exit of mun. Bucharest (km10+000) and to the border point Vama Siret, Suceava County (km 482+230).

## 2. Proactive road infrastructure assessment method (NWA-p)

Proactive road assessment uses a tool in the form of an Excel file No. 1 [3] which contains spreadsheets and calculation formulas. Each assessed road segment is allocated a spreadsheet containing several analysis parameters targeting the built-in safety of road infrastructure, as follows: 1 - lane width, 2 - size of roadside road risk depending on distance to obstacle (RHR)/ obstacle type/slope size, 3 - curve radius, 4 - density of access points to properties, 5 - number and configuration of intersections, 6 - conflicts between pedestrians / cyclists and motorized traffic, 7 - type and width of the shoulder, 8 - presence of additional crossing lanes, 9 - presence and quality of signs and markings, 10 - lighting. The latter was used only in the test version of the integrated assessment methodology.

The Excel file includes for each parameter a calculation formula or algorithm that evaluates the impact on the safety of the analysed road segment. Based on the data entered in the spreadsheet, the proactive tool provides a specific value for the accident change factor (CMF) associated with each parameter [7]. The crash reduction factor (RF) is then automatically calculated by the Excel tool using the formula in equation (1):

$$RF = \frac{1}{CMF} \tag{1}$$

RF is a positive number and always less than or equal to 1. For each segment assessed, the product of the reduction factors, expressed as a percentage, represents the 'estimated road risk level score'. On national roads, depending on the value of the score obtained, the following colour coding is allocated:

- between 0 and less than 50, corresponds to a „High Risk" road, red colour,
- between 50 and less than 80, corresponds to an „Intermediate Risk" road, yellow colour,
- between 80 – 100, corresponds to a „Low Risk" road, green colour, pdf file, [3].

### *2.1. Analysis of proactive assessment tool parameters*

Analysis of the structure, the proactive assessment tool, shows that 2 parameters are not adapted for the evaluation of roads in Romania. The values used in the calculation formulas for assessing the safety of the road infrastructure, in terms of the average width of all traffic lanes as well as the widths of the consolidated/unconsolidated roadsides, do not reflect the approach and requirements of the Romanian norms for the design and construction of roads.

*2.1.1. Parameter No. 1 – „lane width".* For undivided roads, equation (2) provides the formula for calculating the crash modification factor (CMF) associated with parameter no. 1. The value written in cell E6 of the proactive tool is taken by the formula and the CMF is calculated, which is displayed in cell M6. Equation (3) provides the formula for calculating the CMF associated with parameter no. 1, used by the proactive tool in case of divided roads.

$$CMF{=}IF(E6{>=}3{,}4;\ 1;\ IF(E6{>=}3{,}15;\ 1{,}05;\ IF(E6{>=}2{,}7;\ 1{,}12;\ 1{,}19))) \tag{2}$$

$$CMF{=}IF(E6{>=}3{,}4;\ 1;\ IF(E6{>=}3{,}15;\ 1{,}021;\ IF(E6{>=}2{,}7;\ 1{,}08;\ 1{,}12))) \tag{3}$$

Considering the recommendations set out in the report „Crash Modification Factors in Practice" [10] and the absence of consistent data in the CMFs database [7], approximate values like those incorporated in the formulas of the proactive tool of European Commission, are chosen, as there are no significant differences in bandwidths.

For proactive evaluation, depending on parameter no. 1, of roads in Romania belonging to different technical classes, incorporate the appropriate values from Table 2 into the corresponding calculation formulas of the proactive Excel file, according to the equations presented below.

Equation (4) contains the calculation formula for motorways:

$$CMF{=}IF(E6{>=}3{,}75;\ 1;\ 1{,}025) \tag{4}$$

Equation (5) contains the calculation formula for express roads and divided European national roads of technical class II, with 4 lanes:

$$CMF{=}IF(E6{>=}3{,}5;\ 1;\ 1{,}12) \tag{5}$$

Equation (6) contains the calculation formula for European national roads and undivided national roads of technical class II, with 4 lanes:

$$CMF=IF(E6>=3,5; 1; IF(E6>=3; 1,12; IF(E6>=2,75; 1,19; 1,25))) \quad (6)$$

Equation (7) contains the calculation formula for undivided national roads of technical classes III, IV and V, with 2 lanes:

$$CMF=IF(E6>=3,5; 1,025) \quad (7)$$

Table 2 supplies information on the correlation between lane widths and estimated CMF values for various lane widths, categorized by technical classes, following Romanian standards for road design and construction.

**Table 2.** CMF values for lane widths depending on the technical class of the road.

| **Technique** | **Class (RO)** | | | **I** | **II** | **III** | **IV** | **V** | |
|---|---|---|---|---|---|---|---|---|---|
| | Road category | Road structure | Country of the normative | | | Lane width LW (m) | | | |
| | Primary | Undivided | UK | | | 3.4 | 3.15 | 2.7 | |
| ***CMF*** | | | | | | ***1*** | ***1.05*** | ***1.12*** | ***1.19*** |
| | National (DN) | Undivided | RO | | 3.5 | 3.5 | 3 | 2.75 | |
| ***CMF*** | | | | | ***1*** | ***1.025*** | ***1.12*** | ***1.19*** | ***1.25*** |
| | Primary | Divided | UK | | 3.4 | | 3.15 | 2.7 | |
| ***CMF*** | | | | | ***1*** | | ***1.021*** | ***1.08*** | ***1.12*** |
| | National (DN) | Divided | RO | | 3.5 | | | | |
| ***CMF*** | | | | | ***1*** | | | | ***1.021*** |
| | Motorway | Urban | UK | 3.25 | 3 | | | | |
| ***CMF*** | | | | ***1*** | ***1.025*** | | | | ***1.05*** |
| | Motorway | Rural | UK | 3.4 | 3.15 | | | | |
| ***CMF*** | | | | ***1*** | ***1.025*** | | | | ***1.05*** |
| | Motorway | Rural | RO | 3.75 | | | | | |
| ***CMF*** | | | | ***1*** | | | | | ***1.25*** |

*2.1.2. Parameter No. 7 – „type of shoulders" and „shoulders width".* CMF value for parameter No. 7 is displayed in cell M22 of the proactive instrument according to the choice of shoulder type (consolidated OR unconsolidated) and the width of the shoulder. For undivided roads, the assessment is made simultaneously in both directions of travel, and for divided roads, the assessment is made separately in each direction.

The approach of assessment by choosing a single type of shoulder is not applicable for the assessment of national roads or European national roads which, in Romania, are designed and constructed with two types of shoulder, consolidated shoulder AND unconsolidated shoulder. On the other hand, the proactive tool uses values of the widths of the shoulders that do not correspond to the values of the Romanian standards.

The types of shoulder and the values of shoulder widths for national and European national roads are set out in "ANNEX NO. 1 – Widths of platforms and roadways in alignments" from the Annex to UNWTO no. 1296/2017 [6].

For parameter No. 7, the proactive instrument allowed only one type of verge to be selected. This selection is modified and currently the proactive Excel file allows the assessment to be carried out for each type of verge and on each direction of travel.

The Table 3 shows the standardised values for the widths of the two types of roadsides, both for divided and undivided roads.

**Table 3.** Data for the "Lookup Table" sheet and CMFs values for road shoulder types.

| Shoulder width paved. (m) | For VLOOKUP | CMF | Shoulder width unpaved. (m) | For VLOOKUP | CMF |
|---|---|---|---|---|---|
| **Undivided NRs** | | | | | |
| SW = 0 | 0 | 1.3 | SW = 0 | 0 | 1.3 |
| 0 < SW < 0.25 | 0.01 | 1.25 | 0 < SW < 0,5 | 0.01 | 1.25 |
| 0.25 < SW < 0.5 | 0.25 | 1.211 | 0,5 ≤ SW < 0,75 | 0.5 | 1.211 |
| 0.5 ≤ SW < 0.75 | 0.5 | 1.063 | 0.75 ≤ SW < 1.75 | 0.75 | 1.063 |
| SW ≥ 0.75 | 0.75 | 1 | SW ≥ 1.75 | 1.75 | 1 |
| **Divided NRs** | | | | | |
| SW = 0 | 0 | 1.3 | SW = 0 | 0 | 1.3 |
| 0 < SW < 0.5 | 0.01 | 1.211 | 0 < SW < 0.75 | 0.01 | 1.211 |
| 0.5 ≤ SW < 0.75 | 0.5 | 1.063 | 0.75 ≤ SW < 1.5 | 0.75 | 1.063 |
| SW ≥ 0.75 | 0.75 | 1 | SW ≥ 1.5 | 1.5 | 1 |

*2.1.3. Parameter No 10 — „road lighting".* The research aims to find out to what extent the results of proactive evaluation are influenced by removing parameter no. 10. For this purpose, the results of the evaluation of a 132 km segment of the DN 2 road are compared with the test variant of the proactive tool and the results obtained with the variant of the instrument from which parameter no. 2 was removed. 10. The results of the assessment of the DN 2 road by testing the variant of the methodology were presented at the EAEC MVT Congress in 2022 [11].

The results presented in Table 4 show in percentage terms how the results of the proactive assessment of the 132 km segment of the DN 2 national road are influenced due to the elimination of the infrastructure safety assessment based on the influence of parameter no. 10. Thus, for split road segments, initially assessed as „High Risk", the changes are insignificant. For undivided road segments, it is found that 6.29%, representing 8.3km, changed the risk classification from „Intermediate Risk" to „Low Risk" and 1.74%, representing 2.6 km, changed the classification from „High Risk" to „Intermediate Risk".

**Table 4.** Influence of „ lighting" parameter on proactiv assessment status.

| NWA -proactivă | Divided - ascending | | Divided - descending | | Undivided | |
|---|---|---|---|---|---|---|
| **DN 2 - 132 km** | **%** | **km** | **%** | **km** | **%** | **km** |
| ***10 Parameters*** | | | | | | |
| High Risk (Red colour) | 11.48 | 15.15 | 11.48 | 15.15 | 32.58 | 43 |
| Intermediate Risk (Yellow colour) | 1.74 | 2.3 | 1.74 | 2.3 | 25.95 | 34.25 |
| Low Risk (Green colour) | | | | | 28.26 | 37.3 |
| ***9 Parameters*** | | | | | | |
| High Risk (Red colour) | 11.48 | 15.15 | 11.48 | 15.15 | 30.83 | 40.7 |
| Intermediate Risk (Yellow colour) | 1.36 | 1.8 | 1.36 | 1.8 | 21.40 | 28.25 |
| Low Risk (Green colour) | 0.38 | 0.5 | 0.38 | 0.5 | 34.55 | 45.6 |
| ***Influence*** | | | | | | |
| High Risk (Red colour) | | | | | -1.74 | -2.3 |
| Intermediate Risk (Yellow colour) | -0.38 | | -0.38 | | -4.55 | -6 |
| Low Risk (Green colour) | 0.38 | | 0.38 | | 6.29 | 8.3 |

For parameter No. 10, which is the average of the reduction factors for all road segments, expressed as a percentage, recorded a value of 95.4%. This indicates that the final proactive safety score can be influenced by approximately 4.6% for each road segment.

It has been established that each score, obtained through the product of reduction factors, corresponds to a specific colour indicating a certain level of risk. The thresholds for switching from red to yellow and yellow to green are set at 80 and 50 respectively. To identify road segments where scores may be affected by multiplying by the mean value of parameter reduction factor no. 10 (0.954), two numeric filters are applied, with limits from 50 to 52.41 and from 80 to 83.856 on the proactive score column.

To get a detailed insight into the impact of removing parameter no. 10, the study investigated the results of the evaluation using the integrated method, which consists of combining the results of proactive and reactive assessment. The data presented in Table 5 show that the influences are insignificant and do not manifest themselves around interest for carrying out specific road safety inspection procedures. There are no changes in the statistical data of the road segments classified "Very High Priority" and "High Priority".

**Table 5.** Influence of „ lighting" parameter on integrated assessment status.

| **NWA - integrated** | **10 Parameters** | | **9 Parameters** | | **Influence** | |
|---|---|---|---|---|---|---|
| **DN 2 - 132 km** | **%** | **km** | **%** | **km** | **%** | **km** |
| Class 1 (Very Low Priority) | 4.39 | 5.8 | 4.77 | 6.3 | 0.38 | 0.5 |
| Class 2 (Low Priority) | 14.55 | 19.2 | 17.58 | 23.2 | 3.03 | 4.0 |
| Class 3 (Intermediate Priority) | 11.44 | 15.1 | 8.03 | 10.6 | -3.41 | -4.5 |
| Class 4 (High Priority) | 6.14 | 8.1 | 6.14 | 8.1 | - | - |
| Class 5 (Very High Priority) | 63.48 | 83.8 | 63.48 | 83.8 | - | - |

### 2.2. *Results of road assessment by proactive method*

The results obtained from the proactive assessment of DN 2 road, using both the initial testing method and the current method approved by the European Commission, are presented in Table 6. Comparing the results shows that the removal of parameter 10 from the evaluation led to a 1.86% change in the statistical distribution of segment lengths across road risk classes.

**Table 6**. Results of proactive assessment using two working methods (NWA-P).

| **NWA - proactiv** | **Variant 3- test** | | **Variant 5 - work** | | **Comparison** | |
|---|---|---|---|---|---|---|
| **DN 2 - 472.230 km** | **km** | **%** | **km** | **%** | **km** | **%** |
| Low Risk (Green) | 78.935 | 16.72 | 87.735 | 18.58 | 8.8 | 1.86 |
| Intermediate Risk (Yellow) | 328.195 | 69.50 | 321.695 | 68.12 | -6.5 | -1.38 |
| High Risk (Red) | 65.1 | 13.79 | 62.8 | 13.30 | -2.3 | -0.49 |

## 3. Reactive Road Safety Assessment Method (NWA-r)

To apply the reactive assessment method, the DN2 road was divided into homogeneous segments with approximately the same criteria as for the pro-active assessment: physical separation of directions, terrain landform, changes in road configuration (number of lanes, segments in alignment/curve), changes in values for daily average vehicles (AADT), traffic regime (urban/extra-urban). If in the proactive assessment the recommended length of road segments was 2 km, in the reactive assessment the recommended length of road segments can reach up to 7 km in the case of undivided roads and up to 15 km in the case of split roads - pdf file [3].

Of the three approaches to reactive assessment proposed by the European Commission, for the evaluation of DN 2 road, the formation of road segments that include alignments and intersections was chosen as a working method. Each of the road segments is evaluated according to the number of serious accidents occurring on them and the AADT value allocated. Depending on the data entered in the calculation tables of the reagent instrument, the level of associated road risk is determined.

### 3.1. *Statistical data of reactive evaluation*

The reactive tool is presented on the specialized site in the Excel file No. 2, (3). In the „Preliminary info" sheet, for the family of national roads, data taken from official sources were entered. The total length of national roads is 17,913 km, value taken from the Excel file „TEMPO TRN 139A" [12]. For the average daily average value of vehicles (AADT), corresponding to the family of national roads, the value of 8,000 vehicles/ 24 hours is used [8]. Between 2017 and 2020, 10,316 serious deaths and injuries were registered on national roads. The data come from the database of the Romanian Police – Road Department [9].

### *3.2. Results of DN 2 road evaluation by reactive method*

In the Excel spreadsheet, „1. Methodology (S)" the lengths of road segments resulting from the subdivision of DN 2 based on the specific criteria of the method are entered. For each segment, serious accidents are allocated. The average daily traffic (AADT) value recorded on the road section to which the evaluated segment belongs is considered.

The test variant of the reactive method, using the calculation formulas of the reactive instrument calculates the parameters of each segment: „Accident density" and „Accident rate". Based on the accident density thresholds and accident rate thresholds, two classifications are obtained for each section, „accident density classification" and „accident rate classification". The final ranking, according to the test variant approach, is determined considering the more critical classification between the two rankings. The final class to which each section belongs is obtained. „Low risk", „High risk" or „Unsure".

The official working method, agreed by the European Commission, uses the Poisson method. Using the Poisson method, upper and lower thresholds are defined for the observed number of accidents on each section (and junction). These thresholds are then converted into accident density thresholds and accident rate thresholds for each section. For the final ranking of the section is recommended to rely on the comparison of the accident rate if they are available. Otherwise, the ranking is based on the comparison of accident density. (processing after „Deliverable D.4.2 - Network-wide safety assessment tool and guidance" document version no.14, November 2022, European Commission)

The results obtained from the reactive assessment of the DN 2 road, using both the initial test method and the agreed working method, are presented in Table 7.

**Table 7.** Comparative results of DN 2 road reactive assessment (NWA-R).

| **NWA - reactive** | **Variant 3- test** | | **Variant 5 - work** | | **Comparison** | |
|---|---|---|---|---|---|---|
| **DN 2 - 472.230 km** | **km** | **%** | **km** | **%** | **km** | **%** |
| Low Risk (Green) | 45.274 | 9.59 | 27.70 | 5.87 | -17.574 | -3.72 |
| Unsure (Gray) | 153.990 | 32.61 | 287.369 | 60.85 | 133.379 | 28.24 |
| Hight Risk (Red) | 272.966 | 57.80 | 157.161 | 33.28 | -115.810 | -24.52 |

The reactive evaluation through the two working variants reveals significant differences. There is a notable 24.5% decrease in road segments classified as High Risk. The data also indicate that the uncertainty zone in road risk assessment has increased from 32.61% to 60.85%.

## 4. Integrated method of road infrastructure safety assessment (NWA-I)

The method consists of graphically combining the results of proactive assessment and reactive assessment based on a color scheme and results in the classification of road segments into 5 risk classes, according to **Table 8** - Class 1 (Dark green), Class 2 (Green), Class 3 (Yellow), Class 4 (Orange), Class 5 (Red). Data on the statistics of segments classified as Class 5 shall be used by the competent authority for planning specific road safety inspections.

**Table 8.** Results of the safety assessment of DN 2 road infrastructure by the integrated method.

| **NWA - integrated** | **Variant 3- test** | | **Variant 5 - work** | | **Comparison** | |
|---|---|---|---|---|---|---|
| DN 2 - 472.230 km | km | % | km | % | km | % |
| Class 1 (Very Low Priority) | 5.8 | 1.23 | 0 | 0 | -5.8 | -1.23 |
| Class 2 (Low Priority) | 59.564 | 12.61 | 86.435 | 18.30 | 26.871 | 5.69 |
| Class 3 (Intermediate Priority) | 125.8 | 26.64 | 181.934 | 38.53 | 56.134 | 11.89 |
| Class 4 (High Priority) | 8.1 | 1.72 | 46.7 | 9.89 | 38.6 | 8.17 |
| Class 5 (Very High Priority) | 272.966 | 57.80 | 157.161 | 33.28 | -115.805 | -24.52 |

The data show that the results of the classification of road infrastructure by risk classes are significantly influenced by the change in the working method for reactive assessment.

## 5. Conclusions

The study makes significant contributions to the implementation of the road infrastructure safety assessment procedure in Romania. It supports the competent authority in streamlining the planning and monitoring of road safety inspections, as well as in the training of specialized personnel responsible for carrying out the procedures outlined in the RISM Directive.

Following the two working methods of the European Commission, the research conducts a comparative analysis between the results obtained from proactive evaluation and reactive evaluation on the national road, which is the most dangerous in terms of road fatalities, DN 2.

The research findings lead to the adjustment of the parameters of the proactive evaluation tool, and the analysis provides insights into the correlation between the widths of traffic lanes and road shoulders with the estimated CMF values for various technical classes, in accordance with Romanian standards for road design and construction.

Considering that the results of the reactive evaluation significantly impact the outcome of the integrated road network safety classification, it is crucial to pay increased attention when identifying segments based on areas with a high concentration of severe accidents.

## Author's ORCID ID

C I Dimitriu: https://orcid.org/0000-0001-9433-8165

## References

[1] European Parliament, Council of the European Union. https://eur-lex.europa.eu. [Online].; 2019 [cited 2022 July 10. Available from: http://data.europa.eu/eli/dir/2019/1936/oj.

[2] European Parliament, Council of the European Union. https://eur-lex.europa.eu. [Online].; 2008. Available from: http://data.europa.eu/eli/dir/2008/96/oj.

[3] The European Commission. https://road-safety.transport.ec.europa.eu/eu-road-safety-policy/priorities/infrastructure/road-infrastructure-guidelines_en. [Online].; 2023 [cited 29 06 29. Available from: https://road-safety.transport.ec.europa.eu/.

[4] The Romanian Parliament. Law no. 265/2008, republished and updated with subsequent amendments, regarding the management of traffic safety on the road infrastructure. The Official Monitor of Romania. 2012 August 23 07: p. 6-18.

[5] The Government of Romania. Ordinance no. 3/2022 for the amendment and completion of Law no. 265/2008 regarding the management of traffic safety on the road infrastructure. The Official Monitor of Romania. 2022 January 28: p. 1-8.

[6] Ministry of Transport from Romania. Order no. 1296 as of August 30, 2017, for the approval of the Technical Norms regarding the design, construction, and modernization of roads. The Official Monitor of Romania no. 746. 2017 Sep 18: p. 2-15.

[7] U.S. Department of Transportation - Federal Highway Administration. https://www.cmfclearinghouse.org/. [Online].; 2023 [cited 2023 06 30.

[8] C.E.S.T.R.I.N.- Center for Technical Road and Computer Studies. www.cestrin.ro. [Online].; 2022 [cited 2023 Jun 30. Available from: https://cestrin.ro/.

[9] Romanian Police. www.politiaromana.ro. [Online].; 2022 [cited 2023 July 07. Available from: https://www.politiaromana.ro/ro/utile/statistici-evaluari/statistici.

[10] Federal Highway Administration. United States. Crash Modification Factors in Practice: Using CMS to Quantify the Safety in the Value Engineering Process. Road safety. WASHINGTON, DC 20590: Federal Highway Administration, U.S. Department of Transportation; 2013. Report No.: FHWA-SA-13-014.

[11] Dimitriu C I and Bujor C R 2022 The integrated methodology for assessing the safety of the road network infrastructure in Romania. In IOP Conference Series - Materials Science and Engineering, Timisoara, EAEC MVT: IOP Publishing Ltd. p. 8.
[12] National Institute of Statistics - INS, Romania. https://insse.ro/cms/ro. [Online].; 2023 [cited 2023 August 10. Available from: http://statistici.insse.ro:8077/tempo-online/#/pages/tables/insse-table.

# An analysis of emissions and carbon footprints for the various types of vehicle propulsion

**A Oprica[1], B Florescu[1], M Bică[2], I Dumitru[1], D Tutunea[1], L Racila[1] and N D Gencărau[2]**

[1] University of Craiova, Faculty of Mechanics, Department of Vehicle, Transportation and Industrial Engineering, Craiova, Romania
[2] University of Craiova, Faculty of Mechanics, Department of Applied Mechanics and Civil Constructions, Craiova, Romania

E-mail: racila_laurentiu@yahoo.com

**Abstract.** Plans, predictions, and strategies to decrease emissions from vehicles are now being implemented as a direct result of anti-pollution legislation, which is currently in full swing. Countries that have achieved a high level of industrialization have enacted a variety of policies designed to encourage the production of clean (or less polluting) vehicles. These policies include the reduction of taxes as well as the imposition of penalties (such as green surcharges or even a ban on driving) on vehicles that emit high levels of pollutants that are in excess of what is considered to be acceptable levels. The public at large is coming around to the idea that human activity is responsible for the majority of climate change, as well as the emission of carbon dioxide and other greenhouse gases, and this theory has a solid foundation in the scientific community. The paper aims to explore findings in the context of the disproportionately carbon footprint for different means of transportation of goods.

## 1. Introduction

The carbon footprint, also known as the greenhouse gas footprint, is an indicator that may be used to compare the total quantity of greenhouse gases that are released by an activity, product, business, or nation [1]. It is common practice to describe carbon footprints in terms of tons of emissions ($CO_2$-equivalent) per unit of comparison, such as per year, person, kilogram of protein, kilometer traveled, and other similar measures.

The total emissions produced by a product during its entire life cycle, from manufacture all the way through the supply chain to ultimate consumption and disposal, are included in the product's "carbon footprint." The direct and indirect emissions that an organization is responsible for are referred to as Scope 1, 2, and 3 in the Greenhouse Gas Protocol, which is used for the accounting of an organization's carbon footprint. In the same way, an organization's total carbon footprint comprises both of these types of emissions [2].

There are a number of approaches that may be taken and tools that can be used on the internet in order to calculate a person's carbon footprint; this number can also be determined for a nation, a company, a product, or an individual. If a customer wants to be more environmentally conscious about their consumption habits, for instance, considering a product's carbon footprint before making a purchase might be helpful. In the context of efforts that aim to mitigate the effects of climate change, the carbon footprint may be used to differentiate between economic activities that leave a large footprint and those that leave a small impact [3]. To put it another way, the notion of carbon footprint makes it possible for anybody to draw comparisons between the climate-relevant effects that are caused by people, goods, corporations, and governments. When this is done, it is helpful in developing strategies and goals for minimizing the footprint left by carbon emissions.

## 2. Vehicle performances relative to carbon footprint

The carbon footprint is a representation of the emissions that are associated with a certain activity, such as the generation of electricity, transportation, the cooling of space, the construction of structures, the manufacturing of industrial goods, or agricultural goods [4]. The practice of lowering, eliminating, or preventing emissions of greenhouse gases is referred to as carbon offsetting. Carbon offsetting is quantified in $CO_2$-eq equivalent. The $CO_2$-eq is a unit that is used to compare the radiative power of a greenhouse gas (GHG) with the radiative force of carbon dioxide.

The mass of that chemical multiplied by its global warming potential (GWP) will be used in the calculation to determine the carbon dioxide equivalent of that substance. For instance, the global warming potential of methane is 23, but the GWP of nitrous oxide is 296. Carbon dioxide ($CO_2$), methane (CH4), nitrous oxide (N2O), and fluorinated gases are the most significant contributors to global warming. The quantities of greenhouse gases (GHGs) in the atmosphere are already there naturally, but human actions like the combustion of fossil fuels are increasing those levels, which leads to climate change.

Emissions of greenhouse gases are often measured in terms of $CO_2$-equivalents, which is a measurement metric that is used to compare different types of gases that have an effect on the atmosphere with regard to their potential to contribute to global warming in comparison to $CO_2$'s contribution [4, 5]. In general, the total $CO_2$ equivalent emissions made by a vehicle over the course of its lifetime may be calculated as:

$$E_{life} = E_{prod,tot} + E_{util,tot} + E_{recyc,tot} \tag{1}$$

where:

- $E_{prod,tot}, E_{util,tot}, E_{recyc,tot}$ are the emission for production, use and recycling

Even before a vehicle can be used to carry products or people, a significant quantity of greenhouse gas must be generated so that the raw materials can be extracted and the vehicle can be manufactured [6]. Estimates of the total $CO_2$ equivalent that is produced as a result of the manufacturing of a vehicle, $E_{prod,tot}$, may still be derived from the following sources [5]:

$$E_{prod,tot} = m_{veh,body} \times e_{prod,body} + C_{bat} \times e_{prod,bat} \tag{2}$$

where:

- $m_{veh,body}$ and $C_{bat}$, are the weight of the vehicle measured in kilograms and battery capacity measured in kilowatt-hours (only for electric vehicle types)
- $e_{prod,body}$ and $e_{prod,bat}$ are the relative $CO_2$ equivalents of the vehicle body and battery and are represented in kilograms of $CO_2$ equivalents per kilogram and kilograms of $CO_2$ equivalents per kilowatt-hour.

The total weight of the vehicle, denoted by $m_{veh}$, may be broken down into the vehicle's body mass, denoted by $m_{veh,body}$, and its battery mass, denoted by $m_{veh,bat}$ [5, 6, 7], based on the following formula:

$$m_{veh} = m_{veh,body} + m_{veh,bat} = m_{veh,body} + \frac{C_{bat}}{\xi_{bat}} \tag{3}$$

where:

- $C_{bat}$ is the nominal battery capacity measured in kWh
- $\xi_{bat}$ is the specific energy measured in kWh/kg.

The total equivalent of emissions [5, 6] using $E_{util,tot}$ $CO_2$ may still be calculated as follows:

$$E_{util,tot} = c_{WLTP} \times e_{WTW} \times S_{lifetime} \quad (4)$$

where:

- $c_{WLTP}$ is the relative energy/fuel consumption based on the Worldwide Harmonised Light Vehicle Test Procedure (WLTP)
- $S_{lifetime}$ is the expected lifetime of the vehicle in km

Even if the emissions related with the manufacturing of spare parts and the maintenance of vehicles are ignored (which is inconsequential in the grand scheme of things), the total equivalent of emissions can still be determined [5, 7]. $WTW$ relative emissions can be separated according to the following formula:

$$e_{WTW} = e_{WTT} + e_{TTW} \quad (5)$$

This is because WTT emissions typically take into account the extraction processes of the energy source, the production of energy carriers, and the distribution of energy carriers, whereas TTW emissions only take into consideration the energy conversion process required to actually propel the vehicle [8]. There is a clear correlation between the exhaust gases released by traditional ICEVs and the pollutants they produce [5, 10]. Consequently, the relative $TTW\ CO_2$ equivalent emissions of ICEVs may be estimated easily using the following formula:

$$e_{TTW} = \frac{E_{CO_2tot}}{D_{WLTP}} \quad (6)$$

where:

- $E_{CO_2tot}$ is the total $CO_2$ emissions assessed
- $D_{WLTP}$ is the WLTP mileage

In contrast, assessing the emissions produced by PHEVs is more challenging due to the fact that PHEVs may be driven by both an electric motor and an internal combustion engine. Electric motors do not produce any local $CO_2$ emissions [10, 11]. It is required to apply a utility factor (UF), which is defined as the ratio of the distance traveled when the battery charge was reduced to the total distance of the WLTP [4, 5, 6]. The quantity of total $CO_2$ emissions may be computed as follows:

$$E_{CO_2tot} = UF \times E_{CO_2CD} + (1 - UF) \times E_{CO_2CS} \quad (7)$$

where:

- $E_{CO_2CD}$ is the $CO_2$ emissions associated with the 'charge depletion' mode
- $E_{CO_2CS}$ is the $CO_2$ emissions associated with the 'charge support' mode.

When compared, the $CO_2$ emissions produced by BEVs and FCEVs while being driven are considered to be zero. This is due to the fact that these vehicles do not produce $CO_2$ in the immediate environment. However, their use of fuel or energy is something that has to be monitored. With this in mind it is also possible to determine the total $CO_2$ equivalent of recycling emissions using the following formula [5, 10, 12]:

$$E_{recyc,tot} = m_{veh,body} \times e_{recyc,body} + C_{bat} \times e_{recyc,tot} \quad (8)$$

where:

- $E_{recyc,tot}$ are the total $CO_2$ recycling emissions

## 3. WTW emissions evaluation for hybrid vehicle

The evaluation for the hybrid vehicle with regards to WTW emissions was done in the software platform GREET, which stands for "Greenhouse gases, regulated emissions, and energy use in transport". This is a platform that analyzes the implications that vehicle technology, fuels, energy products, and energy systems have across their entire lifecycles. It offers a transparent platform that producers of energy and vehicle systems, researchers, and regulators may use to evaluate the energy and environmental consequences of the technology and products that are used in car and energy system systems.

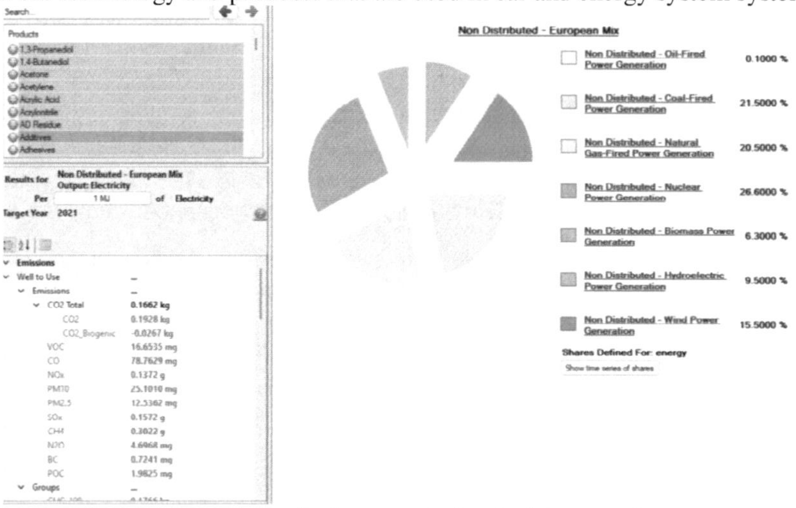

**Figure 1.** The methodical approach for devising new instances.

With the help of the application, we were able to determine overall energy consumption (including non-renewable and renewable forms), air pollutant emissions, greenhouse gas emissions, and water consumption for any energy system or vehicle.

With the use of this modeling tool, we carried out a comparison of the WTW emissions produced by a PHEV E10 (a hybrid vehicle powered by benzene containing 10% ethanol) in conditions involving a variety of different mixtures of electric power generation.

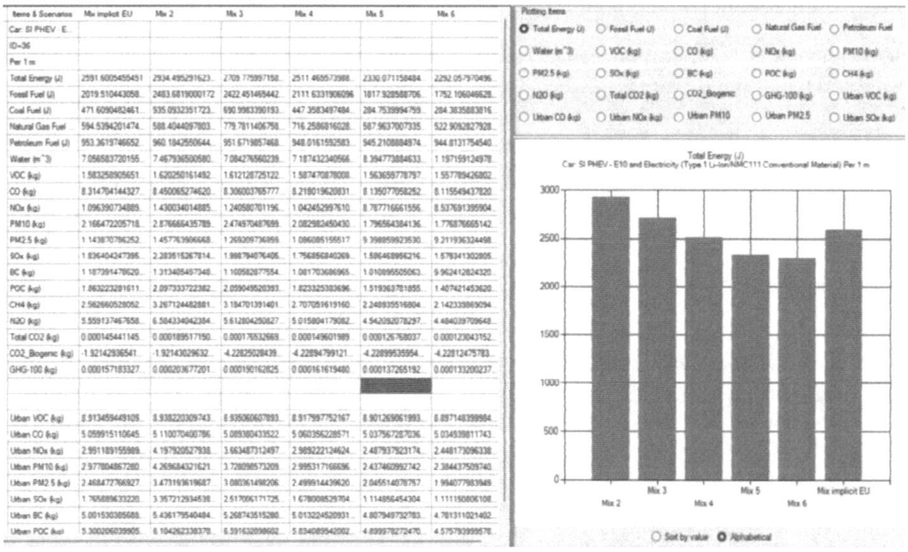

**Figure 2.** Partial results for the energy consumption.

For the study to be consistent and the outputs error to be minimal, we considered five cases:

- The implied mix of energy sources in Europe is as follows: coal (21.5%), natural gas (20.5%), nuclear (26.6%), hydro (9.5%), eolian (15.5%), and other (6.4%).
- Mixul 2 consists of carbon (50%), natural gas (20%), nuclear (10%), hydro (10%), and eolian (10%) sources of energy.
- Mixul 3 includes biomass (35%), natural gas (35%), nuclear (10%), hydro (10%), and eolian (10%) sources of energy.
- Mixul 4 includes carbon (20%), natural gas (30%), nuclear (30%), hydro (10%), and eolian (10%) sources of energy.
- Mixul 5: carbon (ten percent), natural gas (twenty percent), nuclear (forty percent), hydro (15 percent), and eolian (15 percent)
- Mixul 6: carbon dioxide (10%), natural gas (15%), hydro (30%), nuclear (30%), and eolian (15%)

The results of the analysis may be seen in the figures that are farther up. Because all values are expressed in grams per kilometer, it is possible to evaluate all environmental indicators beginning at the same point because they all use the same unit of measurement.

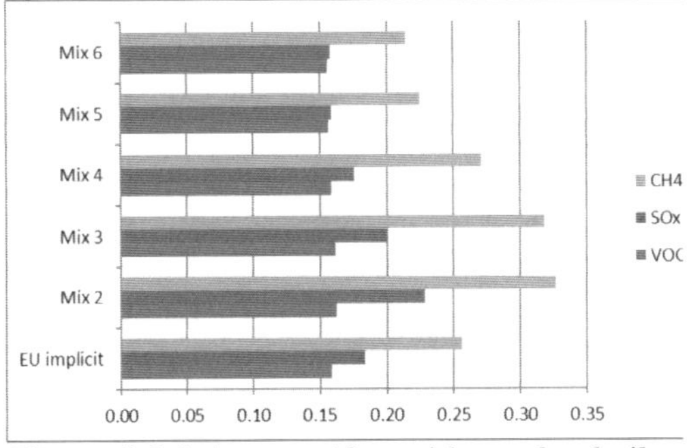


**Figure 3.** Methane, sulfur oxides and volatile compounds emissions.

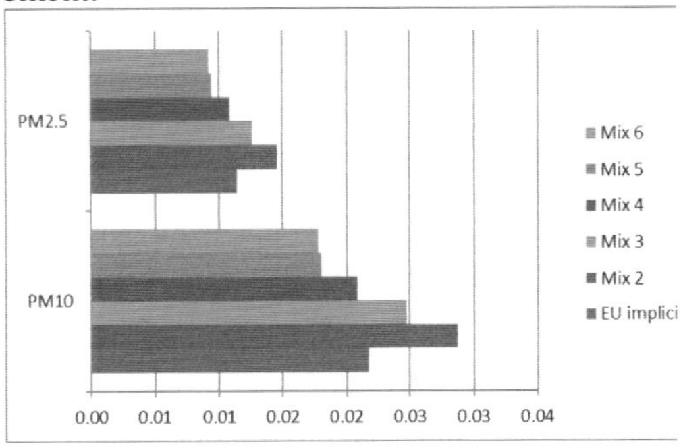


**Figure 4.** Particles emissions.

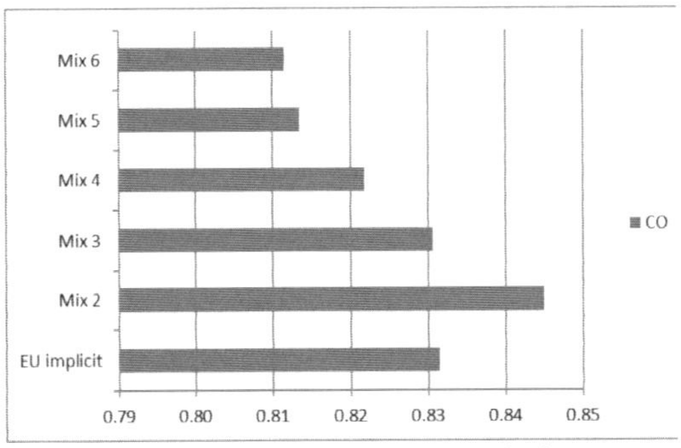


**Figure 5.** Carbon monoxide emissions.

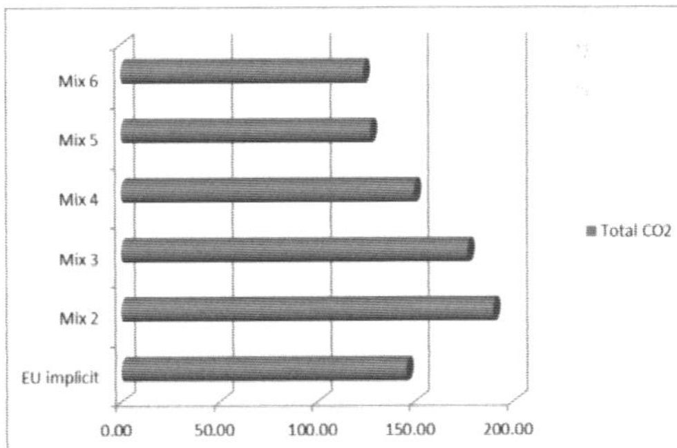


**Figure 6.** Carbon dioxide emissions.

## 4. Conclusion

When compared to conventional cars powered by internal combustion engines, electric vehicles have a number of benefits over their more traditional counterparts, including higher powertrain efficiency, lower maintenance needs, and zero emissions from the exhaust. These factors all contribute to a reduction in urban air pollution. Nevertheless, in contrast to their effects on the environment, reality is more complicated, necessitating a more comprehensive accounting of the consequences that are documented throughout the vehicle's life cycle.

Switching away from electric cars results in large overall savings in $CO_2$ emissions when a major portion of the power is produced from sources that do not produce carbon. In contrast, nations in which a significant amount of power is produced by the burning of coal may find that switching to electric cars results in an increase in the total $CO_2$ emissions related with transportation.

It is important to highlight that when switching to renewable sources for the generation of electricity, considerable cuts may be made in the quantities of pollutants and greenhouse gases produced across all of the categories that were taken into consideration in the WTW study above. In the worst-case scenario, when there is a production rate of 50% coal, total $CO_2$ emissions reach 190g/km traveled. However, if a portion of this proportion migrates to natural gas, which has lower greenhouse gas emissions, gains may be seen. While Mix 4 based primarily on natural gas and nuclear power but with low values for hydropower and wind shows similar results to the current situation, Mixes 5 and 6, where the percentage of nuclear, hydro, and wind energy constitute 3 quarters of production sources, reduce carbon emissions by a third in comparison to the worst case scenario. Mix 5 is based primarily on natural gas and nuclear power while Mix 6 is based primarily on wind and hydropower.

## 5. References

[1] Lajunen A and Lipman T 2016 Lifecycle cost assessment and carbon dioxide emissions of diesel, natural gas, hybrid electric, fuel cell hybrid and electric transit buses, Energy, Volume **106**, Pages 329-342, ISSN 0360-5442.

[2] European Environmental Agency 2018 Electric Vehicles from Life Cycle and Circular Economy Perspectives, TERM 2018: Transport and Environment Reporting Mechanism (TERM) Report, European Environmental Agency: Copenhagen, Denmark

[3] Senecal P K and Leach F 2019 Diversity in transportation: Why a mix of propulsion technologies is the way forward for the future fleet, Results in Engineering, Volume **4**, 100060, ISSN 2590-1230

[4] Correa G, Muñoz P, Falaguerra T and Rodriguez C R 2017 Performance comparison of conventional, hybrid, hydrogen and electric urban buses using well to wheel analysis, Energy, Volume **141**, Pages 537-549, ISSN 0360-5442,

[5] Buberger J, Kersten A, Kuder M, Eckerle R, Weyh T and Thiringer T 2022 Total $CO_2$-equivalent life-cycle emissions from commercially available passenger cars, Renewable and Sustainable Energy Reviews, Volume **159**, 112158, ISSN 1364-0321

[6] Piotr F and Dorota B 2017 Environmental assessment of road transport in a passenger car using the life cycle approach. Transport Problems **12**, 147-153. 10.20858/tp.2017.12.2.14.

[7] Helmers E, Leitão J, Tietge U and Butler T 2019 $CO_2$-equivalent emissions from European passenger vehicles in the years 1995–2015 based on real-world use: Assessing the climate benefit of the European "diesel boom", Atmospheric Environment, Volume **198**, Pages 122-132, ISSN 1352-2310

[8] Pavlovic J, Tansini A, Fontaras G, Ciuffo B, Otura M, Trentadue G, Suarez-Bertoa R and Millo F 2017 The Impact of WLTP on the Official Fuel Consumption and Electric Range of Plug-in Hybrid Electric Vehicles in Europe. 10.4271/2017-24-0133.

[9] Aidin T, Kaveh Z K, Sina C, Pouria A and Mehdi M 2022 Comparative lifecycle assessment of hydrogen fuel cell, electric, CNG, and gasoline-powered vehicles under real driving conditions, International Journal of Hydrogen Energy, Volume **47**, Issue 89, Pages 37990-38002, ISSN 0360-3199

[10] Dumitru I, Dumitru N, Matei L and Racila L 2016 Saturation Flow Mathematical Model Based on Multiple Combinations of Lane Groups, Transportation Research Procedia, Volume **18**, Pages 399-404, ISSN 2352-1465

[11] Pengwei L, Xiaoning X and Jia G 2022 A review of the life cycle carbon footprint of electric vehicle batteries, Separation and Purification Technology, Volume **296**, 121389, ISSN 1383-5866

[12] Hawkins T, Singh B, Majeau-Bettez G and Strømman A 2013 Comparative Environmental Life Cycle Assessment of Conventional and Electric Vehicles. Journal of Industrial Ecology, **17**, 10.1111/j.1530-9290.2012.00532.x.

IOP Conf. Series: Materials Science and Engineering 1303 (2024) 012054 doi:10.1088/1757-899X/1303/1/012054

# Evaluation of different vehicle types by using modelling algorithms in road traffic virtual networks platforms

**N D Gencărău[1], L Matei[1], I Dumitru[1], O V Stănciuc-Oțăt[1], M Racilă[2], A Oprica[1] and Racila L[3]**

[1]University of Craiova, Faculty of Mechanics, Department of Vehicle, Transportation and Industrial Engineering, Craiova, Romania
[3] University of Craiova, Faculty of Mechanics, Department of Applied Mechanics and Civil Constructions, Craiova, Romania
[2] University of Craiova, Faculty of Science, Department of Applied Mathematics, Craiova, Romania

E-mail: racila_laurentiu@yahoo.com

**Abstract.** With the growing demand for simulating and analyzing road traffic, the scope of project planning, modelling, and execution will no longer be restricted to domain experts. In the future, a significantly larger portion of engineering and traffic management professionals will be required to carry out these activities. In this context, it is necessary to have several modelling and analytic algorithms that are based on the current state of affairs. Currently, the research in this subject is facing issues related to driving style variables and the integration of vehicles with the Intelligent Transportation System (ITS). Recently, there has been a focus on studying the significance of driving style for vehicle classes in connection to traffic flow, enhanced comfort, road safety, and the decrease of chemical and noise pollution. The configuration of traffic patterns is constantly changing, leading to the increasing prevalence of the idea of micro-mobility in recent years. The authors provide a comprehensive overview of the primary mathematical models employed in specific applications, as discussed before.

## 1. Introduction

The algorithms that are used to simulate driving styles make use of physical models and algorithms that have been calibrated using data from the actual world. However, due to the high level of complexity, there are currently far too few of these physical models that are accessible for widespread usage. The usage of simulated road networks is one approach that may be taken to address this issue; by doing so, one can also address the more basic issue of knowing how various road traffic situations and modifications could behave in the actual world [1]. Simulation models may also assist us in conceptualizing systems that are far more complicated than what we are able to represent by only employing straightforward mathematical calculations and minute alterations to the flow of real-time traffic.

The use of driving style for different vehicle classes in is an essential component of many of the driving style algorithms that are incorporated into virtual road traffic modelling [2, 3]. Using this parameter, cars are able to enter gaps that are too small to respect the stability of the car following algorithm (how vehicles respond to the vehicles in front of and surrounding them).

Either the vehicle that is changing lanes or the car that is behind it will have to use the brakes at a pace that is up to twice as fast as their maximum rate of deceleration.

IOP Conf. Series: Materials Science and Engineering 1303 (2024) 012054 doi:10.1088/1757-899X/1303/1/012054

## 2. Theoretical considerations with reference to model-building methods

There are multiple considerations with regards to driving style for vehicle classes algorithms used for virtual networks simulations [4]. If the vehicle in the network is considering the decision to make a alter the path, then the decision it is used in the virtual network. If the vehicle in the network is making the actual execution to go to the new path, then the network and the path of the vehicle is adapted to satisfy the virtual network as whole.

For a virtual simulation of road traffic, the driver comportment is assimilated with virtual vehicle within the network. In a virtual simulation of road traffic, the behavior of the driver is represented by a virtual car that operates inside the network. The virtual car is programmed to follow traffic rules and respond to various scenarios encountered on the road. It mimics the actions of a real driver, such as accelerating, braking, and changing lanes.

The algorithms that decide the behavior of a vehicle can be divided into one of two primary categories: tactical or operational [3, 4]. When using tactical category, the virtual cars (virtual drivers) make judgments depending on the existing and predicted characteristics of the traffic. At the same time, the operational choice is solely dependent on the knowledge, which may be restricted depending on the circumstances, on the characteristics of the neighboring traffic.

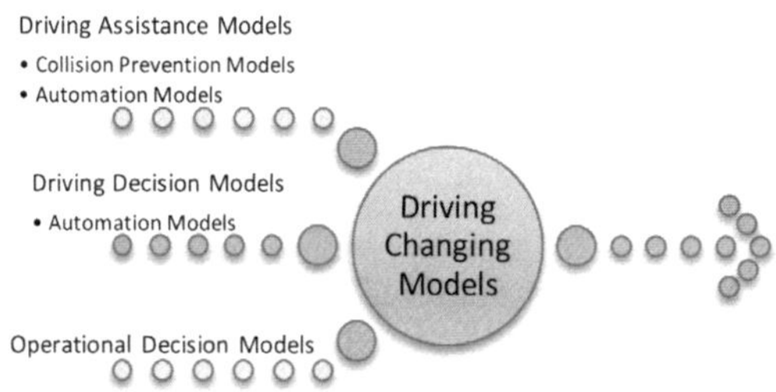


**Figure 1.** Lane change, a classification of the possible models

The traits and actions of the driver have a significant impact on whether or not they have a desire to adjust the driving style, and whether or not they are required to do so. The changing in driving style may be necessary for a variety of reasons, including slower traffic, standing in a line, or being forced to slow down by the car in front of you, among other things. The driver may choose then to change the behavior if the car that they are driving has to slow down because of the vehicle in front of them [5]. The acceleration of the vehicle in the lane that it is now in can be obtained by using the equation 1, bellow:

$$n_{i+1} = \frac{\sigma v^s \Delta v}{\Delta x^d} \tag{1}$$

where:

- $\sigma$ is the sensitivity coefficient
- $v$ is the velocity of the vehicle in question
- $s$ is the speed exponent (ranging from -2 to +2)
- $\Delta v$ is the velocity of the leading vehicle subtracted from the velocity of the vehicle in question
- $x$ is the distance gap between the leading vehicle and the vehicle in question
- $d$ is the distance headway exponent (ranging from +4 to -1).

If $n_{i+1}$ is less than zero, it indicates that the driver will need to slow down his vehicle, which is something that no driver wants to do [6]. In such scenario, the driver wants to switch lanes (behavior change), but if they don't provide their approval, the vehicle will stay in the same lane.

The model incorporates a series of choices that are representative of the algorithm shown in the picture below.

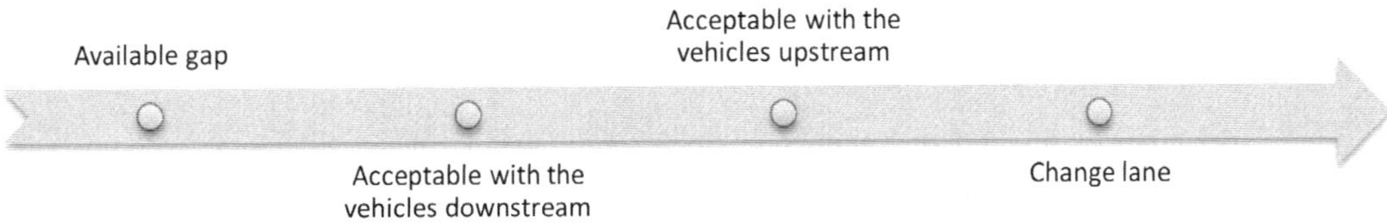


**Figure 2.** Gap acceptance schematics for lane change

Following the schematics shown above, the initial stage of the model involves determining whether or not a change is absolutely necessary [7]. The second stage is to determine whether or not it would be desirable to make a change based on the existing situation in the context of the anticipated improvement in traffic circumstances. The last stage of the model involves determining whether or not the behavior change is possible, after which the switch is actually made.

The decision model developed by Gipps takes into account both the targeted speed of the network vehicle and the maximum speed at which it is safe to go [8, 9]. These speeds are examined simultaneously in order to avoid the effect that may be caused by slow-moving network cars or congestions that are located a significant distance from the vehicle [10].

## 3. Modelling of a virtual network

### *3.1. Network design*

The behavior algorithms that are included into the Aimsun platform are being used for the road traffic evaluation that is being done in connection to driving styles using modelling algorithms. For the purpose of this evaluation, we used a digital model of the road traffic network in Slatina city.

A modeling of the whole region analyzed was carried out in the city of Slatina, and it was based on research that was done on the present traffic. The data on the current traffic was obtained by employing virtual inductive loops that were based on Quercus SmartLoop cameras and also human counters done in different key points. Several calibration components have been included into the modeling as a consequence of the early simulations as well as the need to calibrate the virtual network. These calibration elements are as follows:

- Nicolae Titulescu Boulevard was introduced into the virtual network, beginning at the intersection with Oituz Street and continuing in a clockwise direction toward Vederii Street
- Creating different centroids for secondary and tertiary roads that junctions with the main arteries
- Created were two scenarios lasting two hours each, one for morning and one for afternoon.

The created virtual system is made up of the following components:

- 9 km length sections
- 12 km lane traffic lanes
- 106 sections
- 30 intersections (intersections are all the connections between 2 adjacent sections)
- 17 centroids

**Figure 3.** Virtual network overview

### *3.2. Calibration of current traffic conditions*

After the virtual system of the street frame in the examined region was built, the next step was to fill the virtual network with types of vehicles based on traffic measures as well as the data collected from inductive loops and counting. This was done so that the network could be as close to the real traffic conditions.

For this part of the process, there were a total of 17 centroids employed, which are sites from which vehicles enter and depart the network. A destination origin matrix representing the complete model was constructed using these centroids as the foundation.

| | 3217 | 3218 | 3219 | 3220 | 3226 | 3227 | 3228 | 3229 | 3230 | 3231 | 3232 | 3233 | 3236 | 3237 | 3239 | 3240 | 3241 | Total |
|---|---|---|---|---|---|---|---|---|---|---|---|---|---|---|---|---|---|---|
| 3217 | | 59.2 | 4 | 4 | 4 | 9.6 | 20 | 8 | 9.6 | 20 | 123.2 | 24.8 | 9.6 | | | | 98.4 | 394.4 |
| 3218 | 85.6 | | 16.8 | 11.2 | 21.6 | 11.2 | 24.8 | 24 | | 37.6 | 68 | 56.8 | 11.2 | | | | 199.2 | 568 |
| 3219 | 4.8 | 9.6 | 0.8 | 0.8 | 0.8 | 3.2 | 1.6 | 1.6 | | 0.8 | 4.8 | 1.6 | 0.8 | | | | 1.6 | 32.8 |
| 3220 | 4 | 4 | 0.8 | 0.8 | 1.6 | 3.2 | 1.6 | 1.6 | | 0.8 | 4.8 | 1.6 | 0.8 | | | | 1.6 | 27.2 |
| 3226 | 5.6 | 5.6 | 2.4 | 2.4 | | 27.2 | 8 | 4.8 | 2.4 | 4.8 | 16.8 | 5.6 | 5.6 | | | | 27.2 | 113.4 |
| 3227 | 13.6 | 4 | 1.6 | 1.6 | 27.2 | | 9.6 | 8 | 8 | 18.4 | 8.8 | 16 | 8 | 8 | | | 63.2 | 196 |
| 3228 | 4 | 1.6 | 1.6 | 1.6 | 12 | 4 | | 2.4 | | 2.4 | 4 | 2.4 | | | | | 4 | 40 |
| 3229 | 14.4 | 4 | 4 | 4 | 24.8 | 15.2 | 10.4 | | | 4.8 | 19.2 | 4.8 | | | | | | 105.6 |
| 3230 | 5.6 | 3.2 | 3.2 | 3.2 | 12 | 5.6 | 3.2 | 3.2 | | 3.2 | 3.2 | 5.6 | | | | | 5.6 | 56.8 |
| 3231 | 15.2 | 4 | 4 | 4 | 25.6 | 15.2 | 10.4 | 4.8 | 15.2 | | 40 | 4.8 | | | | | | 143.2 |
| 3232 | 35.2 | 6.4 | 8 | 7.2 | 52.8 | 52.8 | 44 | 29.6 | 4.8 | 31.2 | | 29.6 | | 15.2 | | | 123.2 | 440 |
| 3233 | 10.4 | 5.6 | 5.6 | 5.6 | 27.2 | 16 | 10.4 | | | | 10.4 | | | | | | 43.2 | 134.4 |
| 3236 | 11.2 | 5.6 | 3.2 | 4 | 11.2 | 11.2 | 11.2 | | | | 5.6 | | | | | | 224.8 | 288 |
| 3237 | 1.6 | 1.6 | 1.6 | 1.6 | 1.6 | 1.6 | 1.6 | 1.6 | | 1.6 | 1.6 | 1.6 | 8.8 | | 36 | | 12.8 | 75.2 |
| 3239 | 11.2 | 11.2 | 11.2 | 11.2 | 11.2 | 11.2 | 11.2 | | | | | | | | | 11.2 | 484 | 573.6 |
| 3240 | 5.6 | | | | | 7.2 | | | | | | | | | 14.4 | | | 27.2 |
| 3241 | 96.8 | 13.6 | 12 | 12 | 12 | 423.2 | 48 | 24 | 24 | 36 | 181.6 | 48 | 24 | 12 | 363.2 | 12 | | 1342.4 |
| Total | 374.8 | 139.2 | 98.8 | 75.2 | 245.6 | 617.6 | 236 | 113.6 | 64 | 161.6 | 492 | 203.2 | 68.8 | 35.2 | 413.6 | 23.2 | 1788.8 | 4563.2 |

**Figure 4.** O/D matrix from 07:30 to 09:30 for the small vehicles class

The Aimsun platform creates various characteristics for each of these three kinds of vehicles by basing those characteristics on general aspects like the vehicle's length, speed, consumption, acceleration, and response time, among other things.

Main | Articulated and Doors | Engine | London Emission Model | Experiment Defaults

| | Mean | Deviation | Minimum | Maximum |
|---|---|---|---|---|
| Speed Limit Acceptance | 1.10 | 0.10 | 0.90 | 1.30 |
| Clearance | 1.00 m | 0.30 m | 0.50 m | 1.50 m |
| Maximum Yield Time | 10.00 secs | 2.50 secs | 5.00 secs | 15.00 secs |
| Guidance Acceptance Level | 100.00 % | 0.00 % | 100.00 % | 100.00 % |
| Value of Time | 0.00 $/h | 0.00 $/h | 0.00 $/h | 0.00 $/h |
| Weight | 1300 kg | 400 kg | 800 kg | 2550 kg |

☐ Speed on Slopes Affected by Weight (TWOPAS Model)

| | Mean | Deviation | Minimum | Maximum |
|---|---|---|---|---|
| Weight/Horse Power | 13 kg/hp | 2 kg/hp | 5 kg/hp | 18 kg/hp |
| Weight/Frontal Area | 449 kg/m2 | 244 kg/m2 | 125 kg/m2 | 732 kg/m2 |

**Figure 5.** Individual vehicle characteristics

## 4. Modelling of a virtual network

It is required to employ a dynamic simulation of road traffic in order to get a high degree of detail of the system that is being simulated in order for us to be able to provide a wide variety of reports. The amount to which the model's quality is determined by the availability and precision of the input data is a significant factor. Therefore, the following information must be provided.

- Road traffic network
- Current traffic demand in the region being investigated
- Current plans for road traffic signaling;

In order to analyze the road traffic assessment in connection to driving style and vehicle class by using modeling algorithms, we have used three distinct scenarios based on three different types of vehicles on which the variations are connected to the lane shifting behavior. The following circumstances have been altered from the default scenario: remaining in the lane designated for overtaking, undertaking, and reckless lane shifting. Following the implementation of the aforementioned changes, the virtual network received the following three categories of vehicles:

- cars (light vehicles that are calibrated to be similar to personal vehicle in traffic)
- van (vehicles that used for transport goods in the city and are usual around 3.5 tones)
- truck (heavy vehicles that are used to transport a higher number of goods)

To arrive at an approximation of the road traffic assessment in connection to driving style and vehicle class, we analyze the scenario using the virtual vehicle settings described above, taking into account the fact that all cars entering the network are distributed evenly across the vehicle types used. Under these circumstances, we do an analysis of the situation based on the following criteria: time spent waiting, driving speed, fuel consumption, carbon dioxide emissions, particulate matter emissions during intercity travel.

### *4.1. Simulation results*

The simulation was carried out using the genuine data set, which was derived from the measurements that were taken in the field. The research is carried out in the span of two hours, beginning at 7:30 and ending at 9:30. Following the completion of the simulation, the findings are shown in a graphical format, which may be seen in the figures that are provided below.

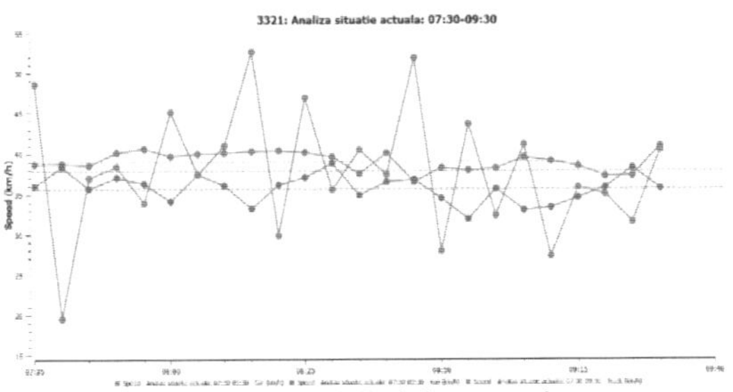


**Figure 6.** Driving speed

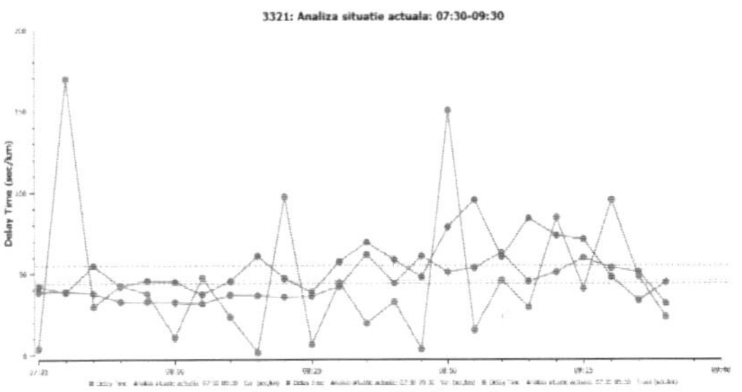


**Figure 7.** Delay time

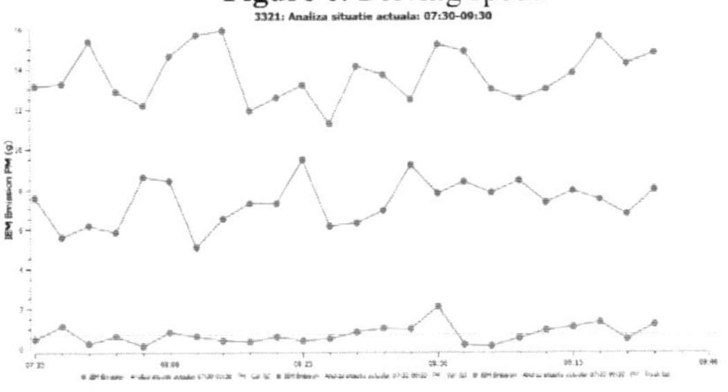


**Figure 8.** PM pollution

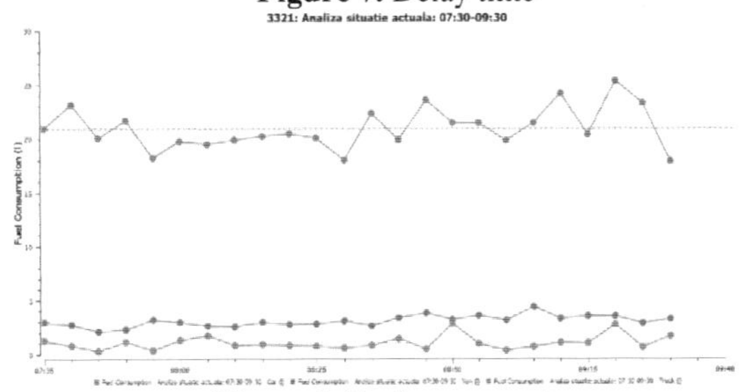


**Figure 9.** Fuel consumption

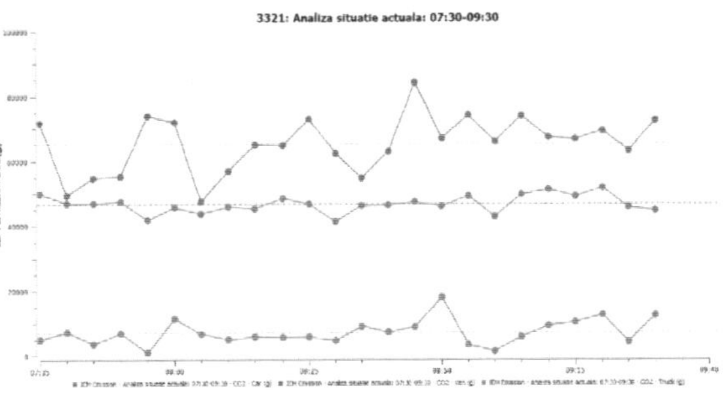


**Figure 10.** CO2 pollution

## 5. Conclusion

As a general rule, vehicle class and driving stile have a major impact on the overall features of traffic flows in an area because of the influence that the different maneuvers have on the traffic that is around them. This is because the flow induces the surrounding traffic to behave in a certain way. Because they may be simply added as an additional resource to already existing lane-changing control and prediction systems, the lane-changing decision models that are based on the characteristics of the traffic are, in many circumstances, the best option. For instance, they may be effective in a project aimed at easing traffic in an area that has a high concentration of automobiles packed into a relatively small space. This quality may be detected in the virtual network that was used to assess the driving style in respect to the road traffic.

The findings of the simulation demonstrate very clearly that the vehicles in the network behave differently based on the driving style of each vehicle class. This suggests that the driving style of a vehicle class has a significant impact on the overall behavior and performance of the network. Furthermore, understanding these differences can help in developing more efficient and tailored strategies for managing traffic flow in mixed vehicle environments.

By analysing the data collected from the simulation, it becomes evident that certain vehicle classes tend to exhibit more aggressive driving behaviours, while others adopt a more cautious approach. This knowledge can be leveraged to create targeted interventions and regulations that encourage safer and smoother traffic flow. Additionally, the findings emphasize the importance of considering the diversity of vehicle classes when designing transportation systems, as the driving style of each class directly influences the overall efficiency and effectiveness of the network.

## References

[1] Moridpour S, Majid S and Geoff R 2010 Lane changing models: A critical review. Transportation Letters: *The International Journal of Transportation Research.* **2**, 10.3328/TL.2010.02.03.157-173, pp. 157-173.

[2] Virginia Tech Transportation Institute Cambridge Systematics Inc 2009 Microscopic Analysis of Traffic Flow in Inclement Weather. Publication No. FHWA- JPO-09-066

[3] Khan Z H and Aaron T 2018 A macroscopic traffic model for traffic flow harmonization *European Transport Research Review* **10**, 10.1186/s12544-018-0291-y

[4] Gipps P G 1986 A model for the structure of lane-changing decisions *Transportation Research* B **20(5)**, pp 403-414

[5] Choudhury C F 2007 *Modeling driving decisions with Latent plans.* PhD thesis, Department of Civil and Environmental Engineering, MIT

[6] Dumitru I, Racilă L, Tutunea D, Matei L, Dima A and Oprica A 2016 *Experimental investigation of a vehicle behaviour using different complex data acquisition systems*, CONAT 2016

[7] Matei L 2020 *Algoritmi de optimizare a circulației la nivel de intersecții rutiere*, Universitaria ISBN: 978-606-14-1590-2

[8] Stewart J A and Aerde M V 1998 An Assessment of Adaptive Co-ordination of Traffic Signal offsets within integration, *Traffic Engineering and Control*, Volume **39**.

[9] Matei L, Dumitru I, Racilă L and Vînatoru M 2016 Adaptive traffic signal control on a national road intrsection *Applied Mechanics and Materials*, ISSN: 1662-7482, Vol. **822**, pp 455-460.

[10] Treiber M and Helbing D 2001 *Microsimulations of Freeway Traffic Including Control Measures*, at Automatisierungstechnik, vol. **49**, no. 11, pp. 478.

IOP Conf. Series: Materials Science and Engineering 1303 (2024) 012055 doi:10.1088/1757-899X/1303/1/012055

# Particularities regarding the design and conception of a bodywork superstructure intended for the transportation of perishable goods

**H Beles, F B Scurt, B A Tolea, T Mitran and V I Tent**

University of Oradea, Romania

E-mail: tent_vlad@yahoo.com

**Abstract.** The aim of this paper is to present the characteristics of the design and construction of a vehicle body superstructure intended for the transportation of perishable goods. In the first phase, the design of the superstructure was conceived, followed by its design. Then, the calculation of the mass distribution on the truck axles, as well as the calculation of the truck's cornering, was carried out, and finally, the superstructure was made, which was mounted on the truck.

## 1. Introduction

At present, the transportation of perishable goods is one of the most demanding sectors in freight transportation, but also with a few regulations that must be followed to ensure that goods arrive at their destination in the best possible condition and do not endanger the health of the final consumer. According to statistical data, in 2010, there were 4 million refrigerated vehicles worldwide, and this number is expected to grow by 2.5% per year by 2030. To limit the growth in the number of refrigerated vehicles, and especially the pollution they generate, it is necessary to improve the superstructure or the materials of the insulation walls, which will lead to a reduction in energy consumption for achieving low temperatures.

Perishable food goods represent a very large proportion of the total goods transported nationally and internationally, making this segment one of the most attractive in transportation.

In Romania and Europe, this matter is generally carried out by road, as delivery times are shorter. In addition, the goods arrive from the loading point to the destination without changing means of transport, so perishable products such as meat, fish, vegetables, fruits, and dairy products must be kept cold or frozen throughout the supply chain [1].

Refrigeration or freezing, food quality and food waste are closely linked [2].

Due to the fact that perishable goods must reach the final consumer safely, it is very important that the vehicle body superstructures used for transportation be constructed in such a way that the heat transfer between the interior of the vehicle body superstructure and the external environment is as small as possible.

## 2. Aspects regarding "Agreement on the International Carriage of Perishable Foodstuffs (ATP)"

The United Nations estimates that approximately 1/3 of all food produced for human consumption is wasted annually, while other sources indicate that food waste is as high as 40% of production [1].

In 1970, the United Nations adopted the *"Agreement on the International Transport of Perishable Foodstuffs (ATP)"*. This agreement contains all regulations for almost all types of refrigerated vehicles used in the transport of perishable goods [3].

Annex 1 of the ATP Agreement provides definitions and standards for special means of transport for the transport of perishable products:

*2.1. Isotherm transport means*

An isotherm transport means is a vehicle with a bodywork superstructure that is constructed with insulating walls, including doors, floors, and roofs, to limit heat exchange between the interior and exterior of the bodywork superstructure. This is achieved by ensuring that the global coefficient of heat transfer (K coefficient) allows the transport means to be classified into one of the following categories:

- $I_N$ = normally insulated equipment specified by a global coefficient of heat transfer (K coefficient) equal to or less than 0.7 $W/m^2{\cdot}K$;
- $I_R$ = heavily insulated equipment specified by a K coefficient equal to or less than 0.4 $W/m^2{\cdot}K$ and by side-walls with a thickness of at least 45 mm for transport equipment of a width greater than 2.50 m.

*2.2. Aspects regarding Annex 1, Appendix 2 of "Agreement on the International Carriage of Perishable Foodstuffs (ATP)"*

In Annex 1, Appendix 2, of the ATP Agreement, are presented the methods and procedures for measuring and verifying the insulation capacity (isothermia) and the efficiency of cooling or heating devices of special means of transport for the transport of perishable goods.

The global coefficient of heat transfer (K coefficient), which characterizes the isothermia of special means of transport, is defined by the equation: (1):

$$K=\frac{W}{S\cdot\Delta T}\left[\frac{W}{m^2\cdot K}\right] \tag{1}$$

where:

- W [W] – the heating power or the cooling capacity, as the case may be, required to maintain a constant absolute temperature difference $\Delta T$ between the mean inside temperature $T_i$ and the mean outside temperature $T_e$, during continuous operation, when the mean outside temperature $T_e$ is constant for a body of mean surface area S;
- S [$m^2$] – the mean surface of the bodywork superstructure = geometric mean of the inside surface area $S_i$ and the outside surface area $S_e$ of the bodywork superstructure.

$$S=\sqrt{S_i\cdot S_e}\left[m^2\right] \tag{2}$$

With respect to the classification of bodywork superstructures, the superstructure in the category of mechanically refrigerated superstructures can be considered (Annex 1):

- thermal insulated bodywork superstructure, equipped either with its own refrigeration unit, or served, together with other transport equipment, by such a unit (compressor, absorption unit, etc.). The unit must be capable, at an average external temperature of +30°C, to lower the temperature inside the bodywork to a value $T_i$, and to maintain it continuously.

## 3. Specifics of the design and construction of a bodywork superstructure for a vehicle intended for the transport of perishable goods.

For illustrative purposes, the specifics of the design and construction of a bodywork superstructure for a vehicle intended for the transport of perishable goods, namely, for the transport of meat carcasses on hooks, will be presented. For the cooling and heating of the interior of the bodywork superstructure, a refrigeration unit was chosen, which was placed in the upper part of the front wall.

The bodywork superstructure for a truck with a maximum authorized total weight of 19000 kg was designed and constructed.

Taking into account the optimal distribution of masses on the truck's axles, the dimensional characteristics of the refrigeration unit, and the maximum total height of the truck (4 m - maximum total height allowed by current legislation [4]), a bodywork superstructure was designed with an external length of 7.207 m and an internal length of 7.03 m. The external width is 2.6 m (this being the maximum width allowed by current legislation), and the useful internal width is 2.47 m. The internal height of the bodywork superstructure is 2.505 m, and the total external height is 2.765 m.

In Figure 1, the main dimensions of the bodywork superstructure are presented:

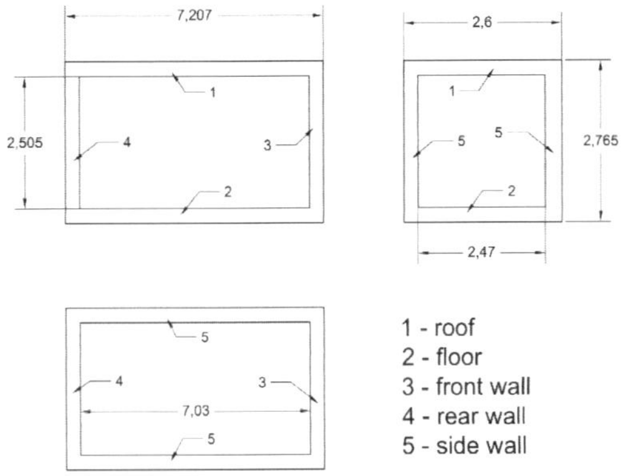


**Figure 1**. The main dimensions of the bodywork superstructure.

From the point of view of construction, the bodywork superstructure able to carry a payload of 9,500 kg was designed and constructed. It consists of the following components:

- floor;
- roof;
- front wall;
- side wall;
- rear wall (rear doors);
- perimeter frame.

The bodywork superstructure was modeled in 3D using the SolidWorks design software [5]. The Figures 2, 3, 4 and 5 show some isometric 3D models of the bodywork superstructure:

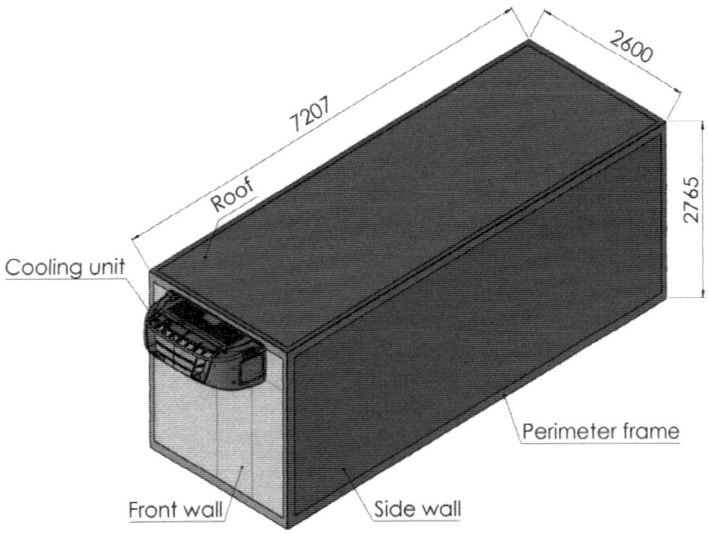


**Figure 2**. Isometric view of the bodywork superstructure from the left front side.

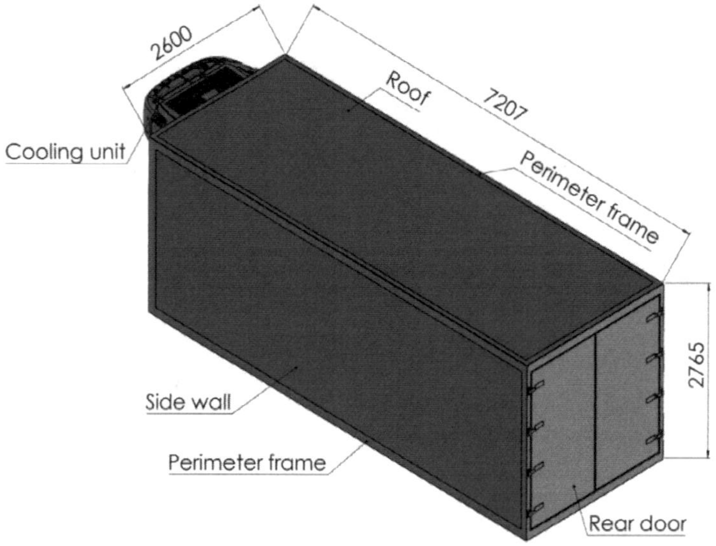


**Figure 3.** Isometric view of the bodywork superstructure with a left rear view.

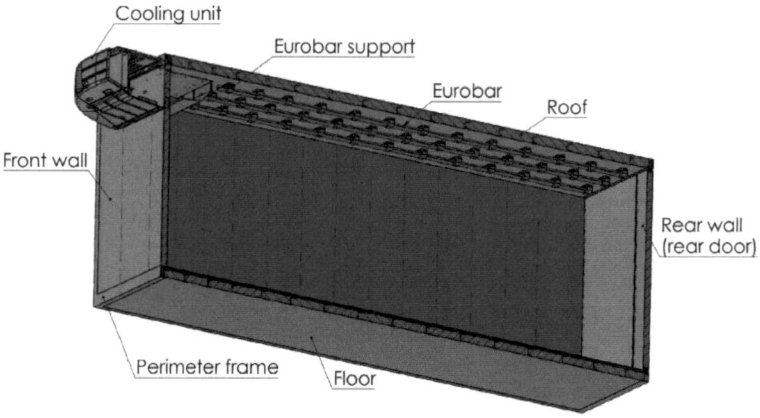


**Figure 4**. Isometric view of the bodywork superstructure in longitudinal section.

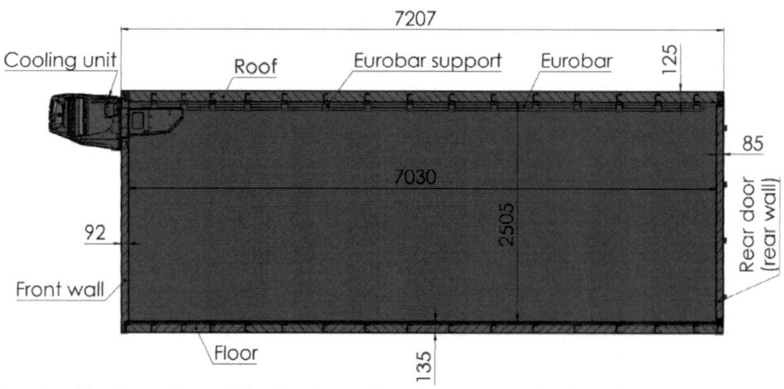


**Figure 5**. A longitudinal section of the bodywork superstructure, showing the clearance dimensions.

IOP Conf. Series: Materials Science and Engineering 1303 (2024) 012055 doi:10.1088/1757-899X/1303/1/012055

In order for this bodywork superstructure to be able to transport meat carcasses, it is necessary to install hooks that slide on 5 eurobars with a diameter of Ø60 mm. These are fixed to the roof by means of 69 supports, which are made of steel and have a load capacity of 500 kg each (according to [6]). This is shown in Figure 6, which is a drawing of the supports.

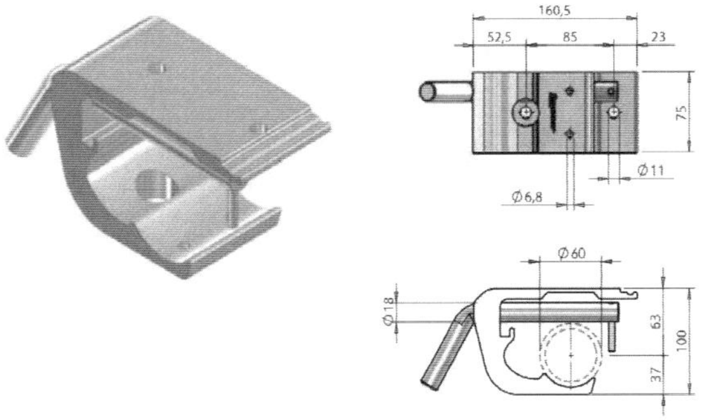


**Figure 6.** Eurobar hook.

The eurobar supports are fixed to the roof with M10 screws, which have a grade of 10.9. This is shown in Figure 7, which is a drawing of the assembly of the eurobar supports and eurobars.

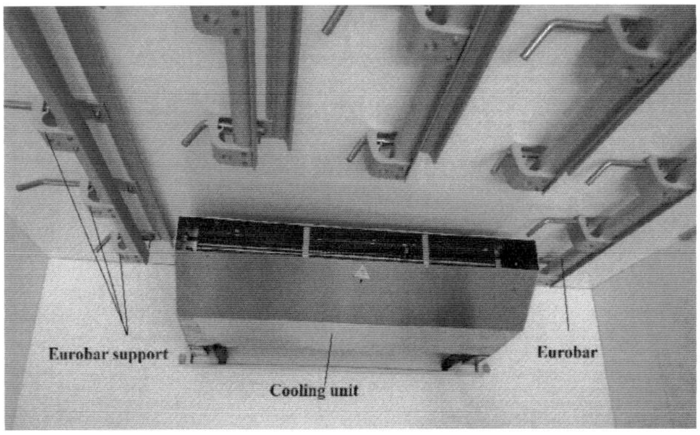


**Figure 7**. Assembly of eurobars together with their supports.

According to [6], a hook has a load capacity of 750 kg of meat carcass, and its main dimensions are shown in Figure 8.

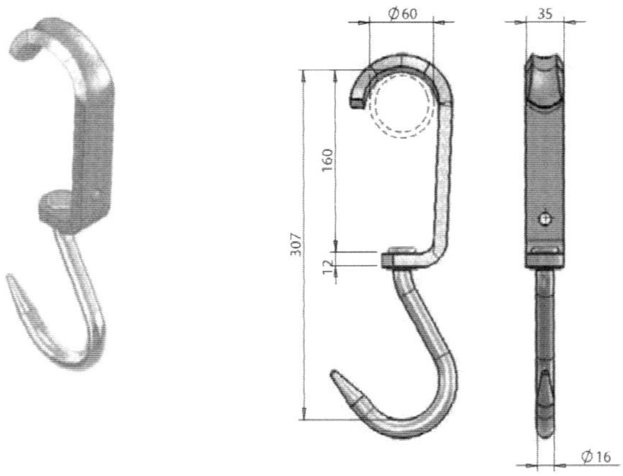


**Figure 8**. Dimensions of a hook destinated for transporting meat carcasses.

The hooks used for transporting meat carcasses are shown in Figure 9.

**Figure 9**. Hooks used for transporting meat carcasses.

## 4. Calculation of load distribution on truck axles

The weight distribution simulation was performed using the Trailer Win software [7].

For the calculation of weight distribution, the useful load is considered to be uniformly distributed over the entire length of the bodywork superstructure. In Figure 10 are presented the overall dimensions of the truck, taken from the simulation performed using the Trailer Win software.

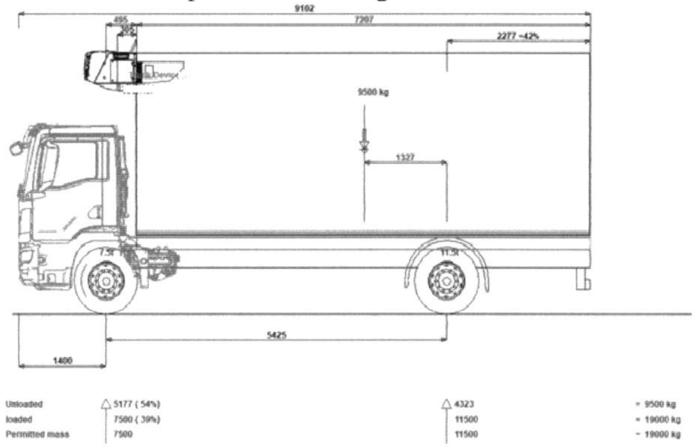


**Figure 10**. The overall dimensions of the truck taken from the simulation performed using the Trailer Win software.

The weight distribution diagram is presented in Figure 11.

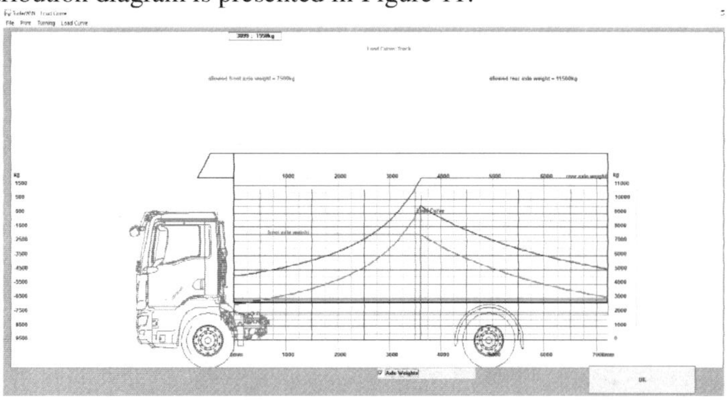

**Figure 11.** Weight distribution diagram.

The calculation of the carrying capacity distribution is described below:

- Interior length of the bodywork superstructure= 7030 mm; (3)
- Rear overhang = 2277 mm; (4)
- $M_{payload}=M_{max}-M_{truck}=19000-9500=9500$ kg. (5)

where:

- $M_{payload}$ – truck payload; (6)
- $M_{max}=19000$ kg; (7)
- $M_{truck}=M_{truck\ front\ axle}+M_{truck\ rear\ axle}=5177+4323=9500$ kg; (8)

The truck payload on the front axle is determined by the following relationship:

$$M_{truck\ front\ axle}=0.34\cdot M_{payload}=0.34\cdot 9500=3230 \text{ kg} \quad (9)$$

The truck payload on the rear axle is determined by the below relationship:

$$M_{truck\ rear\ axle}=M_{payload}-M_{truck\ front\ axle}=9500-3230=6270 \text{ kg} \quad (10)$$

In Table 1 is described the weight distribution on the axles of the truck:

**Table 1.**

| | | Weight on front axle [kg] | Weight on rear axle [kg] | Total weight [kg] |
|---|---|---|---|---|
| 1 | Weight of the truck | 5008 | 4992 | 9500 |
| 2 | Carrying capacity | 3230 | 6270 | 9500 |
| 3 | Weight loaded (1+2) | 7500 | 11500 | 19000 |
| 4 | Gross truck weight | 7500 | 11500 | 19000 |

The verification of the weight distribution is performed using the following relationships:

$$M_{truck\ front\ axle}=7500 \text{ kg} \leq M_{max.\ front\ axle}=7500 \text{ kg} \quad (11)$$

$$M_{truck\ rear\ axle}=11500 \text{ kg} \leq M_{max.\ rear\ axle}=11500 \text{ kg} \quad (12)$$

The authorized weight on the driving axle must be 25% greater than the maximum authorized total weight, which is determined by the following relationship:

$$M_{max\ rear\ axle}=11500 > 25\%M_{max\ truck}=4750 \text{ kg} \quad (13)$$

The authorized weight on the steering axle must be 20% greater than the weight of the truck when the rear axle is loaded to the maximum permissible weight, which is determined with the help of the relationship below:

$$M_{max\ front\ axle}=7500 > 20\%M_{max\ truck}=3800 \text{ kg} \quad (14)$$

## 5. Calculation of truck turning radius

Trucks must be built in such a way that, when they complete a full circle with a radius of 12.5 m, the covered circular crown must not be wider than 7.2 m, when the extreme outer point of the front console moves on the circle with a radius of 12.5 m [4]. When entering the tangent on the circle, no part of the truck must exceed this tangent, outwards, by more than 0.8 m, as shown in Figure 12.

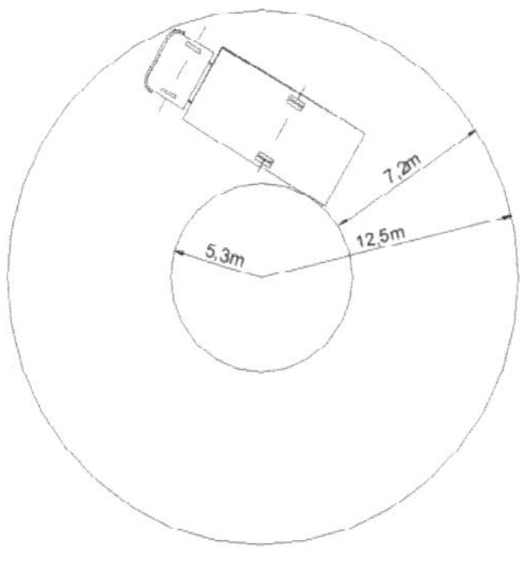


**Figure 12**. Truck turning.

The intermediate radii of the curve entry are represented in Figure 13.

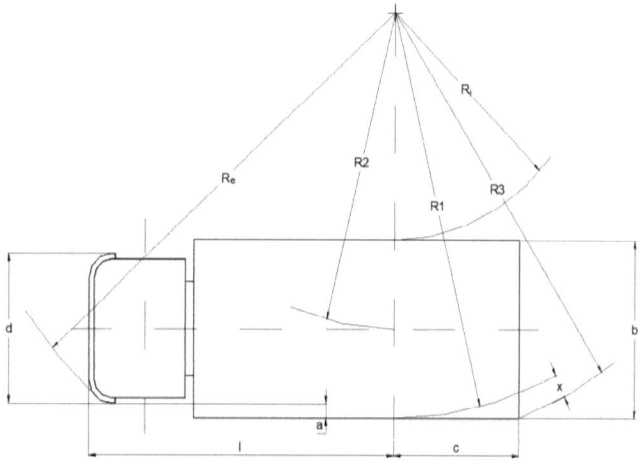


**Figure 13**. The intermediate radii of the curve entry.

where:

- l = 6825 mm - distance from the front of the cabin to the rear axle; (15)
- c = 2277 mm - theoretical rear overhang; (16)
- d = 2488 mm - cabin width; (17)
- b = 2600 mm - bodywork superstructure width; (18)
- $R_{emax}$ = 12500 mm - maximum outer turning radius; (19)
- $R_{imin}$ = 5300 mm - minimum inner turning radius. (20)

The determination of the intermediate radii from Figure 13, is done using the following relationships:

$$R_e=R_{e\ max}=12.5\ \text{m} \tag{21}$$

$$R_1=\sqrt{R_e^2-l^2}+(b\text{-}d)/2=\sqrt{12.5^2+6.825^2}+(2.6\text{-}2.488)/2=14.9\ \text{m} \tag{22}$$

$$R_2=R_1-b/2=14.9-2.6/2=13.6\ \text{m} \tag{23}$$

$$R_3=\sqrt{R_1^2+c^2}=\sqrt{14.9^2+2.277^2}=15.07\ \text{m} \tag{24}$$

The inner turning radius is determined with the following relationship:

$$R_i=R_2-b/2=13.6-2.6/2=12.3\ \text{m} \tag{25}$$

Result:

$$R_i=12.3\ m>R_{imin}=5.3\ m \tag{26}$$

$$X=R_3-R_1=15.07-14.9=0.17\ m \tag{27}$$

X=0.17 m <0.8 m, the truck meets the conditions for entering the curve. (28)

## 6. Conclusion

The bodywork superstructure presented in the paper was designed, engineered, and built, falling within category $I_N$, class C, being a refrigerated and heat-insulating transport vehicle.

The truck with the bodywork superstructure mounted is shown in Figure 14.

**Figure 14.** Truck with the bodywork superstructure mounted.

Respecting [8], [9], [10] the truck was individually approved by the Romanian Road Authority for circulation on public roads.

## 7. Acknowledgments

The research has been funded by the University of Oradea, within the Grants Competition "Scientific Research of Excellence Related to Priority Areas with Capitalization through Technology Transfer: INO - TRANSFER - UO", Project No. 313/21.12.2021.

## 8. References

[1] Mercier S, Villeneuve S, Mondor and Uysal I 2017 Time Temperature Management Along the Food Cold Chain A Review of Recent Developments. Comprehensive Reviews in Food Science and Food Safety Vol. 00.

[2] Paúl B, Torres J, Juan J, Aguirre R, Carlos E, Buenaño M, Efrén V S and Abad Farfán G 2018 Thermal behavior of a refrigerated vehicle: Process simulation, International Journal of Refrigeration.

[3] UNECE Transport Division, United Nationsm 1970 Agreement on the international carriage of perishable foodstuffs and on the special equipment to be used for such carriage (ATP), Geneva, Switzerland
[4] COMMISSION REGULATION (EU) No 1230/2012 of 12 December 2012 implementing Regulation (EC) No 661/2009 of the European Parliament and of the Council with regard to type-approval requirements for masses and dimensions of motor vehicles and their trailers and amending Directive 2007/46/EC of the European Parliament and of the Council
[5] ***SolidWorks Premium 2023 software
[6] ***Pomier – Meat hanging system
[7] ***Trailer Win software
[8] ***Order for the Approval of Regulations on Individual Type Approval, Issuance of Vehicle Registration Document, and Certification of Authenticity of Road Vehicles - RNTR 7
[9] ***Order for the Approval of Regulations on Type Approval and Issuance of Vehicle Registration Document, as well as Type Approval of Products Used in Them - RNTR 2
[10] REGULATION (EU) 2018/858 OF THE EUROPEAN PARLIAMENT AND OF THE COUNCIL of 30 May 2018 on the approval and market surveillance of motor vehicles and their trailers, and of systems, components and separate technical units intended for such vehicles, amending Regulations (EC) No 715/2007 and (EC) No 595/2009 and repealing Directive 2007/46/EC

# AUTHOR INDEX

**Institute of Physics**
Dirac House, Temple Back
Bristol BS1 6BE UK

ISSN: 1757-8981
ISBN 978-1-7138-9567-1

CURRAN ASSOCIATES INC.
proceedings.com